Y0-AIG-647

INDUSTRIAL FASTENERS HANDBOOK

1st Edition

TRADE & TECHNICAL PRESS LTD.,
MORDEN, SURREY,
ENGLAND.

PREFACE

The mode and means of joining metal to metal is of paramount importance in industrial manufacture. The stresses and strains on, and around, the jointing are aggravated by the demands for modern machinery to produce greater power, higher speeds, with intensified torque and its inherent problems. Add to this the many new materials in vogue, the necessity of joining unlike materials, the legion of fastening methods available, and the problem facing the designer, production or chief engineer, r and d personnel, etc, is enormous.

The INDUSTRIAL FASTENERS HANDBOOK provides the answer for the selection and application of an appropriate and preferable fastener for every type of jointing or bonding required. Numbering over 700 pages, it is packed full of useful and practical technical information and data. It is, at time of publication, the most comprehensive and up-to-date reference work on the extensive subject of industrial fastening available anywhere in the world.

Following in logical sequence, two further reference works, 'Handbook of Industrial Materials' and 'Handbook of Industrial Fabrication' will be published shortly after this handbook to complete a vital trio for the engineer's bookshelf.

THE PUBLISHERS

85461 062 6

Printed and published by TRADE & TECHNICAL PRESS LTD., CROWN HOUSE, MORDEN, SURREY, SM4 5EW. ENGLAND.

Paging Glynwed

"or where to find the fastenings you are looking for…fast"

High Tensile Bolts in all steel qualities – Pre-Assembled Lock Washer and Screw Fastener Units – Paint Removing and Self Locking Screws.

See Page 50A

Glynwed Fastenings.
Midland Road · Darlaston · Wednesbury · West Midlands
WS10 8JN
Tel: 021 (B'ham) 526-2895
Telex: 33508

High Grade and General Purpose Studs, 'U' Bolts and Screwed Rods.

See Page 70B

Yarwood Ingram & Co.
Ledsam Street · Ladywood · Birmingham B16 8DW
Tel: 021 (B'ham) 454-3607

Tubular and Semi-Tubular Rivets – Masonry Nails – Special Non-Ferrous Nails – Electrical Cable Clips – Gripfast Barbed Nails.

See Page 254A

Tower Manufacturing.
Navigation Road · Diglis · Worcester WR5 3DE
Tel: 0905 (Worcs.) 356012
Telex: 338880

Glynwed

Glynwed Screws and Fastenings Limited

CONTENTS

SECTION 1 – Sub-section (a) – Mechanical Fasteners

SECTION 1 – Sub-section (b)

SECTION 2 – Sub-section (a) – Adhesives

SECTION 2 – Sub-section (b)

ring round for
Fastenings FAST
or
Telex
your
require~
ments

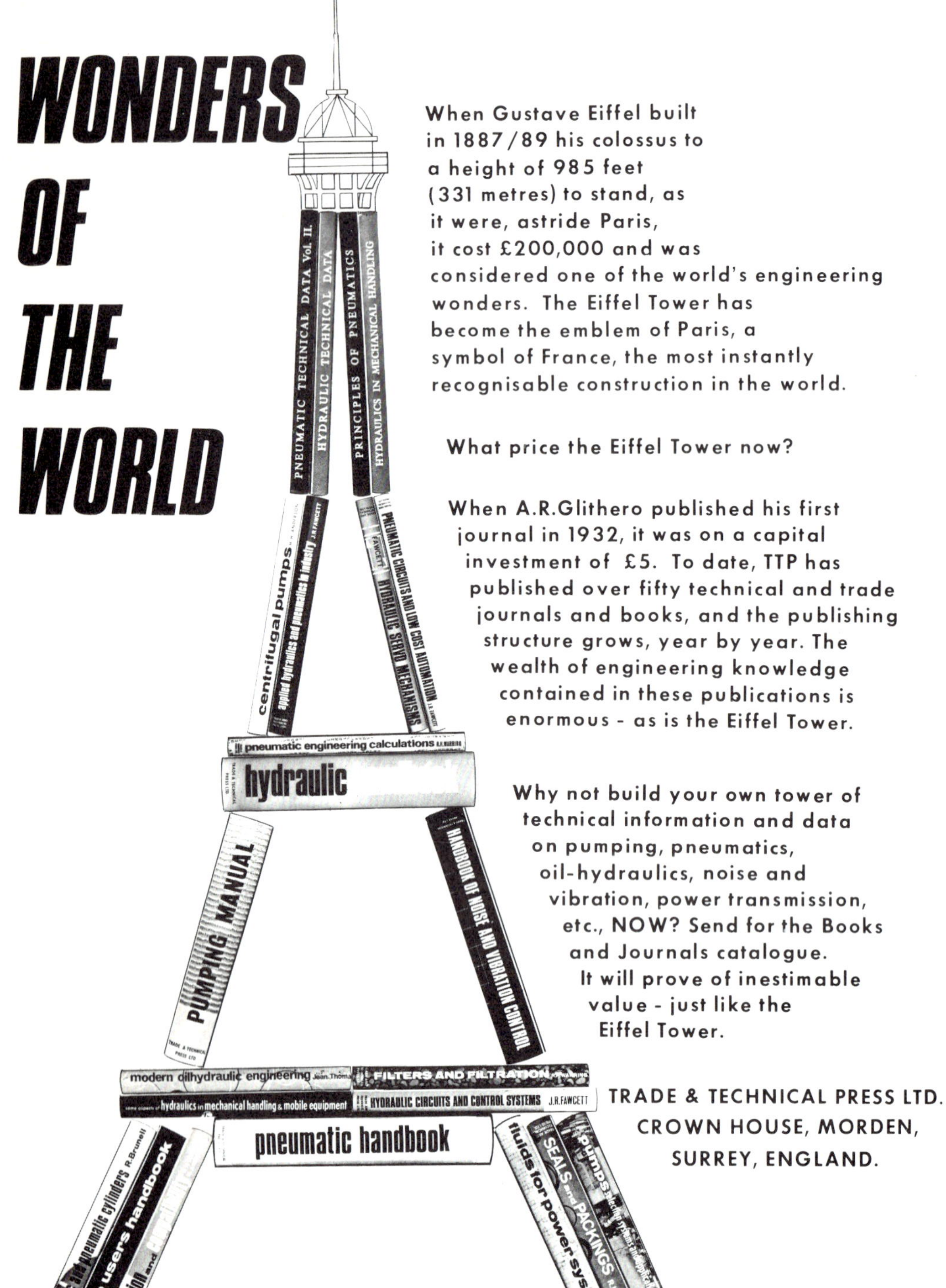
WONDERS OF THE WORLD
When Gustave Eiffel built in 1887/89 his colossus to a height of 985 feet (331 metres) to stand, as it were, astride Paris, it cost £200,000 and was considered one of the world's engineering wonders. The Eiffel Tower has become the emblem of Paris, a symbol of France, the most instantly recognisable construction in the world.
What price the Eiffel Tower now?
When A.R.Glithero published his first journal in 1932, it was on a capital investment of £5. To date, TTP has published over fifty technical and trade journals and books, and the publishing structure grows, year by year. The wealth of engineering knowledge contained in these publications is enormous - as is the Eiffel Tower.
Why not build your own tower of technical information and data on pumping, pneumatics, oil-hydraulics, noise and vibration, power transmission, etc., NOW? Send for the Books and Journals catalogue. It will prove of inestimable value - just like the Eiffel Tower.
TRADE & TECHNICAL PRESS LTD., CROWN HOUSE, MORDEN, SURREY, ENGLAND.
PNEUMATIC TECHNICAL DATA Vol. II.
HYDRAULIC TECHNICAL DATA
PRINCIPLES OF PNEUMATICS
HYDRAULICS IN MECHANICAL HANDLING
centrifugal pumps
applied hydraulics and pneumatics in industry
PNEUMATIC CIRCUITS AND LOW COST AUTOMATION
HYDRAULIC SERVO MECHANISMS
pneumatic engineering calculations
hydraulic
PUMPING MANUAL
HANDBOOK OF NOISE AND VIBRATION CONTROL
modern oilhydraulic engineering
FILTERS AND FILTRATION
hydraulics in mechanical handling & mobile equipment
HYDRAULIC CIRCUITS AND CONTROL SYSTEMS
J.R.FAWCETT
pneumatic handbook
pump users handbook
corrosion and
fluids for power systems
SEALS and PACKINGS
pumps

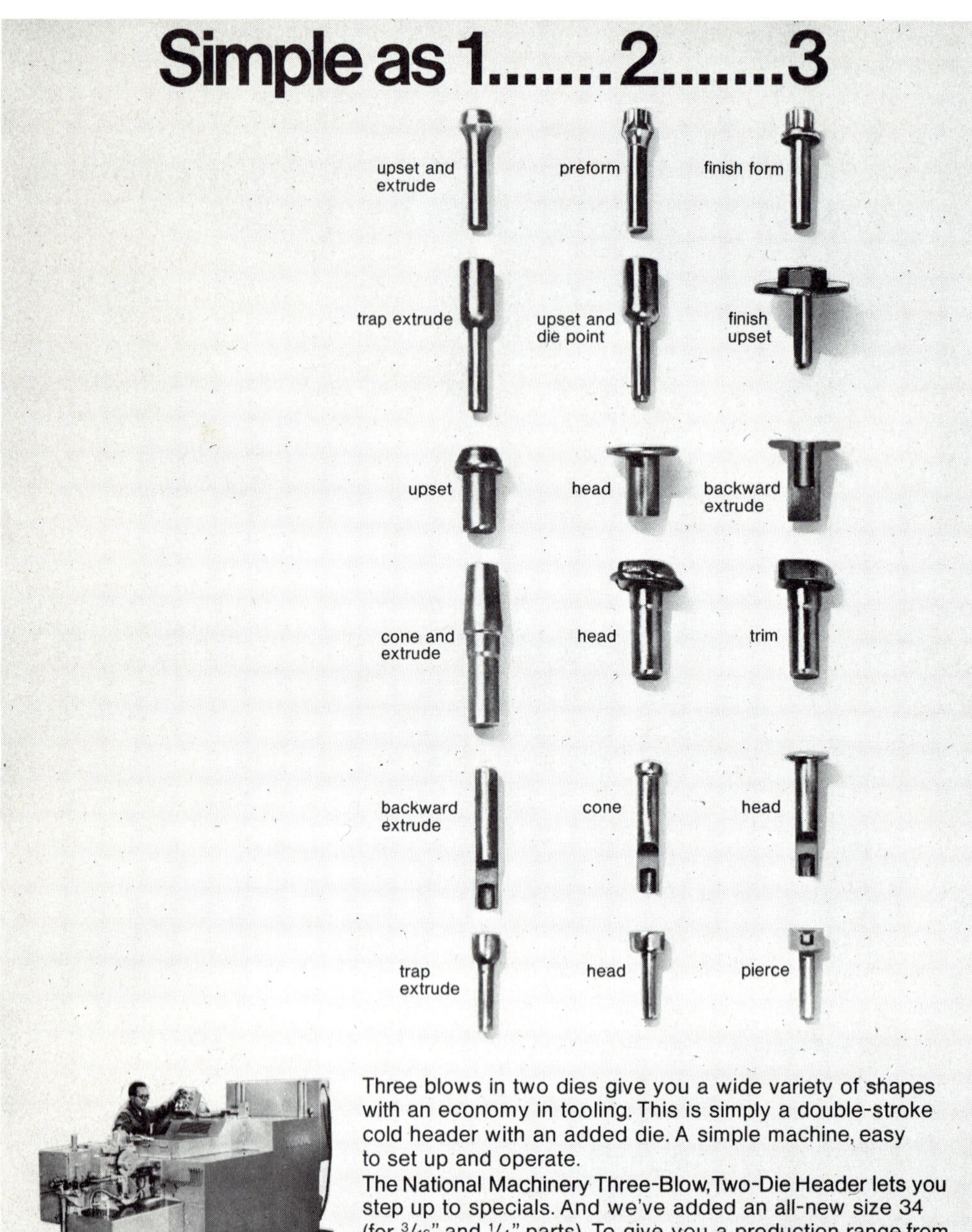

Simple as 1.......2.......3
upset and extrude
preform
finish form
trap extrude
upset and die point
finish upset
upset
head
backward extrude
cone and extrude
head
trim
backward extrude
cone
head
trap extrude
head
pierce
Three blows in two dies give you a wide variety of shapes with an economy in tooling. This is simply a double-stroke cold header with an added die. A simple machine, easy to set up and operate.
The National Machinery Three-Blow, Two-Die Header lets you step up to specials. And we've added an all-new size 34 (for 3/16" and 1/4" parts). To give you a production range from 3/16" to 9/16". All this at speeds up to 200 pieces per minute. We'll be glad to tell you more about it.

National Machinery

NATIONAL MACHINERY CO., TIFFIN, OHIO, U.S.A. 44883 NATIONAL MACHINERY G.m.b.H., 8500 NUERNBERG GERMANY
U.K. OFFICE: 30 BOLDMERE ROAD, SUTTON COLDFIELD, WARWICKSHIRE, ENGLAND. TELEPHONE 021-355 1717
DESIGNERS AND BUILDERS OF HOT AND COLD FORGING MACHINERY

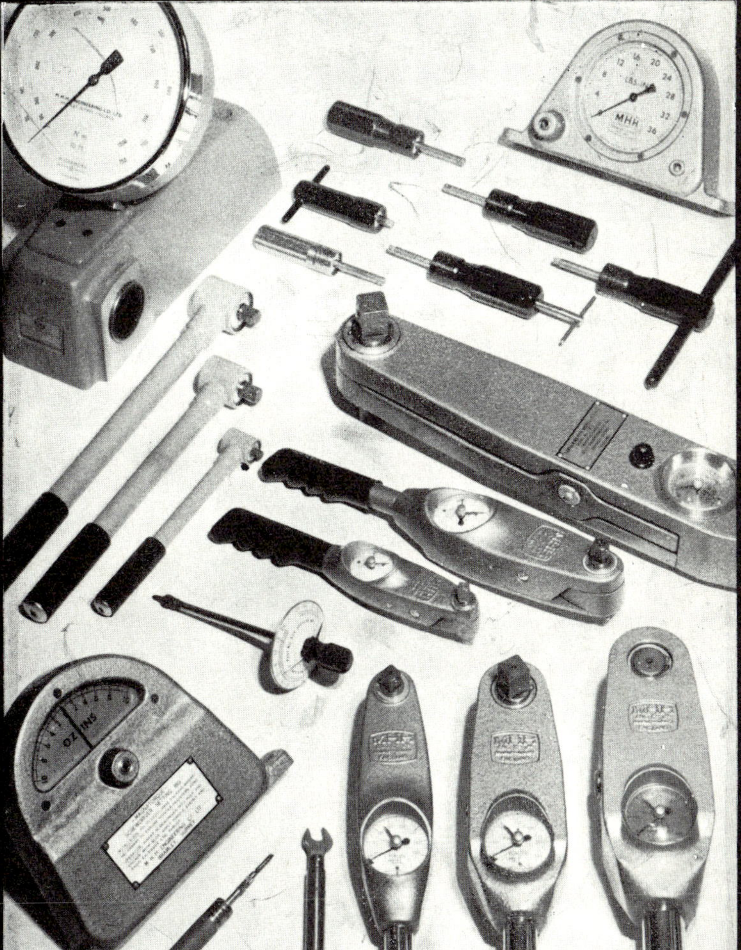

BRIGHT STEEL & BRASS PLAIN WASHERS.
BRIGHT STEEL & BRASS TURNED & CHAMFERED WASHERS.
ENGINEERS BLACK STEEL WASHERS.
NEOPRENE BACKED STEEL WASHERS.
NON-STANDARD WASHERS IN ALL METALS A SPECIALITY.
Washers by the million – with a million uses
CHARLES (WEDNESBURY) LTD.
BRIDGE WORKS, WEDNESBURY, STAFFORDSHIRE, ENGLAND, WS10 ONU.
Telephone: 021·556 1921. Telex: 338083.
Telegrams: Chamfer, Wednesbury.

THOMAS WILLIAM
LENCH
LIMITED

Quality fasteners in quantity

For Agricultural, Automotive, Constructional & Petrochemical Industries

Nearly every type of fastening need is covered by our Standard Range backed by a fast delivery service from our distribution centres. To meet new and unusual fastening requirements we have a range of Special Fasteners and we will also design and develop fasteners to suit specific needs. Most of our fasteners can be supplied with any type of protective coating and we have the most modern Hot Dipped Spun Galvanizing plant to give maximum protection.

Thomas William Lench Ltd

P.O. Box 31, Excelsior Works, Rowley Regis, Warley, West Midlands B65 8BZ. Telephone : 021-559 1530. Telex : 338735

and Our Distributing Companies

Gladstone Ltd.
Portrack Lane, Stockton-on-Tees,
TS18 2HE
Tel. 0642 64481 or 65654
Telex 58169

Warley Fasteners Ltd.
Waterfall Trading Estate,
Cradley Heath, Warley,
West Midlands B64 6PU
Tel. 021-559 4214
Telex 336308

Gladstone Fasteners Ltd.
Harlington Works, Old Newton Road,
Heathfield,
Devon TQ12 6RN
Tel. Bovey Tracey 3777
Telex 42934

Carr & Nichols Ltd.
The Lancashire Screw Bolt Works,
Atherton,
Manchester M29 9LB
Tel. 05-234 2431
Telex 67104

Gwent Fasteners Ltd.
Felnex Trading Estate,
Stephenson Street, Newport,
Mon. NPT OWZ
Tel. 0633 73371
Telex 49160

SECTION 1

Sub-section A

Mechanical Fasteners

Introduction

Fastener technology started with the invention of the nail — first in the form of carved wooden pegs and then as hand forged nails in copper and bronze. Nobody can be sure how many thousands of years the history of the nail goes back, but certainly the Romans were the first large 'industrial' producers and users of iron nails, and the hand forging of nails remained an important industry in Britain all through the Middle Ages and beyond.

Individual variations of nail forms appeared aimed at improving the gripping power — a practice which has become even more widely exploited in recent years — leading eventually to the woodscrew which was *turned* rather than *driven.* But as man came to assemble harder, stronger and thinner materials a fastener which 'held' by virtue of its own form — not its grip in the material — had to evolve, and the original solution — the nut and bolt — has equally well stood the test of time.

This established a different type of fastener — one which had to be *assembled.* The solid rivet emerged as an alternative for 'strong' fastening, and also where a permanent fastening was desirable (although the technique of riveting had been practiced centuries earlier by clenching over nails). And there, apart from material and production improvements and relatively small detail changes, fastener technology virtually remained until comparatively recent years. Nails and screws were for fastening wood. Bolts and rivets for fastening metals. And there were other minor types of fasteners such as eyelets and staples for fastening other materials, but the range of fastener materials available was relatively small.

The three things that have been responsible for a major revolution in fastener technology within the last few decades are speed, cost and the appearance of so many new materials. Speed and cost are inter-related — any technique that speeds up production saves manhours or labour cost. Speed can be achieved by using power drivers — increasing screw, nut and washer assembly rate by a factor of 1.5 or greater for example. (Then somebody had the bright idea of a pre-located washer on the bolt, or 'SEM', to save a little more labour time).

Speed of fastening reduces cost, but that it not the complete answer. The cost of the fastener unit, preparation, applicator, handling and even storage must be taken into account. Speed is the criterion for labour cost. The main variables are assembler labour rate, unit fastener cost, and hole preparation, in that order. On this assumption, speed of completing the fastening *is* the major factor for minimum fastening cost.

(photo GKN)

Most fastening and assembly jobs *could* be done with conventional nuts and bolts, screws or nails, and in a vast number of industries they are cost-favourable. (There are also industries where fastening cost is not important). But there are potentially dozens of types of fastener which could be competitive or even more cost-favourable for many assembly requirements, as well as many alternatives on the 'standard' forms of conventional fasteners. That they exist, and remain competitive, is proof of the success of their original design-intention. The classic example that can be quoted is the self-tapping screw – but this is only the tip of the iceberg.

This Handbook of Industrial Fasteners sets out to do the impossible – cover all the known types of fasteners available (which means that some will inevitably have been missed out), and also present them in logical categories or groups, or under separate chapter headings for ease of reference. This seems absolutely *necessary* to make the Handbook as useful a reference as possible. But unfortunately there is no completely clear cut category for many of the designs of proprietary fasteners which have appeared, so some of the categorisation is of our own origination, designed to make reference to specific sources of information as direct and simple as possible.

This may appear to leave some sections and chapters sparse. *Special Fasteners*, for example, could have embraced types of fasteners categorised and described in perhaps a dozen other individual chapters – and as a consequence made finding *specific* information on separate types more difficult. As it is we have tried to keep them separate, and cross referenced where appropriate.

All categorisation of fasteners is arguable, of course – even nails, screws and bolts have their 'special' and individual developments and sub-classifications. A long way, in fact, from the original classification of fasteners as 'direct' (nails); 'semi-direct' (woodscrews); and 'indirect' (bolts and nuts). Even then people argued about whether woodscrews were 'semi-direct' or 'direct'. Perhaps supporters of the latter condoned the use of a hammer as a screwdriver!

But the primary intent is to present a practical Handbook which 'supplies the answers', not 'debates the question'. And some of the answers may be quite surprising, especially to practical engineers and designers who still think mainly in terms of nuts and bolts (and self-tappers). Modern riveting, for example, using semi-tubular rivets and automatic rivet setting machines, can achieve production rates of 100 rivets per minute or better. This places the potential of riveting in a separate category from engineering or structural fastening. It is equally available as a low-cost, high speed production fastening technique when the primary requirement is 'fastening', not 'structural'.

There are also dozens of different new types of fastener, many designed originally for a specific application field but subsequently proving to be equally acceptable, and competitive, in others.

Then there are the direct alternatives to traditional fasteners. To pick one out (although this cannot help but be unfair to others), the 'Spat' can be regarded as a direct alternative to screws and nuts and self tapping screws for sheet metal assembly work. It is slightly more expensive than the alternatives, its average installation rate on typical work is inferior to self tapping screws, (though superior to screw, washer and nut), but its total in-place cost can work out substantially less. This is achieved by virtue of the fact that the 'Spat' is self-piercing, ie a design which has been influenced by cost-consciousness.

The Handbook attempts to include all types of fastener outside the traditional types which have emerged – and survived. Fastening techniques also extend beyond mechanical fasteners. Adhesive bonding is a recognised – and proven – alternative in many applications. This particular section of the Handbook is much shorter, but covers all modern types of adhesive warranting an industrial application rating.

Compilation of information, data, etc on such a wide range of fastening products would not have been possible without the ready co-operation of the many manufacturers involved, and their help in this matter is gratefully acknowledged. It is to be hoped that there are no significant omissions, and if there are, these are purely omissions of error, not intent. The field of fastening and fasteners is, after all, mainly about proprietary products.

Screw Threads

Historical

The first serious attempt to rationalise screw thread production was undertaken by Sir Joseph Whitworth about 1841, the result of this study being the original Whitworth thread as the standard throughout the British engineering industry. Over the following twenty years it replaced the miscellaneous collection of threads in general use and together with a fine thread series of the same form which became the British Standard Fine (BSF); and also Whitworth watch and instrument threads. All these threads were based on a 55° angle between the flanks with a specified rounding at crest and root, and differed only in the number of threads per inch.

In 1864 Sellers, in America, quite independently proposed another National thread standard based on a 60° angle between the flanks and with cut-off rather than rounded crests and roots (mainly because this was easier to produce by machining). The original Sellers thread was adopted in the American SAE and ASME standards.

In continental Europe, rational threads were also developed with a variety of flank angles, but most again using cut-off V shapes rather than rounded crests and roots. These included the (German) Loeuenhertz thread with an angle of 53°8′, and the (Swiss) Thury thread with an angle of 47½^{o}. Again in an attempt at standardisation the SI (Systems International) thread was originated in 1898 based on a 60° angle between flanks with flat crests and rounded roots. A basic disadvantage of this SI metric thread was that the small root radius led to inferior fatigue characteristics. As a result various countries modified the thread form and a variety of different SI threads came into use. For example, the French SI metric threads followed the US (Sellers) standard and differed from the metric system at about the 100mm size. The German metric thread followed the SI standard between 6 and 12mm, but had slightly different proportions in other sizes.

Additionally, a number of other thread forms were developed for specialised applications – eg British Association (BA) thread based on metric dimensions and a 47½^{o} angle with rounded crests and roots; the British Standard Cycle thread (BSC) and American Cycle Engineers Thread (CIE), both with 60° angle and rounded crests and roots; Acme and Square threads; British, American and continental Europe pipe threads; and a variety of other special thread sizes for watchmaking, model engineering, etc.

No further serious attempts at National or International standardisation were undertaken until after World War II. Then, in 1948, the United States, Great Britain and Canada reached agreement to establish two new Unified Screw Thread Standard Series to be known as UNIFIED COARSE (UNC) and UNIFIED FINE (UNF). Unified threads came to be adopted as a standard for new productions in a number of major industries in these three countries – notably agricultural, automobile and chemical engineering, in the oil industry and to a lesser extent in the aircraft industry – from about 1963 on. The ultimate effect, however, was merely to add another two thread series to existing threads. The two UN series threads never became universally adopted.

In continental Europe the International Standards Organisation attempted to rationalise by having one thread series only (and BS3643 issued in 1963 was based on this ISO metric standard). The general plan was for ISO metric threads to embrace ISO Metric Coarse and ISO Metric Fine (or 'standard') threads with graded pitches, together with other series with constant pitches. Ultimately it was forced to accept two thread series, known as ISO Unified Inch and ISO Metric, adopting the Unified thread profile as the common one for all ISO screw threads. Thus, ISO Metric and ISO Unified thread profiles are identical and their tolerance systems are derived from similar formulas. British Standard BS3692 issued in 1966 as the new Metric fastener standard (subsequently revised) covered hexagon bolts, screws and nuts.

The United States, meanwhile, remains outside the influence of metrication and is in fact planning a new standard thread for their metrication programme, scheduled to be completed by 1980. This differs slightly from the ISO standard; these differences are claimed to give functional advantages.

Metrication

Originally, International standardisation of screw threads merely led to the feed-in of further 'standard' thread forms, especially as British industry in particular was slow to accept metric threads – with some industries continuing to specify older British standard threads such as BSW and BSF. This has complicated, rather than helped, the supply position, with seven different thread forms in use and despite the fact that British fastener manufacturers have phased out the BSF thread and UNF thread, which in many cases are available only on special order. BSF and BSW high tensile precision bolts,screws and nuts to BS1083 have been declared non-preferred since January 1974, and BSW Black Hexagon bolts, screws and nuts to BS916 have been declared non-preferred since January 1975. The current demand for BSW and UNC threads in Britain, however, remains substantially as large as before. The same is probably true of BA threads, although BSW, BSF and BA are essentially now obsolete. Unified threads are not likely to become obsolete until America adopts its new (or International) metric thread standard.

In practice, the ISO Metric Coarse thread provides a replacement for general engineering, superseding BSW, BSF, BA, UNC and UNF, with the advantage of a substantial reduction in the number of sizes of threaded fasteners required. Thus twelve metric sizes will cover the requirements currently met by sixty-nine Imperial sizes, – See Table A.

Substitution

In general, ISO Metric Coarse threads can be regarded as a direct substitute for BSW and BSF. The pitch differs marginally for BSW: and although BSF is a finer thread, this difference is balanced by the higher angle (60° as against 55°); also the metric thread has a large root radius. Comparative data are summarised in Tables I,II and III. Recommended metric diameters to replace inch diameters are given in Table IV.

TABLE A

ISO Metric*	BSW	BSF	UNC	UNF	BA
2			2	2	9, 10
2.5			3	3	7, 8, 6
3	1/8"	1/8"	4, 5	4, 5	6, 5, 4 †
4			6, 8	6, 8	4, 3 †
5	3/16"	3/16"	10, 12	10, 12	2, 1
6	1/4"	1/4"	1/4"	1/4"	0
8	5/16"	5/16"	5/16"	5/16"	
10	3/8", 7/16"	3/8", 7/16"	3/8", 7/16"	3/8", 7/16"	
12	1/2", 9/16"	1/2", 9/16"	1/2", 9/16"	1/2", 9/16"	
16	5/8"	5/8"	5/8"	5/8"	
20	3/4", 7/8",	3/4", 7/8"	3/4", 7/8"	3/4", 7/8"	
24	1"	1"	1"	1"	

* Coarse series generally recommended (thread slightly finer than BSW)
Fine series has more limited application (thread slightly finer than BSF)

† Preferred if maximum strength required

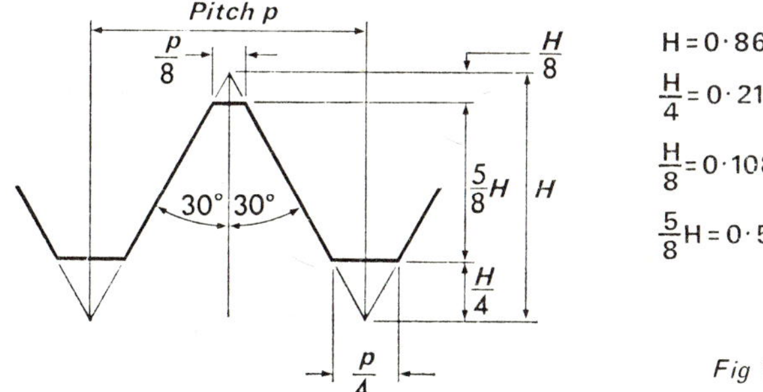

$$H = 0{\cdot}86603p$$

$$\frac{H}{4} = 0{\cdot}21651p$$

$$\frac{H}{8} = 0{\cdot}10825p$$

$$\frac{5}{8}H = 0{\cdot}54127p$$

Fig 1 ISO Metric screw thread.

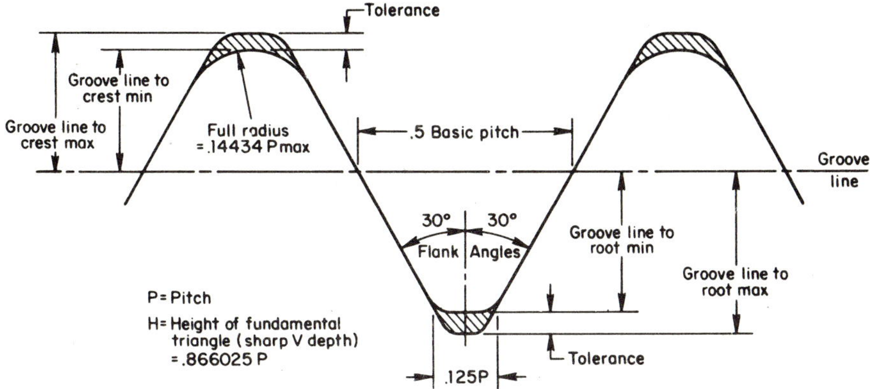

Fig 2 Unified screw thread

TABLE I – COMPARISON OF IMPERIAL AND METRIC THREADS

	BS Whitworth			BS Fine			ISO Metric Coarse			
Diameter	TPI	Stress Area	Basic Minor Diameter	TPI	Stress Area	Basic Minor Diameter	Metric Diameter	TPI (approx)	Stress Area	Basic Minor Diameter
in		in^2	in		in^2	in	mm (in)		mm^2 (in^2)	mm (in)
¼ (0.2500)	20	0.0320	0.1860	26	0.0356	0.2008	6 (0.2362)	25.4	20.1 (0.0312)	4.773 (0.1879)
5/16 (0.3125)	18	0.0527	0.2413	22	0.0567	0.2543	8 (0.3149)	20.3	36.6 (0.0567)	6.466 (0.2546)
3/8 (0.3750)	16	0.0779	0.2950	20	0.0839	0.3110	10 (0.3937)	16.9	58.0 (0.0899)	8.160 (0.3212)
7/16 (0.4375)	14	0.1069	0.3461	18	0.1158	0.3663	12 (0.4724)	14.5	84.3 (0.1307)	9.853 (0.3879)
½ (0.5000)	12	0.1385	0.3932	16	0.152	0.4200	12 (0.4724)	14.5	84.3 (0.1307)	9.853 (0.3879)
9/16 (0.5625)	12	0.183	0.4557	16	0.198	0.4825	14 (0.5512)	12.7	115 (0.1783)	11.546 (0.4546)
5/8 (0.6250)	11	0.227	0.5086	14	0.243	0.5336	16 (0.6299)	12.7	157 (0.2433)	13.546 (0.5333)
¾ (0.7500)	10	0.336	0.6220	12	0.352	0.6432	20 (0.7874)	10.2	245 (0.3798)	16.933 (0.6666)
7/8 (0.8750)	9	0.463	0.7328	11	0.487	0.7586	22 (0.8661)	10.2	303 (0.4697)	18.933 (0.7454)
1 (1.000)	8	0.608	0.8400	10	0.642	0.8720	24 (0.9448)	8.5	353 (0.5520)	20.319 (0.7999)
1.1/8 (1.125)	7	0.767	0.9420	9	0.8147	0.9828	*M27 (1.0629)	8.5	459 (0.7114)	23.319 (0.9181)
							M30 (1.1811)	7.3	561 (0.8695)	25.706 (1.0120)
1¼ (1.250)	7	0.973	1.0670	9	1.027	1.0780	*M33 (1.2992)	7.3	694 (1.0757)	28.706 (1.1302)
*1.3/8 (1.375)	7	0.973	1.0670	8	1.237	1.2150	M36 (1.4173)	6.4	817 (1.2663)	31.093 (1.2241)
1½ (1.500)	6	1.409	1.2866	8	1.496	1.3400	*M39 (1.5354)	6.4	976 (1.5128)	34.093 (1.3422)

Figures in brackets are inch equivalents. **Non preferred sizes.*

TABLE II – COMPARISON OF UN AND METRIC THREADS

Diameter	ISO Metric Fine			UN Coarse			UN Fine		
in	TPI (approx)	Stress Area mm^2 (in^2)	Basic Minor Diameter mm (in)	TPI	Stress Area in^2	Basic Minor Diameter in	TPI	Stress Area in^2	Basic Minor Diameter in
¼ (0.2500)	No ISO Metric Fine for this Diameter			20	0.0324	0.1887	28	0.0368	0.2062
5/16 (0.3125)	25.4	39.2 (0.0608)	6.773 (0.2666)	18	0.0532	0.2443	24	0.0587	0.2614
3/8 (0.3750)	20.3	61.2. (0.0949)	8.466 (0.3333)	16	0.0786	0.2983	24	0.0886	0.3239
7/16 (0.4375)	20.3	92.1 (0.1428)	10.466 (0.4120)	14	0.1078	0.3499	20	0.1198	0.3762
½ (0.5000)				13	0.1438	0.4056	20	0.1612	0.4387
5/8 (0.6250)	17.0	167 (0.2589)	14.160 (0.5575)	11	0.229	0.5135	18	0.258	0.5568
¾ (0.7500)	17.0	272 (0.4216)	18.160 (0.7150)	10	0.338	0.6273	16	0.375	0.6733
7/8 (0.8750)	17.0	333 (0.5162)	20.160 (0.7937)	9	0.467	0.7387	14	0.513	0.7874
1 (1.000)	12.7	384 (0.5952)	21.546 (0.8483)	8	0.612	0.8466	12	0.667	0.8978

TABLE III – COMPARISON OF MM, INCH AND BA SIZES

mm	in	Fractions	BA equiv	mm	in	Fractions	mm	in	Fractions
1.59	0.062	1/16	–	6.35	0.250	¼	17.00	0.670	–
2.00	0.079	–	–	7.00	0.276	–	17.46	0.688	11/16
2.20	0.087	–	8	7.14	0.281	9/32	18.00	0.709	–
2.38	0.094	3/32	–	7.94	0.312	5/16	19.00	0.748	–
2.50	0.098	–	7	8.00	0.315	–	19.05	0.750	¾
2.80	0.110	–	6	8.73	0.344	11/32	20.00	0.787	–
3.00	0.118	–	–	9.00	0.354	–	20.64	0.813	13/16
3.17	0.125	1/8	–	9.52	0.375	3/8	22.22	0.875	7/8
3.20	0.126	–	5	10.00	0.394	–	23.81	0.938	15/16
3.50	0.138	–	–	10.32	0.406	13/32	25.00	0.984	–
3.60	0.142	–	4	11.00	0.433	–	25.40	1.000	1
3.97	0.156	5/32	–	11.11	0.438	7/16	30.00	1.181	–
4.00	0.158	–	–	11.91	0.468	15/32	35.00	1.378	–
4.10	0.161	–	3	12.00	0.472	–	40.00	1.575	–
4.50	0.177	–	–	12.70	0.500	½	45.00	1.772	–
4.70	0.185	–	2	13.00	0.512	–	50.00	1.969	–
4.76	0.187	3/16	–	14.00	0.551	–	100.00	3.937	–
5.00	0.197	–	–	14.29	0.563	9/16	200.00	7.874	–
5.30	0.209	–	1	15.00	0.591	–			
5.56	0.219	7/32	–	15.87	0.625	5/8			
6.00	0.236	–	0	16.00	0.630	–			

TABLE IV- RECOMMENDED METRIC DIAMETERS TO REPLACE INCH DIAMETERS

Inch Diameter		Metric Diameter	Inch
Fraction	Decimal		Equivalent
3/16	0.1875	M5	0.1968
¼	0.2500	M6	0.2362
5/16	0.3125	M8	0.3149
3/8	0.3750	M10	0.3937
7/16	0.4375	M12	0.4724
½	0.5000	—	—
5/8	0.6250	M16	0.6299
¾	0.7500	M20	0.7874
1	1.000	M24	0.9448
1.1/8	1.1250	M30	1.1811
1¼	1.2500	M36	1.4173
1.3/8	1.3750	—	—
1½	1.5000	M42	1.6535
1¾	1.7500	M48	1.8897
2	2.0000	—	—
2¼	2.2500	M56	2.2047
2½	2.5000	M64	2.5196

BRITISH STANDARDS FOR METRIC THREADS AND FASTENERS

BS3643	Screw Threads
Part 1	Thread data and standard thread series
Part 2	Limits and tolerances for coarse pitch series
Part 3	Limits and tolerances for fine pitch series
BS3692	Precision hexagon bolts, screws and nuts
BS4168	Hexagon socket screws and wrench keys
BS4183	Machine screws and machine screw nuts
BS4190	Black hexagon bolts, screws and nuts
BS4219	Slotted grub screws
BS4320	Metal washers for general engineering purposes
BS4439	Screwed studs
BS4395	High strength friction grip bolts
BS4186	Clearance holes
BS4278	Eye bolts
BS4463	Crinkle washers
BS4464	Spring washers

ISO METRIC SCREW THREADS

The ISO Metric thread profile is shown in Fig 1. Dimension data are summarised in Tables VA and VB. Size is normally expressed as nominal diameter prefixed by 'M', eg M2, M2.5. Preferred diameter sizes are first choice. Non-preferred sizes are basically those included to give close equivalents to Imperial sizes — see Table I.

TABLE VA – ISO METRIC THREADS – BASIC DIMENSIONS – COARSE SERIES

Nominal diameter			Pitch	Major diameter	Effective diameter	Minor diameter external thread *	Minor diameter internal thread †	Section at minor diameter mm^2	Tensile stress area mm^2
1st choice	2nd choice	3rd choice							
1	–	–	0.25	1.000	0.894	0.693	0.785	0.377	0.460
–	1.1	–	0.25	1.100	0.994	0.793	0.855	0.494	0.588
1.2	–	–	0.25	1.200	1.094	0.893	0.985	0.626	0.737
–	1.4	–	0.30	1.400	1.265	1.032	1.142	0.837	0.982
1.6	–	–	0.35	1.600	1.373	1.171	1.221	1.07	1.27
–	1.8	–	0.35	1.800	1.573	1.371	1.421	1.48	1.70
2	–	–	0.4	2.000	1.740	1.509	1.567	1.79	2.07
–	2.2	–	0.45	2.200	1.908	1.648	1.713	2.13	2.48
2.5	–	–	0.45	2.500	2.208	1.948	2.013	2.98	3.39
3	–	–	0.5	3.000	2.675	2.387	2.459	4.47	5.03
–	3.5	–	0.6	3.500	3.110	2.764	2.850	6.00	6.78
4	–	–	0.7	4.000	3.545	3.141	3.242	7.75	8.78
–	4.5	–	0.75	4.500	4.013	3.580	3.688	10.1	11.3
5	–	–	0.8	5.000	4.480	4.019	4.134	12.7	14.2
6	–	–	1.0	6.000	5.350	4.773	4.917	17.9	20.1
–	–	7	1.0	7.000	6.350	5.773	5.917	26.2	28.9
8	–	–	1.25	8.000	7.188	6.466	6.647	32.8	36.6
–	–	9	1.25	9.000	8.188	7.466	7.647	43.8	48.1
10	–	–	1.5	10.000	9.026	8.160	8.376	52.3	58.0
–	–	11	1.5	11.000	10.026	9.160	9.376	65.9	72.3
12	–	–	1.75	12.000	10.863	9.853	10.106	76.2	84.3
–	14	–	2.0	14.000	12.701	11.546	11.835	105	115
16	–	–	2.0	16.000	14.701	13.546	13.835	144	157
–	18	–	2.5	18.000	16.376	14.933	15.294	175	192
20	–	–	2.5	20.000	18.376	16.933	17.294	225	245
–	22	–	2.5	22.000	20.376	18.933	19.295	282	303
24	–	–	3.0	24.000	22.051	20.319	20.752	324	353
–	27	–	3.0	27.000	25.051	23.319	23.752	427	459
30	–	–	3.5	30.000	27.727	25.706	26.211	519	561
–	33	–	3.5	33.000	30.727	28.706	29.211	647	694
36	–	–	4.0	36.000	33.402	31.092	31.670	759	817
–	39	–	4.0	39.000	36.402	34.092	34.670	913	976
42	–	–	4.5	42.000	39.077	36.479	37.129	1 050	1 120
–	45	–	4.5	45.000	42.077	39.479	41.029	1 220	1 300
48	–	–	5.0	48.000	44.752	41.866	42.587	1 380	1 470
–	52	–	5.0	52.000	48.752	45.866	46.587	1 650	1 760
56	–	–	5.5	56.000	52.428	49.252	50.046	1 910	2 230
–	60	–	5.5	60.000	56.428	53.252	54.046	2 230	2 360
64	–	–	6.0	64.000	60.103	56.369	57.505	2 250	2 680
–	68	–	6.0	68.000	64.103	60.639	61.505	2 890	3 060

* *Tolerance class 4h* † *Tolerance class 5H*

TABLE VB–ISO METRIC THREADS – BASIC DIMENSIONS – FINE SERIES (contd.)

Nominal diameter 1st choice	Nominal diameter 2nd choice	Nominal diameter 3rd choice	Pitch	Section at minor diameter	Tensile stress area	Nominal diameter 1st choice	Nominal diameter 2nd choice	Nominal diameter 3rd choice	Pitch	Section at minor diameter	Tensile stress area
mm	mm	mm	mm	mm^2	mm^2	mm	mm	mm	mm	mm^2	mm^2
1	–	–	0.2	0.48	0.53	–	14	–	1.25	122	129
–	1.1	–	0.2	0.57	0.67	–	14	–	1.50	116	125
1.2	–	–	0.2	0.72	0.82	–	–	15	1.00	149	155
–	1.4	–	0.2	1.05	1.17	–	–	15	1.50	136	145
1.6	–	–	0.2	1.44	1.59	16	–	–	1.00	171	178
–	1.8	–	0.2	1.90	2.06	16	–	–	1.50	157	167
2	–	–	0.25	2.25	2.45	–	–	17	1.00	195	203
–	2.2	–	0.25	2.82	3.03	–	–	17	1.50	180	191
2.5	–	–	0.35	3.37	3.70	–	18	–	1.0	221	229
3	–	–	0.35	5.19	5.61	–	18	–	1.5	205	216
–	3.5	–	0.35	7.41	7.90	–	18	–	2.0	190	204
4	–	–	0.50	9.00	9.79	20	–	–	1.0	277	285
–	4.5	–	0.50	11.90	12.80	20	–	–	1.5	259	272
5	–	–	0.50	15.10	16.10	20	–	–	2.0	242	258
–	–	5.5	0.50	18.80	19.90	–	22	–	1.0	339	348
6	–	–	0.75	20.30	22.00	–	22	–	1.5	319	333
–	–	7	0.75	29.00	31.30	–	22	–	2.0	300	318
8	–	–	0.75	39.40	41.80	24	–	–	1.0	407	418
8	–	–	1.00	36.00	39.20	24	–	–	1.5	386	401
–	–	9	0.75	51.30	54.10	24	–	–	2.0	365	384
–	–	9	1.00	47.50	51.10	–	–	25	1.0	444	455
10	–	–	0.75	64.80	67.90	–	–	25	1.5	421	437
10	–	–	1.00	60.50	64.50	–	–	25	2.0	399	420
10	–	–	1.25	56.30	61.20	–	–	26	1.5	458	475
–	–	11	0.75	79.80	83.30	–	27	–	1.0	522	533
–	–	11	1.00	75.00	79.50	–	27	–	1.5	497	514
12	–	–	1.00	91.20	96.10	–	27	–	2.0	473	496
12	–	–	1.25	86.00	92.10	–	–	28	1.0	563	575
12	–	–	1.50	81.10	88.10	–	–	28	1.5	537	555
–	14	–	1.00	128	134	–	–	28	2.0	513	536

cont....

TABLE VB—ISO METRIC THREADS — BASIC DIMENSIONS — FINE SERIES (contd.)

Nominal diameter 1st choice	Nominal diameter 2nd choice	Nominal diameter 3rd choice	Pitch	Section at minor diameter	Tensile stress area	Nominal diameter 1st choice	Nominal diameter 2nd choice	Nominal diameter 3rd choice	Pitch	Section at minor diameter	Tensile stress area
mm	mm	mm	mm	mm²	mm²	mm	mm	mm	mm	mm²	mm²
30	—	—	1.0	650	663	48	—	—	3.0	1 540	1 600
30	—	—	1.5	623	642	48	—	—	4.0	1 460	1 540
30	—	—	2.0	596	621	—	—	50	1.5	1 820	1 860
30	—	—	3.0	544	581	—	—	50	2.0	1 780	1 820
—	—	32	1.5	714	735	—	—	50	3.0	1 690	1 750
—	—	32	2.0	686	713	—	52	—	1.5	1 980	2 010
—	33	—	1.5	763	785	—	52	—	2.0	1 930	1 970
—	33	—	2.0	733	761	—	52	—	3.0	1 830	1 900
—	33	—	3.0	675	716	—	52	—	4.0	1 740	1 830
—	—	35	1.5	864	886	—	—	55	1.5	2 220	2 260
36	—	—	1.5	916	940	—	—	55	2.0	2 170	2 220
36	—	—	2.0	844	915	—	—	55	3.0	2 070	2 140
36	—	—	3.0	820	865	—	—	55	4.0	1 970	2 060
—	—	38	1.5	1 030	1 050	56	—	—	1.5	2 300	2 340
—	39	—	1.5	1 080	1 110	56	—	—	2.0	2 250	2 300
—	39	—	2.0	1 050	1 080	56	—	—	3.0	2 150	2 220
—	39	—	3.0	980	1 030	56	—	—	4.0	2 050	2 140
—	—	40	1.5	1 140	1 170	—	—	58	1.5	2 480	2 520
—	—	40	2.0	1 110	1 140	—	—	58	2.0	2 420	2 470
—	—	40	3.0	1 040	1 090	—	—	58	3.0	2 320	2 390
42	—	—	1.5	1 270	1 290	—	—	58	4.0	2 210	2 310
42	—	—	2.0	1 230	1 260	—	60	—	1.5	2 660	2 700
42	—	—	3.0	1 150	1 210	—	60	—	2.0	2 600	2 650
42	—	—	4.0	1 080	1 150	—	60	—	3.0	2 490	2 570
—	45	—	1.5	1 460	1 490	—	60	—	4.0	2 380	2 490
—	45	—	2.0	1 420	1 460	—	—	62	1.5	2 840	2 880
—	45	—	3.0	1 340	1 400	—	—	62	2.0	2 790	2 840
—	45	—	4.0	1 260	1 340	—	—	62	3.0	2 670	2 750
48	—	—	1.5	1 670	1 710	—	—	62	4.0	2 560	2 670
48	—	—	2.0	1 630	1 670	64	—	—	1.5	3 030	3 080

All dimensions in millimetres

NOTE: Nominal diameters extend up to 300mm (BS 3643:Part 3)

In the tolerancing system there are three classes of fits – Close, Medium and Free. Tolerance is fully designated by letter/number symbols, a number and capital letter specifying the tolerance grade and diameter of the nut, and a number and a lower case letter the tolerance grade and diameter of the bolt (see Table VI). Most production fasteners are based on medium class threads.

TABLE VI – COMPARISON OF TOLERANCE SPECIFICATIONS

ISO			BSF and BSW Designation		UNF and UNC Designation	
Class	Bolts	Nuts	Bolts	Nuts	Bolts	Nuts
Close	4h	5H	Close	Medium	3A	3B
Medium	6g	6H	Medium	Normal or Medium	2A	2B
Free	8g	7H	Free	Normal	1A	1B

TABLE VII–ISO METRIC HEXAGON PRECISION BOLTS, SCREWS & NUTS (BS 3692)

Nom size	Pitch of thread mm		Diameter of unthreaded shank mm		Width across flats mm		Width across corners mm	Depth of washer face mm	Radius under head mm		Height of head mm	Thickness of nut mm
	Coarse	Fine	Max	Min	Max	Min	Min		Max	Min	Nom	Nom
M3	0.50	–	3.00	2.86	5.50	5.38	6.08	0.1	0.30	0.10	2.00	2.40
M4	0.70	–	4.00	3.82	7.00	6.85	7.74	0.1	0.35	0.20	2.80	3.20
M5	0.80	–	5.00	4.82	8.00	7.85	8.87	0.2	0.35	0.20	3.50	4.00
M6	1.00	–	6.00	5.82	10.00	9.78	11.05	0.3	0.40	0.25	4.00	5.00
M8	1.25	1.00	8.00	7.78	13.00	12.73	14.38	0.4	0.60	0.40	5.50	6.50
M10	1.50	1.25	10.00	9.78	17.00	16.73	18.90	0.4	0.60	0.40	7.00	8.00
M12	1.75	1.25	12.00	11.73	19.00	18.67	21.10	0.4	1.10	0.60	8.00	10.00
*M14	2.00	1.50	14.00	13.73	22.00	21.67	24.49	0.4	1.10	0.60	9.00	11.00
M16	2.00	1.50	16.00	15.73	24.00	23.67	26.75	0.4	1.10	0.60	10.00	13.00
*M18	2.50	1.50	18.00	17.73	27.00	26.67	30.14	0.4	1.10	0.60	12.00	15.00
M20	2.50	1.50	20.00	19.67	30.00	29.67	33.53	0.4	1.20	0.80	13.00	16.00
*M22	2.50	1.50	22.00	21.67	32.00	31.61	35.72	0.4	1.20	0.80	14.00	18.00
M24	3.00	2.00	24.00	23.67	36.00	35.38	39.98	0.5	1.20	0.80	15.00	19.00
*M27	3.00	–	27.00	26.67	41.00	40.38	45.63	0.5	1.7	1.0	17.00	22.00
M30	3.50	–	30.00	29.67	46.00	45.38	51.28	0.5	1.7	1.0	19.00	24.00
*M33	3.50	–	33.00	32.61	50.00	49.38	55.80	0.5	1.7	1.0	21.00	26.00
M36	4.00	–	36.00	35.61	55.00	54.26	61.31	0.5	1.7	1.0	23.00	29.00
*M39	4.00	–	39.00	38.61	60.00	59.26	66.96	0.6	1.7	1.0	25.00	31.00
M42	4.50	–	42.00	41.61	65.00	64.26	72.61	0.6	1.8	1.2	26.00	34.00

**These are non-preferred diameters to be dispensed with wherever possible.*

ISO Metric Bolts, Screws and Nuts

Standard sizes for ISO Metric Hexagon bolts, screws and nuts are summarised in Table VII (according to BS3592).

The *length of thread* is determined as (2D + l), where D is the bolt diameter (mm) and l is given by:-

normal length up to 125mm, l = 6

normal length 126–200mm, l = 12

normal length over 200mm, l = 25

Recommended standard lengths to replace inch lengths are given in Table VIII.

TABLE VIII – RECOMMENDED METRIC LENGTHS TO REPLACE INCH LENGTHS

Metric length		Nearest inch length	
10mm	(0.3937)	3/8	(0.3750)
12mm	(0.4724)	½	(0.5000)
14mm	(0.5512)	9/16	(0.5625)
16mm	(0.6299)	5/8	(0.6250)
18mm	(0.7086)		
20mm	(0.7874)	¾	(0.7500)
22mm	(0.8661)	7/8	(0.8750)
25mm	(0.9842)	1	(1.0000)
28mm	(1.1023)	1.1/8	(1.1250)
30mm	(1.1811)	1¼	(1.2500)
35mm	(1.3779)	1.3/8	(1.3750)
40mm	(1.5748)	1½	(1.5000)
45mm	(1.7716)	1¾	(1.7500)
50mm	(1.9685)	2	(2.0000)
55mm	(2.1654)	2¼	(2.2500)
60mm	(2.3622)		
65mm	(2.5590)	2½	(2.5000)
70mm	(2.7559)	2¾	(2.7500)
75mm	(2.9527)	3	(3.0000)
80mm	(3.1496)	3¼	(3.2500)
85mm	(3.3464)		
90mm	(3.5433)	3½	(3.5000)
95mm	(3.7401)	3¾	(3.7500)
100mm	(3.9370)	4	(4.0000)
110mm	(4.3307)	4½	(4.5000)
120mm	(4.7244)	4¾	(4.7500)
130mm	(5.1181)	5	(5.0000)
140mm	(5.5118)	5½	(5.5000)
150mm	(5.9055)	6	(6.0000)
160mm	(6.2992)	6¼	(6.2500)
170mm	(6.6929)	6½	(6.5000)
180mm	(7.0866)	7	(7.0000)
190mm	(7.4803)	7½	(7.5000)
200mm	(7.8740)	8	(8.0000)

If lengths over 200 are required then increments of 20mm (0.7874) are the ISO preferred lengths.

Bolts or Screws

Certain *minimum* thread lengths are specified for bolts. Thread fasteners with less than this length are designated as screws.

Diameter	Minimum length to qualify as bolt	Diameter	Minimum length to ·ıalify as bolt
M5	20mm	M24	65mm
M8	25mm	M27	70mm
M10	30mm	M30	80mm
M12	35mm	M33	85mm
M14	40mm	M36	90mm
M16	45mm	M39	100mm
M18	50mm	M42	105mm
M20	55mm		
M22	60mm		

Head Sizes

Bolt head and nut dimensions are given in Table VII, whilst Tables IXA and IXB show a comparison between standard ISO metric sizes and those of Imperial and Unified bolts. It should be noted that a different spanner is required for each thread type (as well as thread size).

TABLE IXA – COMPARISON OF HEAD SIZES ACROSS FLATS (inch sizes in brackets)

ISO Metric			ISO Unified inch			BS Whitworth	
Dia(in)	Max	Min	Dia(in)	Max	Min	Max	Min
M6 (0.2362)	10.0 (0.3937)	9.78 (0.385)	¼	0.4375	0.4305	0.445	0.438
M8 (0.3149)	13.0 (0.5118)	12.73 (0.501)	5/16	0.5000	0.4930	0.525	0.518
M10 (0.3937)	17.0 (0.6692)	16.73 (0.658)	3/8	0.5625	0.5545	0.600	0.592
M12 (0.4724)	19.0 (0.7480)	18.67 (0.735)	7/16	0.6250	0.6170	0.710	0.702
			½	0.7500	0.7420	0.820	0.812
M16 (0.6299)	24.0 (0.9448)	23.67 (0.931)	5/8	0.9375	0.9295	1.010	1.000
M20 (0.7874)	30.0 (1.1811)	29.67 (1.168)	¾	1.1250	1.1150	1.200	1.190
M24 (0.9448)	36.0 (1.41732)	35.38 (1.393)	1	1.5000	1.4880	1.480	1.468
M30 (1.1811)	46.00 (1.8110)	45.38 (1.7866)	1.1/8	1.6875	1.6575	1.670	1.640
M36 (1.4173)	55.00 (2.1653)	54.26 (2.1362)	1¼	1.8750	1.8300	1.860	1.815
			1.3/8	2.0625	2.0175	2.050	2.005
M42 (1.6535)	65.00 (2.5590)	64.26 (2.5299)	1½	2.2500	2.2050	2.220	2.175

TABLE IXB

ISO Metric	Equivalent diameter in inches	Width across flats max in inches	Nearest equivalent inch size Diameter in	Width across flats (inches) UNC Normal in	UNC Heavy in	BA in
M 2.5	0.098	0.197	7BA (0.098)	–	–	0.172
M 3	0.118	0.216	6BA (0.110)	–	–	0.193
M 4	0.158	0.276	4BA (0.142)	–	–	0.248
M 5	0.197	0.315	2BA (0.185)	–	–	0.324
M 6	0.236	0.394	¼ (0.250)	0.438	–	–
M 8	0.315	0.512	5/16 (0.312)	0.500	–	–
M10	0.393	0.669	3/8 (0.375)	0.562	–	–
M12	0.472	0.748	½ (0.500)	0.750	0.875	–
M16	0.629	0.945	5/8 (0.625)	0.938	1.062	–
M20	0.787	1.181	¾ (0.750)	1.125	1.250	–

ISO metric pressed nuts (square and hexagon) differ in size from Table VII dimensions and are shown in Table X.

TABLE X – DIMENSIONS OF ISO METRIC PRESSED NUTS

Size	Width across flats Maximum	Minimum	Thickness Maximum	Minimum
M2.5	5.00	4.82	1.60	1.35
M3	5.50	5.32	1.60	1.35
M4	7.00	6.78	2.00	1.75
M5	8.00	7.78	2.50	2.25
M6	10.00	9.78	3.00	2.75
M8	13.00	12.73	4.00	3.70
M10	17.00	16.73	5.00	4.70

All dimensions in millimetres

UNIFIED THREADS

The Unified Screw thread form is shown in Fig 2. Basic sizes of Fine Thread Series (UNF) are given in Table XI, and sizes of Coarse Thread Series (UNC) are given in Table XII.

Three classes of bolts and nuts are provided.

Class 1A bolt, Class 1B nut – apply to the bulk of screw threads of ordinary commercial standard, corresponding to the 'free' class of BS84.

Class 2A bolt, Class 2B nut – apply to higher grades of interchangeable screw threads, corresponding to 'Medium' class of BS84.

Class 3A bolt, Class 3B nut – for screw threads requiring close fits, corresponding to 'Close' class of BS84. This class is normally only used where refined accuracy of pitch and thread form is particularly required.

ISO METRIC AND UNIFIED THREADS

ISO Metric screw threads are designated by 'M' followed by two numbers separated by 'x' – eg M6 x 1. The 'M' defines 'metric'. The first number is the basic thread diameter in millimetres. The second number is the thread pitch in millimetres. Thus M6 x 1 designates an ISO metric screw thread 6mm dia x 1mm pitch.

Unified screw threads (as specified in USA) are designated by nominal size (diameter), in inches separated by a – from the figure giving number of threads per inch: eg 7/8 – 10 designates a 7/8in nominal size screw with a pitch of 10 threads per inch. The full designation would also include the 'UN' designation – eg 7/8–10 UNC.

Equivalents can be calculated directly from either designation by direct conversion of the diameter sizes and finding the equivalent pitch.

For example 7/8 – 10 UNC becomes

diameter 7/8in = 22.2mm

10 threads per inch = pitch of 0.1in = 2.5mm

Thus equivalent is M22.2 x 2.5

TABLE XI – UNIFIED SCREW THREADS – BASIC SIZES
FINE THREAD SERIES UNF TO BRITISH STANDARD 1580

Designation	Pitch	Major diameter	Effective diameter *	Minor diameter		Approx. diameter of unthreaded shank on rolled thread screws
		Nut and bolt	Nut and bolt	Nut	Bolt (design size)	
	in	in	in	in	in	in
No.10 (0.190)–32UNF	0.03125	0.1900	0.1697	0.1562	0.1517	0.167
¼–28UNF	0.03571	0.2500	0.2268	0.2113	0.2062	0.224
5/16–24UNF	0.04167	0.3125	0.2854	0.2674	0.2614	0.282
3/8–24UNF	0.04167	0.3750	0.3479	0.3299	0.3239	0.345
7/16–20UNF	0.05000	0.4375	0.4050	0.3834	0.3762	0.401
½–20UNF	0.05000	0.5000	0.4675	0.4459	0.4387	0.464
9/16–18UNF	0.05556	0.5625	0.5264	0.5024	0.4943	0.523
5/8–18UNF	0.05556	0.6250	0.5889	0.5649	0.5568	0.585
¾–16UNF	0.06250	0.7500	0.7094	0.6823	0.6733	0.706
7/8–14 UNF	0.07143	0.8750	0.8286	0.7977	0.7874	0.825
1–12UNF	0.08333	1.0000	0.9459	0.9098	0.8978	0.942
1.1/8–12UNF	0.08333	1.1250	1.0709	1.0348	1.0228	
1¼–12UNF	0.08333	1.2500	1.1959	1.1598	1.1478	
1.3/8–12UNF	0.08333	1.3750	1.3209	1.2848	1.2728	
1½–12UNF	0.08333	1.5000	1.4459	1.4098	1.3978	

* Pitch diameter

Similarly, finding an equivalent to M6 x 1 thread in Unified thread
diameter 6mm = 0.236in = 15/64
1mm pitch = 0.039in = 25.6 threads per inch
Thus equivalent is 15/64 – 25.6

Note: It does not follow that equivalents determined on this basis represent available thread sizes, but such conversions enable the nearest available size to be selected.

TABLE XII – UNIFIED SCREW THREADS – BASIC SIZES
COARSE THREAD SERIES, UNC TO BRITISH STANDARD 1580

Designation	Pitch	Major diameter	Effective diameter *	Minor diameter		Approx. diameter of unthreaded shank on rolled thread screws
		Nut and bolt	Nut and bolt	Nut	Bolt (design size)	
	in	in	in	in	in	in
No.4 (0.112)–40UNC	0.02500	0.1120	0.0958	0.0849	0.0813	0.094
No.6 (0.138)–32UNC	0.03125	0.1380	0.1177	0.1042	0.0997	0.115
No.8 (0.164)–32UNC	0.03125	0.1640	0.1437	0.1302	0.1257	0.141
No.10 (0.190)–24UNC (non-preferred)	0.04167	0.1900	0.1629	0.1449	0.1389	0.160
¼–20UNC	0.05000	0.2500	0.2175	0.1959	0.1887	0.215
5/16–18UNC	0.05556	0.3125	0.2764	0.2524	0.2443	0.273
3/8–16UNC	0.06250	0.3750	0.3344	0.3073	0.2983	0.331
7/16–14UNC	0.07143	0.4375	0.3911	0.3602	0.3499	0.388
½–13UNC	0.07692	0.5000	0.4500	0.4167	0.4056	0.446
9/16–12UNC	0.08333	0.5625	0.5084	0.4723	0.4603	0.504
5/8–11UNC	0.09091	0.6250	0.5660	0.5266	0.5135	0.562
¾–10UNC	0.10000	0.7500	0.6850	0.6417	0.6273	0.681
7/8–9UNC	0.11111	0.8750	0.8028	0.7547	0.7387	0.798
1–8UNC	0.12500	1.0000	0.9188	0.8647	0.8466	0.914
1.1/8–7UNC	0.14286	1.1250	1.0322	0.9704	0.9497	
1¼–7UNC	0.14286	1.2500	1.1572	1.0954	1.0747	
1.3/8–6UNC	0.16667	1.3750	1.2667	1.1946	1.1705	
1½–6UNC	0.16667	1.5000	1.3917	1.3196	1.2955	
1¾–5UNC	0.20000	1.7500	1.6201	1.5335	1.5046	
2–4½UNC	0.22222	2.0000	1.8557	1.7594	1.7274	
2¼–4½UNC	0.22222	2.2500	2.1057	2.0094	1.9774	
2½–4UNC	0.25000	2.5000	2.3376	2.2294	2.1933	
2¾–4UNC	0.25000	2.7500	2.5876	2.4794	2.4433	
3–4UNC	0.25000	3.0000	2.8376	2.7294	2.6933	
3¼–4UNC	0.25000	3.2500	3.0876	2.9794	2.9433	
3½–4UNC	0.25000	3.5000	3.3376	3.2294	3.1933	
3¾–4UNC	0.25000	3.7500	3.5876	3.4794	3.4433	
4–4UNC	0.25000	4.0000	3.8376	3.7294	3.6933	

* Pitch diameter

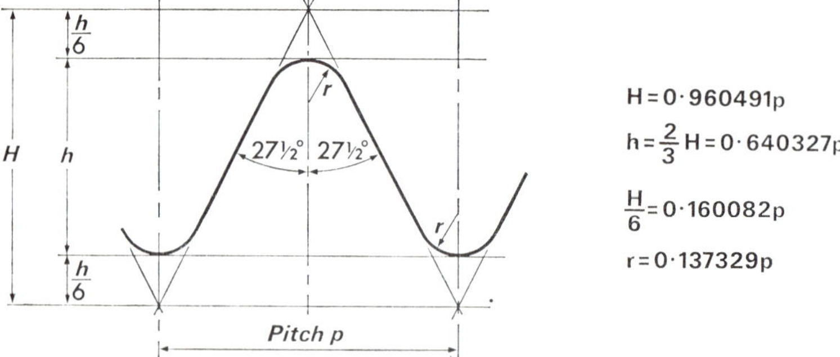

Fig 3 Whitworth screw thread.

BRITISH STANDARD WHITWORTH

The Whitworth thread form is shown in Fig 3. The coarse series is designated BSW and the fine series British Standard Fine (BSF). Both are now considered obsolete, although the BSW continues in constant use and demand. Dimensional data, as given in BS84:1956, are summarised in Tables XIII and XIV.

TABLE XIII

WHITWORTH FORM SCREW THREADS

COARSE THREAD SERIES, BSW TO BRITISH STANDARD 84:1956

Nominal size in.	Number of threads per inch	Pitch in.	Depth of thread in.	Major diameter in.	Effective diameter in.	Minor diameter in.	Approx. diameter of unthreaded shank on rolled thread screws in.
1/8	40	.02500	.0160	.1250	.1090	.0930	.107
3/16	24	.04167	.0267	.1875	.1608	.1341	.158
1/4	20	.05000	.0320	.2500	.2180	.1860	.215
5/16	18	.05556	.0356	.3125	.2769	.2413	.273
3/8	16	.06250	.0400	.3750	.3350	.2950	.331
7/16	14	.07143	.0457	.4375	.3918	.3461	.388
1/2	12	.08333	.0534	.5000	.4466	.3932	.442
9/16	12	.08333	.0534	.5625	.5091	.4557	.505
5/8	11	.09091	.0582	.6250	.5668	.5086	.563
11/16	11	.09091	.0582	.6875	.6293	.5711	.626
3/4	10	.10000	.0640	.7500	.6860	.6220	.682
7/8	9	.11111	.0711	.8750	.8039	.7328	.800
1	8	.12500	.0800	1.0000	.9200	.8400	.916
1 1/8	7	.14286	.0915	1.1250	1.0335	.9420	
1 1/4	7	.14286	.0915	1.2500	1.1585	1.0670	
1 1/2	6	.16667	.1067	1.5000	1.3933	1.2866	
1 3/4	5	.20000	.1281	1.7500	1.6219	1.4938	
2	4.5	.22222	.1423	2.0000	1.8577	1.7154	
2 1/4	4	.25000	.1601	2.2500	2.0899	1.9298	
2 1/2	4	.25000	.1601	2.5000	2.3399	2.1798	
2 3/4	3.5	.28571	.1830	2.7500	2.5670	2.3840	
3	3.5	.28571	.1830	3.0000	2.8170	2.6340	
3 1/4	3.25	.30769	.1970	3.2500	3.0530	2.8560	
3 1/2	3.25	.30769	.1970	3.5000	3.3030	3.1060	
3 3/4	3	.33333	.2134	3.7500	3.5366	3.3232	
4	3	.33333	.2134	4.0000	3.7866	3.5732	
4 1/2	2.875	.34783	.2227	4.5000	4.2773	4.0546	
5	2.75	.36364	.2328	5.0000	4.7672	4.5344	
5 1/2	2.625	.38095	.2439	5.5000	5.2561	5.0122	
6	2.5	.40000	.2561	6.0000	5.7439	5.4878	

TABLE XIV
WHITWORTH FORM SCREW THREADS
FINE THREAD SERIES, BSF TO BRITISH STANDARD 84:1956

Nominal Size	Number of threads per inch	Pitch	Depth of thread	Major diameter	Effective diameter	Minor diameter	Approx. diameter of unthreaded shank on rolled thread screws
in.		in.	in.	in.	in.	in.	in.
3/16	32	.03125	.0200	.1875	.1675	.1475	.165
7/32	28	.03571	.0229	.2188	.1959	.1730	.193
1/4	26	.03846	.0246	.2500	.2254	.2008	.222
9/32	26	.03846	.0246	.2812	.2566	.2320	.253
5/16	22	.04545	.0291	.3125	.2834	.2543	.280
3/8	20	.05000	.0320	.3750	.3430	.3110	.340
7/16	18	.05556	.0356	.4375	.4019	.3663	.398
1/2	16	.06250	.0400	.5000	.4600	.4200	.456
9/16	16	.06250	.0400	.5625	.5225	.4825	.519
5/8	14	.07143	.0457	.6250	.5793	.5336	.576
11/16	14	.07143	.0457	.6875	.6418	.5961	.638
3/4	12	.08333	.0534	.7500	.6966	.6432	.693
7/8	11	.09091	.0582	.8750	.8168	.7586	.813
1	10	.10000	.0640	1.0000	.9360	.8720	.932
1 1/8	9	.11111	.0711	1.1250	1.0539	.9828	
1 1/4	9	.11111	.0711	1.2500	1.1789	1.1078	
1 3/8	8	.12500	.0800	1.3750	1.2950	1.2150	
1 1/2	8	.12500	.0800	1.5000	1.4200	1.3400	
1 5/8	8	.12500	.0800	1.6250	1.5450	1.4650	
1 3/4	7	.14286	.0915	1.7500	1.6585	1.5670	
2	7	.14286	.0915	2.0000	1.9085	1.8170	
2 1/4	6	.16667	.1067	2.2500	2.1433	2.0366	
2 1/2	6	.16667	.1067	2.5000	2.3933	2.2866	
2 3/4	6	.16667	.1067	2.7500	2.6433	2.5366	
3	5	.20000	.1281	3.0000	2.8719	2.7438	
3 1/4	5	.20000	.1281	3.2500	3.1219	2.9938	
3 1/2	4.5	.22222	.1423	3.5000	3.3577	3.2154	
3 3/4	4.5	.22222	.1423	3.7500	3.6077	3.4654	
4	4.5	.22222	.1423	4.0000	3.8577	3.7154	
4 1/4	4	.25000	.1601	4.2500	4.0899	3.9298	

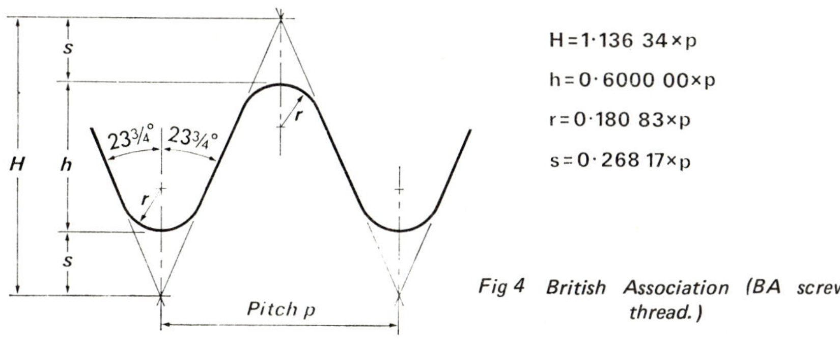

H=1·136 34×p

h=0·6000 00×p

r=0·180 83×p

s=0·268 17×p

Fig 4 British Association (BA screw thread.)

BRITISH ASSOCIATION (BA) SCREW THREADS

The full range extends from 0 (6mm) diameter to 16 (0.8mm) diameter, although sizes smaller than 10BA are little used. The BA thread form is illustrated in Fig 4. Apart from continued use in specific applications (particularly the electrical industry and model making), the original main use of BA screw threads was as a preferred choice to BSW or BSF for all screw sizes smaller than ¼in diameter.

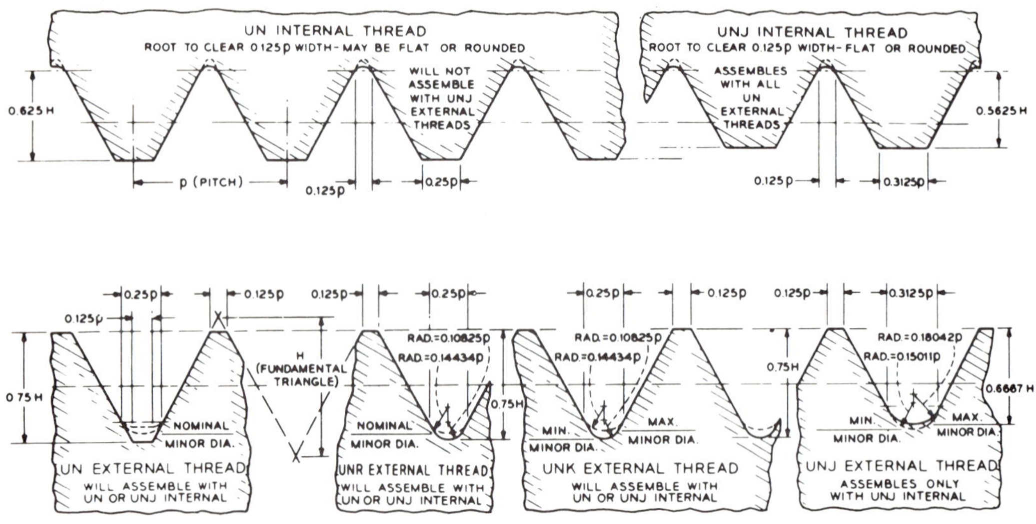

Fig 5 UN screw threads (nominal forms).

Dimensional data are summarised in Table XV. Provision is made for only one class of fit for nuts, and two classes for bolts (close for sizes 0–10BA and normal for sizes 0–16BA). Close class bolts are intended for precision parts subject to stress, no allowance being provided between maximum bolt and minimum nut. Normal class fits provide an allowance of 0.001in between maximum bolt and minimum nut.

TABLE XV

BRITISH ASSOCIATION (BA) SCREW THREADS TO BRITISH STANDARD 93:1951
APPROXIMATE INCH EQUIVALENTS OF METRIC VALUES

BA Number	Approx. pitch	Approx. No.of T.P.I	Depth of thread	Major diameter Bolt and nut	Effective diameter Bolt and nut	Minor diameter Bolt and nut	Approx. diameter of unthreaded shank on rolled thread screws	BA Number
	in.		in.	in.	in.	in.	in.	
0	.03937	25.4	.0236	.2362	.2126	.1890	.209	0
1	.03543	28.2	.0213	.2087	.1874	.1661	.184	1
2	.03189	31.4	.0191	.1850	.1659	.1468	.163	2
3	.02874	34.8	.0173	.1614	.1441	.1268	.141	3
4	.02598	38.5	.0156	.1417	.1262	.1106	.123	4
5	.02323	43.1	.0140	.1260	.1120	.0980	.109	5
6	.02087	47.9	.0126	.1102	.0976	.0850	.095	6
7	.01890	52.9	.0114	.0984	.0870	.0756	.085	7
8	.01693	59.1	.0102	.0866	.0764	.0661	.074	8
9	.01535	65.1	.0093	.0748	.0656	.0563	.064	9
10	.01378	72.6	.0083	.0669	.0587	.0504	.057	10
11	.01220	81.9	.0073	.0591	.0518	.0445		11
12	.01102	90.7	.0067	.0512	.0445	.0378		12
13	.00984	102.0	.0059	.0472	.0413	.0354		13
14	.00906	110.0	.0055	.0394	.0339	.0283		14
15	.00827	121.0	.0049	.0354	.0305	.0256		15
16	.00748	134.0	.0045	.0311	.0266	.0220		16

AMERICAN STANDARD THREADS

Unified and *American National* threads have substantially the same thread form with the same diameter and pitch and are therefore interchangeable, Fig 5. Both series are normally specified by diameter – threads per inch (eg 1/4 – 20) for sizes ¼in and upwards, and by numbers below ¼in diameter. American Standard ASA B1.1 – 1960 approved the designation of Unified Screw Thread Series from sizes 0–80 up to 6in diameter. In addition, the class of screw thread is normally designated, eg ¼ – 20UNC – 2A. Classes are as follows:-

Class 1A and Class 1B – with liberal clearances.

Class 2A and Class 2B – covering the bulk of production threaded fasteners. Class 2A can accommodate plated finishes.

Class 3A and Class 3B – for closer tolerances.

The above replace American National Class 1, Class 2, and Class 3, respectively.

Utilisation of the Unified threads is as follows:-

UNC – (coarse-thread series) for bulk production and general engineering applications, generally in lower tensile materials. (Table XVI).

UNF – (fine thread screws) for use where a coarse-series thread is not applicable, where greater tensile stress area is required in external threads, or where length of engagement or limited wall thickness demands a finer thread. (Table XVII).

UNEF – (extra-fine series) where very fine pitches are desirable, eg for short threaded lengths, thin walled tubing, ferrules, etc.

8UN – offering a compromise between UNF and UNC and consisting of a uniform pitch for larger diameters. (8 threads per inch). (Table XVIII).

12UN – uniform pitch screws for larger diameters (eg about 1.11/16in), requiring threads of medium fine pitch. (12 threads per inch).

16UN – uniform pitch screws for large diameters requiring fine pitch threads (16 threads per inch).

The above three (8UN, 12UN and 16UN) are collectively referred to as CONSTANT PITCH SERIES. Other threads per inch may be specified – eg 4UN, 6UN, 20UN, 28UN or 32UN.

Modified Thread Series – comprising Unified threads with the limits of the major or minor diameter modified to meet special requirements. These are thus basically UNC or UNF threads – eg ¼–20UNC – 2A MOD.

UNS – special combinations of diameter and pitch with pitch allowance based on a length of threaded engagement of 9 times the pitch. Pitch diameter limits are applicable to a length of engagement of from 5 to 15 times the pitch.

UNM – Unified Miniature Thread – see also Table XVIX.

UNR – specifies a Unified external thread having a mandatory radius root to provide higher resistance to fatigue than UN threads. This specification applies to external threads only as they are designed to assemble with UN internal threads.

TABLE XVI
AMERICAN COARSE THREAD SERIES
(ANSI B1.1)

Coarse Thread Series—UNC			
Nominal Size and Threads Per In.	Basic Pitch Dia *	Section at Minor Dia	Tensile Stress Area
	in.	in^2	in^2
— —	—	—	—
1 -64	0.0629	0.00218	0.00263
2 -56	0.0744	0.00310	0.00370
3 -48	0.0855	0.00406	0.00487
4 -40	0.0958	0.00496	0.00604
5 -40	0.1088	0.00672	0.00796
6 -32	0.1177	0.00745	0.00909
8 -32	0.1437	0.01196	0.0140
10 -24	0.1629	0.01450	0.0175
12 -24	0.1889	0.0206	0.0242
1/4 -20	0.2175	0.0269	0.0318
5/16-18	0.2764	0.0454	0.0524
3/8 -16	0.3344	0.0678	0.0775
7/16-14	0.3911	0.0933	0.1063
1/2 -13	0.4500	0.1257	0.1419
9/16-12	0.5084	0.162	0.182
5/8 -11	0.5660	0.202	0.226
3/4 -10	0.6850	0.302	0.334
7/8 - 9	0.8028	0.419	0.462
1 - 8	0.9188	0.551	0.606
1-1/8 - 7	1.0322	0.693	0.763
1-1/4 - 7	1.1572	0.890	0.969
1-3/8 - 6	1.2667	1.054	1.155
1-1/2 - 6	1.3917	1.294	1.405
— —	—	—	—
1-3/4 - 5	1.6201	1.74	1.90
— —	—	—	—
2 - 4 1/2	1.8557	2.30	2.50
2-1/4 - 4 1/2	2.1057	3.02	3.25
2-1/2 - 4	2.3376	3.72	4.00
2-3/4 - 4	2.5876	4.62	4.93
3 - 4	2.8376	5.62	5.97
3-1/4 - 4	3.0876	6.72	7.10
3-1/2 - 4	3.3376	7.92	8.33
3-3/4 - 4	3.5876	9.21	9.66
4 - 4	3.8376	10.61	11.08

TABLE XVII
AMERICAN FINE THREAD SERIES
(ANSI B1.1)

Fine Thread Series—UNF			
Nominal Size and Threads Per In.	Basic Pitch Dia *	Section at Minor Dia	Tensile Stress Area
	in.	in^2	in^2
0 -80	0.0519	0.00151	0.00180
1 -72	0.0640	0.00237	0.00278
2 -64	0.0759	0.00339	0.00394
3 -56	0.0874	0.00451	0.00523
4 -48	0.0985	0.00566	0.00661
5 -44	0.1102	0.00716	0.00830
6 -40	0.1218	0.00874	0.01015
8 -36	0.1460	0.01285	0.01474
10 -32	0.1697	0.0175	0.0200
12 -28	0.1928	0.0226	0.0258
1/4 -28	0.2268	0.0326	0.0364
5/16-24	0.2854	0.0524	0.0580
3/8 -24	0.3479	0.0809	0.0878
7/16-20	0.4050	0.1090	0.1187
1/2 -20	0.4675	0.1486	0.1599
9/16-18	0.5264	0.189	0.203
5/8 -18	0.5889	0.240	0.256
3/4 -16	0.7094	0.351	0.373
7/8 -14	0.8286	0.480	0.509
1 -12	0.9459	0.625	0.663
1-1/8 -12	1.0709	0.812	0.856
1-1/4 -12	1.1959	1.024	1.073
1-3/8 -12	1.3209	1.260	1.315
1-1/2 -12	1.4459	1.521	1.581

TABLE XVIII
AMERICAN CONSTANT PITCH SERIES
(ANSI B1.1)

8-Thread Series—8UN			
Nominal Size and Threads Per In.	Basic Pitch Dia V	Section at Minor Dia	Tensile Stress Area
	in.	in^2	in^2
1 -8	0.9188	0.551	0.606
1-1/8-8	1.0438	0.728	0.790
1-1/4-8	1.1688	0.929	1.000
1-3/8-8	1.2938	1.155	1.233
1-1/2-8	1.4188	1.405	1.492
1-5/8-8	1.5438	1.68	1.78
1-3/4-8	1.6688	1.98	2.08
1-7/8-8	1.7938	2.30	2.41
2 -8	1.9188	2.65	2.77
2-1/4-8	2.1688	3.42	3.56
2-1/2-8	2.4188	4.29	4.44
2-3/4-8	2.6688	5.26	5.43
3 -8	2.9188	6.32	6.51
3-1/4-8	3.1688	7.49	7.69
3-1/2-8	3.4188	8.75	8.96
3-3/4-8	3.6688	10.11	10.34
4 -8	3.9188	11.57	11.81

* Effective diameter

TABLE XIX

UNIFIED MINIATURE SCREW THREADS (Dimensions in inches)

	Size Designation			Pref. 0.30UNM	0.35UNM	0.40UNM	0.45UNM	0.50UNM	0.55UNM	0.60UNM	0.70UNM	0.80UNM	0.90UNM	1.00UNM	1.10UNM	1.20UNM	1.40UNM
Basic Dimensions	Basic major diameter –D			0.0118	0.0138	0.0157	0.0177	0.0197	0.0217	0.0236	0.0276	0.0315	0.0354	0.0394	0.0433	0.0472	0.0551
	Basic pitch diameter –E			0.0098	0.0115	0.0132	0.0152	0.0165	0.0185	0.0198	0.0231	0.0264	0.0297	0.0330	0.0369	0.0409	0.0474
	Basic minor diameter –K			0.0[illegible]5	0.0101	0.0117	0.0136	0.0146	0.0165	0.0175	0.0204	0.0233	0.0262	0.0291	0.0331	0.0370	0.0428
	Threads per inch			318	282	254	254	203	203	169	165	127	113	102	102	102	85
	Lead angle at basic P.D. –(λ)			5° 52′	5° 37′	5° 26′	4° 44′	5° 26′	4° 51′	5° 26′	5° 26′	5° 26′	5°26′	5°26′	4° 51′	4° 23′	4° 32′
External Threads	Major diameter		Max	0.0118	0.0138	0.0157	0.0177	0.0197	0.0217	0.0236	0.0276	0.0315	0.0354	0.0394	0.0433	0.0472	0.0551
			Min	0.0112	0.0131	0.0150	0.0170	0.0189	0.0208	0.0227	0.0265	0.0303	0.0341	0.0380	0.0419	0.0458	0.0535
	Pitch diameter		Max	0.0098	0.0115	0.0132	0.0152	0.0165	0.0185	0.0198	0.0231	0.0264	0.0297	0.0330	0.0369	0.0409	0.0474
			Min	0.0092	0.0109	0.0126	0.0145	0.0158	0.0177	0.0190	0.0222	0.0254	0.0287	0.0319	0.0358	0.0397	0.0462
	Minor diameter		Max	0.0080	0.0095	0.0110	0.0130	0.0138	0.0157	0.0165	0.0193	0.0220	0.0248	0.0276	0.0315	0.0354	0.0409
			Min	0.0072	0.0086	0.0101	0.0120	0.0127	0.146	0.0153	0.0179	0.0205	0.0231	0.0257	0.0296	0.0335	0.0387
Internal Threads	Minor diameter		Max	0.0100	0.0117	0.0134	0.0154	0.0166	0.0186	0.0198	0.0231	0.0263	0.0295	0.0327	0.0367	0.0406	0.0471
			Min	0.0085	0.0101	0.0117	0.0136	0.0146	0.0165	0.0175	0.0204	0.0233	0.0262	0.0291	0.0331	0.0370	0.0428
	Pitch diameter		Max	0.0104	0.0121	0.0138	0.0158	0.0172	0.0192	0.0206	0.0240	0.0273	0.0307	0.0341	0.0380	0.0420	0.0487
			Min	0.0098	0.0115	0.0132	0.0152	0.0165	0.0185	0.0198	0.0231	0.0264	0.0297	0.0330	0.0369	0.0409	0.0474
	Major diameter		Max	0.0129	0.0149	0.0170	0.0190	0.0212	0.0231	0.0254	0.0295	0.0337	0.0379	0.0420	0.0460	0.0499	0.0583
			Min	0.0120	0.0140	0.0160	0.0180	0.0200	0.0220	0.0240	0.0281	0.0321	0.0361	0.0401	0.0440	0.0480	0.0560
Hole Size Before Tapping	Length of Engagement	2/3D	Max	0.0095	0.0111	0.0127	0.0147	0.0158	0.0178	0.0190	0.0221	0.0252	0.0283	0.0314	0.0354	0.0393	0.0455
			Min	0.0089	0.0105	0.0121	0.0141	0.0150	0.0170	0.0181	0.0211	0.0241	0.0270	0.0300	0.0340	0.0379	0.0439
		2/3–1½D	Max	0.0100	0.0117	0.0134	0.0154	0.0166	0.0186	0.0198	0.0231	0.0263	0.0295	0.0327	0.0367	0.0406	0.0471
			Min	0.0093	0.0109	0.0125	0.0145	0.0156	0.0176	0.0187	0.0217	0.0248	0.0279	0.0309	0.0349	0.0388	0.0450
		1½–3D	Max	0.0104	0.0121	0.0138	0.0158	0.0171	0.0191	0.0204	0.0237	0.0270	0.0304	0.0337	0.0376	0.0415	0.0481
			Min	0.0096	0.0113	0.0130	0.0149	0.0161	0.0181	0.0193	0.0224	0.0256	0.0287	0.0319	0.0358	0.0397	0.0460

UNK – is another external thread having a radius root form with greater control than that used for UNR threads. They can be assembled with UN internal threads, but maximum performance is achieved mating with UNJ internal threads.

UNJ – external threads with enlarged radius root form and closer lead and angle tolerances for developing maximum strength in the most critical applications. These are designed for assembling only with UNJ internal threads.

MISCELLANEOUS SCREW THREADS

ACME

The *Acme* screw thread was developed to replace the square thread, used chiefly for providing transverse motions. It utilises a 29° angle between faces with the basic geometry

depth of thread = 0.5 x pitch (+ 0.010in)

width of root = 0.37 x pitch (–0.0052in)

width of crest = 0.37 x pitch

TABLE XX – ACME STANDARD SCREW THREADS

Pitch	No. of Threads per in.	Depth of Thread	Width at Top of Thread	Width at Bottom of Thread	Space at Top of Thread	Thickness at Root of Thread
2	1/2	1.010	.7414	.7362	1.2586	1.2637
1 7/8	8/15	0.9475	.6950	.6897	1.1799	1.1850
1 3/4	4/7	0.8850	.6487	.6435	1.1012	1.1064
1 5/8	8/13	0.8225	.6025	.5973	1.0226	1.0277
1 1/2	2/3	0.7600	.5560	.5508	0.9439	0.9491
1 7/16	16/23	0.7287	.5329	.5277	0.9046	0.9097
1 3/8	8/11	0.6975	.5097	.5045	0.8652	0.8704
1 5/16	16/21	0.6662	.4865	.4813	0.8259	0.8311
1 1/4	4/5	0.635	.4633	.4581	0.7866	0.7918
1 3/16	16/19	0.6037	.4402	.4350	0.7472	0.7525
1 1/8	8/9	0.5725	.4170	.4118	0.7079	0.7131
1 1/16	16/17	0.5412	.3938	.3886	0.6686	0.6739
1	1	0.510	.3707	.3655	0.6293	0.6345
15/16	1 1/15	0.4787	.3476	.3424	0.5898	0.5950
7/8	1 1/7	0.4475	.3243	.3191	0.5506	0.5558
13/16	1 3/13	0.4162	.3012	.2960	0.5112	0.5164
3/4	1 1/3	0.385	.2780	.2728	0.4720	0.4772
11/16	1 5/11	0.3537	.2548	.2496	0.4327	0.4379
2/3	1 1/2	0.3433	.2471	.2419	0.4194	0.4246
5/8	1 3/5	0.3225	.2316	.2264	0.3934	0.3986
9/16	1 7/9	0.2912	.2085	.2033	0.3539	0.3591
1/2	2	0.260	.1853	.1801	0.3147	0.3199
7/16	2 2/7	0.2287	.1622	.1570	0.2752	0.2804
2/5	2 1/2	0.210	.1482	.1430	0.2518	0.2570
3/8	2 2/3	0.1975	.1390	.1338	0.2359	0.2411
1/3	3	0.1766	.1235	.1183	0.2098	0.2150
5/16	3 1/5	0.1662	.1158	.1106	0.1966	0.2018
2/7	3 1/2	0.1528	.1059	.1007	0.1797	0.1849
1/4	4	0.1350	.0927	.0875	0.1573	0.1625
2/9	4 1/2	0.1211	.0824	.0772	0.1398	0.1450
1/5	5	0.110	.0741	.0689	0.1259	0.1311
3/16	5 1/3	0.1037	.0695	.0643	0.1179	0.1232
1/6	6	0.0933	.0617	.0565	0.1049	0.1101
1/7	7	0.0814	.0530	.0478	0.0899	0.0951
1/8	8	0.0725	.0463	.0411	0.0787	0.0839
1/9	9	0.0655	.0413	.0361	0.0699	0.0751
1/10	10	0.060	.0371	.0319	0.0629	0.0681
1/16	16	0.0412	.0232	.0180	0.0392	0.0444

There are also Stub forms of the Acme thread, where a coarse pitch is required with a shallow thread depth. Acme threads are not normally used for fasteners although they may be used in some types of unions (see Table XX). The true STUB THREAD has a 60° angle between faces, with lineated crests and roots. Basic height is 0.433 x pitch and basic thickness 0.5 x pitch.

CYCLE THREADS

The British Standard Cycle (BSC) is a fine thread with a 60° angle between faces and rounded crests and roots — Fig 6. Diameter sizes are based on a match with swg wire sizes — see Table XXI.

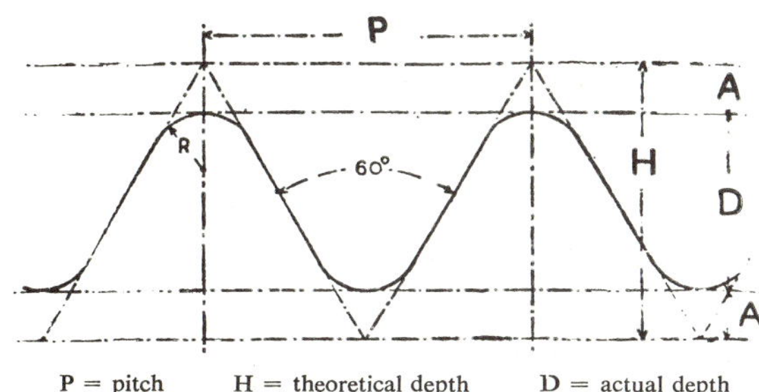

Fig 6 British cycle thread.

TABLE XXI – BRITISH STANDARD CYCLE THREADS
Angle of Thread = 60° included

Size	Diameter inches	Threads per inch	Pitch inches	Core Diameter (ins)	Tapping Size Drill
swg:					
17	0.0560	62	0.0161	0.0388	61
16	0.0640	62	0.0161	0.0468	56
15	0.0720	62	0.0161	0.0548	54
14	0.0800	62	0.0161	0.0628	1/16
13	0.0920	56	0.0178	0.0730	49
12	0.0104	44	0.0227	0.0798	47
inches					
1/8	0.1250	40	0.025	0.0984	2.5mm
0.154	0.1540	40	0.025	0.1274	30
0.175	0.1750	32	0.03125	0.1417	27
3/16	0.1875	32	0.03125	0.1542	23
¼	0.2500	26	0.0384	0.2090	4
0.266	0.2660	26	0.0384	0.2250	1
0.281	0.2810	26	0.0384	0.2400	C
5/16	0.3125	26	0.0384	0.2715	1
3/8	0.3750	26	0.0384	0.3340	8.5mm
9/16	0.5625	20	0.0500	0.5092	13mm
1	1.0000	26	0.0384	0.9590	24.5mm
1.29	1.2900	24	0.04167	1.2456	1¼
1.37	1.3700	24	0.04167	1.3256	1.21/64
1.7/16	1.4375	24	0.04167	1.3931	35.5mm
1½	1.5000	24	0.04167	1.4556	37mm

SPARK PLUG THREADS

Spark plug threads in use include the following types:-

18mm dia x 1.5mm pitch
14mm dia x 1.25mm pitch — metric threads
12mm dia x 1.25mm pitch
10mm dia x 1.0mm pitch

Note: 14mm x 1.25mm is the International standard for sparking plugs.

½in American NTPT (and other pipe threads)
7/8in x 18 tpi SAE
5/8in x 24 tpi SAE
1/4in x 24 tpi SAE

BUTTRESS THREADS

Buttress threads were originally developed with a 45° slope and further modified for specific applications. They are designed to withstand high stresses in the direction parallel to the axis of the screw. Crests and roots of internal and external threads may be truncated or rounded. Standardised forms are given in the American ASA B1.9–1953 specifications.

MACHINE SCREW THREADS

In general, coarse threads (BSW, UNC or M Coarse) are used as machine screws for general engineering applications requiring high assembly strength, and fine threads (BA, BSF and UNF) for more delicate assemblies or where short lengths of thread engagement are involved, particularly in brittle materials. Choice of thread has largely become traditional or accepted practice in specific industries in the United Kingdom, eg

Automobile engineering – UNC and UNF.

Domestic appliances – BA and BSW.

Garden tools and agricultural equipment – BSW and BSF.

Gas cookers – BSW

Electrical switchgear – BA

Radio and TV sets – BA.

Steel shelving and office equipment – BSW.

Sewing machines – UNC and UNF.

Telecommunications equipment – BA.

Bolt and Screw Design Parameters

Standard proportions of threaded bolts and screws are normally based on a shank diameter equal to the full or normal diameter of the thread. There are, however, occasions when an undersize or oversize shank diameter may be preferred.

An *undersize* shank diameter can offer advantages under the following headings:-

(i) Reduction in fatigue under conditions of severe dynamic loading.

(ii) Ability to accommodate a certain degree of misalignment due to the greater flexibility of the shank.

(iii) Installations where the bolt is subject to bending loads: bending stresses are transferred from the thread area to the shank.

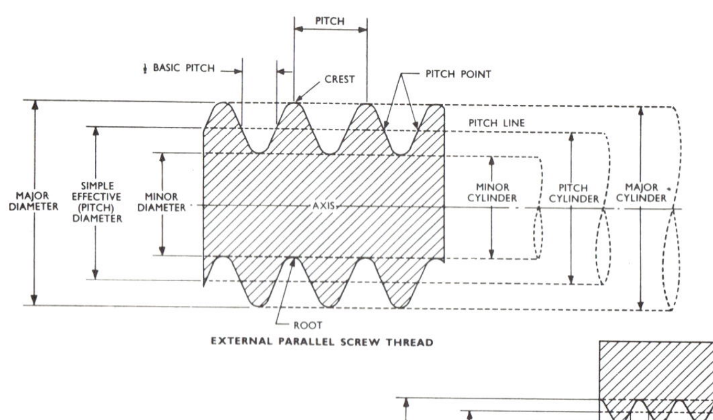

Fig 1 Screw turned definitions.

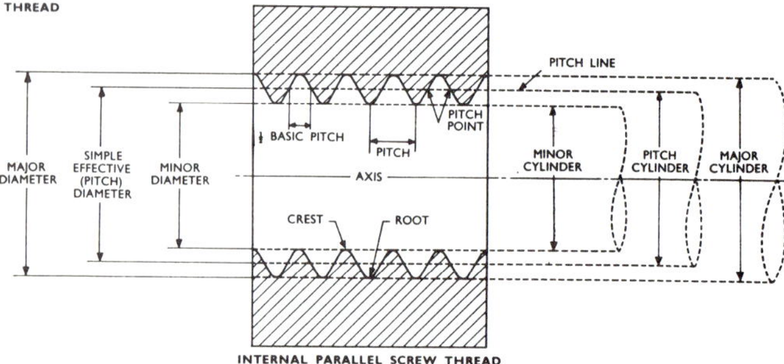

The only standard for undersize bolts is that for aerospace engine bolts where the shank diameter is specified as equal to the pitch diameter of the thread.

An *oversize* shank diameter may be specified where a closer fitting bolt is needed in an already-formed hole or to eliminate a sloppy fit in a worn hole. Shank sizes for oversize shank bolts are usually 0.4 to 0.8mm (1/64in to 1/32in) greater than the normal bolt size (thread overall diameter).

Limitations of Thread Forms

Thread forms with sharp edges and non-continuous geometry can be subject to high stress concentrations at corners and discontinuities, considerably reducing their fatigue performance under dynamic loading. Continuity can be obtained by using a radiused rather than a flat root – Fig 2. Optimum form is a continuous (uniform) radius and the larger the radius the better the fatigue strength of the thread form. MIL aerospace standards accommodate two uniformly radiused threads, MIL-5-8879 with 75% depth radius being mandatory on high strength structural bolts. It should be noted that a radiused root does not appreciably affect the fit of nuts – eg the 75% depth radius thread of MIL-5-8879 can be used with conventional nuts.

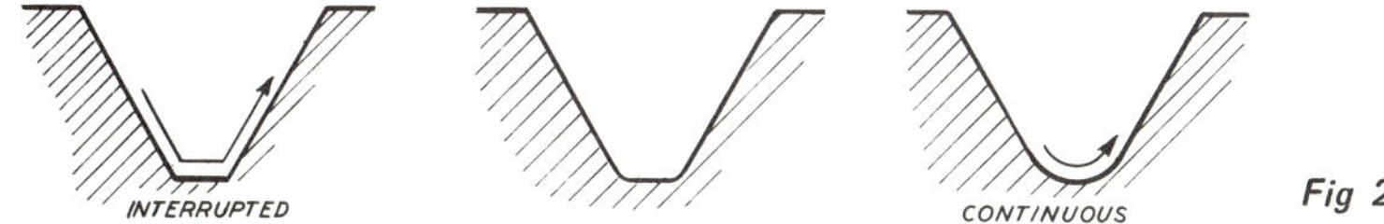

Fig 2

Even larger root radii may be used on notch-sensitive bolt materials – eg beryllium and 300 000lb/in^2 steel aerospace bolts. The 55% radius form shown in Fig 3 is designed for maximum fatigue performance with these materials and has virtually the same thread-stripping strength as a conventional Unified thread form.

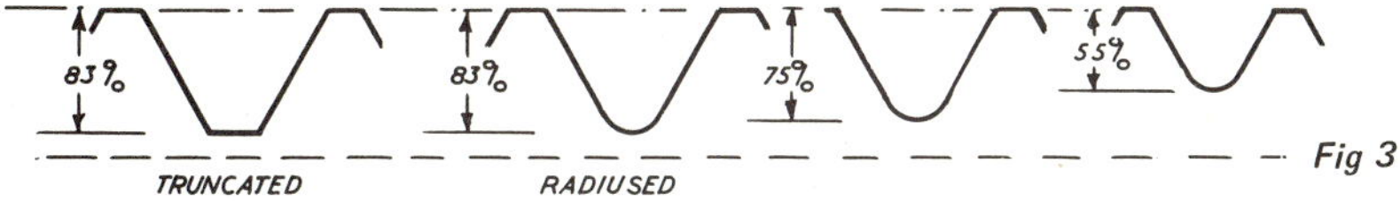

Fig 3

Head Form

Localised high stresses will also be created at sharp corners between the head and shank of a bolt – Fig 4. Again these can be relieved by radiusing to produce a fillet. Two forms of filleting are used on high-strength bolts – a uniform fillet and a two-radius fillet where the smaller radius occurs under the head, blending into a larger fillet running to the shank. The latter geometry yields the higher fatigue strength. Further improvement in fatigue strength can result from cold working the fillet to induce residual compensating stresses in the fillet area. Recommended minimum radii for bolt heads is not less than 0.08D for small bolts (up to 19mm or ¾in diameter); and 0.1D for larger bolts.

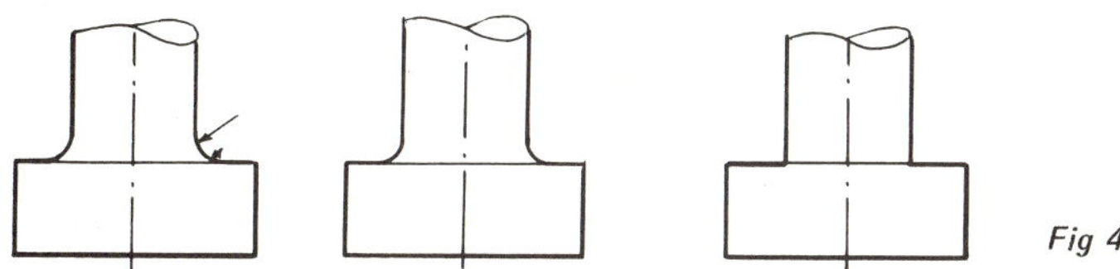

Fig 4

Commercial quality bolts may be radiused by 'necking' or undercutting.

Bolt Thread Relief

Bolt thread relief may be used in the case of bolts (and studs) subject to high dynamic loads to improve fatigue performance, and also for stress relief on bolts subject to bending loads. Optimum proportions for such reliefs are shown in Fig 5, ie

(i) Radius of relief 0.2D
(ii) Length of relief 0.5D
(iii) Depth of relief equal to or slightly less than the root diameter of the thread.

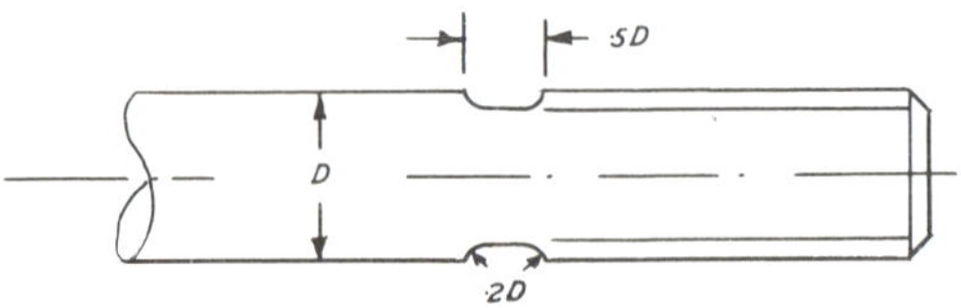

Fig 5

Bolt Body Relief

Relieved body bolts are used for locating or centring members where shear and bending forces are present. In such cases the minimum shank area is determined by tensile strength requirements. The shoulder areas which then act as guide surfaces should be separated from both head and thread – see Fig 6.

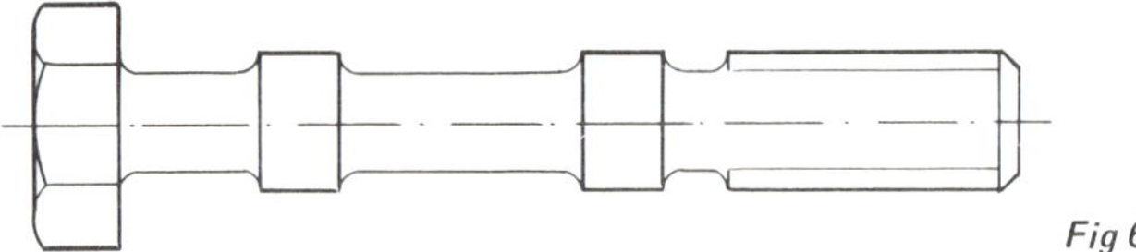

Fig 6

Bolt Proportions

For optimum performance in a bolted-up assembly the unengaged length of thread should be at least equal to one bolt diameter – see Fig 7. In the case of studs under dynamic loading, fatigue strength can be greatly improved by grooving for stress relief under a locating shoulder, as shown. The stress reducing fillet in this case should extend beyond the first mating thread.

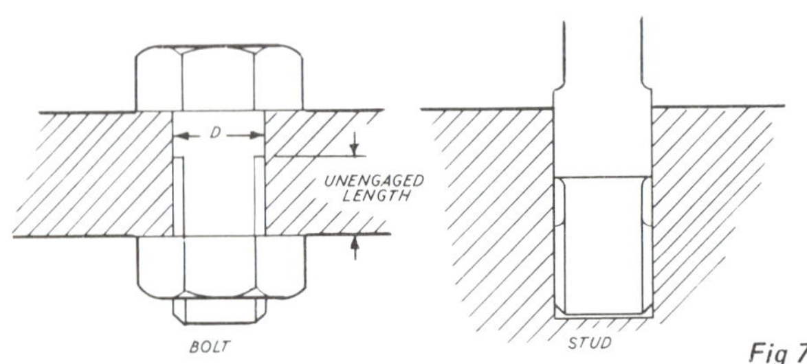

Fig 7

Fine versus Coarse Threads

Theoretically fine threads have a greater stress area and thus higher strength than coarse threads of the same diameter. However, this is not always realisable in practice since thread stripping is more likely on a fine thread than a coarse one. This is because load and stress considerations are lower in a coarse thread and flank engagement is deeper.

For a given torque, higher clamping loads can be achieved with fine threads, and the thread once tightened has less tendency to vibrate loose than a coarse one. Final tightening torque is also more readily controlled (ie less critical). Assembly time is slightly greater than for a coarse thread fastener because of the greater number of turns to tighten.

Coarse threads are generally more robust and less liable to damage, as well as being easier to assemble. They are also a better choice where plated bolts are used.

Clamping Loads

Basically the clamping load of a bolted assembly must be greater than that of subsequent external loads if the assembly is to remain rigid. This means that the tension or preload produced in the bolt by tightening must at least equal the external load. The greater the excess of preload over external load, the greater the resistance to fatigue.

Preloading also provides a locking effect to resist unscrewing due to the friction between bolt or nut and the parent materials. Under dynamic loading conditions, however, additional loads are transmitted to the fastener which can modify its performance. Thus any dynamic loading (eg vibration) which tends to provide movement of the fastener relative to the members it is clamping produces a loss of effective tension in the bolt and a reduction in clamping load. This in turn reduces the locking friction, which can result in the fastener loosening – a condition which can progressively worsen. In such cases where sufficient preload cannot be induced (eg due to lack of bolt strength), locking nuts or some form of thread locking will need to be used to maintain a tight fastening.

There is also a special case of frictional grip used in friction grip bolting which is largely replacing riveting for structural steelwork assemblies. Here high strength bolts are used, tightened to very large preloads and thus applying very high clamping pressures. Working loads are then carried in frictional shear.

See also – *High Strength Friction Grip Bolting, Thread-Locking Systems, Torque-Tension Relationship for Bolts.*

MANUFACTURE

External threads are formed by machining (thread cutting or thread grinding) or rolling. Nut blanks are made by cold forming, hot forming, cold punching or milling from bar stock. Blanks are then tapped in automatic machines for sizes up to about 20mm (¾in). Larger blanks are tapped or milled in hand-fed machines. Additional operations such as drilling, sawing, punching, pressing, etc, as well as thread tapping, may also be included in the production of special nuts.

Operations which may be used in the production of bolts, screws and nuts include drilling, sawing, slotting, shaping, trimming, milling, grinding, turning, pointing, polishing and plating, as well as heat treatment.

THREAD-FIT FAULTS AND CAUSES

Problem	Possible Cause
Thread will not start	(i) Mating parts not same class or size of thread. (ii) Distorted or burred over lead thread. (iii) Damaged or indented lead thread.
Loose or sloppy fit	(i) Mating parts not same class or size of thread. (ii) Undersize pitch diameter. (iii) Pitch diameter of hole is oversize. (iv) Tapered thread.
Galled threads	(i) Burrs. (ii) Excessive lead errors. (iii) Material prone to galling – eg not properly plated. (iv) Poor surface finish on threads. (v) Lapped threads.
Low torque when used with locknut	(i) Unsatisfactory type of nut. (ii) Pitch diameter of bolt is undersize.
Excessive torque when used with locknut	(i) Unsatisfactory type of nut (ii) Pitch diameter of bolt is oversize.
Bolts loosen in tapped hole	(i) Pitch diameter of external thread is too small. (ii) Internal hole oversize. (iii) Incorrect preload. (iv) Tapered threads.
Binding in a tapped hole	(i) Pitch diameter or major diameter oversize. (ii) Excessive lead. (iii) Improper thread form. (iv) Tapered thread.

Heat Treatment

Any of seven different heat treatment methods may be used to enhance the properties of steel fasteners. These are:-

Quenching – to refine the structure of steel and to harden it to a consistent quality. Quenching is normally applied to bolts, screws, pins and washers, and to a lesser extent to medium carbon nuts.

Annealing – to soften the steel. Primarily applied to studs which are to be cold-hardened, but also to special bolts to improve their ductility. Various degrees of softness can be produced by annealing.

Tempering – to control the properties of the steel, and also to produce a black corrosion-resistant surface (by cooling in soluble oil).

Stress Relieving – to relieve high stresses generated during cold forming, again giving a black corrosion-resistant coating.

Carburising or Case Hardening – to produce a hard, wear-resistant surface with a softer interior. Commonly applied to bolts, pins, nuts where resistance to wear is important.

Cyanide Hardening – producing an extremely hard surface of shallow depth for high wear resistance, etc, on bolts, pins and nuts.

Dry Cyaniding – an alternative case-hardening process for bolts, pins, etc.

ROLLED versus CUT THREADS

Thread cutting or grooving by machine produces cuts across the grain lines of the metal, resulting in a more uniform distribution of stress when the threads are loaded in tension or compression. Thread rolling produces plastic deformation of the metal with the grain generally following the contour of the thread. There is also a compacting of the grain at the root of the thread, resulting in increased strength at this critical area because of compression face stressing. A rolled thread, therefore, is inherently stronger than a cut thread, especially under conditions of dynamic loading. Other advantages offered by thread rolling are:-

(i) Better surface finish – reducing the incidence of stress raisers.

(ii) Complete absence of chips or swarf.

(iii) Faster production times.

(iv) Suitability for many materials which are difficult to machine economically.

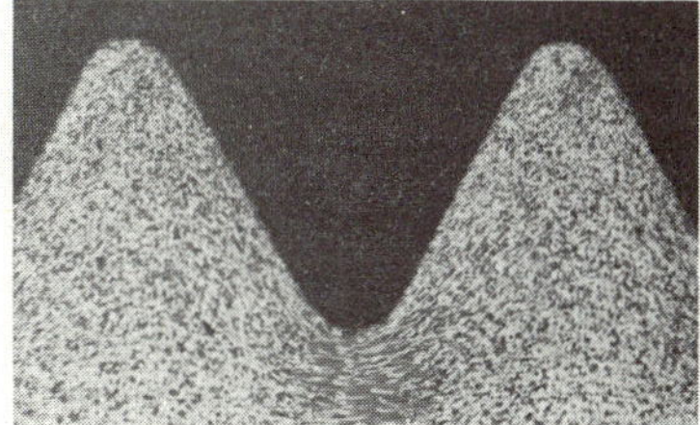

Fig 8 Grain pattern of cut and/or ground thread (left), and rolled thread (right).

The chief limitation with thread rolling is that the material to be formed must be sufficiently ductile to permit plastic deformation without cracking, incipient failure or embrittlement. This eliminates thread rolling as a possibility in such metals as cast iron, some free-cutting steels, certain non-ferrous alloys, and (generally) high-strength alloys with a tensile strength in excess of 100bar (64tonf/in^2). There are also some limitations on the thread forms which can be rolled successfully. Standard parallel threads present no particular problem, but taper threads are generally limited to a maximum taper of 1 in 16.

Short threads are generally formed by 'plunge' rolling. Longer threads are formed by through-rolling. Threads may be rolled after heat treatment, when the full benefits of compressive re-stressing are realised. If heat treatment is applied after thread rolling, then much of the advantage of rolling is lost because heat treatment will produce stress relief.

STRENGTH OF SCREW THREADS

Bolts or screws are normally stressed under tension when tightened. With increasing tensile loading ultimate failure will occur either by fracture of the bolt or by the thread stripping in either the bolt or the nut.

The fracture strength of bolts and screws can be determined with reasonable accuracy on the basis of *stress area loading,* the stress area being the area of a circle with a diameter equal to the mean of the core diameter and effective diameter, ie

$$\text{Bolt tensile stress area} = \frac{\pi}{4}\left(\frac{\text{effective diameter} + \text{minimum diameter}}{2}\right)^2$$

Thus it follows that the tensile loading of the bolt to cause fracture is

Load to fracture = bolt tensile stress area x UTS

$$= 0.7854\ \text{UTS}\left(\frac{\text{effective diameter} + \text{minor diameter}}{2}\right)^2$$

when UTS = ultimate tensile stress of the bolt material

Similarly, the load to cause the bolt material to be stressed to the yield point can be found by multiplying the yield point stress by UTS.

Thread stripping stresses can be determined by relationship to the *basic shear area.* This can be obtained from the following formulas:-

Bolts: basic shear area $= \pi\ L_e \cdot K_n{}^{½} + \dfrac{0.577\ (E_s - K_n)}{P}$

Nuts: basic shear area $= \pi\ L_e \cdot D_s{}^{½} + \dfrac{0.577\ (D_s - E_n)}{P}$

where L_e = axial length of thread engagement
K_n = nut minor diameter
E_s = bolt thread effective diameter
D_s = bolt thread major diameter
E_n = nut thread effective diameter
P = pitch

The stripping strength of a bolt or nut can thus be assessed as

Stripping strength = basic shear area x S_s

where S_s = shear strength of bolt or nut material.

For most practical purposes the shear strength (S_s) is of the order of 0.7 x UTS, so that the formula can be rewritten:

Stripping strength = 0.7 x basic shear area x UTS

This is generally applicable, and gives reasonably close results, for thick walled internally threaded components (eg bearing nuts). It is less accurate for standard or thin nuts, or for bolt threads, where the stripping strength is calculated as

Stripping strength = K x 0.7 x basic shear area x S_s

where K is an empirical constant less than unity.

Typical values for K generally lie between 0.6 and 0.8

In practice, a rather more realistic empirical formula (particularly for calculating the stripping loads on bolt threads) is:

$$1/L = 0.0000027 + \left(\frac{1.050}{Asn\ UTS_n}\right) + \left(\frac{1.190}{Asb\ UTS_b}\right)$$

where L = axial load at which stripping occurs, Kgf

Asn = basic shear area of nut, mm^2

Asb = basic shear area of bolt, mm^2

UTS_n = ultimate tensile stress of nut material, Kgf/mm^2

UTS_b = ultimate tensile stress of bolt material, Kgf/mm^2

Heads and Points

The common form of head for bolts and screws is a hexagon, although square and round heads are also standard. The standard square head offers a larger bearing surface for wrench tightening, with sharp corners for positive grip. A round head is normally associated with a countersunk square shoulder to provide a locking feature, but may have a plain shoulder. There are also various forms of round head styles with multiple projections or lugs for *welding screws* (ie screws designed to be permanently mounted by welding in position), the lugs making them suitable for projection welding.

Examples of basic head forms are shown in Fig 1.

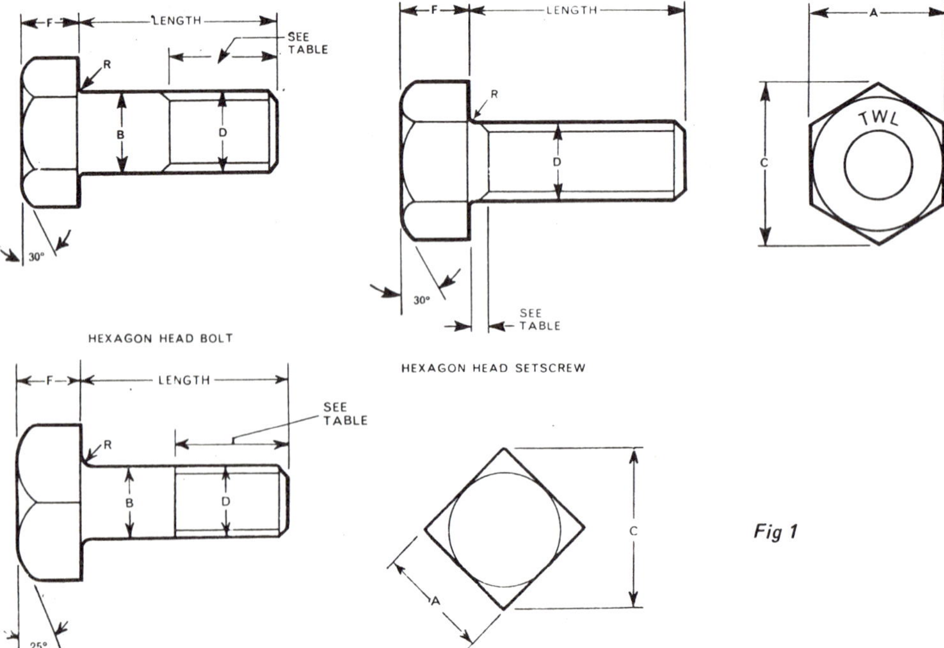

Fig 1

HEADS AND POINTS

Variations on the hexagon head are shown in Fig 2. The undercut head (also known as a finished and waster faced head) features a washer surface cut in the underside of the head with a diameter slightly less than the width across flats. The purpose of this is to prevent possible disfigurement of the surface by the sharp corners of the hexagon when the head is tightened in place. The washer base also applies equal pressure over the seating surface.

The washer-head hexagon has a larger washer surface on the base of the head, again serving to protect the surface against disfigurement, with a larger seating area than the undercut head. It also has a neat appearance. The base of the washer surface may be flat or undercut. The indented hexagon head is used for general identification. Specific grade or property markings for identification purposes are normally symbols.

Full bearing | Washer faced (undercut) | Double chamfered | Washer head

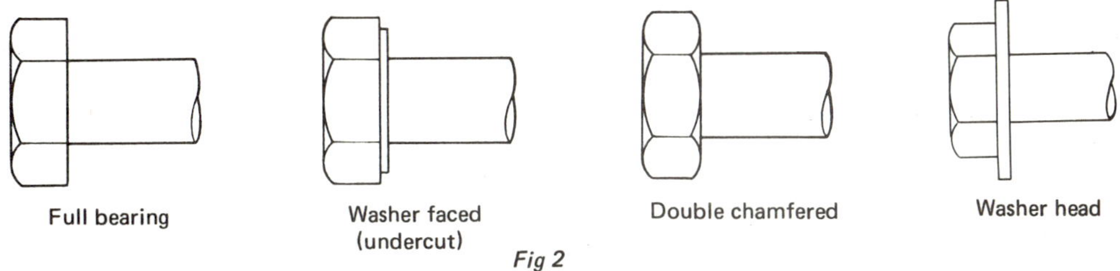

Fig 2

Typical markings for an ISO metric bolt head, and corresponding markings for a nut, are shown in Fig 3. On the bolt head, the letter 'M' designates an ISO metric thread. The grade mark is given as a double number without the decimal point. The manufacturer's name or mark may also appear. The nut marking is based on a dot and dash. The dot indicates 'twelve o'clock' and the position of the dash around the clock face the corresponding grade number.

Head Marking

This mark indicates that bolts are to ISO metric dimensions including ISO metric threads.

This mark indicates that bolts are to ISO grade 8.8 steel and replaces previous mark 8 G used on metric bolts to DIN standards

Nut Marking

This marking is in the form of a code symbol based on a clock face with a single dot indicating 12 o'clock. The second mark, a bar, indicates the grade, i.e. in the case of Grade 8 nut the bar is at the eight o'clock position on the top of the nut. The marks on nuts are indented.

(Tensile Strength min. 80 kgf/mm^2
Yield strength min. 64/kgf/mm^2.)

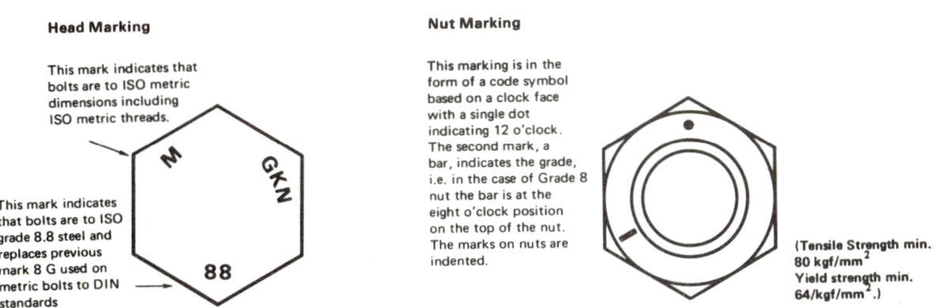

Fig 3 Markings for ISO metric bolt heads and nuts.

Grade markings for (American) ASTM and SAE steel bolts and screws are shown in **Fig 4**. These identify both the specification and the material. ISO grade markings are given by number only. Other American grade markings based on symbols are summarised in Table I.

Variations on the hexagon head for wrenching (spanner-tightening — include the double-hex (8-point), 12-point and spline. All three were developed for the aerospace industry and are applicable to high strength bolts requiring high tightening torques and to shear bolts. The double-hex and 12-point heads provide an appreciably higher wrenching grip than the comparative hexagon head. The spline head permits even higher torque values to be achieved without slipping, the splines providing right angle contact with the wrench surfaces.

Identification Grade Mark	Specification	Fastener Description	Material
	SAE J429 Grade 5 ASTM A449	Bolts, Screws, Studs	Medium carbon steel, quenched and tempered
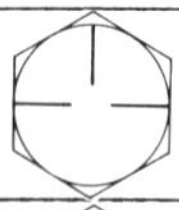	SAE J429 Grade 5.1	Sems	Low or medium carbon steel, quenched and tempered
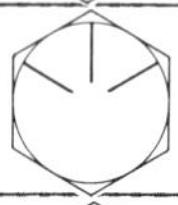	SAE J429 Grade 5.2	Bolts, Screws, Studs	Low carbon martensitic steel, quenched and tempered
 A325	ASTM A325 Type 2	High Strength Structural Bolts	Low carbon martensitic steel, quenched and tempered
	SAE J429 Grade 7	Bolts, Screws	Medium carbon alloy steel, quenched and tempered
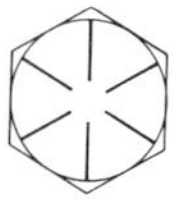	SAE J429 Grade 8 ASTM A354 Grade BD	Bolts, Screws Studs	Medium carbon alloy steel, quenched and tempered Alloy steel, quenched and tempered

Fig 4

TABLE I – ASTM AND ISO GRADE MARKINGS

Mark	Specification	Material
B5	ASTM A193 Grade B5	AISI 501
B6	ASTM A193 Grade B6	AISI 410
B7	ASTM A193 Grade 7	AISI 4140, 4142 or 4145
B8	ASTM A193 Grade 8	AISI 304
B8C	ASTM A193 Grade B8C	AISI 347
B8M	ASTM A193 Grade B8M	AISI 316
B8T	ASTM A193 Grade B8T	AISI 321
B8	ASTM A193 Grade B8	AISI 304 strain hardened
B8C	ASTM A193 Grade B8C	AISI 347 strain hardened
B8M	ASTM A193 Grade B8M	AISI 316 strain hardened
B8T	ASTM A193 Grade B8T	AISI 321 strain hardened
B16	ASTM A193 Grade 16	Chrome-molybdenum-vanadium alloy steel
L7	ASTM A320 Grade L7	AISI 4140, 4142 or 4145
L7A	ASTM A320 Grade L7A	AISI 4037
L7B	ASTM A320 Grade L7B	AISI 4137
L7C	ASTM A320 Grade L7C	AISI 8740
L43	ASTM A320 Grade L43	AISI 4340
A325	ASTM A325 Type 1	Medium carbon steel
A325	ASTM A325, Type 3	Corrosion resistant steel
BB	ASTM A354 Grade BB	Alloy steel
BC	ASTM A354 Grade BC	Alloy steel
A490	ASTM A490	Alloy steel

Coming into general use are variations on the multiple-hex head based on hex-lobular forms rather than sharply defined points – see Fig 5.

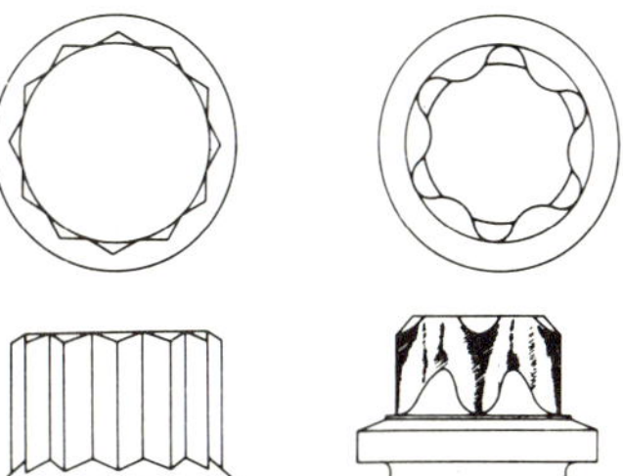

Fig 5 Double hexagon (12 point) and hex-lobular heads for external wrenching.

All the aforementioned can be classified as designs for external wrenching, the same configurations being applicable to bolt heads and nuts.

The other main classification of heads is that designed for internal wrenching, and in this case applicable only to bolt or screw heads. The slotted head is the simplest form, but recessed heads are now preferred for many applications, particularly for the internal wrenching of high-strength fasteners and/or where high speed or automated driving is required.

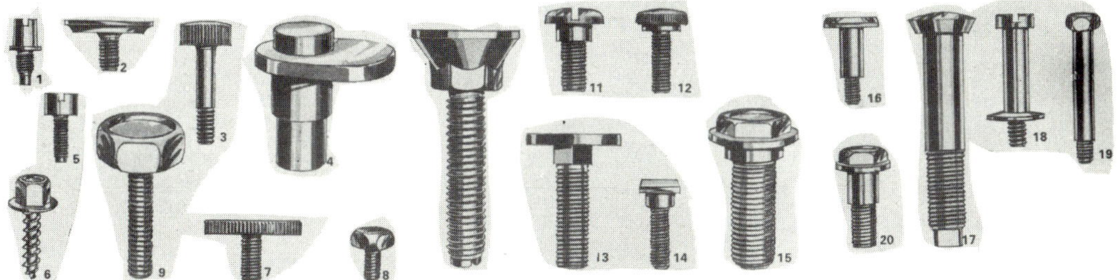

Special screw heads by Linread-Phipard.

1. High carbon steel slotted deep cheese head with integral flange, dog pointed. Heat treated to V quality.
2. NB4 aluminium special head screw.
3. Mild steel cheese head bolt with rolled knurl.
4. Mild steel forged eccentric head.
5. Mild steel deep cheese head screw. Dog pointed.
6. Mild steel washer hexagon head (Hi-Hex).
7. Mild steel large diameter flat head with rolled knurl.
8. Carbon steel forged square head with Taptite thread.
9. Mild steel forged indented hexagon head levelling screw.
10. Mild steel levelling foot with hexagon collar for adjustment.
11. Mild steel slotted pan head collar screw.
12. Mild steel collar screw with forged knurl.
13. High carbon steel square neck bolt, heat treated to P quality.
14. Phosphor bronze square head collar screw.
15. Carbon steel washer hexagon head collar screw.
16. Mild steel Pozi-flat head shoulder screw.
17. Mild steel special head shouldered bolt with spigot end.
18. Mild steel cheese head shoulder screw – captive Belville washer.
19. Steel BS 3111 Type 2 indented hexagon head shoulder bolt, heat treated to W quality.
20. Mild steel washer hexagon head shoulder bolt.

MACHINE SCREW HEADS

Head forms used on machine screws are illustrated in Fig 6. The majority (but not all) may be associated with a slotted head or a recessed head of various forms (eg Phillips, Pozidriv, etc). The most common forms of head are:-

Hexagon – general purpose head for spanner tightening.

Cheese Head – cylindrical head with a flat seating.

Raised Cheese Head – smaller diameter head with a shallow domed top and deeper slot also known as a *filister* head. The same head without doming is known as a *flat filister.*

Connection Head – basically a larger diameter cheese head.

Round Head – roughly semi-spherical in form with a flat seating. This head is also known variously as *cup, snap* and *button head.*

Mushroom Head – a larger diameter round head with the same height as a round head, but offering increased flat seating area. Shallower forms of even larger diameter are known as *truss heads* (also called *oval head, stove head* and *oval binding head).*

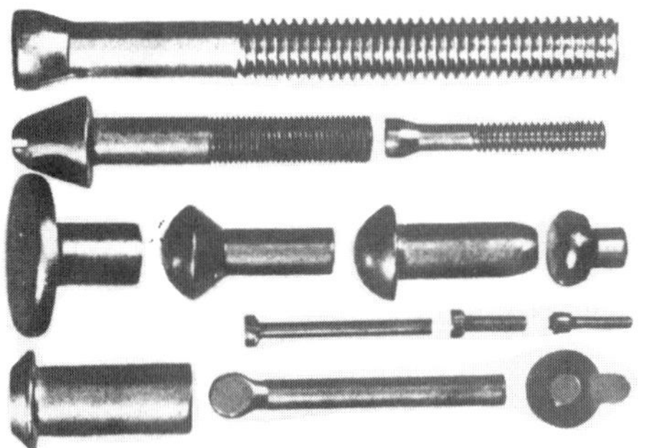

Examples of special heads. (Barton Cold-Form Ltd).

Pan Head – basically a cheese head with tapering sides; it is available in English or metric sizes. The *American pan head* is of larger diameter and shallower, with a rounded top edge. The top is domed in the case of a recessed head. A somewhat similar form of shallow head with slightly tapered sides and undercut base is known as a *binding head* (American).

Countersunk Head – tapered head with flat top, the taper normally being 80–82° to provide flush seating in a countersunk hole, or 120° in the case of aircraft screws. The taper gives the head self-centring properties. It is also known as a *flat head.* Other angles of countersinking may also be used in special flat heads.

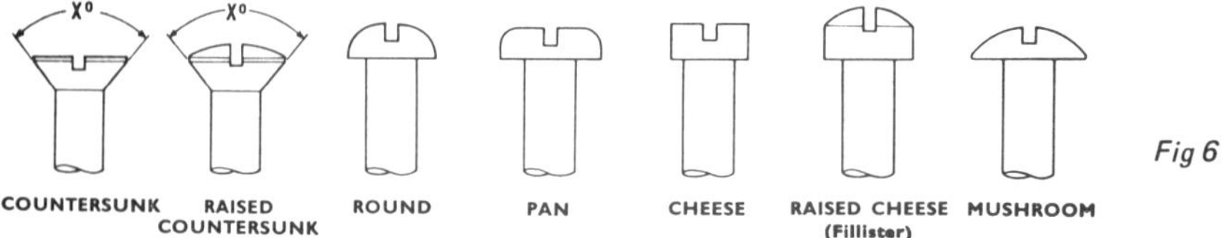

Fig 6

Raised Countersunk – a countersunk head with a shallow domed top. Also known as an *oval head, French head* or *instrument head.* Another form with a smaller head and greater degree of countersink is known as a *lentil head* (American).

Lesser used machine screw heads are:-

Washer Head – either in the form of a round head with larger, integral washer section base, or a washer section with a shallow domed top. The former can be made with either a slot or recessed head, the latter with a recessed head only.

Shoulder Head – similar to a mushroom head, but with a square shank (to provide locking), or a plain round shoulder slightly larger than the thread diameter. These are known as square shoulder and round shoulder heads, respectively. Either type may be produced with a plain (unslotted) or recessed head.

Knurled Shoulder Head – another form of shoulder head where the shoulder is knurled to provide locking.

Fin Head – similar to a shoulder head but with fillets or 'fins' on the underside of the head to provide locking when drawn down tight on a softer material.

Bugle Head – a countersunk head form designed to sink neatly and uniformly into soft materials without crushing or tearing the surface; this is little used on machine screws – mainly on self-tapping and similar screws used for building construction.

TYPICAL MECHANICAL PROPERTIES AND APPLICATION TORQUES FOR POZIDRIV RECESS MILD STEEL MACHINE SCREWS (Steel screws to metric grade 4.8 or 25tonf/in^2 minimum tensile stress)

Thread	*Major dia (in)*	*Minor dia (in)*	*Tensile breaking load*		*Breaking torque*		*Tightening torque*	
			lbf.min	*kgf.min*	*lbf.ft*	*kgf.m*	*lbf.ft*	*kgf.m*
6-32 UNC	0·138	0·0997	509	231	1·30	0·180	0·75	0·104
6-40 UNF	0·138	0·1073	569	258	1·45	0·200	0·91	0·126
8-32 UNC	0·164	0·1257	784	355	2·21	0·361	1·48	0·205
8-36 UNF	0·164	0·1299	826	374	2·52	0·348	1·65	0·228
10-24 UNC	0·190	0·1389	980	444	3·48	0·481	1·98	0·274
10-32 UNF	0·190	0·1517	1190	539	3·95	0·544	2·47	0·341
¼-20 UNC	0·250	0·1887	1781	807	8·30	1·150	5·42	0·750
¼-28 UNF	0·250	0·2062	2039	923	9·40	1·300	6·25	0·865
5/16-18 UNC	0·3125	0·2443	2935	1330	17·30	2·392	10·8	1·492
5/16-24 UNF	0·3125	0·2614	3247	1472	18·30	2·530	11·6	1·605
⅜-16 UNC	0·375	0·2993	4340	1965	30·00	4·150	17·7	2·445
⅜-24 UNF	0·375	0·3239	4919	2230	34·00	4·700	19·5	2·700
	mm	*mm*						
M3	3·000	2·387	447	202	0·90	0·172	0·70	0·096
M4	4·000	3·141	765	347	1·90	0·262	1·25	0·172
M5	5·000	4·019	1236	559	4·00	0·664	3·00	0·415
M6	6·000	4·773	1750	793	7·00	1·07	5·00	0·690
M8	8·000	6·466	3184	1445	18·80	2·60	11·80	1·63
M10	10·000	8·160	4760	2160	35·00	5·60	23·50	3·25

RECESSED HEADS

Recessed heads fall into four main groups:-

(i) Cross recessed
(ii) Internal socket
(iii) High torque (aircraft types)
(iv) High torque (general types)

The main cross recessed heads in general use are the Frearson, Phillips and Pozidriv– Fig 7. All are suitable for low cost production, but require the use of special drivers. The Frearson recess is the least used. The Phillips was virtually the standard cross recess head for many years, but is now largely superseded by the Pozidriv which provides better driver engagement and higher torque transmission with higher driving speeds possible.

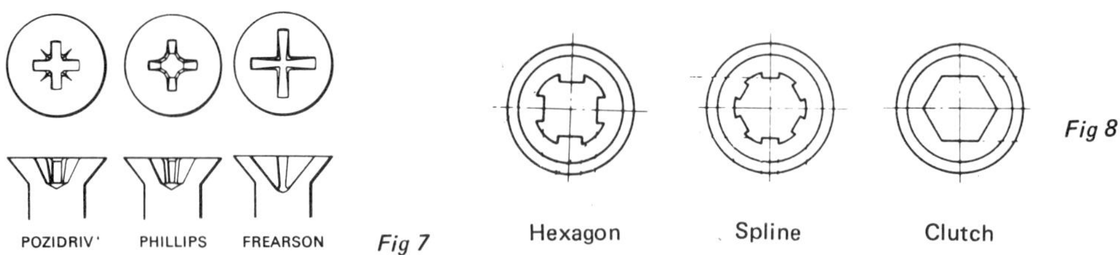

Fig 7

Fig 8

The two main types of internal socket are the hexagon and spline – Fig 8. The clutch recess is also included in this category as being geometrically an internal socket. The hexagon and splined socket are particularly suitable for high strength fasteners requiring high tightening torques and are also generally very versatile. Various other forms of lobular hexagon and splined sockets have also been derived from these.

High torque (aircraft type) recessed heads are shown in Fig 9, all of these being specifically developed to be suitable for use with the shallow (100°) countersunk heads used on screws for the aircraft industry. The Hi-torque and Torq-set are the two main types. The tri-wing recess has a similar performance with the advantage of being tamper-proof (ie it cannot be unscrewed with a conventional screwdriver or a four-wing Armaform driver). A disadvantage is that each type of head requires a different driver for each size of screw (this limitation also applies generally to socket-recessed head screws). Also, all have relatively shallow recesses, demanding careful operator technique and reducing driver speeds.

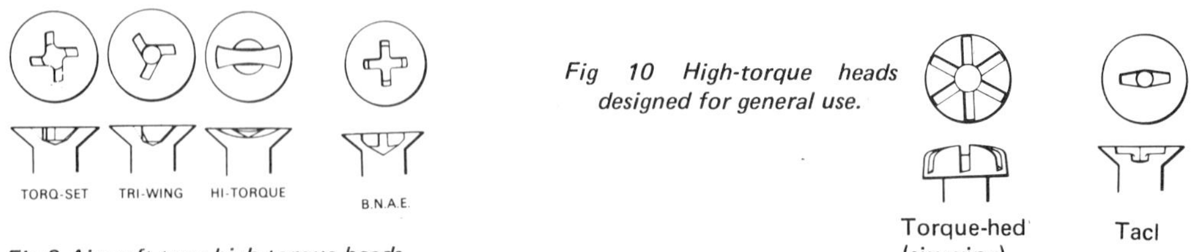

Fig 9 Aircraft-type high-torque heads.

Fig 10 High-torque heads designed for general use.

Two high torque heads intended for general use are shown in Fig 10. The shallow recess of the TACL recess again makes driver engagement and control fairly critical. Also a wide range of driver

TABLE II – MACHINE SCREW HEADS

Type of Head	Application	Torque Rating	Versatility	Suitability for Automation	Remarks
Slotted	Low strength, low torque fasteners	Poor	Good	Poor	Danger of driver slipping. Not suitable for high speed driving
Clutch	Low to moderate strength fasteners	Moderate to very good	Very good to excellent	Moderate	
Hexagon socket	High strength fasteners	Very good	Excellent	Moderate	Maximum possible tightening torques realisable
Splined socket	High strength fasteners	Very good	Excellent	Moderate	Maximum possible tightening torques realisable
Frearson	Low to moderate strength fasteners	Good	Good to very good	Moderate to good	Sharp taper can lead to driver disengagement unless high pressure is applied continuously
Philips	Low to moderate strength fasteners	Good	Moderate to very good good	Moderate to good	Needs bearing end pressure to maintain engagement
Pozidriv	Low to moderate strength fasteners	Good to very good	Good to very good	Moderate to excellent	Vertical driving faces eliminate axial reaction during driving
Torq-set	Moderate high strength fasteners	Good	Very good	Poor	Used on shallow countersunk aircraft screws
Hi-torq	Moderate high strength fasteners	Good	Good to very good	Very poor	Used on shallow countersunk aircraft screws
Tri-wing	Moderate to high strength fasteners	Good	Good to very good	Poor	Tamper proof
BNAE	Moderate to high strength fasteners	Good	Very good	Fair to moderate	Used on shallow countersunk aircraft screws
Torque-hed	Moderate to high strength fasteners	Good	Very good	Moderate to good	
TACL	Low to moderate strength fasteners	Moderate	Moderate to good	Poor	

sizes are needed to cover all screw sizes. The Torque-Hed provides much better control but cannot be used with a flat countersunk head. Both types can be manipulated or tampered with by using an ordinary screwdriver.

Comparative performances, etc, are summarised in Table II.

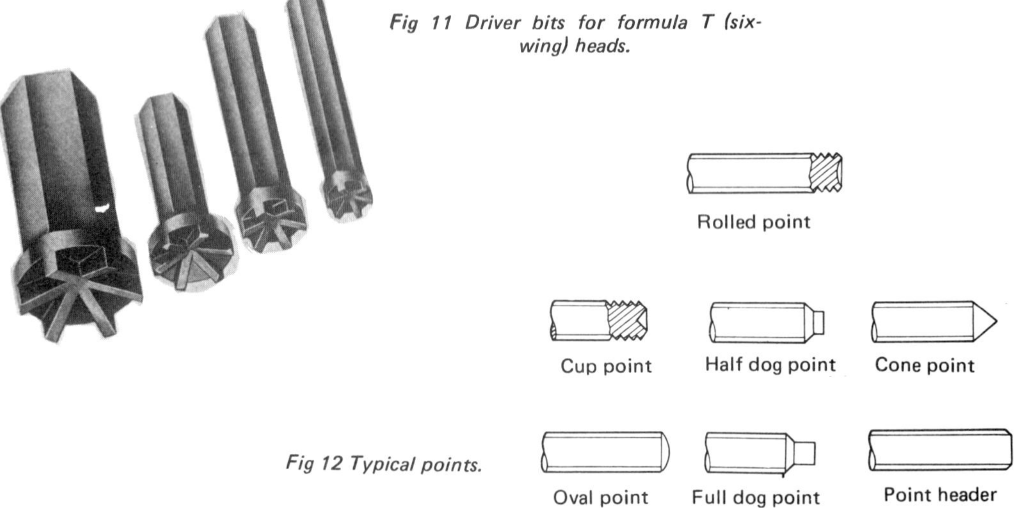

Fig 11 Driver bits for formula T (six-wing) heads.

Fig 12 Typical points.

POINTS

Various point styles for machine screws are illustrated in Fig 12. These include the following:-

Die Point (Header Point) – produced at the time of heading or rolling to provide an end chamfer starting with a diameter smaller than the root diameter of the thread.

Chamfer Point – similar to the die point but with a 45° chamfer regardless of the depth of the thread.

Rolled Point – produced by a rolling operation on long screws, the last 1½ threads being slightly affected by the operation. The point may have a chamfer *(rolled point)*, or not *(rolled end)*.

Dog Point – a parallel point section with a diameter slightly less than the root diameter of the thread. Length may be quite short *(half dog point)*, or extend to approximately two-thirds of the thread diameter *(full dog point)*. This type is also sometimes known as a *pilot point*.

Cone Point – a sharply cut point intended to provide a perforation or aligning function on assembly. If the length is extended to facilitate piercing, it is known as a *needle point*.

Round Point – roughly hemispherical and designed to offer pressure without marking on a hard surface. If the domed shape is truly hemispherical then it is called a *spherical point*.

Cupped Point – a round point with a cupped identation to reduce the contact area of the end surface.

Pinch Point – a round point with a 60° bend in and pinch-off marks on the surface. This is correctly called a *rounded pinch point* to distinguish from a nail point.

Nail Point – a sharp point of approximately 45° angle with squared surfaces produced by 'pinching'. This is also called a *pinch point* and is usually limited in application to screws of 6mm (¼in) diameter or smaller.

Flat Point – a squared off end substantially flat and proportioned to the axis of the screw. It is also known as a *plain point.*

Gimlet Point – a threaded cone point with an angle of approximately 45°, which is used on special screws – eg self-tapping screws.

Cut Points – these include a variety of points similar to the dog point but may incorporate a groove (for a locking ring), cupping (for riveting over), etc.

See also chapter on *Self-Tapping Screws.*

HEAD DIMENSIONS

ISO metric head dimensions for screws are directly related to the basis screw diameter. All head forms (except cheese head) have a maximum diameter of 2 x shank diameter.

HEAD DIMENSIONS FOR POZIDRIV RECESS HEADS (BS4183)

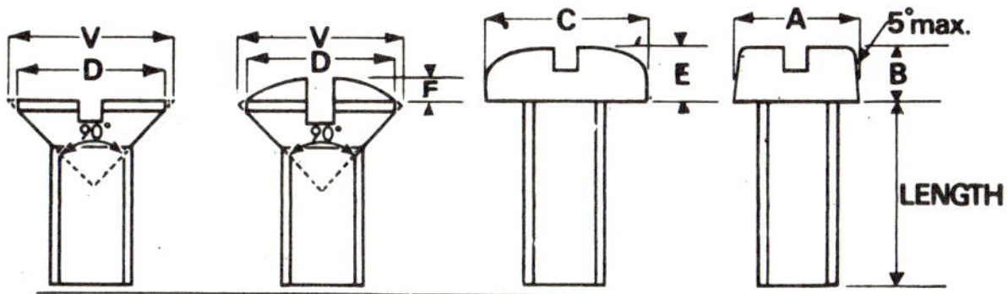

Nominal Size	Pitch	CSK and RSD CSK Head				Pan Head				Recess No
		Dia of Head		Ht. of Raise F	Depth of Head E Max	Dia of Head C		Depth of Head E		
		"V" Max	"D" Min			Max	Min	Max	Min	
M2·5	0·45	5·00	4·38	0·60	1·25	5·00	4·70	1·75	1·61	1
M3	0·50	6·00	5·25	0·75	1·50	6·00	5·70	2·10	1·96	1
M3·5	*0·60*	*7·00*	*6·10*	*0·90*	*1·75*	*7·00*	*6·64*	*2·42*	*2·31*	*2*
M4	0·70	8·00	7·00	1·00	2·00	8·00	7·64	2·80	2·66	2
M4·5	*0·75*	*9·00*	*7·85*	*1·10*	*2·25*	*9·00*	*8·64*	*3·15*	*2·97*	*2*
M5	0·80	10·00	8·75	1·25	2·50	10·00	9·64	3·50	3·32	2
M6	1·00	12·00	10·50	1·50	3·00	12·00	11·57	4·20	3·98	3
M8	1·25	16·00	14·00	2·00	4·00	16·00	15·57	5·60	5·42	4
M10	1·50	20·00	17·50	2·50	5·00	20·00	19·48	7·00	6·78	4
M12	1·75	24·00	21·00	3·00	6·00	—	—	—	—	4

HEAD DIMENSIONS FOR SLOTTED HEADS (BS4183)

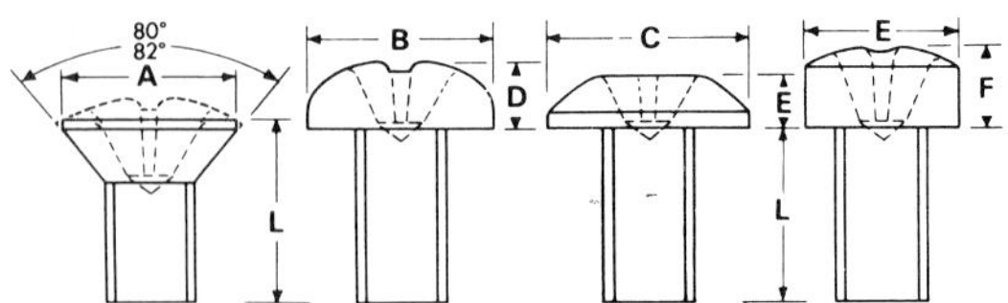

Nominal Size	Pitch	CSK and RSD CSK HEADS				PAN HEADS				CHEESE HEADS			
		Dia of Head		Ht of Raise F	Max Land on Min Dia	Dia of Head C		Depth E		Dia of Head A		Depth B	
		"V" Max (Sharp)	"D" Min			Max	Min	Max	Min	Max	Min	Max	Min
M2	0·40	4·00	3·50	0·50	0·25	—	—	—	—	3·80	3·50	1·30	1·16
M2·2	0·45	4·40	3·85	0·55	0·27	—	—	—	—	4·00	3·70	1·50	1·36
M2·5	0·45	5·00	4·38	0·60	0·31	5·00	4·70	1·50	1·36	4·50	4·20	1·60	1·46
M3	0·50	6·00	5·25	0·75	0·37	6·00	5·70	1·80	1·66	5·50	5·20	2·00	1·86
M3·5	*0·60*	*7·00*	*6·10*	*0·90*	*0·44*	*7·00*	*6·64*	*2·10*	*1·96*	*6·00*	*5·70*	*2·40*	*2·26*
M4	0·70	8·00	7·00	1·00	0·50	8·00	7·64	2·40	2·26	7·00	6·64	2·60	2·46
M4·5	*0·75*	*9·00*	*7·85*	*1·10*	*0·56*	*9·00*	*8·64*	*2·70*	*2·56*	*8·00.*	*7·64*	*3·10*	*2·92*
M5	0·80	10·00	8·75	1·25	0·62	10·00	9·64	3·00	2·86	8·50	8·14	3·30	3·12
M6	1·00	12·00	10·50	1·50	0·75	12·00	11·57	3·60	3·42	10·00	9·64	3·90	3·72
M8	1·25	16·00	14·00	2·00	1·00	16·00	15·57	4·80	4·62	13·00	12·57	5·00	4·82
M10	1·50	20·00	17·50	2·50	1·25	20·00	19·48	6·00	5·82	16·00	15·57	6·00	5·82
M12	1·75	24·00	21·00	3·00	1·50	—	—	—	—	18·00	17·57	7·00	6·78

All dimensions in millimetres

Bolts

BOLTS

For convenience of reference, this chapter is sub-divided into the following sub-sections:-

(i) Black Fasteners
(ii) Bright Bolts
(iii) High Tensile Bolts
(iv) Friction-Grip Bolts
(v) Studs
(vi) Socket Head Screws (Cap Screws)
(vii) Set Screws and Grub Screws
(viii) Thumb Screws and Wing Screws

(i) BLACK FASTENERS

'Black bolts' represent the lowest-cost productions of forged bolts with generous manufacturing tolerances and an average tensile strength of the order of 22–28tonf/in^2 (35–44kg/mm^2). The following standards apply:-

BS916 – hexagon head black bolts with BSW thread.

BS1769 and BS2708 – hexagon head black bolts with UNC thread.

BS4190 – hexagon head black bolts with ISO metric coarse thread.

Hot forged bolts are generally black in appearance, but cold forged black bolts may be partially bright.

The following gives standards for other types of 'black' fastenings.

TABLE I – BLACK FASTENERS

Standard	Description	Applications
BS325 or American standard ASA B185	Cup sq Carriage bolts with sq or hex nuts BSW or UNC	Used by carriage and coach builders, packing case makers, fencing and furniture makers
BS325	Countersunk nibbed and cup nibbed bolts with sq or hex nuts BSW	Used in general and agricultural engineering. The nib locates in a slot in the steelwork, preventing head rotation

cont...

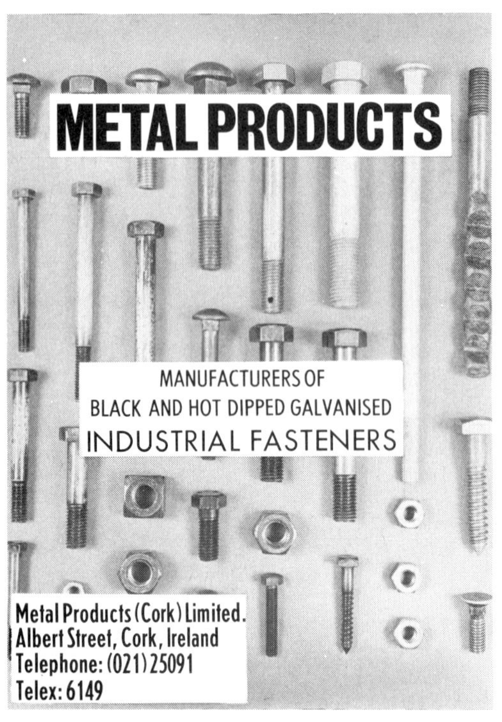
METAL PRODUCTS
MANUFACTURERS OF
BLACK AND HOT DIPPED GALVANISED
INDUSTRIAL FASTENERS
Metal Products (Cork) Limited.
Albert Street, Cork, Ireland
Telephone: (021) 25091
Telex: 6149

Industrial Fastenings
Black, galvanised or
plated bolts & nuts,
tie rods & studs,
manufactured to
metric & imperial
specification
For catalogue and prices
write or phone our Sales Office
Telephone: 021-557 7925
The Bolt & Nut Co.|Tipton|Ltd
PARK LANE EAST, TIPTON, WEST MIDLANDS DY4 8RF
Telephone: 021-557 4731 (3 lines) Grams: "Nuts" Tipton

TABLE I – BLACK FASTENERS (contd.)

Standard	Description	Applications
BS4190	ISO metric hexagon bolts, screws and nuts	All branches of engineering
BS325	Countersunk sq for iron with sq or hex nuts BSW	Mainly used in agricultural industry; the square neck prevents head rotation
BS325	Countersunk sq for wood with sq or hex nuts BSW	Used by coach and carriage builders. They have larger heads than countersunk sq for iron to prevent them pulling through wood.
BS325	Countersunk rd bolts with sq or hex nuts BSW	Used mainly by structural industry where projection of the head is undesirable
BS916 BS2708	Sq or rd bolts and nuts BSW and UNC	All branches of engineering
BS227	Standard large and small tee hd bolted with rd or sq necks hex or sq nuts	Used by valve manufacturers; also on machine tools in 'T' slots for clamping
	Studs	Usually made to customers' requirements
BS1494	Sq hd coach screws (gimlet or plain points)	Used for fastening components to wood, eg telegraph fittings, etc.
BS3410 BS4320	Round washers	General engineering
BS3410	Taper washers	Used in structural industry on beams and channel assemblies to ensure bolt head rests squarely when used on angled surfaces.
BS1802	Sq section spring washers	Locking washer

(ii) BRIGHT BOLTS

Bright bolts are machined from bar (for smaller sizes) or hot forgings (for larger sizes) and have a 'bright' appearance. Average tensile stress is of the order of 25tonf/in^2 (40kg/mm^2) and this is directly comparable with black bolts. The appearance is considerably more attractive, however.

Both black bolts and bright bolts are produced in the widest range of sizes: black bolts from ¼in (6mm) to 2in (50mm) diameter, in lengths up to 75in (1.9m), and bright bolts from 2BA/10UN up to 2½in (62.5mm) diameter, in lengths up to 15in (380mm).

Specific grade (quality) ratings which apply to other than high tensile bolts are:-

'A' quality – minimum tensile strength – 25–28tonf/in^2 (4 000–4 500kgf/cm^2), not stress relieved.

'B' quality – minimum tensile strength 28tonf/in^2 (4 500kgf/cm^2) stress relieved.

(iii) HIGH TENSILE BOLTS

High tensile bolts have a minimum tensile strength of 45–50tonf/in^2 (7 000–8 000kgf/cm^2). Five grades are specified.

BS Grade	Minimum tensile strength tonf/in^2	kgf/cm^2
R (Imperial sizes)	45	7 000
S (Unified and metric sizes)	50	7 875
T	55	8 660
V	65	10 237
X	75	11 810

Corresponding standards to R and S grades are:-

ISO Grade 8.8 (formerly German DIN267 Grade 8G) applicable to all metric sizes – minimum tensile strength 50.8tonf/in^2 (8 000kgf/cm^2).

American Grade 5 – minimum tensile strength 53.6tonf/in^2 (8 441kgf/cm^2) for Unified bolts ¼in to 1in (6mm to 25mm) diameter; and 46.9tonf/in^2 (7.386kgf/cm^2) for bolts over 1in to 1½in (25mm to 37mm) diameter.

Diagramatic illustration of pad deformation characteristic against axial load on an M20 general grade load indicating bolt

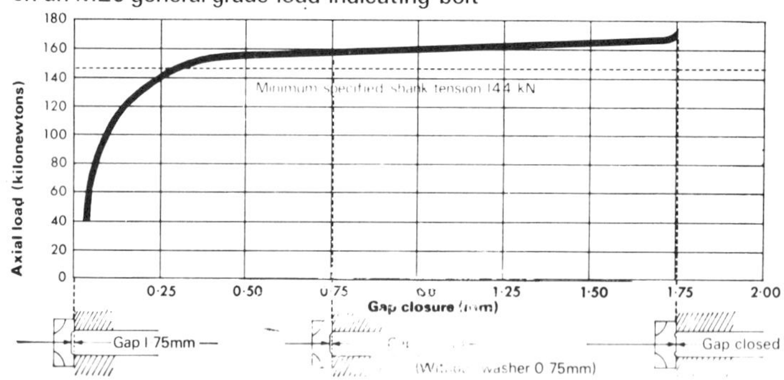

GKN load-indicating bolt. Minimum bolt tension is achieved when the gap width has been reduced to 1mm (0.040inch).

(iv) FRICTION-GRIP BOLTING

High strength bolts together with one or two through hardened washers are used for *friction-grip bolting* in structural steelwork, etc, carrying the loads in frictional shear instead of bending or bearing. Thus the bolts are stressed only in tension and the degree of frictional-grip is determined by the bolt preload. Minimum basic tensions for friction-grip bolting are specified in BS3139, and correspond approximately to the minimum proof load of the bolt. – see also Table II.

TABLE II – BASIC MINIMUM TENSIONS FOR FRICTION-GRIP BOLTING

BOLT SIZE		MINIMUM BOLT TENSION		
in	nearest standard metric	tons	lbs	kg
½	M12	5.37	12 050	5 470
5/8	M16	8.56	19 200	8 720
¾	M20	12.67	28 400	12 900
7/8	M22	17.50	39 300	17 840
1	M24	23.00	51 500	23 400
1.1/8	M30	25.70	56 450	25 630
1¼	M36	32.00	71 700	32 550
1½	M42	46.04	103 950	47 200

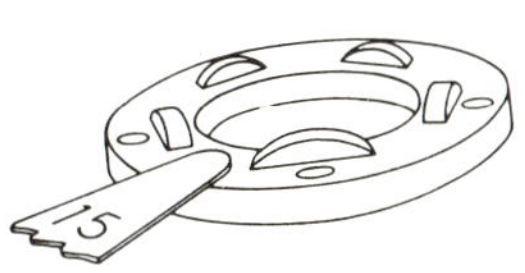

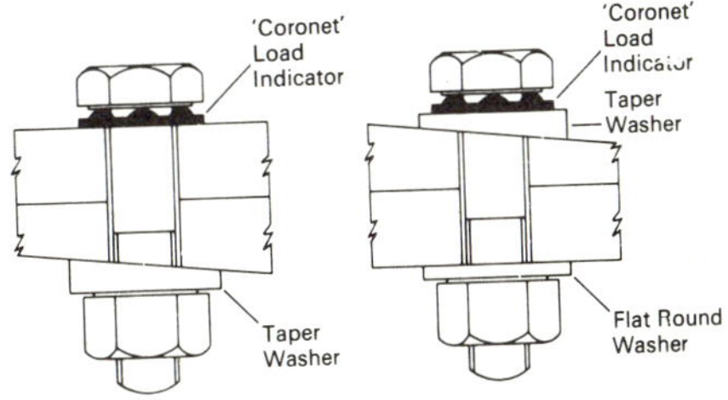

'Coronet' load indicating washers used with friction grip bolting. (Cooper & Turner Ltd).

GKN 'waisted' high-strength bolt with special (Gilbert Rebaris) nut and washer. Stress concentration is removed from the first and second threads of the nut when tightened.

GKN High strength bolts.

HIGH STRENGTH FRICTION – GRIP BOLTS GENERAL GRADE:(BS3139)

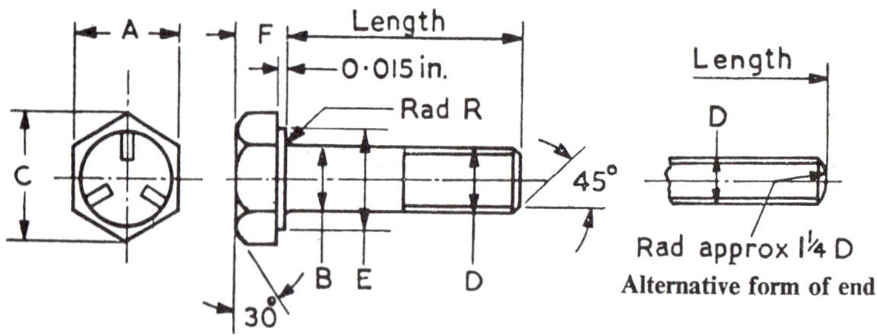

Nom dia D	Diameter of Unthreaded Shank B =		No of Threads per in	Width Across Flats A		Diameter of Washer Face E		Width Across Corners C	Radius Under Head R		Thickness of head F		Nom dia D
	max	min	UNC	max	min	max	min	max	max	min	max	min	
½	0.530	0.496	13	0.8750	0.8550	0.841	0.831	1.010	0.045	0.02	0.323	0.302	½
5/8	0.655	0.619	11	1.0625	1.0375	1.019	1.009	1.227	0.045	0.02	0.403	0.378	5/8
¾	0.780	0.744	10	1.2500	1.2250	1.203	1.188	1.443	0.045	0.02	0.483	0.455	¾
7/8	0.905	0.867	9	1.4375	1.4075	1.380	1.365	1.660	0.045	0.02	0.563	0.533	7/8
1	1.030	0.992	8	1.6250	1.5950	1.559	1.544	1.876	0.045	0.02	0.627	0.597	1
1.1/8	1.165	1.117	7	1.8125	1.7825	1.736	1.721	2.093	0.060	0.03	0.718	0.678	1.1/8
1¼	1.290	1.242	7	2.000	1.9550	1.915	1.900	2.309	0.060	0.03	0.813	0.773	1¼
1½	1.540	1.490	6	2.3750	2.3300	2.271	2.256	2.742	0.060	0.03	0.974	0.934	1½

All dimensions in inches

(v) STUDS (Fig 1)

Studs are headless bolts which may be either double-ended (threaded each end with a plain centre section) or continuously threaded. Both types are also known as stud bolts, although continuously threaded studs are also known as studding. Studs are used as fasteners with a nut on each end, or, alternatively, with one end screwed into a tapped hole in a component with a nut on the other end of the stud holding a second component in position (ie permitting blind or semi-blind assembly).

Fig 1 Stud nomenclature.

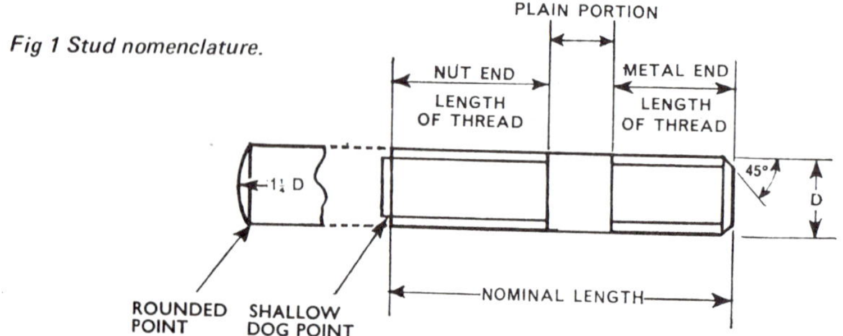

Punch STUD

Length of Stud

A 45 C

Sheet thickness ·060" (1·5 mm) and less

Assembly Anvil

Sheet thickness greater than ·060" (1·5 mm)

Self-aligning captive stud — PSM

Studs are cut with standard threads in steel, alloy steel, brass, etc. For high strength working, alloy steel is normally specified and preferably continuous thread studs (to ensure even distribution of load along the length of the stud). Specific material recommendations are summarised in Table III.

TABLE III — MATERIAL RECOMMENDATIONS FOR STUDS

Duty	Stud	Nut(s)
Small, low strength assemblies	Brass or mild steel	Brass, mild steel
Higher strength assemblies:		
Low temperature services	Alloy steel BS Grade L7 or ASTM A320	Hot forged alloy steel to BS Grade L4 or ASTM A194 Grade A
High temperature services	Alloy steel Grade B7 or B16	Hot forged carbon steel Grade 2 and 2H; or hot forged alloy steel Grade 4
General corrosion-resistant	Stainless steel	Stainless steel

Studding (Screwed rods)

Studding is produced in all standard screw threads and in lengths up to 10ft. Standard sizes in common production are:-

Imperial — ¼in to 2in diameter

Metric — M6 to M36

BA — 0 to 10

Larger diameter studding is also known as threaded rods – eg BA sizes are normally referred to as 'studding', whilst Imperial and Metric sizes may be referred to as 'threaded rods'

ISO Metric Studs

Standard sizes of ISO metric studs for general engineering purposes are summarised in Table IV. Preferred (nominal) stud lengths are (in millimetres).

14, 16, 20 then in 5mm steps up to 90mm.

100, then in 10mm steps up to 200mm.

220, 240. 260, 280, 300, 325, 350, 375, 400, 425, 450, 475, 500.

In the case of double-ended studs for blind assemblies, the length of thread at the top end can be 1 or 1½ diameters; and that at the nut end –

2 diameters + 6mm (up to 125mm length)

2 diameters + 12mm (125–200mm length)

2 diameters + 25mm (over 200mm length)

The plain portion in each case should not be less than ½ diameter.

See also *Bolt and Screw Design Parameters.*

TABLE IV – STANDARD THREAD LENGTHS FOR THE NUT END OF SCREWED STUDS (PREFERRED SIZES) BS4439

Nominal size and thread diameter d	Length of thread for nominal stud length		
	up to 125mm	over 125mm and up to 200mm	above 200mm
	2d + 6mm	2d + 12mm	2d + 25mm
M3	12	—	—
M4	14	—	—
M5	16	—	—
M6	18	—	—
M8	22	—	—
M10	26	32	—
M12	30	36	—
M16	38	44	57
M20	46	52	65
M24	54	60	73
M30	66	72	85
M36	78	84	97

TABLE V – INTERFERENCE THREAD STUDS – UNIFIED THREADS (IFI:1970) TAP END

Nominal Size or Basic Stud Diameter		Thread Length: NC 5 HF, NC 5 CSF – Full Thread	Thread Length: NC 5 HF, NC 5 CSF – Total Thread	Thread Length: NC 5 ONF – Full Thread	Thread Length: NC 5 ONF – Total Thread	Point Diameter	Point Length	
		Min	Max	Min	Max	Max	Max	Min
1/4	0.2500	0.375	0.500	0.687	0.812	0.18	0.150	0.100
5/16	0.3125	0.469	0.625	0.859	1.015	0.23	0.167	0.112
3/8	0.3750	0.562	0.750	1.032	1.219	0.28	0.187	0.125
7/16	0.4375	0.656	0.875	1.203	1.422	0.33	0.214	0.143
1/2	0.5000	0.750	1.000	1.375	1.625	0.39	0.231	0.154
9/16	0.5625	0.843	1.125	1.547	1.828	0.44	0.250	0.167
5/8	0.6250	0.937	1.250	1.718	2.031	0.49	0.273	0.182
3/4	0.7500	1.125	1.500	2.062	2.438	0.60	0.300	0.200
7/8	0.8750	1.312	1.750	2.406	2.844	0.72	0.333	0.222
1	1.0000	1.500	2.000	2.750	3.250	0.82	0.375	0.250
1.1/8	1.1250	1.687	2.250	3.093	3.655	0.92	0.429	0.286
1.1/4	1.2500	1.875	2.500	3.438	4.062	1.04	0.429	0.286
1.3/8	1.3750	2.062	2.750	3.781	4.469	1.14	0.500	0.334
1.1/2	1.5000	2.250	3.000	4.125	4.875	1.27	0.500	0.334

TABLE V – INTERFERENCE THREAD STUDS – UNIFIED THREADS (IFI:1970) NUT END

Nominal Size or Basic Stud Diameter		For Studs Of 3D Length Or Less – Thread Length – Full Thread	Total Thread	For Studs Less Than 4.75D Length and Over 3D Length – Thread Length – Full Thread	Total Thread	For Studs Over 4.75D And Up To And Including 6 in Length – Thread Length – Full Thread	Total Thread	For Studs Over 6 in Length – Thread Length – Full Thread	Total Thread
		Min	Max	Min	Max	Min	Max	Min	Max
1/4	0.2500	–	–	0.375	0.500	0.750	0.875	1.000	1.125
5/16	0.3125	–	–	0.469	0.625	0.875	1.031	1.125	1.281
3/8	0.3750	–	–	0.562	0.750	1.000	1.188	1.250	1.438
7/16	0.4375	–	–	0.656	0.875	1.125	1.344	1.375	1.594
1/2	0.5000	–	–	0.750	1.000	1.250	1.500	1.500	1.750
9/16	0.5625	–	–	0.844	1.125	1.375	1.656	1.625	1.906
5/8	0.6250	–	–	0.938	1.250	1.500	1.812	1.750	2.062
3/4	0.7500	–	–	1.125	1.500	1.750	2.125	2.000	2.375
7/8	0.8750	–	–	1.312	1.750	2.000	2.438	2.250	2.688
1	1.0000	–	–	1.500	2.000	2.250	2.750	2.500	3.000
1.1/8	1.1250	–	–	1.688	2.250	2.500	3.062	2.750	3.312
1.1/4	1.2500	–	–	1.875	2.500	2.750	3.375	3.000	3.625
1.3/8	1.3750	–	–	2.062	2.750	3.000	3.688	3.250	3.938
1.1/2	1.5000	–	–	2.250	3.000	3.250	4.000	3.500	4.250

TABLE VI – MECHANICAL PROPERTIES OF STEEL STUDS (BS4439)

Property			BS Strength Grade Designation						
			4.8	5.8	6.8	8.8	10.9	12.9	14.9
Tensile strength	min	kgf/mm^2	40	50	60	80	100	120	140
	max	kgf/mm^2	55	70	80	100	120	140	160
Brinell hardness	min	HB	110	140	170	225	280	330	390
	max	HB	170	215	245	300	365	425	–
Yield stress R_e	min	kgf/mm^2	32	40	48	–	–	–	–
Stress at permanent set limit R	min	kgf/mm^2	–	–	–	64	90	108	126
Stress under proof load	Sp/R_e Sp/R	and	0.91	0.91	0.91	0.91	0.88	0.88	0.88
Sp		kgf/mm^2	29.1	36.4	43.7	58.2	79.2	95.0	111
Elongation after fracture	min	%	14	10	8	12	9	8	7

TABLE VII – PROOF LOADS FOR STRESS STUDS – BS4436) (COARSE METRIC THREADS)

Nominal size and thread diameter	Tensile stress area	Strength grade designation						
		4.8	5.8	6.8	8.8	10.9	12.9	14.9
		Stress under proof load kgf/mm^2						
		29·1	36·4	43·7	58·2	79·2	95·0	111
		Proof load tonne-force (1000 kgf)						
M3	5·03	0·146	0·183	0·220	0·293	0·398	0·478	0·558
M4	8·78	0·255	0·320	0·384	0·511	0·695	0·834	0·975
M5	14·2	0·413	0·517	0·621	0·826	1·12	1·35	1·58
M6	20·1	0·585	0·732	0·878	1·17	1·59	1·91	2·23
M8	36·6	1·07	1·33	1·60	2·13	2·90	3·48	4·06
M10	58·0	1·69	2·11	2·53	3·38	4·59	5·51	6·44
M12	84·3	2·45	3·07	3·68	4·91	6·68	8·01	9·36
M14	115	3·35	4·17	5·03	6·69	9·11	10·9	12·8
M16	157	4·57	5·71	6·86	9·14	12·4	14·9	17·4
M18	192	5·59	6·99	8·39	11·2	15·2	18·2	21·3
M20	245	7·13	8·92	10·7	14·3	19·4	23·3	27·2
M22	303	8·82	11·0	13·2	17·6	24·0	28·8	33·6
M24	353	10·3	12·8	15·4	20·5	28·0	33·5	39·2
M27	459	13·4	16·7	20·1	26·7	36·4	43·6	50·9
M30	561	16·3	20·4	24·5	32·7	44·4	53·3	62·3
M33	694	20·2	25·3	30·3	40·4	55·0	65·9	77·0
M36	817	23·8	29·7	35·7	47·5	64·7	77·6	90·7
M39	976	28·4	35·5	42·7	56·8	77·3	92·7	108

NOTE Proof load $= \dfrac{\text{stress under proof load} \times \text{tensile stress area of stud}}{1000}$

(vi) SOCKET HEAD AND SCREWS (CAP SCREWS)

Socket head screws and cap screws have a head form proportioned to accommodate a recess for socket driving – eg specifically hexagon socket and splined socket, or patented sockets. The usual range of head form is shown in Fig 2, ie

(i) *Cap head or Shoulder head* – parallel round head with top chamfer; head plain or knurled. On some types the head may be undercut.

(ii) *Countersunk* – 90° countersunk angle.

(iii) *Button head* – similar to a flattened round head.

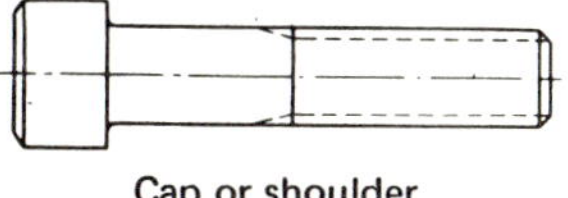

Cap or shoulder

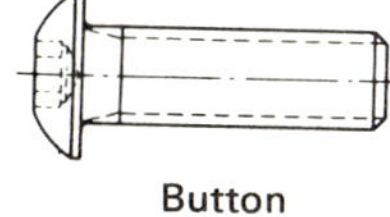

Button

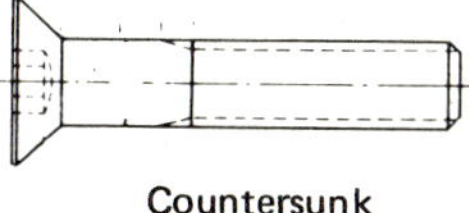

Countersunk

Fig 2. Socket heads.

Standard hexagon socket sizes are detailed in Table I, the matching size of hexagon key or bit being virtually an exact fit (the hexagon width across flats is equal to minimum socket width across flats).

Splined sockets may have 4, 6 or more teeth. A 4-spline socket is used only in small screws, a 6-spline socket being standard for most sizes. Matching spline key sizes are very slightly undersize on the minimum specified value of the socket major diameter.

SOCKET BUTTON HEAD SCREWS
for fastening thin sections when removal or entry is required.

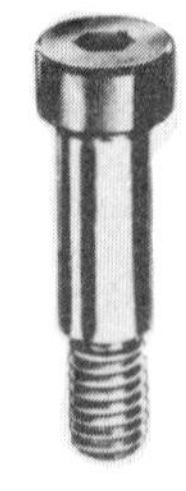

SOCKET SHOULDER SCREWS
for controlled movement of parts on the concentric body.

SOCKET SET SCREWS
for safe flush mounting and positive location on moving parts.

SOCKET HEAD CAP SCREWS
for compact design and positive holding power.

SOCKET COUNTERSUNK SCREWS
for attractive flush mounting of metal sections.

SOCKET PRESSURE (PIPE) PLUGS
for a reliable high pressure seal under hydraulic pressures.

Examples of socket screws (Carbo Engineering Co Ltd).

TABLE VIII – METRIC HEXAGON SOCKET HEAD CAP SCREWS
(Preferred Sizes:BS4168)

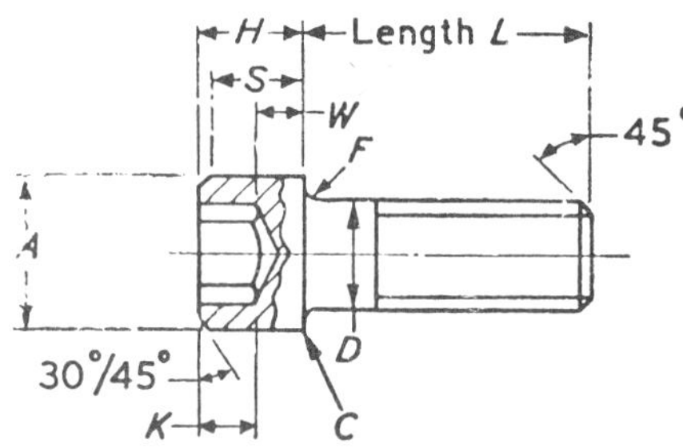

Nominal Size	D Body diameter		A Head diameter		H Head height		S Head side height	Hex socket size	Key engagement		W Wall thickness	C Chamfer or radius
	Max	Min	Max	Min	Max	Min	Min	Nom	Max	Min	Min	Max
M3	3.00	2.86	5.50	5.20	3.00	2.86	2.70	2.50	1.70	1.30	0.96	0.125
M4	4.00	3.82	7.00	6.64	4.00	3.82	3.60	3.00	2.40	2.00	1.28	0.125
M5	5.00	4.82	8.50	8.14	5.00	4.82	4.50	4.00	3.10	2.70	1.60	0.125
M6	6.00	5.82	10.00	9.64	6.00	5.82	5.40	5.00	3.78	3.30	1.92	0.200
M8	8.00	7.78	13.00	12.57	8.00	7.78	7.20	6.00	4.78	4.30	2.56	0.200
M10	10.00	9.78	16.00	15.57	10.00	9.78	9.00	8.00	6.25	5.50	3.20	0.200
M12	12.00	11.73	18.00	17.57	12.00	11.73	10.80	10.00	7.50	6.60	3.84	0.250
M16	16.00	15.73	24.00	23.48	16.00	15.73	14.40	14.00	9.70	8.80	5.12	0.250
M20	20.00	19.67	30.00	29.48	20.00	19.67	18.00	17.00	11.80	10.70	6.40	0.400
M24	24.00	23.67	36.00	35.38	24.00	23.67	21.60	19.00	14.00	12.90	7.68	0.400

All dimensions in millimetres

TABLE IX – COUNTERSUNK SOCKET HEAD CAP SCREWS
(Preferred Sizes:BS4168)

Nominal size	Body diameter		Head diameter		Head height		Hex. socket size	Key engagement
	Max	Min	Theoretical sharp max	Absolute min	Ref	Flushness tolerance	Nom	Min
M3	3·00	2·86	6·72	5·82	1·86	0·20	2·00	1·05
M4	4·00	3·82	8·96	7·78	2·48	0·20	2·50	1·49
M5	5·00	4·82	11·20	9·78	3·10	0·20	3·00	1·86
M6	6·00	5·82	13·44	11·73	3·72	0·20	4·00	2·16
M8	8·00	7·78	17·92	15·73	4·96	0·24	5·00	2·85
M10	10·00	9·78	22·40	19·67	6·20	0·30	6·00	3·60
M12	12·00	11·73	26·88	23·67	7·44	0·36	8·00	4·35
M16	16·00	15·73	33·60	29·67	8·80	0·45	10·00	4·89
M20	20·00	19·67	40·32	35·61	10·16	0·54	12·00	5·45

All dimensions in millimetres

TABLE X – BUTTON SOCKET HEAD CAP SCREWS
(Preferred Sizes:BS4168)

Nominal Size	Head diameter		Head height		Head side height (Nominal)	Key engagement
	Maximum	Minimum	Maximum	Minimum		
M3	5.5	5.32	1.6	1.40	0.38	2.0
M4	7.5	7.28	2.1	1.85	0.38	2.5
M5	9.5	9.28	2.7	2.45	0.50	3
M6	10.5	10.23	3.2	2.95	0.80	4
M8	14	13.75	4.3	3.95	0.80	5
M10	18	17.75	5.3	4.95	0.80	6
M12	21	20.67	6.4	5.90	0.80	8

All dimensions in millimetres

TABLE XI – UNIFIED HEXAGON AND SPLINE SOCKET HEAD CAP SCREWS (ANSI B18.3)

Nominal Size or Basic Screw Diameter	Body Diameter		Head Diameter		Head Height		Head Side Height	Spline Socket Size	Hexagon Socket Size	Key Engagement	Wall Thickness
	Max	Min	Max	Min	Max	Min	Min	Nom	Nom	Min	Min
No. 0 0.0600	0.0600	0.0568	0.096	0.091	0.060	0.057	0.054	0.060	0.050	0.025	0.020
1 0.0730	0.0730	0.0695	0.118	0.112	0.073	0.070	0.066	0.072	1/16 0.062	0.031	0.025
2 0.0860	0.0860	0.0822	0.140	0.134	0.086	0.083	0.077	0.096	5/64 0.078	0.038	0.029
3 0.0990	0.0990	0.0949	0.161	0.154	0.099	0.095	0.089	0.096	5/64 0.078	0.044	0.034
4 0.1120	0.1120	0.1075	0.183	0.176	0.112	0.108	0.101	0.111	3/32 0.094	0.051	0.038
5 0.1250	0.1250	0.1202	0.205	0.198	0.125	0.121	0.112	0.111	3/32 0.094	0.057	0.043
6 0.1380	0.1380	0.1329	0.226	0.218	0.138	0.134	0.124	0.133	7/64 0.109	0.064	0.047
8 0.1640	0.1640	0.1585	0.270	0.262	0.164	0.159	0.148	0.168	9/64 0.141	0.077	0.056
10 0.1900	0.1900	0.1840	0.312	0.303	0.190	0.185	0.171	0.183	5/32 0.156	0.090	0.065
1/4 0.2500	0.2500	0.2435	0.375	0.365	0.250	0.244	0.225	0.216	3/16 0.188	0.120	0.095
5/16 0.3125	0.3125	0.3053	0.469	0.457	0.312	0.306	0.281	0.291	1/4 0.250	0.151	0.119
3/8 0.3750	0.3750	0.3678	0.562	0.550	0.375	0.368	0.337	0.372	5/16 0.312	0.182	0.143
7/16 0.4375	0.4375	0.4294	0.656	0.642	0.438	0.430	0.394	0.454	3/8 0.375	0.213	0.166
1/2 0.5000	0.5000	0.4919	0.750	0.735	0.500	0.492	0.450	0.454	3/8 0.375	0.245	0.190
5/8 0.6250	0.6250	0.6163	0.938	0.921	0.625	0.616	0.562	0.595	1/2 0.500	0.307	0.238
3/4 0.7500	0.7500	0.7406	1.125	1.107	0.750	0.740	0.675	0.620	5/8 0.625	0.370	0.285
7/8 0.8750	0.8750	0.8647	1.312	1.293	0.875	0.864	0.787	0.698	3/4 0.750	0.432	0.333
1 1.0000	1.0000	0.9886	1.500	1.479	1.000	0.988	0.900	0.790	3/4 0.750	0.495	0.380
1.1/8 1.1250	1.1250	1.1086	1.688	1.665	1.125	1.111	1.012	–	7/8 0.875	0.557	0.428
1.1/4 1.2500	1.2500	1.2336	1.875	1.852	1.250	1.236	1.125	–	7/8 0.875	0.620	0.475
1.3/8 1.3750	1.3750	1.3568	2.062	2.038	1.375	1.360	1.237	–	1 1.000	0.682	0.523
1.1/2 1.5000	1.5000	1.4818	2.250	2.224	1.500	1.485	1.350	–	1 1.000	0.745	0.570
1.3/4 1.7500	1.7500	1.7295	2.625	2.597	1.750	1.734	1.575	–	1.1/4 1.250	0.870	0.665
2 2.0000	2.0000	1.9780	3.000	2.970	2.000	1.983	1.800	–	1.1/2 1.500	0.995	0.760

TABLE XII – UNIFIED HEXAGON AND SPLINE SOCKET COUNTERSUNK HEAD CAP SCREWS (AINSI B18.3)

Nominal Size or Basic Screw Diameter	Body Dia		Head Diameter		Head Height		Spline Socket Size	Hexagon Socket Size	Key Engagement
	Max	Min	Theoretical Sharp Max	Abs. Min	Reference	Flushness Tolerance	Nom	Nom	Min
No. 0 0.0600	0.0600	0.0568	0.138	0.117	0.044	0.006	0.048	0.035	0.025
1 0.0730	0.0730	0.0695	0.168	0.143	0.054	0.007	0.060	0.050	0.031
2 0.0860	0.0860	0.0822	0.197	0.168	0.064	0.008	0.060	0.050	0.038
3 0.0990	0.0990	0.0949	0.226	0.193	0.073	0.010	0.072	1/16 0.062	0.044
4 0.1120	0.1120	0.1075	0.255	0.218	0.083	0.011	0.072	1/16 0.062	0.055
5 0.1250	0.1250	0.1202	0.281	0.240	0.090	0.012	0.096	5/64 0.078	0.061
6 0.1380	0.1380	0.1329	0.307	0.263	0.097	0.013	0.096	5/64 0.078	0.066
8 0.1640	0.1640	0.1585	0.359	0.311	0.112	0.014	0.111	3/32 0.094	0.076
10 0.1900	0.1900	0.1840	0.411	0.359	0.127	0.015	0.145	1/8 0.125	0.087
1/4 0.2500	0.2500	0.2435	0.531	0.480	0.161	0.016	0.183	5/32 0.156	0.111
5/16 0.3125	0.3125	0.3053	0.656	0.600	0.198	0.017	0.216	3/16 0.188	0.135
3/8 0.3750	0.3750	0.3678	0.781	0.720	0.234	0.018	0.251	7/32 0.219	0.159
7/16 0.4375	0.4375	0.4294	0.844	0.781	0.234	0.018	0.291	1/4 0.250	0.159
1/2 0.5000	0.5000	0.4919	0.938	0.872	0.251	0.018	0.372	5/16 0.312	0.172
5/8 0.6250	0.6250	0.6163	1.188	1.112	0.324	0.022	0.454	3/8 0.375	0.220
3/4 0.7500	0.7500	0.7406	1.438	1.355	0.396	0.024	0.454	1/2 0.500	0.223
7/8 0.8750	0.8750	0.8647	1.688	1.604	0.468	0.025	–	9/16 0.562	0.248
1 1.0000	1.0000	0.9886	1.938	1.841	0.540	0.028	–	5/8 0.625	0.297
1.1/8 1.1250	1.1250	1.1086	2.188	2.079	0.611	0.031	–	3/4 0.750	0.325
1.1/4 1.2500	1.2500	1.2336	2.438	2.316	0.683	0.035	–	7/8 0.875	0.358
1.3/8 1.3750	1.3750	1.3568	2.688	2.553	0.755	0.038	–	7/8 0.875	0.402
1.1/2 1.5000	1.5000	1.4818	2.938	2.791	0.827	0.042	–	1 1.000	0.435

All dimensions in inches

TABLE XIII – UNIFIED HEXAGON AND SPLINE SOCKET BUTTON HEAD CAP SCREWS (AINSI B18.3)

Nominal Size or Basic Screw Diameter		Head Diameter		Head Height		Head Side Height	Spline Socket Size	Hexagon Socket Size	Key Engagement	Fillet Extension		Maximum Standard Length
		Max	Min	Max	Min	Ref	Nom	Nom	Min	Max	Min	Nom
No. 0	0.0600	0.114	0.104	0.032	0.026	0.010	0.048	0.035	0.020	0.010	0.005	1/2
1	0.0730	0.139	0.129	0.039	0.033	0.010	0.060	0.050	0.028	0.010	0.005	1/2
2	0.0860	0.164	0.154	0.046	0.038	0.010	0.060	0.050	0.028	0.010	0.005	1/2
3	0.0990	0.188	0.176	0.052	0.044	0.010	0.072	1/16 0.062	0.035	0.010	0.005	1/2
4	0.1120	0.213	0.201	0.059	0.051	0.015	0.072	1/16 0.062	0.035	0.010	0.005	1/2
5	0.1250	0.238	0.226	0.066	0.058	0.015	0.096	5/64 0.078	0.044	0.010	0.005	1/2
6	0.1380	0.262	0.250	0.073	0.063	0.015	0.096	5/64 0.078	0.044	0.010	0.005	5/8
8	0.1640	0.312	0.298	0.087	0.077	0.015	0.111	3/32 0.094	0.052	0.015	0.010	3/4
10	0.1900	0.361	0.347	0.101	0.091	0.020	0.145	1/8 0.125	0.070	0.015	0.010	1
1/4	0.2500	0.437	0.419	0.132	0.122	0.031	0.183	5/32 0.156	0.087	0.020	0.015	1
5/16	0.3125	0.547	0.527	0.166	0.152	0.031	0.216	3/16 0.188	0.105	0.020	0.015	1
3/8	0.3750	0.656	0.636	0.199	0.185	0.031	0.251	7/32 0.219	0.122	0.020	0.015	1 1/4
1/2	0.5000	0.875	0.851	0.265	0.245	0.046	0.372	5/16 0.312	0.175	0.030	0.020	2
5/8	0.6250	1.000	0.970	0.331	0.311	0.062	0.454	3/8 0.375	0.210	0.030	0.020	2

All dimensions in inches

TABLE XIV – DIMENSIONS OF METRIC HEXAGON SOCKETS (BS4168)

Nominal socket size	Socket width across flats		Nominal socket size	Socket width across flats	
	Max	Min		Max	Min
1·5	1·61	1·52	6	6·15	6·03
2·0	2·11	2·02	8	8·19	8·04
2·5	2·61	2·52	10	10·19	10·04
3	3·11	3·02	12	12·23	12·05
4	4·15	4·03	14	14·23	14·05
5	5·15	5·03	17	17·23	17·05
			19	19·275	19·065

All dimensions in millimetres

TABLE XV DIMENSIONS OF UNIFIED HEXAGON SOCKETS (ANSI B18.3)

Nominal Socket Size		Socket Width Across Flats		Nominal Socket Size		Socket Width Across Flats		Nominal Socket Size		Socket Width Across Flats	
		Max	Min			Max	Min			Max	Min
	0.028	0.0285	0.0280	3/16	0.188	0.1900	0.1875	7/8	0.875	0.8850	0.8750
	0.035	0.0355	0.0350	7/32	0.219	0.2217	0.2187	1	1.000	1.0100	1.0000
	0.050	0.0510	0.0500	1/4	0.250	0.2530	0.2500	1 1/4	1.250	1.2650	1.2500
1/16	0.062	0.0635	0.0625	5/16	0.312	0.3160	0.3125	1 1/2	1.500	1.5150	1.5000
5/64	0.078	0.0791	0.0781	3/8	0.375	0.3790	0.3750	1 3/4	1.750	1.7650	1.7500
3/32	0.094	0.0952	0.0937	7/16	0.438	0.4420	0.4375	2	2.000	2.0150	2.0000
7/64	0.109	0.1111	0.1094	1/2	0.500	0.5050	0.5000	2 1/4	2.250	2.2650	2.2500
1/8	0.125	0.1270	0.1250	9/16	0.562	0.5680	0.5625	2 3/4	2.750	2.7650	2.7500
9/64	0.141	0.1426	0.1406	5/8	0.625	0.6310	0.6250	3	3.000	3.0150	3.0000
5/32	0.156	0.1587	0.1562	3/4	0.750	0.7570	0.7500				

All dimensions in inches

TABLE XVI – DIMENSIONS OF SPLINE SOCKETS (ANSI B18.3)

Nominal Socket and Key Size	Number of Teeth	Socket Major Diameter		Socket Minor Diameter		Width of Tooth	
		Max	Min	Max	Min	Max	Min
0.033	4	0.035	0.034	0.026	0.0255	0.012	0.0115
0.048	6	0.050	0.049	0.041	0.040	0.011	0.010
0.060	6	0.062	0.061	0.051	0.050	0.014	0.013
0.072	6	0.074	0.073	0.064	0.063	0.016	0.015
0.096	6	0.098	0.097	0.082	0.080	0.022	0.021
0.111	6	0.115	0.113	0.098	0.096	0.025	0.023
0.133	6	0.137	0.135	0.118	0.116	0.030	0.028
0.145	6	0.149	0.147	0.128	0.126	0.032	0.030
0.168	6	0.173	0.171	0.150	0.147	0.036	0.033
0.183	6	0.188	0.186	0.163	0.161	0.039	0.037
0.216	6	0.221	0.219	0.190	0.188	0.050	0.048
0.251	6	0.256	0.254	0.221	0.219	0.060	0.058
0.291	6	0.298	0.296	0.254	0.252	0.068	0.066
0.372	6	0.380	0.377	0.319	0.316	0.092	0.089
0.454	6	0.463	0.460	0.386	0.383	0.112	0.109
0.595	6	0.604	0.601	0.509	0.506	0.138	0.134
0.620	6	0.631	0.627	0.535	0.531	0.149	0.145
0.698	6	0.709	0.705	0.604	0.600	0.168	0.164
0.790	6	0.801	0.797	0.685	0.681	0.189	0.185

All dimensions in inches

TABLE XVII – STANDARD SIZES FOR METRIC HEXAGON WRENCH KEYS (BS4168)

Nominal key size	Hexagon width across flats		Length of short arm ± 2·5 mm	Length of long arm ± 2·5 mm	Radius of bend approx	End chamfer
	Max	Min				Max
1·5	1·5	1·440	14	45	1·5	0·15
2·0	2·0	1·940	16	50	2·0	0·2
2·5	2·5	2·440	18	56	2·5	0·2
3·0	3·0	2·940	20	63	3·0	0·25
4·0	4·0	3·925	25	71	4·0	0·3
5·0	5·0	4·925	28	80	5·0	0·4
6·0	6·0	5·925	32	90	6·0	0·5
8·0	8·0	7·910	36	100	8·0	0·6
10·0	10·0	9·910	40	112	10·0	0·8
12·0	12·0	11·890	45	125	12·0	1·0
14·0	14·0	13·890	55	140	14·0	1·2
17·0	17·0	16·890	60	160	17·0	1·4
19·0	19·0	18·870	70	180	19·0	1·9

All dimensions in millimetres

XVIII – MACHINE HEXAGON WRENCH KEYS (METRIC SIZES)

Nominal key size	Type of screw			
	Cap screw	Countersunk head screw	Button head screw	Set screw
1·5				M3
2·0		M3	M3	M4
2·5	M3	M4	M4	M5
3·0	M4	M5	M5	M6
4·0	M5	M6	M6	M8
5·0	M6	M8	M8	M10
6·0	M8	M10	M10	M12, M14
8·0	M10	M12	M12	M16
10·0	M12	M14, M16		M18, M20
12·0	M14	M18, M20		M22, M24
14·0	M16, M18			
17·0	M20, M22			
19·0	M24			

TABLE XVIIIA – BSW AND BSF THREADS (55° INCLUDED THREAD ANGLE)

Nominal Size	Thread Diameter	No. of Threads per inch		Stress Area in^2		Recommended Tap Drill Size BS1157:1965		Shank Clearance Hole Dia.	Recommended Counterbore Diameter
in	in	BSW	BSF	BSW	BSF	BSW	BSF		in
1/8	0.1250	40		0.0080		2.55mm		9/64 in	17/64in
3/16	0.1875	24	32	0.0170	0.0194	3.70mm	5/32in	5.20mm	23/64
1/4	0.2500	20	26	0.0320	0.0356	5.10mm	5.30mm	17/64in	27/64
5/16	0.3125	18	22	0.0527	0.0567	6.50mm	6.80mm	8.50mm	31/64
3/8	0.3750	16	20	0.0779	0.0839	5/16in	8.30mm	10.20mm	39/64
7/16	0.4375	14	18	0.1069	0.1158	9.30mm	9.70mm	11.80mm	43/64
1/2	0.5000	12	16	0.1385	0.152	10.50mm	7/16in	17/32in	51/64
5/8	0.6250	11	14	0.227	0.243	13.50mm	14.00mm	21/32in	59/64
3/4	0.7500	10	12	0.336	0.352	41/64in	16.75mm	25/32in	1.1/16
7/8	0.8750	9	11	0.463	0.487	19.25mm	25/32in	29/32in	1.3/16
1	1.000	8	10	0.608	0.642	22.00mm	22.75mm	1.1/32in	1.3/8

TABLE XVIIIB – BA THREADS (47½° INCLUDED THREAD ANGLE)

Nominal Size	Thread Diameter	No. of Threads per inch	Stress Area in^2	Rec.Tap Drill Size BS1157:1965	Shank Clearance Hole Dia.	Recommended Counterbore Diameter
	in			mm	mm	in
8BA	0.0866	59.07	0.0040	1.80	2.50	11/64
6BA	0.1102	47.92	0.0065	2.30	3.10	7/32
5BA	0.1260	43.05	0.0087	2.65	3.60	17/64
4BA	0.1417	38.48	0.0110	3.00	4.00	17/64
3BA	0.1614	34.79	0.0144	3.40	4.50	19/64
2BA	0.1850	31.36	0.0192	4.00	5.10	23/64
1BA	0.2087	28.22	0.0245	4.50	5.70	23/64
OBA	0.2362	25.40	0.0317	5.10	6.40	27/64

SOCKET HEAD SCREW DATA (GKN)

TABLE XVIIIC – ISO INCH (UNIFIED) THREADS (60° INCLUDED THREAD ANGLE)

Nominal Size	Thread Diameter	No. of Threads per inch		Stress Area in^2		Recommended Tap Drill Size BS1157:1965		Shank Clearance Hole Dia.	Recommended Counterbore Diameter
	in	UNC	UNF	UNC	UNF	UNC	UNF		
No.4	0.1120	40	48	0.0062	0.0067	2.35mm	2.40mm	1/8in	5.60mm
No.5	0.1250	40	44	0.0081	0.0084	2.65mm	2.70mm	9/64in	1/4in
No.6	0.1380	32	40	0.0093	0.0103	2.85mm	2.95mm	3.90mm	6.90mm
No.8	0.1640	32	36	0.0142	0.0149	3.50mm	9/64in	4.60mm	8.00mm
No.10	0.1900	24	32	0.0179	0.0203	3.90mm	4.10mm	5.20mm	23/64in
1/4	0.2500	20	28	0.0324	0.0368	5.20mm	5.50mm	17/64in	27/64in
5/16	0.3125	18	24	0.0532	0.0587	6.60mm	6.90mm	8.50mm	31/64
3/8	0.3750	16	24	0.0786	0.0886	8.00mm	8.50mm	10.20mm	39/64
7/16	0.4375	14	20	0.1078	0.1198	9.40mm	9.90mm	11.80mm	43/64
1/2	0.5000	13	20	0.1438	0.1612	10.80mm	11.50mm	17/32in	51/64
5/8	0.6250	11	18	0.229	0.258	13.50mm	14.50mm	21/32in	59/64in
3/4	0.7500	10	16	0.338	0.375	16.50mm	11/16in	25/32in	1.1/16in
7/8	0.8750	9	14	0.467	0.513	49/64in	51/64in	29/32in	1.3/16in
1	1.0000	8	12	0.612	0.667	22.25mm	23.25mm	1.1/32in	1.3/8in

TABLE XVIIID – ISO METRIC THREADS (60° INCLUDED THREAD ANGLE)

Nominal Size		Pitch	Stress Area	Recommended Tap Drill Size	Shank Clearance Hole Dia.	Recommended Counterbore Diameter
mm	in	mm	mm^2	mm	mm*	mm
M3	0.118	0.5	5.03	2.5	3.4	6
M4	0.158	0.7	8.78	3.3	4.5	8
M5	0.197	0.8	14.2	4.2	5.5	10
M6	0.236	1.0	20.1	5.0	6.6	11
M8	0.315	1.25	36.6	6.8	9	15
M10	0.394	1.5	58	8.5	11	18
M12	0.472	1.75	84.3	10.2	13	20
M16	0.630	2	157	14.0	17	26
M20	0.787	2.5	245	17.5	21	34
M24	0.945	3	353	21.0	25	40

* It is essential that the sharp edge of the shank clearance hole be suitably chamfered to clear the rounded shoulder under the head diameter without excessive reduction of the underhead bearing surface.

(vii) SET SCREWS AND GRUB SCREWS

Set screws can be classified by their head geometry, viz

(i) Square head – for spanner wrenching

(ii) Headless – (a) square or round, slotted for screwdriver tightening

(b) hexagon or spline socketed, for key or bit tightening.

Headless set screws are known as *grub screws* in British engineering practice.

Each type is usually also available with alternative point types (although square head set screws are most commonly produced with plain or knurled cup points and grub screws almost invariably with cone or cup points) — Fig 3. Alternative point types include:-

(i) Flat
(ii) Cup
(iii) 'W' cup
(iv) External knurled cup
(v) Internal knurled cup
(vi) Diamond cup
(vii) Cone
(viii) Rounded (oval)
(ix) Dog
(x) Half-dog

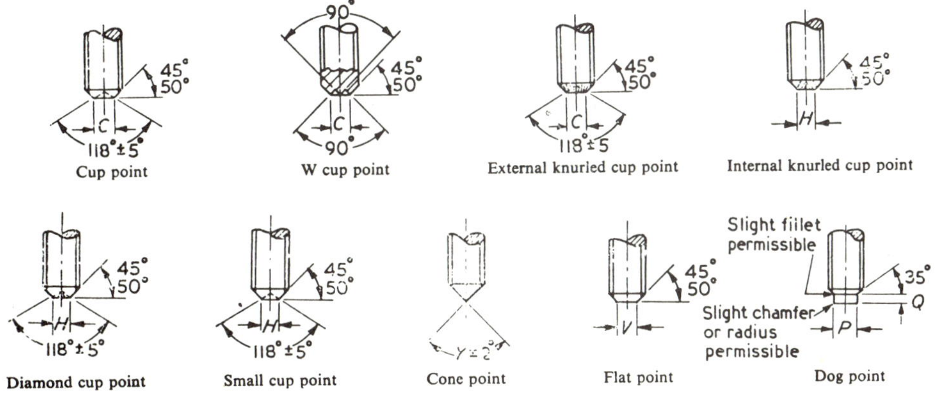

Fig 3 Set screw points (BS4168).

TABLE XIX – HEXAGON SOCKET SET SCREWS (BS4168) (Preferred Metric Sizes)

Nominal size	Hex. socket size	Key engagement	'W' cup point external knurled cup point diameters		Internal knurled diamond and small cup point diameters		Cone point angle 90° for these lengths and over; 118° for shorter lengths	Flat point diameter		Dog Point Diameter		Dog Point Length	
	Nom	Min	Max	Min	Max	Min		Max	Min	Max	Min	Max	Min
M3	1.5	1.2	1.4	1.00	1.5	1.10	5.0	2.0	1.60	2.0	1.86	2.40	2.25
M4	2.0	1.6	2.0	1.60	2.0	1.60	6.0	2.6	2.20	2.5	2.36	2.80	2.60
M5	2.5	2.0	2.5	2.10	2.5	2.10	8.0	2.4	2.92	2.5	3.32	2.80	2.60
M6	3.0	2.4	3.0	2.60	3.0	2.60	9.0	4.0	3.52	4.5	4.32	3.35	3.15
M8	4.0	3.2	5.0	4.52	4.0	3.52	12.0	5.5	5.02	6.0	5.82	4.70	4.50
M10	5.0	4.0	6.0	5.52	5.0	4.52	16.0	7.0	6.42	7.0	6.78	4.90	4.70
M12	6.0	4.8	8.0	7.42	6.0	5.52	18.0	8.5	7.92	9.0	8.78	6.40	6.20
M16	8.0	6.4	10.0	9.42	8.0	7.42	25.0	12.0	11.30	12.0	11.73	8.15	7.95
M20	10.0	8.0	14.0	13.30	10.0	9.42	30.0	15.0	14.30	15.0	14.73	7.85	7.65
M24	12.0	9.6	16.0	15.30	12.0	11.30	38.0	18.0	17.30	18.0	17.73	9.65	9.40

All dimensions in millimetres

TABLE XX – RECOMMENDED TIGHTENING TORQUES FOR SET SCREWS

CUP POINT AND 'W' POINT ISO INCH (UNIFIED) THREADS UNC AND UNF

Nominal Size	Nominal Wrench Size	Tightening Torque		Axial Holding Power
		UNC/UNF	UNC/UNF	UNC/UNF
in	A/Flats in	lbf in	lbf ft	lbf
No.4	0.050	5		143
No.5	1/16	9		240
No.6	1/16	9		217
No.8	5/64	18		366
No.10	3/32	32		561
1/4	1/8	70		933
5/16	5/32	138		1 472
3/8	3/16	234		2 080
7/16	7/32	373		2 834
1/2	1/4	564	47	3 760
5/8	5/16		91	5 813
3/4	3/8		158	8 427

Note: Cup point (Nos.4 to 8); 'W' point (No.10 and larger)

SET SCREWS : BSW AND BSF THREADS

Nominal Size	Nominal Wrench Size	Tightening Torque		Axial Holding Power
		BSW/BSF	BSW/BSF	BSW/BSF
in	A/Flats in	lbf in	lbf ft	lbf
3/16	3/32	32		569
1/4	1/8	70		933
5/16	5/32	138		1 472
3/8	3/16	234		2 080
7/16	7/32	372		2 834
1/2	1/4	564	47	3 760
5/8	5/16		91	5 813
3/4	3/8		158	8 427
7/8	1/2		320	14 629
1	9/16		455	18 187

Note : 'W' Point throughout

SET SCREWS : BA THREADS

Nominal Size	Nominal Wrench Size	Tightening Torque	Axial Holding Power
	A/Flats in	lbf in	lbf
8BA	0.035	1.6	62
6BA	0.050	5	145
5BA	1/16	9	238
4BA	1/16	9	211
3BA	5/64	18	372
2BA	3/32	32	577
OBA	1/8	70	990

Note : Cup Point (8BA to 3BA); 'W' Point (2BA and OBA)

TABLE XXI – STANDARD METRIC SIZE GRUB SCREWS (BS4219)

Nominal size and thread diameter	Pitch of thread (coarse pitch series)	Slot width		Slot depth		Cup point diameter	
		Maximum	Minimum	Maximum	Minimum	Maximum	Minimum
M1.6	0.35	0.50	0.36	1.00	0.80	–	–
M2	0.40	0.50	0.36	1.00	0.80	–	–
M2.5	0.45	0.60	0.46	1.20	1.00	–	–
M3	0.50	0.70	0.56	1.40	1.00	1.40	1.00
M4	0.70	0.80	0.66	1.60	1.20	2.00	1.60
M5	0.80	1.00	0.86	2.00	1.60	2.50	2.10
M6	1.00	1.20	1.06	2.20	1.80	3.00	2.60
M8	1.25	1.51	1.26	2.70	2.30	5.00	4.52
M10	1.50	1.91	1.66	3.20	2.80	6.00	5.52
M12	1.75	2.31	2.06	4.24	3.76	8.00	7.42

All diameters in millimetres

Set screws are normally made from carbon steel and hardened, or from alloy steel and heat treated, in natural finish. They are also available in other materials and with plated or corrosion resistant finishes. All set screws normally have full form threads extending over the whole length of the screw, except for the point. Square head set screws are threaded to the underside of the head, or to a neck relief section immediately under the head.

(viii) THUMB SCREWS AND WING SCREWS

Thumb screws are screws with a flattened head intended to be turned by gripping between thumb and finger. The flattened head form in this case is basically circular or elliptical – Fig 4.

Wing screws have a 'winged' head form, similar to that of a wing nut, enabling greater tightening torque to be obtained by manual turning than can be obtained with a thumb screw. The head may be of integral construction (eg die-cast or hot-formed) or of two-piece construction with the wings stamped from sheet, die-cast, or cold formed. Two-piece constructions may or may not be welded. There are also a variety of wing shapes to be found, but that of Fig 5 is typical.

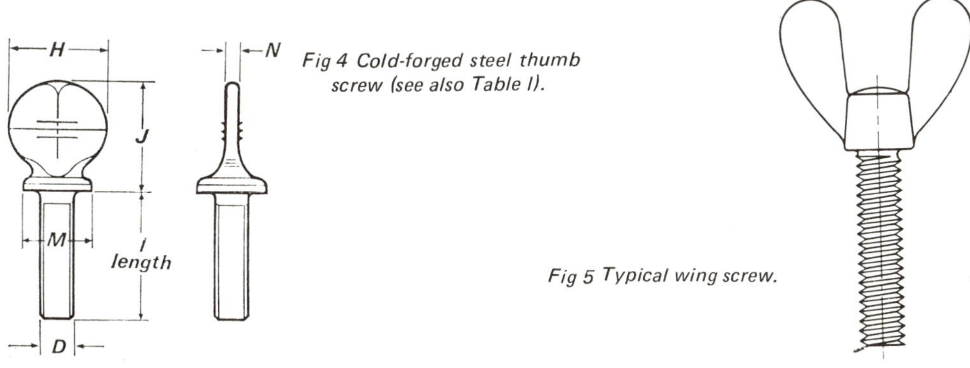

Fig 4 Cold-forged steel thumb screw (see also Table I).

Fig 5 Typical wing screw.

TABLE XXII – TYPICAL SIZE RANGE OF COLD FORGED STEEL THUMB SCREWS (GKN)
(see also Fig 4).

Nominal Size	Shouldered type							
D	H		J		M		N	
	Max	Min	Max	Min	Max	Min	Max	Min
3/16 BSW	0.422	0.398	0.480	0.456	0.347	0.323	0.064	0.052
1/4 BSW	0.551	0.521	0.635	0.605	0.465	0.435	0.072	0.058
5/16 BSW	0.697	0.667	0.780	0.750	0.585	0.555	0.095	0.079
3/8 BSW	0.827	0.797	0.952	0.922	0.705	0.675	0.114	0.098

BSW threads are rolled to BS84 medium class.
Dimensions in inches

Size Range BSW – Article 2486

Length	Diameter			
l	3/16	1/4	5/16	3/8
3/8	BSW			
1/2	BSW	BSW	BSW	
5/8		BSW		
3/4	BSW	BSW	BSW	BSW
1	BSW	BSW	BSW	BSW
1¼		BSW	BSW	
1½		BSW		BSW
2		BSW		

Dimensions in inches

Bent Bolts

The general classification *Bent Bolts* covers threaded rods with a length of plain section bent to a specific shape. The main shapes are (see also Fig 1) –

(i) Hook bolts – (a) right-angle bend (cup bolt)
(b) square bend (square hook)
(c) round bend (round hook)
(d) angled-straight bend, acute or obtuse

(ii) Eyebolts – (a) closed loop (closed eyebolt)
(b) open loop (open eyebolt)

(iii) U-bolts – (a) square bend (square U-bolt)
(b) round bend (round U-bolt)

(iv) J-bolt

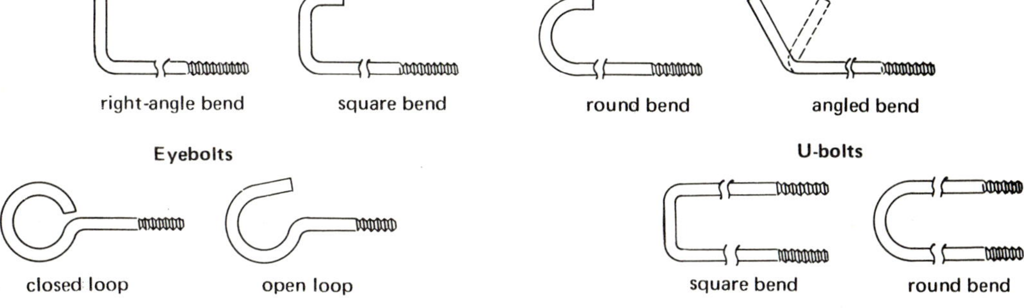

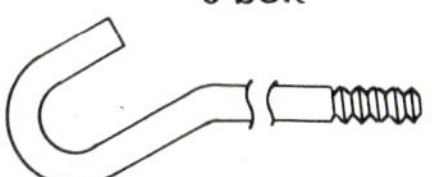

Fig 1

Dimensional nomenclature is standardised as:-

Length – length from end of thread to inside of bend end

Diameter – diameter of rod

Thread length – actual length of thread from threaded end to last complete thread form

Head size – (a) Hook bolts – inside dimension of hook *or* projected height of straight bend.

(b) Eyebolts – inside diameter of eye

(c) U-bolts – distance between parallel legs

(d) J-bolts – inside diameter of eye

Thread form can be Imperial, Unified or Metric matching the diameter of the bolt.

Other types of bent bolt may be produced for specific applications – eg *insulator spindles,* Fig 2.

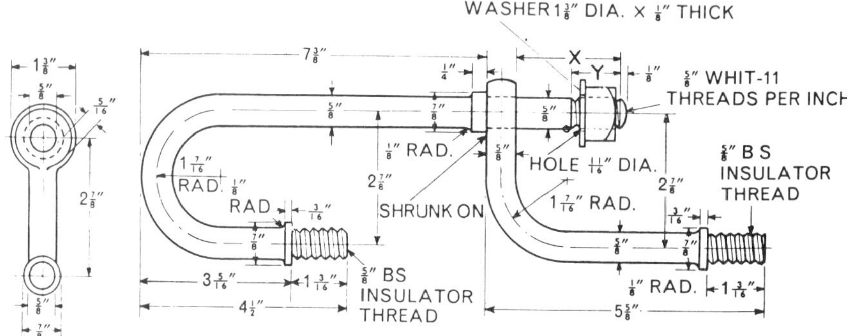

Fig 2 Single and double 'J' insulator spindles. (Thomas William Lench Ltd).

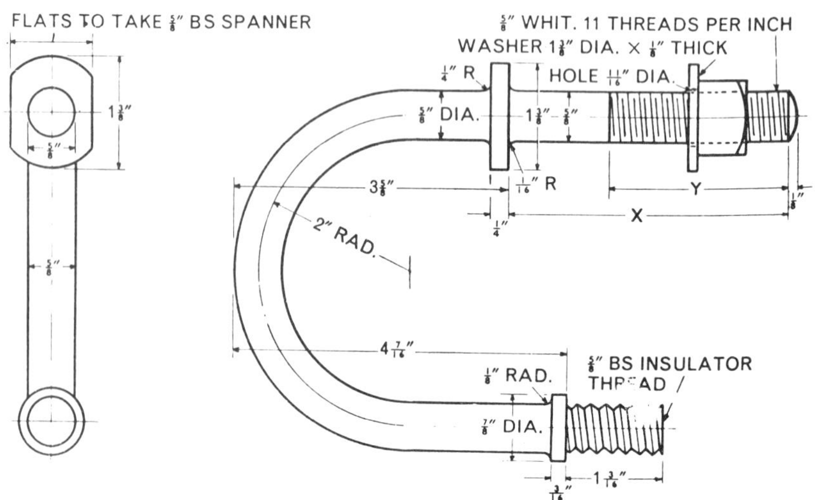

Miscellaneous and Special Bolts

FOUNDATION BOLTS

The original form of foundation bolts was the standard bolt embedded in concrete, or rag bolt (Lewis bolt) with rugged protusions for better grip in the concrete when set – Fig 1. These have largely been replaced by the modern expanded-body or indented foundation bolt – see Fig 2.

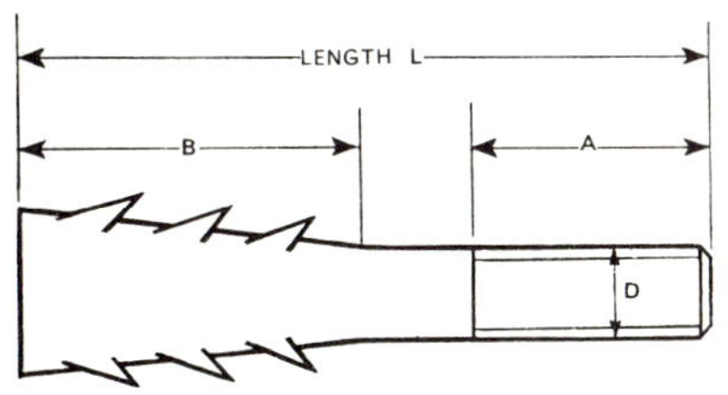

A = length of thread.
B = length of taper = ½ overall length L
D = nominal size (typically 3/8in, ½in, 5/8in and ¾in).
L = overall length (typically 3–12in) in 1in steps, depending on size)

Fig 1 Rag bolt.

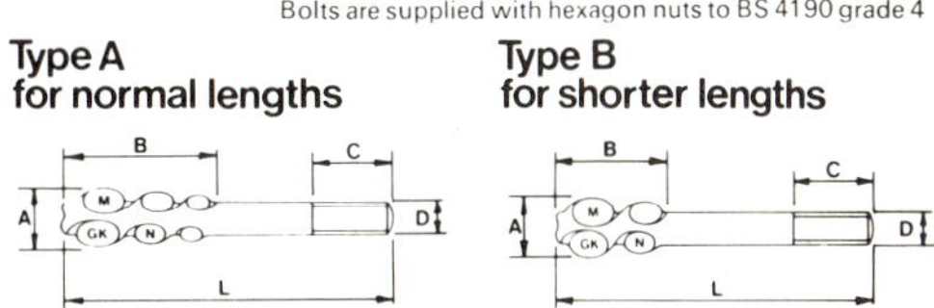

Fig 2 GKN indented foundation bolt. (See also Table I).

Standard bolts of this type are made in diameter sizes from ¼in to 1½in, indented length being five times the diameter. Overall diameter, over the indented section, is approximately twice the bolt diameter. Recommendations for depth of embedding (embedded length) are:-

Indented length + 1½ x bolt diameter for static loads

Indented length + 3 x bolt diameter for live loads

Indented length + 4 x bolt diameter for live loads with heavy vibrations

Foundation bolts are produced in mild steel, alloy steel, stainless steel and non-ferrous metals. They can also be supplied with protective coatings – eg galvanised or cadmium plated.

TABLE I – GKN FOUNDATION BOLTS

				Minimum length which should be embedded in concrete			
				Static loads	Live loads	Heavy vibration	
Dia D	'A'	'B'	'C'	B+ $1\frac{1}{2}$D	B+ 3D	B+ 4D	Shortest bolt length
M6	12	30	15	39	48	54	50
M8	16	40	20	52	64	72	65
M10	20	50	25	65	80	90	80
M12	24	60	30	78	96	108	100
M16	32	80	40	104	128	144	130
M20	40	100	50	130	160	180	160
M24	48	120	60	156	192	216	190
M30	60	150	75	195	240	270	240
M36	72	180	90	234	288	324	280

				Minimum length which should be embedded in concrete			
				Static loads	Live loads	Heavy vibration	
Dia D	'A'	'B'	'C'	B+ $1\frac{1}{2}$D	B+ 3D	B+ 4D	Shortest bolt length
M16	32	64	32	88	112	128	100
M20	40	80	40	110	140	160	130

MINIMUM BREAKING LOADS

DIA	M8	M10	M12	M16	M20	M24	M30	M36
kN	14·4	22·7	33·1	61·6	96·1	138	220	321

The ultimate strength of the bolt is limited only by the stress area of the thread. The holding power of the bolt is limited by the strength of the cement (provided the bolt was installed correctly without air pockets).

An alternative type of ragbolt based on a plain rod with a splayed or bifurcated end is known as a *fishtail bolt*.

See also *Screw Anchors and Bolt Anchors*

HYDRAULICALLY STRETCHED BOLTS

Special designs of high duty bolts are capable of being stretched longitudinally as a preliminary to fitting in 'undersize' holes. On release of hydraulic pressure the bolt diameter expands to provide a predetermined degree of fit. Such bolts can only be removed by hydraulic stretching. The purpose of such a technique is to eliminate the need for hammering or driving bolts in position where particular rigidity or accurately aligned assemblies are required, and to ensure that there is no possibility of scoring the bolt or hole surface.

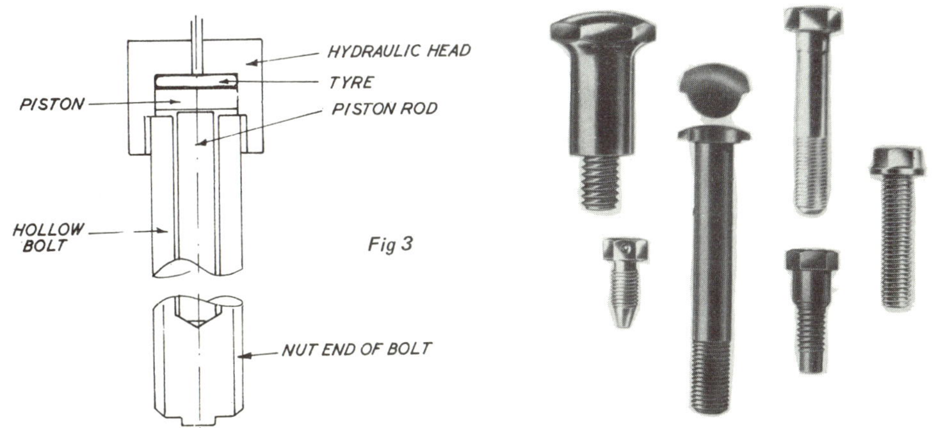

Fig 3

Special and patterned high tensile bolts by Glynwed.

A bolt designed for hydraulic stretching consists of a blind drilled bolt with a reduced diameter screwed head which can be assembled as a unit with a loose piston rod, piston, rubber 'tyre' and a hydraulic head, in that order — Fig 3. Hydraulic pressure is applied directly to the centre of the hydraulic head and is transmitted via the tyre and piston to the piston rod, generating a predetermined force sufficient to stretch the bolt the required amount, but without exceeding the maximum permissible stress in the bolt material. The reactive force is contained jointly by the hydraulic head and the bolt head to which it is attached.

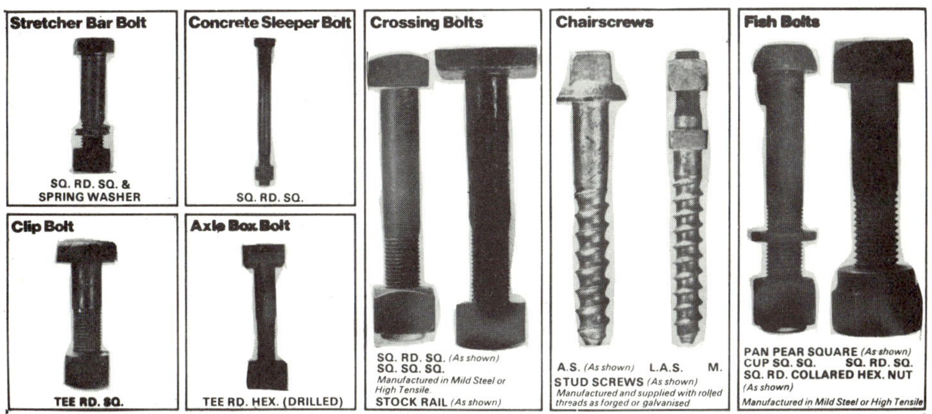

Examples of railway bolts and fastenings. (George Cooper (Sheffield) Ltd).

The total contraction which can be achieved on the bolt is of the order of D/2 500, where D is the bolt diameter. This is generally adequate to provide a radial gripping pressure of the order of 1.4tons per sq in (212bar) between the bolt and hole surface. Axial grip can also be provided, if required by controlling the run-up tightness of the nut. Hydraulic pressures used for stretching are of the order of 30 000–50 000lb/in^2 (2 100–3 500bar).

LARGE DIAMETER BOLTS

Large diameter bolts may require the use of flogging hammers and ring spanners to tighten the nut sufficiently to produce the desired preload. A number of systems have been introduced whereby final tightening can be achieved without nut rotation.

One example is the two-part rubber-filled 'Pilgrim' nut which uses solid nitrile rubber as a jacking medium – Fig 4. After initial tightening with a C-spanner or tommy bar to take up all the

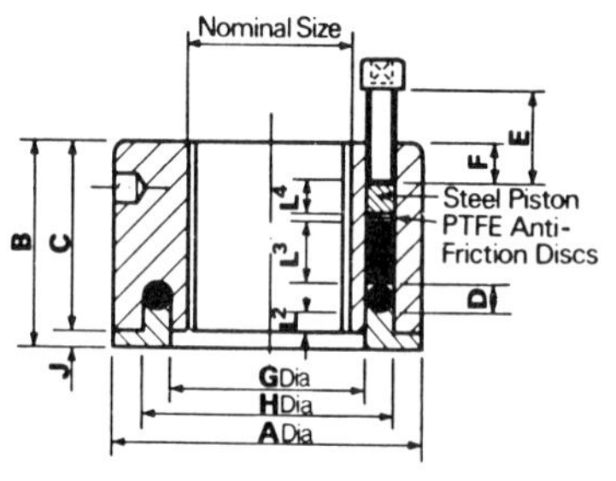

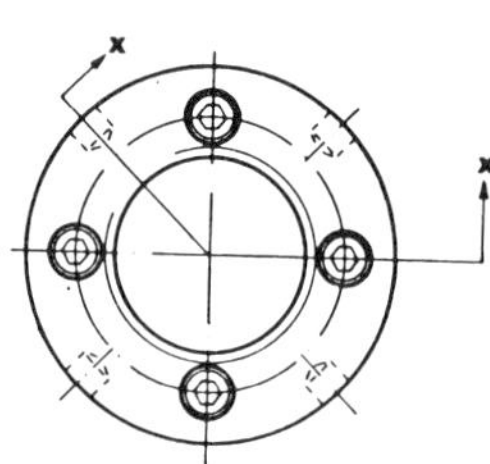

Fig 4 Rubber filled 'Pilgrim' nut (GKN).

A Outside diameter.
B Overall height
C Length of engagement.
D Diameter rubber tyre
E Stressing screw length
F Depth of tapping in nut
G Inside diameter piston
H Outside diameter piston
J Piston flange thickness
K P.C.D. stressing screw
L^1 Stressing screw diameter
L^2 Height of piston groove
L^3 Height of rubber plug
L^4 Height of steel plunger
M Maximum shim
N Ring area in^2 (mm^2)

TABLE II – MAXIMUM SHIM SIZES FOR 'PILGRIM' NUTS

RUBBER-FILLED NUTS				OIL-FILLED NUTS			
Nominal Size		Shim		Nominal Size		Shim	
in	mm	in	mm	in	mm	in	mm
1	M24	0.036	0.914	1–1.5/8	M24–M39	1/8	3.17
1¼	M30	0.036	0.914	1¾–2	M42–M48	5/32	3.97
1.3/8	M36	0.026	0.660	2¼	M56	3/16	4.76
1½	M36	0.023	0.584	2½	M64	7/32	5.56
1.5/8	M39	0.020	0.508	2¾	M68	¼	6.35
1¾	M42	0.031	0.787	3	M72	¼	6.35
2	M48	0.026	0.660	3¼	M80	5/16	7.94
2¼	M56	0.034	0.864	3½	M90	5/16	7.94
2½	M64	0.042	1.067	3¾	M95	5/16	7.94
2¾	M68	0.037	0.940	4	M100	3/8	9.52
3	M72	0.046	1.168	4½	M110	3/8	9.52
3¼	M80	0.068	1.727	5	M125	7/16	11.11
3½	M90	0.062	1.575	5½	M140	½	12.70
3¾	M95	0.057	1.448	6	M150	½	12.70
4	M100	0.080	2.032	6½	M160	9/16	14.29
4½	M110	0.068	1.727	7	M180	5/8	15.87
5	M125	0.084	2.134	7½	M190	11/16	17.46
5½	M140	0.102	2.591	8	M200	¾	19.05
6	M150	0.090	2.286	8½	M210	¾	19.05
				9	M220	13/16	20.64

A Outside diameter
B Overall height
C Length of thread engagement
D Tyre diameter
E Hydraulic connection
F Depth of tapping in nut
G Inside diameter piston
H Outside diameter piston
J Piston flange thickness
K P.C.D. of hydraulic connection
M Maximum shim
N Ring area sq ins (mm^2)
P Maximum load ton f(kN/kilos)

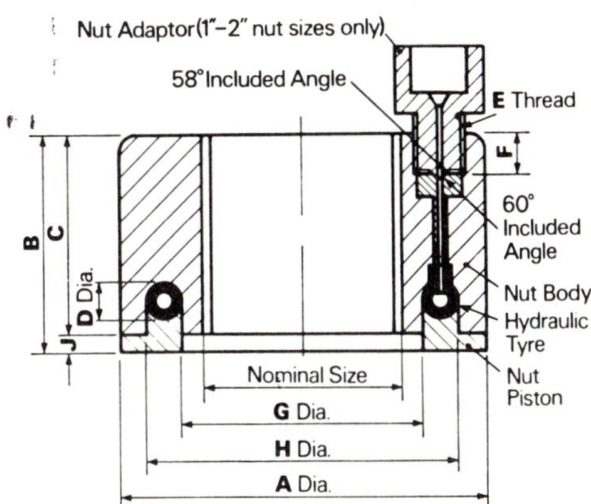

Fig 5 Oil-filled 'Pilgrim' nut (GKN).

Typical stainless steel harness being used to simultaneously pressurise 44 oil filled pilgrim nuts on a test pressure vessel at James Howden & Company Limited.

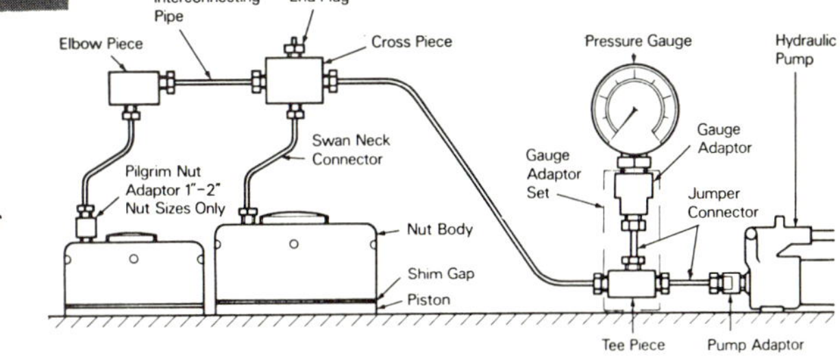

Fig 6

slack in the assembly, stressing screws are inserted, rotated hand tight and then further tightened three turns each in sequence with a socket wrench. The resulting gap between the piston and body of the nut is filled with shims, when the screws can be removed and the holes plugged to exclude dirt. Maximum shim sizes are shown in Table II.

To achieve higher bolt pre-tension (ie larger shim gaps), a similar jacking action can be provided hydraulically, as in the oil-filled 'Pilgrim' nut – Fig 5. After initial tightening hydraulic connections are made to the nut and the nut pressurised to a value which will generate the required bolt load. The resulting gap between piston and body is then filled with shims, when hydraulic pressure can be relieved and the hydraulics disconnected. Apart from the higher preloads possible, a further advantage of this system is that several bolts may be tightened simultaneously with a uniform load – Fig 6.

LIFTING EYEBOLTS

BS4278 specifies three types of eyebolt for lifting purposes:-

(i) *Collar eyebolts* for permanent attachment to massive items, normally fitted in pairs.

(ii) *Eyebolt with link* for general lifting purposes; this is an alternative to collar eyebolts where the loading cannot be confined to a single plane.

(iii) *Dynamo eyebolt* for vertical lifting only.

COLLAR EYEBOLTS TO BS4278

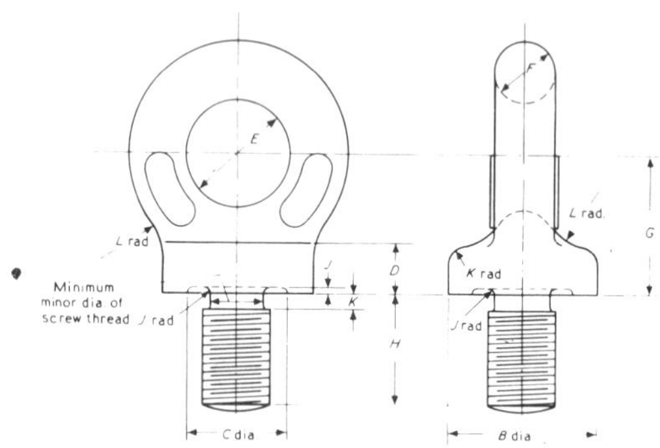

Safe working load (vertical)	Metric thread	Dia.of un-mach.shank A	Rec.forged dia.of shank	B	C	D	E	F	G	H	K	L	J
tonnef	mm	mm	mm										mm
0.32	12	10	14										1
0.63	16	13	18										1
1.00	18	16	21										1
1.25	20	18	23										1
1.60	22	20	25										1
2.00	24	23	28										2
2.50	27	26	31										2
3.20	30	29	34	2.25	1.5A	0.75A	1.5A	0.9A	2A	1.75A	0.25A	0.9A	2
4.00	33	32	38										3
5.00	36	36	41										3
6.30	39	40	44										3
8.00	45	45	50										3
10.00	52	51	55										3
12.50	56	57	60										4
16.00	64	64	68										4
20.00	70	72	74										4
25.00	76	80	80										4

The standard specifically recommends that eyebolts with screwed shank sizes of less than 12mm should not be used for lifting purposes due to the possibility of high torsional stresses or excessive bending stresses being developed by screwing up too tightly or by misalignment respectively.

EYEBOLTS WITH LINK TO BS4278

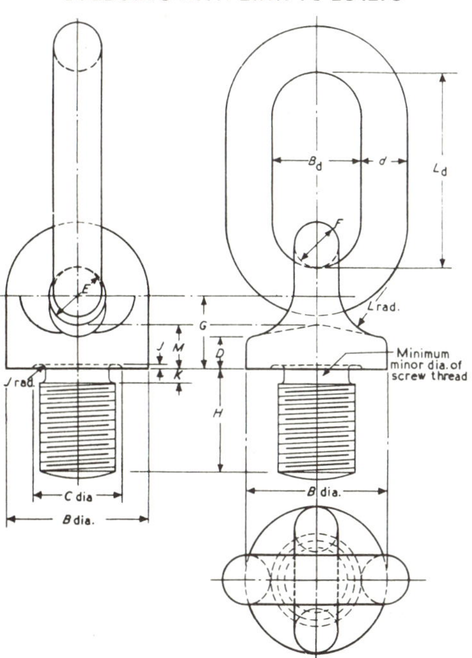

Safe working load	Metric thread	Dia. of unmach shank (A)	B	C	D	E	F	G	H	K	L	M	J
tonnef	mm	mm											mm
0.80	20	15											1
1.25	24	18											1
2.00	30	23	2.6A	1.6A	0.6A	A	0.8A	1.3A	1.8A	0.25A	0.9A	0.8A	2
3.20	36	29											2
5.00	45	36											3

LINKS : d = 0.84A B_d = 1.60A L_d = 3.50A

VINE EYES

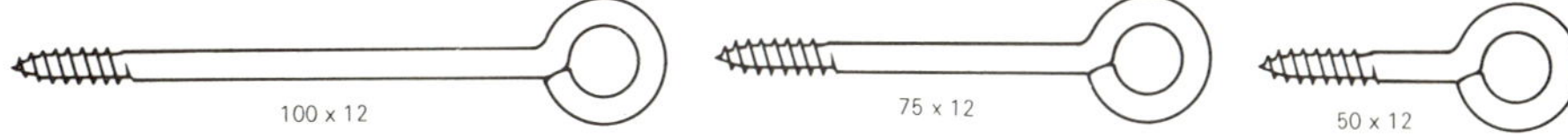

Vine eyes dimensions in mm.
(The British Screw Co Ltd)

DYNAMO EYEBOLTS TO BS4278

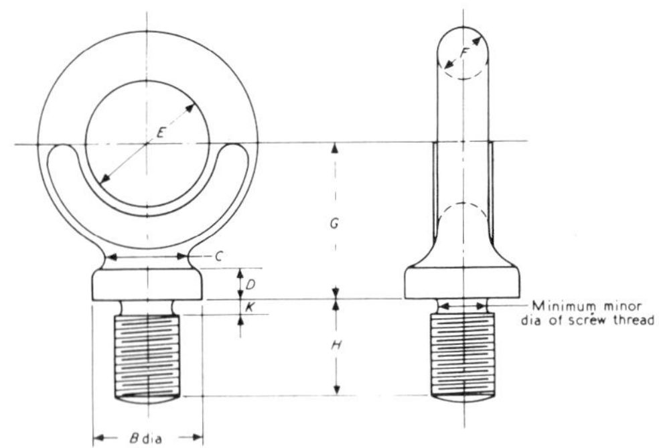

Safe working load (vertical)	Metric thread	Dia.of un-mach.shank (A)	Rec.forged dia. of shank	B	C	D	E	F	G	H	K
tonnef	mm	mm	mm								
0.32	12	10	14								
0.63	16	13	18								
1.00	18	16	21								
1.25	20	18	23								
1.60	22	20	25								
2.00	24	23	28	1.75A	1.4A	0.5A	2.2A	0.83A	2.63A	1.75A	0.25A
2.50	27	26	31								
3.20	30	29	34								
4.00	33	32	38								
5.00	36	36	41								
6.30	39	40	44								
8.00	45	45	50								
10.00	52	51	55								

SCREW HOOKS AND EYES

A screw hook is a bent shape of plain wire with a woodscrew point for inserting into a suitable surface (eg wood). Typical configurations are shown in the diagram below.

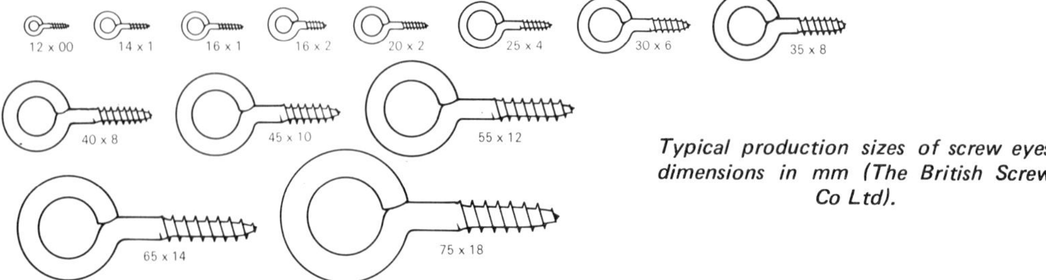

Typical production sizes of screw eyes dimensions in mm (The British Screw Co Ltd).

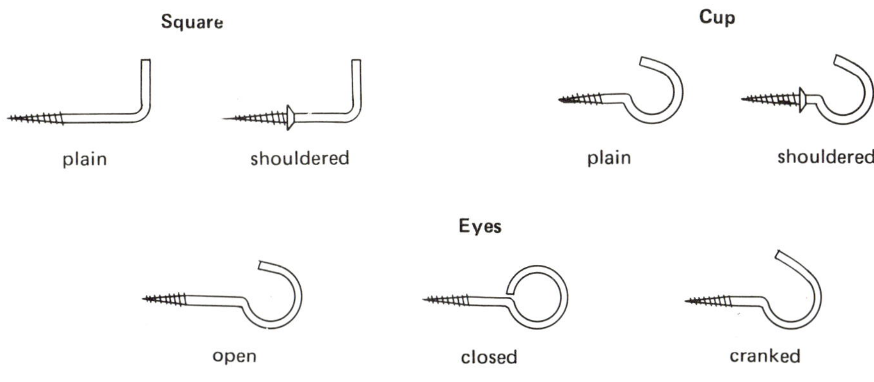

Examples of cup hooks.

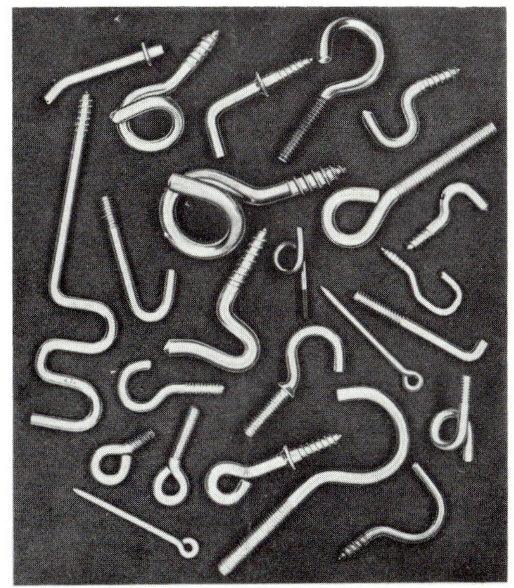

Examples of special hooks and eyes. (The British Screw Co.Ltd.)

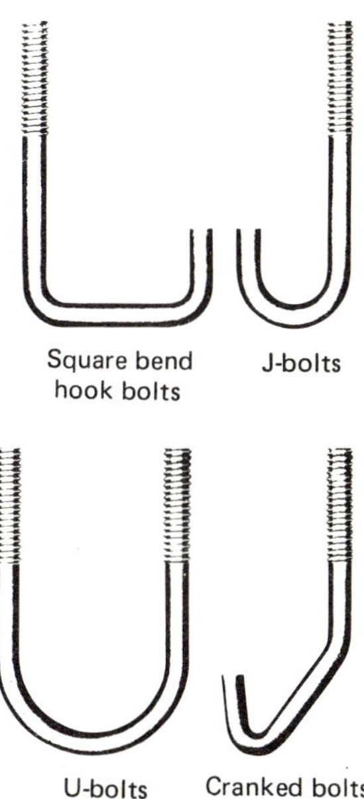

Examples of special hook bolts. (The British Screw Co.Ltd).

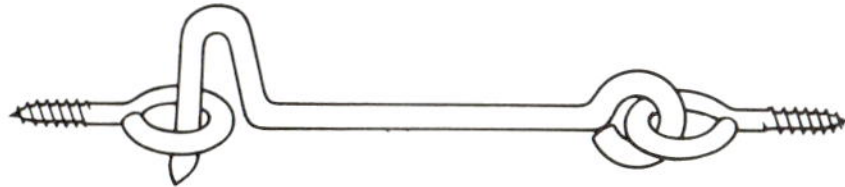

Gate hook and eye.

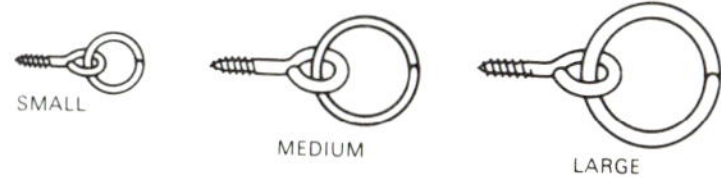

Screw rings (The British Screw Co Ltd)

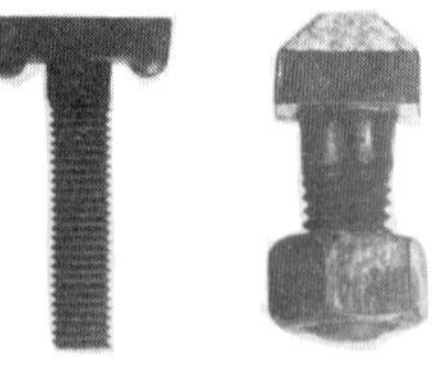

T Head bolt | Spill Plate bolt | Liner bolts

Examples of miscellaneous bolts. (George Cooper (Sheffield) Ltd).

CAST-IN SOCKETS, ETC.

Cast-in sockets are available in a wide range of designs and materials. These sockets are associated with a fixing bolt and are designed to be positioned before pouring the concrete. They may have temporary sealing plates or plugs to keep the threads clear of concrete.

Other types of cast-in fixing include channels, straps and deformed rods and hook bolts, etc. Cast-in channels are normally used for suspended fixings, fixing being by a specially headed bolt which slides in the channel, or a standard bolt screwing into a nut sliding in the channel. Cast-in rods may be used where rigid stud fixings are required.

Explosive Fixings (Cartridge Fixings)

Explosive fixings depend on the detonation of an explosive charge to drive a hardened fixing into the base material. Holding power is produced by compression of the displaced base material against the shank of the fastener, which may be a hardened and tempered steel pin or a hardened, pointed, knurled and threaded bolt. When concrete is the base material plain pins are invariably used. Additional holding strength is obtained by virtue of the conversion of the kinetic energy of the pin into heat, fusing the silicates present in the concrete into glass. The bulk of the holding power, however, is realised over the lower part of the pin – ie below the level of crack propagation.

Pins are formed with an ogive tip. For fixing into concrete the shank is plain. For fixing into steel the shank incorporates a knurled section. A variety of head forms are available.

There are two basic types of tool used for fixing pins:-

(i) Direct-acting pistol where the expanding gases from the charge act directly on the head of the pin to propel it from the barrel with high velocity and high muzzle energy.

(ii) Indirect-firing when the tool incorporates an intermediate captive piston. The expanding charge acts on the piston to hammer it against the head of the pin. Piston inertia force is responsible for ejecting the pin at a comparatively low velocity and low muzzle energy. This type of tool may be pistol form, trigger operated or be hammer-type fired by striking the head of the tool with a hammer.

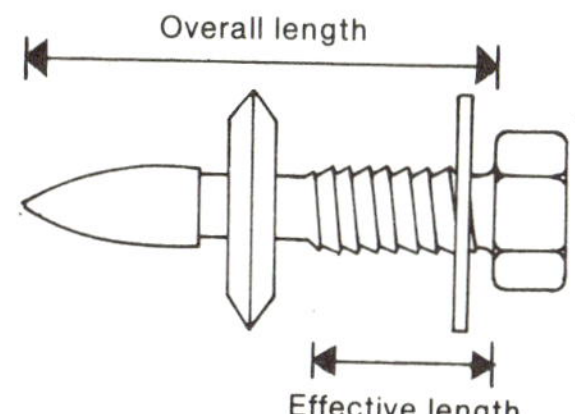

Examples of patent cartridge screw, (The British Screw Co Ltd).

TABLE I – CARTRIDGE COLOUR CODING

Colour	Strength
Green	Low
Yellow	Low–medium
Blue	Medium
Red	Medium–high

Explosive charges or cartridges are available in a range of strengths, colour coded for identification (BS4078) – Table I. Pins and cartridges are specially made for individual tools and are not interchangeable. All tools incorporate a safety device to prevent them being fired unless the tool is pressed against a firm surface and is substantially at right angles to that surface. Care in use at all times is essential for safety. The general recommendation is that the weakest cartridge in the range should be used first for a test shot, increasing the strength gradually on further test shots until the required degree of penetration is achieved.

Performance in concrete may be unsatisfactory because of:-

(i) misapplication of the tool.

(ii) unsuitable base material (concretes containing hard aggregates are generally unsuitable).

(iii) spacing is unsuitable (see Table II).

Poor or unsatisfactory results will also be obtained if the concrete is not hard enough (eg should be at least five days' old).

TABLE II – SPECIAL RECOMMENDATIONS

	General recommendation	Minimum
Spacing	More than 20 x shank diameter	20 x shank diameter
Penetration depth	8 x shank diameter	6 x shank diameter
Edge distance	15 x shank diameter	100mm with direct-acting piston; 50mm with piston-type tool

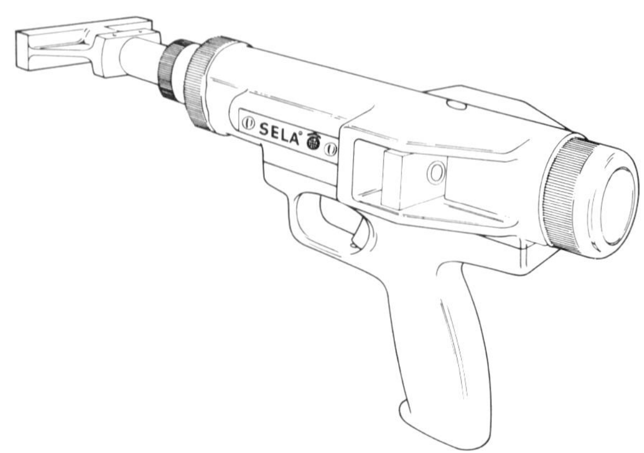

Example of piston-type cartridge tool. (The British Screw Co Ltd).

Results from firing into mild steel are generally more consistent, provided the depth penetration does not drive the pin into the material beyond the smooth length of the shank, when holding strength will be reduced. The presence of lubricants or coatings on the surface of the steel can also reduce the holding strength as these are driven through the steel with the pin. Spacing of pins can be closer than that for concrete and minimum edge distance can be reduced to 15mm.

SELA® PIN

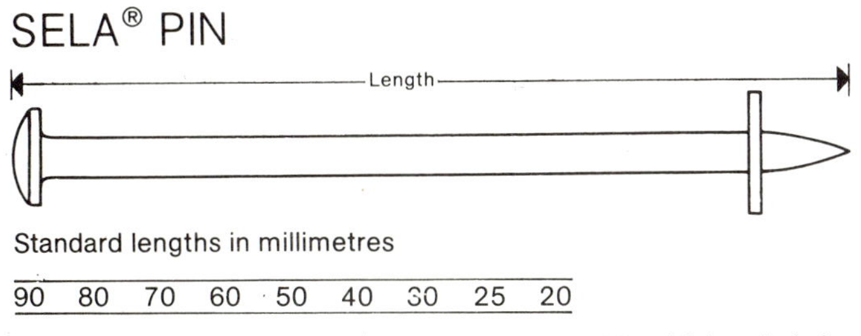

Standard lengths in millimetres

90	80	70	60	50	40	30	25	20

All sizes up to and including 50 mm. also available with knurled shanks as well as plain.

SELA® STUD

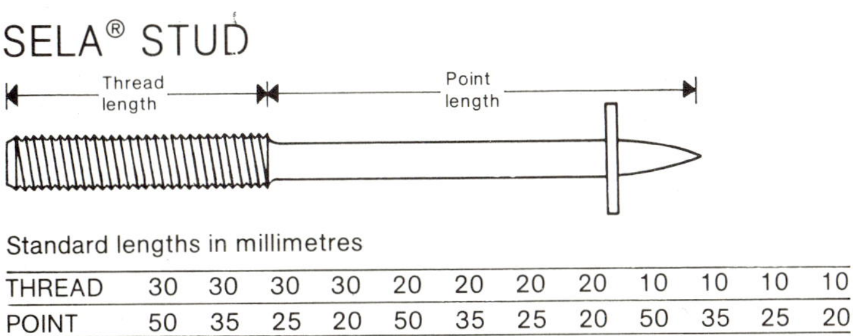

Standard lengths in millimetres

THREAD	30	30	30	30	20	20	20	20	10	10	10	10
POINT	50	35	25	20	50	35	25	20	50	35	25	20

Examples of patent cartridge pin and stud.
(The British Screw Co Ltd).

Nuts

The basic plain nut types are the square and hexagon which are applicable to most industries. Square nuts are for low-cost and/or low-torque applications, hexagonal nuts for general use. Square nuts may be unchamfered or chamfered on one face. Hexagonal nuts may be single or double chamfered – Fig 1.

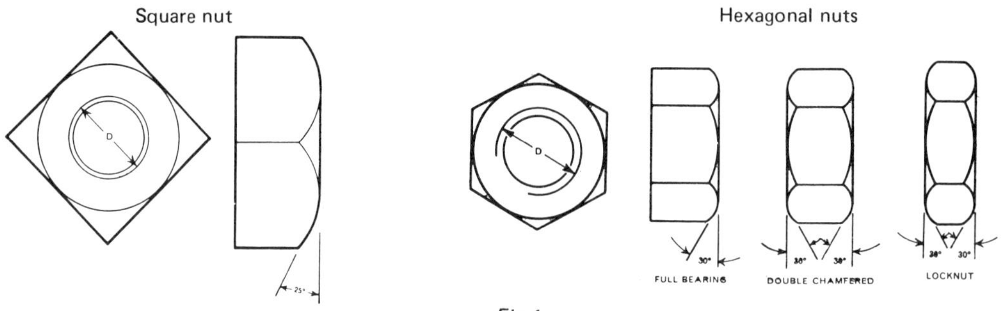

Fig 1

Other standard forms of hexagonal nut include the slotted nut and castellated forms, for use with split pins (cotter pins) for mechanical locking, and the flanged nut – Fig 2. The latter may be used where extra bearing area is required – eg for tightening a softer material or to achieve higher preloads with high-strength bolts.

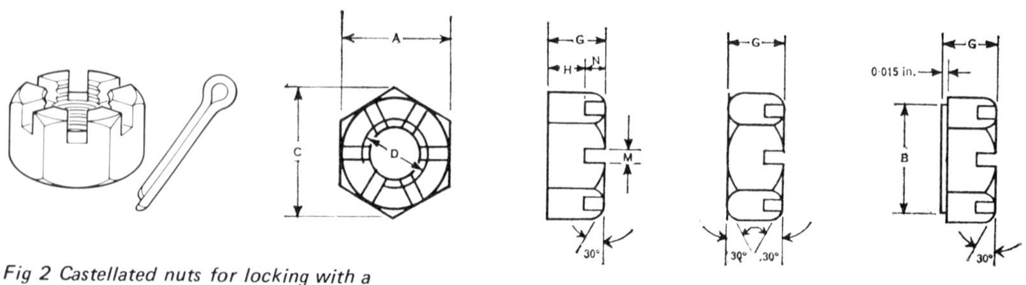

Fig 2 Castellated nuts for locking with a split pin.

Domed nuts (Fig 3) are used where it is desired to hide or cover the end of the screw thread.

The aerospace industry also uses 12-point internal and external nuts and splined nuts where high wrenching torques are required on high-strength bolts and screws. In addition there are literally hundreds of individual and patented nut designs in both non-locking (free-running) and locking configurations.

Nuts may also be described according to their method of manufacture, viz:-

(i) Pressed — brass or steel (square nuts and smaller sizes of hexagonal nuts).

(ii) Machined from bar — aluminium, brass and steel ('bright' hexagon nuts), typical size range in steel ¼in to 1½in (6mm to 375mm).

(iii) Hot forged — all metals (hexagonal nuts) typical size in steel 1½in (37.5mm) and above.

(iv) Cold forged — steel only (hexagonal nuts) — usual production of steel nuts.

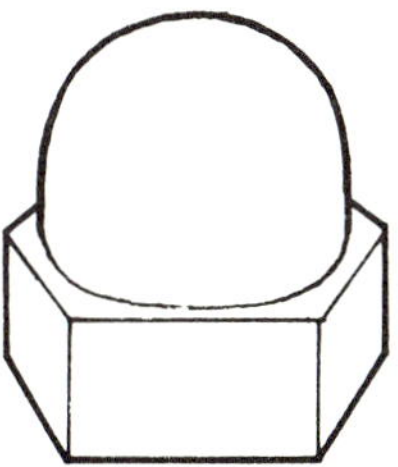

Fig 3 Domed nut.

BS Grades

British Standard grades of nuts are 0, 1, 3 and 5 (BS1768): and A, P, R and T (BS1083). Grade A is the most often used for cold forged nuts (28tonf/in^2) in conjunction with Grade R and Grade S high tensile bolts; and Grade P (35tonf/in^2) for use with Grade V bolts, machined from EN3 or EN6B steel. See also Tables I and II.

TABLE I — GRADES OF NUTS

BOLT GRADE		NUT GRADE		
Metric	Imperial	BS1768	BS1083	BS3692
4.6 or 4.8	A	O	A	4
4.6 or 4.8	B	O	A	4
5.6 or 6.6	P	O	A	5 or 6
8.8	R	—	A	8
8.8	S	1	—	8
8.8 or 10.9	T	3	P	8 or 10
10.9	V	5	R	12
12.9	X	5	T	12

TABLE II – MECHANICAL PROPERTIES OF NUTS

BS1083 : 1951

Code Symbol	Common Application	Ultimate Tensile Stress tons/sq in	Brinell Hardness Numbers
A	Nuts for A, B and R bolts	28 min	120/235
P	Nuts for T bolts	35/48	152/235
R	Nuts for V bolts	45 min	201/271
T	Nuts for X bolts	55 min	248/316

BS1768 : 1963

Grade	Common Application	Brinell Hardness Numbers	
		1in diameter and under	Over 1in diameter
		HB 10/3000	HB 10/3000 min
0	Nuts for A and B bolts	120–235	120
1	Nuts for S bolts	163–240	180
3	Nuts for T bolts	183–300	230
5	Nuts for V and X bolts	270–335	270

ISO Metric

Grade	Proof Load Stress	
	kgf/mm^2	N/mm^2
8	80	784
12	120	1 176
14	140	1 372

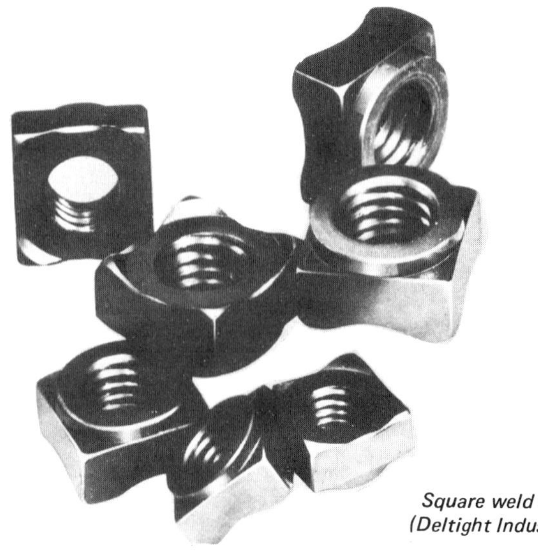

Square weld nuts. (Deltight Industries).

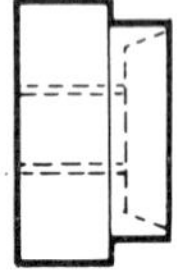
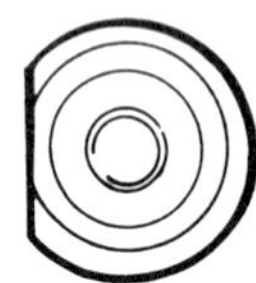
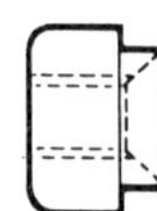

Examples of special nuts. Double flat and single flat 'Fixt-nuts'. (Barton Cold-Form Ltd.)

Standard nut sizes are given in Tables III, IV and V (see also chapter *Dimensional Data*). Nuts may also be described as 'thick' (standard) or 'thin'. The latter are intended for use as lock nuts used in conjunction with a thick (standard) nut, and for this reason may also be described as 'jam nuts'. Contrary to common belief, GKN recommend that the thin (locking) nut should go on first – see chapter on *Locking Nuts.*

TABLE III – PRESSED NUTS TO BS2827

(Imperial Sizes)

Size		Width across flats		Thickness	
Whit BSF	*BA*	*Max*	*Min*	*Max*	*Min*
Inches		*Inches*	*Inches*	*Inches*	*Inches*
3/32	6	0·250	0·244	0·082	0·078
1/8	5	0·250	0·244	0·082	0·078
5/32	4 and 3	0·312	0·306	0·095	0·091
3/16	2	0·375	0·369	0·106	0·102
7/32	1	0·437	0·429	0·128	0·123
1/4	0	0·500	0·492	0·128	0·123
5/16		0·562	0·552	0·160	0·154
3/8		0·625	0·615	0·179	0·173

Note: Light gauge nuts are one gauge thinner than the above

TABLE IV – PRESSED NUTS TO BS3155

(Unified Sizes)

Size	Width across flats		Thickness	
	Max	*Min*	*Max*	*Min*
	Inches	*Inches*	*Inches*	*Inches*
4	0·250	0·241	0·098	0·087
6	0·312	0·302	0·114	0·102
8	0·344	0·332	0·130	0·117
10	0·375	0·362	0·130	0·117
12	0·437	0·423	0·161	0·148
1/4	0·437	0·423	0·193	0·178
5/16	0·562	0·545	0·225	0·208
3/8	0·625	0·607	0·257	0·239

TABLE VA – ISO METRIC PRESSED NUTS

Size	Width Across Flats		Thickness	
	Maximum	Minimum	Maximum	Minimum
M2.5	5.00	4.82	1.60	1.35
M3	5.50	5.32	1.60	1.35
M4	7.00	6.78	2.00	1.75
M5				2.25
M8				2.75
M10				3.70
M12				4.70

Dimensions in millimetres

Weld Nuts

Weld nuts are produced specifically for fixing to the parent metal to produce a permanently anchored nut. The design is to arrange for limited contact area and thus heat concentration at these points for ease of welding (eg by projection welding) and also protect the nut against deformation due to excessive heating during the welding operation.

TABLE V – STANDARD AMERICAN NUT DIMENSIONS (ANSI B18.2.2)

Nut Type	Nut Size	Width Across Flats		Nut Thickness		Width Across Corners
		Basic	Tolerance (Minus)	Basic	Tolerance (Plus or Minus)	Limits
Square	1/4 to 5/8 3/4 to 1.1/2	1.500 D + 0.062 1.500 D	0.050 D 0.050 D	0.875 D 0.875 D	0.016 D + 0.012 0.016 D + 0.012	Maximum = 1.4142 (Max F) Minimum = 1.373 (Min F)
Hex flat	1.1/8 to 1.1/2	1.500 D	0.050 D	0.875 D	0.016 D + 0.012	Maximum = 1.1547 (Max F) Minimum = 1.14 (Min F)
Hex flat jam	1.1/8 1.1/4 to 1.1/2	1.500 D 1.500 D	0.050 D 0.050 D	0.500 D + 0.062 0.500 D + 0.125	0.016 D + 0.012 0.016 D + 0.012	Maximum = 1.1547 (Max F) Minimum = 1.14 (Min F)
Hex and hex slotted	1/4 5/16 to 5/8 3/4 to 1.1/8 1.1/4 to 1.1/2	1.500 D + 0.062 1.500 D 1.500 D 1.500 D	0.015 D + 0.006 0.015 D + 0.006 0.050 D 0.050 D	0.875 D 0.875 D 0.875 D – 0.016 0.875 D – 0.031	0.015 D + 0.003 0.015 D + 0.003 0.016 D + 0.012 0.016 D + 0.012	Maximum = 1.1547 (Max F) Minimum = 1.14 (Min F)
Hex jam	1/4 5/16 to 5/8 3/4 to 1.1/8 1.1/4 to 1.1/2	1.500 D + 0.062 1.500 D 1.500 D 1.500 D	0.015 D + 0.006 0.015 D + 0.006 0.050 D 0.050 D	 0.500 D + 0.047 0.500 D + 0.094	0.015 D + 0.003 0.015 D + 0.003 0.016 D + 0.012 0.016 D + 0.012	Maximum = 1.1547 (Max F) Minimum = 1.14 (Min F)
Hex thick, hex thick slotted and hex castle	1/4 5/16 to 5/8 3/4 to 1.1/2	1.500 D + 0.062 1.500 D 1.500 D	0.015 D + 0.006 0.015 D + 0.006 0.050 D		0.015 D + 0.003 0.015 D + 0.003 0.015 D + 0.003	Maximum = 1.1547 (Max F) Minimum = 1.14 (Min F)
Heavy square	1/4 to 1.1/2	1.500 D + 0.125	0.050 D	1.000 D	0.016 D + 0.012	Maximum = 1.4142 (Max F) Minimum = 1.373 (Min F)
Heavy hex flat	1.1/8 to 4	1.500 D + 0.125	0.050 D	1.000 D	0.016 D + 0.012	Maximum = 1.1547 (Max F) Minimum = 1.14 (Min F)
Heavy hex flat jam	1/4 to 1.1/8 1.1/4 to 2.1/4 2.1/2 to 4	1.500 D + 0.125 1.500 D + 0.125 1.500 D + 0.125	0.050 D 0.050 D 0.050 D	0.500 D + 0.062 0.500 D + 0.125 0.500 D + 0.250	0.016 D + 0.012 0.016 D + 0.012 0.016 D + 0.012	Maximum = 1.1547 (Max F) Minimum = 1.14 (Min F)
Heavy hex and heavy hex slotted	1/4 to 1.1/8 1.1/4 to 2 2.1/4 to 3 3.1/4 to 4	1.500 D + 0.125 1.500 D + 0.125 1.500 D + 0.125 1.500 D + 0.125	0.050 D 0.050 D 0.050 D 0.050 D	1.000 D – 0.016 1.000 D – 0.031 1.000 D – 0.047 1.000 D – 0.062	0.016 D + 0.012 0.016 D + 0.012 0.016 D + 0.012 0.016 D + 0.012	Maximum = 1.1547 (Max F) Minimum = 1.14 (Min F)
Heavy hex jam	1/4 to 1.1/8 1.1/4 to 2 2.1/4 2.1/2 to 3 3.1/4 to 4	1.500 D + 0.125 1.500 D + 0.125 1.500 D + 0.125 1.500 D + 0.125 1.500 D + 0.125	0.050 D 0.050 D 0.050 D 0.050 D 0.050 D	0.500 D + 0.047 0.500 D + 0.094 0.500 D + 0.078 0.500 D + 0.203 0.500 D + 0.188	0.016 D + 0.012 0.016 D + 0.012 0.016 D + 0.012 0.016 D + 0.012 0.016 D + 0.012	Maximum = 1.1547 (Max F) Minimum = 1.14 (Min F)

D = nominal diameter (basic diameter of thread)
F = width across flats

Most weld nuts are of patent design, a typical range being shown in Fig 4 (GKN patent weldnuts). Typical applications of these various types of weldnuts are summarised in Table VI.

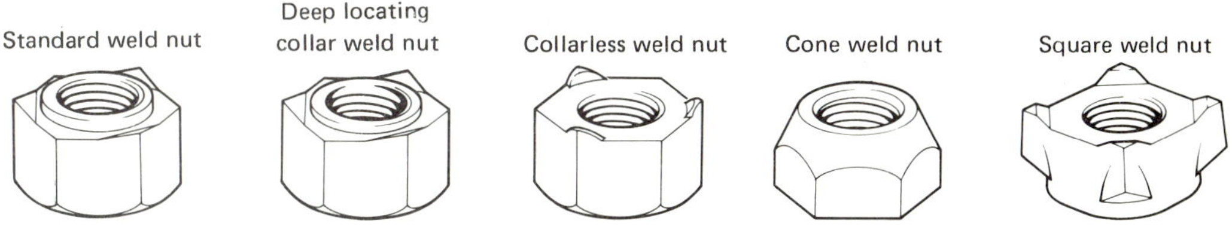

Fig 4

TABLE VI – TYPICAL APPLICATIONS OF WELD NUTS

Type	Application(s)
Standard locating collar	18–20 swg (1.2–0.9mm) sheet
Deep locating collar	8–16 swg (4–1.6mm) sheet
Standard collarless nut	Sheet less than 20 swg (0.9mm) thick
Collarless square weld nut	Most sheet thicknesses
Cone weld nut	Sheet thicker than 8 swg (4mm)
Special weld nuts with heavier projections	Thicker sheet and plate

Wing Nuts

Wing nuts are nuts with 'wings' or diametrically opposed flat arms which make it possible to turn the nut readily by hand, either for tightening or release. They can be classified by method of manufacture – eg stamped, die-cast, cold forged or hot forged; and also by style or wing geometry.

BS856 specifies three types of wing nuts, classified according to the manufacturing process –

Type HS – hot stamped products

Type DC – cast products

Type CF – cold forged products

Standard blank geometry is shown in Fig 5.

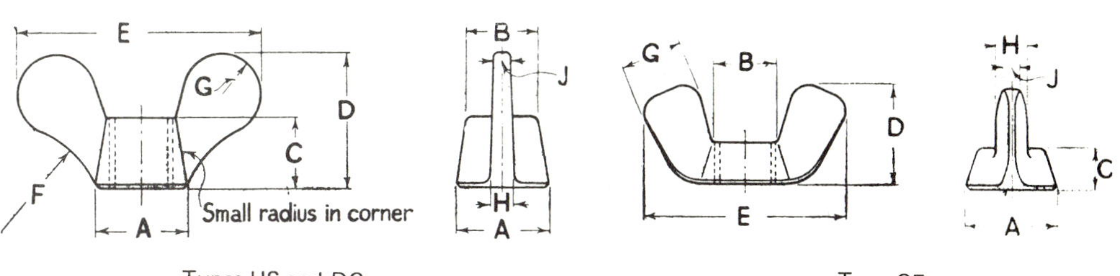

Fig 5

Weld-anchored bolts are an alternative to weld nuts for captive threaded fastenings. 'Weldbolts' by Thomas Haddon & Stokes Ltd.

TABLE VII – NOMINAL DIMENSIONS OF TYPES HS AND DC WING NUTS (BS856)

Blank No.*	Nominal screw thread size				A		B		C		D		E		F		G		H		J	
	ISO metric	UNC UNF	BSW BSF	BA	mm	in	mm	in	mm	in	mm	in	mm	in	mm	in	mm	in	mm	in	mm	in
1	M3	No. 4 & No. 6	1/8 in	5 & 4	9	11/32	6·5	1/4	7	9/32	13·5	17/32	22	7/8	19	3/4	3·5	9/64	2·5	3/32	1·5	1/16
2	M4 & M5	No. 8 & No. 10	3/16 in	3 & 2	10	13/32	8	5/16	9	11/32	15	19/32	25·5	1	19	3/4	4	5/32	2·5	3/32	1·5	1/16
3	M6	No. 12 & 1/4 in	1/4 in	1 & 0	13	1/2	9·5	3/8	11	7/16	18	23/32	30	13/16	19	3/4	5	3/16	2·5	3/32	1·5	1/16
4	M8	5/16 in	5/16 in	—	16	5/8	12	15/32	13	1/2	23	29/32	38	1 1/2	19	3/4	6·5	1/4	3	1/8	2·5	3/32
5	M10	3/8 in	3/8 in	—	17·5	11/16	14	9/16	14	9/16	25·5	1	44·5	1 3/4	19	3/4	7	9/32	5	3/16	3	1/8
6	M12	7/16 in	7/16 in	—	19	3/4	16	5/8	15	19/32	28·5	1 1/8	51	2	25·5	1	8	5/16	5	3/16	3	1/8
7	(M14)	1/2 in	1/2 in	—	22	7/8	17·5	11/16	17	21/32	32	1 1/4	59	2 5/16	25·5	1	9·5	3/8	5·5	7/32	4	5/32
8	M16	5/8 in	5/8 in	—	25·5	1	20·5	13/16	19	3/4	36·5	1 7/16	63·5	2 1/2	32	1 1/4	10	13/32	6·5	1/4	5	3/16
9	(M18)	3/4 in	3/4 in	—	32	1 1/4	27	1 1/16	22	7/8	41	1 5/8	78	3 1/16	32	1 1/4	12	15/32	7	9/32	5·5	7/32

* The numbers given in Column 1 are the customary trade designations for the sizes of the nut blanks, which are common for Type HS and Type DC wing nuts. The values of the metric dimensions given are arithmetic conversions of the basic inch dimensions, rounded-off to the nearest 0·5 mm.

TABLE VIII – NOMINAL DIMENSIONS OF TYPE CF WING NUTS (BS856)

Blank No.*	Nominal screw thread size			BA†	A		B		C		D		E		G		H		J	
	ISO metric	UNC UNF	BSW BSF		mm	in	mm	in	mm	in	mm	in	mm	in	mm	in	mm	in	mm	in
1	M3	No. 4 & No. 6	1/8 in	5 & 4	8	5/16	5	13/64	3	1/8	9	11/32	16·5	21/32	5	13/64	2·5	3/32	1	3/64
2	M4 & M5	No. 8 & No. 10	3/16 in	3 & 2	11	27/64	6	15/64	4·5	11/64	11	7/16	21·5	27/32	7	17/64	3	1/8	2	5/64
3	M6	No. 12 & 1/4 in	1/4 in	1 & 0	12	31/64	8	5/16	5	13/64	13	1/2	27	1 1/16	8	5/16	4	5/32	2·5	3/32
4	M8	5/16 in	5/16 in	—	14	9/16	9·5	3/8	6	15/64	16	5/8	31	1 7/32	9·5	3/8	5	3/16	3	7/64
5	M10	3/8 in	3/8 in	—	17	43/64	12	15/32	7	17/64	18	23/32	36	1 13/32	11	7/16	5·5	7/32	3	1/8
6	M12	7/16 in & 1/2 in	7/16 in & 1/2 in	—	22	7/8	16	5/8	9·5	3/8	23	29/32	47·5	1 7/8	15	19/32	7·5	19/64	4·5	11/64
8	M16 & M20	5/8 in & 3/4 in	5/8 in & 3/4 in	—	29·5	1 5/32	22	7/8	13·5	17/32	35	1 3/8	68	2 11/16	20·5	13/16	9·5	3/8	5	3/16

* The numbers given in Column 1 are the customary trade designations for the sizes of the nut blanks. The values of the metric dimensions given are arithmetic conversions of the basic inch dimensions, rounded-off to the nearest 0·5 mm.

Captive Nuts

Captive nuts or self-retained nuts can be either surface mounted or hole retained. They can be used for blind or semi-blind assemblies according to type, Surface-mounted captive nuts include the following types:-

(i) *Anchor nuts* – basically nuts with an extended base or base 'legs' which can be riveted or screwed to the surface to prevent the nut rotating, once mounted.

(ii) *Caged nuts* – where the nut is contained within a prefabricated 'cage' or retainer secured to the surface.

(iii) *Weld nuts* – where the nut is physically welded to the surface.

Hole-retained captive nuts include the following types:-

(i) *Clinch nuts* – a design of nut which can be partly pushed through a hole and the lower portion then clinched over to retain the nut.

(ii) *Self-clinching nuts* – which can be pressed into a preformed hole, the lower part of the nut being of a form which locks in the hole.

(iii) *Press nuts* – basically self-clinching nuts of shallower form which fit flush with the surface when pressed into a preformed hole.

(iv) *Rivet nuts* – similar to clinch nuts but specifically designed for riveting in position with special tools. Also known as rivet bushes.

(v) *Pierce nuts* – nuts with a serrated length of reduced diameter capable of both piercing a hole in the parent material under pressure and providing a locking-seating in the pierced hole.

There are also numerous individual proprietary designs which do not clearly fall into any of the aforementioned categories.

Anchor Nuts and Caged Nuts

Both anchor and caged nuts may incorporate plain or self-locking nuts. The anchor base may also take various forms, although the usual geometry is an extended base length for two-point fixing — Fig 1. Caged nuts may incorporate square or hexagonal nuts and again either plain or self-locking nuts — Fig 2. Rather than rigid assembly, the cage usually provides a certain amount of clearance for the nut to 'float', making it easier to align the bolt or screw.

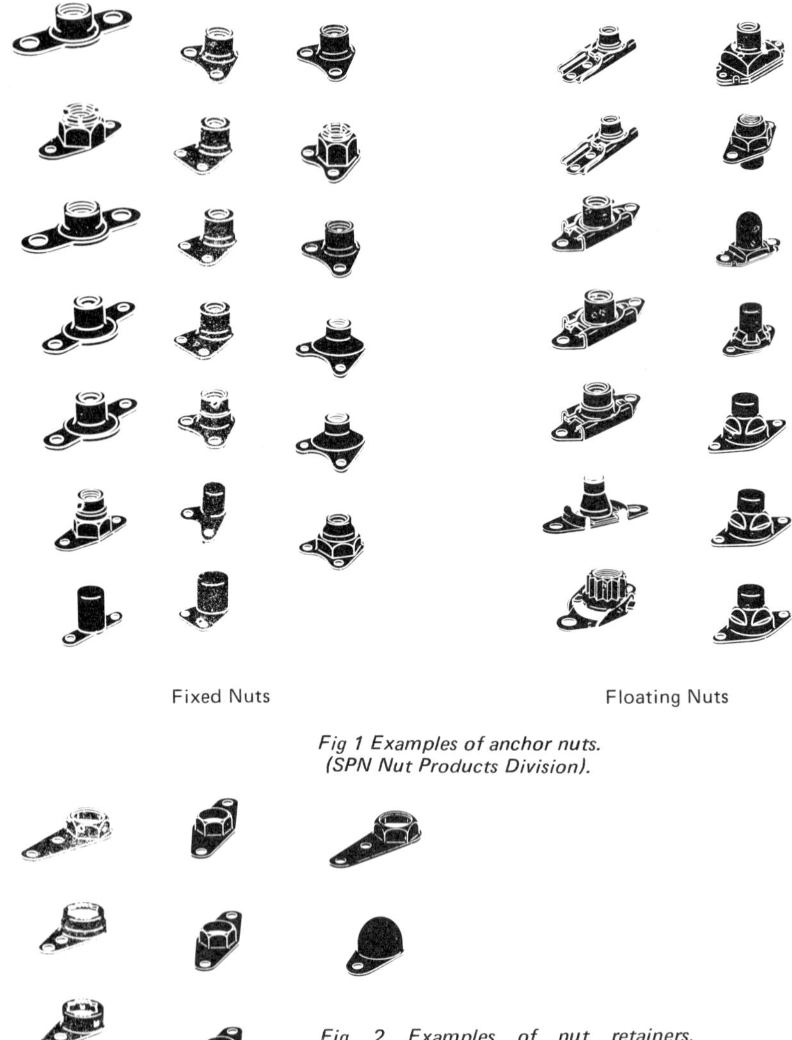

Fig 1 Examples of anchor nuts. (SPN Nut Products Division).

Fig 2 Examples of nut retainers. (SPN) Nut Products Division).

Clinch Nuts

Clinch nuts are of various individual or patented forms, one example being shown in Fig 3, together with a typical assembly sequence (see also Table I). A matching sized hole is drilled or punched in the parent sheet, the spigot end of the nut inserted and the end of the spigot riveted over flush with a hammer or convex punch. This particular type of nut is self-locking. Other types of clinch nut differ in the form and rake-off of the spigot section, and may be based on plain or self-locking nuts.

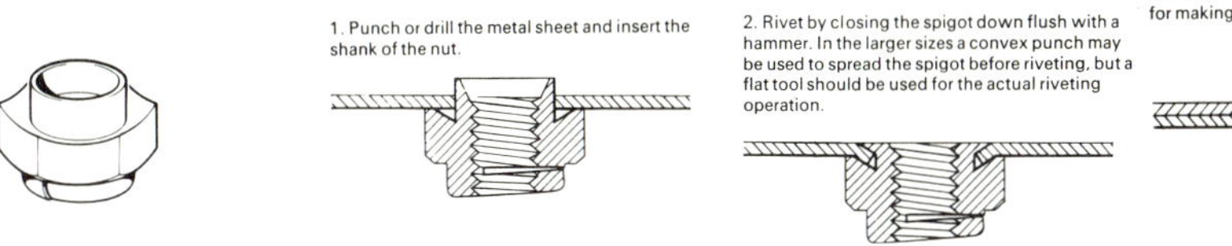

Fig 3 GKN 'Aerotight' clinch nut.

TABLE I – GKN 'AEROTIGHT' CLINCH NUTS

Size	Width Across Flats (in)	Spigot Diameter (Max) (in)	Overall Thickness (Max) (in)	Maximum Thickness of Metal Sheet swg	in	mm
BSW/BSF						
1/4"	0.525	0.375	0.319	16	0.064	1.626
5/16"	0.600	0.444	0.380	14	0.080	2.032
3/8"	0.710	0.625	0.414	12	0.104	2.642
BA						
6	0.248	0.192	0.180	18	0.048	1.219
4	0.342	0.255	0.219	18	0.048	0.219
2	0.413	0.312	0.267	18	0.048	1.219

There is also a wide variety of patented captive nut designs, as well as spring fasteners capable of providing blind or semi-blind assembly and fastening of bolts and screws – see *Threaded Inserts, Spring Nuts* and *Blind Screw Anchors, etc.*

Weld Nuts – see chapter on *Nuts.*

Rivet Nuts

Rivet nuts (or nut bushes) are conventional nut forms with an integral spigot for mounting in a matching hole – see Fig 4 and Table II; also Fig 5. The ends of the spigot are then rounded over to retain the nut in position. Rivet bushes are not normally anchored against rotation although a nylon body ring may be used on, or incorporated in, a collar to increase the amount of torque

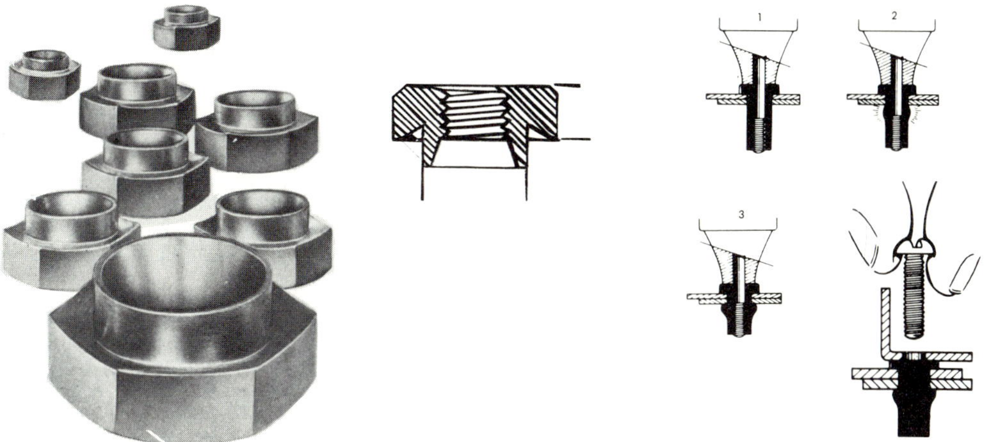

Fig 4 Rivet bushes (Deltight Industries).

Fig 5 Rivet nuts are set with a special tool.

TABLE II – STANDARD SIZES OF DELTIGHT RIVET BUSHES

Metric	Size UNC UNF	BSW BSF	BA	Spigot Dia	Width Across Flats Max	Min	Bush † Thickness	Sheet Metal Thickness swg	in
–	–	–	8	0.125	0.193	0.189	0.078	18	0.048
3	4	1/8	5 or 6	0.219	0.282	0.277	0.125	18	0.048
4	6 or 8	5/32	3 or 4	0.266	0.324	0.319	0.125	18	0.048
5	10	3/16	1 or 2	0.312	0.413	0.408	0.150	18	0.048
6	1/4	1/4	0	0.375	0.525	0.518	0.200	16	0.064
8	5/16	5/16	–	0.500	0.710	0.702	0.250	14	0.080
10	3/8	3/8	–	0.625	0.820	0.812	0.300	12	0.104

† Dimensions in inches

needed to turn the nut. Specific designs of nut bushes may, however, incorporate serrations on the bottom face of the nut or sides of the spigot, to provide locking against rotation. These are generally described as anchor nut bushes – Fig 6.

There are also self-piercing types of rivet nuts, where the spigot is shaped to provide a cutting action through the sheet material, shearing out a slug as the spigot is set.

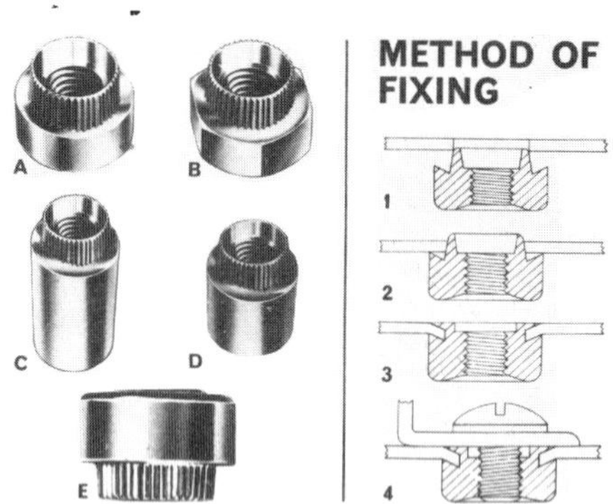

Fig 6 Serrated anchor bushes. (P.S.M.)

Clinch nuts, rivet nuts (rivet bushes) and pierce nuts are thus not clearly defined as individual types although the basic differences are:-

Clinch nuts are not necessarily heavily clinched or riveted over and are mainly for use in soft metals, and non-metals.

Pierce nuts are self-piercing and self-locking in operation.

Rivet nuts are positively riveted in place and thus less suited for use on thin sheets or weak materials where oversetting could cause dimpling or local weakening of the material, and in thicker sheet where there may be insufficient length of spigot left for satisfactory riveting over.

Self-piercing rivet nuts are essentially pierce nuts set in place with a spigot and die tool (usually in a single operation) – see Fig 7 and Tables IIIA, IIIB and IIIC.

TABLE IIIA – GKN HANK RIVET BUSHES

Size	Sheet gauge swg	Sheet thickness (in)	Die cavity 4 (in)	Cavity diameter ±0.0005(in)	C (in)
6·5 BA $\frac{1}{8}$"BSW	18	0·050	0·045	0·222	0·156
No 4 UNC	20	0·038	0·033	0·222	0·156
4,3 BA	18	0·050	0·045	0·269	0·187
$\frac{5}{32}$" BSW	20	0·038	0·033	0·269	0·187
Nos 6 & 8 UNC	22	0·031	0·026	0·269	0·187
2 BA	16	0·065	0·060	0·316	0·250
$\frac{3}{16}$" BSW	18	0·050	0·045	0·316	0·250
No 10 UNC/UNF	20	0·038	0·033	0·316	0·250
0 BA $\frac{1}{4}$" BSW	16	0·065	0·060	0·379	0·250
$\frac{1}{4}$" UNC/UNF	18	0·050	0·045	0·379	0·250
$\frac{5}{16}$" BSW $\frac{5}{16}$" UNC/UNF	14	0·080	0·075	0·505	0·250
$\frac{3}{8}$" BSW	12	0·100	0·095	0·631	0·250
$\frac{3}{8}$" UNF	14	0·080	0·075	0·631	0·250

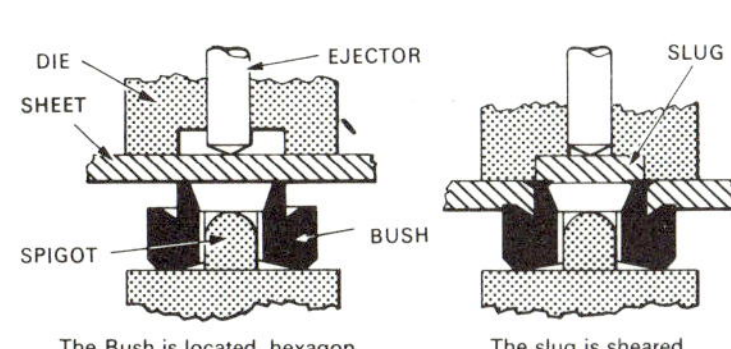

The Bush is located, hexagon downwards, on a spigot. The ram contains a die which forces the sheets onto the Bush.

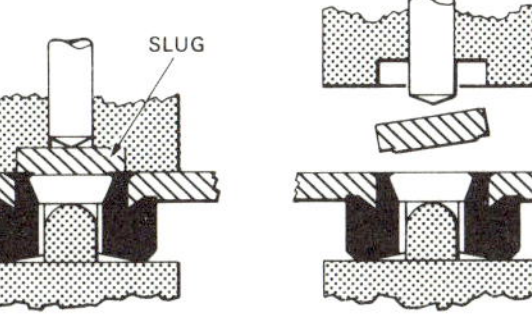

The slug is sheared and the spigot is set.

The ram retracts and the ejector moves down to eject the slug from the die cavity.

Fig 7 Assembly of GKN 'Hank' self-piercing rivet bush.

TABLE IIIB – SETTING LOADS FOR GKN HANK BUSHES

Size	Setting load for thinnest sheet to be used (lbf)
1/8" BSW 4–40 UNC 6 and 5 BA	4 500
5/32" BSW Nos 6 and 8 UNC 4 and 3 BA	5 600
3/16" BSW 10–24 UNC 10–32 UNF 2BA	5 600
1/4" BSW 1/4" UNC/UNF OBA	6 700
5/16" BSW 5/16" UNC/UNF	11 200
3/8" BSW 3/8" UNF	15 700

TABLE IIIC – SIZES AND APPLICATION DATA FOR GKN HANK BUSHES

Stock range	BSW (in)		$\frac{1}{8}$		$\frac{5}{32}$	$\frac{3}{16}$	$\frac{1}{4}$	$\frac{5}{16}$	$\frac{3}{8}$	$\frac{1}{2}$
	BA	8	6, 5	4	3	2	0			
	UNC		4–40	6–32	8–32	10–24	$\frac{1}{4}$″	$\frac{5}{16}$″		
	UNF					10–32	$\frac{1}{4}$″	$\frac{5}{16}$″	$\frac{3}{8}$″	
Suitable for sheets up to a max SWG thickness of		18	18	18	18	16	16	14	12	11
Diameter of spigot (in)		$\frac{1}{8}$	$\frac{7}{32}$	$\frac{17}{64}$	$\frac{17}{64}$	$\frac{5}{16}$	$\frac{3}{8}$	$\frac{1}{2}$	$\frac{5}{8}$	$\frac{3}{4}$
Size across flats max BSW, BA (in)		0·187	0·312	0·312	0·312	0·375	0·437	0·562	0·750	0·875
Size across flats max UNC, UNF (in)			0·312	0·312	0·312	0·375	0·437	0·562	0·750	
Depth of hexagon body (in)		0·078″	0·125	0·125	0·125	0·150	0·200	0·250	0·300	0·400
Max spigot length (in)			0·070	0·070	0·070	0·085	0·090	0·105	0·130	

Tank Bushes

Tank Bushes have a blind hole and are normally used for attaching connections to water tanks and other liquid containers – eg GKN 'Hank' rivet bushes.

TABLE IV – RECOMMENDATIONS FOR RIVET BUSH PERFORMANCE

Size	Torque to turn lb/in	Pull out load lb	Rivet setting load lb
6BA–M3–4UN	15	130	4 500
4BA–M3.5–6UN	20	150	5 600
2BA–M5–10UN	30	200	5 600
1/4in –M6	35	250	6 700
5/16in–M8	150	400	11 200
3/8in–M10	200	500	15 700

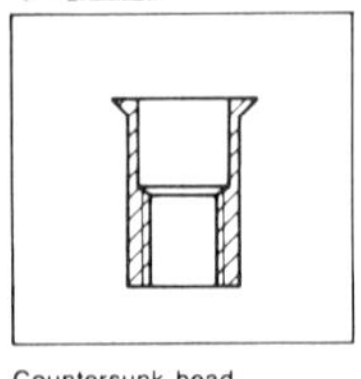

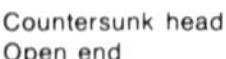

Countersunk head
Open end

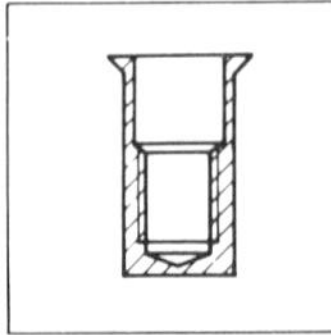

Countersunk head
Closed end

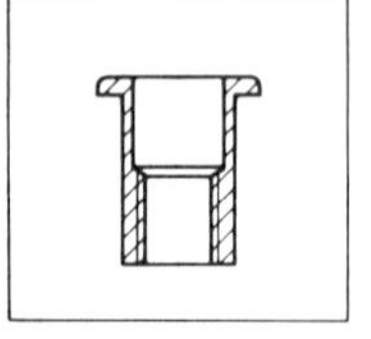

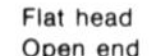

Flat head
Open end

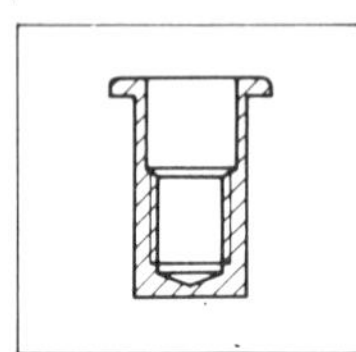

Flat head
Closed end

'Bear-nut' rivet nuts (Tucker Fasteners Ltd).

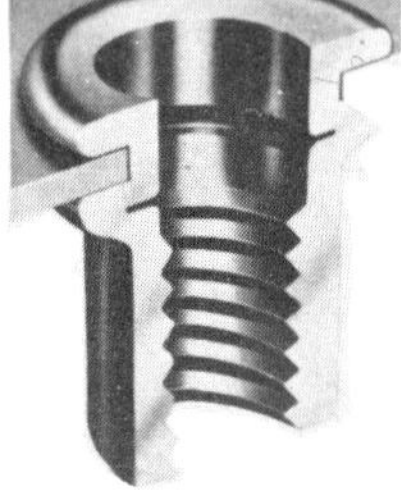

'Pul-sert' rivet nut (blind anchor nut). (P.S.M. Fasteners Ltd).

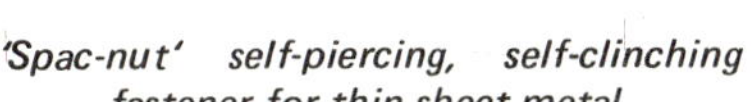

'Spac-nut' self-piercing, self-clinching fastener for thin sheet metal.

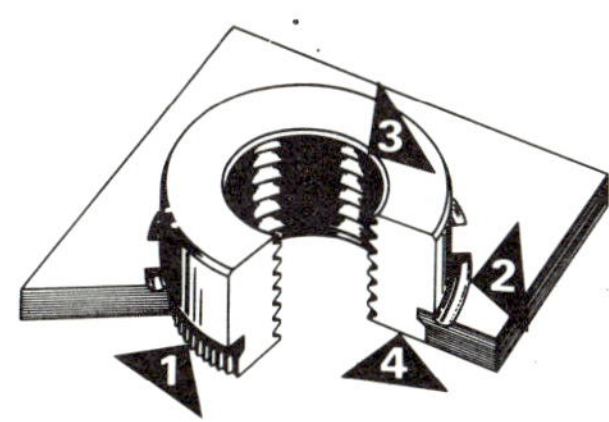

'Prestinsert' steel bushes pierce nut.

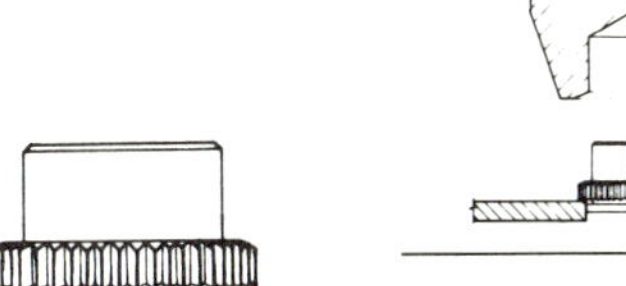

Press nut for assembly by hand punching (Instrument Screw Co Ltd).

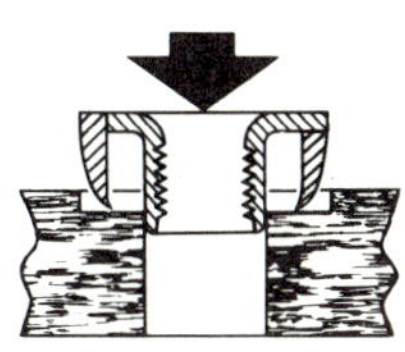

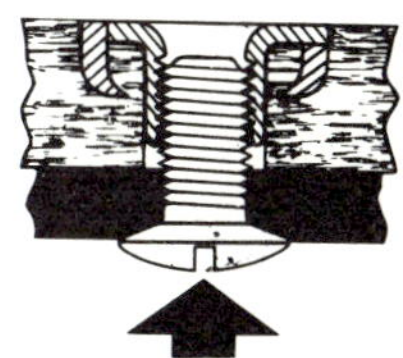

'L' 'Drive-in' nut – a proprietary variation of a self-anchoring press nut for use in wood.

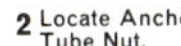

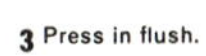

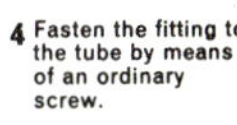

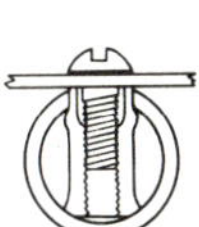

Anchor tube nut (P.S.M. Fasteners Ltd).

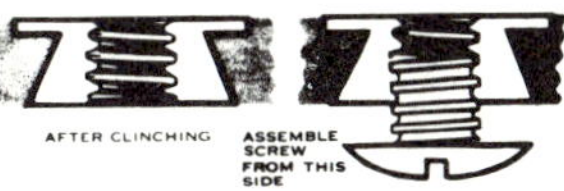

'P.S.M.' flush self-clinching fastener for use in softer materials.

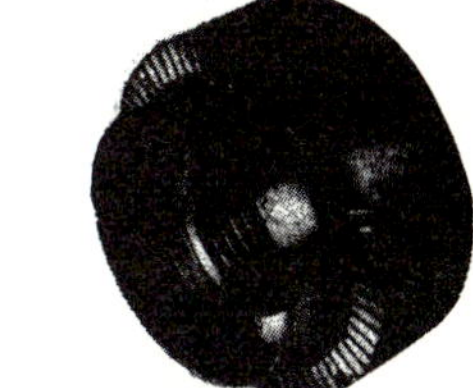

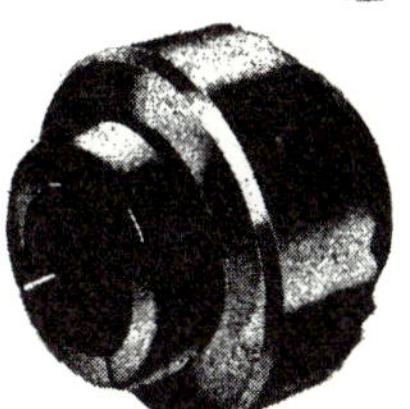

Rivet bushes by Benton Engineering Ltd. A–Wunop. B–Willock. C–Wunlock.

STRENGTH DATA FOR TYPICAL RIVET NUTS

Thread size	Material	Ultimate shear strength		Ultimate tensile strength		Upset load	
		kp	Newton	kp	Newton	kp	Newton
M 4	ALUMINIUM	125	1225	200	1960	235	2305
M 5		175	1715	275	2700	345	3385
M 6		310	3040	485	4760	650	6375
M 8		400	3925	660	6475	765	7505
M 4	STEEL	245	2400	335	3285	525	5150
M 5		360	3530	485	4760	740	7260
M 6		575	5640	950	9320	1240	12165
M 8		585	5740	840	8240	1120	10990
M10		550	5395	790	7750	1230	12070
M 4	BRASS	175	1715	200	1960	340	3335
M 5		260	2550	325	3190	430	4220
M 6		430	4220	485	4760	800	7850
M 8		380	3725	545	5345	720	7065
M10		445	4365	585	5740	750	7360

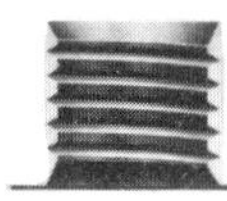

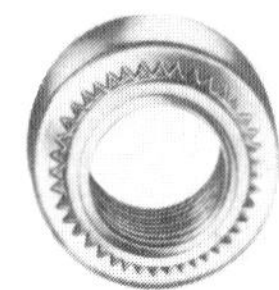

'Flexithread' swage nuts, self-locking types (left) and non-locking (right). (S.P.S. Nut Products Division).

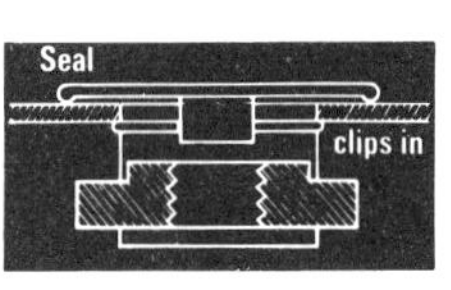

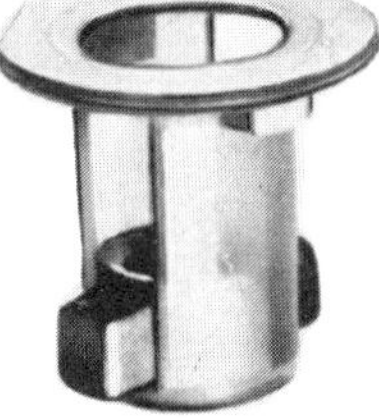

Blind fitting push-fit captive nut

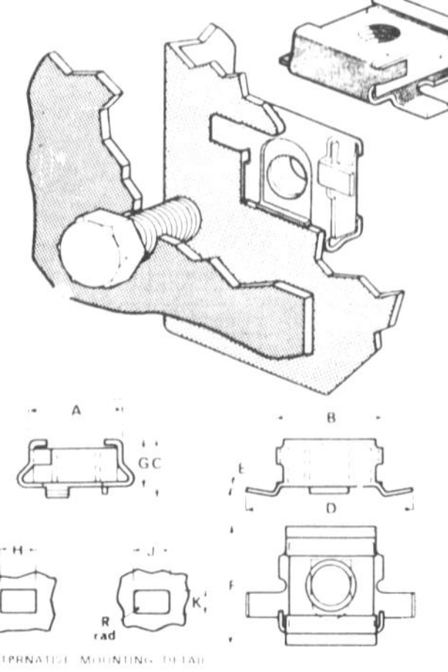
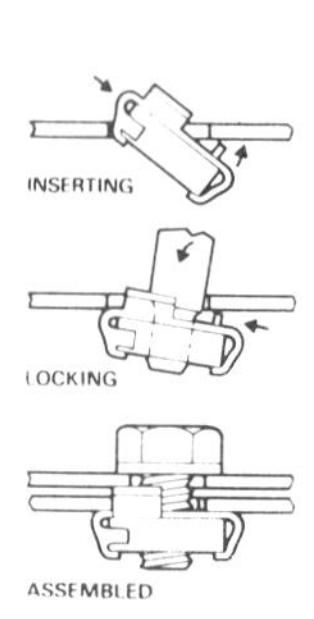

Front mounting cage nut fastener. (GKN).

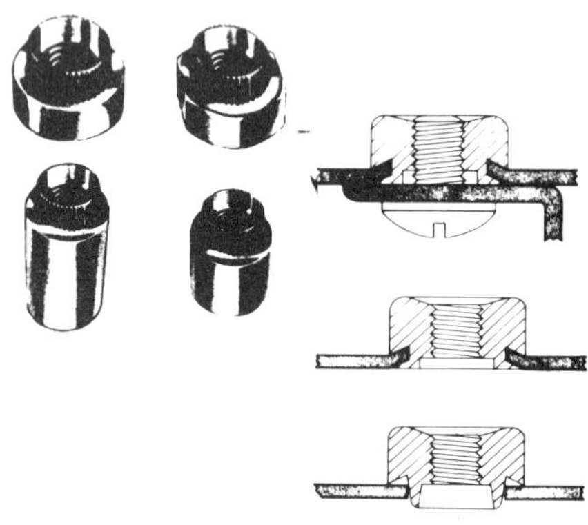

'P.S.M.' anchor right bushes.

Screw Anchors and Bolt Anchors

Blind screw anchors can be divided into blind nuts and blind screw anchors. The former also embrace anchor nuts, captive nuts, rivet nuts, welded nuts, etc, (which see) fitted as part of an assembly. The description 'blind nut' here is confined to types designed for insertion in drilled or pierced holes. Again these can be sub-divided into types for use in thin materials with clear space on the blind side for assembly and types for fitting in blind holes in thicker materials – eg for wall bolt fixing.

A variety of proprietary blind nuts for use in thin materials are two-piece, based on blind wrenching principles, the nut being captive in a hollow body. When the screw is engaged and tightened the nut is drawn up inside the body to deform it and lock the body in place.

An example of a lightweight blind nut of this type is shown in Fig 1, the body being of nylon and the internal nut aluminium. The outer diameter of the nut is knurled to prevent rotation within the body. The advantage of a blind nut of this type is that the fastener has only to be pressed into its hole until the shoulder locates against the surface and the screw engaged and tightened.

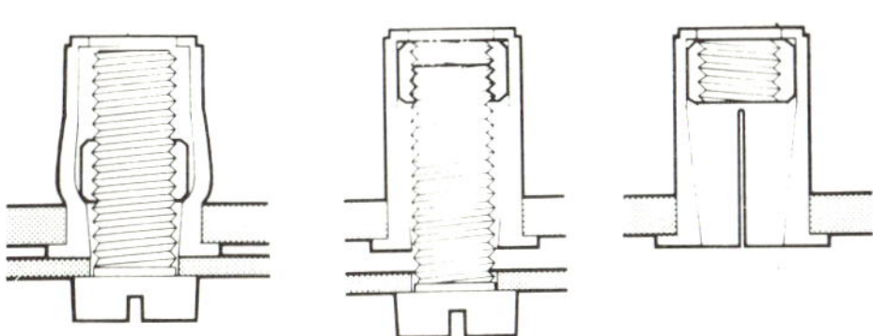

Fig 1 Spirol blind nut.

Blind screw anchors are normally based on a captive nut retained in a flexible cage (usually metal). The working principle is similar. As the screw is tightened the nut is pulled upwards, deforming the cage to grip securely in the original hole. Depending on the design of the cage, the hole size is normally less critical than in the case of a blind nut; also greater material thicknesses can be accommodated merely by increasing the length of the cage.

Simple screw anchors require no special tools for fitting and are deformed by insertion and tightening of the screw (eg Fig 2). Other types may require special tools to form the cage once the fastener has been inserted to ensure design deformation (eg Fig 3).

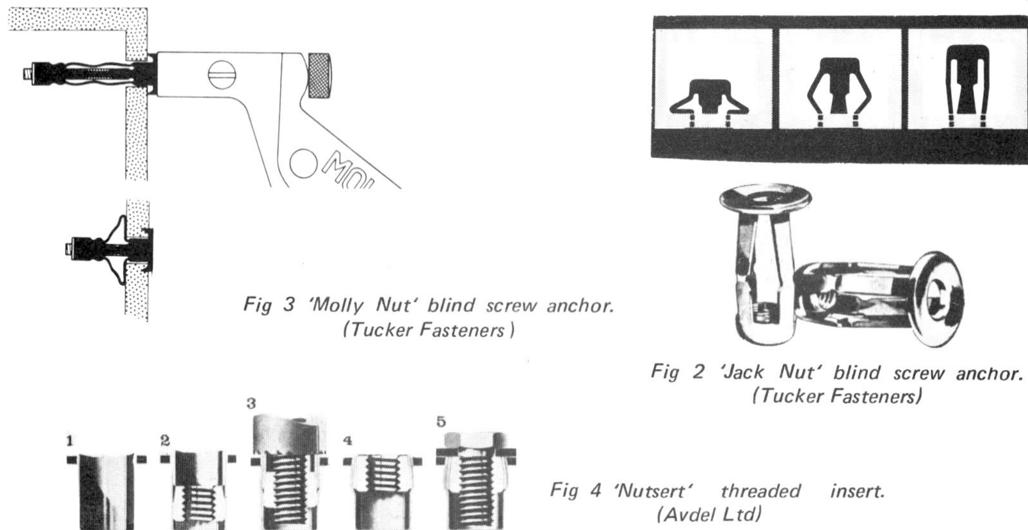

Fig 3 'Molly Nut' blind screw anchor. (Tucker Fasteners)

Fig 2 'Jack Nut' blind screw anchor. (Tucker Fasteners)

Fig 4 'Nutsert' threaded insert. (Avdel Ltd)

There are numerous types of threaded blind fasteners which are more generally classified as threaded inserts, although the distinction between a blind screw fastener and a threaded insert is not always clear. Basically, a blind screw fastener is used in thinner material and protrudes on the blind side of the assembly, whilst a threaded insert is more properly considered as a blind fastener fitting a blind hole. Manufacturers' descriptions are not consistent on this point, however. For example, the 'formed nut' type of insert shown in Fig 4 is called a threaded insert although by the aforementioned definition it would be classified as a blind screw anchor.

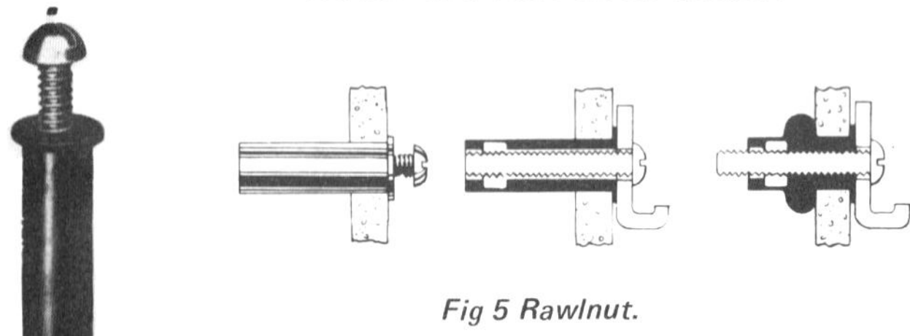

Fig 5 Rawlnut.

The 'Rawlnut' – Fig 5 – is a similar type of blind screw anchor for use in cavities and light-weight solid materials. This consists of a tough rubber sleeve with a metal nut bonded in one end and an external flange on the other end. As the screw is turned the nut travels up the screw thread compressing the rubber sleeve to form a 'rivet head' in cavities, or compressing the sleeve in the hole in solid materials. It is intended for use as a demountable fitting in plastic, glass, plywood, cellular lightweight building blocks, etc, and also for use in sheet metals to mount a fixing screw which is electrically insulated from the metal.

See also chapters on *Threaded Inserts, Spring Fasteners and Captive Nuts.*

Non-Threaded Blind Nuts

Non-threaded blind 'nuts' may be used with self tapping screws to provide superior holding power in thin sheet materials. The 'nut' in this case is normally a plastic moulding (eg nylon) which can be pressed into a matching hole on one piece of the assembly. The assembly is then completed with a thread-forming self-tapping screw yielding a high resistance to pull-out and a high clamping load on the joint – see Fig 6 and Tables IA and IB. This type of nut can be described as an expansion nut – see also chapter on *Plastic Fasteners.*

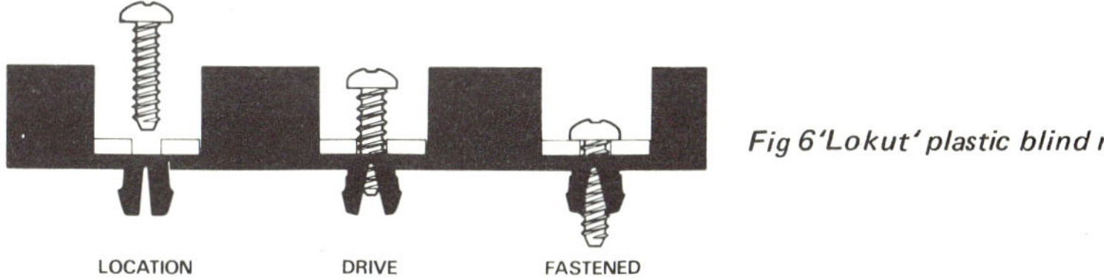

Fig 6 'Lokut' plastic blind nut.

TABLE IA – TYPICAL PERFORMANCE OF FASTEX NYLON 'LOKUT' NUT

Screw Size	Driving Torque lb/in	Driving Torque kg/cm	Stripping Torque lb/in	Stripping Torque kg/cm	Pull Out Load lb	Pull Out Load kg
6	5	5.8	10	11.5	40	18
8	8	9.2	15	17.5	100	45
8	8	9.2	25	29	120	54
10	10	11.5	40	46	120	54
12	15	17.5	55	63	500	226

TABLE IB – TYPICAL PERFORMANCE OF FASTEX NYLON REVERSE 'LOKUT' NUT

Screw Size	Driving Torque lb/in	Driving Torque kg/cm	Stripping Torque lb/in	Stripping Torque kg/cm	Pull Out Load lb	Pull Out Load kg
8	5	5.8	15	17.5	150	68
8	5	5.8	50	57.5	400	180
10	8	9.2	20	23	150	68
10	10	11.5	60	69	400	180
12	10	11.5	65	75	400	180
14	10	11.5	85	98	400	180

BOLT ANCHORS

Bolt anchors may be considered as more robust forms of blind screw anchors, designed for blind fastening bolts or studs. Such anchors are normally based on an expansion principle whereby tightening the bolt expands the anchor body or a section of the body to grip the sides of the hole into which it is fitted.

'Biff' expansion anchors illustrated in Fig 7 are available in both nut-retaining and bolt-retaining types. In the former the nut retained in the anchor is drawn up on tightening to expand the body. In the projection bolt type the bolt is contained within the anchor and the bolt head drawn up when a nut is tightened on the thread to draw the head up and expand the anchor. Pull out characteristics are summarised in Table II.

An example of a sleeve anchor is shown in Fig 8. This is designed to anchor in building blocks, brick, concrete blocks and similar materials with a relatively low compressive strength. The sleeve anchor is tapped into a hole of the same diameter and the screwhead run on and tightened a few turns to set the anchor. Subsequently heads can be removed and replaced without affecting the grip of the anchor.

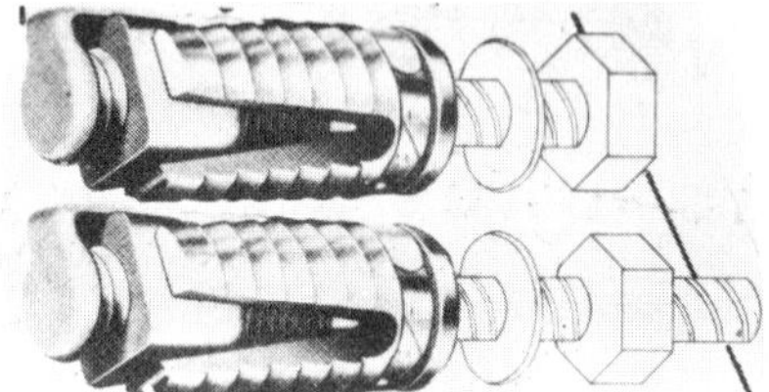

Fig 7 'Biff' expansion anchor.

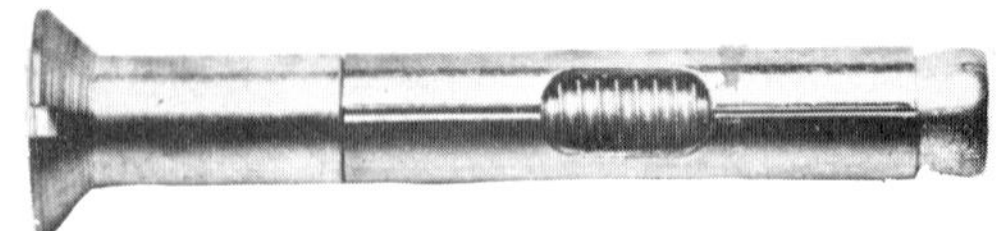

Fig 8 Parabolt sleeve anchor – set by tightening the screw head.

TABLE II – PULL OUT CHARACTERISTICS OF BIFF EXPANSION ANCHORS

R. Quality Bolt Size	Bolt Tightening Torque	Pull Out Load	Remarks
1/4	5lbs/ft	2 875lb	Shield pulled out of concrete
1/4	8lb/ft	2 660lb	Concrete cracked and shield pulled out of slab
5/16	11lb/ft	2 875lb	Cracking of concrete
5/16	13lb/ft	2 660lb	Cracking of concrete
All above tests carried out with shield flush to surface of concrete slab			
5/16	11lb/ft	3 160lb	Cracking of concrete
Above test carried out with shield 5/8in below surface of concrete slab			
3/8	19lb/ft	3 575lb	Cracking of concrete
3/8	30lb/ft	4 930lb	Cracking of concrete
All above tests carried out with shield 1/4in below surface of concrete slab			
3/8	33lb/ft	4 030lb	Cracking of concrete
Above test carried out with shield 1/2in below surface of concrete slab			
1/2	46lb/ft	5 450lb	Concrete slab broke. No movement of shield
Above test carried out with shield 1/8in below surface of concrete			
1/2	46lb/ft	7 450lb	Concrete slab broke. No movement of shield
Above test carried out with shield 1/4in below surface of concrete			
1/2	46lb/ft	7 990lb	Concrete slab broke. No movement of shield
Above test carried out with shield 1/2in below surface of concrete			

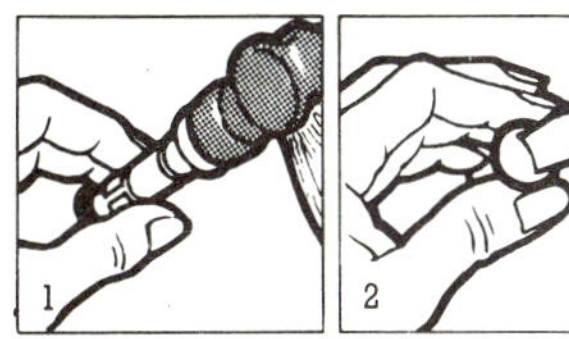

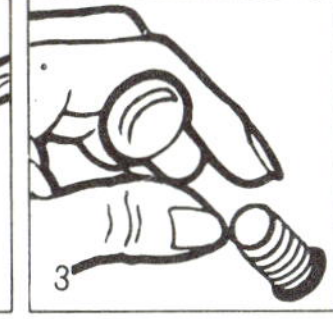

Fixing

1. Tap a Screwhead Anchor into a hole of the same diameter. Over-deep holes will not affect holding power, although ideal embedment should be at least 4½ times the anchor diameter.
2. Run on the Screwhead and tighten 3 or 4 turns to set the anchor, or until sufficient resistance is felt.
3. Installation completed. Screwheads can be removed and replaced without affecting the anchor's grip.

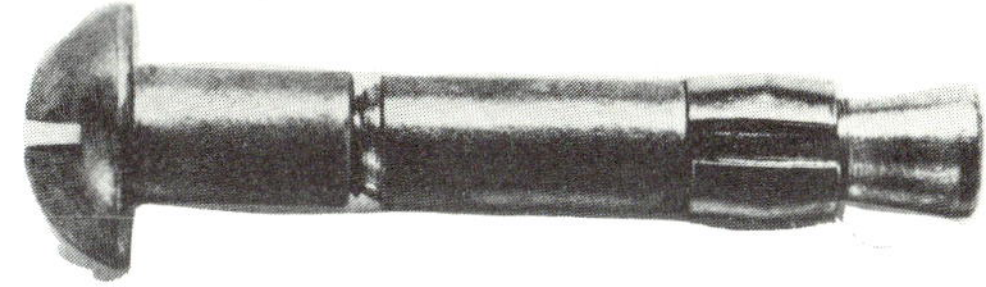

Fig 9 Screwhead anchor.

Collet-type anchors are shown in Figs 9 and 10, the anchoring device being a split sleeve immediately above a taper section on the fastener. This fits into a hole of the same diameter as the anchor. As the anchor is tightened the split sleeve is expanded to bite hard into the surrounding material. The use of a parabolic taper gives rapid sleeve expansion.

Fig 10 'Parabolt' drop-in anchor — set by tightening the nut.

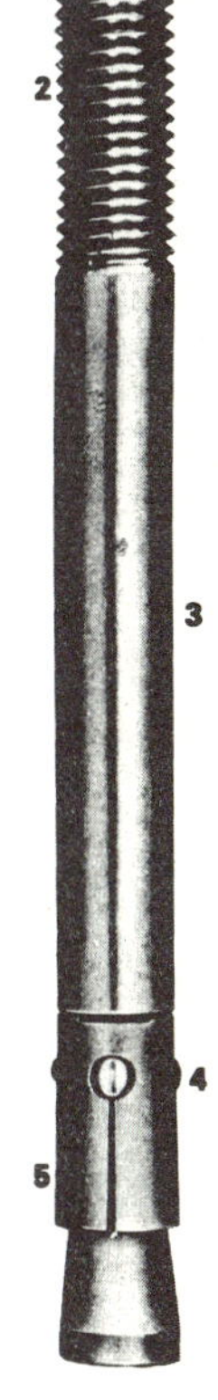

Fig 11 'Spitfix' wedging anchor bolt. Diameter sizes 6, 8, 10, 12, 16 and 20mm. Lengths 10–20mm to 110–160mm, depending on diameter size.
Note: A similar 'dovetail locking' principle is employed in the Rotospit F anchor for fixing in concrete in 2BA, ¼in, 3/8in, ½in and 5/8in diameter sizes. (Spit Fixings).

1—The chamfer protects the first few threads when the anchor is hammered into place.
2—The Spitfix is made in standard metric sizes.
3—The anchor is made of a steel and lead alloy, zinc plated and passivated.
4—The bosses on the sleeve prevent rotation in the hole.
5—The sleeve is made to elongate by 30%, this high elasticity ensuring a high pull-out strength.

Another collet-type anchor is shown in Fig 11, whilst Figs 12 and 13 show anchors which obtain their grip by the spread of a split end, the expansive end being also serrated or shaped for grip. The Rawlplug 'Loden' anchor is similar in anchoring principle, but is designed for hammering home – Fig 14. It is particularly suitable for use in aerated concrete blocks to BS2028,1364:1968, also softwoods and plywoods. It can also be used in pre-drilled holes in hardwoods, plastics or metals.

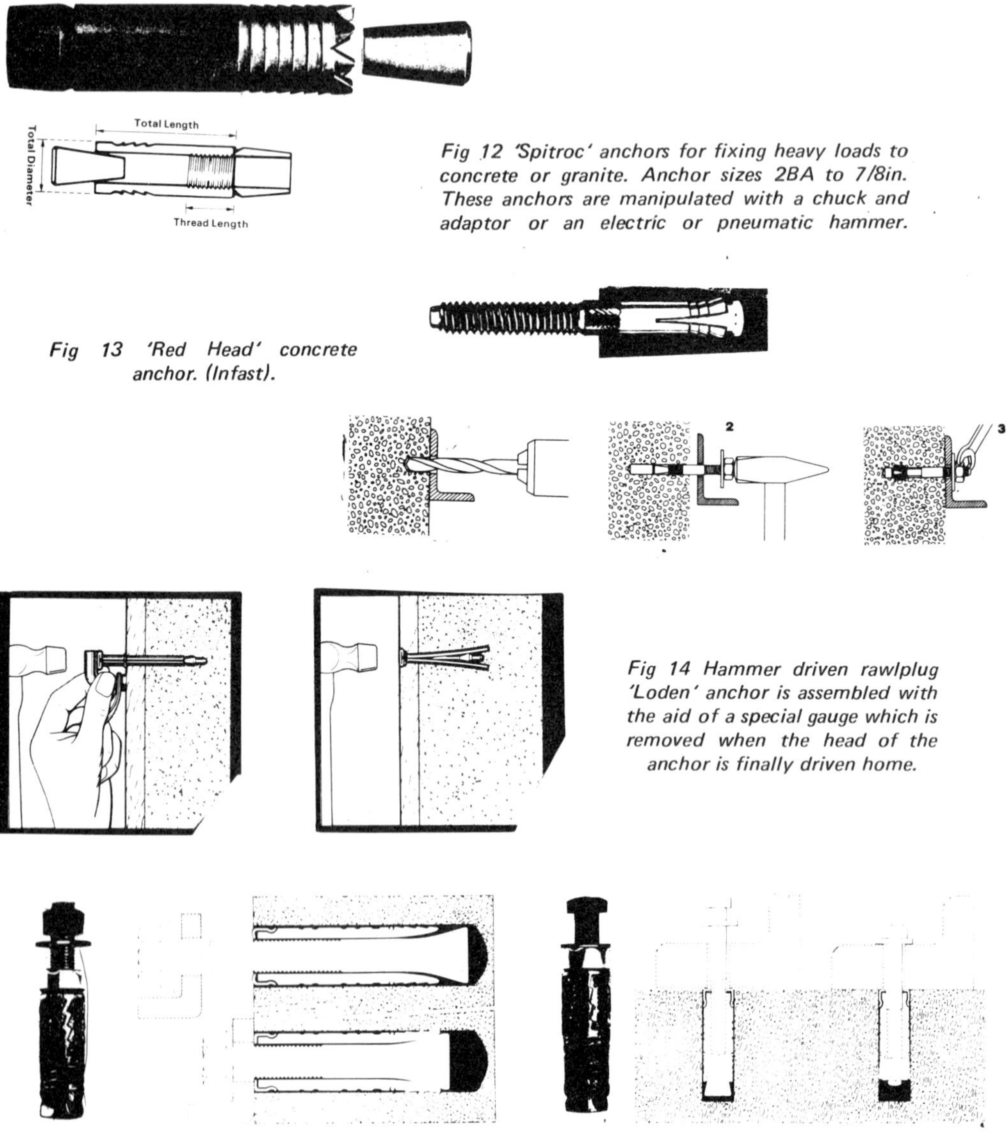

Fig 12 'Spitroc' anchors for fixing heavy loads to concrete or granite. Anchor sizes 2BA to 7/8in. These anchors are manipulated with a chuck and adaptor or an electric or pneumatic hammer.

Fig 13 'Red Head' concrete anchor. (Infast).

Fig 14 Hammer driven rawlplug 'Loden' anchor is assembled with the aid of a special gauge which is removed when the head of the anchor is finally driven home.

Fig 15 Wall and ceiling fixing rawlbolt (left) and floor fixing rawlbolt, (right).

Rawlbolt metal expansion bolts — Fig 15 — are designed for floor fixings, or wall and ceiling fixings. The shell in both cases is segmental, in malleable iron, containing either a bolt shaped base (wall fixing type) or an internal expander nut (floor fixing type). Tightening the fixing bolt forces the segments of the shell apart to grip the sides of the hole. Floor-type Rawlbolts can also be of loose bolt type to accommodate either bolts or studding.

Recommended loads for Rawlbolts are given in Table III.

TABLE III — RECOMMENDED LOADS FOR STANDARD RAWLBOLTS

Nominal Size of Bolt	Code Letter	Tensile Loads				Shear Loads			
		Safe Steady Load		Safe Shock Load		Safe Steady Load		Safe Shock Load	
ins		**lb**	**kg**	**lb**	**kg**	**lb**	**kg**	**lb**	**kg**
$\frac{3}{16}$	A	280	127	140	63·5	210	95	105	48
$\frac{1}{4}$	C	480	218	240	109	360	163	180	82
$\frac{5}{16}$	D	940	426	470	213	700	317·5	350	159
$\frac{3}{8}$	E	1 260	571	630	286	940	426	470	213
$\frac{1}{2}$	G	2 510	1 138	1 255	569	1 880	853	940	426
$\frac{5}{8}$	H	3 360	1 524	1 680	762	2 500	1 134	1 250	567
$\frac{3}{4}$	J	4 220	1 914	2 110	957	3 200	1 451	1 600	726
1	K	8 500	3 855	4 250	1 927	6 400	2 902	3 200	1 451

Bonded Anchors

Bonding is an alternative form of anchoring studs. In this case a stud is fitted in an oversize hole partially filled with a suitable adhesive or bonding medium — eg thermosetting or chemically hardening synthetic resin. Bonding has proved suitable for heavy duty applications and is particularly attractive where fixings have to be placed near the edges of forms, in softer building materials and also where vibratory or shock loads are anticipated.

In one proprietary type of bonded anchor the bonding medium (epoxy resin and hardener) is contained in a capsule which is inserted into the prepared hole. The stud is then inserted on a special driver to the end of the capsule and the stud driven to the bottom of the hole with a rotating action, providing complete mixing of resin and hardener.

An example of a *caulking anchor* is shown in Fig 16. Designed for use in hard or brittle materials the sleeve is of antimonial lead with controlled wall thickness to provide even expansion throughout. It is secured by caulking with a setting tool which action causes the lead to flow and fill the hole and any irregularities in it. This type of anchor is particularly suitable for shallow hole installations for light and medium loads — see also Table IV.

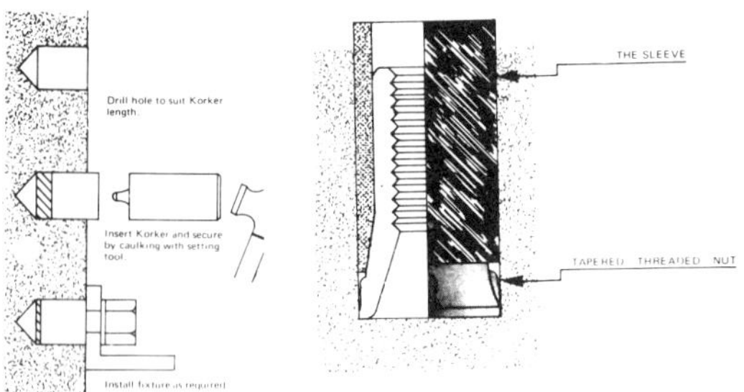

Fig 16 'Biff' caulking anchor.

TABLE IV – BIFF CAULKING ANCHOR DATA

Korker Size (in Whit)	Korker Length (in)	Drill Size (in)	Holding Power in average concrete (lb)
1/4	7/8	1/2	3 000
5/16	1	5/8	3 700
3/8	1¼	3/4	6 500
1/2	1½	7/8	7 200

Washers

Plain washers are now normally specified by diameter size, although data are still quoted for nominal washer sizes related to specific or nominal bolt and screw sizes. In Britain the range of standard washer sizes originally covered by BS57, BS916, BS1082, BS1768, BS1769, BS1981 and BS2708 are now collated under one standard, BS3410, and manufacturers' stock sizes are normally based on this standard (see Table II).This covers washers suitable for both Imperial and Unified

TABLE I – SELECTION GUIDE TO LOCK WASHERS

Type	Applications and Remarks
Helical spring	A general purpose lock washer with spring take-up to compensate for any looseness or loss of bolt preload which may occur in the assembly. Generally poor locking action.
Spring ring	As for helical spring.
Internal tooth (light duty)	Used with small screw heads; also when it is desirable that the washer teeth do not show.
Internal tooth (heavy duty)	Used with larger bolts and nuts – particularly on castings, etc.
External tooth	Maximum locking action under bolt heads or nuts. Heads (or nuts) must be large enough to cover and flatten teeth when tightened.
External tooth, countersunk	Used with countersunk screw heads.
External-internal	Used where a large bearing area is required with maximum locking effect; also with oversize or elongated holes.
Domed	Used with thin or weak materials; and/or oversize or elongated holes.
Dished	An alternative to domed washers, particularly if joint resilience is more desirable than joint rigidity.
Pyramidal	Used for applications requiring very high tightening torques, good rigidity of joint and some resilience under very high loads. Used under fully threaded screw heads or nuts.
Wave	Widely used as an alternative choice to spring washers.
Tab	Provide a positive locking action.
Belleville (conical spring)	Widely used as a spring/locking washer for screwed connections.

TABLE II – ABRIDGED LIST OF PLAIN AND BEVELLED WASHERS TO BS3410

BA Unified Whit BSF		10		8 2	7 3 $\frac{3}{32}$	6 4	5 5 $\frac{1}{8}$	4 6	3 8 $\frac{5}{32}$	2 10 $\frac{3}{16}$	1 12 $\frac{7}{32}$	0
Table 1	**i.d.**	—	—	0·099/0·094	0·111/0·106	0·123/0·118	0·140/0·135	0·157/0·152	0·177/0·172	0·202/0·197	0·228/0·223	0·256/0·251
	o.d.	—	—	0·185/0·180	0·208/0·203	0·233/0·228	0·268/0·263	0·301/0·296	0·341/0·336	0·391/0·386	0·443/0·438	0·500/0·495
Thickness		—	—	0·020	0·022	0·024	0·024	0·028	0·028	0·032	0·036	0·040
Table 2	**i.d.**	0·078/0·073	0·086/0 081	0·099/0·094	0·111/0·106	0·123/0·118	0·140/0·135	0·157/0·152	0·177/0·172	0·202/0·197	0·228/0·223	0·256/0·251
	o.d.	0·176/0·171	0·197/0·192	0·228/0·223	0·257/0·252	0·288/0·283	0·335/0·330	0·378/0·373	0·432/0·427	0·500/0·495	0·565/0·560	0·625/0·620
Thickness		0·016	0·016	0·020	0·028	0·036	0·036	0·040	0·040	0·048	0·048	0·056
Unified Whit BSF		$\frac{1}{4}$	$\frac{5}{16}$	$\frac{3}{8}$	$\frac{7}{16}$	$\frac{1}{2}$	$\frac{9}{16}$	$\frac{5}{8}$	$\frac{3}{4}$	$\frac{7}{8}$	1	$1\frac{1}{4}$
Table 3	**i.d.**	0·270/0·265	0·333/0·328	0·395/0·390	0·458/C·453	0·520/0·515	0·593/0·588	0·656/0·651	0·781/0·776	0·906/0·901	1·031/1·026	1·281/1·276
	o.d.	0·562/0·557	0·625/0·620	0·750/0·745	0·875/0·870	1·000/0·995	1·125/1·115	1·250/1·240	1·500/1·490	1·625/1·615	1·875/1·865	2·375/2·355
Heavy gauge		0·560	0·072	0·072	0·092	0·092	0·104	0·116	0·128	0·144	0·160	0·176
Light gauge		0·036	0·040	0·048	0·048	0·056	0·056	0·072	0·072	0·072	0·072	0·116
Table 4	**i.d.**	0·270/0·265	0·333/0·328	0·395/0·390	0·458/0·453	0·520/0·515	0·593/0·588	0·656/0·651	0·781/0·776	0·906/0·901	1·031/1·026	1·281/1·276
	o.d.	0·625/0·620	0·750/0·745	0·875/0·870	1·000/0·995	1·125/1·115	1·250/1·240	1·375/1·365	1·625/1·615	1·875/1·865	2·125/2·115	2·625/2·605
Heavy gauge		0·056	0·072	0·072	0·092	0·092	0·104	0·116	0·128	0·144	0·160	0·176
Light gauge		0·040	0·048	0·048	0·056	0·056	0·056	0·072	0·072	0·072	0·092	0·116
Table 5	**i.d.**	0·270/0·265	0·333/0·328	0·395/0·390	0·458/0·453	0·520/0·515	0·593/0·588	0·656/0·651	0·781/0·776	0·906/0·901	1·031/1·026	—
	o.d.	0·750/0·745	0·875/0·870	1·000/0·995	1·125/1·115	1·250/1·240	1·375/1·365	1·625/1·615	1·875/1·865	2·125/2·115	2·375/2·355	—
Thickness		0·064	0·072	0·080	0·092	0·104	0·116	0·128	0·144	0·160	0·192	—
Table 6	**i.d.**	0·270/0·265	0·333/0·328	0·395/0·390	0·458/0·453	0·520/0·515	0·593/0·588	0·656/0·651	0·781/0·776	0·906/0·901	1·031/1·026	
	o.d.	1·000/0·995	1·250/1·240	1·375/1·365	1·500/1·490	1·625/1·615	1·750/1·740	2·000/1·990	2·250/2·240	2·500/2·480	2·750/2·730	—
Thickness		0·064	0·080	0·092	0·104	0·116	0·128	0·128	0·144	0·160	0·192	—

bolts and screws. ISO metric washers are specified in BS4320 for normal and large diameters – see Table III. American production standards are based on ANSI B27.2 for Type A and Type B, equivalent to ISO metric normal and large diameters, see Table IV.

TABLE III – ISO METRIC WASHERS TO BS4320

Normal diameters **Normal and light range**

Nominal size of bolt	A		B		C		D	
	Max	*Min*	*Max*	*Min*	*Max*	*Min*	*Max*	*Min*
M2·5	2·85	2·7	6·5	6·2	0·6	0·4	—	—
M3	3·4	3·2	7·0	6·7	0·6	0·4	—	—
M3·5	3·9	3·7	7·0	6·7	0·6	0·4	—	—
M4	4·5	4·3	9·0	8·7	0·9	0·7	—	—
M5	5·5	5·3	10·0	9·7	1·1	0·9	—	—
M6	6·7	6·4	12·5	12·1	1·8	1·4	0·9	0·7
M8	8·7	8·4	17·0	16·6	1·8	1·4	1·1	0·9
M10	10·9	10·5	21·0	20·5	2·2	1·8	1·45	1·05
M12	13·4	13·0	24·0	23·5	2·7	2·3	1·8	1·4
M14	15·4	15·0	28·0	27·5	2·7	2·3	1·8	1·4
M16	17·4	17·0	30·0	29·5	3·3	2·7	2·2	1·8
M18	19·5	19·0	34·0	33·2	3·3	2·7	2·2	1·8
M20	21·5	21·0	37·0	36·2	3·3	2·7	2·2	1·8
M22	23·5	23·0	39·0	38·2	3·3	2·7	2·2	1·8
M24	25·5	25·0	44·0	43·2	4·3	3·7	2·7	2·3

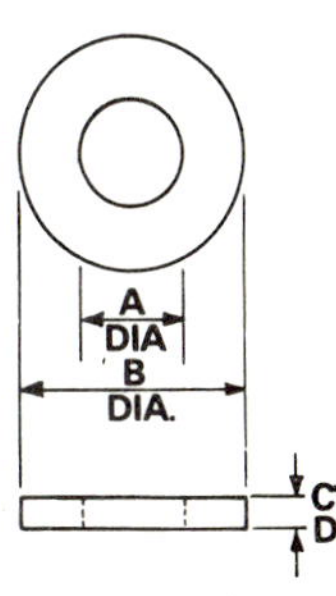

C = heavy gauge
D = light gauge

Large diameters **Normal and light range**

Nominal size of bolt	A		B		C		D	
	Max	*Min*	*Max*	*Min*	*Max*	*Min*	*Max*	*Min*
M4	4·5	4·3	10.0	9·7	0·9	0·7	—	—
M5	5·5	5·3	12.5	12·1	1·1	0·9	—	—
M6	6·7	6·4	14.0	13·6	1·8	1·4	0·90	0·70
M8	8·7	8·4	21.0	20·5	1·8	1·4	1·10	0·90
M10	10·9	10·5	24.0	23·5	2·2	1·8	1·45	1·05
M12	13·4	13·0	28.0	27·5	2·7	2·3	1·80	1·40
M14	15·4	15·0	30.0	29·5	2·7	2·3	1·80	1·40
M16	17·4	17·0	34.0	33·2	3·3	2·7	2·20	1·80
M18	19·5	19·0	37.0	36·2	3·3	2·7	2·20	1·80
M20	21·5	21·0	39.0	38·2	3·3	2·7	2·20	1·80
M22	23·5	23·0	44.0	43·2	3·3	2·7	2·20	1·80
M24	25·5	25·0	50.0	49·2	4·3	3·7	2·70	2·30

Standard washer production is in steel or brass as well as plastic materials. Tolerances specified apply only to metallic washers.

Lock Washers

Lock washers may be grouped under the following classifications:-

(i) Spring washers (helical spring and spring ring)
(ii) Toothed washers
(iii) Domed washers
(iv) Dished washers
(v) Pyramidal washers
(vi) Wave washers
(vii) Tab washers
(viii) Special friction-grip washers

Helical spring washers feature spring take-up for locking action and hardened thrust bearing surfaces reducing the friction between the bearing faces to facilitate tightening and/or disassembly. The standard form of cutting (Fig 1) results in a free height of washer approximately twice the thickness of the washer section. When compressed flat and then released, the recovered free height should be at least two thirds of the original free height if the temper is satisfactory. They are not true locking washers.

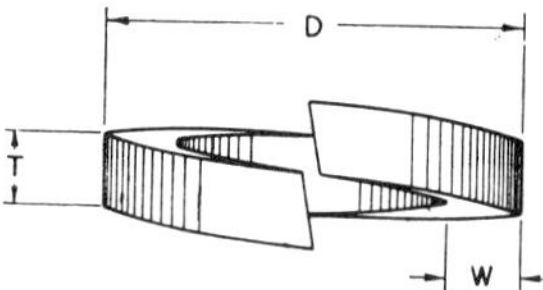

Fig 1 Square or rectangular section spring washer. (BS 1802).

A 'twist test' is also applicable to spring washers. If held in the vice as shown, a 90° segment of the free end is gripped in a wrench, it should be possible to rotate the unclamped washer section through 90° without fracturing in the case of carbon steel washers (45° with non-ferrous washers and corrosion resistant steel washers).

Materials used for spring lock washers include:-

(i) *Carbon steel* – heat treated to a hardness of 454–598 (Vickers)
(ii) *Corrosion resistant steel* – types 302, 305 and 420.
(iii) *Phosphor bronze* – minimum hardness 180 (Vickers)
(iv) *Silicon bronze* – minimum hardness 180 (Vickers)

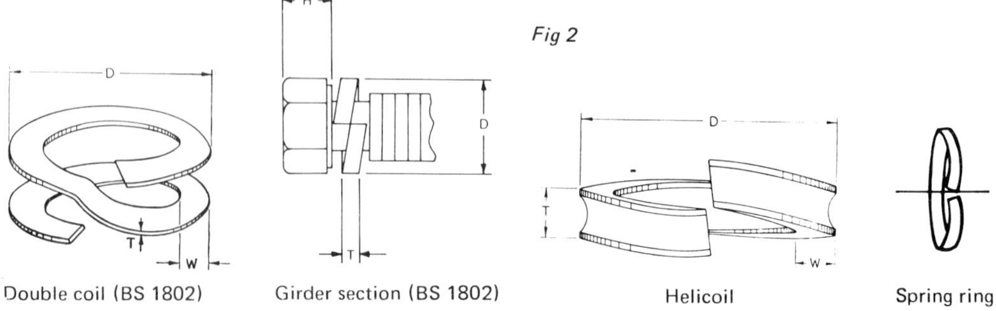

Fig 2

Double coil (BS 1802) Girder section (BS 1802) Helicoil Spring ring

(v) *K-Monel* – age-hardened to 316–389 (Vickers)

(vi) *Aluminium-zinc alloy* – heat treated to a minimum hardness of 138 (Vickers)

There are also variations on the simple form of helical spring washers – see Fig 2.

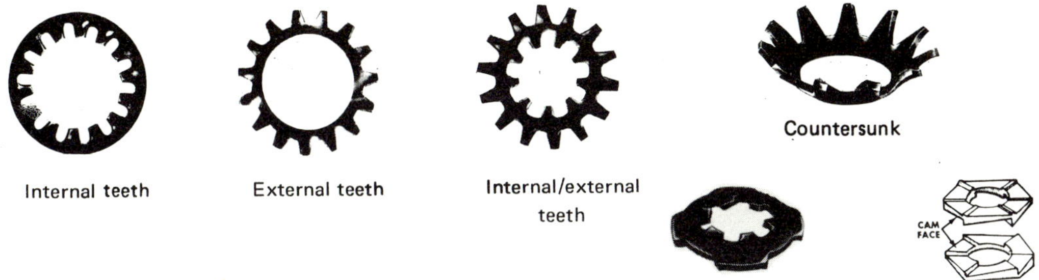

Fig 3 Toothed washers.

Toothed Lock Washers

Toothed washers include the following basic types (see also Fig 3).

(i) *Internal tooth* – standard and heavy duty

(ii) *External tooth* –

(iii) *External tooth, countersunk* – (external-internal countersunk also produced).

(iv) *External-internal* –

Toothed lock washers are normally made from carbon steel of Rockwell 40-50 hardness or equivalent. The number and form of the teeth, the width of the rim and the free length of the washer is largely at the option of the individual manufacturer concerned.

Domed washers are like cup washers with a saucer-shaped rim and toothed hole. The periphery may be plain or toothed, the latter providing enhanced grip on the parent surface.

Dished washers are similar, except that the top of the domed shape is not flattened. They are rather more elastic than domed washers and thus produce a less rigid assembly. Again the periphery may be plain or toothed.

Pyramidal washers are basically a dished type but of square or hexagonal planform with corner protusions on the periphery for gripping the parent surface. The hole is in the form of a single thread spring grip. This type of locking washer is capable of taking very high tightening torques to produce a rigid assembly whilst retaining a certain amount of resilience under very heavy loads.

Cam-faced washers or nesting washers used in pairs to produce a cam-locking action · eg Oldenkott.

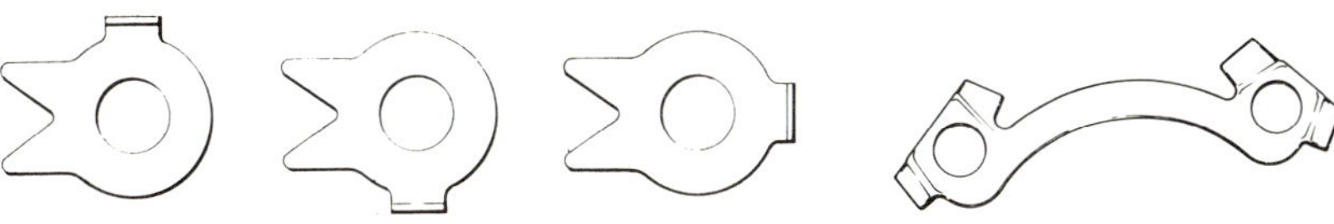

Fig 4a Simple tab washers.

Fig 4b Tab plate.

Tab Washers

Tab washers take the form of plain washers with one or more projections or tabs which can be bent to provide a mechanical lock between a bolt or nut and the component being clamped up. In some cases one (or more) of the tabs are pre-bent, the remaining tab or tabs being bent into position after assembly. There are also forms of tab plates locating on two (or more) bolt positions, and thus automatically locked to the assembly, leaving only the nut or bolt head locking tabs to be bent after assembly – eg see Fig 4b. These are produced in a variety of configurations to suit specific applications.

Properly used, tab washers provide positive locking, but may call for special designs of components to accommodate them. They can fail if badly assembled and should not be regarded as re-usable.

Wave washers are spring washers with an undulating surface, which flattens under pressure. The form and number of 'waves' can vary. Although most washers of this type are simple stampings, some are split (and virtually equivalent to the spring ring form of spring washer). British Standard 4463 specifies a particular geometry consisting of three equi-spaced radial corrugations under the description *crinkle* washers – Table VIII. Metric specifications (DIN128 and DIN137) detail a different form of corrugation and split rings.

Examples of some special designs of *friction-grip* washers are shown in Fig 5.

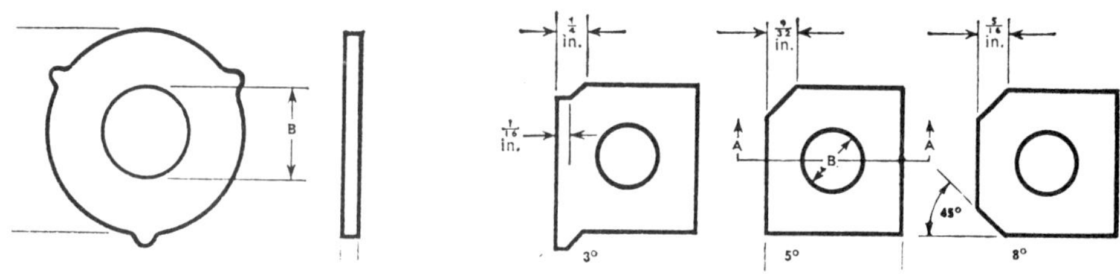

Fig 5 High strength friction-grip washers.

Taper Washers

Taper washers are so called because they taper in thickness – Fig 6. They are also known as wedge washers and as bevelled washers (eg the latter in America), but are not to be confused with conventional bevelled washers (ie plain washers bevelled on the outer edge only). They are mainly intended for structural applications involving standard channels and bearers to compensate for lack of alignment in assemblies – eg where the outer face of the bolted parts has a slope of 1:20 or more with respect to a plane normal to the bolt axis. Bevelled washers thus have a compensating slope of specific proportions, eg 3°, 5°, or 8° for general use. They are normally of square form, and fabricated from malleable iron or steel. Steel washers may be hardened or non-hardened, depending on the strength of the bolts with which they are to be used.

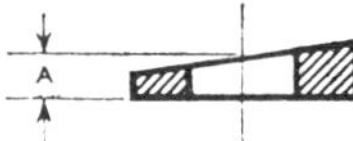

Fig 6 Tapered washer. A–mean thickness.

Vossloh Screw Security Devices

High Duty Lock Washers and other Spring Washers

We place detailed brochures at your disposal

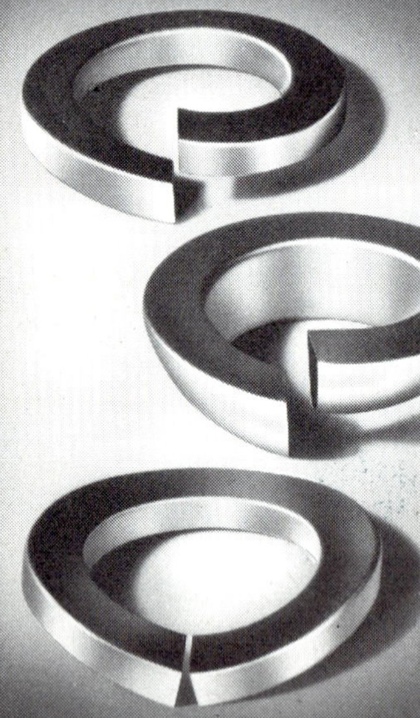

Vossloh-Werke GmbH.
D-598 Werdohl, Germany
P.O. Box 36
Telephone (2392) 521
Telex 826444

this way
comme ceci
oder so
así

or that
ou comme cela
so
o así

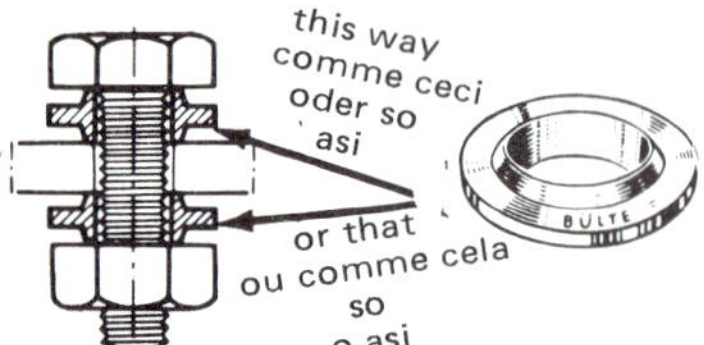

Plastic Washers with Triple Effect

Rondelle en matière plastique triplement efficace

Diese Plastic-Unterlegscheibe ist dreifach wirksam:

Esta arandela de plástico es de triple efecto:

Washers which act as: screw and nut lock-screw gasket – protection from corrosion and damage. For all screw connections in mechanical engineering, auto-motive and marine construction industries, for metal constructions and outside facings, in the construction industry, in mining etc.
A most reasonably priced ring of super-polyamide.
Supplied in millimetres and inches.

Schraubensicherung – Schraubendichtung Rost- und Materialschutz für alle Schraubverbindungen im Maschinen- und Metallbau, Auto- und Schiffsbau, für Aussenverkleidungen und Bauindustrie, im Bergbau usw. Ein preiswerter Ring aus Superpolyamid. Lieferbar in Metrisch und Zoll.

Write for free samples!
Echantillon gratuit! Veuillez nous contacter.
Muster kostenlos! Bitte anfordern.
¡Muestra gratuita! ¡ Pidase!

En tant que bague de sécurité pour vis, joint d'étanchéité pour vis et comme protection du matériel contre la rouille et la corrosion. Particulièrement appropriée pour tous les raccords à vis dans la construction mécanique et métal lique, la construction d'automobiles et de navires, l'industrie de construction et l'exploitation des mines, etc.
Une bague en superpolyamide très avantageuse du point de vue prix.
Livrable en dimensions métriques et en pouces.

Asegura el tomillo – Hermetiza la atornilladura – Protege contra el óxido y al material. Para toda clase de atomilladuras en las construcciónes metálicas y de maquinaria, industrias del automóvil y astilleros, para revestimientos exteriores e industria de la construcción, para minería, etc. Una arandela económica hecha de superpoliamida.
Se sirve en medidas métricas y en pulgadas.

NORBERT BÜLTE KG
D-47 10 Lüdinghausen/West Germany,
Sendener Str. 14–16, Telex 08 9 803

*Responsable pour la France: M. PELTIER, 15b, route d'Ingersheim, 6800 Colmar/Ht. Rhin

THE FIRST!

HANDBOOK OF POWER DRIVES

Over 600 pages containing thousands of diagrams, tables, charts, illustrations, etc., stiff board bound and gold blocked.

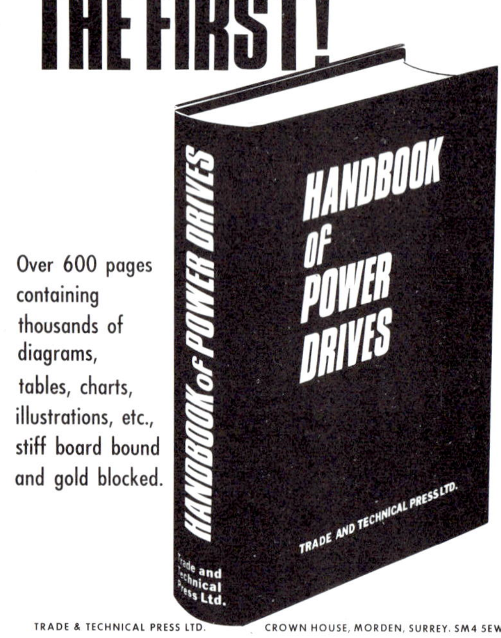

TRADE & TECHNICAL PRESS LTD. CROWN HOUSE, MORDEN, SURREY. SM4 5EW

TABLE IV – AMERICAN TYPE A* PLAIN WASHERS (ANSI B27.2)

Nominal Washer Size			Inside Diameter			Outside Diameter			Thickness		
			Basic	Tolerance		Basic	Tolerance		Basic	Max	Min
				Plus	Minus		Plus	Minus			
–	–		0.078	0.000	0.005	0.188	0.000	0.005	0.020	0.025	0.016
–	–		0.094	0.000	0.005	0.250	0.000	0.005	0.020	0.025	0.016
–	–		0.125	0.008	0.005	0.312	0.008	0.005	0.032	0.040	0.025
No. 6	0.138		0.156	0.008	0.005	0.375	0.015	0.005	0.049	0.065	0.036
8	0.164		0.188	0.008	0.005	0.438	0.015	0.005	0.049	0.065	0.036
10	0.190		0.219	0.008	0.005	0.500	0.015	0.005	0.049	0.065	0.036
3/16	0.188		0.250	0.015	0.005	0.562	0.015	0.005	0.049	0.065	0.036
12	0.216		0.250	0.015	0.005	0.562	0.015	0.005	0.065	0.080	0.051
1/4	0.250	N	0.281	0.015	0.005	0.625	0.015	0.005	0.065	0.080	0.051
1/4	0.250	W	0.312	0.015	0.005	0.734	0.015	0.007	0.065	0.080	0.051
5/16	0.312	N	0.344	0.015	0.005	0.688	0.015	0.007	0.065	0.080	0.051
5/16	0.312	W	0.375	0.015	0.005	0.875	0.030	0.007	0.083	0.104	0.064
3/8	0.375	N	0.406	0.015	0.005	0.812	0.015	0.007	0.065	0.080	0.051
3/8	0.375	W	0.438	0.015	0.005	1.000	0.030	0.007	0.083	0.104	0.064
7/16	0.438	N	0.469	0.015	0.005	0.922	0.015	0.007	0.065	0.080	0.051
7/16	0.438	W	0.500	0.015	0.005	1.250	0.030	0.007	0.083	0.104	0.064
1/2	0.500	N	0.531	0.015	0.005	1.062	0.030	0.007	0.095	0.121	0.074
1/2	0.500	W	0.562	0.015	0.005	1.375	0.030	0.007	0.109	0.132	0.086
9/16	0.562	N	0.594	0.015	0.005	1.156	0.030	0.007	0.095	0.121	0.074
9/16	0.562	W	0.625	0.015	0.005	1.469	0.030	0.007	0.109	0.132	0.086
5/8	0.625	N	0.656	0.030	0.007	1.312	0.030	0.007	0.095	0.121	0.074
5/8	0.625	W	0.688	0.030	0.007	1.750	0.030	0.007	0.134	0.160	0.108
3/4	0.750	N	0.812	0.030	0.007	1.469	0.030	0.007	0.134	0.160	0.108
3/4	0.750	W	0.812	0.030	0.007	2.000	0.030	0.007	0.148	0.177	0.122
7/8	0.875	N	0.938	0.030	0.007	1.750	0.030	0.007	0.134	0.160	0.108
7/8	0.875	W	0.938	0.030	0.007	2.250	0.030	0.007	0.165	0.192	0.136
1	1.000	N	1.062	0.030	0.007	2.000	0.030	0.007	0.134	0.160	0.108
1	1.000	W	1.062	0.030	0.007	2.500	0.030	0.007	0.165	0.192	0.136
1 1/8	1.125	N	1.250	0.030	0.007	2.250	0.030	0.007	0.134	0.160	0.108
1 1/8	1.125	W	1.250	0.030	0.007	2.750	0.030	0.007	0.165	0.192	0.136
1 1/4	1.250	N	1.375	0.030	0.007	2.500	0.030	0.007	0.165	0.192	0.136
1 1/4	1.250	W	1.375	0.030	0.007	3.000	0.030	0.007	0.165	0.192	0.136
1 3/8	1.375	N	1.500	0.030	0.007	2.750	0.030	0.007	0.165	0.192	0.136
1 3/8	1.375	W	1.500	0.045	0.010	3.250	0.045	0.010	0.180	0.213	0.153
1 1/2	1.500	N	1.625	0.030	0.007	3.000	0.030	0.007	0.165	0.192	0.136
1 1/2	1.500	W	1.625	0.045	0.010	3.500	0.045	0.010	0.180	0.213	0.153
1 5/8	1.625		1.750	0.045	0.010	3.750	0.045	0.010	0.180	0.213	0.153
1 3/4	1.750		1.875	0.045	0.010	4.000	0.045	0.010	0.180	0.213	0.153
1 7/8	1.875		2.000	0.045	0.010	4.250	0.045	0.010	0.180	0.213	0.153
2	2.000		2.125	0.045	0.010	4.500	0.045	0.010	0.180	0.213	0.153
2 1/4	2.250		2.375	0.045	0.010	4.750	0.045	0.010	0.220	0.248	0.193
2 1/2	2.500		2.625	0.045	0.010	5.000	0.045	0.010	0.238	0.280	0.210
2 3/4	2.750		2.875	0.065	0.010	5.250	0.065	0.010	0.259	0.310	0.228
3	3.000		3.125	0.065	0.010	5.500	0.065	0.010	0.284	0.327	0.249

N = narrow or SAE size
W = wide or 'plate' size

*Note : Type B washers are of larger diameter to distribute loads over a wider bearing area. These are specified in narrow, regular and wide series (ANSI B27.2)

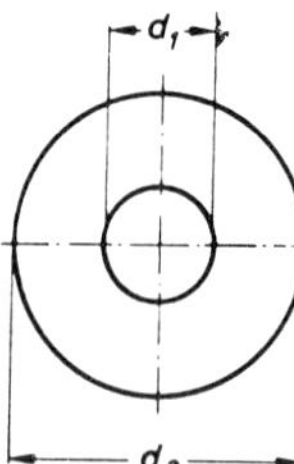

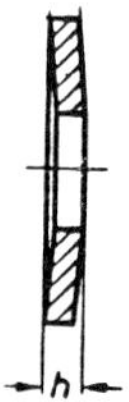

Fig 7 Belleville washer or conical spring washer.

Belleville Washers

Belleville washers are cup-shaped springs, with the basic geometry defined in Fig 7. Their performance as springs can be calculated from first principles, but the process is tedious and complicated. The basic formulas are:-

$$\text{load (P)} = \frac{Ef}{(1-\delta^2)\,M\,a^2}\left[\left(h-\frac{f}{2}\right)(h-f)\,t+t^3\right]$$

$$\text{stress (S)} = \frac{Ef}{(1-\delta^2)\,M\,a^2}\left[C_1\left(h-\frac{f}{2}\right)+C_2\,t\right]$$

when E = modulus of elasticity of material
f = deflection
a = half o.d.
h = free height of washer, less thickness
t = thickness of washer material
δ = Poisson's ratio (= 0.3 for steel)

$$M = \frac{6}{\pi \log_e A} \cdot \frac{(A-1)^2}{A2}$$

where A = ratio o.d./i.d.

$$C_1 = \frac{6}{\pi \log_e A} \cdot \left[\frac{A-1}{\log_e A} - 1\right]$$

$$C_2 = \frac{6}{\pi \log_e A} \cdot \frac{(A-1)}{2}$$

Conical spring washers are also produced in curved form — see Fig 8.

Sealing Washers

Soft plain washers can act both as a locking device and a seal, used under bolt or screw heads. Sealing washers, as such, are designed primarily to act as seals. The complete fastener may also be designed with a sealing head. Fig 9 shows some proprietary examples.

An example of a washer designed to provide both sealing and locking action is shown in Fig 10. See also chapter on *Locking Nuts* for other examples.

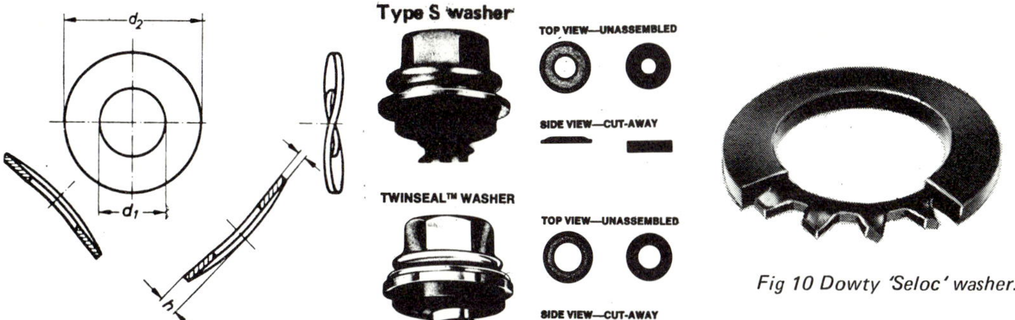

Fig 8 Curved conical spring washer.

Fig 10 Dowty 'Seloc' washer.

Fig 9 Examples of fasteners with sealed heads. ('Buildex').

TABLE VA – DIMENSIONS OF BS SINGLE COIL RECTANGULAR SPRING WASHERS (BS1802)

Nominal size of washer (dia. of bolt)	Nominal outside diameter D	Nominal width W		Nominal thickness T	
in.	in.	in.	S.W.G.	in.	S.W.G.
⅛ (and 6 B.A.)	0·237	0·048	18	0·032	21
5/32 (and 4 B.A.)	0·300	0·064	16	0·032	21
2 B.A. (and 3/16)	0·387	0·092	13	0·048	18
¼	0·465	0·092	13	0·064	16
5/16	0·600	0·128	10	0·064	16
⅜	0·694	0·144	9	0·064	16
7/16	0·789	0·160	8	0·064	16
½	0·851	0·160	8	0·092	13
9/16	1·009	0·192	6	0·092	13
⅝	1·072	0·192	6	0·128	10
¾	1·276	0·232	4	0·128	10
⅞	1·442	0·252	3	0·128	10
1	1·624	0·281	—	0·187	—
1⅛	1·750	0·281	—	0·187	—
1¼	1·936	0·312	—	0·187	—
1⅜	2·062	0·312	—	0·187	—
1½	2·312	0·375	—	0·187	—
1¾	2·686	0·437	—	0·250	—
2	3·062	0·500	—	0·250	—

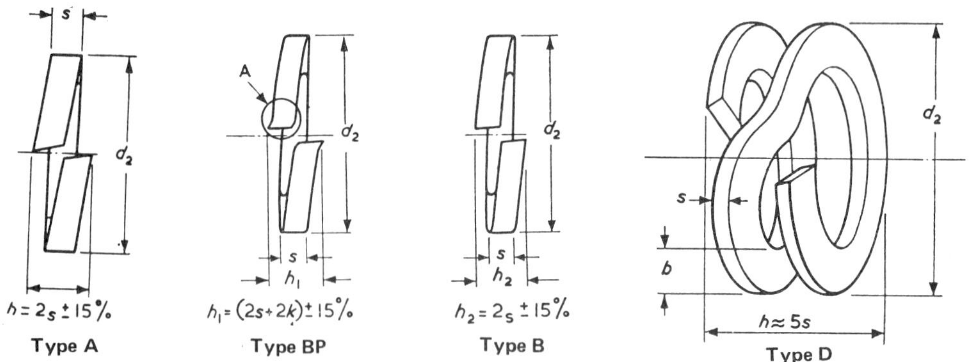

TABLE VB – SPRING WASHERS FOR GENERAL ENGINEERING PURPOSES (BS4464)

TYPE A: Nominal Size	Inside diameter d_1 Maximum	Inside diameter d_1 Minimum	S	d_2 Maximum
M3	3.3	3.1	1 ± 0.1	5.5
M4	4.35	4.1	1.2 ± 0.1	6.95
M5	5.35	5.1	1.5 ± 0.1	8.55
M6	6.4	6.1	1.5 ± 0.1	9.6
M8	8.55	8.2	2 ± 0.1	12.75
M10	10.6	10.2	2.5 ± 0.15	15.9
M12	12.6	12.2	2.5 ± 0.15	17.9
M16	16.9	16.3	3.5 ± 0.2	24.3
M20	21.1	20.3	4.5 ± 0.2	30.5
M24	25.3	24.4	5 ± 0.2	35.7
M30	31.5	30.5	6 ± 0.2	43.9
M36	37.6	36.5	7 ± 0.25	52.1
M42	43.8	42.6	8 ± 0.25	60.3
M48	50.0	48.8	8 ± 0.25	66.5

TYPES BP AND B Nominal Size	Inside diameter d_1 Maximum	Inside diameter d_1 Minimum	b (width)	s (thickness)	d_2 Maximum
M1.6	1.9	1.7	0.7 ± 0.1	0.4 ± 0.1	3.5
M2	2.3	2.1	0.9 ± 0.1	0.5 ± 0.1	4.3
M2.5	2.8	2.6	1.0 ± 0.1	0.6 ± 0.1	5.0
M3	3.3	3.1	1.3 ± 0.1	0.8 ± 0.1	6.1
M4	4.35	4.1	1.5 ± 0.1	0.9 ± 0.1	7.55
M5	5.35	5.1	1.8 ± 0.1	1.2 ± 0.1	9.15
M6	6.4	6.1	2.5 ± 0.15	1.6 ± 0.1	11.7
M8	8.55	8.2	3 ± 0.15	2 ± 0.1	14.85
M10	10.6	10.2	3.5 ± 0.2	2.2 ± 0.15	18.0
M12	12.6	12.2	4 ± 0.2	2.5 ± 0.15	21.0
M16	16.9	16.3	5 ± 0.2	3.5 ± 0.2	27.3
M20	21.1	20.3	6 ± 0.2	4 ± 0.2	33.5
M24	25.3	24.4	7 ± 0.25	5 ± 0.2	39.8
M30	31.5	30.5	8 ± 0.25	6 ± 0.25	48.0
M36	37.6	36.5	10 ± 0.25	6 ± 0.25	58.1

cont...

TABLE VB – SPRING WASHERS FOR GENERAL ENGINEERING PURPOSES (BS4464) (contd.)

M42	43.8	42.6	12 ±0.25	7 ±0.25	68.3
M48	50.0	48.8	12 ±0.25	7 ±0.25	74.5
M56	58.1	56.8	14 ±0.25	8 ±0.25	86.6
M64	66.3	64.9	14 ±0.25	8 ±0.25	93.8
TYPE D DOUBLE COIL SPRING WASHERS					
M2	2.4	2.1	0.9 ±0.1	0.5 ±0.05	4.4
M2.5	2.9	2.6	1.2 ±0.1	0.7 ±0.1	5.5
M3	3.6	3.3	1.2 ±0.1	0.8 ±0.1	6.2
M4	4.6	4.3	1.6 ±0.1	0.8 ±0.1	8.0
M5	5.6	5.3	2 ±0.1	0.9 ±0.1	9.8
M6	6.6	6.3	3 ±0.15	1 ± 0.1	12.9
M8	8.8	8.4	3 ±0.15	1.2 ±0.1	15.1
M10	10.8	10.4	3.5 ±0.2	1.2 ±0.1	18.2
M12	12.8	12.4	3.5 ±0.2	1.6 ±0.1	20.2
M16	17.0	16.5	5 ±0.2	2 ±0.1	27.4
M20	21.5	20.8	5 ±0.2	2 ±0.1	31.9
M24	26.0	25.0	6.5 ±0.2	3.25 ±0.15	39.4
M30	33.0	31.5	8 ±0.25	3.25 ±0.15	49.5
M36	40.0	38.0	10 ±0.25	3.25 ±0.15	60.5
M42	46.0	44.0	10 ±0.25	4.5 ±0.2	66.5
M48	52.0	50.0	10 ±0.25	4.5 ±0.2	72.5
M56	60.0	58.0	12 ±0.25	4.5 ±0.2	84.5
M64	70.0	67.0	12 ±0.25	4.5 ±0.2	94.5

All dimensions in millimetres

TABLE VI – DIMENSIONS OF HELICOIL WASHERS

HELICOIL WASHERS—I.S.O. METRIC THREADS

Screw Size	F Across Flats	G Across Corners	H Head Height	T Material Thickness	D Washer Diameter
I.S.O.	Min. Max.	Min. Max.	Min. Max.	Min. Max.	Max.
M6	9·78 10·00	11·05 11·50	3·85 4·15	1·50 1·70	10·6
M8	12·73 13·00	14·38 15·00	5·35 5·65	1·90 2·10	13·45
M10	16·73 17·00	18·90 19·60	6·82 7·18	2·35 2·65	17·20

HELICOIL WASHERS—UNIFIED THREADS

Screw Size	F Across Flats	G Across Corners	H Head Height	T Material Thickness	D Washer Diameter
UNF UNC	Min. Max.	Max.	Min. Max.	Min. Max.	Min. Max.
$\frac{1}{4}$	·4305 ·4375	·505	·153 ·163	·062 ·066	·448 ·464
$\frac{5}{16}$	·493 ·500	·577	·201 ·211	·078 ·082	·540 ·556
$\frac{3}{8}$	·5545 ·5625	·650	·233 ·243	·094 ·098	·635 ·641

TABLE VII – AMERICAN REGULAR* SPRING WASHERS (ANSI B27.1)

Nominal Washer Size		Inside Diameter		Outside Diameter	Washer Section	
					Width	Thickness
		Min	Max	Max	Min	Min
No. 2	0.086	0.088	0.094	0.172	0.035	0.020
3	0.099	0.101	0.107	0.195	0.040	0.025
4	0.112	0.115	0.121	0.209	0.040	0.025
5	0.125	0.128	0.134	0.236	0.047	0.031
6	0.138	0.141	0.148	0.250	0.047	0.031
8	0.164	0.168	0.175	0.293	0.055	0.040
10	0.190	0.194	0.202	0.334	0.062	0.047
12	0.216	0.221	0.229	0.377	0.070	0.056
1/4	0.250	0.255	0.263	0.489	0.109	0.062
5/16	0.312	0.318	0.328	0.586	0.125	0.078
3/8	0.375	0.382	0.393	0.683	0.141	0.094
7/16	0.438	0.446	0.459	0.779	0.156	0.109
1/2	0.500	0.509	0.523	0.873	0.171	0.125
9/16	0.562	0.572	0.587	0.971	0.188	0.141
5/8	0.625	0.636	0.653	1.079	0.203	0.156
11/16	0.688	0.700	0.718	1.176	0.219	0.172
3/4	0.750	0.763	0.783	1.271	0.234	0.188
13/16	0.812	0.826	0.847	1.367	0.250	0.203
7/8	0.875	0.890	0.912	1.464	0.266	0.219
15/16	0.938	0.954	0.978	1.560	0.281	0.234

cont....

TABLE VII – AMERICAN REGULAR* SPRING WASHERS (ANSI B27.1)

1	1.000	1.017	1.042	1.661	0.297	0.250
1 1/16	1.062	1.080	1.107	1.756	0.312	0.266
1 1/8	1.125	1.144	1.172	1.853	0.328	0.281
1 3/16	1.188	1.208	1.237	1.950	0.344	0.297
1 1/4	1.250	1.271	1.302	2.045	0.359	0.312
1 5/16	1.312	1.334	1.366	2.141	0.375	0.328
1 3/8	1.375	1.398	1.432	2.239	0.391	0.344
1 7/16	1.438	1.462	1.497	2.334	0.406	0.359
1 1/2	1.500	1.525	1.561	2.430	0.422	0.375

*Note : Heavy and extra duty washers have same inside diameter sizes but increased washer sections
Separate specifications apply to Hi-collar washers for use on socket head cap screws.

TABLE VIII – CRINKLE WASHERS TO BS4463

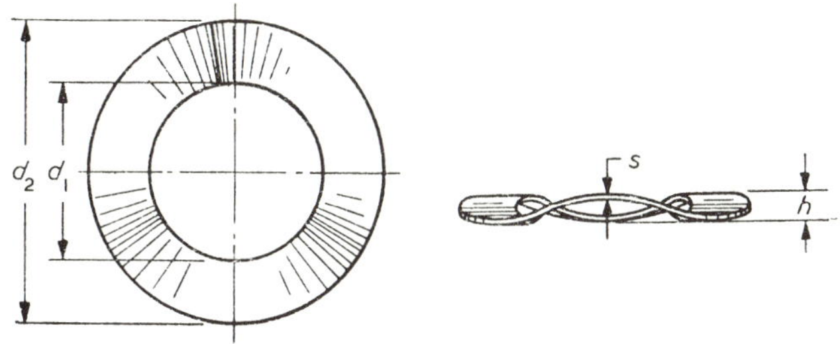

Nominal size and thread diameter d	Inside diameter d_1		Outside diameter d_2		Height h		Thickness s
	max	min	max	min	max	min	nominal
M1·6	1·8	1·7	3·7	3·52	0·51	0·36	0·16
M2	2·3	2·2	4·6	4·42	0·53	0·38	0·16
(M2·2)	2·5	2·4	5·2	5·02	0·53	0·38	0·16
M2·5	2·8	2·7	5·8	5·62	0·53	0·38	0·16
M3	3·32	3·2	6·4	6·18	0·61	0·46	0·16
(M3·5)	3·82	3·7	6·9	6·68	0·79	0·63	0·20
M4	4·42	4·3	8·1	7·88	0·84	0·69	0·28
M5	5·42	5·3	9·2	8·98	0·89	0·74	0·30
M6	6·55	6·4	11·5	11·23	1·14	0·99	0·40
M8	8·55	8·4	15·0	14·73	1·40	1·25	0·40
M10	10·68	10·5	19·6	19·27	1·70	1·55	0·55
M12	13·18	13·0	22·0	21·67	1·90	1·65	0·55
(M14)	15·18	15·0	25·5	25·17	2·06	1·80	0·55
M16	17·18	17·0	27·8	27·47	2·41	2·16	0·70
(M18)	19·21	19·0	31·3	30·91	2·41	2·16	0·70
M20	21·21	21·0	34·7	34·31	2·66	2·16	0·70

NOTE. Sizes shown in brackets are non-preferred, and are not usually stock sizes.

All dimensions in millimetres

TABLE IX DIMENSIONS OF FLAT ROUND FRICTION GRIP WASHERS (BS3139)

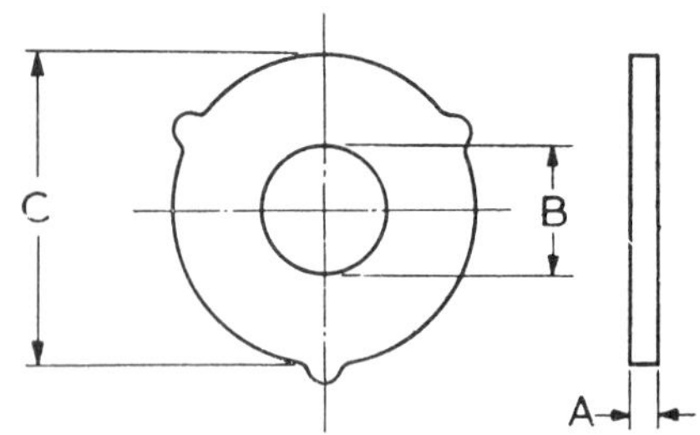

Bolt size	Inside diameter B	Outside diameter C	Thickness A	
in.	in.	in.	S.W.G.	in. (approx)
½	9/16	1⅜	12	0·104
⅝	11/16	1½	10	0·128
¾	13/16	1¾	9	0·144
⅞	15/16	2	8	0·160
1	1 1/16	2¼	8	0·160
1⅛	1¼	2½	8	0·160
1¼	1⅜	2¾	8	0·160
1½	1⅝	3¼	7	0·176

TABLE X – DIMENSIONS OF METRIC TAPER WASHERS (BS3139)

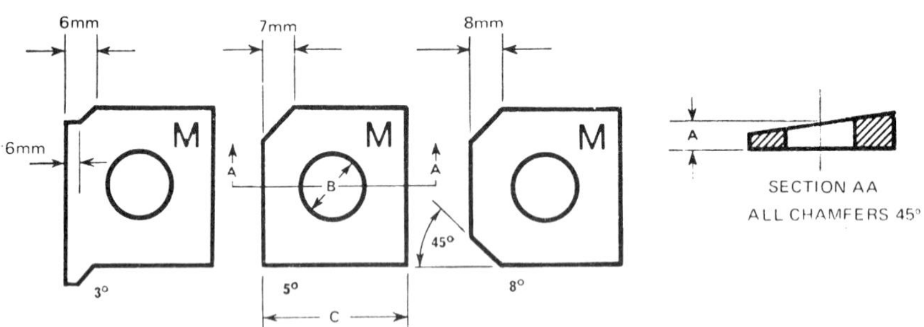

Nom dia	Inside diameter B max	Inside diameter B min	Overall size C	Mean thickness A 3° and 5° Taper	Mean thickness A 8° Taper
M16	18·2	17·4	38·10	4·76	6·35
M20	21·9	21·1	38·10	4·76	6·35
M22	23·8	23·0	44·45	4·76	6·35
M24	26·8	26·0	57·15	4·76	6·35
M27	29·8	29·0	57·15	4·76	6·35
M30	33·2	32·4	57·15	4·76	6·35
M36	39·2	38·4	57·15	4·76	6·35

Self-Sealing Fasteners

Fasteners may be made to provide sealing as well as fastening by a variety of methods, eg

(i) *Interference fit* – as on Dryseal threads for pipe couplings, sealing rivets which expand to an interference fit on setting, etc. Interference fits may also be provided by rubber or nylon sleeves, or special designs of fastener incorporating elastically deformed sleeve sections.

(ii) *Liquid/adhesive thread coatings* – applied to screws and/or nut threads prior to assembly. These may be pipe thread sealants (eg as used on pipe couplings) or provide extra thread locking.

(iii) *Sealing washers* – these are of resilient materials such as lead, soft aluminium, synthetic rubber, etc, designed to provide a seal against the bearing surface under a bolt head or nut when the fastener is tightened. They may range from simple washer forms or rubber rings to composite bonded or laminated metal and elastomer or plastic constructions.

(iv) *Moulded-in seals* – more generally applied to washers, but may also be incorporated in other designs of fasteners or fastener elements. The moulded-in sealing element is usually either rubber or nylon.

(v) *Preassembled washers* – these are virtually the same as sealing washers, but are usually of rubber, plastic or composite construction preassembled on the bolt or screw under the head.

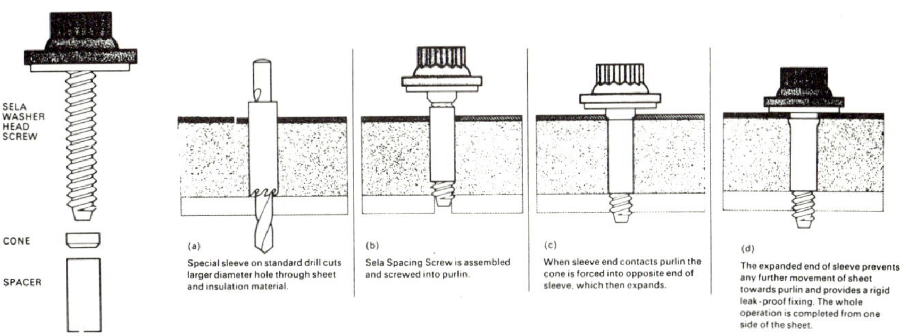

Example of a special design of self-sealing fastener to secure cladding sheets in building construction. (The British Screw Co Ltd).

(vi) *Sealing nuts* – these include a variety of proprietary designs of nut incorporating an element which is compressed to produce a seal when the nut is tightened. The sealing element may be of rubber, plastic or soft metal.

(vii) *Special head designs* – applicable to bolts where the underside of the head is further flared or shaped to accommodate a particular type of sealing ring or sealing washer. The combination may or may not be preassembled.

(viii) *Flared-in sealants* – liquid or semi-liquid sealant compounds for use with washers where a more conventional type of sealing washer may not be suitable or is less readily fitted. Basically it is the modern equivalent of (ii) above applied to washers.

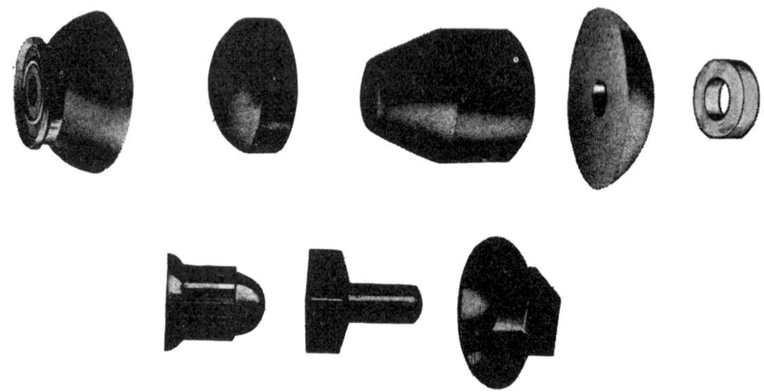

Examples of proprietary sealing washers, covers, caps and sealing nuts for building construction. (The British Screw Co Ltd).

Sems

Sems is the generic term for screw and washer assemblies, their purpose being to reduce time in hand assembly and inspection, to ensure that the washer is the right type and size, and to eliminate washer losses.

Sems are produced in all standard thread forms and heads, (eg machine screws and self-tapping screws), pre-assembled with flat washers, single-coil spring washers, toothed locking washers, or special assemblies to an individual customer's specifications. The maximum diameter of the unthreaded shank retaining the washer is less than the maximum major diameter of the thread by an amount sufficient to prevent disassembly of the washer from the screw. Heat treated Sems may be heat treated after assembly, or screws and washers heat treated prior to assembly.

Examples of Sems are shown in the diagram.

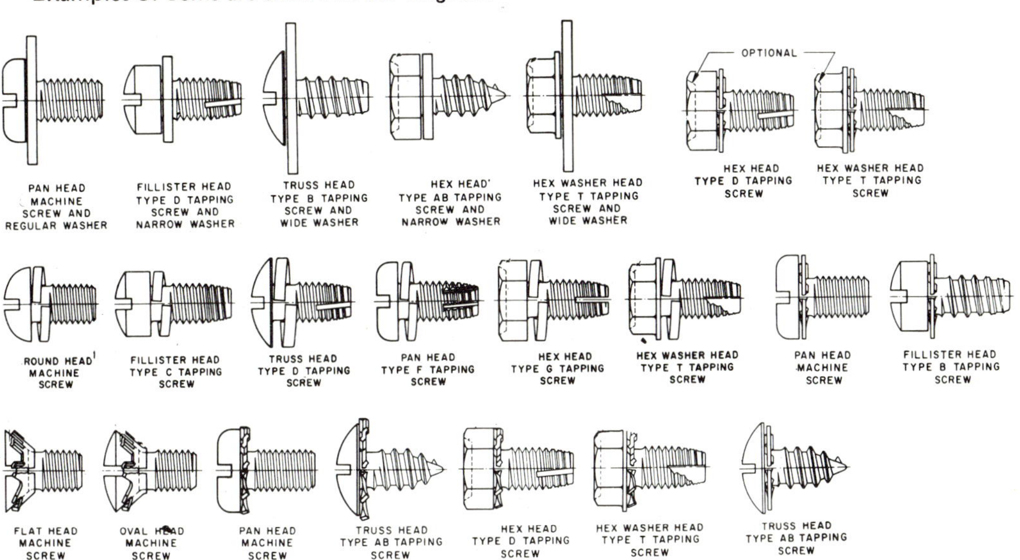

Examples of sems.

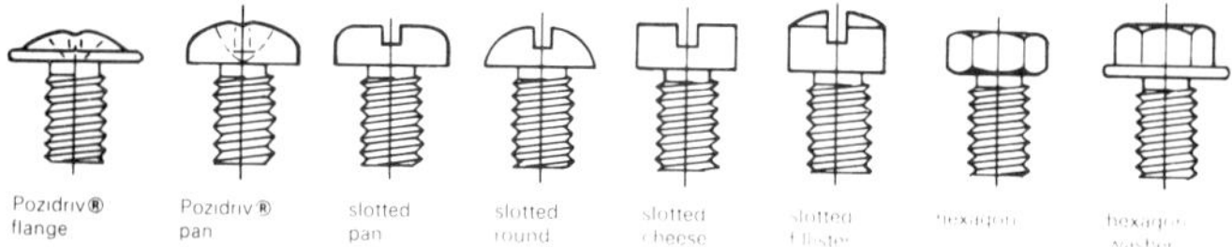

Head styles for sems.

Examples of Sems ITW (Fastex).

Dimensional Data

MANUFACTURERS' RECOMMENDED RANGES OF METRIC SIZES
PRECISION HEXAGON HEAD BOLTS, STEEL, GRADE 8.8 BS3693

Nominal Length mm	Nominal Diameter M5	M6	M8	M10	M12	M16	M20	M24
25	●	●						
30	●	●	●					
35	●	●	●	●				
40	●	●	●	●	●			
45	●	●	●	●	●	●		
50	●	●	●	●	●	●		
55		●	●	●	●	●	●	
60		●	●	●	●	●	●	
65		●	●	●	●	●	●	●
70		●	●	●	●	●	●	●
75		●	●	●	●	●	●	●
80			●	●	●	●	●	●
90			●	●	●	●	●	●
100			●	●	●	●	●	●
110				●	●	●	●	●
120				●	●	●	●	●
130					●	●	●	●
140						●	●	●
150						●	●	●
160						●	●	●

PRECISION HEXAGON HEAD SCREWS, STEEL, GRADE 8.8 BS3692

Nominal Length mm	Nominal Diameter M5	M6	M8	M10	M12	M16	M20	M24
10	●							
12	●	●						
16	●	●	●					
20	●	●	●	●				
25	●	●	●	●	●			
30	●	●	●	●	●	●		
35		●	●	●	●	●		
40		●	●	●	●	●	●	
45			●	●	●	●	●	
50		●	●	●	●	●	●	●
55				●	●	●	●	●
60			●	●	●	●	●	●
70				●	●	●	●	
80				●	●	●	●	

cont...

BLACK HEXAGON HEAD BOLTS,STEEL, GRADE 4.6, BS4190

Nominal Length mm	Nominal Diameter						
	M6	M8	M10	M12	M16	M20	M24
20	●						
25	●	●	●	●			
30	●	●	●	●	○		
35	●	●	●	●	○		
40	●	●	●	●	○	○	
45	●	●	●	●	○	○	
50	●	●	●	●	⊕	○	
55			●	●	⊕	○	
60	●	●	●	●	⊕	⊕	○
65			●	●	⊕	⊕	
70	●	●	●	●	⊕	⊕	⊕
75			●	●	⊕	⊕	
80	●	●	●	●	⊕	⊕	⊕
90	●	●	●	●	⊕	⊕	⊕
100	●	●	●	●	⊕	⊕	⊕
110				●	●	●	●
120		●	●	●	●	●	●
130					●	●	
140			●	●	●	●	●
150						●	
160				●	●	●	●
180				●	●	●	●
200				●	●	●	●
220				●	●	●	●
260				●	●	●	●
300				●	●	●	●

● Standard thread lengths
○ Short thread lengths
⊕ Available in standard and short thread lengths

BLACK HEXAGON HEAD SCREWS, STEEL, GRADE 4.6,BS4190

Nominal Length mm	Nominal Diameter					
	M6	M8	M10	M12	M16	M20
16	●	●				
20	●	●	●			
25	●	●	●	●		
30	●	●	●	●	●	
35	●	●	●	●	●	
40	●	●	●	●	●	●
45	●	●	●	●	●	●
50	●	●	●	●	●	●
60	●	●	●	●	●	●
70	●	●	●	●	●	●
80			●	●	●	●
100			●	●	●	●

BLACK BOLTS, NUTS AND LOCK NUTS – IMPERIAL SIZES:BS916

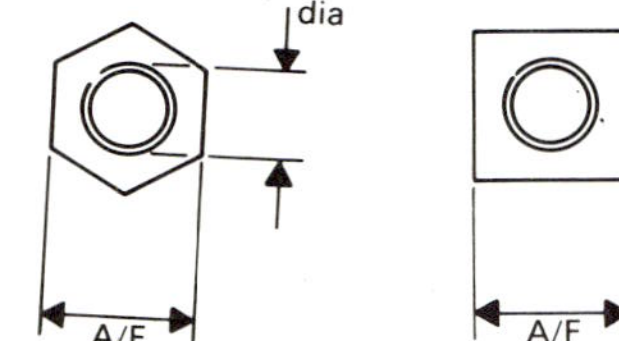

Nominal Size and Maximum Diameter of Bolt	Number of Threads per inch		Dimension Across Flats of Bolt Heads and Nuts (hexagon and square)		Approximate Maximum Dimension Across Corners		Thickness of Bolt Heads (hexagon and square)		Thickness of Nuts (hexagon and square)		Thickness of Lock Nuts (hexagon)	
—	BS Whit	BS Whit	Max in	Min in	Hexagon in	Square in	Max in	Min in	Max in	Min in	Max in	Min in
$\frac{1}{4}$″	20	26	·445	·435	·51	·63	·20	·18	·22	·20	·14	·12
$\frac{5}{16}$″	18	22	·525	·515	·61	·74	·23	·21	·27	·25	·18	·16
$\frac{3}{8}$″	16	20	·600	·585	·69	·85	·28	·26	·33	·31	·22	·20
$\frac{7}{16}$″	14	18	·710	·695	·82	1·00	·34	·32	·39	·37	·26	·24
$\frac{1}{2}$″	12	16	·820	·800	·95	1·16	·40	·37	·46	·43	·31	·28
$\frac{9}{16}$″	12	16	·920	·900	1·06	1·30	·46	·43	·53	·50	·35	·32
$\frac{5}{8}$″	11	14	1·010	·985	1·17	1·43	·51	·48	·60	·56	·41	·37
$\frac{3}{4}$″	10	12	1·200	1·175	1·39	1·70	·62	·59	·72	·68	·49	·45
$\frac{7}{8}$″	9	11	1·300	1·270	1·50	1·84	·69	·65	·81	·75	·55	·49
1″	8	10	1·480	1·450	1·71	2·09	·80	·76	·93	·87	·63	·57
1$\frac{1}{8}$″	7	9	1·670	1·640	1·93	2·36	·91	·87	1·06	1·00	·72	·65
1$\frac{1}{4}$″	7	9	1·860	1·815	2·15	2·63	1·02	·96	1·20	1·12	·81	·73
1$\frac{1}{2}$″	6	8	2·220	2·175	2·56	3·14	1·24	1·18	1·45	1·37	·98	·90
1$\frac{3}{4}$″	5	7	2·580	2·520	2·98	3·65	1·50	1·40	1·72	1·62	1·16	1·06
2″	4·5	7	2·760	2·700	3·19	3·90	1·61	1·51	1·85	1·75	1·25	1·15

BRASS PAN HEAD MACHINE SCREWS SLOTTED AND RECESSED, BRASS, BS4183

Nominal Length	Nominal Diameter						
mm	M1.6	M2	M2.5	M3	M4	M5	M6
4	●*						
5	●*	●	●	●	●		
6	●*	●	●	●	●		
8	●*	●	●	●	●		
10		●	●	●	●	●	
12		●	●	●	●	●	●
16				●	●	●	●
20				●	●	●	●
25				●	●	●	●
30					●	●	●
40						●	●
50						●	●

* Available only as turned screws with slotted cheese heads

BRASS COUNTERSUNK MACHINE SCREWS SLOTTED AND RECESSED, BRASS,BS4183

Nominal Length	Nominal Diameter						
mm	M1.6	M2	M2.5	M3	M4	M5	M6
4	●*	●					
5	●*	●	●	●			
6	●*	●	●	●	●		
8	●*	●	●	●	●		
10		●	●	●	●	●	
12		●	●	●	●	●	●
16				●	●	●	●
20				●	●	●	●
25				●	●	●	●
30					●	●	●
40						●	●
50						●	●

* Available only as turned screws with slotted heads

ISO METRIC BLACK HEXAGON BOLTS AND SCREWS (Preferred Sizes: BS4190)

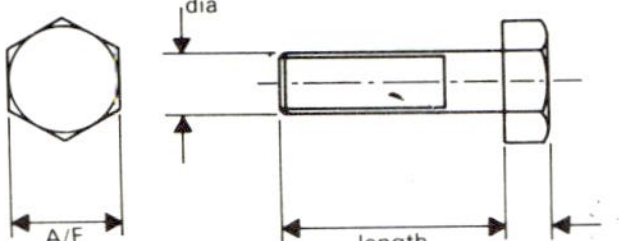

Nominal Size and thread diameter	Pitch of thread	Diameter of unthreaded shank		Width across flats		Width across corners		Height of head	
d	(Coarse pitch series)	Max	Min	Max	Min	Max	Min	Max	Min
M5	0.80	5.48	4.52	8.00	7.64	9.2	8.63	3.875	3.125
M6	1.00	6.48	5.52	10.00	9.64	11.5	10.89	4.375	3.625
M8	1.25	8.58	7.42	13.00	12.57	15.0	14.20	5.875	5.125
M10	1.50	10.58	9.42	17.00	16.57	19.6	18.72	7.45	6.55
M12	1.75	12.70	11.30	19.00	18.48	21.9	20.88	8.45	7.55
M16	2.0	16.70	15.30	24.00	23.16	27.7	26.17	10.45	9.55
M20	2.5	20.84	19.16	30.00	29.16	34.6	32.95	13.90	12.10
M24	3.0	24.84	23.16	36.00	35.00	41.6	39.55	15.90	14.10
M30	3.5	30.84	29.16	46.00	45.00	53.1	50.85	20.05	17.95
M36	4.0	37.00	35.00	55.00	53.80	63.5	60.79	24.05	21.95
M42	4.5	43.00	41.00	65.00	63.80	75.1	72.09	27.05	24.95
M48	5.0	49.00	47.00	75.00	73.80	86.6	83.39	31.05	28.95
M56	5.5	57.20	54.80	85.00	83.60	98.1	94.47	36.25	33.75
M64	6.0	65.20	62.80	95.00	93.60	109.7	105.77	41.25	38.75

All dimensions in millimetres   Note other sizes up to M150

ISO METRIC PRECISION HEXAGON BOLTS AND SCREWS (Preferred Sizes:BS3692)

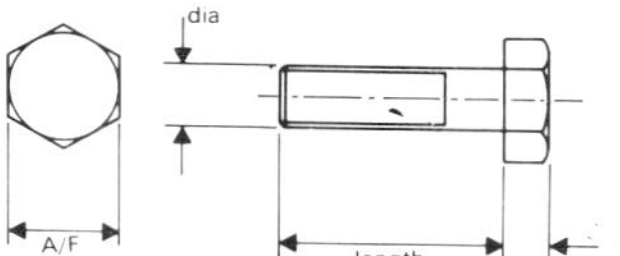

Nominal size and thread dia	Pitch of thread (coarse pitch series)	Thread runout	Diameter of unthreaded shank		Width across flats		Width across corners		Diameter of washer face		Depth of washer face	Transition diameter	Radius under head		Height of Head	
d		max	max	min	max	min	max	min	max	min		max	max	min	max	min
M1.6	0.35	0.8	1.6	1.46	3.2	3.08	3.7	3.48	–	–	–	2.0	0.2	0.1	1.225	0.975
M2	0.4	1.0	2.0	1.86	4.0	3.88	4.6	4.38	–	–	–	2.6	0.3	0.1	1.525	1.275
M2.5	0.45	1.0	2.5	2.36	5.0	4.88	5.8	5.51	–	–	–	3.1	0.3	0.1	1.825	1.575
M3	0.5	1.2	3.0	2.86	5.5	5.38	6.4	6.08	5.08	4.83	0.1	3.6	0.3	0.1	2.125	1.875
M4	0.7	1.6	4.0	3.82	7.0	6.85	8.1	7.74	6.55	6.30	0.1	4.7	0.35	0.2	2.925	2.675
M5	0.8	2.0	5.0	4.82	8.0	7.85	9.2	8.87	7.55	7.30	0.2	5.7	0.35	0.2	3.650	3.350
M6	1	2.5	6.0	5.82	10.0	9.78	11.5	11.05	9.48	9.23	0.3	6.8	0.4	0.25	4.150	3.850
M8	1.25	3.0	8.0	7.78	13.0	12.73	15.0	14.38	12.43	12.18	0.4	9.2	0.6	0.4	5.650	5.350
M10	1.5	3.5	10.0	9.78	17.0	16.73	19.6	18.90	16.43	16.18	0.4	11.2	0.6	0.4	7.180	6.820
M12	1.75	4.0	12.0	11.73	19.0	18.67	21.9	21.10	18.37	18.12	0.4	14.2	1.1	0.6	8.180	7.820
M16	2	5.0	16.0	15.73	24.0	23.67	27.7	26.75	23.27	23.02	0.4	18.2	1.1	0.6	10.180	9.820
M20	2.5	6.0	20.0	19.67	30.0	29.67	34.6	33.53	29.27	28.80	0.4	22.4	1.2	0.8	13.215	12.785
M24	3	7.0	24.0	23.67	36.0	35.38	41.6	39.98	34.98	34.51	0.5	26.4	1.2	0.8	15.215	14.785
M30	3.5	8.0	30.0	29.67	46.0	45.38	53.1	51.28	44.98	44.36	0.5	33.4	1.7	1.0	19.260	18.740
M36	4	10.0	36.0	35.61	55.0	54.26	63.5	61.31	53.86	53.24	0.5	39.4	1.7	1.0	23.260	22.740
M42	4.5	11.0	42.0	41.61	65.0	64.26	75.1	72.61	63.76	63.04	0.6	45.6	1.8	1.2	26.260	25.740
M48	5	12.0	48.0	47.61	75.0	74.26	86.6	83.91	73.76	73.04	0.6	52.6	2.3	1.6	30.260	29.740
M56	5.5	19.0	56.0	55.54	85.0	84.13	98.1	95.07	–	–	–	63.0	3.5	2.0	35.310	34.690
M64	6	21.0	64.0	63.54	95.0	94.13	109.7	106.37	–	–	–	71.0	3.5	2.0	40.310	39.690

All dimensions in millimetres

ISO METRIC PRECISION HEXAGON NUTS AND THIN NUTS – (Preferred Sizes – BS3692)

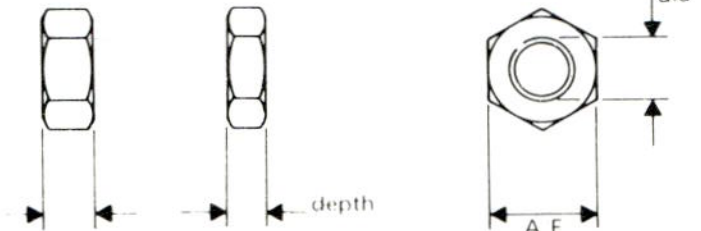

Nominal size and thread diameter	Pitch thread (coarse pitch series)	Width across flats		Width across corners		Thickness of normal nut		Tolerance on squareness of thread to face of nut	Eccentricity of hexagon	Thickness of thin nut	
		max	min	max	min	max	min	max	max	max	min
M1.6	0.35	3.2	3.08	3.7	3.48	1.3	1.05	0.05	0.14	–	–
M2	0.40	4.0	3.88	4.6	4.38	1.6	1.35	0.06	0.14	–	–
M2.5	0.45	5.0	4.88	5.8	5.51	2.0	1.75	0.08	0.14	–	–
M3	0.50	5.5	5.38	6.4	6.08	2.4	2.15	0.09	0.14	–	–
M4	0.70	7.0	6.85	8.1	7.74	3.2	2.90	0.11	0.18	–	–
M5	0.80	8.0	7.85	9.2	8.87	4.0	3.70	0.13	0.18	–	–
M6	1.00	10.0	9.78	11.5	11.05	5.0	4.70	0.17	0.18	–	–
M8	1.25	13.0	12.73	15.0	14.38	6.5	6.14	0.22	0.22	5	4.70
M10	1.50	17.0	16.73	19.6	18.90	8.0	7.64	0.29	0.22	6	5.70
M12	1.75	19.0	18.67	21.9	21.10	10.0	9.64	0.32	0.27	7	6.64
M16	2.0	24.0	23.67	27.7	26.75	13.0	12.57	0.41	0.27	8	7.64
M20	2.5	30.0	29.67	34.6	33.53	16.0	15.57	0.51	0.33	9	8.64
M24	3.0	36.0	35.38	41.6	39.98	19.0	18.48	0.61	0.33	10	9.64
M30	3.5	46.0	45.38	53.1	51.28	24.0	23.48	0.78	0.33	12	11.57
M36	4.0	55.0	54.26	63.5	61.31	29.0	28.48	0.94	0.39	14	13.57
M42	4.5	65.0	64.26	75.1	72.61	34.0	33.38	1.11	0.39	16	15.57
M48	5.0	75.0	74.26	86.6	83.91	38.0	37.38	1.29	0.39	18	17.57
M56	5.5	85.0	84.13	98.1	95.07	45.0	44.38	1.46	0.46	–	–
M64	6.0	95.0	94.13	109.7	106.37	51.0	50.26	1.63	0.46	–	–

All dimensions in millimetres   Larger sizes up to M150

COLD FORMED HEXAGON NUTS – METRIC SERIES (BS3692)

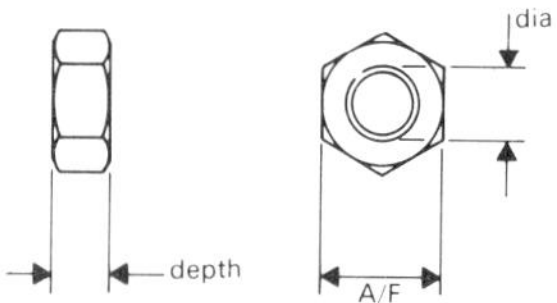

Nominal size and thread diameter	Width across flats		Width across corners		Thickness of ordinary nut		Thickness of locknut	
	max	min	max	min	max	min	max	min
M5	8	7.85	9.2	8.87	4.0	3.70	—	—
M6	10	9.78	11.5	11.05	5.0	4.70	—	—
M8	13.	12.73	15.0	14.38	6.5	6.14	5	4.70
M10	17	16.73	19.6	18.90	8.0	7.64	6	5.70
M12	19	18.67	21.9	21.10	10.0	9.64	7	6.64
M16	24	23.67	27.7	26.75	13.0	12.57	8	7.64
M20	30	29.67	34.6	33.53	16.0	15.57	9	8.64

Dimensions in millimetres

HEXAGON NUTS – UNIFIED SERIES (BS2708)

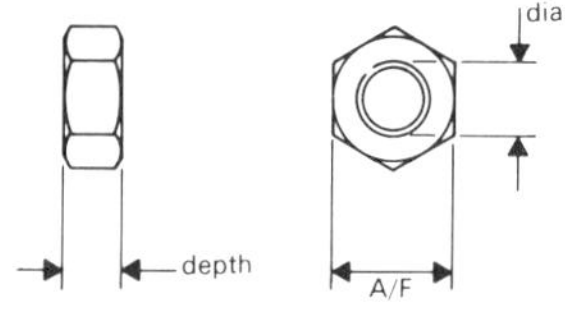

Nominal size and thread diameter	Width across flats		Width across corners	Thickness of ordinary nut				Thickness of locknut			
				Black		Faced one side		Black		Faced both sides	
	max	*min*	*max*	*max*	*min*	*max*	*min*	*max*	*min*	*max*	*min*
$\frac{1}{4}$	0·4375	0·428	0·505	0·226	0·212	—	—	0·163	0·150	—	—
$\frac{5}{16}$	0·5000	0·490	0·577	0·273	0·258	—	—	0·195	0·180	—	—
$\frac{3}{8}$	0·5625	0·551	0·650	0·337	0·320	—	—	0·227	0·210	—	—
$\frac{7}{16}$	0·6875	0·675	0·794	0·385	0·365	—	—	0·260	0·240	—	—
$\frac{1}{2}$	0·7500	0·736	0·866	0·448	0·427	—	—	0·323	0·302	—	—
$\frac{5}{8}$	0·9375	0·912	1·083	0·559	0·535	—	—	0·387	0·363	—	—
$\frac{3}{4}$	1·1250	1·100	1·299	0·665	0·625	—	—	0·446	0·414	—	—
$\frac{7}{8}$	1·3125	1·282	1·516	0·776	0·724	0·754	0·724	0·510	0·458	0·478	0·458
1	1·5000	1·470	1·732	0·887	0·831	0·871	0·831	0·575	0·519	0·539	0·519
$1\frac{1}{8}$	1·6875	1·657	1·949	0·999	0·939	0·979	0·939	0·639	0·579	0·599	0·579
$1\frac{1}{4}$	1·8750	1·830	2·165	1·094	1·030	1·080	1·030	0·751	0·687	0·707	0·687
$1\frac{3}{8}$	2·0625	2·017	2·382	1·206	1·138	1·188	1·138	0·815	0·752	0·767	0·747
$1\frac{1}{2}$	2·2500	2·205	2·598	1·317	1·245	1·295	1·245	0·880	0·816	0·828	0·808
$1\frac{3}{4}$	2·6250	2·565	3·031	1·540	1·460	1·510	1·460	1·009	0·932	0·949	0·929
2	3·0000	2·940	3·464	1·763	1·675	1·735	1·675	1·138	1·050	1·070	1·050

Sizes up to 8½in nominal thread diameter
Dimensions in inches

BSW COUNTERSUNK NIB BOLTS (BS325)

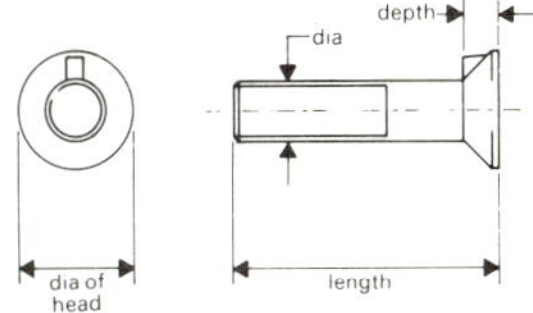

Nominal size and thread diameter	Diameter of head	Maximum depth of flash	Nib	
			Depth	Width
$\frac{1}{4}$	0·459	$\frac{1}{32}$	0·218	0·062
$\frac{5}{16}$	0·574	$\frac{1}{32}$	0·234	0·078
$\frac{3}{8}$	0·690	$\frac{1}{32}$	0·280	0·094
$\frac{7}{16}$	0·805	$\frac{3}{64}$	0·327	0·109
$\frac{1}{2}$	0·920	$\frac{3}{64}$	0·358	0·125
$\frac{5}{8}$	1·149	$\frac{3}{64}$	0·420	0·156
$\frac{3}{4}$	1·379	$\frac{1}{16}$	0·509	0·188
$\frac{7}{8}$	1·609	$\frac{1}{16}$	0·561	0·219
1	1·839	$\frac{1}{16}$	0·625	0·250

Dimensions in inches

Length of thread

Length of bolt	Length of thread
Up to and including 8 in	2 dias
Over 8 in	2½ dias

BSW CUP OVAL FISH BOLTS (BS536)

Nominal size and thread diameter	Diameter of head	Depth of head	Oval neck	
			Depth	Major axis
$\frac{3}{8}$	0·750	0·187	0·250	0·562
$\frac{1}{2}$	0·937	0·312	0·312	0·687
$\frac{5}{8}$	1·125	0·500	0·375	0·875

Dimensions in inches

BSW CUP NIB BOLTS (BS325)

Nominal size and thread diameter	Diameter of head	Depth of head	Max depth of flash	Max radius under head	Nib		
					Depth	Projection	Width
1/4	0·500	0·188	1/32	1/32	0·125	0·062	0·062
5/16	0·625	0·234	1/32	1/32	0·156	0·078	0·078
3/8	0·750	0·281	1/32	1/32	0·188	0·094	0·094
7/16	0·875	0·328	1/32	1/32	0·219	0·109	0·109
1/2	1·000	0·375	1/32	1/32	0·250	0·125	0·125
5/8	1·250	0·469	1/16	3/64	0·312	0·156	0·156
3/4	1·500	0·562	1/16	3/64	0·375	0·188	0·188
7/8	1·750	0·656	3/32	1/16	0·438	0·219	0·219
1	2·000	0·750	3/32	1/16	0·500	0·250	0·250

Dimensions in inches

Length of thread

Length of bolt	Length of thread
Up to and including 8 in	2 dias
Over 8 in	2½ dias

BSW CUP ROUND BOLTS (BS325)

Nominal size and thread diameter	Diameter of head	Depth of head	Max depth of flash	Max radius under head
1/4	0·500	0·188	1/32	1/32
5/16	0·625	0·234	1/32	1/32
3/8	0·750	0·281	1/32	1/32
7/16	0·875	0·328	1/32	1/32
1/2	1·000	0·375	1/32	1/32
5/8	1·250	0·469	1/16	3/64
3/4	1·500	0·562	1/16	3/64
7/8	1·750	0·656	3/32	1/16
1	2·000	0·750	3/32	1/16

Dimensions in inches

Length of thread

Length of bolt	Length of thread
Up to and including 8 in	2 dias
Over 8 in	2½ dias

BSW CUP SQUARE BOLTS (BS325)

Nominal size and thread diameter	Diameter of head	Depth of head	Maximum depth of flash	Depth of square neck	Maximum radius under head
1/4	0·562	0·125	1/32	0·188	1/32
5/16	0·703	0·156	1/32	0·234	1/32
3/8	0·844	0·188	1/32	0·281	1/32
7/16	0·984	0·250	1/32	0·328	1/32
1/2	1·125	0·281	1/32	0·375	1/32
5/8	1·406	0·375	1/16	0·469	3/64
3/4	1·688	0·438	1/16	0·562	3/64
7/8	1·969	0·500	3/32	0·656	1/16
1	2·250	0·562	3/32	0·750	1/16

Dimensions in inches

Length of thread

Length of bolt	Length of thread
Up to and including 8 in	2 dias
Over 8 in	2½ dias

BSW COUNTERSUNK ROUND BOLTS (BS325)

Nominal size and thread diameter	Diameter of head	Maximum depth of flash
1/4	0·459	1/32
5/16	0·574	1/32
3/8	0·690	1/32
7/16	0·805	3/64
1/2	0·920	3/64
5/8	1·149	3/64
3/4	1·379	1/16
7/8	1·609	1/16
1	1·839	1/16

Dimensions in inches

Length of thread

Length of bolt	Length of thread
Up to and including 8 in	2 dias
Over 8 in	2½ dias

BSW IRON COUNTERSUNK SQUARE BOLTS (BS325)

Nominal size and thread diameter	Diameter of head	Maximum depth of flash	Total depth of head and square
3/8	0·690	1/32	0·313
7/16	0·805	3/64	0·375
1/2	0·920	3/64	0·422
5/8	1·149	3/64	0·515
3/4	1·379	1/16	0·625
7/8	0·609	1/16	0·719
1	1·839	1/16	0·812

Dimensions in inches

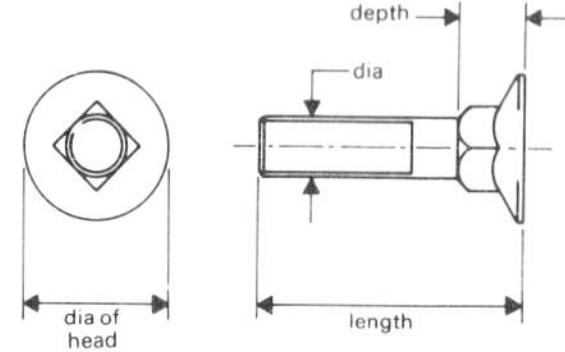

Length of thread

Length of bolt	Length of thread
Up to and including 8 in	2 dias
Over 8 in	2½ dias

BSW WOOD COUNTERSUNK SQUARE BOLTS (BS325)

Nominal size and thread diameter	Diameter of head	Maximum depth of flash	Total depth of head and square
1/4	0·625	1/32	0·350
5/16	0·781	1/32	0·429
3/8	0·938	1/32	0·509
7/16	1·094	1/32	0·589
1/2	1·250	1/16	0·699
5/8	1·562	1/16	0·859
3/4	1·875	1/16	1·018
7/8	2·188	3/32	1·209
1	2·500	3/32	1·369

Dimensions in inches

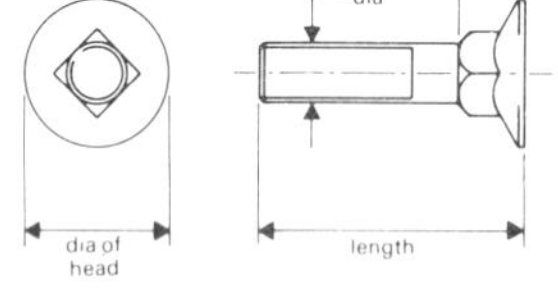

Length of thread

Length of bolt	Length of thread
Up to and including 8 in	2 dias
Over 8 in	2½ dias

BSW SQUARE/SQUARE BOLTS (PITCH ARCH BOLTS) (BS3916)

Nominal size and thread diameter	Diameter of unthreaded shank	Width across flats		Width across corners	Radius under head	Thickness of head		Depth of square
	max	*max*	*min*	*max*	*max*	*max*	*min*	*basic*
$\frac{1}{2}$	0·530	0·820	0·800	1·16	$\frac{1}{32}$	0·363	0·333	0·375
$\frac{9}{16}$	0·592	0·920	0·900	1·30	$\frac{3}{64}$	0·405	0·375	0·375
$\frac{5}{8}$	0·665	1·010	0·985	1·43	$\frac{3}{64}$	0·447	0·417	0·437
$\frac{3}{4}$	0·790	1·200	1·175	1·70	$\frac{3}{64}$	0·530	0·500	0·500
$\frac{7}{8}$	0·915	1·300	1·270	1·84	$\frac{1}{16}$	0·623	0·583	0·562
1	1·040	1·480	1·450	2·09	$\frac{1}{16}$	0·706	0·666	0·687
$1\frac{1}{8}$	1·175	1·670	1·640	2·36	$\frac{1}{8}$	0·79	0·75	0·750
$1\frac{1}{4}$	1·300	1·860	1·815	2·63	$\frac{1}{8}$	0·89	0·83	0·875
$1\frac{3}{8}$	1·425	2·050	2·005	2·90	$\frac{1}{8}$	0·98	0·92	0·875
$1\frac{1}{2}$	1·550	2·220	2·175	3·14	$\frac{1}{8}$	1·06	1·00	1·000
$1\frac{5}{8}$	1·685	2·410	2·365	3·41	$\frac{1}{8}$	1·18	1·08	1·375
$1\frac{3}{4}$	1·810	2·580	2·520	3·65	$\frac{1}{8}$	1·27	1·17	1·375
2	2·060	2·760	2·700	3·90	$\frac{1}{8}$	1·43	1·33	1·625
$2\frac{1}{4}$	—	3·150	3·090	4·45	$\frac{3}{16}$	1·60	1·50	2·000
$2\frac{1}{2}$	—	3·550	3·490	5·02	$\frac{3}{16}$	1·77	1·67	2·125

Dimensions in inches

Pitch Arch Bolts

M16	16·70	24·0	23·16	27·7	1·0	10·45	9·55	10·0
M20	20·84	30·0	29·16	34·6	1·0	13·90	12·10	10·0

Dimensions in millimetres

Length of thread — inch square/square

Length of bolt	All dias up to and including $\frac{1}{2}$ in	All dias over $\frac{1}{2}$ in
Up to and including 4 in	2 dias	$1\frac{1}{2}$ dias
Over 4 in up to and including 8 in	2 dias	2 dias
Over 8 in	$2\frac{1}{2}$ dias	$2\frac{1}{2}$ dias

Length of thread — metric Pitch Arch

Length of bolt	Dia M16	Dias M20 and over
Up to and including 70 mm	$1\frac{1}{2}$ dias	
Over 70 mm up to and including 200 mm		2 dias

COMBINER BOLTS (BSW THREAD TO BS16)

Heads and Nuts (BS 916)										Washers		
BS Size No.	Dia of bolt	Length of bolt	Length of thread	Width across flats, heads and nuts		Thickness				Dia of hole after galvg	Outside diameter	Thick-ness
						Heads		Nuts				
				max	min	max	min	max	min			swg
320		2	1									
325		$2\frac{1}{2}$	$1\frac{1}{2}$									
330		3	$1\frac{1}{2}$									
335	$\frac{3}{8}$	$3\frac{1}{2}$	$1\frac{1}{2}$	0·600	0·585	0·28	0·26	0·33	0·31	$\frac{7}{16}$	$\frac{7}{8}$	14
340		4	2									
345		$4\frac{1}{2}$	2									
350		5	2									
355		$5\frac{1}{2}$	2									
420		2	1									
425		$2\frac{1}{2}$	$1\frac{1}{2}$									
430		3	$1\frac{1}{2}$									
435	$\frac{1}{2}$	$3\frac{1}{2}$	$1\frac{1}{2}$	0·820	0·800	0·40	0·37	0·46	0·43	$\frac{9}{16}$	$1\frac{1}{8}$	12
440		4	2									
445		$4\frac{1}{2}$	2									
450		5	2									
455		$5\frac{1}{2}$	2									
520		2	1									
525		$2\frac{1}{2}$	$1\frac{1}{2}$									
530		3	$1\frac{1}{2}$									
535	$\frac{5}{8}$	$3\frac{1}{2}$	$1\frac{1}{2}$	1·010	0·985	0·51	0·48	0·60	0·56	$\frac{11}{16}$	$1\frac{3}{8}$	10
540		4	2									
545		$4\frac{1}{2}$	2									
550		5	2									
555		$5\frac{1}{2}$	2									

Dimensions in inches

ARMBOLTS (BSW THREAD TO BS16)

Heads and Nuts (BS 916)									Washers					
Dia of bolt	Width across flats, heads and nuts		Thickness				Length of thread	Dia of hole after galvg	Matching washers are either neck washers or nut washers.					
			Heads		Nuts									
	max	*min*	*max*	*min*	*max*	*min*								
½	0·820	0·800	0·40	0·37	0·46	0·43	2¼	9/16	2	2	⅛	3	2	¼
⅝	1·010	0·985	0·51	0·48	0·60	0·56	2¾	11/16	3	2	¼	3	2	¼

Dimensions in inches

HEXAGON NUTS – UNIFIED HEAVY SERIES (BS1769)

dia
depth
A/F

Nominal size and thread diameter	Width across flats		Width across corners	Thickness of ordinary nut				Thickness of lock nut			
				As forged		Faced one side		As forged		Faced both sides	
	max	*min*	*max*	*max*	*min*	*max*	*min*	*max*	*min*	*max*	*min*
½	0·8750	0·8550	1·01	0·520	0·480	0·504	0·464	0·332	0·292	0·317	0·277
⅝	1·0625	1·0375	1·23	0·647	0·603	0·631	0·587	0·397	0·353	0·381	0·337
¾	1·2500	1·2250	1·44	0·774	0·726	0·758	0·710	0·462	0·414	0·446	0·398
⅞	1·4375	1·4075	1·66	0·901	0·849	0·885	0·833	0·526	0·474	0·510	0·458
1	1·6250	1·5950	1·88	1·028	0·972	1·012	0·956	0·590	0·534	0·575	0·519
1⅛	1·8125	1·7825	2·09	1·155	1·095	1·139	1·079	0·655	0·595	0·639	0·579
1¼	2·0000	1·9550	2·31	1·282	1·218	1·251	1·187	0·782	0·718	0·751	0·687
1½	2·3750	2·3300	2·74	1·536	1·464	1·505	1·433	0·911	0·839	0·880	0·808
1¾	2·7500	2·6900	3·18	1·790	1·710	1·759	1·679	1·040	0·960	1·009	0·929
2	3·1250	3·0650	3·61	2·044	1·956	2·013	1·925	1·169	1·081	1·138	1·050

Dimensions in inches. Sizes up to 8½in nominal thread diameter

THICK HEXAGON NUTS – UNIFIED NORMAL SERIES (BS1768)

Nominal size and thread diameter	Width across flats		Width across corners	Diameter of washer face		Thickness	
	max	*min*	*max*	*max*	*min*	*max*	*min*
$\frac{1}{4}$	0·437	0·430	0·505	0·421	0·411	0·286	0·276
$\frac{5}{16}$	0·500	0·493	0·577	0·483	0·473	0·333	0·323
$\frac{3}{8}$	0·562	0·554	0·650	0·545	0·535	0·411	0·401
$\frac{7}{16}$	0·687	0·679	0·794	0·668	0·658	0·458	0·448
$\frac{1}{2}$	0·750	0·742	0·866	0·730	0·720	0·567	0·557
$\frac{9}{16}$	0·875	0·867	1·010	0·855	0·845	0·614	0·604
$\frac{5}{8}$	0·937	0·929	1·083	0·918	0·908	0·724	0·714

Dimensions in inches

SLOTTED HEXAGON NUTS – IMPERIAL SERIES (BS1083)

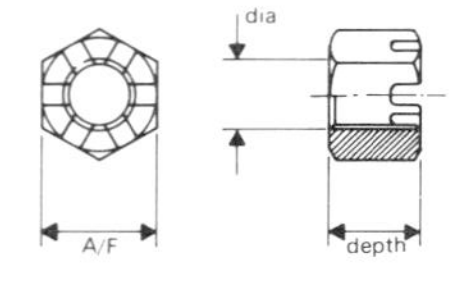

Nominal size and thread diameter	Width across flats		Width across corners	Thickness		Lower face of nut to bottom of slots		Slots: Width		Slots: Depth
	max	*min*	*max*	*max*	*min*	*max*	*min*	*max*	*min*	*approx*
$\frac{1}{4}$	0·445	0·438	0·51	0·260	0·250	0·170	0·160	0·100	0·090	0·090
$\frac{5}{16}$	0·525	0·518	0·61	0·280	0·270	0·190	0·180	0·100	0·090	0·090
$\frac{3}{8}$	0·600	0·592	0·69	0·312	0·302	0·222	0·212	0·100	0·090	0·090
$\frac{7}{16}$	0·710	0·702	0·82	0·375	0·365	0·235	0·225	0·135	0·125	0·140
$\frac{1}{2}$	0·820	0·812	0·95	0·437	0·427	0·297	0·287	0·135	0·125	0·140
$\frac{9}{16}$	0·920	0·912	1·06	0·500	0·490	0·313	0·303	0·175	0·165	0·187
$\frac{5}{8}$	1·010	1·000	1·17	0·562	0·552	0·375	0·365	0·175	0·165	0·187
$\frac{3}{4}$	1·200	1·190	1·39	0·687	0·677	0·453	0·443	0·218	0·208	0·234

Dimensions in inches

HIGH TENSILE HEXAGON ROUND BOLTS AND SETSCREWS – UNIFIED NORMAL SERIES (BS1768)

Nominal size and thread diameter	Diameter of unthreaded shank		Width across flats		Width across corners	Diameter of washer face		Thickness of head		Radius under head	
	max	*min*	*max*	*min*	*max*	*max*	*min*	*max*	*min*	*max*	*min*
$\frac{1}{4}$	0·250	0·246	0·437	0·430	0·505	0·421	0·411	0·163	0·153	0·025	0·015
$\frac{5}{16}$	0·312	0·309	0·500	0·493	0·577	0·483	0·473	0·211	0·201	0·025	0·015
$\frac{3}{8}$	0·375	0·371	0·562	0·554	0·650	0·545	0·535	0·243	0·233	0·025	0·015
$\frac{7}{16}$	0·437	0·433	0·625	0·617	0·722	0·605	0·595	0·291	0·281	0·025	0·015
$\frac{1}{2}$	0·500	0·496	0·750	0·742	0·866	0·730	0·720	0·323	0·313	0·025	0·015
$\frac{9}{16}$	0·562	0·558	0·812	0·804	0·938	0·792	0·782	0·371	0·361	0·045	0·020
$\frac{5}{8}$	0·625	0·619	0·937	0·929	1·083	0·918	0·908	0·403	0·393	0·045	0·020
$\frac{3}{4}$	0·750	0·744	1·125	1·115	1·300	1·100	1·090	0·483	0·463	0·045	0·020
$\frac{7}{8}$	0·875	0·867	1·312	1·300	1·515	1·285	1·275	0·563	0·543	0·065	0·040
1	1·000	0·992	1·500	1·488	1·732	1·473	1·463	0·627	0·597	0·095	0·060
$1\frac{1}{8}$	1·125	1·117	1·687	1·657	1·948	1·641	1·625	0·718	0·678	0·095	0·060
$1\frac{1}{4}$	1·250	1·242	1·875	1·830	2·165	1·813	1·797	0·813	0·773	0·095	0·060
$1\frac{3}{8}$	1·375	1·365	2·062	2·017	2·382	2·001	1·985	0·878	0·838	0·095	0·060
$1\frac{1}{2}$	1·500	1·490	2·250	2·205	2·598	2·188	2·172	0·974	0·934	0·095	0·060
$1\frac{3}{4}$	1·750	1·740	2·625	2·565	3·031	2·543	2·527	1·134	1·074	0·095	0·060
2	2·000	1·990	3·000	2·940	3·464	2·918	2·902	1·263	1·203	0·095	0·060

Dimensions in inches

HEXAGON ROUND BOLTS AND SETSCREWS – UNIFIED HEAVY SERIES (BS1769)

Nominal size and thread diameter	Diameter of unthreaded shank	Width across flats		Width across corners	Thickness of head		Radius under head
		max	*min*	*max*	*max*	*min*	*max*
½	0·530	0·875	0·855	1·01	0·353	0·323	$\frac{1}{32}$
⅝	0·675	1·062	1·037	1·23	0·433	0·403	$\frac{1}{16}$
¾	0·800	1·250	1·225	1·44	0·513	0·483	$\frac{1}{16}$
⅞	0·938	1·437	1·407	1·66	0·605	0·563	$\frac{1}{16}$
1	1·063	1·625	1·595	1·88	0·667	0·627	$\frac{1}{16}$
1⅛	1·188	1·812	1·782	2·09	0·758	0·718	⅛
1¼	1·313	2·000	1·955	2·31	0·873	0·813	⅛
1⅜	1·469	2·187	2·142	2·53	0·938	0·878	⅛
1½	1·594	2·375	2·330	2·74	1·034	0·974	⅛
1¾	1·844	2·750	2·690	3·18	1·234	1·134	⅛
2	2·094	3·125	3·065	3·61	1·363	1·263	⅛

Dimensions in inches

Length of thread

Length of bolt	Length of thread
Up to and including 6 in	2 dias + ¼ in
Above 6 in	2 dias + ½in

HIGH STRENGTH FRICTION GRIP HEXAGON HEAD BOLTS – METRIC SERIES (BS4395)

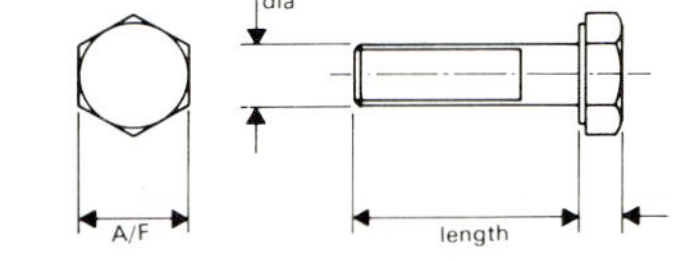

Nominal size and thread diameter	Diameter of unthreaded shank		Width across flats		Width across corners		Diameter of washer face		Depth of washer face	Radius under head		Thickness of head	
	max	*min*	*max*	*min*	*max*	*min*	*max*	*min*	*max*	*max*	*min*	*max*	*min*
M16	16·70	15·30	27	26·16	31·2	29·30	27	24·91	0·4	1·0	0·6	10·45	9·55
M20	20·84	19·16	32	31·00	36·9	35·03	32	29·75	0·4	1·2	0·8	13·90	12·10
M22	22·84	21·16	36	35·00	41·6	39·55	36	33·75	0·4	1·2	0·8	14·90	13·10
M24	24·84	23·16	41	40·00	47·3	45·20	41	38·75	0·5	1·2	0·8	15·90	14·10
M27	27·84	26·16	46	45·00	53·1	50·85	46	43·75	0·5	1·5	1·0	17·90	16·10
M30	30·84	29·16	50	49·00	57·7	55·37	50	47·75	0·5	1·5	1·0	20·05	17·95
M36	37·0	35·0	60	58·80	69·3	66·44	60	57·75	0·5	1·5	1·0	24·05	21·95

Dimensions in millimetres

Length of thread

Nominal length of bolt	Length of thread
Up to and including 125 mm	2 dias + 6mm
Over 125 mm up to and including 200 mm	2 dias + 12mm
Over 200 mm	2 dias + 25mm

HIGH STRENGTH FRICTION GRIP HEXAGON NUTS – METRIC SERIES (BS4395)

Nominal size and thread diameter	Width across flats		Width across corners		Diameter of washer face		Depth of washer face	Thickness of nut	
	max	*min*	*max*	*min*	*max*	*min*	*max*	*max*	*min*
M16	27·00	26·16	31·20	29·30	27·00	24·91	0·4	15·55	14·45
M20	32·00	31·00	36·90	35·03	32·00	29·75	0·4	18·55	17·45
M22	36·00	35·00	41·60	39·55	36·00	33·75	0·4	19·65	18·35
M24	41·00	40·00	47·30	45·20	41·00	38·75	0·5	22·65	21·35
M27	46·00	45·00	53·10	50·85	46·00	43·75	0·5	24·65	23·35
M30	50·00	49·00	57·70	55·37	50·00	47·75	0·5	26·65	25·35
M36	60·00	58·80	69·30	66·44	60·00	57·75	0·5	31·80	30·20

Dimensions in millimetres

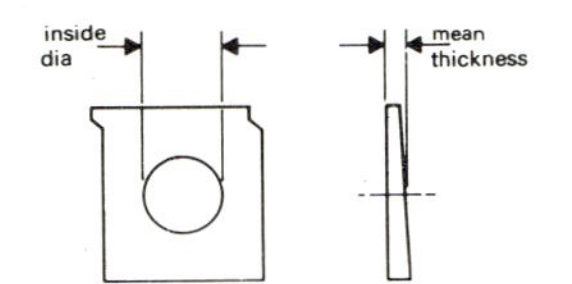

SQUARE ROUND BOLTS UNIFIED NORMAL SERIES (BS2708)

Nominal size and thread diameter	Dia of un-threaded shank	Width across flats		Width across corners	Radius under head	Thickness of head		
	max	*max*	*min*	*max*	*max*	*nominal*	*max*	*min*
$\frac{1}{4}$	0·280	0·3750	0·3650	0·530	$\frac{1}{32}$	$\frac{11}{64}$	0·188	0·168
$\frac{5}{16}$	0·342	0·5000	0·4900	0·707	$\frac{1}{32}$	$\frac{13}{64}$	0·220	0·200
$\frac{3}{8}$	0·405	0·5625	0·5475	0·795	$\frac{1}{32}$	$\frac{1}{4}$	0·268	0·248
$\frac{7}{16}$	0·468	0·6250	0·6100	0·884	$\frac{1}{32}$	$\frac{19}{64}$	0·316	0·296
$\frac{1}{2}$	0·530	0·7500	0·7300	1·061	$\frac{1}{32}$	$\frac{21}{64}$	0·348	0·318
$\frac{5}{8}$	0·665	0·9375	0·9125	1·326	$\frac{3}{64}$	$\frac{27}{64}$	0·444	0·414
$\frac{3}{4}$	0·790	1·1250	1·1000	1·591	$\frac{3}{64}$	$\frac{1}{2}$	0·524	0·494
$\frac{7}{8}$	0·915	1·3125	1·2825	1·856	$\frac{1}{16}$	$\frac{19}{32}$	0·620	0·580
1	1·040	1·5000	1·4700	2·121	$\frac{1}{16}$	$\frac{21}{32}$	0·684	0·644
$1\frac{1}{8}$	1·175	1·6875	1·6575	2·386	$\frac{1}{8}$	$\frac{3}{4}$	0·780	0·740
$1\frac{1}{4}$	1·300	1·8750	1·8300	2·652	$\frac{1}{8}$	$\frac{27}{32}$	0·876	0·816
$1\frac{3}{8}$	1·425	2·0625	2·0175	2·917	$\frac{1}{8}$	$\frac{29}{32}$	0·940	0·880
$1\frac{1}{2}$	1·550	2·2500	2·2050	3·182	$\frac{1}{8}$	1	1·036	0·976

Dimensions in inches

Length of thread

Length of bolt	All dias up to and including ½in All dias over 1in	5/8in, ¾in, 7/8in and 1in diameters
Up to and including 6in	2 dias + ¼in	1½ dias
Over 6in	2 dias + ½in	2½ dias

BSW HEXAGON ROUND BOLTS – HIGH YIELD STRESS (BS4360 and BS916)

Nominal size and thread diameter	Diameter of unthreaded shank	Width across flats		Width across corners	Radius under head	Thickness of head	
	max	*max*	*min*	*max*	*max*	*max*	*basic*
$\frac{1}{2}$	0·530	0·820	0·800	0·95	$\frac{1}{32}$	0·363	0·333
$\frac{9}{16}$	0·592	0·920	0·900	1·06	$\frac{3}{64}$	0·405	0·375
$\frac{5}{8}$	0·665	1·010	0·985	1·17	$\frac{3}{64}$	0·447	0·417
$\frac{3}{4}$	0·790	1·200	1·175	1·39	$\frac{3}{64}$	0·530	0·500
$\frac{7}{8}$	0·915	1·300	1·270	1·50	$\frac{1}{16}$	0·623	0·583
1	1·040	1·480	1·450	1·71	$\frac{1}{16}$	0·706	0·666
$1\frac{1}{8}$	1·175	1·670	1·640	1·93	$\frac{1}{8}$	0·79	0·75
$1\frac{1}{4}$	1·300	1·860	1·815	2·15	$\frac{1}{8}$	0·89	0·83
$1\frac{3}{8}$	1·425	2·050	2·005	2·37	$\frac{1}{8}$	0·98	0·92
$1\frac{1}{2}$	1·550	2·220	2·175	2·56	$\frac{1}{8}$	1·06	1·00
$1\frac{5}{8}$	1·685	2·410	2·365	2·78	$\frac{1}{8}$	1·18	1·08
$1\frac{3}{4}$	1·810	2·580	2·520	2·98	$\frac{1}{8}$	1·27	1·17
2	2·060	2·760	2·700	3·19	$\frac{1}{8}$	1·43	1·33

Dimensions in inches

Length of thread

Length of Bolt	All dias up to and including $\frac{1}{2}$ in	All dias over $\frac{1}{2}$ in
Up to and including 4 in	2 dias	$1\frac{1}{2}$ dias
Over 4 in up to and including 8 in	2 dias.	2 dias.
Over 8 in	$2\frac{1}{2}$ dias.	$2\frac{1}{2}$ dias.

SLOTTED RAISED COUNTERSUNK HEAD MACHINE SCREWS – METRIC SERIES (Preferred Sizes – BS4183)

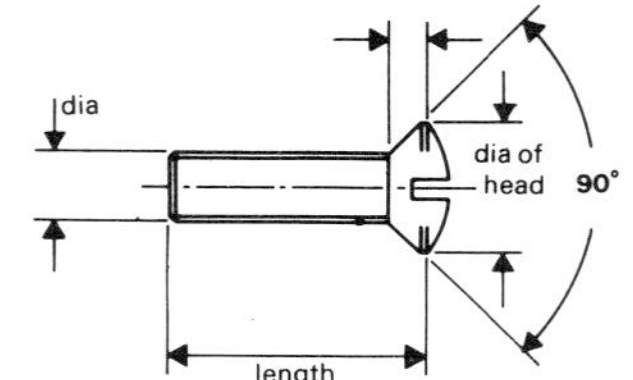

Nominal size and thread diameter	Head diameter		Head height		Radius under head	Thread length	Thread run-out	Height of raised portion	Head radius	Width of slot		Depth of Slot	
	Max	Min	Max	Min		Min	Max	Nom	Nom	Max	Min	Max	Min
M1	2.20	1.76	0.60	0.48	0.1		0.5	0.25	2.0	0.45	0.31	0.50	0.40
M1.2	2.64	2.14	0.72	0.60	0.1		0.5	0.30	2.5	0.50	0.36	0.60	0.48
M1.6	3.52	2.86	0.96	0.84	0.1	15	0.7	0.40	3.0	0.60	0.46	0.80	0.64
M2	4.40	3.50	1.20	1.08	0.1	16	0.8	0.50	4.0	0.70	0.56	1.00	0.80
M2.5	5.50	4.45	1.50	1.38	0.1	18	0.9	0.60	5.0	0.80	0.66	1.25	1.00
M3	6.30	5.25	1.65	1.50	0.1	19	1.0	0.75	6.0	1.00	0.86	1.50	1.20
M4	8.40	7.04	2.20	2.00	0.2	22	1.4	1.00	8.0	1.20	1.06	2.00	1.60
M5	10.00	8.75	2.50	2.25	0.2	25	1.6	1.25	10.0	1.51	1.26	2.50	2.00
M6	12.00	10.50	3.00	2.70	0.25	28	2.0	1.50	12.0	1.91	1.66	3.00	2.40
M8	16.00	14.00	4.00	3.60	0.4	34	2.5	2.00	16.0	2.31	2.06	4.00	3.20
M10	20.00	17.50	5.00	4.50	0.4	40	3.0	2.50	20.0	2.81	2.56	5.00	4.00
M12	24.00	21.00	6.00	5.40	0.6	46	3.5	3.00	25.0	3.31	3.06	6.00	4.80
M16	32.00	28.00	8.00	7.20	0.6	58	4.0	4.00	32.0	4.37	4.07	8.00	6.40
M20	40.00	35.00	10.00	9.00	0.8	70	5.0	5.00	40.0	5.37	5.07	10.00	8.00

All dimensions in millimetres

SLOTTED COUNTERSUNK HEAD MACHINE SCREWS – METRIC SERIES
(Preferred Sizes – BS4183)

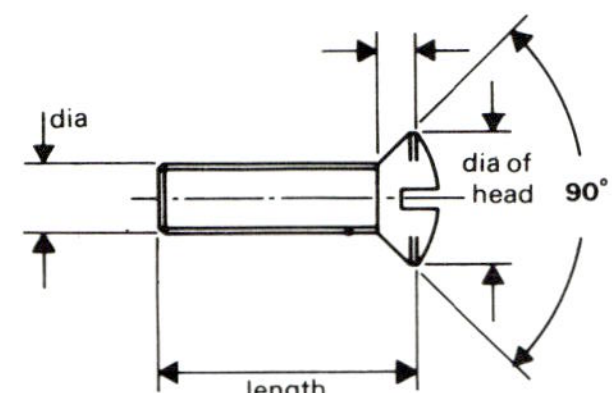

Nominal size and thread diameter	Head diameter		Head height		Radius	Thread	Thread	Width of slot		Depth of slot	
	Max	Min	Max	Min		Min	Max	Max	Min	Max	Min
M1	2.20	1.76	0.60	0.48	0.1		0.50	0.45	0.31	0.30	0.20
M1.2	2.64	2.14	0.72	0.60	0.1		0.50	0.50	0.36	0.36	0.24
M1.6	3.52	2.86	0.96	0.84	0.1	15.0	0.70	0.60	0.46	0.48	0.32
M2.0	4.40	3.50	1.20	1.08	0.1	16.0	0.80	0.70	0.56	0.60	0.40
M2.5	5.50	4.45	1.50	1.38	0.1	18.0	0.90	0.80	0.66	0.75	0.50
M3	6.30	5.25	1.65	1.50	0.1	19.0	1.00	1.00	0.86	0.90	0.60
M4	8.40	7.04	2.20	2.00	0.2	22.0	1.40	1.20	1.06	1.20	0.80
M5	10.0	8.75	2.50	2.25	0.2	25.0	1.60	1.51	1.26	1.50	1.00
M6	12.00	10.50	3.00	2.70	0.25	28.0	2.00	1.91	1.66	1.80	1.20
M8	16.00	14.00	4.00	3.60	0.4	34.0	2.50	2.31	2.06	2.40	1.60
M10	20.00	17.50	5.00	4.50	0.4	40.0	3.00	2.81	2.56	3.00	2.00
M12	24.00	21.00	6.00	5.40	0.6	46.0	3.50	3.31	3.06	3.60	2.40
M16	32.00	28.00	8.00	7.20	0.6	58.0	4.00	4.37	4.07	4.80	3.20
M20	40.00	35.00	10.00	9.00	0.8	70.0	5.00	5.37	5.07	6.00	4.00

All dimensions in millimetres

SLOTTED CHEESE HEAD MACHINE SCREWS – METRIC SERIES (Preferred Sizes – BS4183)

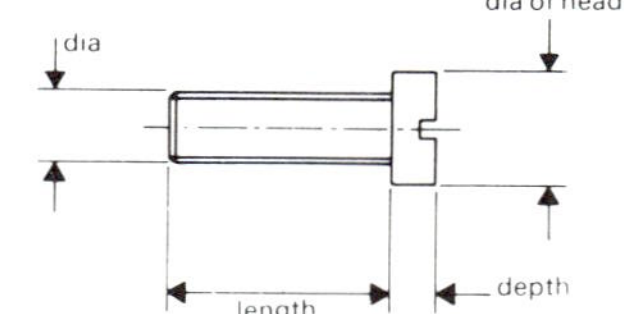

Nominal size and thread diameter	Head diameter		Head height		Radius	Transition diameter	Thread length	Thread run-out	Width of slot		Depth of slot	
	Max	Min	Max	Min	Min	Max	Min	Max	Max	Min	Max	Min
M1	2.0	1.75	0.7	0.56	0.10	1.3		0.5	0.45	0.31	0.44	0.30
M1.2	2.3	2.05	0.8	0.66	0.10	1.5		0.5	0.50	0.36	0.49	0.35
M1.6	3.0	2.75	1.0	0.86	0.10	2.0	15	0.7	0.60	0.46	0.65	0.45
M2	3.8	3.50	1.3	1.16	0.10	2.6	16	0.8	0.70	0.56	0.85	0.60
M2.5	4.5	4.20	1.6	1.46	0.10	3.1	18	0.9	0.80	0.66	1.00	0.70
M3	5.5	5.20	2.0	1.86	0.10	3.6	19	1.0	1.00	0.86	1.30	0.90
M4	7.0	6.64	2.6	2.46	0.20	4.7	22	1.4	1.20	1.06	1.60	1.20
M5	8.5	8.14	3.3	3.12	0.20	5.7	25	1.6	1.51	1.26	2.00	1.50
M6	10.0	9.64	3.9	3.72	0.25	6.8	28	2.0	1.91	1.66	2.30	1.80
M8	13.0	12.57	5.0	4.82	0.40	9.2	34	2.5	2.31	2.06	2.80	2.30
M10	16.0	15.57	6.0	5.82	0.40	11.2	40	3.0	2.81	2.56	3.20	2.70
M12	18.0	17.57	7.0	6.78	0.60	14.2	46	3.5	3.31	3.06	3.80	3.20
M16	24.0	23.48	9.0	8.78	0.60	18.2	58	4.0	4.37	4.07	4.60	4.00
M20	30.0	29.48	11.0	10.73	0.80	22.4	70	5.0	5.27	5.07	5.60	5.00

All dimensions in millimetres

SLOTTED AND RECESSED PAN HEAD MACHINE SCREWS – METRIC SERIES – (Preferred Sizes – BS4183)

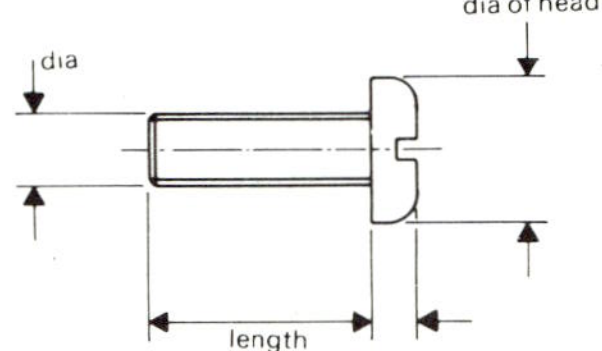

Nominal size and thread diameter	Head diameter		Head height slotted		Head height recessed		Head radius	Head radius	Radius under	Transition diameter	Thread length	Thread run-out	Slot width		Slot depth	
	Max	Min	Max	Min	Max	Min	Max	Nom	Min	Max	Min	Max	Max	Min	Max	Min
M2.5	5.00	4.70	1.50	1.36	1.75	1.61	1.00	3.8	0.10	3.1	18.	0.90	0.80	0.66	0.90	0.60
M3	6.00	5.70	1.80	1.66	2.10	1.96	1.20	4.4	0.10	3.6	19	1.0	1.00	0.86	1.08	0.72
M4	8.00	7.64	2.40	2.26	2.80	2.66	1.60	6.2	0.20	4.7	22	1.4	1.20	1.06	1.44	0.96
M5	10.00	9.64	3.00	2.86	3.50	3.32	2.00	7.2	0.20	5.7	25.	1.6	1.51	1.26	1.80	1.20
M6	12.00	11.57	3.60	3.42	4.20	3.98	2.50	9.5	0.25	6.8	28.	2.0	1.91	1.66	2.16	1.44
M8	16.00	15.57	4.80	4.62	5.60	5.42	3.20	13.4	0.40	9.2	34	2.5	2.31	2.06	2.88	1.92
M10	20.00	19.48	6.00	5.82	7.00	6.78	4.00	19.8	0.40	11.2	40	3.0	2.81	2.56	3.60	2.40

All dimensions in millimetres

RECESSED COUNTERSUNK HEAD MACHINE SCREWS – METRIC SERIES – (Preferred Sizes BS4183)

Nominal size and thread diameter	Depth of recess		Penetration		Recess diameter	Recess and driver point number
	Maximum	Minimum	Maximum	Minimum	Nominal	
M2.5	1.60	1.19	1.35	0.94	2.39	1
M3	1.73	1.32	1.47	1.06	2.51	1
M4	2.18	1.72	1.80	1.34	3.71	2
M5	2.90	2.44	2.51	2.05	4.42	2
M6	3.45	2.99	2.92	2.46	6.10	3
M8	4.27	3.81	3.68	3.22	7.85	4
M10	5.84	5.38	5.26	4.80	9.42	4
M12	6.63	6.17	6.04	5.58	10.18	4

All dimensions in millimetres

RECESSED PIN HEAD MACHINE SCREWS – METRIC SERIES – (Preferred Sizes – BS4183)

Nominal size and thread diameter	Depth of recess		Penetration		Recess diameter	Recess and driver point number
	Maximum	Minimum	Maximum	Minimum	Nominal	
M2.5	1.85	1.44	1.57	1.16	2.64	1
M3	2.11	1.70	1.83	1.42	2.89	1
M4	2.72	2.26	2.31	1.85	4.27	2
M5	3.10	2.64	2.72	2.26	4.67	2
M6	4.06	3.60	3.51	3.05	6.76	3
M8	4.85	4.39	4.17	3.71	8.46	4
M10	6.40	5.94	5.72	5.26	9.96	4

All dimensions in millimetres

RECESSED RAISED COUNTERSUNK HEAD MACHINE SCREWS – METRIC SERIES – (Preferred Sizes – BS4183)

Nominal size and thread diameter	Depth of recess		Penetration		Recess diameter	Recess and driver point number
	Maximum	Minimum	Maximum	Minimum	Nominal	
M2.5	1.98	1.57	1.70	1.29	2.77	1
M3	2.18	1.78	1.93	1.52	2.97	1
M4	2.77	2.31	2.36	1.90	4.32	2
M5	3.81	3.35	3.38	2.92	5.28	2
M6	4.47	4.01	3.86	3.40	7.11	3
M8	5.21	4.75	4.50	4.04	8.79	4
M10	7.37	6.91	6.68	6.22	10.92	4
M12	8.23	7.77	7.54	7.08	11.76	4

All dimensions in millimetres

RAISED COUNTERSUNK, ROUND AND CHEESE HEAD SCREWS – IMPERIAL SIZES (BS450)

Nominal Size and Max Dia of Screw	Number of Threads per inch		Countersunk and Raised Csk Head (Max) 90° Csk			Round Head (Max)			Cheese Head (Max)	
			Diameter of Head	Depth of Csk Portion	Height of Raise, Rsd Csk Hds	Diameter of Head	Depth of Head	Radius of Head	Diameter of Head	Depth of Head
	B S W	B S F								
$\frac{3}{32}$	48	—	·164	·043	·021	·164	·066	·164	·140	·066
$\frac{1}{8}$	40	—	·219	·058	·026	·219	·087	·219	·187	·087
$\frac{5}{32}$	32	—	·273	·072	·034	·274	·109	·274	·234	·109
$\frac{3}{16}$	24	32	·328	·087	·041	·328	·131	·328	·281	·131
$\frac{7}{32}$	24	28	·383	·102	·048	·383	·153	·383	·328	·153
$\frac{1}{4}$	20	26	·437	·116	·055	·437	·175	·437	·375	·175
$\frac{5}{16}$	18	22	·547	·146	·069	·547	·218	·547	·469	·218
$\frac{3}{8}$	16	20	·656	·174	·082	·656	·262	·656	·562	·262
$\frac{7}{16}$	14	18	·766	·204	—	·760	·300	·766	·656	·306
$\frac{1}{2}$	12	16	·875	·233	—	·875	·350	·875	·750	·350
$\frac{5}{8}$	11	14	1·094	·291	—	1·094	·437	1·094	·937	·437
$\frac{3}{4}$	10	12	1·312	·349	—	1·312	·525	1·312	1·125	·525

All dimensions in inches

BA MACHINE SCREWS AND NUTS (BS57 (1951))

B A Desig- nating No	Nominal Diameter		Threads		Diameter of Heads (Csk , Rsd Csk , Rnd and Cheese) 90° Csk		Width Across Flats (Hex Heads and Nuts)		Depth of Heads (maximum)			* Thickness of Full Nuts		* Thickness of Lock Nuts	
	in	mm	No per inch	Pitch in mm	in	in	in	in	Csk in	Round and Cheese in	Hex in	in	in	in	in
					Max	Min	Max	Min				Max	Min	Max	Min
0	·236	6·0	25·4	1·00	·413	·403	·413	·408	·099	·167	·177	·213	·203	·157	·147
1	·209	5·3	28·2	·90	·366	·356	·365	·360	·089	·148	·156	·188	·178	·139	·129
2	·185	4·7	31·4	·81	·319	·309	·324	·319	·077	·130	·139	·167	·157	·123	·113
3	·161	4·1	34·8	·73	·283	·273	·282	·277	·071	·113	·121	·153	·143	·108	·098
4	·142	3·6	38·5	·66	·252	·242	·248	·243	·065	·101	·106	·135	·125	·094	·084
5	·126	3·2	43·0	·59	·221	·211	·220	·216	·058	·088	·094	·120	·110	·084	·074
6	·110	2·8	47·9	·53	·194	·184	·193	·189	·051	·078	·083	·105	·095	·073	·063
7	·098	2·5	52·9	·48	·173	·163	·172	·169	·047	·069	·074	·094	·087	—	—
8	·087	2·2	59·1	·43	·157	·147	·152	·149	·043	·063	·065	·082	·075	·058	·051
9	·075	1·9	65·1	·39	·128	·123	·131	·128	·035	·052	·056	·071	·064	—	—
10	·067	1·7	72·6	·35	·112	·107	·117	·114	·030	·045	·050	·064	·057	—	—

*Both full and lock nuts are double chamfered

HIGH STRENGTH FRICTION GRIP NUTS – UNC SERIES (BS3139)

Nominal size and thread diameter	Width across flats		Diameter of washer face		Width across corners	Thickness of nut	
	max	*min*	*max*	*min*	*max*	*max*	*min*
½	0·875	0·855	0·841	0·831	1·010	0·504	0·464
⅝	1·062	1·037	1·019	1·009	1·227	0·631	0·587
¾	1·250	1·225	1·203	1·188	1·443	0·758	0·710
⅞	1·437	1·407	1·380	1·365	1·660	0·885	0·833
1	1·625	1·595	1·559	1·544	1·876	1·012	0·956
1⅛	1·812	1·782	1·736	1·721	2·093	1·139	1·079
1¼	2·000	1·955	1·915	1·900	2·309	1·251	1·187
1½	2·375	2·330	2·271	2·256	2·742	1·505	1·433

Dimensions in inches

FRICTION GRIP WASHERS – INCH SERIES

Round Washers

Bolt size	Inside diameter	Outside diameter	Thickness swg	(approx)
½	9/16	1⅜	12	0·104
⅝	11/16	1½	10	0·128
¾	13/16	1¾	9	0·144
⅞	15/16	2	8	0·160
1	1 1/16	2¼	8	0·160
1⅛	1¼	2½	8	0·160
1¼	1⅜	2¾	8	0·160
1½	1⅝	3¼	7	0·176

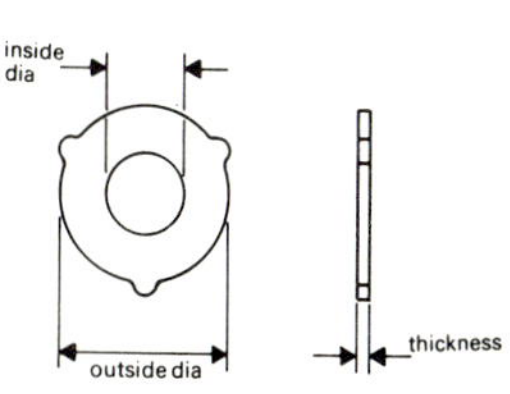

Square Taper Washers

Nominal diameter	Inside diameter	Overall size	Mean thickness 3° and 5° taper	Mean thickness 8° taper
½	9/16	1¼	3/16	¼
⅝	11/16	1½	3/16	¼
¾	13/16	1½	3/16	¼
⅞	15/16	1¾	3/16	¼
1	1 1/16	2	3/16	¼
1⅛	1¼	2¼	3/16	¼
1¼	1⅜	2¼	3/16	¼
1½	1⅝	2½	3/16	¼

Dimensions in inches

HIGH STRENGTH FRICTION GRIP HEXAGON HEAD BOLTS – UNC SERIES (BS3139)

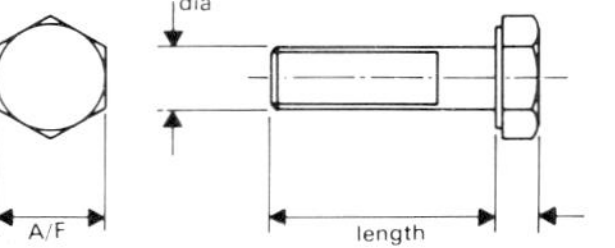

Nominal size and thread diameter	Diameter of unthreaded shank		Width across flats		Diameter of washer face		Width across corners	Radius under head		Thickness of head	
	max	*min*	*max*	*min*	*max*	*min*	*max*	*max*	*min*	*max*	*min*
$\frac{1}{2}$	0·530	0·496	0·875	0·855	0·841	0·831	1·010	0·045	0·020	0·323	0·302
$\frac{5}{8}$	0·655	0·619	1·062	1·037	1·019	1·009	1·227	0·045	0·020	0·403	0·378
$\frac{3}{4}$	0·780	0·744	1·250	1·225	1·203	1·188	1·443	0·045	0·020	0·483	0·455
$\frac{7}{8}$	0·905	0·867	1·437	1·407	1·380	1·365	1·660	0·045	0·020	0·563	0·533
1	1·030	0·992	1·625	1·595	1·559	1·544	1·876	0·045	0·020	0·627	0·597
$1\frac{1}{8}$	1·165	1·117	1·812	1·782	1·736	1·721	2·093	0·060	0·030	0·718	0·678
$1\frac{1}{4}$	1·290	1·242	2·000	1·955	1·915	1·900	2·309	0·060	0·030	0·813	0·773
$1\frac{1}{2}$	1·540	1·490	2·375	2·330	2·271	2·256	2·742	0·060	0·030	0·974	0·934

Dimensions in inches

Length of thread

Length of bolt	Length of thread
Up to and including 6 in	2 dias + ¼in
Above 6 in	2 dias + ½in

FRICTION GRIP WASHERS – METRIC SERIES

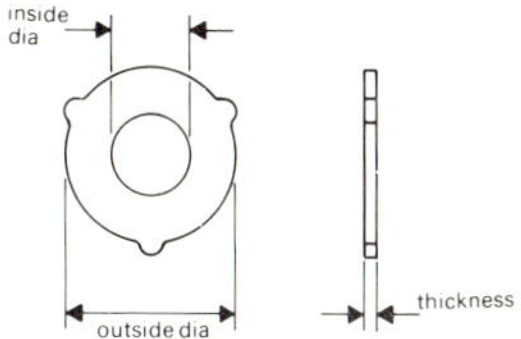

Round Washers

Nom dia	Inside diameter		Outside diameter		Thickness	
	max	*min*	*max*	*min*	*max*	*min*
M16	17·8	17·4	37	36	3·4	3·0
M20	21·5	21·1	44	43	3·7	3·3
M22	23·4	23·0	50	48·5	4·2	3·8
M24	26·4	26·0	56	54·5	4·2	3·8
M27	29·4	29·0	60	58·5	4·2	3·8
M30	32·8	32·4	66	64·5	4·2	3·8
M36	38·8	38·4	85	83·5	4·6	4·2

Dimensions in millimetres

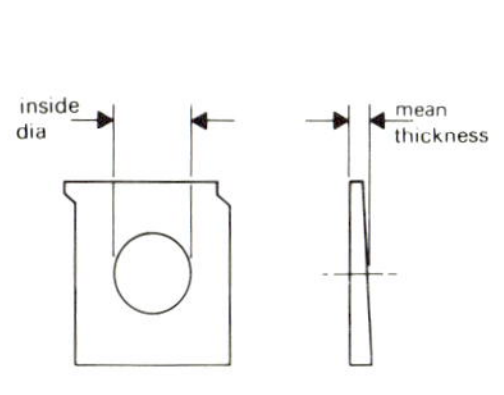

Square Taper Washers

Nom dia	Inside diameter		Overall size	Mean thickness	
	max	*min*		3° and 5° Taper	8° Taper
M16	18·2	17·4	38·10	4·76	6·35
M20	21·9	21·1	38·10	4·76	6·35
M22	23·8	23·0	44·45	4·76	6·35
M24	26·8	26·0	57·15	4·76	6·35
M27	29·8	29·0	57·15	4·76	6·35
M30	33·2	32·4	57·15	4·76	6·35
M36	39·2	38·4	57·15	4·76	6·35

ISO METRIC HEAD STYLES AND DIMENSIONS

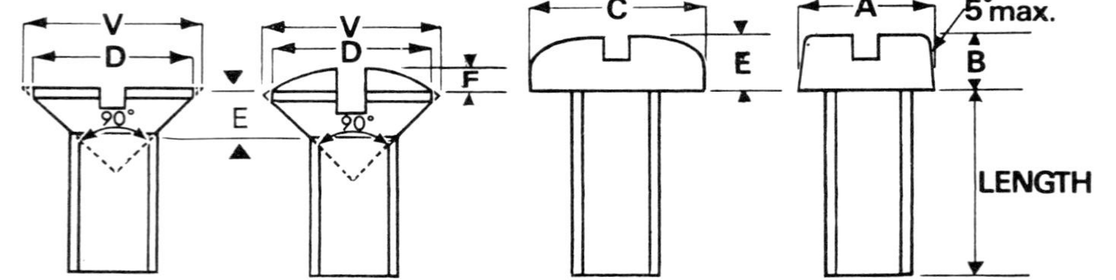

Head dimensions

Slotted Heads to BS4183:1967

Nominal Size	Pitch	CSK and RSD CSK				PAN HEADS				CHEESE HEADS			
		Dia of Head		Ht of Raise F	Depth of Head E Max	Dia of Head C		Depth E		Dia of Head A		Depth B	
		"V" * Max (Sharp)	"D" Min			Max	Min	Max	Min	Max	Min	Max	Min
M2	0·40	4·40	3·50	0·50	1·20	4·00	3·70	1·20	1·06	3·80	3·50	1·30	1·16
M2·5	0·45	5·50	4·45	0·60	1·50	5·00	4·70	1·50	1·36	4·50	4·20	1·60	1·46
M3	0·50	6·30	5·25	0·75	1·65	6·00	5·70	1·80	1·66	5·50	5·20	2·00	1·86
M3·5	*0·60*	*7·35*	*6·12*	*0·90*	*1·93*	*7·00*	*6·64*	*2·10*	*1·96*	*6·00*	*5·70*	*2·40*	*2·26*
M4	0·70	8·40	7·04	1·00	2·20	8·00	7·64	2·40	2·26	7·00	6·64	2·60	2·46
M5	0·80	10·00	8·75	1·25	2·50	10·00	9·64	3·00	2·86	8·50	8·14	3·30	3·12
M6	1·00	12·00	10·50	1·50	3·00	12·00	11·57	3·60	3·42	10·00	9·64	3·90	3·72
M8	1·25	16·00	14·00	2·00	4·00	16·00	15·57	4·80	4·62	13·00	12·57	5·00	4·82
M10	1·50	20·00	17·50	2·50	5·00	20·00	19·48	6·00	5·82	16·00	15·57	6·00	5·82
M12	1·75	24·00	21·00	3·00	6·00	—	—	—	—	18·00	17·57	7·00	6·78

All dimensions in millimetres

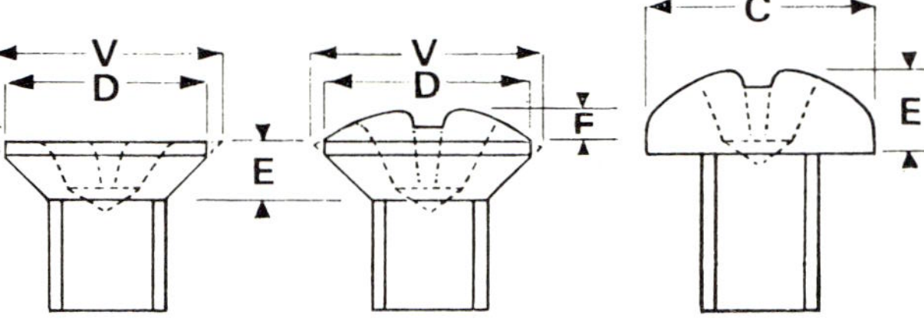

Head dimensions **Pozidriv Recess Heads to BS4183:1967**

Nominal Size	Pitch	CSK and RSD CSK Head				Pan Head				Recess No
		Dia of Head		Ht of Raise F	Depth of Head E Max	Dia of Head C		Depth of Head E		
		"V" Max *	"D" Min			Max	Min	Max	Min	
M2	0·40	4·40	3·50	0·50	1·20	4·00	3·70	1·60	1·46	0
M2·5	0·45	5·50	4·45	0·60	1·50	5·00	4·70	1·95	1·80	1
M3	0·50	6·30	5·25	0·75	1·65	6·00	5·70	2·30	2·16	1
M3·5	*0·60*	*7·35*	*6·12*	*0·90*	*1·93*	*7·00*	*6·64*	*2·45*	*2·31*	2
M4	0·70	8·40	7·04	1·00	2·20	8·00	7·64	2·80	2·66	2
M5	0·80	10·00	8·75	1·25	2·50	10·00	9·64	3·50	3·32	2
M6	1·00	12·00	10·50	1·50	3·00	12·00	11·57	4·20	3·98	3
M8	1·25	16·00	14·00	2·00	4·00	16·00	15·57	5·60	5·42	4
M10	1·50	20·00	17·50	2·50	5·00	20·00	19·48	7·00	6·78	4
M12	1·75	24·00	21·00	3·00	6·00	—	—	—	—	4

** "V" Max is the theoretical sharp diameter, and is the diameter to which holes should be countersunk to enable the screw heads to fix flush with the surface.*

All dimensions in millimetres

HEXAGON ROUND BOLTS AND SET SCREWS – UN SERIES (BS7708)

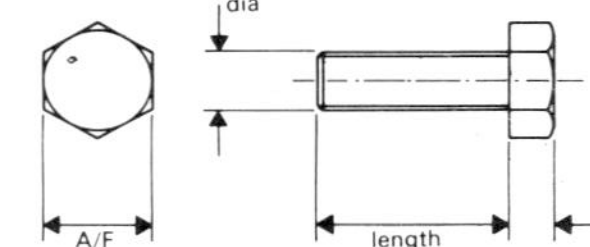

Nominal size and thread diameter	Diameter of unthreaded shank	Width across flats		Width across corners	Radius under head	Thickness of head		
	max	*max*	*min*	*max*	*max*	*nominal*	*max*	*min*
$\frac{1}{4}$	0·280	0·437	0·427	0·505	$\frac{1}{32}$	$\frac{11}{64}$	0·183	0·163
$\frac{5}{16}$	0·342	0·500	0·490	0·577	$\frac{1}{32}$	$\frac{7}{32}$	0·231	0·211
$\frac{3}{8}$	0·405	0·562	0·547	0·650	$\frac{1}{32}$	$\frac{1}{4}$	0·263	0·243
$\frac{7}{16}$	0·468	0·625	0·610	0·722	$\frac{1}{32}$	$\frac{19}{64}$	0·311	0·291
$\frac{1}{2}$	0·530	0·750	0·730	0·866	$\frac{1}{32}$	$\frac{11}{32}$	0·353	0·323
$\frac{5}{8}$	0·665	0·937	0·912	1·083	$\frac{3}{64}$	$\frac{27}{64}$	0·433	0·403
$\frac{3}{4}$	0·790	1·125	1·100	1·299	$\frac{3}{64}$	$\frac{1}{2}$	0·513	0·483
$\frac{7}{8}$	0·915	1·312	1·282	1·516	$\frac{1}{16}$	$\frac{37}{64}$	0·605	0·563
1	1·040	1·500	1·470	1·732	$\frac{1}{16}$	$\frac{43}{64}$	0·667	0·627
$1\frac{1}{8}$	1·175	1·687	1·657	1·949	$\frac{1}{8}$	$\frac{3}{4}$	0·758	0·718
$1\frac{1}{4}$	1·300	1·875	1·830	2·165	$\frac{1}{8}$	$\frac{27}{32}$	0·873	0·813
$1\frac{3}{8}$	1·425	2·062	2·017	2·382	$\frac{1}{8}$	$\frac{29}{32}$	0·938	0·878
$1\frac{1}{2}$	1·550	2·250	2·205	2·598	$\frac{1}{8}$	1	1·034	0·974
$1\frac{3}{4}$	1·810	2·625	2·565	3·031	$\frac{1}{8}$	$1\frac{5}{32}$	1·234	1·134
2	2·060	3·000	2·940	3·464	$\frac{1}{8}$	$1\frac{11}{32}$	1·363	1·263

Dimensions in inches

HEXAGON HEAD SET SCREWS – METRIC SERIES (BS4190)

Nominal size and thread diameter	Width across flats max	Width across flats min	Width across corners max	Width across corners min	Thickness of head max	Thickness of head min	Radius under head max	Length of thread max
M6	10·00	9·64	11·5	10·89	4·375	3·625	0·40	4
M8	13·00	12·57	15·0	14·20	5·875	5·125	0·8	4·5
M10	17·00	16·57	19·6	18·72	7·45	6·55	0·8	5
M12	19·00	18·48	21·9	20·88	8·45	7·55	1·25	6
M16	24·00	23·16	27·7	26·17	10·45	9·55	1·25	7·5
M20	30·00	29·16	34·6	32·95	13·90	12·10	1·78	9
M24	36·00	35·00	41·6	39·55	15·90	14·10	1·78	11
M30	46·00	45·00	53·1	50·85	20·05	17·95	2·28	12
M36	55·00	53·80	63·5	60·79	24·05	21·95	2·7	15
M42	65·00	63·80	75·1	72·09	27·05	24·95	2·8	16
M48	75·00	73·80	86·6	83·39	31·05	28·95	3·8	18
M56	85·00	83·60	98·1	94·47	36·25	33·75	4·9	20
M64	95·00	93·60	109·7	105·77	41·25	38·75	4·9	22

Dimensions in millimetres

HEXAGON SLOTTED NUTS – UNIFIED HEAVY SERIES (BS1769)

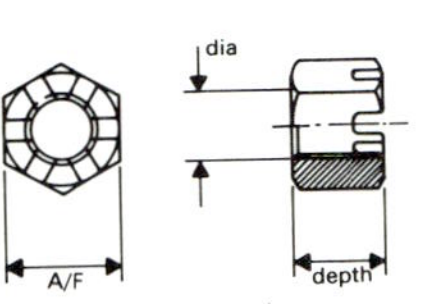

Nominal size and thread diameter	Width across flats max	Width across flats min	Width across corners max	Thickness of nut max	Thickness of nut min	Slot Width	Slot Depth
$\frac{1}{2}$	0·875	0·855	1·01	0·504	0·464	$\frac{5}{32}$	$\frac{5}{32}$
$\frac{5}{8}$	1·062	1·037	1·23	0·631	0·587	$\frac{3}{16}$	$\frac{7}{32}$
$\frac{3}{4}$	1·250	1·225	1·44	0·758	0·710	$\frac{3}{16}$	$\frac{1}{4}$
$\frac{7}{8}$	1·437	1·407	1·66	0·885	0·833	$\frac{3}{16}$	$\frac{1}{4}$
1	1·625	1·595	1·88	1·012	0·956	$\frac{1}{4}$	$\frac{9}{32}$
$1\frac{1}{8}$	1·812	1·782	2·09	1·139	1·079	$\frac{1}{4}$	$\frac{11}{32}$
$1\frac{1}{4}$	2·000	1·955	2·31	1·251	1·187	$\frac{5}{16}$	$\frac{3}{8}$
$1\frac{3}{8}$	2·187	2·142	2·53	1·378	1·310	$\frac{5}{16}$	$\frac{3}{8}$
$1\frac{1}{2}$	2·375	2·330	2·74	1·505	1·433	$\frac{3}{8}$	$\frac{7}{16}$
$1\frac{3}{4}$	2·750	2·690	3·18	1·759	1·679	$\frac{7}{16}$	$\frac{1}{2}$
2	3·125	3·065	3·61	2·013	1·925	$\frac{7}{16}$	$\frac{9}{16}$

Dimensions in inches

Sizes up to 8½in nominal thread diameter

SQUARE NUTS – UNIFIED NORMAL SERIES (BS2708)

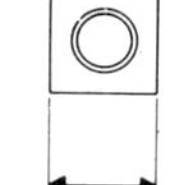

Nominal size and thread diameter	Width across flats		Width across corners	Thickness of nut		
	max	*min*	*max*	*nominal*	*max*	*min*
$\frac{1}{4}$	0·4375	0·4275	0·619	$\frac{7}{32}$	0·235	0·215
$\frac{5}{16}$	0·5625	0·5525	0·795	$\frac{17}{64}$	0·283	0·263
$\frac{3}{8}$	0·6250	0·6100	0·884	$\frac{21}{64}$	0·346	0·326
$\frac{7}{16}$	0·7500	0·7350	1·061	$\frac{3}{8}$	0·394	0·374
$\frac{1}{2}$	0·8125	0·7925	1·149	$\frac{7}{16}$	0·458	0·428
$\frac{5}{8}$	1·0000	0·9750	1·414	$\frac{35}{64}$	0·569	0·529
$\frac{3}{4}$	1·1250	1·1000	1·591	$\frac{21}{32}$	0·680	0·640
$\frac{7}{8}$	1·3125	1·2825	1·856	$\frac{49}{64}$	0·792	0·742
1	1·5000	1·4700	2·121	$\frac{7}{8}$	0·903	0·847
$1\frac{1}{8}$	1·6875	1·6575	2·386	1	1·030	0·970
$1\frac{1}{4}$	1·8750	1·8300	2·652	$1\frac{3}{32}$	1·126	1·062
$1\frac{3}{8}$	2·0625	2·0175	2·917	$1\frac{13}{64}$	1·237	1·169
$1\frac{1}{2}$	2·2500	2·2050	3·182	$1\frac{5}{16}$	1·348	1·276

Dimensions in inches

Sizes up to 8½in nominal thread diameter

BSW HEXAGON AND SQUARE HEAD BOLTS (BS916)

Nominal size and thread diameter	Diameter of unthreaded shank	Width across flats hex and square		Width across corners hex and square		Radius under head	Thickness of head hex and square	
	max	*max*	*min*	*hex*	*square*	*max*	*max*	*basic*
$\frac{1}{4}$	0·280	0·445	0·435	0·51	0·63	$\frac{1}{32}$	0·186	0·166
$\frac{5}{16}$	0·342	0·525	0·515	0·61	0·74	$\frac{1}{32}$	0·228	0·208
$\frac{3}{8}$	0·405	0·600	0·585	0·69	0·85	$\frac{1}{32}$	0·270	0·250
$\frac{7}{16}$	0·468	0·710	0·695	0·82	1·00	$\frac{1}{32}$	0·312	0·292
$\frac{1}{2}$	0·530	0·820	0·800	0·95	1·16	$\frac{1}{32}$	0·363	0·333
$\frac{9}{16}$	0·592	0·920	0·900	1·06	1·30	$\frac{3}{64}$	0·405	0·375
$\frac{5}{8}$	0·665	1·010	0·985	1·17	1·43	$\frac{3}{64}$	0·447	0·417
$\frac{3}{4}$	0·790	1·200	1·175	1·39	1·70	$\frac{3}{64}$	0·530	0·500
$\frac{7}{8}$	0·915	1·300	1·270	1·50	1·84	$\frac{1}{16}$	0·623	0·583
1	1·040	1·480	1·450	1·71	2·09	$\frac{1}{16}$	0·706	0·666
$1\frac{1}{8}$	1·175	1·670	1·640	1·93	2·36	$\frac{1}{8}$	0·79	0·75
$1\frac{1}{4}$	1·300	1·860	1·815	2·15	2·63	$\frac{1}{8}$	0·89	0·83
$1\frac{1}{2}$	1·550	2·220	2·175	2·56	3·14	$\frac{1}{8}$	1·06	1·00
$1\frac{3}{4}$	1·810	2·580	2·520	2·98	3·65	$\frac{1}{8}$	1·27	1·17
2	2·060	2·760	2·700	3·19	3·90	$\frac{1}{8}$	1·43	1·33

Dimensions in inches

Length of thread — bolts

Length of Bolt	All dias up to and including $\frac{1}{2}$ in	All dias over $\frac{1}{2}$ in
Up to and including 4 in	2 dias	$1\frac{1}{2}$ dias
Over 4 in up to and including 8 in	2 dias	2 dias
Over 8 in	$2\frac{1}{2}$ dias	$2\frac{1}{2}$ dias

Length of thread — setscrews

Threads per inch	Distance from underside of head
16 and under	2 × pitch
Over 16, up to and including 20	$2\frac{1}{2}$ × pitch
Over 20	3 × pitch

ISO METRIC HEXAGON NUTS AND HEXAGON THIN NUTS – (Preferred Sizes – BS4190)

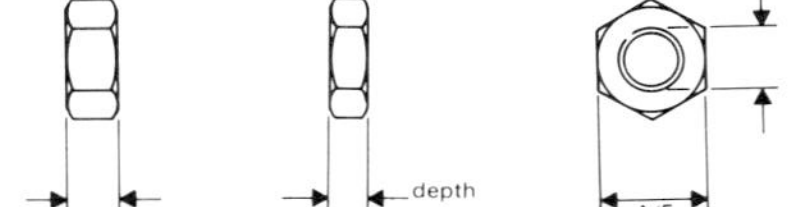

Nominal size and thread diameter	Pitch of thread (Coarse pitch series)	Width across flats		Width across corners		Thickness of nut: Black		Thickness of nut: Faced one side		Thickness of thin nut (faced both sides)	
		max	min	max	min	max	min	max	min	max	min
M5	0.8	8.00	7.64	9.2	8.63	4.375	3.625	4.0	3.52	–	–
M6	1.0	10.00	9.64	11.5	10.89	5.375	4.625	5.0	4.52	–	–
M8	1.25	13.00	12.57	15.0	14.20	6.875	6.125	6.5	5.92	5	4.52
M10	1.50	17.00	16.57	19.6	18.72	8.45	7.55	8	7.42	6.	5.52
M12	1.75	19.00	18.48	21.9	20.88	10.45	9.55	10	9.42	7.	6.42
M16	2.0	24.00	23.16	27.7	26.17	13.55	12.45	13	12.30	9	8.42
M20	2.5	30.00	29.16	34.6	32.95	16.55	15.45	16	15.30	9	8.42
M24	3.0	36.00	35.00	41.6	39.55	19.65	18.35	19	18.16	10	9.42
M30	3.5	46.00	45.00	53.1	50.85	24.65	23.35	24	23.16	12	11.30
M36	4.0	55.00	53.80	63.5	60.79	29.65	28.35	29	28.16	14	13.30
M42	4.5	65.00	63.80	75.1	72.09	34.80	33.20	34	33.0	16	15.30
M48	5.0	75.00	73.80	86.6	83.39	38.80	37.20	38	37.0	18	17.30
M56	5.5	85.00	83.60	98.1	94.47	45.80	44.20	45	44.0	–	–
M64	6.0	95.00	93.60	109.7	105.77	51.95	50.05	51	49.8	–	–

All dimensions in millimetres

COLD FORMED HEXAGON NUTS – UNIFIED NORMAL SERIES (BS1768)

Nominal size and thread diameter	Width across flats		Width across corners	Diameter of washer face		Thickness			
						Ordinary nut		Lock nut	
	max	*min*	*max*	*max*	*min*	*max*	*min*	*max*	*min*
$\frac{1}{4}$	0·437	0·430	0·505	0·421	0·411	0·224	0·214	0·161	0·151
$\frac{5}{16}$	0·500	0·493	0·577	0·483	0·473	0·271	0·261	0·192	0·182
$\frac{3}{8}$	0·562	0·554	0·650	0·545	0·535	0·333	0·323	0·224	0·214
$\frac{7}{16}$	0·687	0·679	0·794	0·668	0·658	0·380	0·370	0·255	0·245
$\frac{1}{2}$	0·750	0·742	0·866	0·730	0·720	0·442	0·432	0·317	0·307
$\frac{9}{16}$	0·875	0·867	1·010	0·855	0·845	0·489	0·479	0·349	0·339
$\frac{5}{8}$	0·937	0·929	1·083	0·918	0·908	0·552	0·542	0·380	0·370
$\frac{3}{4}$	1·125	1·115	1·300	1·100	1·090	0·651	0·631	0·432	0·412

Dimensions in inches

depth
dia
A/F

COLD FORMED SLOTTED HEXAGON NUTS – UNIFIED NORMAL SERIES (BS1768)

Nominal size and thread diameter	Width across flats		Width across corners	Diameter of washer face		Thickness		Lower face of nut to bottom of slot		Width of slot		Depth of slot
	max	*min*	*max*	*max*	*min*	*max*	*min*	*max*	*min*	*max*	*min*	*nom*
$\frac{1}{4}$	0·437	0·430	0·505	0·421	0·411	0·224	0·214	0·161	0·151	0·088	0·078	0·062
$\frac{5}{16}$	0·500	0·493	0·577	0·483	0·473	0·271	0·261	0·177	0·167	0·104	0·094	0·094
$\frac{3}{8}$	0·562	0·554	0·650	0·545	0·535	0·333	0·323	0·224	0·214	0·135	0·125	0·109
$\frac{7}{16}$	0·687	0·679	0·794	0·668	0·658	0·380	0·370	0·255	0·245	0·135	0·125	0·125
$\frac{1}{2}$	0·750	0·742	0·866	0·730	0·720	0·442	0·432	0·302	0·292	0·166	0·156	0·141
$\frac{9}{16}$	0·875	0·867	1·010	0·855	0·845	0·489	0·479	0·333	0·323	0·166	0·156	0·156
$\frac{5}{8}$	0·937	0·929	1·083	0·918	0·908	0·552	0·542	0·364	0·354	0·198	0·188	0·187
$\frac{3}{4}$	1·125	1·115	1·300	1·100	1·090	0·651	0·631	0·432	0·412	0·198	0·188	0·219

Dimensions in inches

dia
A/F
depth

HEXAGON WRENCHES (ALLEN KEYS)

Size across flats in	To fit (by nominal diameter)					
	Cap	Set	Shoulder in	Countersunk	Pipe Plugs (Taper) in	Button
·035		8BA				
·050		6BA, 4UN		6BA		
$\frac{1}{16}$	8BA	5, 4BA, 5, 6UN, $\frac{1}{8}''$		4BA, 4UN		
$\frac{5}{64}$	6BA, 4UN	3BA, 8UN		3BA, 6UN		6UN
$\frac{3}{32}$	5, 4BA, 5, 6UN, $\frac{1}{8}''$	2BA, 10UN, $\frac{3}{16}''$		2BA, 8UN, $\frac{3}{16}''$		8UN
$\frac{1}{8}$	3BA, 8UN	0BA, $\frac{1}{4}''$	$\frac{1}{4}$	0BA, 10UN		2BA, 10UN, $\frac{3}{16}''$
$\frac{5}{32}$	2, 1BA, 10UN, $\frac{3}{16}''$	$\frac{5}{16}''$	$\frac{5}{16}$	$\frac{1}{4}''$		$\frac{1}{4}''$
$\frac{3}{16}$	0BA, $\frac{1}{4}''$	$\frac{3}{8}''$, $\frac{1}{8}''$ Gas	$\frac{3}{8}$	$\frac{5}{16}''$	$\frac{1}{8}$	$\frac{5}{16}''$
$\frac{7}{32}$	$\frac{5}{16}''$	$\frac{7}{16}''$		$\frac{3}{8}''$		$\frac{3}{8}''$
$\frac{1}{4}$		$\frac{1}{2}''$, $\frac{1}{4}''$ Gas	$\frac{1}{2}$	$\frac{7}{16}''$	$\frac{1}{4}$	
$\frac{5}{16}$	$\frac{3}{8}''$, $\frac{7}{16}''$	$\frac{5}{8}''$, $\frac{3}{8}''$ Gas	$\frac{5}{8}$	$\frac{1}{2}''$	$\frac{3}{8}$	$\frac{1}{2}''$
$\frac{3}{8}$	$\frac{1}{2}''$	$\frac{3}{4}''$	$\frac{3}{4}$	$\frac{5}{8}''$, $\frac{3}{4}''$	$\frac{1}{2}$	
$\frac{1}{2}$	$\frac{5}{8}''$	$\frac{7}{8}''$			$\frac{5}{8}$	
$\frac{9}{16}$	$\frac{3}{4}''$, $\frac{7}{8}''$	1″			$\frac{3}{4}$	
$\frac{5}{8}$	1″				1	
$\frac{3}{4}$	$1\frac{1}{4}''$					

Size across flats mm	To fit				
	Cap	Set	Shoulder	Csk	Button
1·5		M3			
2		M4		M3	M3
2·5	M3	M5		M4	M4
3	M4	M6	M5	M5	M5
4	M5	M8	M6	M6	M6
5	M6	M10	M8	M8	M8
6	M8	M12	M10	M10	M10
8	M10	M16	M12	M12	M12
10	M12	M20	M16	M16	
12		M24			
14	M16				
17	M20				
19	M24				

Gauging

ISO Metric Threads

Gauges for ISO metric threads are covered by BS919 : Part 3. There are three tolerance qualities designated –

fine
medium
coarse

There are seven tolerance grades for external threads and five for internal threads. Provision is also made for the introduction of an allowance through different tolerance positions. A combination of a tolerance grade and a tolerance position designates a tolerance class.

For general productions the recommended tolerance is 6g for bolts and screws, and 6H for nuts see Table I on page 168.

American screw thread practice adopts five classes of fit (Class 5 being an interference fit). Class 1A and Class 2A are clearance fits, with an allowance. Class 3A does not have an allowance, thus theoretically under true maximum material condition provides complete contact of mating threads.

Classes 1A and 1B are used on threaded components where quick and easy assembly is required, or where threads are likely to be dirty or slightly bruised.

Classes 2A and 2B are recommended for general applications, the tolerances and allowances of these classes also being best suited to threaded fastener production.

Class 3A or Class 3B is specified where closeness of fit and accuracy of lead and thread angle are important. These classes correspond to high quality production, with efficient inspection and gauging.

TABLE I – LIMITING DIMENSIONS FOR ISO METRIC BOLTS AND NUTS (contd.)

Nominal Size	Pitch P	Basic Thread Designation	External Thread (Bolt) Tol Class	Allowance	Major Diameter Max	Major Diameter Min	Pitch Diameter Max	Pitch Diameter Min	Pitch Diameter Tol	Minor Diameter Max	Minor Diameter Min	Internal Thread (Nut) Tol Class	Minor Diameter Min	Minor Diameter Max	Pitch Diameter Min	Pitch Diameter Max	Pitch Diameter Tol	Major Dia Min
1.6	0.35	M1.6	6g	0.019	1.581	1.496	1.354	1.291	0.063	1.151	1.063	6H	1.221	1.321	1.373	1.458	0.085	1.6
1.8	0.35	M1.8	6g	0.019	1.781	1.696	1.554	1.491	0.063	1.351	1.263	6H	1.421	1.521	1.573	1.658	0.085	1.8
2	0.4	M2	6g	0.019	1.981	1.886	1.721	1.654	0.067	1.490	1.394	6H	1.567	1.679	1.740	1.830	0.090	2.0
2.2	0.45	M2.2	6g	0.020	2.180	2.080	1.888	1.817	0.071	1.628	1.525	6H	1.713	1.838	1.908	2.003	0.095	2.2
2.5	0.45	M2.5	6g	0.020	2.480	2.380	2.188	2.117	0.071	1.928	1.825	6H	2.013	2.138	2.208	2.303	0.095	2.5
3	0.5	M3	6g	0.020	2.980	2.874	2.655	2.580	0.075	2.367	2.256	6H	2.459	2.599	2.675	2.775	0.100	3.0
3.5	0.6	M3.5	6g	0.021	3.479	3.354	3.089	3.004	0.085	2.742	2.614	6H	2.850	3.010	3.110	3.222	0.112	3.5
4	0.7	M4	6g	0.022	3.978	3.838	3.523	3.433	0.090	3.119	2.979	6H	3.242	3.422	3.545	3.663	0.118	4.0
4.5	0.75	M4.5	6g	0.022	4.478	4.338	3.991	3.901	0.090	3.558	3.414	6H	3.688	3.878	4.013	4.131	0.118	4.5
5	0.8	M5	6g	0.024	4.976	4.826	4.456	4.361	0.095	3.994	3.841	6H	4.134	4.334	4.480	4.605	0.125	5.0
6	1	M6	6g	0.026	5.974	5.794	5.324	5.212	0.112	4.747	4.563	6H	4.917	5.153	5.350	5.500	0.150	6.0
7	1	M7	6g	0.026	6.974	6.794	6.324	6.212	0.112	5.747	5.563	6H	5.917	6.153	6.350	6.500	0.150	7.0
8	1.25	M8	6g	0.028	7.972	7.760	7.160	7.042	0.118	6.439	6.231	6H	6.647	6.912	7.188	7.348	0.160	8.0
10	1.5	M10	6g	0.032	9.968	9.732	8.994	8.862	0.132	8.127	7.879	6H	8.376	8.676	9.026	9.206	0.180	10.0
12	1.75	M12	6g	0.034	11.966	11.701	10.829	10.679	0.150	9.819	9.543	6H	10.106	10.441	10.863	11.063	0.200	12.0
14	2	M14	6g	0.038	13.962	13.682	12.663	12.503	0.160	11.508	11.204	6H	11.835	12.210	12.701	12.913	0.212	14.0
16	2	M16	6g	0.038	15.962	15.682	14.663	14.503	0.160	13.508	13.204	6H	13.385	14.210	14.701	14.913	0.212	16.0
18	2.5	M18	6g	0.042	17.958	17.623	16.334	16.164	0.170	14.891	14.541	6H	15.294	15.744	16.376	16.600	0.224	18.0
20	2.5	M20	6g	0.042	19.958	19.623	18.334	18.164	0.170	16.891	16.541	6H	17.294	17.744	18.376	18.600	0.224	20.0
22	2.5	M22	6g	0.042	21.958	21.623	20.334	20.164	0.170	18.891	18.541	6H	19.294	19.744	20.376	20.600	0.224	22.0
24	3	M24	6g	0.048	23.952	23.577	22.003	21.803	0.200	20.271	19.855	6H	20.752	21.252	22.051	22.316	0.265	24.0
27	3	M27	6g	0.048	26.952	26.577	25.003	24.803	0.200	23.271	22.855	6H	23.752	24.252	25.051	25.316	0.265	27.0
30	3.5	M30	6g	0.053	29.947	29.522	27.674	27.462	0.212	25.653	25.189	6H	26.211	26.771	27.727	28.007	0.280	30.0
33	3.5	M33	6g	0.053	32.947	32.522	30.674	30.462	0.212	28.653	28.189	6H	29.211	29.771	30.727	31.007	0.280	33.0
36	4	M36	6g	0.060	35.940	35.465	33.342	33.118	0.224	31.033	30.521	6H	31.670	32.270	33.402	33.702	0.300	36.0
39	4	M39	6g	0.060	38.940	38.465	36.342	36.118	0.224	34.033	33.521	6H	34.670	35.270	36.402	36.702	0.300	39.0

*Required for high strength applications where rounded root is specified

All dimensions in millimetres

Thread-Locking Systems

Types of fasteners and thread locking systems capable of retaining their clamping force and resisting vibration or anything which could cause them to loosen can be classified as follows:-

(i) Locking nuts

(ii) Locking washers

(iii) Captive nuts

(iv) Self-locking nuts and fasteners

(v) Prevailing torque fasteners (stiffnuts)

(vi) Thread inserts and thread locks

(vii) Chemical thread-locking (see *Adhesives* section)

Locking Nuts – include the use of two nuts (one standard and one thin nut) tightened separately to butt against each other, castellated nuts and drilled nuts. The last two types are secured (locked) by a split pin passing through a hole drilled in the bolt, preferable *after* the nut has been tightened to the required degree of preload – Fig 1. See also chapter on *Nuts.*

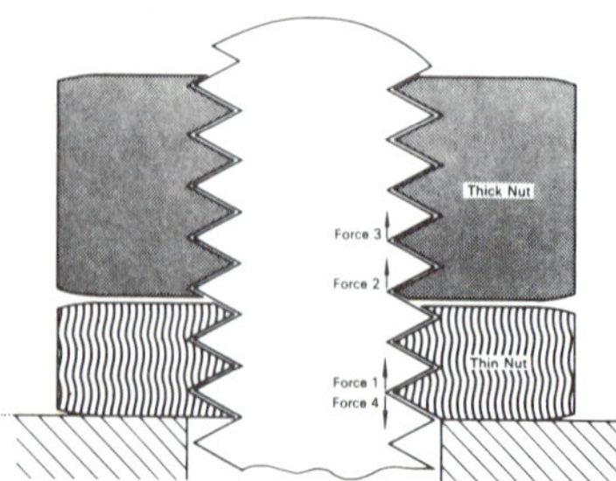

Fig 1 Correct use of a lock nut (GKN). Force 1 is generated by tightening the thin nut. Force 2 is generated by tightening the thick nut. Force 3 generated in the thick nut changes force 1 into a downward force (Force 4).

Locking Washers include spring washers (limited locking ability) and various forms of serrated washers, toothed washers and tab washers – see chapter on *Washers*. There is also a variety of spring nut fasteners capable of providing comparable locking action on light duty assemblies – see chapter on *Spring Nuts.*

TOMORROWS FASTENERS TO-DAY FROM P.A. !
Peter Abbott
MANUFACTURERS AND DISTRIBUTORS OF
INDUSTRIAL FASTENERS
SELF-LOCKING NUTS • HELI-COIL INSERTS AND EQUIPMENT •
HIGH TENSILE BOLTS AND SETSCREWS • WING NUTS •
SOCKET SCREWS • THREADRIV BITS • BLIND RIVETS •
RIVET BUSHES AND SHEET METAL FIXINGS •
1 Peter Abbott & Co. Ltd.,
Bridge Close Industrial Estate,
Romford, Essex
RM7 0AB.
Phone: Romford 25111 (10 Lines)
Telex: 897072
Answerback: Threadfast Rmfd.
2 Peter Abbott & Co. (Central) Ltd.,
Morley Street,
Daybrook,
Nottingham NG5 6JX.
Phone: (0602) 264222/3/4
Telex: 377216
Answerback: Abbottco Nottm.
3 Peter Abbott & Co. (Northern) Ltd.,
Beza Street,
Industrial Estate,
Hunslet,
Leeds LS10 2TB.
Phone: Leeds (0532) 700681/2
Telex: 557145
Answerback: Abbottco Leeds.
4 Peter Abbott & Co. (Western) Ltd.,
Manor Road,
Marston Trading Estate,
Frome,
Somerset.
Phone: (0373) 4509/5084.
Telex 449997
Answerback Abbottco Frome
3
2
1
4

WONDERS OF THE WORLD

When Gustave Eiffel built in 1887/89 his colossus to a height of 985 feet (331 metres) to stand, as it were, astride Paris, it cost £200,000 and was considered one of the world's engineering wonders. The Eiffel Tower has become the emblem of Paris, a symbol of France, the most instantly recognisable construction in the world.

What price the Eiffel Tower now?

When A.R.Glithero published his first journal in 1932, it was on a capital investment of £5. To date, TTP has published over fifty technical and trade journals and books, and the publishing structure grows, year by year. The wealth of engineering knowledge contained in these publications is enormous - as is the Eiffel Tower.

Why not build your own tower of technical information and data on pumping, pneumatics, oil-hydraulics, noise and vibration, power transmission, etc., NOW? Send for the Books and Journals catalogue. It will prove of inestimable value - just like the Eiffel Tower.

TRADE & TECHNICAL PRESS LTD.,
CROWN HOUSE, MORDEN,
SURREY, ENGLAND.

Captive Nuts are basically conventional nuts and stiffnuts enclosed in a cage or anchor which is itself rigidly secured to the component. These may be categorised separately as caged nuts and anchor nuts and also described by other names. See chapter on *Anchor Nuts.*

Self-locking Fasteners are free-running nuts (or bolts) incorporating geometric design features which when finally tightened provide a locking action against subsequent rotation. These are described in the chapter *Locking Nuts.*

All the aforementioned categories of fasteners provide locking only when finally tightened. Locking action can be fully or partially lost with relaxation of preload.

Prevailing Torque Fasteners are nuts with a thread-binding feature, or some form of thread interference, providing a continuous locking torque on thread whether tightened or not. This particular torque can account for up to 50% of the tightening torque on the smaller sizes of fastener, but is usually of the order of 10% on larger sizes.

Stiffnuts of this type are largely individual and/or patented designs, but the American IFI ('Industrial Fasteners Institute) Standard for prevailing torque requirements is widely adopted by manufacturers.

See also chapter on *Prevailing Torque Nuts (Stiffnuts).*

IFI STANDARD FOR PREVAILING TORQUE REQUIREMENTS

Size (UNC)	Prevailing torque (max) inch-lb	Prevailing torque (min) Inch-lb Unhardened Nuts		Hardened Nuts	
		1st removal*	5th removal	1st removal	5th removal
1/4	50	5	3.5	6	4.5
5/16	60	8	5.5	10.5	7.5
3/8	80	12	8.5	16	11.5
7/16	100	17	12	23	16
1/2	150	22	15	30	20
9/16	200	30	21	40	28
5/8	300	39	27	57	36
3/4	400	58	41	78	54
7/8	600	88	62	117	82
1	800	120	84	160	112
1.1/8	900	150	105	200	140
1.1/4	1 000	188	132	250	176
1.3/8	1 200	220	154	293	205
1.1/2	1 350	260	182	346	242

* after tightening to 75% of proof load

Adhesive Systems

The most recent developments in this category of thread fastening are the use of anaerobic adhesives curing when air is excluded by the tightening of the fastener; and microencapsulated adhesives ('Scotch-Grip' Adhesive 2353) which can be preapplied and remains dormant until the fastener is tightened, the twisting action causing the adhesive to cure — see *Adhesive Section.*

Self-Locking Threads

This category of thread locking system embraces locking devices applied to the bolt or screw thread rather than the nut. They are now most commonly based on resilient inserts, a bonded resilient patch or coating or some alternative means of trapping a resilient interface between the mating threads. There are also all-metal systems based on 'sprung' deformation of the bolt or screw threads during manufacture, or the use of interference fit threads.

Thread inserts apply thread-interference locking to the bolt (or screw) threads rather than to nuts, and again are normally a prevailing torque type of fastener. The original 'Wedglok' system incorporated a small nylon fillet inserted in the thread parallel to the bolt axis with a sufficient degree of protrusion to interfere elastically with the mating nut thread. The later 'Longlok' system incorporates a strip of nylon embedded in a groove milled into the thread parallel to the bolt axis, thus giving a longer 'locking' length – Fig 1.

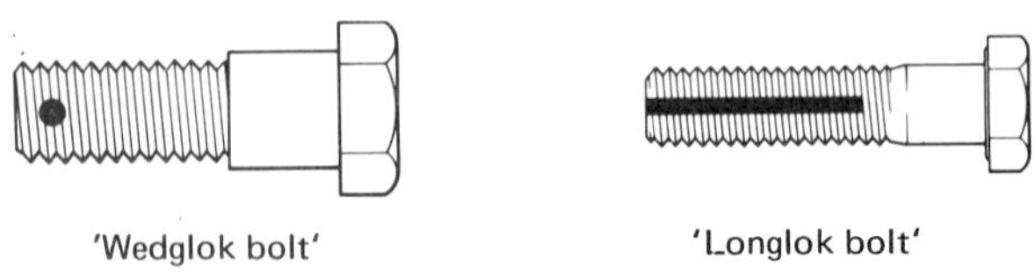

'Wedglok bolt' 'Longlok bolt'

Fig 1

Both types typically yield a primary torque of the order of 7–9kg-cm (6–8lb-in) for 6mm (¼in) diameter screws and up to 11-14kg (10-12lb) for 12.5mm (½in) screws, ie substantially less than that normally provided by stiffnuts. There is also the chance that in the absence of a smooth lead-in on the mating thread the nylon insert can be sheared rather than deformed elastically, with partial or complete loss of interference.

Variations on the 'Longlok' principle include the 'Strip-lok' as a lower cost general purpose fastener; and the 'T-sert' for specialised applications (both by Trutite).

Thread inserts are now rivalled, or even superceded by later types of locking threads where the elastic member is applied to the threads in the form of a coating, patch or spiral winding bonded to the threads. The 'Eslok' bolt and 'Tuf-lok' bolt both employ a patch of powered nylon fused to the threads – Figs 2 and 3. Other types include the 'Locwel' (nylon strip bonded to the threads) and the 'Circloc' (spiral band of nylon wrapped around the threads).

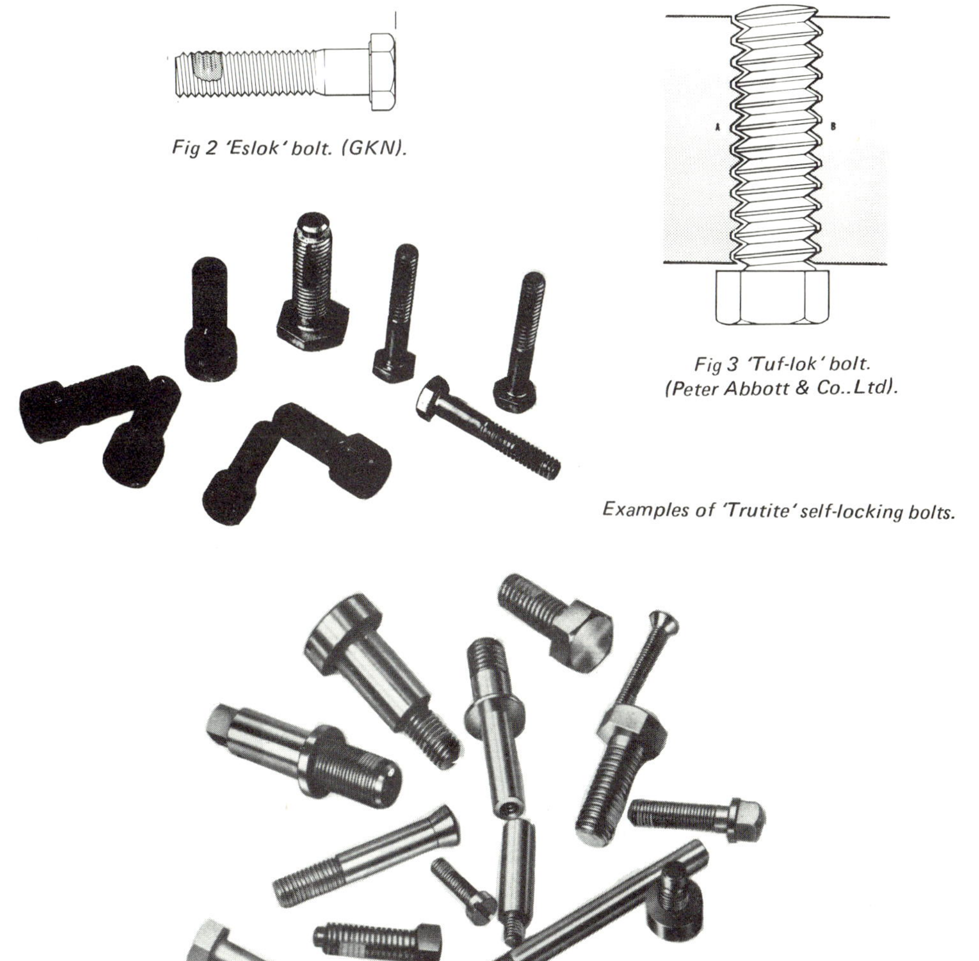

Fig 2 'Eslok' bolt. (GKN).

Fig 3 'Tuf-lok' bolt. (Peter Abbott & Co..Ltd).

Examples of 'Trutite' self-locking bolts.

A selection of 'Tuf-lok' self-locking screws and fasteners. (Peter Abbott & Co Ltd).

All-metal thread-locking systems are based on a special form of thread with an 'interference area' which reforms metal from the crest of the female thread along the helix of the male, filling the voids between the flanks of the mating threads. The special thread form is basically a modified form of the standard thread, carefully proportioned to avoid overstressing the mating thread when engaged – eg Lok-thread – see Fig 4 and Table I. Essentially this provides a prevailing torque locking system. Other all-metal systems of this type are described in the chapter on *Prevailing Torque Nuts.*

TABLE I – 'LOK-THREAD' DIMENSIONS

60% COARSE THREAD SERIES

Thread Size	Root Dia.	Root Width T	Range Tap Drill
6-32	.1136	.0156	.114 - .110
8-32	.1396	.0156	.140 - .136
10-24	.1576	.0208	.158 - .154
12-24	.1836	.0208	.184 - .180
1/4-20	.2110	.0250	.212 - .208
5/16-18	.2693	.0278	.270 - .265
3/8-16	.3262	.0313	.327 - .322
7/16-14	.3819	.0357	.383 - .377
1/2-13	.4400	.0385	.441 - .435
9/16-12	.4975	.0417	.498 - .492
5/8-11	.5542	.0455	.556 - .550
3/4-10	.6720	.0500	.673 - .666
7/8-9	.7884	.0556	.789 - .782
1-8	.9026	.0625	.904 - .896

60% FINE THREAD SERIES

Thread Size	Root Dia.	Root Width T	Range Tap Drill
6-40	.1186	.0125	.119 - .115
8-36	.1424	.0139	.143 - .139
10-32	.1656	.0156	.166 - .162
12-28	.1882	.0179	.190 - .186
1/4-28	.2222	.0179	.223 - .219
5/16-24	.2801	.0208	.281 - .276
3/8-24	.3426	.0208	.343 - .338
7/16-20	.3985	.0250	.399 - .393
1/2-20	.4610	.0250	.462 - .456
9/16-18	.5193	.0278	.520 - .514
5/8-18	.5818	.0278	.583 - .576
3/4-16	.7012	.0313	.704 - .695
7/8-14	.8194	.0357	.820 - .813
1-14	.9444	.0357	.945 - .938

70% COARSE THREAD SERIES

Thread Sze	Root Dia.	Root Width T	Range Tap Drill
6-32	.1096	.0133	.110 - .106
8-32	.1356	.0133	.136 - .132
10-24	.1521	.0177	.153 - .149
12-24	.1781	.0177	.179 - .175
1/4-20	.2046	.0213	.205 - .199
5/16-18	.2619	.0236	.263 - .257
3/8-16	.3182	.0266	.319 - .312
7/16-14	.3725	.0304	.373 - .366
1/2-13	.4300	.0327	.430 - .422
9/16-12	.4867	.0354	.488 - .480
5/8-11	.5424	.0386	.544 - .536
3/4-10	.6590	.0425	.661 - .653
7/8-9	.7740	.0472	.776 - .768
1-8	.8864	.0531	.888 - .880

70% FINE THREAD SERIES

Thread Sze	Root Dia.	Root Width T	Range Tap Drill
6-40	.1153	.0106	.116 - .112
8-36	.1387	.0118	.139 - .135
10-32	.1616	.0133	.162 - .158
12-28	.1835	.0152	.184 - .180
1/4-28	.2176	.0152	.218 - .212
5/16-24	.2747	.0177	.275 - .269
3/8-24	.3372	.0177	.338 - .331
7/16-20	.3921	.0213	.393 - .386
1/2-20	.4546	.0213	.455 - .447
9/16-18	.5119	.0236	.513 - .505
5/8-18	.5744	.0236	.575 - .567
3/4-16	.6932	.0266	.694 - .686
7/8-14	.8100	.0304	.811 - .803
1-14	.9350	.0304	.936 - .928

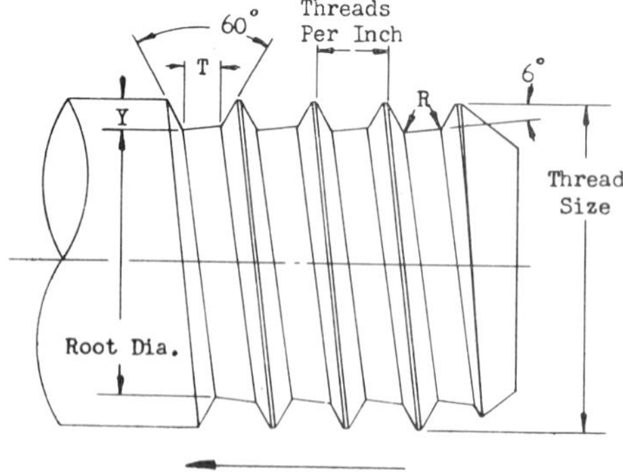

'Lok-Thred' studs are tapered in direction indicated.

Enlarged thread form

Fig 4 Basic dimensions of 'Lok-Thred' stud specifications 60% (when maximum strength of part is desired) and standard 70% for general use. Y = 60% or 70% of American National Form Thread.

Chemical thread locking is similar in principle to a coated thread lock, the coating being applied in the form of a liquid to the threads immediately prior to assembly. The chemical coating can be in the form of a metal adhesive for permanent, rigid assembly, or yield a semi-resilient or resilient bond which enables the nut to be removed. Both one-part and two-part chemical thread locking compounds are used in a variety of combinations – see *Adhesives* section, *Thread Locking Compounds.*

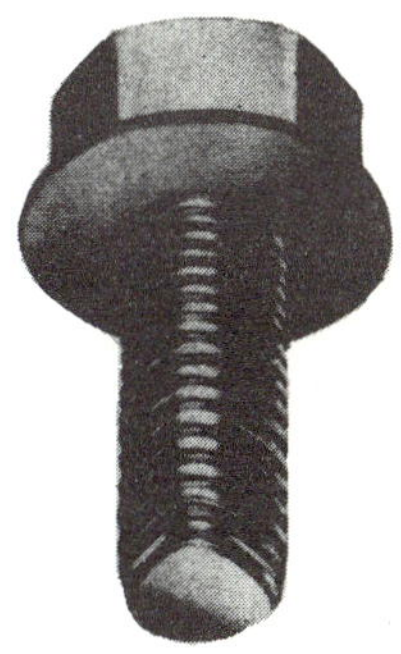

'Powerlok' tri-lobular thread locking screw. (Linread Ltd).

Nominal Size	Prevailing Torque During 1st Installation in-lbs max	Breakaway Torque in lbs-min	
		1st Removal	5th Removal
4-40	3.0	1.5	1.0
6-32	6.0	3.0	1.5
8-32	10.0	7.0	5.0
10-32	12.0	10.0	6.0
1/4-20	30.0	18.0	11.0
5/16-18	65.0	45.0	20.0
3/8-16	95.0	50.0	30.0

Examples of GKN 'Eslok' patch fasteners.

Gamp-Tork.

The *Gamp-Tork* (Glynwed Screws and Fastenings) is a very recent example of the 'deformed thread' principle. The (reusable) locking action is developed by a carefully predetermined pitch change in the standard screw thread which is produced during the thread rolling operation. It can be applied to any standard male thread, in such a position that it can be adjusted to suit special needs (provided at least two full threads are left between the pitch change and the point of the screw). Prevailing torque values can also be varied from standard (which is better than MIL-F-18240D requirements) by recalculation of the pitch change or increased penetration of engaged thread.

Locking Nuts and Free Running Systems

Locking nuts are defined as those which rely on being torqued tight to produce a locking action to hold the nut against subsequent loosening and loss of the bolt or screw preload under vibration, etc. They include the following types:-

(i) *Jam nuts* – thin nuts used in combination with a standard nut – see chapter on *Nuts.*

(ii) *Flexible nuts* – standard nut forms with a concave bottom surface in a material ductile enough for the upper section to flex and tighten on the screw thread when torqued tight.

(iii) *Base-insert nuts* – where a non-metallic insert in the base of the nut is deformed on tightening and expands to grip the bolt or screw threads to provide frictional locking.

(iv) *'Grip' nuts* – where the base of the nut is knurled or serrated or incorporates small 'teeth' which bite into the parent surface when the nut is tightened. The same principle is applied to bolt heads.

(v) *Spring nuts* – spring steel pressings which engage a single turn of the bolt or screw thread and are drawn tight against the thread to provide a locking grip when the bolt is fully tightened.

(vi) *Jamming nuts* – patented types.

(vii) *Place bolts*

The design of nuts within each type may vary in detail, except in the case of conventional jam nuts which are standard thin nuts – Fig 1. All have the characteristic of being free-running up to the final tightening turns. All are re-usable, but the base-insert and spring nut may have limited re-usability. Simple spring nuts are suitable only for low loads (ie low bolt or screw preloads). There are a vast number of individual and patented designs of spring nuts and blind screw anchors – see chapters on *Anchor Nuts, Spring Nuts, Blind Fasteners,* etc.

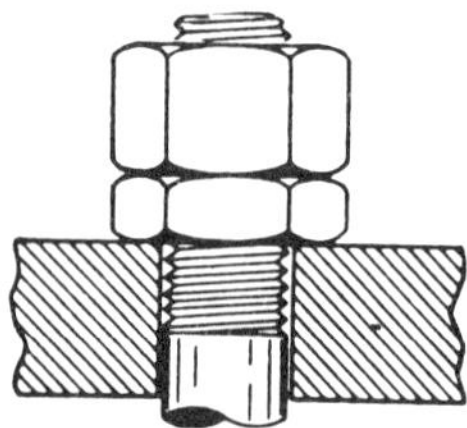

Fig 1 Locking nut or jam nut.

All the aforementioned types of locking nut are resistant to the 'unscrewing' effect of vibration once the nut has been torqued tight.

Base-insert Nuts

An example of a proprietary base-insert nut is shown in Fig 2. An elastomeric washer type insert in the base of the nut is expanded when the nut is tightened so as both to grip the serrated base of the nut and be compressed against the bolt or screw thread. Under this condition it provides similar locking action to the type of primary-torque locknut with an elastomeric insert.

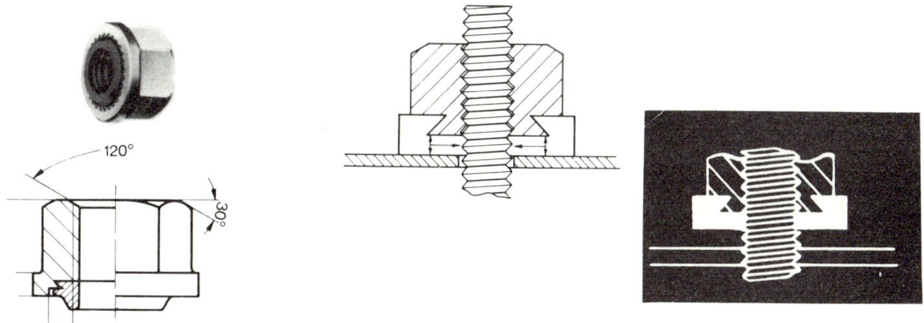

Fig 2 Otalu base insert nut.

Fig 3 'Deltight' nut.

Another example of a similar type of nut is shown in Fig 3. This differs in that the 'washer' section is in acetal resin with a concave bearing surface, (ie forming a composite flexible nut). When compressed on tightening this produces a self-sealing assembly as well as one that is self-locking. It is particularly suitable for use on vitreous enamel, glass, plastic and other hard or brittle surfaces which might be cracked by the use of hard-surface locking nuts – see also Table I.

TABLE I – 'DELTIGHT' SELF-LOCKING NUTS

Thread	Washer Thickness	Washer Diameter	Height of Nut
6 BA	0.064	0.141	0.233
4 BA	0.071	0.175	0.301
2 BA	0.088	0.215	0.391
0 BA	0.106	0.269	0.500
Whit and BSF			
3/16	0.088	0.215	0.391
1/4	0.122	0.272	0.561
5/16	0.122	0.322	0.625
3/8	0.125	0.387	0.725
1/2	0.158	0.530	1.00
5/8	0.214	0.640	1.25
3/4	0.214	0.801	1.50
1	0.215	0.990	1.87

All dimensions in inches

Washer Systems

The 'Nylite' fastener is another proprietary type of self-locking, self-sealing washer which is applied under the head of a screw rather than combined with the nut. The sealing element is a specially shaped nylon washer with a rolled edge. When the fastener is tightened, this washer is forced to cold flow to the final form shown in Fig 4, providing both sealing and thread locking action This type of sealing lock is also used on self-tapping screws.

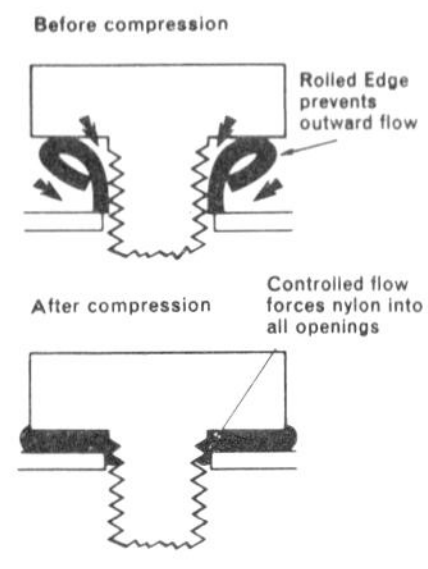

Fig 4 'Nylite' fastener.

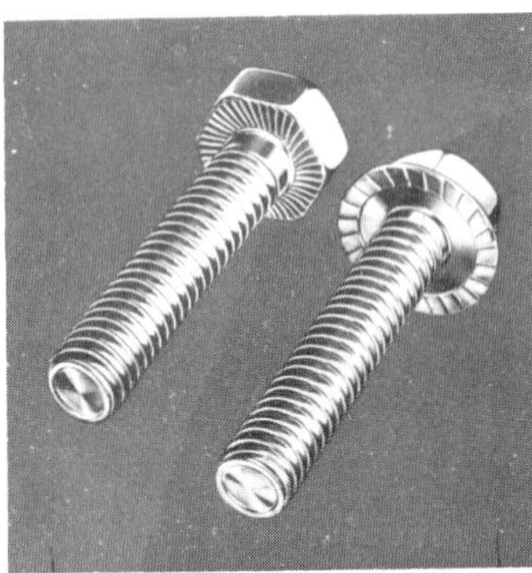

Fig 5 'Whiz-Tite' system.

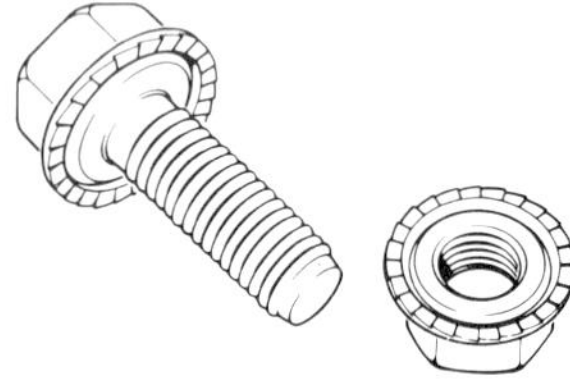

Fig 5b 'Grip' bolt and washer.

'Grip' Nuts

The 'Whiz-Tite' system shown in Fig 5 is a typical example of a 'grip' bolt where the locking action on final tightening is provided by the spiralling serrations or teeth on the underside of the head. There is also a convex angle on the base to improve the locking effectiveness of the teeth. The serrations are so formed that they do not grip or oppose rotation to any marked degree on

TABLE II – SUGGESTED TIGHTENING TORQUES FOR 'WHIZ-TITE' SCREWS

Unplated Parts

The chart gives suggested tightening torque figures in respect of unplated fasteners on plain steel bearing surfaces

Size	Grade B lb/ft	Grade T lb/ft
1/4 – 20	8	12
1/4 – 28	10	15
5/16 – 18	14	24
5/16 – 24	16	28
3/8 – 16	30	40
3/8 – 24	35	45
7/16 – 14	50	60
7/16 – 20	58	69
1/2 – 13	65	90
1/2 – 20	75	100

final tightening. Teeth and bottom surface of the nut are case hardened to Rockwell C40 (minimum) and 'Whiz-Tite' fasteners can be re-used up to fifteen times or more with no loss of performance – see also Table II.

In a similar system, the 'Tensilok', the teeth are formed on a flexible skirt around the periphery of the bolt head or nut.

Place Bolts

Place bolts are cold formed hexagon headed bolts with slots in the upper face forming a number of uniform segments, and a circular recess in the underside of the head adjacent to the shank – Fig 6. The slots and recess impart a certain amount of flexibility to the head. They thus have a locking action similar to that of flexible nuts as well as a high resistance to fatigue (particularly in shorter bolt lengths).

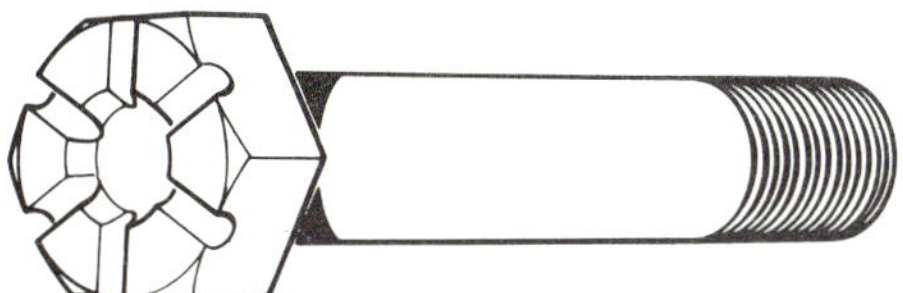

Fig 6 Place bolt.

Place bolts are used mainly in heavily loaded applications.

Other Types

Crimp nuts are modified forms of standard nuts which are locked after tightening by a crimping tool. In the 'Huckrimp' system (Fig 7) the nut is of elongated form with one end threaded, the other consisting of a smooth collar which clears the bolt thread. After running up snug a special crimping tool is used to apply pressure to the collar, forming new threads over this section. This not only provides locking action on the threads, but a predetermined amount of bolt tension.

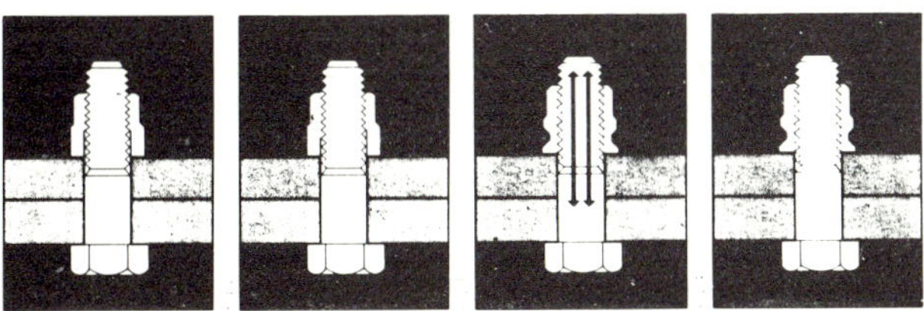

Fig 7 Huckrimp fastener system.

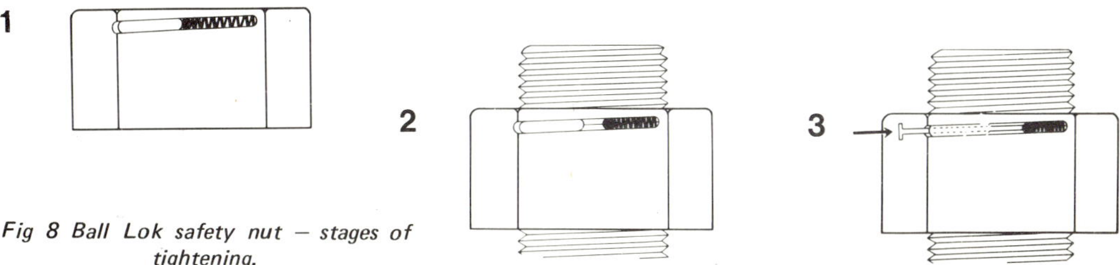

Fig 8 Ball Lok safety nut – stages of tightening.

A patented *jamming nut* is shown in Fig 8. This is the 'Ball-Lok' safety nut which is free running onto a male thread. If the direction is reversed, the spring loaded ball jams into the thread for locking action. To remove the nut a thin pin is inserted to press the ball against the spring, when the nut will turn easily.

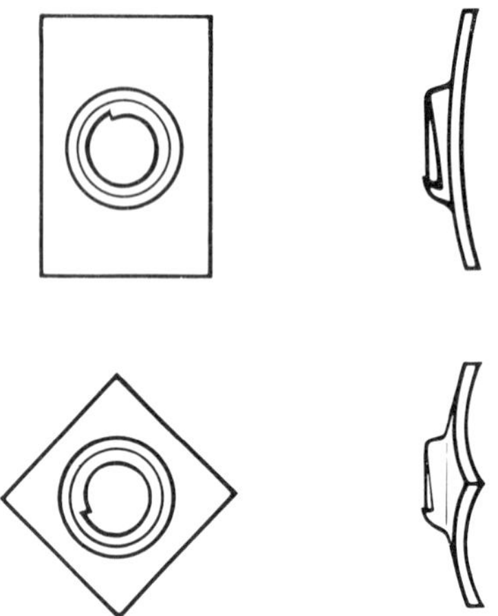

TABLE III – SIZES AND DATA ON 'SPIROL' TENSIONUTS (SPRING NUTS)

Size	Flange Size			Thickness	Rec Holes	Recommended Inst Torque in-lb	Maximum Tensile Strength lb	Weight lb/1000 pcs	
	Square	Rectangular						Sq	Rect
4BA and 6 UNF	11/32	9/32	7/16	0.017	5/32	4	120*	0.5	0.5
2BA and 10 UNF	7/16	3/8	5/8	0.017	7/32	8	300*	1.0	1.2
6 UNC	11/32	9/32	7/16	0.017	5/32	5	150*	0.5	0.5
8 UNC	3/8	5/16	1/2	0.017	3/16	7	250*	0.7	0.8
3/16 BSW and 10 UNC	7/16	3/8	5/8	0.022	7/32	12	400*	1.3	1.5
1/4 BSW and 1/4 UNC	1/2	1/2	3/4	0.025	9/32	30	900*	1.9	2.9
5/16 BSW and 5/16 UNC	19/32	9/16	1	0.028	11/32	30	1 100*	2.7	4.3
6Z	3/8	5/16	1/2	0.025	5/32	11	500	1.1	1.2
8Z	7/16	13/32	5/8	0.028	3/16	17	750	1.5	2.0
10Z	1/2	1/2	3/4	0.031	7/32	30	1 300	2.3	3.5

Standard finishes : Plain, zinc, plated, phosphate coated.
Materials: Spring steel, heat treated unless otherwise specified.

*Machine screw fails

Locking Washers

Typical data on Spring Nuts are given in Table III.

Locking action with free-running (standard) nuts can also be provided by locking washers, eg

(i) *Spring washers* – which may be used in conjunction with a plain washer.

(ii) *Elastic washers* – providing a grip similar to that of a base-insert nut but with less control of spread and more limited grip since the bulk of flow of washer material will be outwards.

(iii) *Toothed washers* – various designs of locking and 'shakeproof' washer.

(iv) *Serrated or kinked face washers* – providing a similar locking action to a 'grip' nut. One or both faces of the washer may be given a gripping surface, or the whole washer section may have a dimpled or serrated form.

See also *Washers* and *Thread Locking Systems.*

Prevailing Torque Nuts (Stiff Nuts)

Self-locking nuts or stiffnuts are defined as nuts with inherent stiffness (high frictional torque) or prevailing torque action when run on to a mating thread. Locking action does not depend on the nut being fully tightened and is thus largely independent of bolt or screw preload. They are correctly classified as prevailing torque nuts. Basic types are:-

(i) Non-metallic insert stiffnuts.
(ii) Deformed thread nuts.
(iii) Integral spring nuts.
(iv) Deflected body nuts.
(v) Deflected thread nuts.
(vi) Spring insert nuts.
(vii) Jamming nuts.

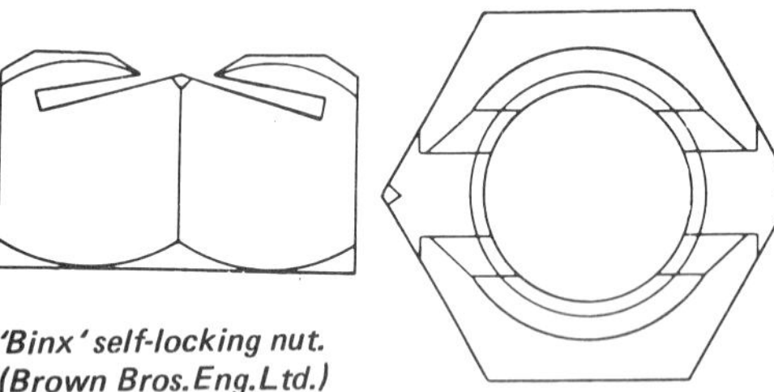

'Binx' self-locking nut.
(Brown Bros.Eng.Ltd.)

A.

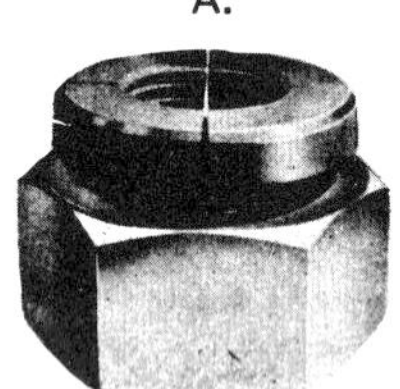

B.

Locknuts by Benton Engineering Ltd.
A—'Slipnot' collet —type grip.
B—'Benton Loc' all metal.

TABLE I — TORQUE PERFORMANCE OF STIFFNUTS
(Industrial Fasteners Institute of America)

Nut Size UN	¼in	5/16in	3/8in	7/16in	½in
First installation	30 max	60 max	80 max	100 max	150 max
First removal	5 min	8 min	12 min	17 min	22 min
Fifth removal	3.5 min	5.5 min	8.5 min	12 min	15 min

Torque values in lbf-in

Note: these recommendations are widely accepted in Europe

The A to Z of Nyloc and Cleveloc

LONDON & HOME COUNTIES
Peter Abbott & Co Ltd
Bridge Close Industrial Estate Romford Essex
Tel: Romford 25111 (10 lines) Telex: 897072

Alder Miles Druce Ltd
Alderberry Works Beaconsfield Road Hayes Middlesex
Tel: 01-573 7766 (10 lines)

Delson & Co Ltd
Wessex Road Bourne End Bucks
Tel: 0628-522711 (3 lines)

Alder Miles Druce Ltd
Station Road Sutton-at-Hone Dartford Kent
Tel: Dartford 862611 Telex: 896577

Millerservice Ltd
PO Box 19 Slough Bucks SL1 4SG
Tel: Slough 25511 (10 lines)

Pillar Engineering Supplies Ltd
99 Waldegrave Road Teddington Middlesex
Tel: 01-977 8844 (20 lines) Telex: 928138

Tern Screw Co Glynwed Distribution Ltd
500 Old Kent Road London SE1
Tel: 01-237 8221 Telex: 884807

SOUTH COAST
J. R. Smith & Sons (Structural) Ltd
Hamworthy Poole Dorset BH15 4LG
Tel: 020-135115 (15 lines)

Alder Miles Druce Ltd
Anglesea Road Shirley Southampton
Tel: 0703 775352

Pillar Engineering Supplies Ltd
42 Aston Road Waterlooville Portsmouth Hants
Tel: Waterlooville 54341/3 Telex: 86153

SOUTH WESTERN
Alder Miles Druce Ltd
New Queen Street Whitehouse Street
Bedminster Trading Estate Bristol BS3 4BY
Tel: Bristol 6604217 Telex: 44814

Woodberry Chillcott & Co Ltd
Atlas Street, Feeder Road Bristol 2 BS99 7UW
Tel: 0272 770407 Telex: 44328

WALES
G. N. Hunter & Co Ltd
The Airport Tremorfa Cardiff CF2 2XG
Tel: 44631 (10 lines)

Miller Bridges Fastenings,
Glynwed Distribution Limited,
Trecenydd Industrial Estate,
Caerphilly, Glamorgan.
Tel: 0222 868411 (5 lines) 0222 867573

EAST ANGLIA
Alder Miles Druce Ltd
Fengate Drove Brandon Suffolk
Tel: 084-281 0771

Anglian Nuts & Bolts Ltd
Warehouse 10C6 Elstow Storage Depot
Kempston Hardwick Bedford
Tel: Bedford 740650

MIDLANDS
Alder Miles Druce Ltd
Stourvale Works Clensmore Street
Kidderminster Worcs
Tel: 0562 2748

Walford Manufacturing Co Ltd
Excelsior Works Great Lister Street
Birmingham B7 4LR
Tel: 021-359 2681 (10 lines)

Delson & Co Ltd
Studley Road Redditch Worcs
Tel: 0527 25148

Miller Bridges Fastenings
Glynwed Distribution Ltd
Humpage Road Bordesley Green
Birmingham B9 5HP
Tel: 021-773 1222 (10 lines) Telex: 338768

Alder Miles Druce Ltd
Evelyn Street (Off Queens Road) Beeston Notts
Tel: Nottingham 252612 Telex: 37287

Peter Abbott & Co (Central) Ltd
Morley Street Daybrook Notts
Tel: 0602 264222 Telex: 377216

Miller Bridges Fastenings
Glynwed Distribution Ltd
Portway Road Wednesbury Staffs
Tel: 021-556 1748

Miller Bridges Fastenings
Glynwed Distribution Ltd
Middlemore Lane West Aldridge Staffs WS9 8DS
Tel: Aldridge 55121

NORTHERN ENGLAND
Alder Miles Druce Ltd
Midland Road Balm Road Leeds 10
Tel: 0532 700671

J. A. Challiner & Co. Ltd.,
Midland Street, Manchester M12 6LB
Tel: 061-273 3221 (4 lines)

Alder Miles Druce Ltd
Tenax Road Trafford Park Manchester 17
Tel: 061-872 3562 Telex: 669424

C. Walters & Son Ltd.,
Blackbrook Estate,
Wellington Road North,
Manchester M19 2QS
Tel: 061-432 8167

Delson & Co Ltd
Newby Road Hazel Grove Industrial Estate
Stockport Cheshire
Tel: 061-483 9731 (10 lines)

Dudley & Green Glynwed Distribution Ltd
Mersey Bolt & Screw Works Simpson Street
Liverpool 1
Tel: 051-709 9666 (10 lines) Telex: 627033

John Heaton Fastenings Ltd
Mount Pleasant Leyland Preston PR5 3BS
Tel: Leyland 22281 Telex: 67487

Williams Bros of Sheffield
Green Lane Sheffield S3 8SF
Tel: 0742 77365

Industrial Fasteners Supplies (South Yorkshire) Ltd.,
237 Edmund Road,
Sheffield S2 4EL
Tel: Sheffield 24613

The Hull Factoring Co Ltd
Bourne House Anlaby Road Hull HU3 2RN
Tel: 224683 (3 lines)

Macnays Ltd
GPO Box No 14, 48/50 West Street Middlesbrough
Teesside
Tel: 0642 48144 (12 lines)
Direct line to stock control 0642 2125/6 Telex: 58514

Alder Miles Druce Ltd
Swindon Street Hebburn Co Durham
Tel: 0632 834169

Delson Screwparts
First Avenue Tyne Tunnel Trading Estate
North Shields NE29 Northumberland
Tel: 089-45 78244/5/6

Peter Abbott & Co. (Northern) Ltd.,
Beza Street, Industrial Estate,
Hunslet, Leeds, LS10 2TP
Tel: 0532 700681-2

Components & Engineering Ltd.,
Birchill Road, Kirkby, Nr. Liverpool, Lancs.
Tel: 051-548 4004 (10 lines)

Miller Bridges Fastenings,
Glynwed Distribution Limited,
Pym Street, Hunslet Road, Leeds 10
Tel: 0532 448331 Telex: 557125

SCOTLAND
George Boyd & Co Ltd
24 Annandale Street Edinburgh EH7 4AR
Tel: 031-556 4771 (5 lines) Telex: 72483

D. F. Wishart & Co Ltd
GPO Box No 208 St Clair Street Edinburgh EH6 8LJ
Tel: 031-554 4393 (10 lines)

George Boyd & Co Ltd
13-15 Bell Street Dundee DD1 9NN
Tel: 0382 21591 (4 lines) Telex: 76211

Alder Miles Druce Ltd
3 Graham Street Dundee
Tel: Dundee 89005 Telex: 76383

George Boyd & Co Ltd
300 Crownpoint Road Glasgow G40 2UP
Tel: 041-554 1844 (20 lines) Telex: 77315

Alder Miles Druce Ltd
Caledonian Steel Works Rutherglen Glasgow
Tel: 041-647 9181 Telex: 77114

Stephens & Smith (Power Transmission) Ltd
13 York Street Glasgow C2
Tel: 041-248 4051/2

NORTHERN IRELAND
Kennedy & Morrison Limited,
Boucher Road, Belfast BT12 6QF
Tel: 0232 663621 (5 lines)

As you can see there's a distributor of **Nyloc** and **Cleveloc*** self locking nuts near you.

That's the great thing about buying Firth Cleveland products. Not only are they manufactured to a high standard but across the country there are high calibre distributors holding large stocks for you to draw on. They are your guarantee of faster fastenings service.

But that's not all you get with Nyloc and Cleveloc nuts. They're backed by an enormous technical organisation which can offer you professional advice or practical help on any fastening problem.

To be really up-to-date on fastenings you should read the technical booklet "Helpful hints on threaded fastenings". Like our service, it is freely available.

For the latest information on self locking nuts write to the address below.

Firth Cleveland Fastenings Limited

Self Locking Nut Division

Treforest, Pontypridd, Glamorgan, Great Britain.

A member of The GKN Group of Companies 

*Nyloc and Cleveloc are Registered Trade Marks

There are also several individual and patented designs of prevailing torque nuts.

Non-metallic insert stiffnuts employ an insert (originally fibre but now replaced almost universally by nylon) in the form of a compression collar – see Fig 1 and Table II. The mating thread

Fig 1 GKN 'polystop' locking nuts with nylon inserts (see also Table I).

'Nyloc' self-locking nuts. (Peter Abbott Group).

TABLE II – GKN 'POLYSTOP' NYLON INSERT LOCKNUTS

		Type P (thick)			Type T (thin)		
Dia	Thread	H (max) ±·020 *in*	S *in*	M (min) ±·010 *in*	H (max) ±·020 *in*	S *in*	M (min) ±·010 *in*
0	BA	0·302	0·413	0·204	0·236	0·413	0·138
2	BA	0·253	0·324	0·160	0·199	0·324	0·106
4	BA	0·212	0·248	0·118	0·180	0·248	0·086
6	BA	0·161	0·193	0·087	0·139	0·193	0·065
3/16″	BSF WHIT	0·271	0·324	0·179	0·206	0·324	0·114
1/4″	BSF WHIT	0·311	0·445	0·193	0·249	0·445	0·126
5/16″	BSF WHIT	0·366	0·525	0·244	0·282	0·525	0·160
3/8″	BSF WHIT	0·445	0·600	0·306	0·341	0·600	0·202
7/16″	BSF WHIT	0·532	0·710	0·369	0·407	0·710	0·244
1/2″	BSF WHIT	0·595	0·820	0·432	0·451	0·820	0·286
9/16″	BSF WHIT	0·678	0·920	0·495	0·512	0·920	0·328
5/8″	BSF WHIT	0·757	1·010	0·558	0·573	1·010	0·371
3/4″	BSF WHIT	0·928	1·200	0·684	0·702	1·200	0·455
7/8″	BSF WHIT	1·000	1·300	0·746	0·753	1·300	0·496
1″	BSF WHIT	1·171	1·480	0·871	0·883	1·480	0·579

is an interference fit with this insert so that when the nut is run on the thread a prevailing torque is created and the insert further compressed by a predetermined amount. Thus the degree of thread locking produced depends primarily on the thread profile and major diameter of the mating thread. The use of a resilient insert material (eg nylon) makes these nuts frequently re-usable. The main limitation of such types is the maximum service temperature as determined by the point at which nylon starts to soften and flow – typically 110°C (230°F), or up to 130°C (266°F) with superior grades of nylon.

A similar principle of 'elastic locking' can be applied to threads themselves, by incorporating a resilient insert or patch in the length of the bolt or screw thread – see *Prevailing Torque Threads.*

Modified or deflected thread forms rely on an interference fit to provide prevailing torque and locking action. Examples are 'Lok-Thread' (see chapter on *Thread Locking Systems*); 'E-Lok' (Fig 2); 'Vibresist (Fig 3); and 'Powerlok' (Fig 4).

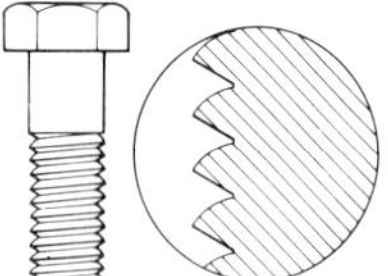
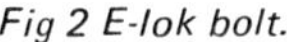

Fig 2 E-lok bolt.

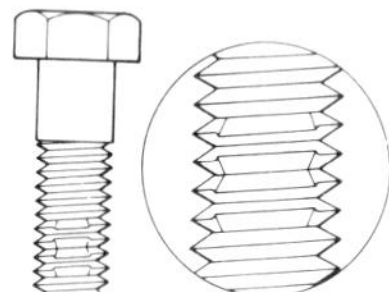

Fig 3 Vibresist bolt.

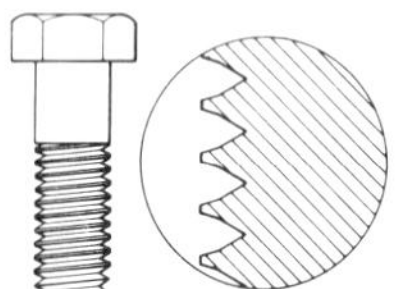

Fig 4 Powerlok bolt.

Deformed thread nuts have basically a standard tapped thread, subsequently deformed over part of the length to provide an interference fit with the mating bolt around a proportion of the circumference – Figs 5, 6, 7, and 8. When the nut is assembled the interference section is 'sprung', thus providing high frictional grip via prevailing torque. Such nuts are normally re-usable, the nut recovering to its original deformed thread shape when removed from the bolt.

There are a number of possible variations on this principle. The most satisfactory is to incorporate the deformation in a collar section on the nut so that its body remains undisturbed when it is assembled, thus reducing the risk of the spanner or wrench slipping. Characteristics of various all-metal prevailing torque nuts are summarised in Tables III and IV. Being all-metal, they have no temperature limitation for general use, other than that imposed by the material.

TABLE III – RECOMMENDED TIGHTENING TORQUES AND BOLT TENSION GKN 'AUTOLOK' NUTS

	'S' quality bolt		'T' quality bolt		'V' quality bolt	
Size	*75% proof load (lbf)*	*Tightening torque (lbf ins)*	*75% proof load (lbf)*	*Tightening torque (lbf ins)*	*75% proof load (lbf)*	*Tightening torque (lbf ins)*
¼" UNC	2020	120	2225	130	2700	150
¼" UNF	2320	130	2530	145	3060	165
5/16" UNC	3340	180	3670	200	4420	230
5/16" UNF	3700	230	4050	250	4870	300
⅜" UNC	4940	375	5400	400	6530	480
⅜" UNF	5600	425	6100	430	7350	500
7/16" UNC	6800	44 lbf ft	7450	50 lbf ft	8980	60 lbf ft
7/16" UNF	7650	48 lbf ft	8250	55 lbf ft	9970	65 lbf ft
½" UNC	9050	75 lbf ft	9900	80 lbf ft	11920	90 lbf ft
½" UNF	10700	85 lbf ft	11100	90 lbf ft	13400	100 lbf ft

TABLE IIIA – GKN AEROTIGHT STIFF NUTS – RECOMMENDED TIGHTENING TORQUES AND BOLT TENSIONS

Size		'A' quality bolt 75% proof load (lbf)	'A' quality bolt Tightening torque (lbf in)	'S' quality bolt 75% proof load (lbf)	'S' quality bolt Tightening torque (lbf in)
UNC	4-40	248	5·56	414	9·28
	6-32	374	10·3	623	17·2
	1/4	1300	65·3	2180	109
	5/16	2140	134	3570	223
	3/8	3170	19·8 lbf ft	5280	33 lbf ft
	1/2	5800	48 lbf ft	9660	80 lbf ft
UNF	10-32	818	31	1360	52
	1/4	1480	74	2470	123
	5/16	2370	147	3940	20·5 lbf ft
	3/8	3570	22 lbf ft	5950	37 lbf ft
	7/16	4717	34·4 lbf ft	7875	57 lbf ft
	1/2	6450	54 lbf ft	10830	90 lbf ft
	5/8	10400	108 lbf ft	17340	180 lbf ft
BSW	3/16	689	25·8	1150	43
	1/4	1290	64·7	2160	108
	5/16	2120	133	3540	18·4 lbf ft
	3/8	3140	19·6 lbf ft	5230	32·7 lbf ft
	1/2	5580	46·5 lbf ft	9300	77·5 lbf ft
	5/8	9150	95 lbf ft	15250	159 lbf ft
	3/4	13550	169 lbf ft	22600	282 lbf ft
	7/8	18700	272 lbf ft	31200	454 lbf ft
	1	24500	387 lbf ft	40850	705 lbf ft
BSF	1/4	1440	72	2400	120
	5/16	2290	143	3820	19·8 lbf ft
	3/8	3390	21 lbf ft	5880	36·7 lbf ft
	1/2	6130	51 lbf ft	10200	85 lbf ft
BA	6	264	5·8	440	9·7
	4	444	12·6	740	21
	2	770	28·6	1290	47·6

4·6/4·8 grade bolts

Sizes metric	75% proof load (lbf)	75% proof load (N)	Tightening torque (lbf in)	Tightening torque (Nm)
M3	262	1165	6·2	0·07
M4	456	2030	14·36	1·62
M5	740	3290	29·08	3·28
M6	1050	4660	49·50	5·59
M8	2040	9070	128·4	14·51
M10	3020	13400	19·8 lbf ft	26·8
M12	4390	19500	34·5 lbf ft	46·8
M16	8160	36300	84·7 lbf ft	115
M20	12770	56800	167 lbf ft	227
M24	17875	79500	280 lbf ft	382

8·8 grade bolts

Size metric	75% proof load (lbf)	75% proof load (N)	Tightening torque (lbf in)	Tightening torque (Nm)
M3	524	2330	12·4	1·40
M4	914	4060	28·8	3·25
M5	1480	6580	58·2	6·57
M6	2100	9330	99	11·18
M8	3800	16940	20 lbf ft	27·12
M10	6030	26800	39·5 lbf ft	53·57
M12	6780	30140	69 lbf ft	93·5
M16	16330	72600	171 lbf ft	232
M20	25500	113600	335 lbf ft	454
M24	35750	159000	563 lbf ft	763

Note: The figures quoted above give approximate tensions and recommended tightening torques for self colour, lightly oiled nuts, applied to self colour bolts. Any changes in assembly conditions will require re-adjustment of these recommendations.

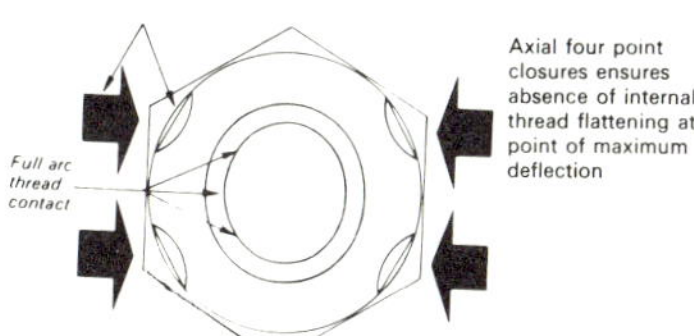

Fig 5 GKN 'Autolok' stiffnut (deformed thread type). The collar on top of the hexogan body being deformed into an eclipse. Standard sizes ¼in, 5/16in, 3/8in, 7/16in and ½in UNC and UNF.

Integral spring nuts incorporate spring 'arms' or a similar thread-gripping feature on the top of the nut. These arms are themselves threaded and deflected downwards and inwards. When the nut is assembled on a mating thread they are sprung apart, causing them to grip the thread tightly. Again, these are re-usable nuts. Their only particular limitation is that at least two threads should project through the arms in the tightened position to ensure a satisfactory level of prevailing torque.

TABLE IV – LESTER STAR STIFFNUTS

Nom size	Width across flats	Normal series thickness	Thick series thickness*	Weight per 100 nuts normal series	
	max	*max*	*max*	*lb*	*kg*
in	in	in	in	UNF	UNF
$\frac{1}{4}$	0·4375	0·224	0·270	0·61	0·28
$\frac{5}{16}$	0·500	0·271	0·325	0·94	0·43
$\frac{3}{8}$	0·5625	0·333	0·385	1·38	0·63
$\frac{7}{16}$	0·6875	0·380	0·440	2·52	1·15
$\frac{1}{2}$	0·750	0·442	0·510	3·26	1·48
$\frac{5}{8}$	0·9375	0·552	0·638	6·40	2·90
$\frac{3}{4}$	1·125	0·651	0·752	11·19	5·08
Standard Hexagon—BS 1083: Whit /BSF Thread					
in	in	in	in	BSF	BSF
$\frac{1}{4}$	0·445	0·200	0·242	0·62	0·28
$\frac{5}{16}$	0·525	0·250	0·302	1·03	0·45
$\frac{3}{8}$	0·600	0·312	0·376	1·64	0·73
$\frac{7}{16}$	0·710	0·375	0·434	2·75	1·25
$\frac{1}{2}$	0·820	0·437	0·508	4·18	1·90
$\frac{5}{8}$	1·010	0·562	0·650	8·38	3·81
$\frac{3}{4}$	1·200	0·687	0·795	13·73	6·21
ISO Metric Hexagon—BS 3692: Coarse Pitch					
	mm	mm	mm		
M6	10·00	5·00	6·50	0·53	0·24
M8	13·00	6·50	8·00	0·88	0·40
M10	17·00	8·00	10·00	1·99	0·90
M12	19·00	10·00	11·50	3·09	1·40
M16	24·00	13·00	15·00	6·09	2·76
M20	30·00	16·00	18·50	11·47	5·20

*Differs from Normal Series in thickness and weight only

The basic principle of a deflected body self-locking nut is shown in Fig 6. The body of the nut is so finished – eg with a slot – that when the nut is finally tightened the 'free' end of the body is deflected or sprung out of axial alignment to provide gripping action. These nuts are re-usable, but are not true stiffnuts since locking torque is only provided with the final tightening turns. Once tightened, however, they will retain their locking action even if the preload is lost.

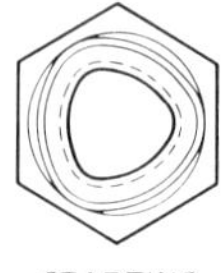

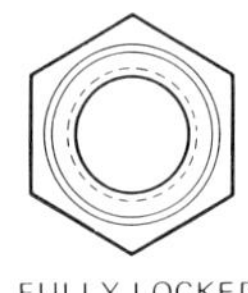

Fig 6 'Flexloc' deformed thread stiffnuts –sizes 0(0.060in) to 4(0.112in).

Deflected thread self-locking nuts work on a similar principle but here the thread form is deflected on assembly to provide the locking action – see Fig 7. The 'spring' of the nut is thus radial rather than axial. The integral spring nut is a form of deflected thread nut, although classified here as a separate type.

Spring-insert nuts are similar in action to integral spring nuts, except that the locking spring section is a separate metal insert. Individual designs vary and some may require a key or special tool to release the spring to remove the nut after assembly.

Some comparative performance data for self-locking nuts are given in Table VIII.

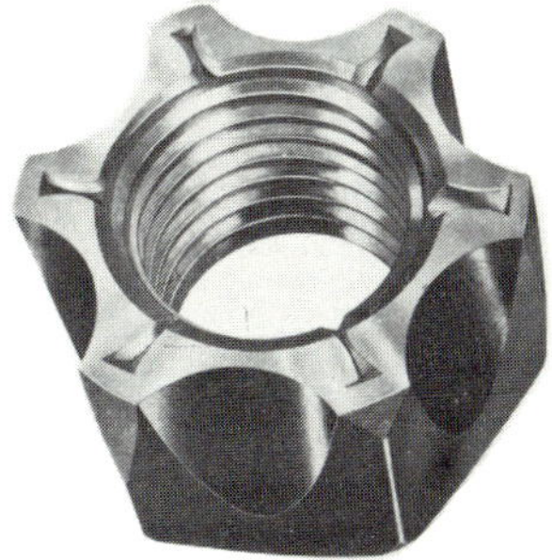

Fig 7 'Lester Star' deformed thread stiffnut features six flutes closed after thread tapping during manufacture to provide predetermined thread diameter reduction.

Fig 8 'Conelok' hi-hex stiffnuts.

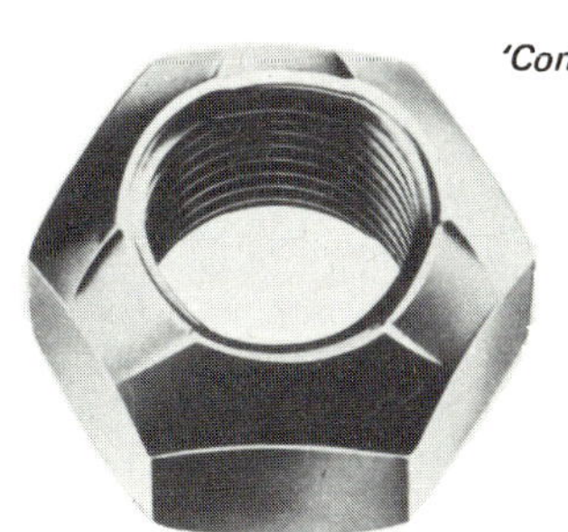

'Conelok' all metal stiffnut.

'Armalok' nylon inserted locknut.

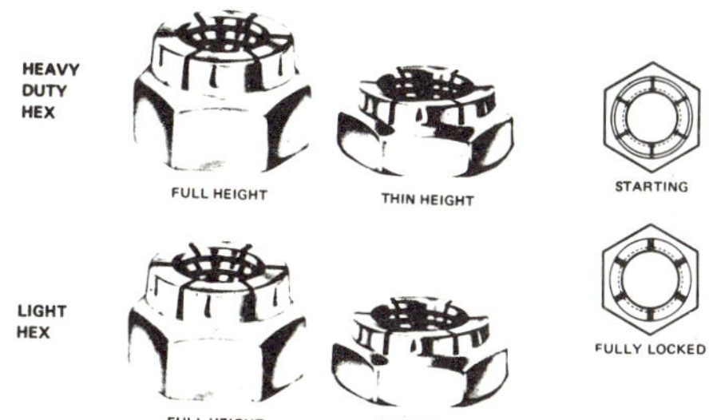

Fig 9 'Flexloc' stiffnuts (fully loaded in 1½ turns).

TABLE V – GKN 'AEROTIGHT' STIFFNUTS (INTERNAL SPRING TYPE)

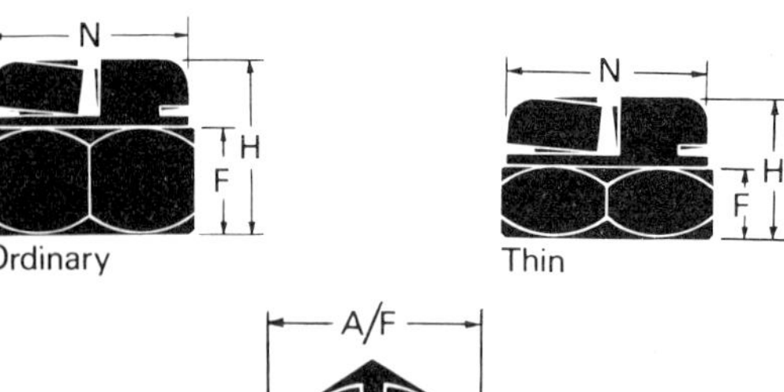

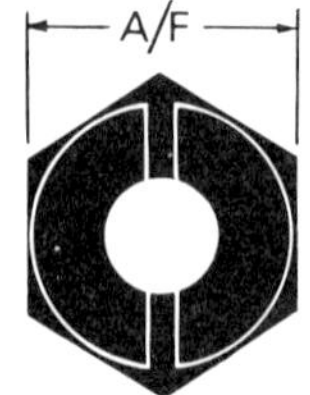

Thread	*Size*	*A/F*	*Dia N max*	*H max*		*F min*	
				Ordinary	*Thin*	*Ordinary*	*Thin*
		in	*in*	*in*	*in*	*in*	*in*
UNC/ UNF	No 4	0·250	0·177	0·172	0·140	0·080	0·053
	No 6	0·312	0·238	0·203	0·167	0·100	0·064
	No 8	0·344	0·301	0·243	0·201	0·130	0·088
	No 10	0·375	0·332	0·266	0·222	0·142	0·098
	$\frac{1}{4}$	0·438	0·425	0·312	0·245	0·175	0·108
	$\frac{5}{16}$	0·500	0·488	0·408	0·299	0·250	0·141
	$\frac{3}{8}$	0·563	0·550	0·459	0·355	0·287	0·183
	$\frac{7}{16}$	0·688	0·675	0·539	0·414	0·350	0·225
	$\frac{1}{2}$	0·750	0·738	0·621	0·475	0·412	0·266
	$\frac{5}{8}$	0·938	0·922	0·773	0·586	0·537	0·350
	$\frac{3}{4}$	1·125	1·110	0·934	0·705	0·662	0·433
	$\frac{7}{8}$	1·313	1·297	1·020	0·770	0·720	0·470
	1	1·500	1·480	1·172	0·880	0·845	0·553
BA	8	0·152	0·149	0·125	—	0·065	—
	6	0·193	0·183	0·157	0·125	0·080	0·053
	4	0·248	0·238	0·203	0·162	0·110	0·069
	3	0·282	0·272	0·231	0·186	0·128	0·083
	2	0·324	0·314	0·256	0·212	0·142	0·098
	1	0·365	0·355	0·289	0·240	0·163	0·114
	0	0·413	0·403	0·328	0·272	0·188	0·132
BSW/ BSF	$\frac{3}{16}$	0·324	0·314	0·256	0·212	0·142	0·098
	$\frac{1}{4}$	0·445	0·433	0·312	0·245	0·175	0·108
	$\frac{5}{16}$	0·525	0·513	0·383	0·299	0·225	0·141
	$\frac{3}{8}$	0·600	0·588	0·459	0·355	0·287	0·183
	$\frac{7}{16}$	0·710	0·698	0·539	0·414	0·350	0·225
	$\frac{1}{2}$	0·820	0·808	0·621	0·475	0·412	0·266
	$\frac{5}{8}$	1·010	0·995	0·773	0·586	0·537	0·350
	$\frac{3}{4}$	1·200	1·185	0·934	0·705	0·662	0·433
	$\frac{7}{8}$	1·300	1·285	1·020	0·770	0·720	0·470
	1	1·480	1·460	1·172	0·880	0·845	0·553

TABLE VI – TORQUE & LOAD FIGURES FOR 'FLEXLOC' STIFFNUTS

Size	Threads per in	40 000lb/in^2 preload* Tightening torque (in-lb)	40 000lb/in^2 preload* Load (in lbs)	90 000lb/in^2 preload* Tightening torque (in-lb)	90 000lb/in^2 preload* Load (in lbs)
1/4	20 28	50 60	1 270 1 450	115 130	2 860 3 275
5/16	18 24	100 110	2 100 2 300	230 250	4 720 5 200
3/8	16 24	170 190	3 100 3 500	385 425	7 000 7 900
7/16	14 20	260 280	4 270 4 760	575 630	9 600 10 700
1/2	13 20	375 420	5 700 6 400	840 940	12 800 14 400
9/16	12 18	510 580	7 300 8 100	1 150 1 300	16 400 18 300
5/8	11 18	690 780	9 000 10 200	1 550 1 750	20 300 23 000
3/4	10 16	1 100 1 200	13 400 14 900	2 400 2 650	30 100 33 600
7/8	9 14	1 500 1 600	18 500 20 200	3 300 3 650	41 600 45 900
1	8 12 14	2 000 2 150 2 200	24 200 26 500 26 700	4 400 4 800 4 850	54 500 59 700 60 200
1.1/8	7 12	2 500 2 800	30 500 34 200	5 700 6 300	68 700 77 000
1.1/4	7 12	3 400 3 700	38 800 42 900	7 600 8 300	87 300 96 600
1.3/8	6 12	4 200 4 800	46 200 52 500	9 500 10 800	103 900 118 300
1.1/2	6 12	5 600 6 300	55 700 63 100	12 600 14 200	126 500 142 300
1.5/8	5½	7 200	63 000	16 200	141 700
1.3/4	5	10 000	76 000	22 400	171 000
1.7/8	5	13 200	88 700	29 600	199 300
2	4½	17 300	100 000	39 000	225 000

*Preloads – Unless limited by strength of screw, 40 000lb/in^2 applies to Thin Height; 90 000lb/in^2 applies to Full Height 'Flexloc' Nuts

TABLE VII – 'CONELOK' LOCKNUTS

UNC and UNF thread series

Size (in)	1/4	5/16	3/8	7/16	1/2	9/16	5/8	3/4
Width across flats	0.437	0.500	0.562	0.687	0.750	0.875	0.937	1.125
Width across cones	0.489	0.557	0.628	0.768	0.840	0.982	1.051	1.240
Height	0.226	0.273	0.337	0.385	0.448	0.496	0.559	0.665
Body length	0.152	0.190	0.250	0.265	0.326	0.350	0.412	0.463

All dimensions in inches

BSW and BSF thread series

Width across flats	0.437	0.516	0.594	0.703	0.812	0.906	1.000
Width across cones	0.505	0.598	0.683	0.810	0.937	1.053	1.154
Height	0.257	0.309	0.360	0.416	0.508	0.538	0.617
Body length	0.180	0.216	0.255	0.286	0.352	0.371	0.435

All dimensions in inches

ISO metric thread series

Size	M3	M4	M5	M6	M8	M10	M12	M14	M16	M18	M20	M24
Width across flats	5.5	7.0	8.0	10.0	13.0	17.0	19.0	22.0	24.0	27.0	30.0	36.0
Width across cones	6.08	7.74	8.87	11.05	14.38	18.90	21.10	24.49	26.75	30.14	33.53	39.98
Height	3.0	4.0	5.0	6.0	8.0	10.0	12.0	14.0	16.0	18.0	20.0	24.0
Body length	2.56	3.02	3.32	4.32	5.82	6.78	8.78	9.78	10.73	13.73	14.73	15.73

All dimensions in millimetres

TABLE VIII – 'CLEVELOC' PREVAILING TORQUE LOCKNUTS

'CLEVELOC' Prevailing Torque Locknuts
Grade A – mild steel for locks up to 90 000 lb/in^2
Grade B – mild steel for locks up to 120 000 lb/in^2
Grade C – heat treated steel for locks up to 150 000 lb/in^2
Grade 8 – ISO metric locknuts

GRADES A & B Thread Size	4BA 6 UNC	8 UNC	2BA 10 UNF	¼	5/16	3/8	7/16	½	9/16	5/8	¾	7/8	1
Installation torque lb/in (max)	8.0	12	13	30	60	80	100	150	204	300	420	600	840
Prevailing torque, 1st removal, highest reading min lb/in	1.5	2.0	2.5	5.0	8.0	12	17	22	30	39	58	88	120
Prevailing torque, 1st removal, lowest reading min lb/in	0.5	1.0	1.5	2.5	4.0	5.0	7.5	10	15	17.5	25	40	60
Prevailing torque, 5th removal, highest reading min lb/in	1.0	1.5	2.0	3.5	5.5	8.5	12	15	21	27	41	62	84
Prevailing torque, 5th removal, lowest reading min lb/in	0.5	0.5	1.0	1.5	2.5	4.0	5.0	7.5	10	12.5	20	30	40

GRADE C Thread Size	¼	5/16	3/8	7/16	½
Installation torque lb/in (max)	40	80	110	135	204
Prevailing torque, 1st removal, highest reading min lb/in	6.0	10.5	16	23	30
Prevailing torque, 1st removal, lowest reading min lb/in	3.0	5.0	7.5	10	15
Prevailing torque, 5th removal, highest reading min lb/in	4.5	7.5	11.5	16	20
Prevailing torque, 5th removal, lowest reading min lb/in	2.0	3.0	5.0	7.5	10

PREVAILING TORQUE is measured with a minimum of two bolt threads protruding, the nut in motion and no load on the bolt.

FIRST AND FIFTH REMOVAL torque figures are taken after one loaded cycle to 75% bolt proof load utilising a conventional load measuring device.

GRADE 8 Thread Size	M3	M4	M5	M6	M8	M10	M12	M14	M16	M18	M20	M24
Installation torque Nm (max)	0.45	0.9	1.0	3.0	6.0	10.5	15.5	23.5	31.5	42.0	54.0	80.0
Prevailing torque, 1st removal, highest reading min Nm	0.12	0.18	0.29	0.45	0.85	1.5	2.3	3.3	4.5	6.0	7.5	11.5
Prevailing torque, 1st removal, lowest reading min Nm	0.06	0.09	0.14	0.2	0.4	0.7	1.0	1.5	2.0	3.0	3.5	5.5
Prevailing torque, 5th removal, highest reading min Nm	0.08	0.12	0.23	0.3	0.6	1.0	1.6	2.3	3.0	4.2	5.3	8.0
Prevailing torque, 5th removal, lowest reading min Nm	0.04	0.06	0.1	0.15	0.3	0.5	0.8	1.0	1.5	2.0	2.5	4.0

'CLEVELOC' ISO METRIC PREVAILING TORQUE LOCKNUTS

Size	Width across Flats Max	Width across Flats Min	Thickness Max	Thickness Min
M3 x 0.5	5.5	5.38	4.0	3.7
M4 x 0.7	7.0	6.85	4.0	3.7
M5 x 0.8	8.0	7.85	5.0	4.7
M6 x 1.0	10.0	9.78	6.0	5.5
M8 x 1.25	13.0	12.73	7.5	7.0
M10 x 1.5	17.0	16.73	9.0	8.5
M12 x 1.75	19.0	18.67	11.0	10.5
M14 x 2.0	22.0	21.67	12.0	11.5
M16 x 2.0	24.0	23.67	14.0	13.5
M18 x 2.5	27.0	26.67	18.0	17.5
M20 x 2.5	30.0	29.67	20.0	19.5
M24 x 2.5	36.0	35.38	24.0	23.5

TABLE IX – 'TORQLOK' PREVAILING TORQUE LOCKNUTS

Imperial

P Type SIZE	Thickness ±·010	Approx. weight lbs/100	ACROSS FLATS	T Type SIZE	Thickness ±·010	Approx. weight lbs/100
$\frac{1}{4}$	·305	0·87	·437	—	—	—
$\frac{5}{16}$	·370	1·50	·500	$\frac{5}{16}$	·300	1·20
$\frac{3}{8}$	·435	2·00	·562	$\frac{3}{8}$	·330	1·40
$\frac{7}{16}$	·500	3·50	·687	$\frac{7}{16}$	·390	2·40

dimensions in inches

Metric

P Type SIZE	Thickness ±·25	Approx. weight kg/100	Approx. weight lbs/100	ACROSS FLATS	T Type SIZE	Thickness ±·25	Approx. weight kg/100	Approx. weight lbs/100
M 2·5	4·10	0·02	0·05	5·00	M 2·5	3·40	0·02	0·05
M 3	4·10	0·03	0·07	5·50	M 3	3·40	0·02	0·06
M 4	5·60	0·07	0·16	7·00	M 4	4·40	0·06	0·13
M 5	7·10	0·16	0·35	8·00	M 5	5·80	0·15	0·33
M 6	7·50	0·25	0·55	10·00	M 6	6·35	0·24	0·53
M 8	9·90	0·57	1·26	13·00	M 8	8·30	0·52	1·15
M 10	11·60	1·25	2·75	17·00	M 10	9·60	1·05	2·31
M 12	14·40	1·88	4·15	19·00	M 12	12·00	1·72	3·79

All dimensions in millimetres

also in BA, UNF and UNC threads

TABLE X – TORQUE-TENSION REQUIREMENTS FOR AMERICAN UNIFIED THREAD PREVAILING TORQUE NUTS (Grade C) (Industrial Fasteners Institute of America)

Coarse Thread Series

Nom Size and Threads per Inch	Proof Load	Clamp[1] Load	Prevailing Torque					Locknut Tightening Torque ft lb	
			First Installation in lb	First Removal		Fifth Removal			
				Highest Reading min	Lowest Reading min	Highest Reading min	Lowest Reading min		
	lb	lb	Max	in lb	in lb	in lb	in lb	Max	Min
1/4–20	4750	2850	40	6.0	3.0	4.5	2.0	10.5	7.0
5/16–18	7850	4700	80	10.5	5.0	7.5	3.0	16.0	11.0
3/8–16	11600	6950	110	16.0	7.5	11.5	5.0	29.0	21.0
7/16–14	16000	9600	135	23.0	10.0	16.0	7.5	44.0	32.0
			ft lb						
1/2 – 13	21300	12800	17	30.0	15.0	20.0	10.0	66.0	49.0
9/16–12	27300	16400	25	40.0	20.0	28.0	12.5	95.0	70.0
5/8–11	33900	20300	35	52.0	25.0	36.0	15.0	122.5	90.0
3/4–10	50100	30100	45	78.0	35.0	54.0	25.0	210.0	155.0
Fine Thread Series									
1/4–28	5450	3250	40	6.0	3.0	4.5	2.0	10.5	7.0
5/16–24	8700	5200	80	10.5	5.0	7.5	3.0	17.0	12.0
3/8–24	13200	7900	110	16.0	7.5	11.5	5.0	30.0	22.0
7/16–20	17800	10700	135	23.0	10.0	16.0	7.5	45.0	33.0
1/2–20	24000	14400	17	30.0	15.0	20.0	10.0	69.0	51.0
9/16–18	30500	18300	25	40.0	20.0	28.0	12.5	100.0	75.0
5/8–18	38400	23000	35	52.0	25.0	36.0	15.0	130.0	95.0
3/4–16	56000	33600	45	78.0	35.0	54.0	25.0	210.0	155.0

Note: [1] *Clamp loads equal 75% of the proof loads specified for SAE J429 Grade 8 and ASTM A354 Grade BD bolts.*

'Spring-Stop' self-locking nut features an integral steel spring disc. When the screw is rotated in the nut the disc tabs are deformed elastically by the thread profile, producing friction locking. A feature is the low first installation torque. These nuts are available in a variety of forms, including caged nuts, anchor nuts, as well as plain and capped nuts. Suitable for temperatures up to 300°C.

TABLE XI – COMPARATIVE FATIGUE PERFORMANCE OF SELF-LOCKING NUTS*
(Nuts power wrenched to 75% bolt proof load then backed off ¼ turn and vibrated 1800 Hz)

Type of Nut	Cycles to produce:- 1 turn unwinding		run nut off bolt	
	max	min	max	min
Nylon insert	110000	8000	182000	10000
Deformed centre threads	10000	–	12000	5000
Deformed body-top				
(3 sides)	80000	12000	130000	16000
(2 sides)	119000	11000	135000	16000
Deformed thread-top				
(3 points)	10000	–	21000	3000
(2 points)	14000	–	20000	–

**Tests on a Sonntag Universal Fatigue machine by GKN.*

Torque Tension Relationship for Bolts

The general relationship between torque (T) applied to tighten a bolt and the resulting bolt load or tension (W) is

$$T = \frac{K\,D\,W}{12}$$

where D = nominal bolt diameter in inches
T = torque in lbf-ft
W = tension in lbf

$$T = \frac{K\,D\,W}{1\,000}$$

where D = nominal bolt diameter in mm
T = torque in kgf-mm
W = tension in kgf

K is a friction factor depending on the bolt/nut material and its condition and is also modified by the form of the thread and length of engagement (see *Detailed Analysis* later).

Typical values are:-

Material	K
Lubricated steel	0.11
Cadmium plated steel	0.15
Plain steel	0.20
Stainless steel	0.30

The maximum permissible bolt tension recommended is 80% of the yield point of the material. Thus for plain (self-colour) steel bolts with a typical yield point of 34 tons/in^2 in R quality material or equivalent working formulas are:-

T (lbf-ft) = 620 x D (inches)

T (kgf-m) = 4 x D (mm)

Factors for other materials are given in Table I. These calculations can be applied to ISO Metric (Coarse), UNC, UNF, BSW and BSF threads with a reasonable degree of accuracy. Average torque figures are also shown in Table II. It should be noted that all such figures derived from the basic formula are average values. If particularly accurate results are required, each batch of bolts should

TABLE I – STRESS CLASSIFICATION FOR BOLT MATERIALS

Material	Grade/ Specification	Minimum UTS		Yield Point	
		lb/in ^{2}	kg/mm^2	lb/in ^{2}	kg/mm^2
ISO Metric	4.6	57000	40.0	34000	24.0
	4.8	57000	40.0	45500	32.0
	5.6	71000	50.0	42500	30.0
	6.6	85000	60.0	51000	36.0
	8.8	114000	80.0	91000	64.0
	10.9	142000	110.0	128000	90.0
	12.9	170000	120.0	153500	108.0
	14.9	200000	140.0	180000	126.0
ISO Unified Inch	B (BS1768) (BS1083)	63000	44.0	–	–
	P (BS1768) (BS1083)	78400	55.0	47000	33.0
	R (BS1083)	100000	70.0	76000	53.0
	S(BS1768)	112000	78.4	90000	63.0
	T (BS1768)	123000	86.0	92000	64.0
	V (BS1768) (BS1083)	146000	102.0	116500	81.5
	X (BS1768) (BS1083)	168000	118.0	141000	99.0
Commercial steel bolts	SAE 1	55000	38.5	44000	30.8
Low carbon steel bolts	SAE 2				
up to ½in		69000	48.0	55000	38.5
½ – ¾in		64000	45.0	51000	35.7
¾ – 1in		55000	38.5	44000	30.8
Medium carbon steel bolts (graded and tempered)	SAE 5				
up to ¾in		120000	84.0	96000	67.2
¾ – 1in		115000	80.5	92000	64.4
Medium carbon alloy steel bolts (graded and tempered) up to 1½in	SAE 8	150000	105.0	120000	94.4

TABLE II – AVERAGE TORQUE FIGURES FOR UNIFIED & IMPERIAL STEEL THREADS *

Nominal diameter	UNC *A & B*	*S*	*V*	UNF *A & B*	*S*	*V*	UN 8TPI *S*	*V*	BSW *A & B*	*R*	*V*	BSF *A & B*	*R*	*V*
in	*lbf.ft*	*lbf.ft*	*lbf.ft*	*lbf.ft*	*lbf.ft*	*lbf.ft*	*lbf.ft*	*lbf.ft*	*lbf.ft*	*lbf.ft*	*lbf.ft*	*lbf.ft*	*lbf.ft*	*lbf.ft*
$\frac{1}{4}$	4	9	12	4	11	14			4	8	12	4	9	14
$\frac{5}{16}$	7	20	25	8	22	28			7	17	25	8	18	27
$\frac{3}{8}$	13	35	45	14	40	50			13	30	45	14	32	49
$\frac{7}{16}$	21	56	73	22	62	81			21	47	72	22	51	78
$\frac{1}{2}$	31	85	111	36	96	125			31	70	107	34	77	118
$\frac{9}{16}$	46	123	160	51	137	180			46	105	160	50	112	173
$\frac{5}{8}$	64	170	212	72	192	238			63	145	210	68	155	224
$\frac{3}{4}$	113	300	393	126	335	435			113	255	390	118	268	410
$\frac{7}{8}$	182	485	632	200	535	695			182	410	628	191	432	660
1	274	730	950	297	795	1 000			272	620	945	288	650	995
$1\frac{1}{8}$	388	1 035	1 345	433	1 160	1 500	1 075	1 390	388	875	1 330	410	930	1 420
$1\frac{1}{4}$	546	1 460	1 900	604	1 610	2 090	1 500	1 950	550	1 235	1 890	575	1 300	1 990
$1\frac{3}{8}$	720	1 920	2 480	815	2 150	2 800	2 040	2 650	–	–	–	765	1 730	2 630
$1\frac{1}{2}$	950	2 540	3 300	1 068	2 840	3 700	2 690	3 500	950	2 150	3 280	1 005	2 270	3 480

**To produce a preload of 80% of the yield stress of the bolt or screw*

be calibrated to take into account possible variations in manufacture affecting the coefficient of friction between thread and nut. Tightening torque can also be affected by the finish and condition of the bolt thread and nut faces, and the faces of the surfaces being clamped.

Mechanical Considerations

When a bolt is being tightened it is subject to torsion and tension, the majority of the applied torque being used to overcome friction. When tightened, the bolt is subject only to residual tension produced by the stretching of the bolt, the value of which is determined by the degree of tightening. To maintain a tight joint the clamping load provided by this tension must be greater than any external loads subsequently applied. Also any subsequent deformation of the surface against which the bolt head (or nut) presses must be less than the stretch of the bolt when tightened, otherwise bolt tension will be lost.

A certain minimum preload is necessary to maintain effective clamping – ie a certain degree of elastic stretch must be maintained in the bolt. The higher the preload the greater the locking effect to resist unscrewing, and also the greater the resistance to fatigue effects.

Once tightened, the bolt will not be subject to any stress other than its preload, unless the external load exceeds this figure. Tensional stresses in excess of the preload will be additional tensile stress on the bolt. The actual preload figure used can be a normal minimum up to a practical maximum represented by 80% of the yield point of the bolt material.

Such conditions apply only in the case of rigid joints, ie metal-to-metal contact where the elastic modulus of the bolt is similar to that of the clamped materials. If the material being bolted is much softer, or the joint is filled with a gasket or certain types of backwashers, the joint is non-rigid and the preload requirements are modified. The preload on non-rigid joints is normally adjusted to be the maximum possible safely without crushing the softer material, or to produce and maintain the required degree of compression of a gasket, etc. Bolts will then be subject to additional torsional stress if the external load exceeds the preload established on tightening. Also, with non-rigid joints, there is the possibility of external forces providing shear loads on the bolt — see chapter on *Rivets* for determining the strength of bolts in shear.

Approximate factors which can be used to establish maximum tightening torques when clamping different materials are given below. These factors apply to the basic torque formulas given previously.

Material	Factor
Brass	1.0 (same as steel)
Copper	0.75
Zinc	0.68
Aluminium & aluminium alloy	0.57
Magnesium & magnesium alloy	0.57
Fibre	0.43

Detailed Analysis

A more detailed evaluation of the parameters affecting the value of the constant 'K' in the basic torque-tension relationship can be undertaken on the basis that this is a torque coefficient rather than a friction coefficient. A complete expression for K is:-

$$K = \mu B \frac{R_B}{D} + \frac{R_T}{D} \left(\frac{\mu T \sec\beta + \tan C}{1 - \mu T \sec\beta \tan C} \right)$$

where μB = coefficient of friction at bearing face of bolt or nut

μT = coefficient of friction at thread contact surfaces

R_B = effective radius of action of friction forces on bearing face

R_T = effective radius of action of friction forces on thread surfaces

β = thread angle/2

C = thread helix angle

Since the quantity $\mu T \sec B \tan C$ is normally of the order of unity (within 2%) in the case of threaded fasteners, this formula simplifies to:-

$$K = \mu B \frac{R_B}{D} + \mu T \frac{R_T}{D} \sec\beta + \frac{R_T}{D} \tan C$$

which is conveniently expressed in the form

$$K = K_1 + K_2 + K_3$$

TABLE III – AVERAGE TORQUE FIGURES FOR METAL THREADS
(Coarse threads except where noted)
Torque in kgf/m

Grade / Nominal dia.	4.6 (hot forged)	4.8 (cold forged)	5.6	6.6	8.8	8.8 (fine)	10.9	12.9	14.9
M6	0.45	0.60	0.56	0.66	1.19	1.31	1.66	2.00	2.4
M8	1.10	1.45	1.36	1.62	2.90	3.10	4.00	5.00	5.8
M10	2.15	2.87	2.70	3.20	5.74	6.05	8.00	9.75	12.0
M12	3.75	5.00	4.70	5.60	10.00	10.93	14.0	17.0	20.0
M14	6.00	8.00	7.50	9.00	16.00	17.00	22.5	27.0	32.0
M16	9.50	12.50	11.75	14.00	25.00	26.00	35.0	42.5	50.0
M20	18.00	24.00	22.50	27.00	48.00	54.00	67.0	81.5	96.0
M22	25.00	33.00	31.00	37.00	66.00	74.00	92.5	112.0	132.0
M24	31.50	42.00	39.50	47.00	84.00	91.00	118.0	143.0	168.0

Basically K_1 represents the bearing friction to be overcome by tightening torque, K_2 the thread friction and K_3 the modifying effect of the actual thread form, or the amount of torque actually producing bolt tension. Respective values are likely to be of the order of 50%, 40% and 10%, showing that only a small proportion of the tightening torque is effective (or useful torque) in producing tension in the assembly.

The effective radii also need defining. In the case of the bearing face, the numerical value of R_B is given by

$$R_B = \frac{2}{3} \left(\frac{R_o^3 - R_i^3}{R_o^2 - R_i^2} \right)$$

where R_o = outer bearing radius
R_i = inner bearing radius

The effective thread surface radius is approximately equal to that of the pitch radius of the thread, provided contact pressure is uniform. If not, R_T can vary between the inner and outer radii of contact. Another characteristic of R_T is that it will differ between fine and coarse threads. It will tend to be larger with fine threads (of the same nominal diameter) and also more subject to variation because the thread angle (β) is more likely to be deformed under load. In practice, however, there is normally little difference between the value of K for fine and coarse threads, although there may be a fairly substantial difference in K_3. This is because of the relatively small contribution of K_3 to the final value of K.

Typical contributed values for American size threads are given in Table IV. These also serve to illustrate the fact that the 'typical' value of K = 0.2, normally used in the simpler empirical formula, is justified for most screw threads.

TABLE IIIA — RECOMMENDED TIGHTENING TORQUES (GKN)

				DIAMETERS													
				1/4″	5/16″	3/8″	7/16″	1/2″	9/16″	5/8″	3/4″	7/8″	1″	1 1/8″	1 1/4″	1 3/8″	1 1/2″
UNC	THREADS PER INCH			20	18	16	14	13	12	11	10	9	8	7	7	6	6
	TENSILE STRESS AREA		in²	·0324	·0532	·0786	·1078	·1438	·184	·229	·338	·467	·612	·771	·978	1·166	1·418
	Bolt Tension Based on 85% of Proof Load	A* (Mild Steel)	ton f	0·39	0·64	0·95	1·31	1·74	2·23	2·77	4·09	5·66	7·41	9·34	11·85	14·12	17·18
		S		1·05	1·72	2·54	3·48	4·65	5·94	7·40	10·91	15·08	19·77	24·90	31·59	37·66	45·81
		T		1·13	1·85	2·74	3·76	5·01	6·41	7·98	11·78	16·27	21·33	26·87	34·09	40·63	49·43
		V		1·36	2·34	3·31	4·53	6·05	7·74	9·63	14·22	19·65	25·76	32·44	41 15	49·06	59·67
	Approx Torque	A (Mild Steel)	lb f ft	3.7	7.5	13	21	32	47	65	115	185	280	390	560	720	960
		S		9.8	20.0	36	55	90	125	175	310	490	740	1050	1450	1950	2550
		T		10.0	22.0	38	60	95	135	185	320	540	800	1150	1600	2100	2750
		V		13.0	26.0	46	75	115	160	220	400	640	960	1350	1900	2500	3350
UNF UP TO 1″ DIA UN–8 TPI FOR 1.1/8″ DIA AND ABOVE	THREADS PER INCH			28	24	24	20	20	18	18	16	14	12	8	8	8	8
	TENSILE STRESS AREA		in²	·0368	·0587	·0886	·1198	·1612	·205	·258	·375	·513	·667	·798	1·008	1·242	1·502
	Bolt Tension Based on 85% of Proof Load	A* (Mild Steel)	ton f	0·45	0·71	1·05	1·45	1·95	2·48	3·13	4·54	6·21	8·08	9·67	12·21	15·04	18·19
		S		1·19	1·89	2·86	3·87	5·21	6·62	8·33	12·11	16·57	21·54	25·77	32·55	40·12	48·51
		T		1·28	2·05	3·09	4·17	5·62	7·15	8·99	13·07	17·88	23·25	27·81	35·13	43·28	52·34
		V		1·55	2·47	3·73	5·04	6·78	8·63	10·85	15·78	21·58	28·06	33·58	42·41	52·26	63·20
	Approx Torque	A (Mild Steel)	lb f ft	4.1	8.3	15	24	36	50	75	125	200	300	410	560	780	1000
		S		11.0	22.0	40	65	95	140	195	340	540	800	1100	1500	2050	2700
		T		12.0	24.0	43	70	105	150	200	360	580	860	1150	1650	2200	2950
		V		14.0	29.0	50	80	125	180	250	440	700	1050	1400	2000	2700	3550
BSW	THREADS PER INCH			20	18	16	14	12	12	11	10	9	8	7	7	–	6
	TENSILE STRESS AREA		in²	·0320	·0527	·0779	·1069	·1385	·183	·227	·336	·463	·608	·767	·973	–	1·409
	Bolt Tension Based on 85% of Proof Load	A* (Mild Steel)	ton f	0·39	0·64	0·94	1·29	1·68	2·22	2·75	4·07	5·61	7·36	9·29	11·79	–	17·07
		R*		0·88	1·45	2·14	2·93	3·80	5·02	6·23	9·22	12·71	16·69	21·06	26·71	–	38·68
		T*		1·12	1·84	2·71	3·73	4·83	6·38	7·91	11·71	16·14	21·19	26·73	33·91	–	49·10
		V*		1·35	2·22	3·28	4·50	5·83	7·70	9·55	14·14	19·48	25·58	32·27	40·94	–	59·28
	Approx Torque	A (Mild Steel)	lb f ft	3.6	7.4	13	21	31	47	65	115	185	270	390	540	–	950
		R		8.4	17.0	30	48	70	105	145	260	415	620	880	1200	–	2175
		T		10.0	21.0	38	60	90	135	185	320	520	800	1100	1600	–	2750
		V		13.0	26.0	46	75	110	160	220	400	640	960	1350	1900	–	3300
BSF	THREADS PER INCH			26	22	20	18	16	16	14	12	11	10	9	9	8	8
	TENSILE STRESS AREA		in²	·0356	·0567	·0839	·1158	·152	·198	·243	·352	·487	·642	·8147	1·027	1·237	1·496
	Bolt Tension Based on 85% of Proof Load	A* (Mild Steel)	ton f	0·43	0·69	1·07	1·40	1·84	2·40	2·94	4·26	5·90	7·78	9·87	12·44	14·98	18·12
		R*		0·98	1·56	2·30	3·18	4·17	5·44	6·67	9·66	13·37	17·63	22·37	28·20	33·96	41·07
		T*		1·24	1·98	2·92	4·04	5·30	6·90	8·47	12·27	16·97	22·37	28·39	35·79	43·11	52·14
		V*		1·50	2·39	3·53	4·87	6·40	8·33	10·22	14·81	20·49	27·01	34·28	43·21	52·05	62·94
	Approx Torque	A (Mild Steel)	lb f ft	4·0	8·0	14	23	34	50	70	120	195	290	410	580	760	1000
		R		9·1	18	32	50	80	115	155	270	435	660	940	1300	1750	2300
		T		12	23	41	65	100	145	200	340	560	840	1200	1650	2200	2900
		V		14	28	50	80	120	170	240	420	660	1000	1450	2000	2650	3550

*THEORETICAL PROOF LOADS BASED ON 95% OF YIELD STRESS

TABLE IV – THEORETICAL TORQUE COEFFICIENTS FOR UNIFIED THREADS

Size	K_1	K_2	K_3	K	Measured K (average) High-point torque	Measured K (average) Mid-point torque
1/4 -20	0.1055	0.0753	0.0318	0.213	0.243	0.267
1/4 -28	0.1055	0.0786	0.0227	0.207	0.216	0.231
5/16-18	0.0993	0.0766	0.0284	0.204	0.206	0.186
5/16-24	0.0993	0.0790	0.0212	0.200	0.194	0.183
3/8 -16	0.0950	0.0772	0.0265	0.199	0.200	0.247
3/8 -24	0.0950	0.0802	0.0176	0.193	0.192	0.234
7/16-14	0.0980	0.0772	0.0260	0.201	0.217	0.224
7/16-20	0.0980	0.0800	0.0181	0.196	0.194	0.190
1/2 -13	0.0950	0.0780	0.0245	0.198	0.205	0.158
1/2 -20	0.0950	0.0811	0.0159	0.192	0.167	0.205
9/16-12	0.0970	0.0781	0.0235	0.199	0.194	0.214
9/16-18	0.0970	0.0811	0.0157	0.194	0.196	0.207
5/8 -11	0.0950	0.0783	0.0231	0.196	0.178	0.196
5/8 -18	0.0950	0.0816	0.0141	0.191	0.183	0.175
3/4 -10	0.0950	0.0790	0.0212	0.195	0.169	0.172
3/4 -16	0.0950	0.0819	0.0132	0.190	0.170	0.180
7/8 - 9	0.0950	0.0793	0.0201	0.194	0.181	0.194
7/8 -14	0.0950	0.0819	0.0130	0.190	0.171	0.178
1 - 8	0.0950	0.0795	0.0199	0.194	0.188	0.204
1 -14	0.0950	0.0826	0.0109	0.189	0.161	0.167
Average				0.197	0.191	0.201

Practical Considerations

Alternative methods which may be used to assess the degree of tightening or preload produced in a bolted assembly are:-

(i) Feel – ie the operator judges when the assembly is tightened to an optimum degree. Accuracy can vary widely and even with an experienced operator there may be differences in preload of the order of up to plus or minus 40%.

(ii) Angle-controlled tightening – the basis of this being that the bolt is tightened to a snug fit and then a final specific amount of rotation is applied to the nut. Techniques can vary from the snug fit being assessed as 'finger tight' to 'wrench tight', with the final nut turns varying accordingly. It will also vary with the size and strength of the bolt and the pitch of the threads. With a properly established technique, this method can, however, produce repeatability of preload within plus or minus 15%. See also (viii) on page 205.

(iii) Torque wrench – to apply a predetermined torque figure. Surprisingly, this can produce variations as high as plus or minus 25% in preload, although this can be reduced substantially if surface finishes and condition of bearing and thread faces are consistent and torque wrenches are accurately calibrated.

(iv) Determination of elongation of the bolt. This is one of the most accurate methods since the preload is directly proportional to the degree of extension produced in the bolt when tightened. Accuracy achieved depends on the method of measuring elongation, and can be better than plus or minus 5%. However the method is tedious and costly.

(v) Measurement of tensile stress by strain gauges. This is also a very accurate method but is even more costly. It would therefore only be considered for especially critical applications. A considerably less expensive — but less accurate — method of measuring tensile stress in the bolt is with special load-indicating washers. Accuracy achieved with these can be of the order of plus or minus 10%.

Methods (iv) and (v) have the advantage that they are independent of the condition of the bolt and nut and bearing faces.

Bolts tightened to produce a tensile stress beyond the yield point of the material are no longer subject to a simple torque-tension relationship. Galling of the contact surfaces is likely to occur as well as deformation of the thread angle, both producing a substantial increase in K as the failure point is approached. This increase can be highly variable and so torque to produce failure cannot be used as a measure of the strength of the bolt. Also failure itself may occur by stripping of the threads or shearing of the bolt.

(vi) Load-indicating bolts — These are bolts with a special head design incorporating bearing pads of pre-determined performance. As the bolt is approaching fine tightness the pads deform elastically and the remaining gap is a measure of the bolt tension. Thus a specific minimum bolt tension can be expressed in terms of residual gap measured by a feeler gauge.

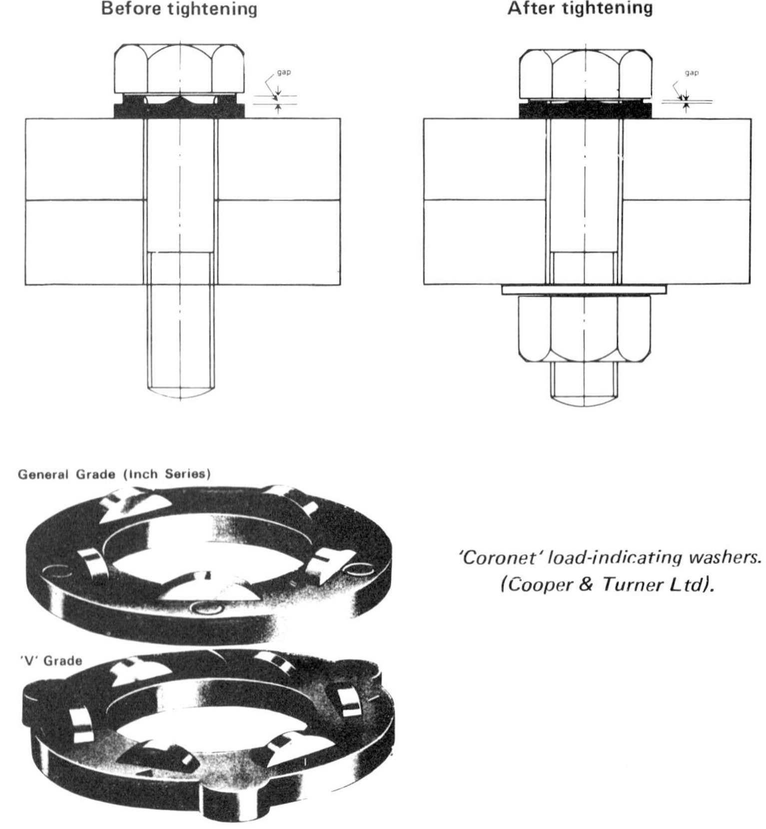

'Coronet' load-indicating washers. (Cooper & Turner Ltd).

Bolts of this type may require tightening on to a hardened washer to avoid false readings due to indentation of the parent metal. Other designs allow for the extent to which the bearing pads will have bitten into the surface at the predetermined bolt tension.

Note: Torque-tightening can be particularly variable or misleading in the case of plated threads.

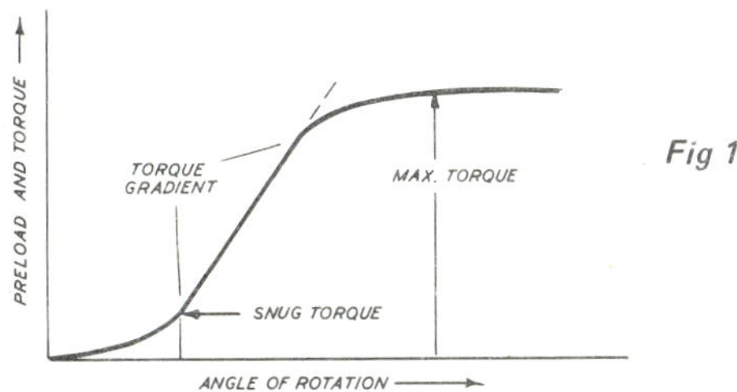

Fig 1

(vii) Gradient controlled tightening is based on the assumption that preload is a function of the angle of rotation of the nut, part of which relationship is linear, ie has a constant gradient. At some point linearity is lost and the gradient flattens out – Fig 1. This corresponds with the approach of the yield point of the bolt or screw.

Basically, in gradient-controlled tightening, torque and angle of nut rotation are read by sensors during the tightening process and the results are fed to an electronic circuit. This circuit differentiates these readings to determine the gradient and sets itself to detect when there is a significant drop in it (ie indicating the approach of the yield point of the material). At this point a control signal is initiated shutting off the power circuit.

Although it is in a relatively early stage of practical development, gradient-controlled tightening is well suited to modern automated and semi-automatic wrenching operations and can achieve very precise preloads as preload sensing is independent of the effects of friction and other variable factors.

(viii) Angle-controlled tightening ('turn of the nut') is dependent on the following theoretical relationship

$$\text{elongation} = \frac{\phi P}{360^{\circ}}$$

where ϕ = angle of rotation of nut
P = pitch of thread

It is thus theoretically possible to calculate the amount of angular rotation of the nut (number of turns) to produce a specific amount of elastic elongation of the bolt, with a given thread pitch, starting with a rigid assembly (snug fit). However, this relationship is modified by elastic deformation of the thread and frictional effects. Also, accuracy is affected by the initial determination of what is a snug fit.

Despite these apparent limitations, very satisfactory specifications for angle-controlled tightening, expressed in terms of turns of the nut after a snug fit, can be established by practical lists and the method is quite widely used in America and in the German automobile engineering industry. It would appear to be most precise and consistent when the final tightening of the bolt stresses the bolt material beyond its yield point (as is common practice with powered drivers).

TABLE V – STANDARD TORQUE VALUES FOR AMERICAN INDUSTRIAL FASTENERS

Fastener	Grade Designation	Tensile Strength Minimum lb/in²	Material	Screw, Stud, or Bolt Shank Size or Diameter																				Torque figures are in foot pounds unless otherwise noted							
				2	3	4	5	6	8	10	1/4	5/16	3/8	7/16	1/2	9/16	5/8	3/4	7/8	1	1-1/8	1-1/4	1-3/8	1-1/2	1-5/8	1-3/4	1-7/8	2	2-1/4	2-1/2	2-3/4	3
CAP SCREWS	SAE 2 ASTM A-307 Steel	64000	Low carbon steel								6	11	19	30	45	66	93	150	202	300	474	659	884	1057	1448	1884	2336	2721	3117	4380	7319	9455
	SAE 3 Steel	100000	Medium carbon steel								9	17	30	47	69	103	145	234	372	551	872	1211	1624	1943	2660	3463	4659	5427	7226	8049	13450	17548
	ASTM A-449 SAE 5 Steel; ASTM 354BB Steel	105000	Medium carbon steel or low alloy heat treated								9	18	31	50	75	110	150	250	378	583	782	1097	1461	1748	2392	3114	4191	4504	6497	7144	12092	15775
	ASTM A-325*														100		200	355	525	790	1060	1495	1960	2600	*Torque specifications are for permanent fastenings on steel structures							
	ASTM A-354-BC steel	125000	Low alloy or med.carb. quenched tempered								11	20	34	54	81	119	167	269	427	644	1002	1392	1868	2234	3059	3982	5457	5749	8308	9255	15465	20176
	SAE 6 Steel; SAE 7 Steel	133000	Med carbon steel, quenched tempered; Med carbon alloy, quenched tempered, roll threaded								12.5	24	43	69	106	150	209	350	550	825	1304	1815	2434	2913	3985	5189	6980	7491	10825	14983	20151	2628[illegible]
	SAE 8 Steel	150000	Med.carbon alloy, quenched tempered								13	28	46	75	115	165	225	370	591	893	1410	1964	2633	3150	4311	5614	7550	8104	11710	16208	22440	28436
	A-354-BD A490*	150000	Med.carbon alloy, quenched tempered										55	90	138	198	270	444	709	1071	1692	2360	3159	3780	5173	6736	*Specifications are for permanent steel structures					
	Socket Head Cap Screw also NAS Aircraft Std.; NAS 144 Aircraft Std MS 20000 Mil Std	160000	High carbon alloy, quenched tempered								14	30	50	81	121	176	240	395	629	964	1523	2120	2843	3402	1655	6063	8154	8751	12645	17503	2354[illegible]	30709

	NAS 624 National Aircraft Std Steel	180000	High carbon alloy, quenched tempered								16	34	56	91	136	198	270	444	708	1085	1713	2385	3198	3827	5237	6821	9173	9845	14226	1969	26484	34548
	Aircraft No Number assigned Steel	220000	High carbon alloy, quenched tempered								19	41	69	111	166	232	330	534	865	1326	2094	2916	4009	4678	6401	8337	11212	11685	17387	24067	32368	42225
Studs	Studs Steel			As for Cap Screw figures – depending on grade																												
				in/lb	in/lb	in/lb	in/lb	in/lb	in/lb	in/lb	in/lb	in/lb	16																			
Set Screws	Socket Set Screws (Steel)	212000 to 225000	High carbon alloy, quenched tempered					9	16	30	70	140	18	29	43	63	100	146														
CAP & MACHINE SCREWS	Stainless Steel		18-8	2.5	4	5.2	8	9.6	20	23	75	132	20	31	43	57	92	124	194	259	390	480		703								
	Stainless Steel		316 Series	2.6	4	5.5	8	10	21	24	79	138	21	33	45	59	97	130	202	271	408	504		732								
	Yellow Brass	60000	CU 63 ZN 37	2	3.2	4.3	6.3	8	16	19	62	107	16	26	35	47	76	102	158	212	318	394		575								
	Silicon Bronze (Low) Type'B'	70000	CU 96 Min SI 1.5-2 ZN 1-5 Min	2.3	3.6	4.8	7.1	9	18	21	69	123	18	29	40	52	86	115	178	240	361	447		651								
	Aluminium 2024-T4	55000	CU 3.8-4.9 1.2-1.8 MN 0.3-0.9 AL Balance	1.4	2.1	2.9	4.2	5.3	11	14	45	80	12	19	26	34	60	80	125	166	251	308		450								
	Monel	82000	NI 67 CU 30 FE 1.4	2.5	4	5.3	7.8	10	20	26	85	149	22	35	49	65	110	149	229	310	470	575		840								
Sems and Mach Screw	Steel	55000	1010 Etc not heat treated	2.5	3.7	6	8	11	20	32	75	140																				
Sems	Heat Treated Steel	120000	1018 1022	4	5	7	10	14	25	35	85	195	in/lb 325																			

Bearing Stresses in Bolted Assemblies

The bearing stress in a bolted assembly is related to the bearing area in contact with the bolt head or nut. Bolt heads are normally flat hex, but nuts may be flat, single-chamfered or double-chamfered, so different considerations may apply.

Basically, the bearing area depends on the diameter of the bolt head (or effective diameter of the nut) and that of the hole – Fig 2. The former is standardised for the size of bolt, but the hole diameter is a variable factor depending on the clearance allowed for assembly. Excessive clearance can reduce the bearing area to such an extent that the preload applied to the reduced area exceeds the yield surface of the material being bolted up, with the result that the head indents the material – Fig 3. This can limit the preload which can be applied, or in extreme cases cause the material to flow to such an extent that the bolt head is pulled right into the hole, especially if the material being bolted is relatively soft.

Normal practice to safeguard against this is to fit a washer under the bolt head (or nut). This will have the immediate effect of distributing the bearing load over a larger surface and so reducing the bearing stress – Fig 4. However, if excessive clearance is present there may still be a concentration of stress at the edges of the hole, especially if the washer is relatively soft. With an oversize hole, therefore, it is always desirable to use a hard washer.

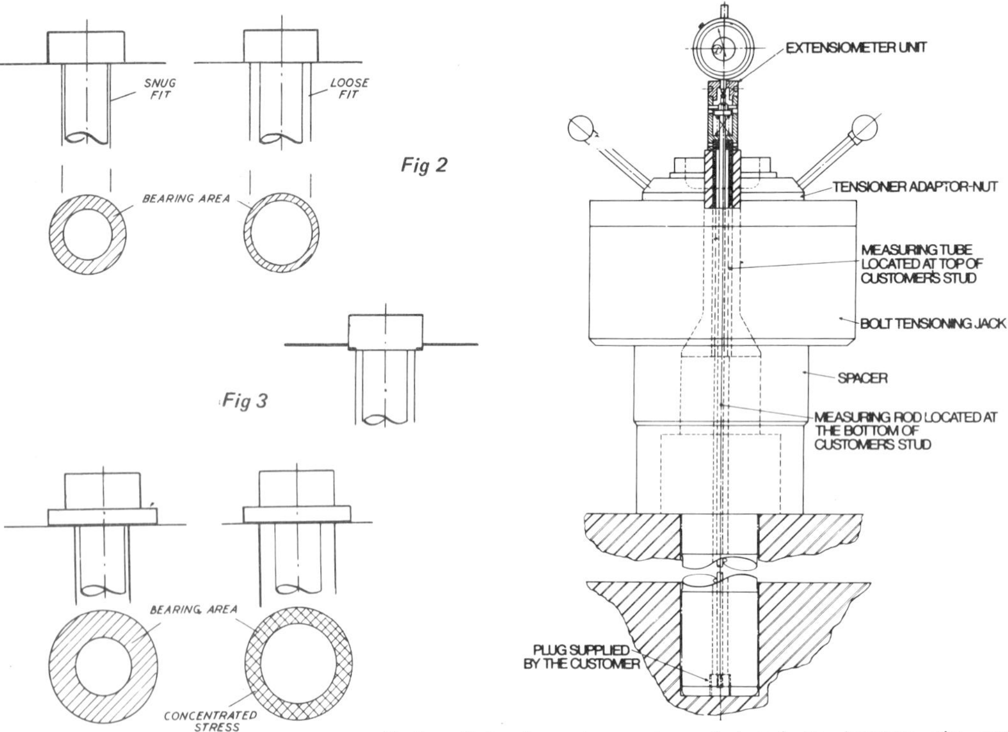

Madan Extensionmeter system designed to measure the actual extension of the stud/bolt while the load is being applied, capable of measuring extension to 0.0005in (0.012mm). It should be noted that it is necessary to use bolts or studs with a hollow centre to allow the measuring rod to be fitted. This is usually allowed for at the design stage.

TORQUE CONVERSION: kgf-m to lbf-ft **TABLE VI**

Kgf Metres	—	·1	·2	·3	·4	·5	·6	·7	·8	·9
—	—	·7233	1·4466	2·1699	2·8932	3·6165	4·3398	5·0631	5·7864	6·5097
1	7·233	7·956	8·679	9·462	10·126	10·849	11·572	12·295	13·019	13·742
2	14·466	15·189	15·912	16·636	17·359	18·082	18·806	19·529	20·252	20·976
3	21·699	22·422	23·145	23·869	24·592	25·315	26·039	26·762	27·485	28·209
4	28·932	29·655	30·378	31·102	31·825	32·548	33·272	33·995	34·718	35·442
5	36·165	36·888	37·611	38·335	39·058	39·781	40·505	41·228	41·991	42·675
6	43·398	44·121	44·844	45·568	46·291	47·014	47·738	48·461	49·184	49·908
7	50·631	51·354	52·077	52·801	53·524	54·247	54·971	55·694	56·417	57·141
8	57·864	58·587	59·310	60·034	60·757	61·480	62·204	62·927	63·650	64·374
9	65·097	65·820	66·543	67·267	67·990	68·713	69·437	70·160	70·883	71·607
10	72·330	—	—	—	—	—	—	—	—	—

TORQUE CONVERSION: lbf-ft to kgf-m

lbf-ft	—	1	2	3	4	5	6	7	8	9
—	—	·13826	·27651	·41477	·55302	·69128	·82953	·96778	1·10604	1·24430
10	1·3826	1·521	1·659	1·797	1·935	2·073	2·211	2·350	2·488	2·626
20	2·7651	2·903	3·041	3·179	3·317	3·456	3·594	3·732	3·870	4·009
30	4·1477	4·286	4·424	4·562	4·700	4·838	4·977	5·115	5·253	5·391
40	5·5302	5·669	5·807	5·945	6·083	6·221	6·360	6·498	6·636	6·774
50	6·9128	7·051	7·189	7·327	7·465	7·604	7·742	7·880	8·018	8·156
60	8·2953	8·434	8.572	8.710	8.848	8.987	9.125	9.263	9.401	9.540
70	9·6778	9·816	9·954	10·092	10·230	10·368	10·507	10·645	10·783	10·921
80	11·0604	11·199	11·337	11·475	11·613	11·752	11·891	12·029	12·167	12·305
90	12·4430	12·581	12·719	12·857	12·998	13·133	13·272	13·410	13·548	13·686
100	13·8255	—	—	—	—	—	—	—	—	—

TORQUE CONVERSION: kgf-m to lbf-in

Kgf Metres	—	·1	·2	·3	·4	·5	·6	·7	·8	·9
—	—	8·680	17·359	26·039	34·719	43·398	52·078	60·757	69·437	78·117
1	86·796	95·48	104·16	112·84	121·52	130·20	138·88	147·56	156·24	164·92
2	173·592	182·27	190·95	199·63	208·31	216·99	225·67	234·35	243·03	251·71
3	260·388	269·07	277·75	286·43	295·11	303·79	312·47	321·15	329·83	338·51
4	347·185	355·87	364·55	373·23	381·91	390·59	399·27	407·95	416·63	425·31
5	433·981	442·66	451·34	460·02	468·70	477·38	486·06	494·74	503·42	512·10
6	520·78	529·46	538·14	546·82	555·50	564·18	572·86	581·54	590·22	598·90
7	607·57	616·25	624·93	633·61	642·29	650·97	659·65	668·33	677·01	685·69
8	694·37	703·05	711·73	720·41	729·09	737·77	746·45	755·13	763·81	772·49
9	781·17	789·85	798·53	807·21	815·89	824·57	833·25	841·93	850·61	859·29
10	867·96	—	—	—	—	—	—	—	—	—

TORQUE CONVERSION: lbf-in to kgf-m

lbf-in	—	1	2	3	4	5	6	7	8	9
—	—	·01152	·02304	·03456	·04609	·05761	·06913	.08065	·09217	·10369
10	·11521	·1267	·1382	·1497	·1612	·1727	·1842	·1957	·2074	.2187
20	·23043	·2419	·2534	·2649	·2764	·2879	·2994	·3110	·3225	·3340
30	·34564	·3571	·3686	·3801	·3916	·4032	·4147	·4262	·4378	·4493
40	·46085	·4724	·4839	·4954	·5069	·5185	·5300	·5415	·5530	·5645
50	·57606	·5876	·5991	·6106	·6222	·6337	·6452	·6567	·6683	·6798
60	·69128	·7028	·7143	·7258	·7374	·7489	·7604	·7720	·7836	·7951
70	·80649	·8180	·8295	·8410	·8526	·8641	·8756	·8871	·8987	·9102
80	·92170	·9332	·9447	·9562	·9678	·9793	·9908	1·0033	1·0139	1·0254
90	1·03691	1·0484	1·0599	1·0714	1·0830	1·0945	1·1060	1·1175	1·1291	1·1403
100	1·15212	—	—	—	—	—	—	—	—	—

Actual bolt head (or nut) diameter and mating hole size are both factors limiting the preload which can be achieved without causing indentation of the surfaces being bolted, especially in the case of softer materials. Washers distribute the bearing load, allowing higher tightening torques to be achieved without indentation of the material, but do not necessarily allow 'full' tightening torques to be achieved.

Bending Stresses in Bolts

When the bearing surfaces are not parallel in a bolted up assembly a bending stress is produced in the bolt. This can be determined directly as:-

$$S_B = \frac{Ea}{2L}$$

where S_B = bending stress

E = modulus of elasticity of bolt material

a = offset or actual gap between bearing surfaces

L = effective length of bolt

TABLE VII – TORQUE DATA UNIFIED HIGH TENSILE HEXAGON HEAD BOLTS & SCREWS

Tensile stress areas					Recommended Bolt Tension (see note 1)		Approximate Torques (see note 2)	
Nominal bolt dia in	UNC		UNF		UNC	UNF	UNC	UNF
	TPI	Stress area in^2	TPI	Stress area in^2	ton f	ton f	lbf/ft	lbf/ft
$\frac{1}{4}$	20	0·0324	28	0·0368	1·05	1·19	10	11
$\frac{5}{16}$	18	0·0532	24	0·0587	1·72	1·89	20	22
$\frac{3}{8}$	16	0·0786	24	0·0886	2·54	2·86	36	40
$\frac{7}{16}$	14	0·1078	20	0·1198	3·48	3·87	55	65
$\frac{1}{2}$	13	0·1438	20	0·1612	4·65	5·21	90	95
$\frac{5}{8}$	11	0·229	18	0·258	7·40	8·33	175	195
$\frac{3}{4}$	10	0·338	16	0·375	10·91	12·11	310	340
$\frac{7}{8}$	9	0·467	14	0·513	15·08	16·57	490	540
1	8	0·612	12	0·667	19·77	21·54	740	800

NOTE 1 Recommended bolt tension is based on 85% of proof load.

NOTE 2 Approximate torque figures given are for bolts in the unplated condition and do not take into account the effect of special lubricants or smooth and hard mating surfaces such as hardened washers.

TABLE VIII – TORQUE DATA METRIC HIGH TENSILE HEXAGON BOLTS

Diameter	Tensile Stress Area (mm²)	PROOF LOAD (kN)			APPROXIMATE TORQUE* (Nm)		
		Grade 8·8	Grade 10·9	Grade 12·9	Grade 8·8	Grade 10·9	Grade 12·9
M5	14·2	8·1	11·0	13·2	6·9	9·4	11·2
M6	20·1	11·5	15·6	18·7	11·7	15·9	19·1
M8	36·6	20·9	28·4	34·1	28	38	46·4
M10	58·0	33·1	45·1	54·1	56	77	92
M12	84·3	48·1	65·5	78·6	98	134	160
M16	157	89·6	122	146	244	332	397
M20	245	140	190	228	476	646	775
M24	353	201	274	329	822	1 120	1 342

*Approximate torque values for self-colour bolts. These have been calculated to induce a theoretical bolt load equal to 85% of the proof load.

TABLE IX – TORQUE DATA METRIC BLACK HEXAGON HEAD BOLTS & SCREWS

Diameter	Tensile Stress Area (mm²)	Proof Load (kN)	Approximate Torque* (Nm)
M6	20·1	4·5	4·54
M8	36·6	8·1	11
M10	58·0	12·8	22
M12	84·3	18·7	38
M16	157	34·8	95
M20	245	54·3	185
M24	353	78·2	320
M30	561	124	633
M36	817	181	1 110

*The torque figures quoted are approximate and are applicable to fasteners in the self-colour (unplated) condition only. They do not take into account the effect of plated finishes, special lubricants or the effect of hard and smooth mating surfaces such as hardened washers, etc. The torque figures quoted have been based on a theoretical bolt load equal to 85% of the proof load of the bolting material.

TABLE X – TORQUE DATA POZIDRIV RECESS MILD STEEL MACHINE SCREWS

Thread	Basic Major dia (in)	Basic Minor dia (in)	Minimum Tensile Breaking Load		Breaking Torque		Tightening Torque	
			lbf	kgf	lbf/ft	kgf/m	lbf/ft	kgf/m
6-32 UNC	·138	·0997	509	231	1·30	0·180	0·75	0·104
6-40 UNF	·138	·1073	569	258	1·45	0·200	0·91	0·126
8-32 UNC	·164	·1257	784	355	2·21	0·361	1·48	0·205
8-36 UNF	·164	·1299	826	374	2·52	0·348	1·65	0·228
10-24 UNC	·190	·1389	980	444	3·48	0·481	1·98	0·274
10-32 UNF	·190	·1517	1190	539	3·95	0·544	2·47	0·341
¼-20 UNC	·250	·1887	1781	807	8·30	1·150	5·42	0·750
¼-28 UNF	·250	·2062	2039	923	9·40	1·300	6·25	0·865
5/16-18 UNC	·3125	·2443	2935	1330	17·3	2·392	10·8	1·492
5/16-24 UNF	·3125	·2614	3247	1472	18·3	2·530	11·6	1·605
⅜-16 UNC	·375	·2993	4340	1965	30·0	4·150	17·7	2·445
⅜-24 UNF	·375	·3239	4919	2230	34·0	4·700	19·5	2·700
	mm	mm						
M2	2·0	1·509	182	83	0·33	0·046	0·17	0·023
M2·5	2·5	1·948	291	133	0·5	0·069	0·30	0·042
M3	3·0	2·387	447	202	1·0	0·138	0·60	0·083
M4	4·0	3·141	765	347	2·4	0·332	1·4	0·194
M5	5·0	4·019	1236	559	4·5	0·624	2·8	0·388
M6	6·0	4·773	1750	793	6·7	0·928	4·8	0·665
M8	8·0	6·466	3184	1445	14·3	1·98	11·7	1·62
M10	10·0	8·160	4760	2160	29·2	4·05	23·2	3·22

**Thomas Haddon & Stokes Ltd*

The properties shown are for steel screws produced to 25tonf/in^2 (39.37kgf/mm^2) minimum tensile strength (BS1981) and ISO metric to 40kgf/mm^2 minimum. Torque values are given as a guide to control application. The information was obtained from tests with self-colour screws assembled with standard hexagon nuts and washers. Applications involving special lubricants, alternative lengths of thread engagement, surface coatings and electroplated finishes will give different values from assemblies with the normal self-colour finish.

Tightening torques for brass screws are approximately 10% less than those quoted for mild steel. Values for high tensile steel fasteners ('S' quality and metric grade 8.8) at 55tonf/in^2 (86.6kgf/mm^2) are approximately 75% greater than those quoted for mild steel. Tensile breaking loads for brass screws are 80% of those given for mild steel. Values for high tensile steel screws are slightly greater than twice those quoted for mild steel.

TABLE XI – RECOMMENDED TIGHTENING TORQUES FOR SELF-COLOUR SOCKET SCREWS (GKN)

Cap Head Screws: ISO Metric Threads

Nominal Size mm	**Nominal Wrench Size** A/Flats mm	**Tightening Torque**			**Nominal Induced Load**	
		lbf in	lbf ft	Nm	lbf	kN
M3	2·5	21		2·4	936	4·16
M4	3	49		5·5	1632	7·26
M5	4	99		11·2	2640	11·74
M6	5	168	14	19·0	3744	16·65
M8	6	407	34	46·0	6804	30·27
M10	8		67	91·2	10788	47·99
M12	10		117	159·1	15684	69·77
M16	14		291	394·8	29196	129·87
M20	17		568	770·3	45576	202·73
M24	19		982	1331·8	65652	292·03

Countersunk Head Screws: ISO Metric Threads

Nominal Size mm	**Nominal Wrench Size** A/Flats mm	**Tightening Torque**			**Nominal Induced Load**	
		lbf in	lbf ft	Nm	lbf	kN
M3	2	18		2	625	2·8
M4	2·5	35		4	937	4·2
M5	3	58		6·6	1236	5·5
M6	4	142	12	16	2500	11·1
M8	5	274	23	31	3630	16·1
M10	6		39	53	4964	22·1
M12	8		94	127	9913	44·1
M16	10		182	247	14460	64·3
M20	12		270	365	17094	76

Button Head Screws: ISO Metric Threads

Nominal Size mm	**Nominal Wrench Size** A/Flats mm	**Tightening Torque**			**Nominal Induced Load**	
		lbf in	lbf ft	Nm	lbf	kN
M3	2	18		2	681	3·0
M4	2·5	35		4	1020	4·5
M5	3	58		6·6	1350	6·0
M6	4	142	12	16	2725	12·1
M8	5		23	31	3960	17·6
M10	6		39	53	5416	24·1
M12	8		94	127	10814	48·1

N.B. The tightening torque values for countersunk head and button head screws are restricted by the related wrench key size.

cont...

TABLE XI (contd.)

Set Screws: Cup Point & 'W' Point ISO Metric Threads

Nominal Size	Nominal Wrench Size	Tightening Torque			Axial Holding Power	
mm	A/Flats mm	lbf in	lbf ft	Nm	lbf	kN
M3	1·5	8		0·9	226	1·0
M4	2	18		2	381	1·7
M5	2·5	35		4	610	2·7
M6	3	58		6·6	833	3·7
M8	4	142	12	16	1480	5·6
M10	5	274	23	31	2320	10·3
M12	6		39	53	3330	14·8
M16	8		94	127	5980	26·6
M20	10		182	247	9270	41·2
M24	12		270	365	11400	50·7

Shoulder Screws

Shoulder Diameter	Nominal Thrd Size	Nominal Wrench Size	Thread Neck		Seating Torque			Nominal Tensile Load	
mm	mm	A/Flats mm	Dia mm	Area mm²	lbf in	lbf ft	Nm	lbf	kN
6	M5	3	3·52	9·731	35		4	1509	6·7
8	M6	4	4·22	13·987	61		6·9	2169	9·6
10	M8	5	5·82	26·603	156		17·6	4120	18·4
12	M10	6	7·48	43·943		27	36	6816	30·3
16	M12	8	9·18	56·187		41	56	8715	38·8

Note: The values are restricted by the thread neck diameter

contd...

TABLE XI (contd.)

Cap Head Screws: ISO Inch (Unified) Threads UNC & UNF

Nominal Size	Nominal Wrench Size	Tightening Torque				Nominal Induced Load lbf	
		UNC		UNF			
in	A/Flats in	lbf in	lbf ft	lbf in	lbf ft	UNC	UNF
No. 4	5/64	13				616	
No. 5	3/32	19				810	
No. 6	3/32	24		27		928	1030
No. 8	1/8	44		—		1420	—
No. 10	5/32	65		73		1790	2030
1/4	3/16	154		175		3240	3680
5/16	7/32	316		349		5320	5870
3/8	5/16	560	47	631	53	7860	8860
7/16	5/16		75		83	10780	11980
1/2	3/8		114		128	14380	16120
5/8	1/2		226		255	22880	25780
3/4	9/16		402		446	33820	37540

Cap Head Screws: BSW & BSF Threads

Nominal Size	Nominal Wrench Size	Tightening Torque				Nominal Induced Load lbf	
		BSW		BSF			
in	A/Flats in	lbf in	lbf ft	lbf in	lbf ft	BSW	BSF
1/8	3/32	19		—		800	—
3/16	5/32	61		69		1710	1950
1/4	3/16	152		169		3200	3560
5/16	7/32	313		337		5270	5680
3/8	5/16	555	46	598	50	7790	8400
7/16	5/16		74		80	10690	11590
1/2	3/8		110		120	13850	15200
5/8	1/2		225		241	22710	24310
3/4	9/16		399		419	33590	35250

Cap Head Screws: BA Threads

Nominal Size	Nominal Wrench Size	Tightening Torque	Nominal Induced Load
in	A/Flats in	lbf in	lbf
8BA	1/16	6	400
6BA	5/64	14	650
5BA	3/32	21	870
4BA	3/32	27	1100
3BA	1/8	44	1440
2BA	5/32	67	1920
1BA	5/32	97	2450
0BA	3/16	142	3170

contd...

TABLE XI (contd.)

Countersunk Head Screws: ISO Inch (Unified) Threads UNC & UNF*

Nominal Size	Nominal Wrench Size	Tightening Torque		Nominal Induced Load
		UNC/UNF	UNC/UNF	UNC/UNF
in	A/Flats in	lbf in	lbf ft	lbf
No. 4	1/16	9		335
No. 6	5/64	18		543
No. 8	3/32	32		813
No. 10	1/8	70		1535
1/4	5/32	138		2300
5/16	3/16	234		3120
3/8	7/32	372	31	4133
1/2	5/16		91	9083
5/8	3/8		158	12640
3/4	1/2		320	21333

Countersunk Head Screws: BSW & BSF Threads

Nominal Size	Nominal Wrench Size	Tightening Torque		Nominal Induced Load
		BSW/BSF	BSW/BSF	BSW/BSF
in	A/Flats in	lbf in	lbf ft	lbf
3/16	3/32	32		711
1/4	5/32	138		2300
5/16	3/16	234		3120
3/8	7/32	372		4133
7/16	1/4	564	47	5371
1/2	5/16		91	9083
5/8	3/8		158	12640
3/4	3/8		158	10533

Countersunk Head Screws: BA Threads

Nominal Size	Nominal Wrench Size	Tightening Torque	Nominal Induced Load
in	A/Flats in	lbf in	lbf
6BA	0·050	5	182
4BA	1/16	9	265
3BA	5/64	18	466
2BA	3/32	32	720
0BA	1/8	70	1240

contd...

TABLE XI (contd.)

Button Head Screws: ISO Inch (Unified) Threads UNC & UNF*

Nominal Size	Nominal Wrench Size	Tightening Torque		Nominal Induced Load
		UNC/UNF	UNC/UNF	UNC/UNF
in	A/Flats in	lbf in	lbf ft	lbf
No. 6	5/64	18		593
No. 8	3/32	32		887
No. 10	1/8	70		1675
1/4	5/32	138		2509
5/16	3/16	234		3404
3/8	7/32	372	31	4509
1/2	5/16		91	9909

Button Head Screws: BSW & BSF Threads

Nominal Size	Nominal Wrench Size	Tightening Torque		Nominal Induced Load
		BSW/BSF	BSW/BSF	BSW/BSF
in	A/Flats in	lbf in	lbf ft	lbf
3/16	1/8	70		1697
1/4	5/32	138		2509
5/16	3/16	234		3404
3/8	7/32	372	31	4509
1/2	5/16		91	9909

Button Head Screws: BA Threads

Nominal Size	Nominal Wrench Size	Tightening Torque	Nominal Induced Load
in	A/Flats in	lbf in	lbf
2BA	1/8	70	1720

*The tightening torque values for countersunk head and button head screws are restricted by the related wrench key size.

cont...

TABLE XI (contd.)

Shoulder Screws: BSW Threads

Shoulder Diameter	Nominal Thread Size	Nominal Wrench Size	Thread Neck		Tightening Torque		Nominal Tensile Load
			Diameter	Area			
in	in	A/Flats in	in	in²	lbf in	lbf ft	lbf
$\frac{1}{4}$	$\frac{3}{16}$	$\frac{1}{8}$	0·117	0·0107	24		1070
$\frac{5}{16}$	$\frac{1}{4}$	$\frac{5}{32}$	0·169	0·0224	67		2240
$\frac{3}{8}$	$\frac{5}{16}$	$\frac{3}{16}$	0·223	0·0391	147		3910
$\frac{1}{2}$	$\frac{3}{8}$	$\frac{1}{4}$	0·275	0·0594	267	22	5940
$\frac{5}{8}$	$\frac{1}{2}$	$\frac{5}{16}$	0·371	0·1081		54	10810
$\frac{3}{4}$	$\frac{5}{8}$	$\frac{3}{8}$	0·486	0·1855		116	18550

Shoulder Screws: BSF Threads

Shoulder Diameter	Nominal Thread Size	Nominal Wrench Size	Thread Neck		Tightening Torque		Nominal Tensile Load
			Diameter	Area			
in	in	A/Flats in	in	in²	lbf in	lbf ft	lbf
$\frac{1}{4}$	$\frac{3}{16}$	$\frac{1}{8}$	0·132	0·0137	31		1370
$\frac{5}{16}$	$\frac{1}{4}$	$\frac{5}{32}$	0·185	0·0269	81		2690
$\frac{3}{8}$	$\frac{5}{16}$	$\frac{3}{16}$	0·236	0·0437	164		4370
$\frac{1}{2}$	$\frac{3}{8}$	$\frac{1}{4}$	0·293	0·0674	303	25	6740
$\frac{5}{8}$	$\frac{1}{2}$	$\frac{5}{16}$	0·399	0·1250		63	12500
$\frac{3}{4}$	$\frac{5}{8}$	$\frac{3}{8}$	0·511	0·2051		128	20510

Note: The values are restricted by the thread neck diameter

Self-tapping Screws

Self-tapping screws are special designs of hardened fasteners, capable of providing a matching thread in the materials being joined. They thus offer a simpler type of fastener than bolted joints or riveting. They fall into three categories:-

(i) *Thread-forming tapping screws* where the mating thread in the material being jointed is produced by displacement of the material around the crests of the tapping screw. Displaced material generates extra thread depth with the core diameter smaller than the drilled hole, providing a strong threaded joint. This type is suitable for use in ductile materials with good strength and rigidity. Thread-forming self-tapping screws can be further classified as conventional (ie for fitting in drilled holes) or self-piercing and self-drilling types which cut their own hole.

(ii) *Thread-cutting tapping screws* where the thread is actually cut in the material being joined in a similar manner to that generated by a tap. The depth of thread is the difference between the diameter of the screw and the hole in the material. This type of self-tapping screw is suitable for use in brittle material.

(iii) *Drive screws* which have multiple threads and a large lead angle with a pilot point. They are intended to be driven into holes under pressure.

In addition there are two basic types of thread form used on self-tapping screws – spaced threads and conventional screw threads. Spaced threads are more suitable for use on thin sheet materials or thin sections, yielding more material between pitches to improve the load bearing characteristics of the fastening.

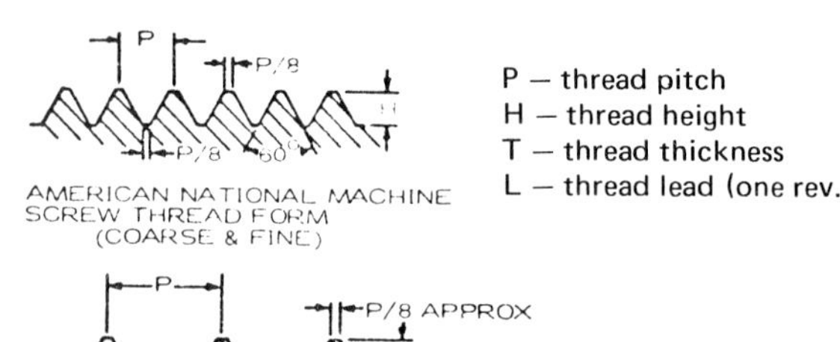

Fig 1 Comparison of conventional self-tapping thread and American National Screw Thread.

Conventional self-tapping screws have a thread form based on the American ASA B 18.6.4 specification (now incorporated in BS 4174) – see Fig 1. The diameter size is expressed by a number, which is the same both in inch and metric systems. Length can be expressed in inches or rounded-off metric equivalents – see Table I.

Some fourteen different basic types of self-tapping screw have been produced, and are designated by letters – see Fig 2. Rationalisation and standardisation of production have reduced this to about eight (varying from country to country and between individual manufacturers) excluding special self-piercing types.

The main types now used are:-

Thread forming – spaced thread – A, AB, B.

conventional thread – C.

Thread cutting – spaced thread – BF, BT.

conventional thread – D. F, T.

Drive screws – Type U.

Type A – coarse threads with a gimlet point, originally specified for thin metals, and still widely employed. Directly alternative types are:-

Type AB – with a less coarse spaced thread and blunter point.

Type B – spaced threads with a blunt point and incomplete entry threads. These are used in sheet metals of all thicknesses up to structural steel sections, also in plastics, etc, and non-ferrous castings. They have the same pitch as Type AB. Type B was formerly known as type 'Z'.

Other standard types include the following:-

Type BF and BT – spaced threads as in Type B but with the addition of one (Type BT) or more (Type BF) cutting edges and chip cavities. Type BT is also known as Type 25 or Type BT/25.

Type BP – same as Type B with a conical point for piercing fabrics or for use in assemblies where holes may be misaligned.

Type C – with Unified thread forms, blunt point and tapered incomplete entering threads.

Type D – as type C but with one thread cutting slot. The tapered entering threads are incomplete. Type D is also known as Type 1 or Type D/1.

Type F – as Type C with more than one chip cavity and cutting edges. The tapered entering threads may be complete or incomplete.

Type G – similar to Type D.

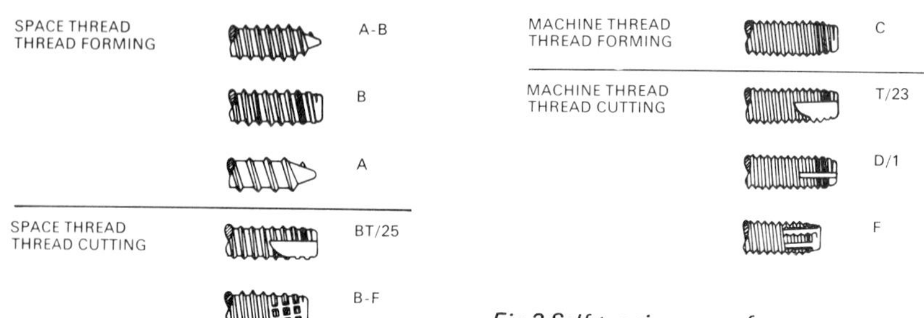

Fig 2 Self-tapping screw forms.

Type T – similar to Type C with one wide thread-cutting slot. Type T is also known as Type 23 or Type T/23.

Type U – metallic drive screws, used mainly in brittle plastics, cast iron and GRP mouldings.

Type Z – now known as Type B.

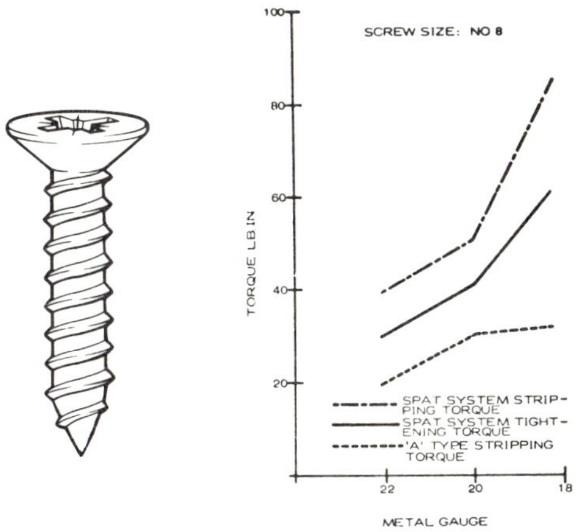

Fig 3 Self-piercing hand tapping screw or 'Spat' and performance compared with type A self-tapping screw.

The *self-piercing* self-tapping screw originated by GKN is called a 'Spat' – Fig 3. The screw carries an unthreaded diamond point for piercing, blending into a two-start fast helix thread with controlled thread non-cut to prevent misalignment when driving. It is used with a high energy impact driver to force the point of the screw through sheet metal, followed by rotary action to drive the screw into the hole with a thread forming action.

Self-drilling screws are self-tapping screws with a drill point at the lead end. These are mostly thread-cutting and the drill point forms the core diameter of the tapped hole.

A proprietary example of a self-drilling, self-tapping screw is shown in Fig 4, the advantage offered being that no separate drilling operation is required. Fasteners of this type are suitable for use in sheet metals as well as other materials (including metal-to-wood fastening). The critical factor involved is that drilling must be completed before the self-tapping threads engage, and thus the length of pilot section used must be related to the total thickness to be drilled. Examples of screw selection are given in Table IV.

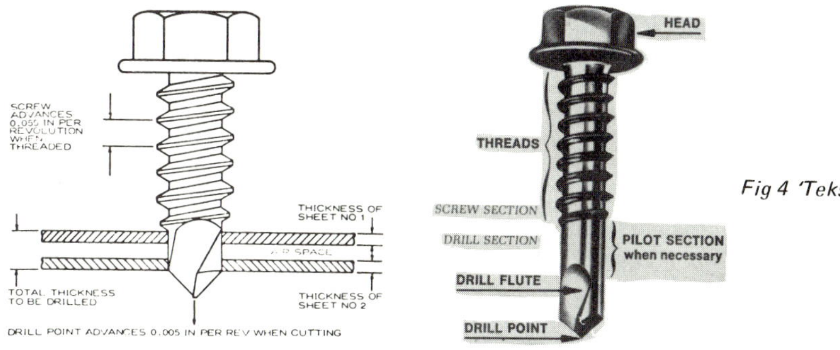

Fig 4 'Teks' self-drilling screw.

In addition a variety of high strength screws with conventional threads may be used as self-tappers (mostly thread-forming) for more general machine assembly use. The main difference between these and standard thread-forming self-tapping screws is the point formation which generates the full thread form. Examples are shown in Fig 5.

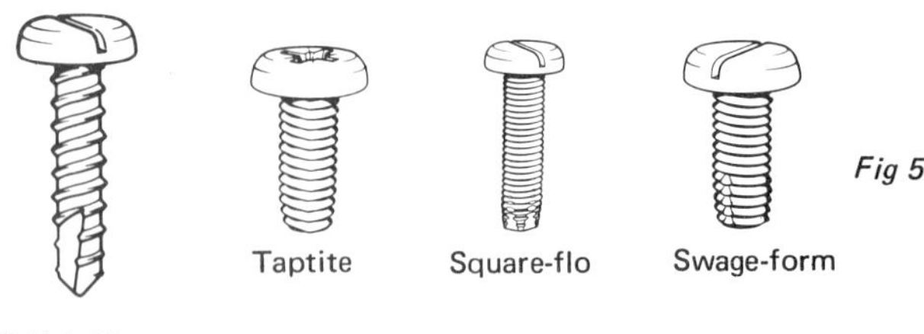

Fig 5

The *'swage-form'* screw has a fully threaded point form with three bands of protuberances carried on the flanks of the thread. The effect of these is to 'swage' the material and so form the thread. This is particularly suitable for use in die-cast and thermoplastic components, providing the equivalent of a machine screw/tapped hole fastening without the necessity of tapping the hole (see also Table XVII).

The *'square-flow'* screw is an orthodox high strength machine screw terminating in a tapered square section point with interrupted threads. This has similar application to the 'swage form' screw (see also Table XVIII).

'Taptite' screws are of less conventional form with a tri-lobular shape throughout the length of the thread to reduce tapping torque. They are particularly suitable for use in more delicate materials. This form of screw is also produced with a fully rolled point ('Taptite 'W' point') with sharp thread crests. Again, its applications are as above.

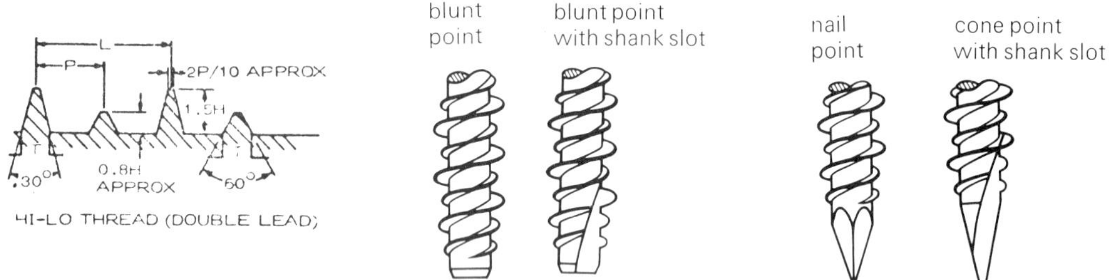

Fig 6 Hi-lo thread form and point forms.

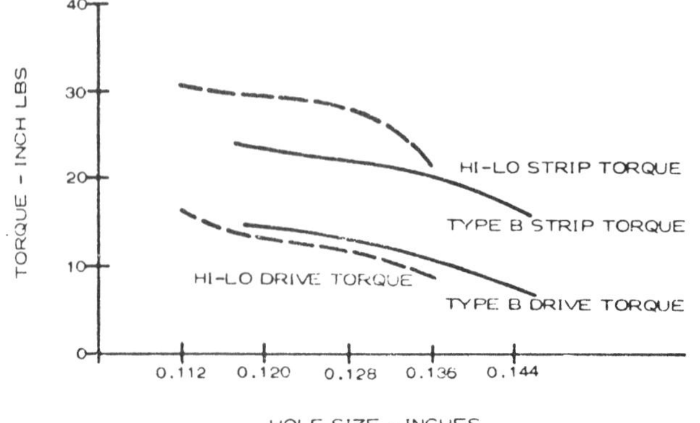

Fig 7 Comparison of drive torque and stripping torque of 'Hi-lo' and type B self-tapping screws.

Modified thread forms may be used on self-tapping fasteners designed for use in brittle materials where thread stripping or thread crumbling can be a problem. The *Hi-Lo* thread screw is an example with a double lead made up of a high and low thread. The high thread is sharp with a 30° included angle form. The low thread is of conventional 60° form with a height of approximately 40% of the high thread – Fig 6. This thread form reduces radial pressures as well as providing an increase in differential between driving and stripping torques. It is particularly suitable for use in plastic materials, including phenolics – see also Table III and Fig 7.

Choice of Self-tapper

Self-tapping screws are produced with a variety of heads, but mainly countersunk, raised countersunk, pan, hexagon and hexagon with integral washer (hexagon washer) – see Fig 8.

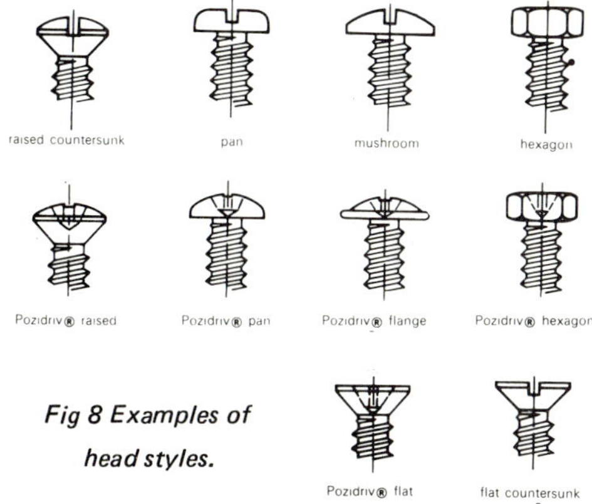

Fig 8 Examples of head styles.

The choice of type of self-tapping screw depends on the thickness of the material and its characteristics (eg whether delicate or brittle), and the method of preparing the hole (eg drilling, punching, extruding, moulding). The method of driving may also need consideration – eg by hand or power screwdriver.

The choice of size of screw depends on strength requirements (affecting the diameter size required) and material thicknesses, (affecting the screw length).

The choice of *type* and *size* of screw, together with the characteristics and dimensions of the materials to be joined, determine the diameter of the *pilot hole* required. This in turn is governed by the fact that the stripping torque should be at least equal to the torsional strength of the screw for full utilisation of its characteristics.

In practice a tapping torque of one-third of the stripping torque is recommended, and a tightening torque of two-thirds. Recommendations for pilot hole sizes are normally based on these parameters, although there can be limitations in the case of brittle or weak materials (eg it may be necessary to reduce the tightening torque considerably to avoid stripping the threads formed in the material). On the other hand, in suitable materials a tightening torque as high as 80% of the stripping torque may be possible for maximum fastening strength – see also Figs 9 and 10.

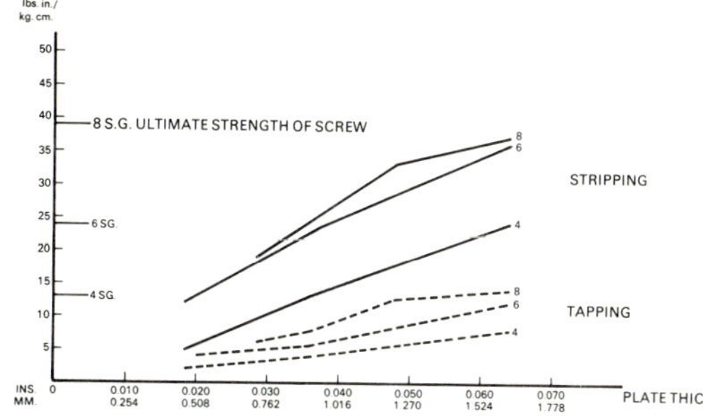

Fig 9 Type A–B tapping/stripping torques (GKN)

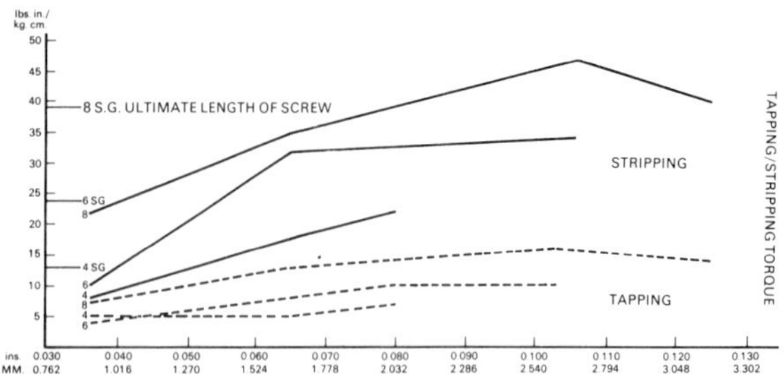

Fig 10 Type B tapping/stripping torques. (GKN).

A general selection guide is given in Table III. Specific recommendations are summarised below:-

(i) The tensile strength of the joint approaches that of the screw only when the effective length of thread produced (or thickness of material) exceeds 75% of the screw diameter (Type AB and Type B screws).

(ii) Plunged rather than drilled holes offer a substantial increase to the tensile strength of the fastener in thin sheet steel.

(iii) In the case of low tensile strength sheet materials – eg aluminium – the maximum strength of the assembly is governed directly by the sheet or plate thickness (Type B screws).

(iv) Variations in hole size can appreciably modify the performance of the fastener. This can also result from tolerances in hole sizes which occur in production.

Power Driving Torque

The tightening torque required with self-tapping screws can vary widely with screw design and strength, the materials being fastened and the screw and hole size used.

For maximum performance, but allowing a suitable factor for safety, a tapping/tightening torque ratio of 2:1 is recommended. Higher ratios may be necessary with thin metal or weak materials in order to avoid any possibility of thread stripping – eg 3:1 or higher, depending on the material concerned.

TABLE I – SIZES OF CONVENTIONAL SELF-TAPPING SCREWS

Nominal Length		**Type AB**				**Types B & Y**			
		Max		Min		Max		Min	
in	mm	in	mm	in	mm	in	mm	in	mm
1/8	**3·20**	0·149	**3·78**	0·102	**2·58**	0·125	**3·18**	0·102	**2·58**
3/16	**4·50**	0·211	**5·36**	0·164	**4·16**	0·188	**4·76**	0·164	**4·16**
1/4	**6·50**	0·280	**7·10**	0·220	**5·60**	0·250	**6·35**	0·220	**5·60**
3/8	**9·50**	0·404	**10·27**	0·345	**8·77**	0·375	**9·52**	0·345	**8·77**
1/2	**13·00**	0·535	**13·60**	0·465	**11·80**	0·500	**12·70**	0·465	**11·80**
5/8	**16·00**	0·661	**16·78**	0·590	**14·98**	0·625	**15·88**	0·590	**14·98**
3/4	**19·00**	0·791	**20·10**	0·709	**18·00**	0·750	**19·05**	0·709	**18·00**
7/8	**22·00**	0·916	**23·27**	0·833	**21·17**	0·875	**22·22**	0·833	**21·17**
1	**25·00**	1·041	**26·45**	0·959	**24·35**	1·000	**25·40**	0·959	**24·35**
1 1/4	**32·00**	1·300	**33·00**	1·201	**30·50**	1·250	**31·75**	1·201	**30·50**
1 1/2	**38·00**	1·549	**39·35**	1·451	**36·85**	1·500	**38·10**	1·451	**36·85**
1 3/4	**45·00**	1·799	**45·70**	1·701	**43·20**	1·750	**44·45**	1·701	**43·20**
2	**50·00**	2·059	**52·30**	1·941	**49·30**	2·000	**50·80**	1·941	**49·30**

For production purposes a suitable value for tightening torque can be determined empirically by measuring the torque applied to a sample run of at least 10 screws to strip or otherwise cause failure of the joint. A specified application torque can then be based on 60–70% of the *lowest* failure torque of the test group.

TABLE II – SELECTION GUIDE FOR CONVENTIONAL SELF-TAPPING SCREWS

Screw	Suitability	Remarks
Type AB	(i) Best for thin sheet metals up to 18SWG (2.5mm) Use on light and heavy gauge sheet metals, soft plastics, high-impact plastics, resin-bonded plywoods, asbestos board, zinc and aluminium diecastings.	Stripping strength approximates to screw strength. Hole size less critical than for Type B.
Type B	(i) More restricted depth of thread in thin sheet metals. (ii) Better than Type A for metal thicknesses above 1.6mm. Use on non-ferrous castings, zinc and aluminium based die castings, high impact plastics.	
Type C	Use where machine screw thread pitch is preferred or needed.	
Type BF, Type BT	Thread cutting – particularly plastics, asbestos, etc.	Onbbrittle plastics the more cutting flanges the better (Type BF).
Type D	Blind holes in square based die castings, cast iron, brittle plastic.	Slot provides clearance for swarf.
Type F	Aluminium, zinc and lead die-castings, steel sheets and sections, cast iron, brass, plastics, etc.	Thread cutting
Type T	Blind holes in zinc-based die castings, cast iron and brittle plastics.	Better than type D for extended thread depths.
Type U	Ferrous and non-ferrous castings, GRP mouldings, plastics.	Press or drive into position.
'Square Flo' 'Swage Form' etc.	Consider as direct alternatives to machine screw/tapped holes for assemblies	

TABLE III – COMPARISON OF HI-LO AND TYPE B SELF-TAPPING SCREWS USED IN PLASTICS

Material	Boss Diameter for Equivalent Stripping Torque (in)		Pull-Out Force (lb)	
	8–18 Type B	8–18 Hi-Lo	8–18 Type B	8–18 Hi-Lo
ABS	3/8	0.228	403	576
Nylon	3/8	0.194	640	812
Delrin	3/8	0.256	610	745

TABLE IV – 'TEKS' SELF-DRILLING SCREWS

'TEKS' Number	Drilling Capacity*	Application(s)
/2 /3	0.35"–0.110" (0.9–2.8mm)	Light duty fastenings on sheet metal, etc.
/3	0.110"–0.250" (2.8–6mm)	Attachment of facia panels, clips, accessories, etc.
/4	0.175"–0.250" (4.5–6mm)	Metal and other assemblies. Modified design reduces driving torque required.
/5	0.250"–0.500" (6–12.5mm)	Metal assemblies

*Depending on screw size

TABLE V – SURFACE HARDNESS

TYPES 'AB', 'B' AND 'U'

Screw size (No.)	Plate thickness (in.)	Hole diameter (in.)
2	0·048	0·076
4	0·048	0·086
6	0·080	0·116
8	0·080	0·136
10	0·125	0·159
12	0·125	0·188
14	0·187	0·217
16	0·187	0·272
$\frac{3}{8}$"	0·187	0·328

TAPTITE THREAD FORMING SCREWS ISO INCH (UNIFIED)

Screw size (No.)	Plate thickness (in.)	Hole size ±0·025 mm	Hole size ±0·001 in.
4-40	0·125	2·55	0·1004
5-40	0·125	2·85	0·1122
6-32	0·187	3·10	0·1220
8-32	0·187	3·80	0·1496
10-24	0·250	4·40	0·1732
10-32	0·250	4·50	0·1772
$\frac{1}{4}$"-20	0·250	5·70	0·2244
$\frac{5}{16}$"-18	0·375	7·40	0·2913
$\frac{3}{8}$"-16	0·375	8·90	0·3504

TAPTITE THREAD FORMING SCREWS ISO METRIC

Screw size (No.)	Plate thickness (mm)	Hole size ±0·25
M2·5	3·15	2·25
M3	3·15	2·75
M3·5	3·15	3·20
M4	5·00	3·60
M5	5·00	4·60
M6	6·30	5·50
M8	10·00	7·30
M10	10·00	9·20

Low carbon steel self-tapping screws are required to have a minimum surface hardness of 450 HV and minimum core of 270 HV. Case depth 0.003"/0.008". The screws should be driven into a low carbon steel test plate of hardness between 125/165 HV30. Hole sizes to be used in this test are specified above

TABLE VI – TORSIONAL STRENGTH

TYPES 'AB', 'B' AND 'Y'

Screw size (No.)	Minimum torsional load lbf in	kgf cm	Nm
2	4	5	0·49
4	13	15	1·47
6	24	28	2·74
8	39	45	4·41
10	56	64	6·27
12	88	101	10·78
14	142	163	15·98
$\frac{5}{16}$"	290	334	34·71
$\frac{3}{8}$"	590	678	70·80

For the torsional strength test, the shank of the screw is clamped so that at least two threads protrude above the clamping device. Using a calibrated torque measuring device, torque is applied until fracture occurs. Screws have to meet the minimum torsional strength shown in the tables

TAPTITE THREAD FORMING SCREWS ISO INCH (UNIFIED)

Size	Minimum torsional strength lbf in	kgf cm	Nm
2-56	6·7	7·7	0·76
4-40	16	18·4	1·81
5-40	22·6	26	2·55
6-32	24	27·6	2·71
8-32	52	60	5·88
10-24	65	75	7·35
10-32	81	93	9·16
$\frac{1}{4}$"-20	176	202	19·9
$\frac{5}{16}$"-18	380	437	43
$\frac{3}{8}$"-16	700	805	79

TAPTITE THREAD FORMING SCREWS ISO METRIC

Size	Minimum torsional strength kgf cm	Nm	lbf in
M2·5	13·82	1·36	12
M3	23·10	2·27	20
M3·5	34·58	3·39	30
M4	54·16	5·31	47
M5	107·18	10·51	93
M6	195·93	19·21	170
M8	507·11	49·73	440
M10	956·54	93·80	830

TABLE VII – RECOMMENDED HOLE SIZES (DRILLED OR CORED HOLES) – TYPE B SELF-TAPPING SCREWS

In non-ferrous castings, aluminium, zinc, brass, bronze, etc.

Screw size (No.) and nominal diameter	Minimum penetration				Normal maximum penetration			
	Hole depth	Hole diameter	Drill size		Hole depth	Hole diameter	Drill size	
	in.	in.	mm	Alternatives	in.	in.	mm	Alternatives
2 (0·086")	$\frac{1}{8}$	0·071	1·80	50	$\frac{1}{4}$	0·079	2·00	47
4 (0·112")	$\frac{5}{32}$	0·096	2·45	41	$\frac{5}{16}$	0·104	2·65	37
6 (0·138")	$\frac{3}{16}$	0·130	3·30	30	$\frac{3}{8}$	0·130	3·30	30
7 (0·151")	$\frac{7}{32}$	0·142	3·60	28	$\frac{7}{16}$	0·142	3·60	27
8 (0·164")	$\frac{7}{32}$	0·153	3·90	24	$\frac{7}{16}$	0·153	3·90	23
10 (0·186")	$\frac{1}{4}$	0·177	4·50	16	$\frac{1}{2}$	0·177	4·50	16
12 (0·212")	$\frac{9}{32}$	0·201	5·10	7	$\frac{9}{16}$	0·201	5·10	7
14 (0·242")	$\frac{5}{16}$	0·236	6·00	A	$\frac{5}{8}$	0·236	6·00	B
16 (0·311")	$\frac{7}{16}$	0·295	7·50	M	$\frac{7}{8}$	0·295	7·50	M
$\frac{3}{8}$"	$\frac{9}{16}$	0·358	9·10	T	1	0·358	9·10	T

Notes:

1 Cored holes. A side taper of 1° 11' is permissible. The diameter of a cored hole should equal the nominal hole size shown in the table above at one half the screw penetration depth.

2 Porous castings may require the use of a smaller hole and/or increased depth of engagement.

TABLE VIII – RECOMMENDED HOLE AND DRILL SIZES – TYPE B SELF-TAPPING SCREWS

Hexagon head in structural steel

Screw size (No.) and nominal diameter	Metal thickness			Cadmium plated or lubricated screws			Self colour, zinc plated or non-lubricated screws		
			SWG or fraction	Hole diameter required	Drill size		Hole diameter required	Drill size	
	in.	mm		in.	mm	Alternatives	in.	mm	Alternatives
6 (0·138″)	0·036	0·91	20	0·102	2·60	38	0·102	2·60	38
	0·064	1·62	16	0·110	2·80	35	0·110	2·80	35
	0·080	2·03	14	0·114	2·90	33	0·114	2·90	33
	0·104	2·64	12	0·122	3·10	$\frac{1}{8}$″	0·122	3·10	$\frac{1}{8}$″
8 (0·164″)	0·064	1·62	16	0·130	3·30	30	0·130	3·30	30
	0·080	2·03	14	0·142	3·60	28	0·142	3·60	28
	0·104	2·64	12	0·142	3·60	28	0·146	3·70	26
	0·125	3·18	$\frac{1}{8}$″	0·146	3·70	26	0·150	3·80	25
10 (0·186″)	0·064	1·62	16	0·150	3·80	25	0·150	3·80	25
	0·104	2·64	12	0·158	4·00	22	0·158	4·00	22
	0·125	3·18	$\frac{1}{8}$″	0·161	4·10	20	0·161	4·10	20
	0·187	4·75	$\frac{3}{16}$″	0·173	4·40	17	0·177	4·50	16
14 (0·242″)	0·125	3·18	$\frac{1}{8}$″	0·220	5·60	2	0·220	5·60	2
	0·187	4·75	$\frac{3}{16}$″	0·232	5·90	A	0·232	5·90	A
	0·250	6·35	$\frac{1}{4}$″	0·232	5·90	A	0·232	5·90	A
	0·312	7·92	$\frac{5}{16}$″	0·232	5·90	A	0·232	5·90	A
$\frac{5}{16}$″	0·125	3·18	$\frac{1}{8}$″	0·283	7·20	K	0·287	7·30	L
	0·187	4·75	$\frac{3}{16}$″	0·291	7·40	L	0·295	7·50	M
	0·250	6·35	$\frac{1}{4}$″	0·291	7·40	L	0·299	7·60	N
	0·312	7·92	$\frac{5}{16}$″	0·295	7·50	M	0·303	7·70	N
	0·375	9·53	$\frac{3}{8}$″	0·295	7·50	M	0·303	7·70	N
	0·500	12·70	$\frac{1}{2}$″	0·299	7·60	N	0·303	7·70	N
$\frac{3}{8}$″	0·125	3·18	$\frac{1}{8}$″	0·339	8·60	R	0·354	9·00	T
	0·187	4·75	$\frac{3}{16}$″	0·350	8·90	S	0·362	9·20	U
	0·250	6·35	$\frac{1}{4}$″	0·354	9·00	T	0·362	9·20	U
	0·312	7·92	$\frac{5}{16}$″	0·358	9·10	T	0·366	9·30	U
	0·375	9·53	$\frac{3}{8}$″	0·362	9·20	U	0·366	9·30	U
	0·500	12·70	$\frac{1}{2}$″	0·362	9·20	U	0·366	9·30	U
$\frac{1}{2}$″	0·250	6·35	$\frac{1}{4}$″	0·468	11·90	$\frac{15}{32}$″	0·484	12·30	$\frac{31}{64}$″
	0·312	7·92	$\frac{5}{16}$″	0·468	11·90	$\frac{15}{32}$″	0·488	12·40	$\frac{31}{64}$″
	0·375	9·53	$\frac{3}{8}$″	0·472	12·00	$\frac{15}{32}$″	0·488	12·40	$\frac{31}{64}$″
	0·500	12·70	$\frac{1}{2}$″	0·472	12·00	$\frac{15}{32}$″	0·488	12·40	$\frac{31}{64}$″

Note: If very hard material is being used a hole size slightly larger may have to be used, and in very soft material a smaller hole may be necessary.

TABLE IX – RECOMMENDED HOLE AND DRILL SIZES FOR THERMOPLASTICS – TYPES AB, B

Screw size (No.) and nominal diameter	Normal penetration			Minimum penetration in blind holes
	Hole diameter required	Drill size		
	in.	mm	Alternatives	
2 (0·086″)	0·070	1·80	50	$\frac{1}{4}$″
4 (0·112″)	0·093	2·35	42	$\frac{1}{4}$″
6 (0·138″)	0·114	2·90	32	$\frac{1}{4}$″
7 (0·151″)	0·125	3·10	$\frac{1}{8}$″	$\frac{5}{16}$″
8 (0·164″)	0·135	3·40	29	$\frac{5}{16}$″
10 (0·186″)	0·154	3·90	23	$\frac{5}{16}$″
12 (0·212″)	0·180	4·60	15	$\frac{3}{8}$″
14 (0·242″)	0·210	5·30	4	$\frac{3}{8}$″

Note: It may be necessary to increase or decrease the recommended hole size to obtain optimum fastening conditions.

TABLE X – RECOMMENDED HOLE AND DRILL SIZES – TYPES AB, B SELF-TAPPING SCREWS
In mild steel, brass, aluminium alloy, stainless steel and monel metal sheet

Screw size (No.) and nominal dia.	Material thickness in.	Material thickness mm	Material thickness SWG or fraction	Pierced or extruded hole dia. in.	Drilled or clean-punched holes: Hole dia required in.	Drill size mm	Drill size Alternatives
2 (0·086″)	0·018	0·45	26	—	0·063	1·60	52
	0·036	0·91	20	—	0·073	1·85	49
	0·064	1·62	16	—	0·077	1·95	48
4 (0·112″)	0·018	0·45	26	—	0·081	2·05	46
	0·036	0·91	20	0·098	0·091	2·30	42
	0·064	1·62	16	—	0·095	2·40	41
	0·080	2·03	14	—	0·102	2·60	38
6 (0·138″)	0·018	0·45	26	—	0·092	2·35	42
	0·036	0·91	20	0·111	0·110	2·80	35
	0·064	1·62	16	—	0·116	2·95	32
	0·080	2·03	14	—	0·122	3·10	31
	0·104	2·64	12	—	0·126	3·20	30
7 (0·151″)	0·036	0·91	20	0·120	0·118	3·00	32
	0·064	1·62	16	—	0·126	3·20	$\frac{1}{8}$″
	0·080	2·03	14	—	0·130	3·30	30
	0·104	2·64	12	—	0·134	3·40	29
8 (0·164″)	0·028	0·71	22	—	0·114	2·90	33
	0·036	0·91	20	0·136	0·122	3·10	$\frac{1}{8}$″
	0·048	1·22	18	—	0·126	3·20	30
	0·064	1·62	16	—	0·134	3·40	29
	0·104	2·64	12	—	0·146	3·70	26
	0·125	3·18	$\frac{1}{8}$″	—	0·150	3·80	25
10 (0·186″)	0·028	0·71	22	—	0·134	3·40	29
	0·048	1·22	18	—	0·142	3·60	28
	0·064	1·62	16	—	0·150	3·80	25
	0·104	2·64	12	—	0·161	4·10	20
	0·125	3·18	$\frac{1}{8}$″	—	0·169	4·30	18
	0·187	4·75	$\frac{3}{16}$″	—	0·177	4·50	16
12 (0·212″)	0·028	0·71	22	—	0·161	4·10	20
	0·048	1·22	18	—	0·169	4·30	18
	0·064	1·62	16	—	0·177	4·50	16
	0·104	2·64	12	—	0·189	4·80	12
	0·125	3·18	$\frac{1}{8}$″	—	0·193	4·90	10
	0·187	4·75	$\frac{3}{16}$″	—	0·201	5·10	7
14 (0·242″)	0·048	1·22	18	—	0·189	4·80	12
	0·064	1·62	16	—	0·205	5·20	6
	0·080	2·03	14	—	0·213	5·40	3
	0·125	3·18	$\frac{1}{8}$″	—	0·224	5·70	1
	0·187	4·75	$\frac{3}{16}$″	—	0·232	5·90	A
	0·250	6·35	$\frac{1}{4}$″	—	0·236	6·00	B

Note: If very hard material is being used a hole size slightly larger may have to be used, and in very soft material a smaller hole may be necessary.

TABLE XI – RECOMMENDED HOLE AND DRILL SIZES – GKN TYPES AB, B SELF-TAPPING SCREWS – 18–8 STAINLESS STEEL

In mild steel, monel metal, brass and aluminium alloy sheet

Screw size (No.) and nominal diameter	Material thickness in.	mm	SWG	Drilled or clean-punched holes: Hole diameter required in.	Drill size mm	Alternatives
4 (0·112″)	0·018	0·45	26	0·087	2·20	44
	0·036	0·91	20	0·091	2·30	43
6 (0·138″)	0·018	0·45	26	0·106	2·70	36
	0·036	0·91	20	0·110	2·80	35
8 (0·164″)	0·028	0·71	22	0·118	3·00	32
	0·048	1·22	18	0·126	3·20	$\frac{1}{8}$″
	0·064	1·62	16	0·134	3·40	29
10 (0·186″)	0·028	0·71	22	0·138	3·50	29
	0·048	1·22	18	0·146	3·70	26
	0·064	1·62	16	0·150	3·80	25
14 (0·242″)	0·048	1·22	18	0·213	5·40	3
	0·064	1·62	16	0·213	5·40	3

Notes:

1. **Because conditions differ widely it may be necessary to vary the hole size to suit a particular application.**
2. **18–8 quality stainless steel self-tapping screws are much softer than case hardened steel screws and therefore care must be exercised in using them. They cannot be used in very hard material. Also due to the galling tendency of stainless steel they should not be used in stainless steel sheet.**

TABLE XII – RECOMMENDED HOLE AND DRILL SIZES – GKN TYPE B SELF-TAPPING SCREWS 18–8 STAINLESS STEEL – **In non-ferrous castings or sections, aluminium, magnesium, zinc, brass, bronze, etc.**

Screw size (No.) and nominal diameter	Minimum penetration: Hole depth in.	Hole diameter in.	Drill size mm	Alternative	Normal maximum penetration: Hole depth in.	Hole diameter in.	Drill size mm	Alternative
4 (0·112″)	$\frac{5}{32}$	0·096	2·45	41	$\frac{5}{16}$	0·104	2·65	37
6 (0·138″)	$\frac{3}{16}$	0·130	3·30	30	$\frac{3}{8}$	0·130	3·30	30
8 (0·164″)	$\frac{7}{32}$	0·153	3·90	24	$\frac{7}{16}$	0·153	3·90	23
10 (0·186″)	$\frac{1}{4}$	0·177	4·50	16	$\frac{1}{2}$	0·177	4·50	16
14 (0·242″)	$\frac{5}{16}$	0·236	6·00	A	$\frac{9}{16}$	0·236	6·00	B

Notes:

1. **Cored Holes. A side taper of 1° 11′ is permissible. The diameter of a cored hole should equal the nominal hole size shown in the table above at one half the screw penetration depth.**
2. **Porous castings may require the use of a smaller hole and/or increased depth of engagement.**
3. **18–8 quality stainless steel self-tapping screws are much softer than case hardened steel screws and therefore care must be exercised in using them. They cannot be used in very hard material.**

TABLE XIII – RECOMMENDED HOLE AND DRILL SIZES – TYPE Y SELF-TAPPING SCREWS
In thermoset plastics and cast iron

Screw size (No.) and nominal diameter	Material thickness in.	Cellulose acetate, acrylic resin, cellulose nitrate (ie, perspex types)			Phenol formaldehyde (ie, bakelite types)			Cast iron		
		Hole diameter required in.	Drill size mm or in.	Drill size No. drill or equiv.	Hole diameter required in.	Drill size mm or in.	Drill size No. drill equiv.	Hole diameter required in.	Drill size mm or in.	No. drill equiv.
4 (0·112″)	$\frac{1}{8}$	0·094	$\frac{3}{32}$	42	0·100	2·55	39	—	—	—
	$\frac{1}{4}$	0·094	$\frac{3}{32}$	42	0·100	2·55	39	—	—	—
	$\frac{1}{2}$	0·095	2·40	42	0·100	2·55	39	0·102	2·60	38
6 (0·138″)	$\frac{1}{8}$	0·118	3·00	31	—	—	—	—	—	—
	$\frac{1}{4}$	0·125	$\frac{1}{8}$	—	0·130	3·30	30	—	—	—
	$\frac{1}{2}$	0·125	$\frac{1}{8}$	—	0·130	3·30	30	0·125	$\frac{1}{8}$	—
8 (0·164″)	$\frac{1}{8}$	0·150	3·80	25	—	—	—	—	—	—
	$\frac{1}{4}$	0·150	3·80	25	0·150	3·80	25	—	—	—
	$\frac{1}{2}$	0·150	3·80	25	0·150	3·80	25	0·153	3·90	23
10 (0·186″)	$\frac{1}{4}$	0·173	4·40	17	0·177	4·50	16	—	—	—
	$\frac{1}{2}$	0·177	4·50	16	0·177	4·50	16	0·177	4·50	16

Notes:
1 Because conditions differ widely it may be necessary to vary the hole size to suit a particular application.
2 Type 'U' screws are not generally suitable in materials other than those listed in the table above.

TABLE XIV – 'TAPTITE SCREW SIZES

ISO INCH (UNIFIED)

Screw Size	Threads per inch	C 0 −0·006	D +0 −0·006
No. 2	56	0·0875	0·084
No. 4	40	0·1145	0·1095
No. 5	40	0·1275	0·1225
No. 6	32	0·141	0·135
No. 8	32	0·167	0·161
No. 10	24	0·194	0·185
No. 10	32	0·193	0·187
$\frac{1}{4}$″	20	0·255	0·246
$\frac{5}{16}$″	18	0·318	0·308
$\frac{3}{8}$″	16	0·381	0·371

Tolerance on screw length

Nominal length	Tolerance
Up to $\frac{3}{4}$	+0 −0·031
Over $\frac{3}{4}$ up to $1\frac{1}{2}$	+0 −0·046
Over $1\frac{1}{2}$	+0 −0·062

All dimensions in inches

ISO METRIC

Screw Size	Pitch	C +0 −0·15	D 0 −0·15
M2·5	0·45	2·55	2·46
M3	0·50	3·05	2·95
(M3·5)	0·60	3·56	3·44
M4	0·70	4·07	3·93
M5	0·80	5·08	4·92
M6	1·00	6·10	5·90
M8	1·25	8·13	7·88
M10	1·50	10·15	9·85

The thread diameter in parentheses is non-preferred

Tolerance on screw length

Nominal length	Tolerance
Up to 10	+0 −0·40
Over 10 up to 20	+0 −0·80
Over 20 up to 40	+0 −1·20
Over 40	+0 −1·50

All dimensions in millimetres

TABLE XV – RECOMMENDED PILOT HOLE SIZES AND PERCENTAGE THREAD ENGAGEMENT FOR GKN 'TAPTITE' SCREWS

Steel Sheet and Bar

Material thickness in	·031–·078			·062–·140			·125–·266			·250–·326			·312–·500		
Screw size	Drill Size mm	in	% thread	Drill Size mm	in	% thread	Drill Size mm	in	% thread	Drill Size mm	in	% thread	Drill Size mm	in	% thread
2–56 UNC	1·90	·074	85	2·00	·079	65	2·05	·081	50						
4–40 UNC	2·50	·098	85	2·60	·102	65	2·65	·104	50						
6–32 UNC				3·10	·122	90	3·10	·122	90	3·20	·126	60			
8–32 UNC				3·70	·146	85	3·80	·150	70	3·90	·154	50			
10–24 UNC				4·20	·165	90	4·40	·173	70	4·50	·177	50			
10–32 UNF				4·40	·173	85	4·50	·177	65	4·60	·181	50			
¼"–20 UNC							5·70	·224	80	5·80	·228	70	6·00	·236	50
5/16"–18 UNC							7·10	·280	85	7·30	·287	75	7·40	·291	60
⅜"–16 UNC							8·70	·343	75	8·90	·350	65	9·00	·354	50

Aluminium Sheet and Bar

Material thickness in	·031–·078			·062–·140			·125–·266			·250–·328			·312–·500		
Screw size	Drill Size mm	in	% thread	Drill Size mm	in	% thread	Drill Size mm	in	% thread	Drill Size mm	in	% thread	Drill Size mm	in	% thread
2–56 UNC	1·90	·074	85	2·00	·079	65	2·00	·079	65						
4–40 UNC	2·50	·098	85	2·55	·100	75	2·60	·102	65	2·60	·102	65			
6–32 UNC	3·10	·118	90	3·10	·122	90	3·10	·122	90	3·20	·126	65	3·20	·126	65
8–32 UNC				3·70	·146	85	3·80	·150	70	3·80	·150	70	3·90	·154	60
10–24 UNC				4·20	·165	90	4·20	·165	90	4·30	·170	75	4·40	·173	65
10–32 UNF				4·40	·173	85	4·40	·173	85	4·50	·177	65	4·60	·181	55
¼"–20 UNC				5·50	·217	95	5·60	·221	90	5·70	·224	80	5·80	·228	70
5/16"–18 UNC				7·10	·280	85	7·10	·280	85	7·20	·284	75	7·30	·287	70
⅜"–16 UNC							8·60	·339	90	8·70	·343	75	8·90	·350	65

TABLE XVI – RECOMMENDED EXTRUDED HOLE DIAMETERS FOR TAPTITE SCREWS (Diameter in inches)

Screw Size	MATERIAL THICKNESS (Inches)											
	0.02	0.03	0.04	0.06	0.09	0.13	0.16	0.19	0.22	0.25	0.31	0.38
6–32 UNC	0.116 0.119	0.117 0.119	0.118 0.121	0.119 0.122	0.122 0.125							
8–32 UNC	0.142 0.145	0.143 0.146	0.143 0.146	0.144 0.147	0.146 0.149	0.149 0.152						
10–24 UNC	0.160 0.164	0.161 0.165	0.162 0.166	0.173 0.167	0.166 0.170	0.169 0.163						
10–32 UNF	0.167 0.170	0.168 0.171	0.169 0.172	0.170 0.173	0.172 0.175	0.174 0.177						
¼in – 20 UNC			0.215 0.219	0.217 0.221	0.220 0.224	0.222 0.226	0.224 0.228	0.227 0.231	0.229 0.233	0.231 0.235		
5/16in 18 UNC				0.271 0.275	0.272 0.276	0.275 0.279	0.277 0.281	0.279 0.283	0.281 0.285	0.284 0.288		
3/8in 16 UNC						0.332 0.336	0.334 0.338	0.336 0.340	0.337 0.341	0.339 0.343	0.342 0.346	0.345 0.349

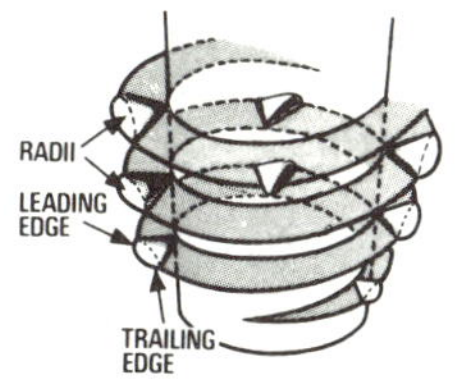

TABLE XVII – RECOMMENDED HOLE SIZES FOR SWAGE FORM

INCH-UNIFIED — MATERIAL THICKNESS

Screw Diam.	Material	.020-.078 Drill	.020-.078 Hole Dia	.063-.140 Drill	.063-.140 Hole Dia	.125-.266 Drill	.125-.266 Hole Dia	.250-.328 Drill	.250-.328 Hole Dia	.313-.500 Drill	.313-.500 Hole Dia
4-40	Steel	40	.098	38	.102	37	.104				
	Alum. Zinc	40	.098	39	.100	38	.102	38	.102		
6-32	Steel			31	.120	⅛	.125	30	.129		
	Alum. Zinc	31	120	31	.120	⅛	.125	⅛	.125	⅛	.125
8-32	Steel			26	.147	25	.150	23	.154		
	Alum. Zinc			26	.147	25	.150	24	.152	24	.152
10-24	Steel			19	.166	11/64	.172	16	.177		
	Alum. Zinc			19	.166	19	.166	18	.170	17	.173
10-32	Steel			17	.173	16	.177	15	.180		
	Alum Zinc			17	.173	17	.173	16	.177	15	.180

METRIC — MATERIAL THICKNESS

Screw Diam.	Material	.50-2.00 Hole Dia Drill Size	1.50-3.50 Hole Dia Drill Size	3.00-6.50 Hole Dia Drill Size	6.00-8.00 Hole Dia Drill Size	8.00-12.00 Hole Dia Drill Size
M3	Steel	2.70	2.75	2.80		
	Alum. Zinc	2.60	2.70	2.75		
M3.5	Steel	3.10	3.20	3.20	3.30	3.30
	Alum. Zinc	3.10	3.10	3.20	3.20	3.20
M4	Steel	3.60	3.70	3.70	3.80	3.80
	Alum. Zinc		3.60	3.70	3.70	3.70
M5	Steel		4.50	4.60	4.70	4.70
	Alum. Zinc		4.50	4.50	4.60	4.70

Ephraim Phillips square-flo screw.

TABLE XVIII – RECOMMENDED HOLE SIZES FOR SQUARE-FLO

Screw Size	Stock Thickness in Inches									
	0.050	0.060	0.083	0.109	0.125	0.140	3/16	1/4	5/16	3/8
IN STEEL										
6-32	0.1100	0.1130	0.1160	0.1160	0.1160	0.1200	0.1250	0.1250	......	......
8-32	0.1360	0.1405	0.1405	0.1440	0.1440	0.1470	0.1495	0.1495	0.1495	......
10-24	0.1520	0.1540	0.1610	0.1610	0.1660	0.1695	0.1730	0.1730	0.1730	0.1730
10-32	0.1590	0.1660	0.1660	0.1695	0.1695	0.1695	0.1770	0.1770	0.1770	0.1770
12-24	......	0.1800	0.1820	0.1875	0.1910	0.1910	0.1990	0.1990	0.1990	0.1990
1/4-20	......	......	0.2130	0.2188	0.2210	0.2210	0.2280	0.2280	0.2280	0.2280
1/4-28	......	......	0.2210	0.2280	0.2280	0.2340	0.2344	0.2344	0.2344	0.2344
5/16-18	......	......	......	0.2770	0.2770	0.2813	0.2900	0.2900	0.2900	0.2900
5/16-24	......	......	......	0.2900	0.2900	0.2900	0.2950	0.2950	0.2950	0.2950
3/8-16	......	......	......	......	0.3390	0.3390	0.3480	0.3580	0.3580	0.3580
3/8-24	......	......	......	......	0.3480	0.3480	0.3580	0.3580	0.3580	0.3580
IN ALUMINIUM										
6-32	0.1094	0.1094	0.1110	0.1130	0.1160	0.1160	0.1200	0.1250	......	......
8-32	0.1360	0.1360	0.1360	0.1405	0.1405	0.1440	0.1470	0.1495	0.1495	......
10-24	0.1495	0.1520	0.1540	0.1570	0.1590	0.1610	0.1660	0.1719	0.1730	0.1730
10-32	0.1610	0.1610	0.1610	0.1660	0.1660	0.1660	0.1719	0.1770	0.1770	0.1770
12-24	......	0.1770	0.1800	0.1820	0.1850	0.1875	0.1910	0.1990	0.1990	0.1990
1/4-20	......	......	0.2055	0.2090	0.2130	0.2130	0.2210	0.2280	0.2280	0.2280
1/4-28	......	......	0.2188	0.2210	0.2210	0.2210	0.2280	0.2344	0.2344	0.2344
5/16-18	......	......	......	0.2660	0.2720	0.2720	0.2810	0.2900	0.2900	0.2900
5/16-24	......	......	......	0.2810	0.2812	0.2812	0.2900	0.2950	0.2950	0.2950
3/8-16	......	......	......	......	0.3281	0.3320	0.3390	0.3480	0.3480	0.3480
3/8-24	......	......	......	......	0.3438	0.3438	0.3480	0.3580	0.3580	0.3580

Note: Because conditions differ widely, it may be necessary to vary the hole size to suit a particular application.

TABLE XVIX — EQUIVALENT SIZES; SELF-TAPPING SCREWS, MACHINE SCREWS AND WOODSCREWS

Inches Decimal	Fraction	Wood Screw Gauge	Self-Tapping Screw Types A & B Gauge	Machine Screws		
				UNC/UNF	BA	ISO Metric
0.060		0				
0.062	1/16					
0.070		1				
0.082		2				
0.086			2		8	
0.092	3/32					
0.094		3				
0.098					7	2.5
0.108		4				
0.110					6	
0.112			4	4		
0.118						3
0.122		5				
0.125	1/8					
0.136		6				
0.138			6	6		
0.141					4	
0.150		7				
0.151			7			
0.156	5/32					
0.158						4
0.164		8	8	8		
0.185					2	
0.187	3/16					
0.190			10	10		
0.192		10				
0.197						5
0.216			12			
0.218	7/32					
0.220		12				
0.236					0	6
0.242			14			
0.248		14				
0.250	1/4			1/4		
0.276		16				
0.281	9/32					
0.304		18				
0.312	5/16			5/16		
0.315						8
0.332		20				
0.375	3/8			3/8		
0.394						10
0.437	7/16			7/16		

Woodscrews

Woodscrews conform to the geometry shown in Fig 1. Length is defined as the length from point to flat (maximum diameter) section of the head; diameter is the diameter of the parallel plain length of shank. The threaded section extends for approximately two-thirds of the length. Three types of head are standard:-

(i) Countersunk, with an angle of approximately 90° (80°–82° in American practice).

(ii) Round head

(iii) Raised head (Industrial head or Raised Countersunk), with an angle of approximately 90° (80°–82° in American practice).

Head diameters follow these proportions:-

	Countersunk	Round	Raised
Full diameter	2D	2D	2D
Head thickness	0.5D*	6.75D	0.5D + 0.333D
Slot width (S)	depends on screw diameter		
Slot depth (N)	0.25D*	0.35D	0.333D

*plus 0.005in to 0.025in, depending on screw size

Heads may be slotted or recessed (cross-slotted, Phillips or Pozidriv) – see Fig 2.

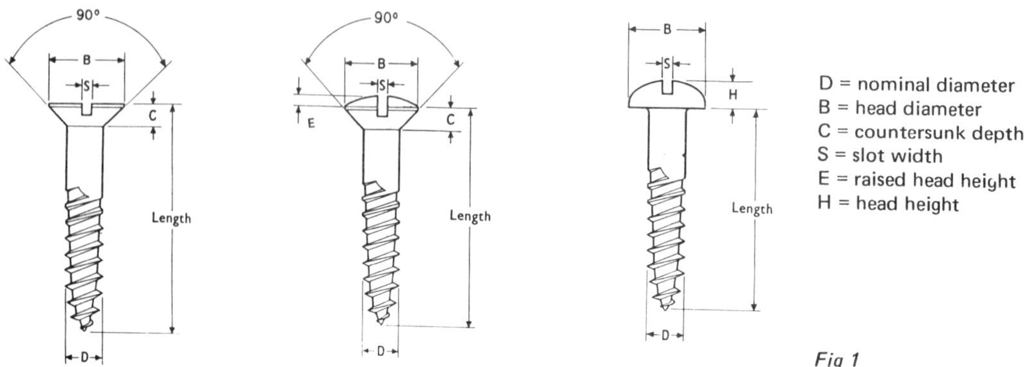

Fig 1

Slotted Heads

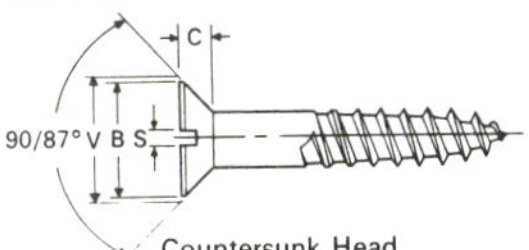

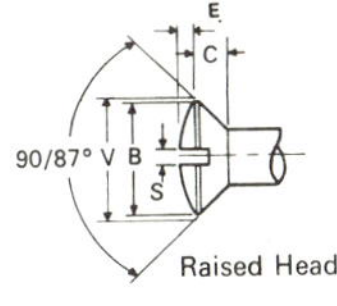

Round Head

Pozidriv Heads

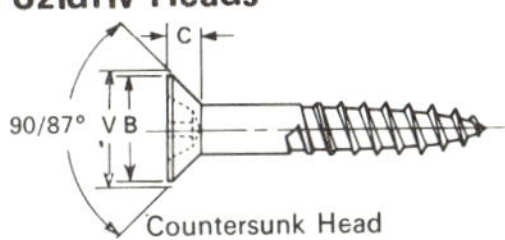

Raised Head

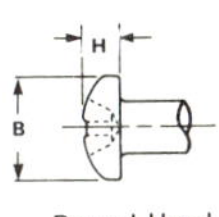

Round Head

Phillips Heads

Fig 2

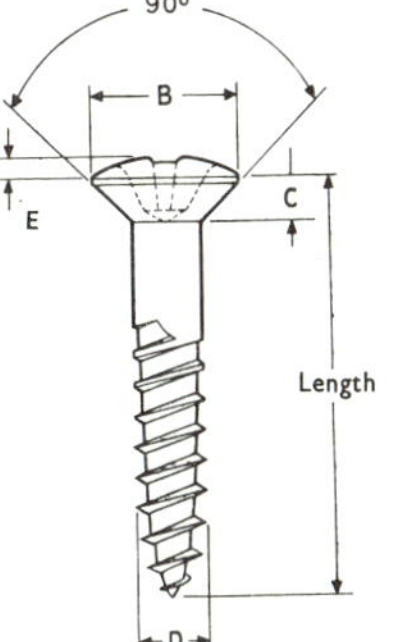

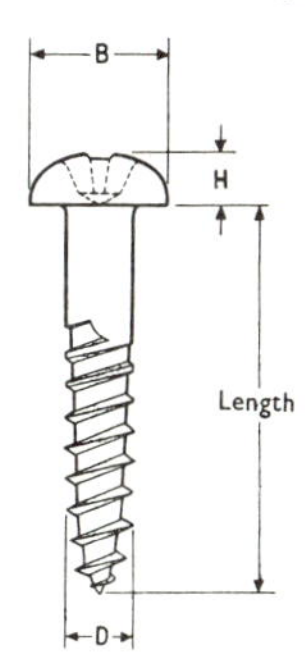

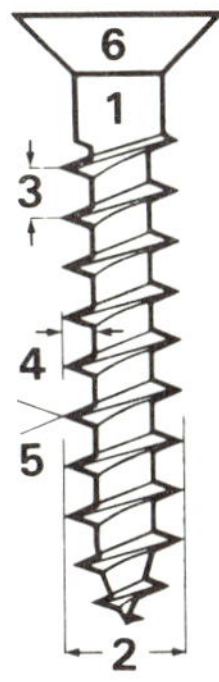

Winglin Autopilot woodscrew with deep, thin threads designed for screwing into wood without a pilot hole. (Linread Fasteners Ltd).

1—Natural, fully heat treated steel.
2—Major diameters comparable to standard wood-screw ranges.
3—Widely spaced thread to prevent stripping out.
4—Extra deep thread to provide full grip.
5—Thin thread, deep cut for easy penetration.
6—Pozidriv or slotted head.

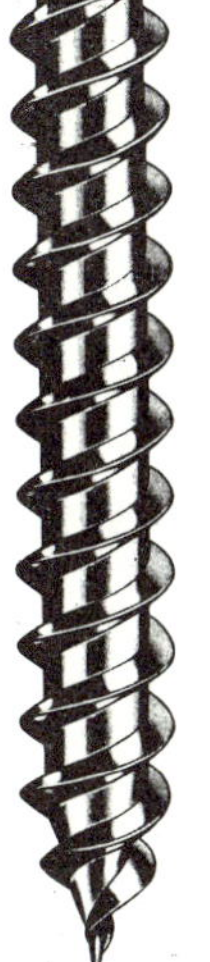

Bierbach woodscrew with twin thread and self-centering point.

Turned single-thread wood screw.

TABLE I – IMPERIAL SIZE WOODSCREWS†

Screw Gauge	Nominal Size in	Threads per inch	Countersunk and Raised Countersunk				Round heads		Slot Width
			*V**	*B* min	*C* max	*H* ☆ nom	*D* max	*E* max	*S* min
0	0.060	30	0.126	0.114	0.035	–	0.116	0.045	0.016
1	0.070	28	0.147	0.133	0.041	–	0.140	0.053	0.021
2	0.082	26	0.174	0.156	0.048	0.020	0.164	0.062	0.026
3	0.094	24	0.199	0.179	0.054	0.024	0.189	0.071	0.030
4	0.108	22	0.230	0.205	0.064	0.027	0.215	0.081	0.032
5	0.122	20	0.261	0.232	0.073	0.030	0.241	0.090	0.035
6	0.136	18	0.291	0.258	0.082	0.034	0.267	0.100	0.040
7	0.150	16	0.323	0.285	0.091	0.038	0.293	0.109	0.040
8	0.164	14	0.353	0.312	0.100	0.041	0.319	0.118	0.045
9	0.178	12	0.384	0.338	0.109	–	0.345	0.127	0.045
10	0.192	12	0.414	0.365	0.117	0.048	0.372	0.136	0.050
12	0.220	10	0.476	0.418	0.135	0.055	0.424	0.154	0.055
14	0.248	9	0.538	0.472	0.153	–	0.476	0.171	0.065
16	0.276	8	0.599	0.524	0.170	–	0.529	0.190	0.065
18	0.304	7½	0.660	0.578	0.188	–	0.580	0.207	0.075
20	0.332	7	0.721	0.631	0.205	–	0.632	0.226	0.075

* The dimensions for *V* are the theoretical diameters of head to sharp corners and are given for design purposes only.

☆ Raised only *(H)*

† Note : This table covers the normal production range.
Sizes up to gauge number 40 (nominal dia 0.612in) are detailed in BS1210

Screw diameter size is designated by number (see Tables I and II). The choice of diameter size and screw length are determined by the width and thickness of the wood into which it is to be screwed. Ideally, the screw diameter should not be greater than one tenth the width of wood involved. (Fig 3). The length of screw should be such that a minimum of seven full turns of thread are engaged – or in the case of hardwoods a minimum of four full turns of thread.

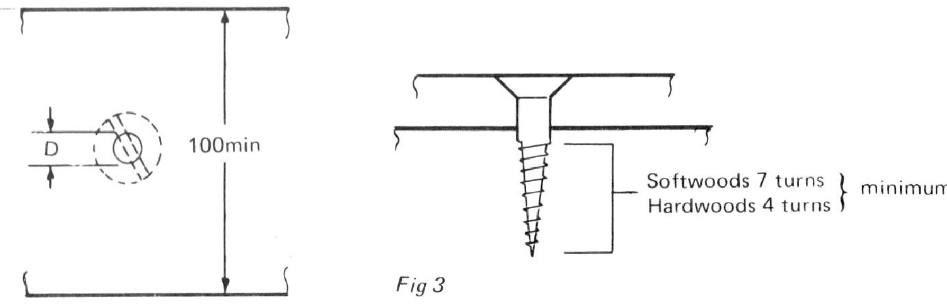

Fig 3

TABLE II – AMERICAN STANDARD WOODSCREWS (ANSI B18.6.1)

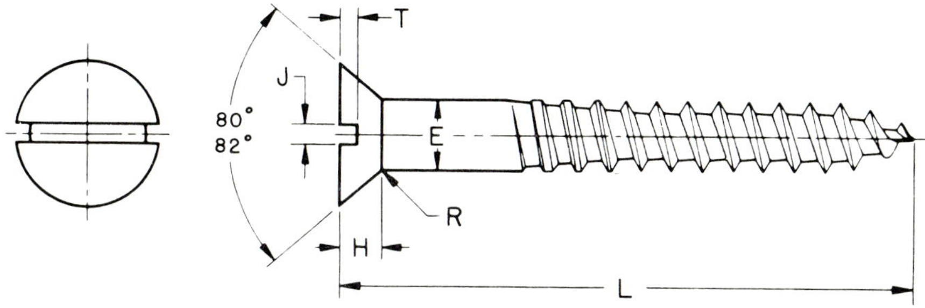

Nominal Size and Basic Screw Diameter (in)		Threads per Inch	E		A		H	J		T		R
			Body Diameter		Head Diameter		Head Height	Slot Width		Slot Depth		Fillet Radius
			Max	Min	Max, Edge Sharp	Min, Edge Rounded or Flat		Max	Min	Max	Min	Max
0	0.060	32	0.064	0.053	0.119	0.099	0.035	0.023	0.016	0.015	0.010	0.031
1	0.073	28	0.077	0.066	0.146	0.123	0.043	0.026	0.019	0.019	0.012	0.031
2	0.086	26	0.090	0.079	0.172	0.147	0.051	0.031	0.023	0.023	0.015	0.031
3	0.099	24	0.103	0.092	0.199	0.171	0.059	0.035	0.027	0.027	0.017	0.031
4	0.112	22	0.116	0.105	0.225	0.195	0.067	0.039	0.031	0.030	0.020	0.031
5	0.125	20	0.129	0.118	0.252	0.220	0.075	0.043	0.035	0.034	0.022	0.062
6	0.138	18	0.142	0.131	0.279	0.244	0.083	0.048	0.039	0.038	0.024	0.062
7	0.151	16	0.155	0.144	0.305	0.268	0.091	0.048	0.039	0.041	0.027	0.062
8	0.164	15	0.168	0.157	0.332	0.292	0.100	0.054	0.045	0.045	0.029	0.062
9	0.177	14	0.181	0.170	0.358	0.316	0.108	0.054	0.045	0.049	0.032	0.062
10	0.190	13	0.194	0.183	0.385	0.340	0.116	0.060	0.050	0.053	0.034	0.062
12	0.216	11	0.220	0.209	0.438	0.389	0.132	0.067	0.056	0.060	0.039	0.062
14	0.242	10	0.246	0.235	0.491	0.437	0.148	0.075	0.064	0.068	0.044	0.093
16	0.268	9	0.272	0.261	0.544	0.485	0.164	0.075	0.064	0.075	0.049	0.093
18	0.294	8	0.298	0.287	0.597	0.534	0.180	0.084	0.072	0.083	0.054	0.093
20	0.320	8	0.324	0.313	0.650	0.582	0.196	0.084	0.072	0.090	0.059	0.093
24	0.372	7	0.376	0.365	0.756	0.679	0.228	0.094	0.081	0.105	0.069	0.093

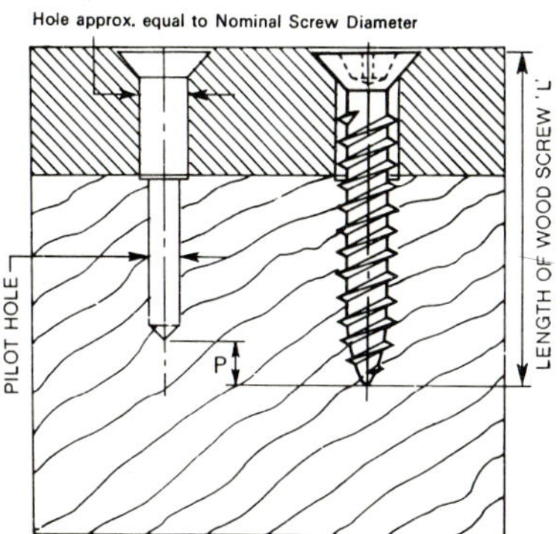

Fig 4 'P' is depth allowance – see Table III.

To eliminate the risk of splitting the wood (and also possibly shearing the screw in the case of brass screws in hardwood), woodscrews should be assembled in pre-drilled clearance and pilot holes – Fig 4. The clearance hole can extend completely through the upper of the two surfaces to be joined (or one-third of the length of the screw, if screwed into a single piece of wood). The pilot hole can extend to some 80% of the length of the screw, to leave a suitable depth allowance (P).

Recommended drill sizes for clearance and pilot holes are given in Tables III and IIIA.

Recommendations for the spacing of woodscrews are:-

Distance from end of wood	10 x D*
Distance from edge of wood	5 x D
Distance between lines of screws	3 x D*
Distance along grain,between screws	10 x D*

where D is the screw diameter

*These distances should be doubled if the screws are assembled without pilot holes.

As a general rule, woodscrews should not be used in end grain wood as their holding capacity will then be poor; and brass screws should not be used on assemblies subject to high shear loads.

Standard materials used for the manufacture of woodscrews are brass, mild steel, stainless steel and silicon bronze. Brass, stainless steel and silicon bronze screws are used for marine assemblies. Chrome plated brass or stainless steel screws are a logical choice where the screw head is exposed and is required to maintain a clean, attractive appearance (in the latter case a recess head is also normally preferred).

TABLE III – RECOMMENDED CLEARANCE AND PILOT HOLE SIZES FOR WOODSCREWS (ABRIDGED LIST)

Screw Gauge No	Clearance Drill			Pilot Hole Size Hardwoods			Pilot Hole Size Softwoods			Depth *Allowance	
	dia	in	mm	dia	in	mm	dia	in	mm	in	mm
Under 3	–	–	–	pointed awl			pointed awl			3/32	2
3	0.094	3/32	2.40	0.057	no 59	1.45	pointed awl			1/8	3
4	0.108	7/64	2.75	0.066	no 51	1.70	pointed awl			1/8	3
5	0.122	1/8	3.10	0.073	no 49	1.85	pointed awl			5/32	4
6	0.136	no 29	3.50	0.082	no 45	2.10	0.059	no 53	1.50	5/32	4
7	0.150	no 24	3.90	0.091	3/32	2.30	0.066	no 51	1.70	3/16	5
8	0.164	no 19	4.20	0.097	no 41	2.50	0.071	no 50	1.80	7/32	5.5
9	0.178	no 16	4.50	0.103	no 38	2.65	0.078	5/64	2.00	1/4	6
10	0.192	no 10	4.90	0.108	7/64	2.75	0.082	no 45	2.15	1/4	7
12	0.220	7/32	5.60	0.124	1/8	3.15	0.097	no 40	2.50	5/16	8
14	0.248	1/4	6.30	0.140	9/64	3.60	0.108	7/64	2.75	11/32	9

*See Fig 4

TABLE IIIA – PILOT HOLES FOR WOOD SCREWS (FULL RANGE)

WOOD SCREW				PILOT HOLE							
Diameter		Core diameter	3× Pitch	Hard Woods				Soft Woods			
				Pilot hole diameter	Drill sizes			Pilot hole diameter	Drill sizes		
Screw Gauge	Nom.	Approx	Approx		Fraction	mm	Letter or number		Fraction	mm	Letter or number
00	·057	·036	$\frac{3}{32}$	·034		·875	65				
0	·063	·040	$\frac{3}{32}$	·038		·975	62				
1	·070	·044	$\frac{3}{32}$	·042		1·05	58	Not required			
2	·082	·051	$\frac{1}{8}$	·049		1·25					
3	·094	·059	$\frac{1}{8}$	·057		1·45					
4	·108	·068	$\frac{1}{8}$	·066		1·70	51				
5	·122	·076	$\frac{5}{32}$	·073		1·85	49				
6	·136	·085	$\frac{5}{32}$	·082		2·10	45	·059		1·50	53
7	·150	·094	$\frac{3}{16}$	·091	$\frac{3}{32}$	2·30		·066		1·70	51
8	·164	·102	$\frac{7}{32}$	·097		2·50	40	·071		1·80	49
9	·178	·111	$\frac{1}{4}$	·103		2·65	37	·078	$\frac{5}{64}$	2·00	47
10	·192	·120	$\frac{1}{4}$	·108	$\frac{7}{64}$	2.75	35	·084		2·15	44
11	·206	·129	$\frac{9}{32}$	·116		2·95	32	·090		2·30	
12	·220	·138	$\frac{5}{16}$	·124	$\frac{1}{8}$	3·15		·097		2·50	40
14	·248	·155	$\frac{11}{32}$	·140	$\frac{9}{64}$	3·60	28	·108	$\frac{7}{64}$	2·75	35
16	·276	·173	$\frac{3}{8}$	·156	$\frac{5}{32}$	4·00	22	·121		3·10	
18	·304	·190	$\frac{13}{32}$	·171	$\frac{11}{64}$	4·35	17	·131		3·35	
20	·332	·208	$\frac{7}{16}$	·187	$\frac{3}{16}$	4·75	12	·145		3·70	26
22	·360	·225	$\frac{15}{32}$	·202	$\frac{13}{64}$	5·20	6	·157	$\frac{5}{32}$	4·00	22
24	·388	·243	$\frac{1}{2}$	·219	$\frac{7}{32}$	5·60	2	·170	$\frac{11}{64}$	4·35	18
28	·444	·278	$\frac{9}{16}$	·250	$\frac{1}{4}$	6·40	E	·195		5·00	9
32	·500	·313	$\frac{5}{8}$	·282	$\frac{9}{32}$	7·20	K	·219	$\frac{7}{32}$	5·60	2

Load Capacity of Woodscrews

Typical 'holding' strengths for woodscrews are given in BS CP 112, Tables 25 and 26. Approximate values of pullout loads for standard woodscrews can be calculated from the following empirical formula

$$\text{Load} = K.f.D.L.(SG)^2$$

where D = screw diameter
L = length of penetration
SG = specific gravity of wood
K = grain factor
= 1.0 for penetration into long grain
= 0.65 for penetration into end grain
f = factor dependent on units employed
= 2375 for load in lbf and D and L in inches
= 1.67 for load in kgf and D and L in millimetres

Twinfast Threads

Woodscrews with twin parallel threads have the advantage of greater holding power and faster driving (ie driving in half the number of turns). They also tend to drive more truly. They are particularly suitable for use in softer materials, such as fibreboards, blockboards and lower density woods.

Pilot hole sizes required are slightly larger than for conventional woodscrews (see Table IV). Smaller screw lengths are threaded to the head. Dimensional data of GKN Twinfast Woodscrews are summarised in Table V.

TABLE IV – RECOMMENDED PILOT HOLE SIZES FOR POZIDRIV TWINFAST

Screw Gauge	Hardwoods			Softwoods			Depth Allowance (Fig 4)
	Pilot Hole Diameter	Drill Sizes		Pilot Hole Diameter	Drill Sizes		
		Fraction	mm		Fraction	mm	
3	0.063	1/16in		0.035		0.90	0.100
4	0.070		1.80	0.049		1.25	0.125
5	0.082		2.10	0.057		1.45	0.136
6	0.093	3/32in		0.062	1/16in		0.150
7	0.106		2.70	0.065		1.65	0.166
8	0.116		2.95	0.076		1.95	0.166
10	0.125	1/8in		0.089		2.25	0.200
12	0.142		3.60	0.102		2.60	0.231

TABLE V – GKN TWINFAST WOODSCREWS

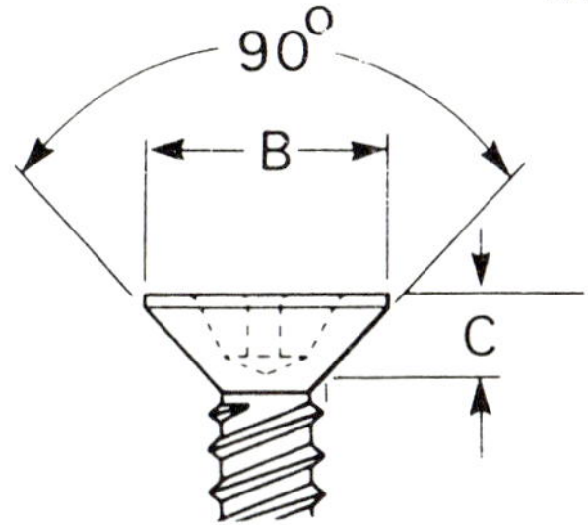

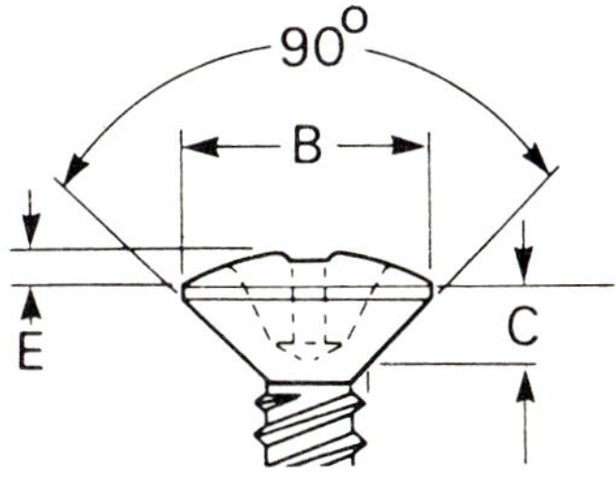

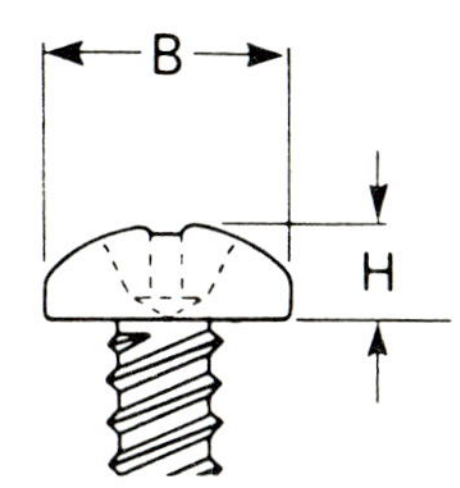

Head dimensions

All heads			Countersunk			Raised Countersunk				Round		
Screw gauge	*Basic thread dia*	*Threads per inch*	*B max*	*C max*	*Driver point no*	*B max*	*C max*	*E nom*	*Driver point no*	*B max*	*H max*	*Driver point no*
	Inches		*Inches*	*Inches*		*Inches*	*Inches*	*Inches*		*Inches*	*Inches*	
3	0·099	26	0·188	0·055	1	0·188	0·055	0·024	1	0·193	0·071	1
4	0·112	24	0·216	0·064	1	0·216	0·064	0·026	1	0·219	0·080	1
5	0·125	22	0·244	0·073	2	0·244	0·073	0·030	2	0·245	0·089	2
6	0·138	20	0·272	0·082	2	0·272	0·082	0·033	2	0·270	0·097	2
7	0·151	18	0·300	0·091	2	0·300	0·091	0·040	2	0·296	0·106	2
8	0·164	18	0·328	0·100	2	0·328	0·100	0·040	2	0·322	0·115	2
10	0·190	15	0·384	0·117	2	0·384	0·117	0·046	2	0·373	0·133	2
12	0·216	13	0·440	0·135	3	0·440	0·135	0·055	3	0·425	0·151	3

Length of thread – all head styles

Nom length \ Dia, screws gauge	4	5	6	7	8	10	12
¾ and under	*Threaded to head*						
⅞	0·780	0·780	0·780	0·765	0·765	0·750	
1	0·874	0·874	0·874	0·859	0·859	0·844	0·844
1¼	1·061	1·061	1·061	1·046	1·046	1·031	1·031
Over 1¼	¾ Length+0·030			¾ Length+0·015		¾ Length	

Dimensions in inches

Dowel Screws

Dowel screws are a special type of woodscrew for invisibly joining T-joints, etc. They are double-ended woodscrews with plain centre sections. Lengths available are usually from 19mm (¾in) to 62.5mm (2½in) in a limited number of gauges.

Coach Screws

The traditional coachscrew is a sturdy (heavy gauge) woodscrew with a square head for turning in the work with a spanner. They are used to provide strong fixings in heavy wood constructions. Sizes may range up to 16in in length.

'Blind' Woodscrews

These are 'headless' woodscrews with a relatively long plain, parallel shank above the woodscrew length. A portion of the plain shank length is cut or formed with a machine screw thread to take a special matching nut with axial grooves. These screws are used for 'hidden' fastenings where the nut can be tightened by a screwdriver.

Special Screws

Drive Screws

Drive screws are special 'jamming' screw forms (usually in hardened steel) for hammer driving into matching holes for fastening heavy gauge metal sheet, structural steelwork, etc, and for fastening to ferrous and non-ferrous castings. Pointed tip drive screws are suitable for blind fastening of insulating panels, wallboards, linings, etc, to sheet metal and similar hard materials. They are basically a coarse pitch, shallow thread form of self-tapping screw for straight driving or pressing into position – see also chapter on *Self-Tapping Screws.*

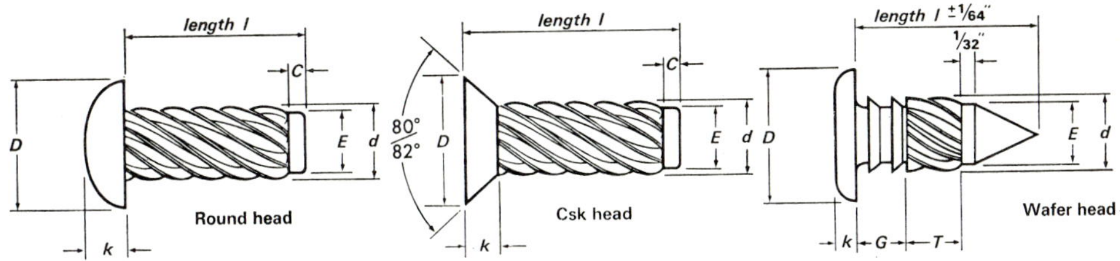

Examples of drive screws

GKN HAMMER DRIVE SCREWS TYPE U21 (WAFER)

Screw Size No	Length		Major diameter Maximum		Point dia Maximum		Head dia Maximum		Head depth Maximum		G Maximum		T Nominal	
	in	mm	in	mm	in	mm	in	mm	in	mm	in	mm	in	mm
4	11/32	8.73	0.116	2.95	0.095	2.41	0.256	6.50	0.027	0.69	0.125	3.18	0.068	1.73
6	7/16	11.11	0.140	3.56	0.115	2.92	0.302	7.67	0.047	1.19	0.125	3.18	0.134	3.40
7	1/2	12.70	0.154	3.91	0.125	3.18	0.302	7.67	0.047	1.19	0.136	3.45	0.173	4.39
8	7/16	11.11	0.167	4.24	0.135	3.43	0.307	7.80	0.052	1.32	0.142	3.61	0.091	2.31

GKN HAMMER DRIVE SCREWS TYPE U (ROUND OR CSK HEAD)

Screw Size No	All Types				Round Head				Countersunk Head			
	Major diameter maximum		Point diameter maximum		Head diameter maximum		Head depth maximum		Head diameter maximum		Head depth maximum	
	in	mm	in	mm	in	mm	in	mm	in	mm	in	mm
00	0.060	1.52	0.049	1.24	0.099	2.51	0.034	0.86	0.110	2.79	0.038	0.97
0	0.075	1.90	0.063	1.60	0.127	3.22	0.049	1.24	0.137	3.48	0.046	1.17
2	0.100	2.54	0.083	2.11	0.162	4.11	0.069	1.75	0.188	4.78	0.064	1.62
4	0.116	2.95	0.096	2.44	0.211	5.36	0.086	2.18	0.214	5.44	0.072	1.83
6	0.140	3.56	0.116	2.95	0.260	6.60	0.103	2.62	0.263	6.68	0.089	2.26
7	0.154	3.91	0.126	3.20	0.285	7.24	0.111	2.82	0.289	7.34	0.099	2.51
8	0.167	4.24	0.136	3.45	0.309	7.85	0.120	3.05	0.316	8.03	0.109	2.77
10	0.182	4.62	0.150	3.81	0.359	9.12	0.137	3.48	0.364	9.25	0.130	3.30
12	0.212	5.38	0.177	4.50	0.408	10.36	0.153	3.89	0.415	10.54	0.144	3.66
14	0.242	6.15	0.202	5.13	0.457	11.61	0.170	4.32	0.466	11.84	0.159	4.05

Length of pilot

Length of screw l	Below 1/8	Below 3.2	1/8 to 5/32	3.2 to 4	3/16 to 5/16	4.8 to 8	3/8 to 1/2	9.5 to 13	5/8 to 7/8	16 to 22	1 and over	25 and over
Length of pilot Min **C**	0.020	0.5	0.035	0.9	0.045	1.2	0.062	1.6	0.078	2.0	0.125	3.2

RECOMMENDED HOLE AND DRILL SIZES FOR HAMMER DRILL SCREWS

Steel Type 'U' GKN Hammer Drive Screws

Screw Gauge and Nominal Diameter	in	Thin Sheet Metal, Non-Ferrous Castings, Plastics, etc. — Drill Size: Hole Dia required in	mm	Number alter-native	Cast Iron and Thick Sheet Metal — Drill Size: Hole Dia required in	mm	Number alter-native	Clearance Hole — Drill Size: Hole Dia required in	mm or in	Alter-natives
00	·059	·051	1·30	55	·055	1·40	54	·067	1·70	51
0	·074	·065	1·65	52	·069	1·75	50	·082	2·1	45
2	·099	·087	2·20	44	·091	2·30	42	·107	2·7	36
4	·114	·100	2·55	39	·106	2·70	36	·125	$\frac{1}{8}$"	—
6	·138	·122	3·10	31	·130	3·30	30	·150	3·8	25
7	·152	·134	3·40	29	·142	3·60	27	·166	4·2	19
8	·165	·146	3·70	27	·154	3·90	23	·181	4·6	15
10	·180	·161	4·10	20	·169	4·30	18	·196	5·0	9
12	·209	·189	4·80	12	·197	5·00	8	·228	5·8	1
14	·239	·217	5·50	2	·228	5·80	1	·261	6·6	G

Notes

1. The material should be thick enough to provide adequate thread engagement, and normally should not be less than the screw diameter.
2. In applications in plastic, the rigidity of the section and the brittleness of the plastic must be considered.

18-8 Stainless Steel Type 'U' GKN Hammer Drive Screws

Screw Gauge and Nominal Diameter	in	In Non-Ferrous Sheet and Castings — Drill Size: Hole Dia. required in	mm or in	Number Alter-natives	In Plastics — Drill Size: Hole Dia required in	mm	Number alter-native
00	·059	·055	1·40	54	·051	1·30	55
0	·074	·071	1·80	50	·067	1·70	51
2	·099	·089	2·25	43	·087	2·20	44
4	·114	·108	2·75	36	·104	2·65	37
6	·138	·125	$\frac{1}{8}$"	—	·118	3·00	31

Note

18-8 quality stainless steel self-tapping screws are softer than case hardened steel screws and care must be exercised in using them. They cannot be used in very hard materials.

Drywall Screws

Hi-Lo drywall screws are based on the Hi-Lo thread form (see chapter on *Self-Tapping Screws*) and have been specifically developed as fasteners for building construction, etc. Examples of drywall screw applications are:-

Type S – Bugle head	Screw length in
Single layer ½in drywall to metal stud	7/8
Steel channels to framing	7/8
¼in backerboard to metal stud	7/8
Single layer ½in or 5/8in drywall to metal stud	1
Steel channels to wood framing	1
5/8in drywall to channel when resiliency is a problem	1.1/8
5/8in drywall ceiling into resilient channel	1¼
Batten strips through board into steel stud	1¼
Trim items to demountable wall systems (cadmium plate)	1¼
Double layer of ½in drywall to metal stud	1.5/16
Double layer of 5/8in drywall to metal stud	1.5/8
½in drywall through coreboard to metal runners in solid partitions	1.7/8
5/8in drywall through coreboard to metal runners in solid partitions	2¼
Double layer of 1in drywall to stud (used around elevator shafts and stair wells)	2.5/8
Type S – Oval Head	
Steel cabinets, base plate brackets or other accessories to walls of ½in or 5/8in wallboard and steel studs	1¼
Cabinets, shelves or wood frames, made from 3/8in or 5/8in hardboard or plywood, through wallboard to steel studs	1.5/8
Cabinets having a ¾in wood frame through double layers of ½in wallboard to steel studs	2¼
Type G – Bugle Head	
Attachment of wallboard to wallboard during laminating process. (Other than double layer of 3/8in)	1½
Type W – Bugle Head	
Single layer wallboard to wood framing	1¼
Channels to wood framing	1¼
Type S – Pan Head	
Metal studs to metal runners, and splicing studs	3/8
Type S – Hex Washer Head	
Metal studs to metal runners	3/8

Type S – Trim Head

Narrow batten to steel stud	1
Wood trim over single layer of wallboard to steel stud	1.5/8
Wood trim over double layer of wallboard to steel stud	2¼

Type S-12 – Bugle Head

Single layer of ½in or 5/8in wallboard or sheathing to metal framing	1
Metal lath to metal studs for curtain wall applications	1¼
Double layer ½in or 5/8in wallboard to metal framing	1.5/8
Multi-layer applications of wallboard to steel framing	1.7/8
Laminated layers of wallboard up to 1.5/8in total thickness to metal framing	2
Double layer of 1in wallboard to metal framing	2.3/8
Laminated layers of wallboard up to 2¼in to metal framing	2.5/8
Double layer of 1in wallboard plus a layer of ½in wallboard to metal framing	3

Type S-12 – Pan Head

Metal studs to runners	3/8
Door frame clips to metal studs	3/8
Any metal attachment not exceeding 12 gauge	3/8
Attachments of stud to runner in curtain wall or heavy door frame applications	½
Any steel combination not exceeding 12 gauge	½

Type S-12 – Trim Head

Narrow batten to metal stud	1
Wood trim over single layer of wallboard to metal framing	1.5/8
Wood trim over double layer of wallboard to metal framing	2¼

Speed Screws

'Speed screws' are similar in appearance to self-tapping screws, but are not self-tapping. They have hardened threads of a semi-buttress form and are designed for use with spring nuts. They have a superior performance to self-tapping screws or machine screws used with this type of nut.

SPIRE SPEED SCREWS (Replacing the former type 'J')

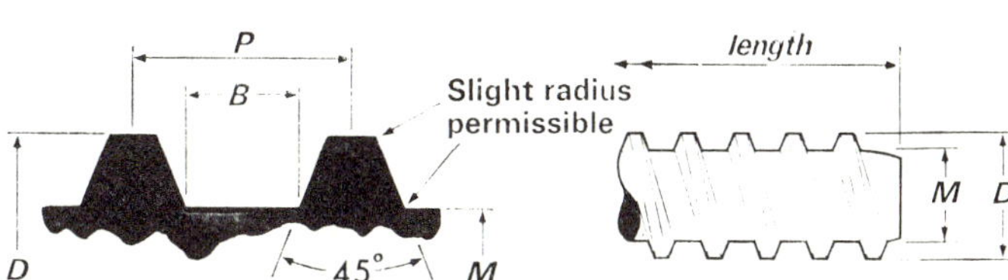

SPIRE SPEED SCREWS (Replacing the former type J)

Screw Size No	Threads per inch	Major diameter D		Minor diameter M		Root width B	
		Maximum	Minimum	Maximum	Minimum	Maximum	Minimum
4	24	0.114	0.110	0.086	0.081	0.022	0.019
6	18	0.141	0.136	0.102	0.096	0.028	0.025
8	15	0.168	0.162	0.123	0.116	0.034	0.030
10	12	0.194	0.188	0.137	0.130	0.042	0.038
12	11	0.221	0.215	0.162	0.155	0.046	0.042

Dimensions in inches

Screw Size No	Pitch P	Major diameter D		Minor diameter M		Root width B	
		Maximum	Minimum	Maximum	Minimum	Maximum	Minimum
4	1.059	2.90	2.80	2.18	2.06	0.56	0.48
6	1.410	3.58	3.45	2.59	2.44	0.71	0.64
8	1.694	4.27	4.11	3.12	2.95	0.86	0.76
10	2.116	4.93	4.78	3.48	3.30	1.07	0.96
12	2.309	5.61	5.46	4.11	3.94	1.17	1.07

Dimensions in millimetres

Paint Removing Screws

Paint removing screws have been developed to overcome the problems presented by the need to re-tap captive nuts prior to screw assembly where the nut has been dipped or sprayed before assembly. In the proprietary design shown the leading threads of the screw incorporate two cutting surfaces formed by two longitudinally parallel grooves which also serve to disperse hard paint blocking the nut.

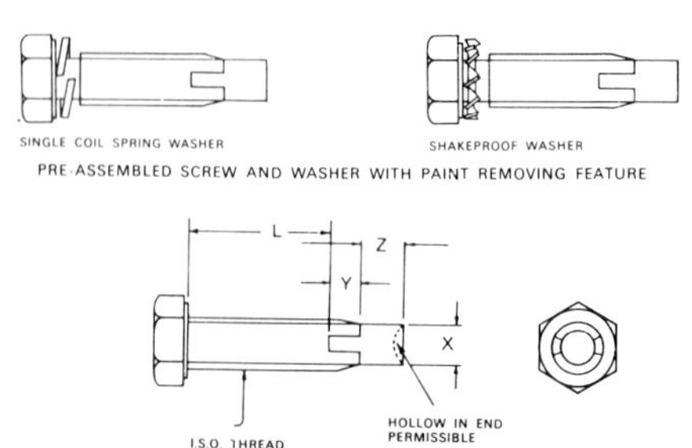

PRE-ASSEMBLED SCREW AND WASHER WITH PAINT REMOVING FEATURE

Size	Pitch	Spigot Length	Spigot Diameter	Flute Length
M5	0.8	3.6 4.3	3.715 3.969	2.0 2.8
M6	1.0	3.9 4.7	4.436 4.690	2.5 3.5
M8	1.25	5.3 6.1	6.103 6.357	3.125 4.375
M10	1.5	7.2 7.9	7.761 8.015	3.75 5.25
M12	1.75	7.7 8.5	9.416 9.670	4.375 6.125

Dimensions in millimetres

G.S.F. paint-removing screws and sems.

Nails

BS1202 now specifies nails in metric dimensions. The following Tables based on this standard also include the (original size) swg equivalents, but these no longer form part of the specification.

Steel nails to BS1202, Part I. *Wire nails* are made from mild steel. *Cut nails* are made from black rolled steel. Finishes for wire nails are:-

(i) bright

(ii) galvanized

(iii) resin-coated (cement coated)

(iv) cadmium plated, lead coated, sheradized, etc, as required

Cut nail finishes are 'as rolled' with bright cut sides, or galvanized.

BS1202:Part 1:1974 Round Plain Head Nails

Length L	Shank Diameter D	swg Size	Approximate Number of Nails per kg
mm	mm		
15	1.40	17	4 400
20	1.40	17	3 750
20	1.60	16	2 710
25	1.60	16	2 120
25	1.80	15	1 720
25	2.00	14	1 430
30	1.80	15	1 410
30	2.00	14	1 170
30	2.36	13	840
40	2.00	14	970
40	2.36	13	750
40	2.65	12	575
45	2.00	14	840
45	2.36	13	640
45	2.65	12	510
50	2.36	13	550
50	2.65	12	440
50	3.00	11	340
50	3.35	10	290
60	2.65	12	385
60	3.00	11	310
60	3.35	10	255
65	2.65	12	350
65	3.00	11	275
65	3.35	10	230
65	3.75	9	175
75	3.00	11	236
75	3.35	10	194
75	3.75	9	154
75	4.00	8	121
90	3.35	10	152
90	3.75	9	123
90	4.00	8	106
90	4.50	7	88
100	3.75	9	110
100	4.00	8	88
100	4.50	7	77
100	5.00	6	66
115	5.00	6	57
125	5.00	6	53
125	5.60	5	42
150	5.60	5	35
150	6.00	4	29
180	6.70	3	22
200	8.00	5/16	13

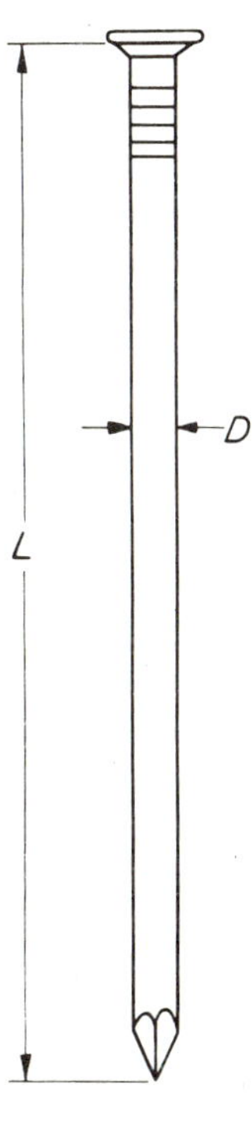

Head diameters :

2.75D for 1.40mm diameter
2.5D for 1.60 to 2.36mm diameter
2.25D for 2.65 to 3.75mm diameter
2D for 4.00mm diameter or thicker

BS1202:Part 1:1974 Round Lost Head Nails

Length L	Shank Diameter D	swg Size	Approximate Number of Nails per kg
mm	mm		
15	1.00	19	9 400
20	1.00	19	8 030
25	1.00	19	6 100
30	2.00	14	1 190
40	2.36	13	760
50	2.65	12	420
50	3.00	11	360
60	3.00	11	330
60	3.35	10	270
65	3.00	11	270
65	3.35	10	240
75	3.75	9	160

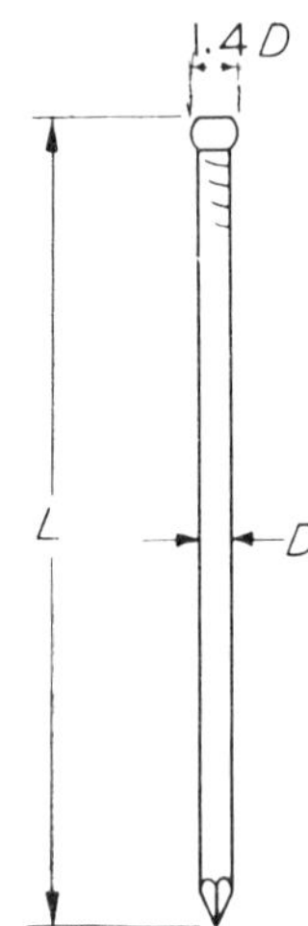

BS1202:Part 1:1974 Clout or Slate Nails

Length L	Shank Diameter D	swg Size	Approximate Number of Nails per kg
mm	mm		
15	2.00	14	2 380
15	2.36	13	1 540
20	2.65	12	1 035
25	2.65	12	815
30	2.36	13	830
30	2.65	12	660
30	3.00	11	540
40	2.36	13	700
40	2.65	12	570
40	3.35	10	350
45	2.65	12	460
45	3.35	10	330
50	2.65	12	430
50	3.00	11	340
50	3.35	10	290
50	3.75	9	230
65	3.75	9	180
75	3.75	9	150
90	4.50	7	85
100	4.50	7	75

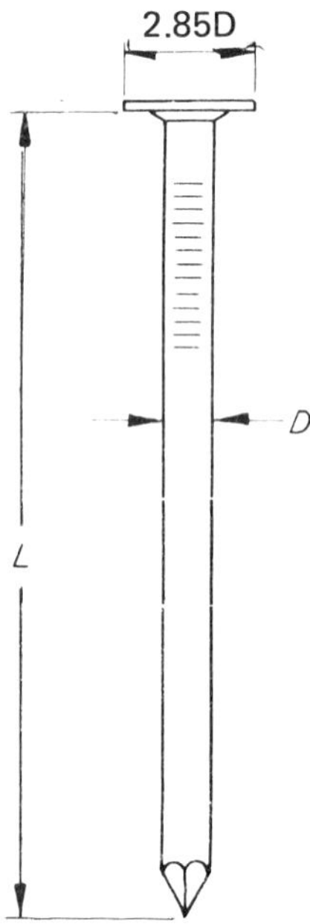

Head diameters:
3.3D for 2.0 to 3.35mm diameter
3.1D for 3.75mm diameter
2.85D for 4.50mm diameter

BS1202:Part 1:1974 Extra Large Head Clout or Felt Nails

Length L	Shank Diameter D	swg Size	Approximate Number of Nails per kg
mm	mm		
13	3.00	11	780
15	3.00	11	650
20	3.00	11	580
25	3.00	11	485
30	3.00	11	420
40	3.00	11	350

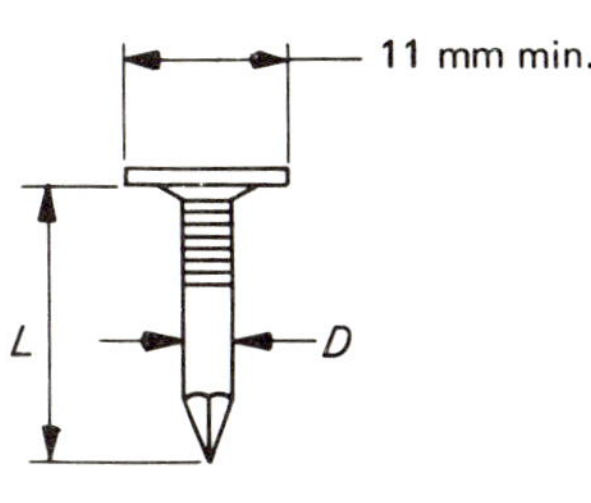

BS1202:Part 1:1974 Convex Head Roofing Nails (Chisel or Diamond Point)

Length L	Shank Diameter D	swg Size	Approximate Number of Nails per kg
mm	mm		
65	5.60	5	79
65	6.00	4	66
75	5.60	5	68

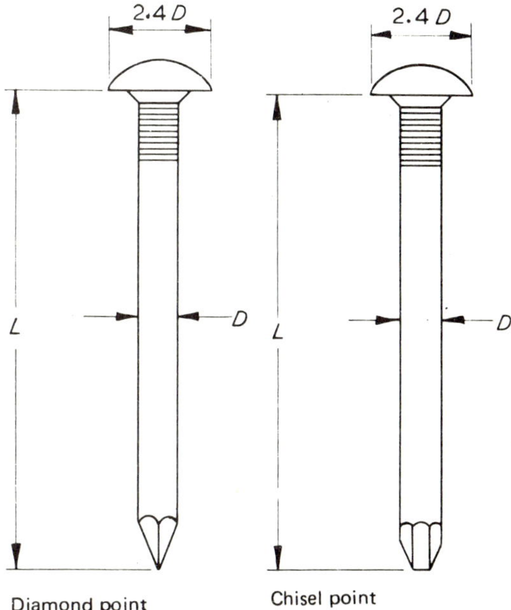

Diamond point
Chisel point

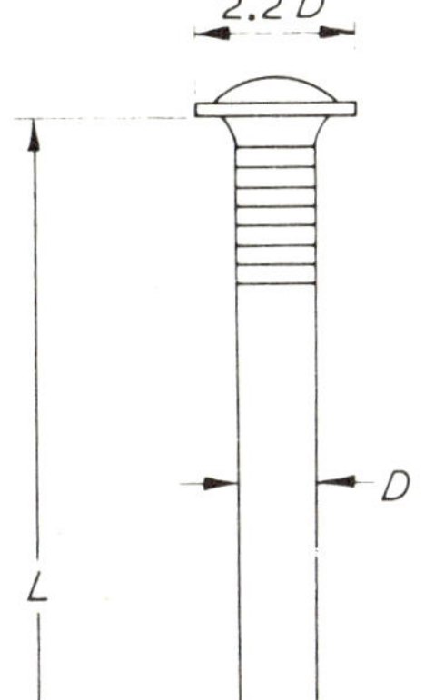

BS1202:Part 1:1974 Pipe Nails (Chisel Point)

Length L	Shank Diameter D	swg Size	Approximate Number of Nails per kg
mm	mm		
50	8.00	5/16	44
65	8.00	5/16	35
75	8.00	5/16	31
90	8.00	5/16	26
100	8.00	5/16	24

BS1202: Part 1: 1974 Panel Pins

Length L	Shank Diameter D	swg Size	Approximate Number of Nails per kg
mm	mm		
15	1.00	19	8 800
15	1,25	18	6 400
20	1,25	18	5 290
20	1.40	17	3 970
20	1.60	16	3 140
25	1.40	17	3 090
25	1.60	16	2 340
30	1.60	16	1 900
40	1.60	16	1 590
50	2.00	14	770
65	2.65	12	345
75	2.65	12	290

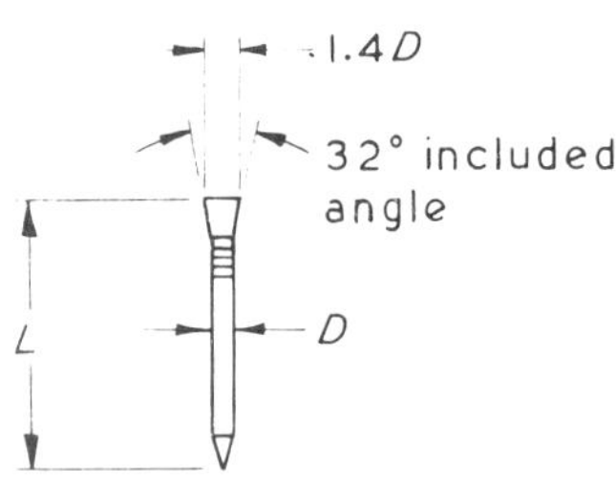

BS1202:Part 1: 1974 Hardboard Panel Pins (Round Shank)

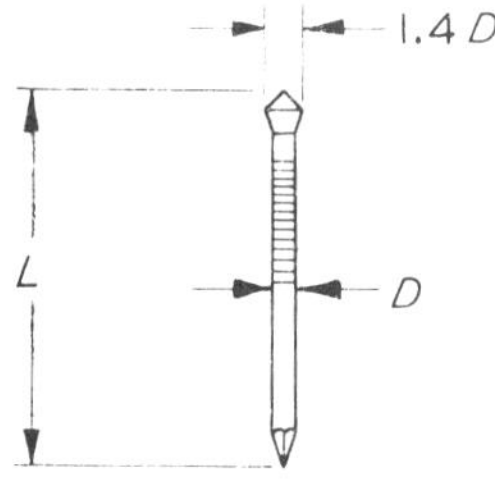

Length L	Shank Diameter D	swg Size	Approximate Number of Nails per kg
mm	mm		
20	1.40	17	3 970
20	1.60	16	3 140
25	1.40	17	3 090
25	1.60	16	2 340

BS1202:Part 1: 1974 Hardboard Panel Pins (Square Shank)

Length L	Shank Diameter D	swg Size	Approximate Number of Nails per kg
mm	mm		
20	1.40	17	3 470

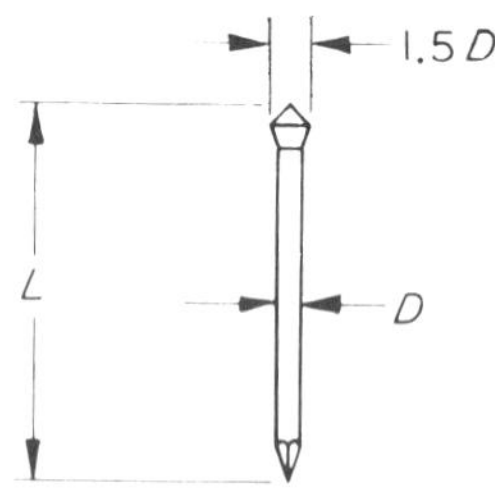

BS1202:Part 1: 1974 Lath Nails

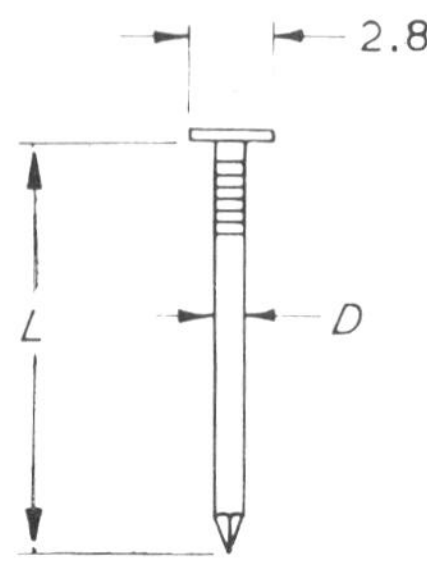

Length L	Shank Diameter D	swg Size	Approximate Number of Nails per kg
mm	mm		
20	1.60	16	2 370
20	1.80	15	1 750
25	1.60	16	2 140
25	1.80	15	1 740
25	2.00	14	1 430
30	2.00	14	1 170
40	2.00	14	970

BS1202:Part 1:1974 Plasterboard Nails (Jagged Shank)

Length L	Shank Diameter D	swg Size	Approximate Number of Nails per kg
mm	mm		
30	2.65	12	700
40	2.65	12	570

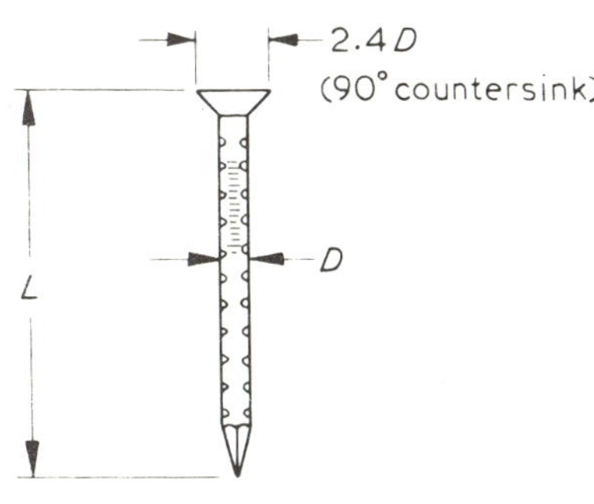

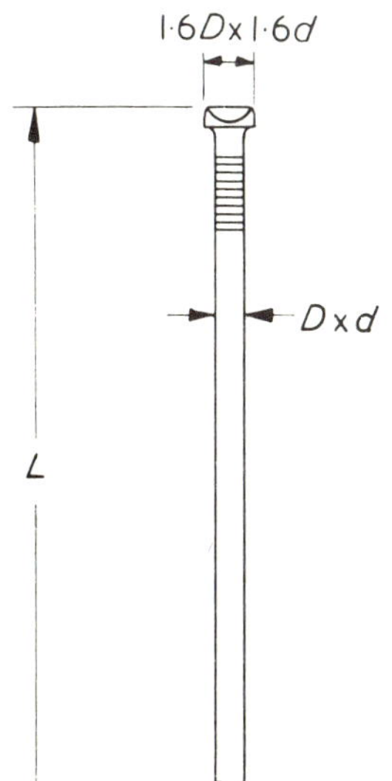

BS1202:Part 1:1974 Oval Brad Head Nails

Length L	Shank Dimensions D x d	swg Size	Approximate Number of Nails per kg
mm	mm		
20	2.00 x 1.25	14 x 18	4 500
25	2.00 x 1.25	14 x 18	2 530
30	2.65 x 1.60	12 x 16	1 480
40	2.65 x 1.60	12 x 16	940
45	3.35 x 2.00	10 x 14	655
50	3.35 x 2.00	10 x 14	470
60	3.75 x 2.36	9 x 13	340
65	4.00 x 2.65	8 x 12	230
75	5.00 x 3.35	6 x 10	125
90	5.60 x 3.75	5 x 9	90
100	6.00 x 4.00	4 x 8	64
125	6.70 x 4.50	3 x 7	44
150	7.10 x 5.00	2 x 6	31

BS1202:Part 1:1974 Oval Lost Head Nails

Length L	Shank Dimensions D x d	swg Size	Approximate Number of Nails per kg
mm	mm		
20	2.00 x 1.25	14 x 18	4 500
25	2.00 x 1.25	14 x 18	2 530
30	2.65 x 1.60	12 x 16	1 480
40	2.65 x 1.60	12 x 16	940
45	3.35 x 2.00	10 x 14	655
50	3.35 x 2.00	10 x 14	470
60	3.75 x 2.36	9 x 13	340
65	4.00 x 2.65	8 x 12	230
75	5.00 x 3.35	6 x 10	125
90	5.60 x 3.75	5 x 9	90
100	6.00 x 4.00	4 x 8	64
125	6.70 x 4.50	3 x 7	44
150	7.10 x 5.00	2 x 6	31

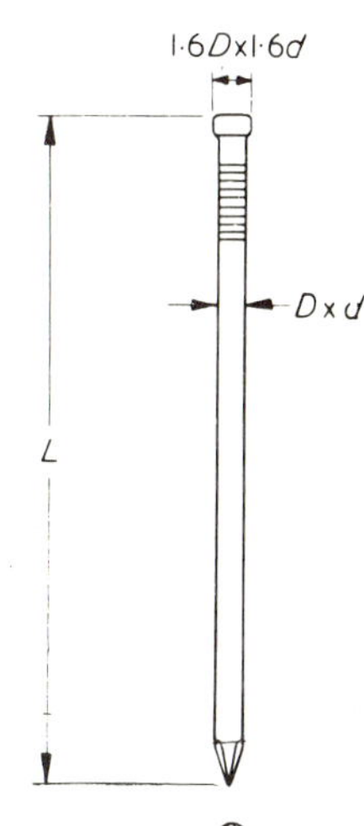

BS1202:Part 1:1974 Tile Pegs

Length L	Shank Diameter D	swg Size	Approximate Number of Nails per kg
mm	mm		
30	6.00	4	106
40	6.00	4	88

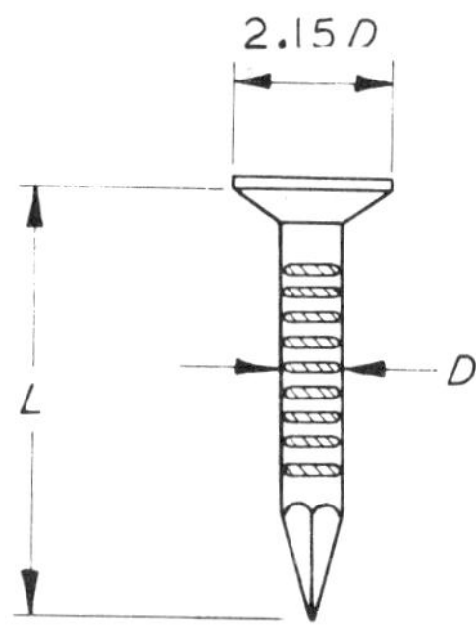

BS1202:Part 1:1974 Tram Nails (Flat or Raised Head and Chisel Point)

Length L	Shank Diameter D	swg Size	Approximate Number of Nails per kg
mm	mm		
65	8.00	5/16	37

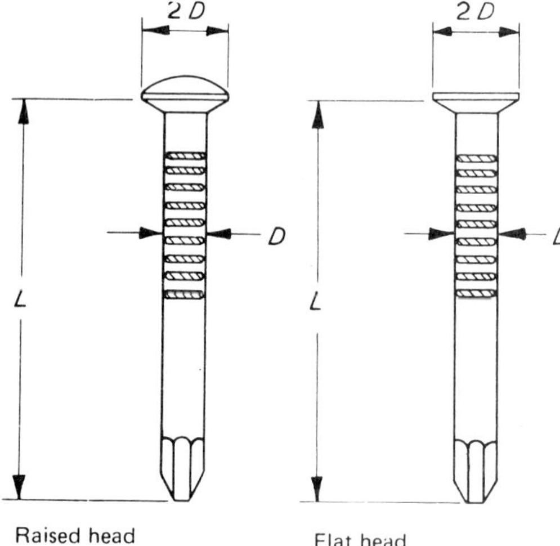

Raised head

Flat head

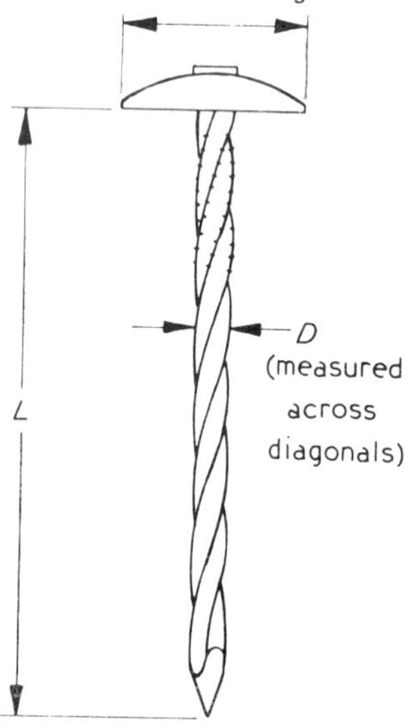

BS1202:Part 1:1974 Spring Head Twisted Shank Nails

Length L	Shank Diameter D	swg Size	Approximate Number of Nails per kg
mm	mm		
65	3.35	10	140

BS1202: Part 1: 1974 Square Twisted Shank Flat Head Nails

Length L	Shank Diameter D	swg Size	Approximate Number of Nails per kg
mm	mm		
40	2.36	13	860
50	2.65	12	515
50	3.00	11	415
65	3.35	10	235

Head diameters:
2.50D for 40mm nail
2.25D for remainder

Note: Because of the square twisted shank, the shape of the head will differ from that of a round wire nail

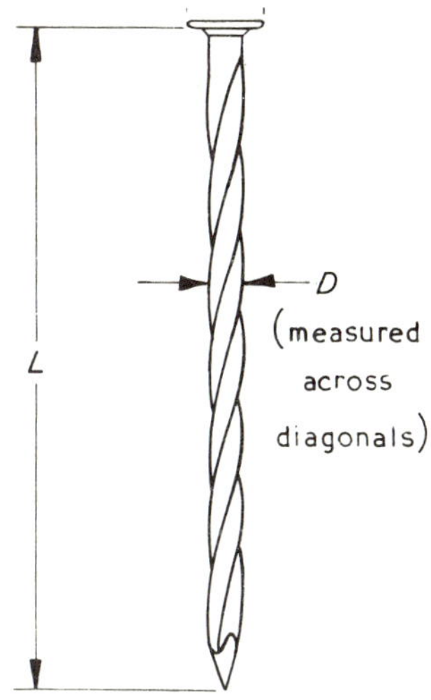

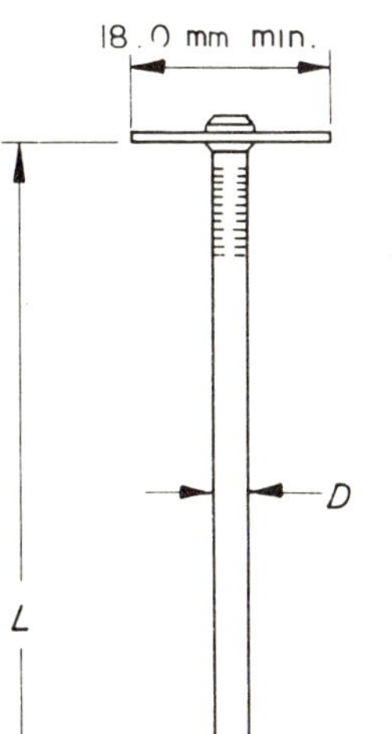

BS1202:Part 1:1974 Washer Head Slab Nails

Length L	Shank Diameter D	swg Size	Approximate Number of Nails per kg
mm	mm		
100	3.35	10	100

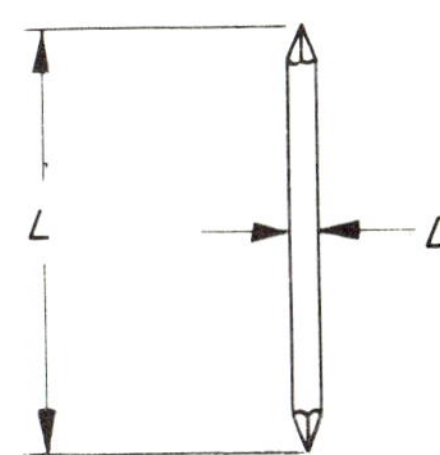

BS1202:Part 1:1974 Dowels

Length L	Shank Diameter D	swg Size	Approximate Number of Nails per kg
mm	mm		
40	2.65	12	610
45	2.65	12	520
50	2.65	12	455

BS1202:Part 1:1974 Tenter Hooks

Length L	Shank Diameter D	swg Size	Approximate Number of Nails per kg
mm	mm		
20	2.36	13	740
25	2.36	13	690

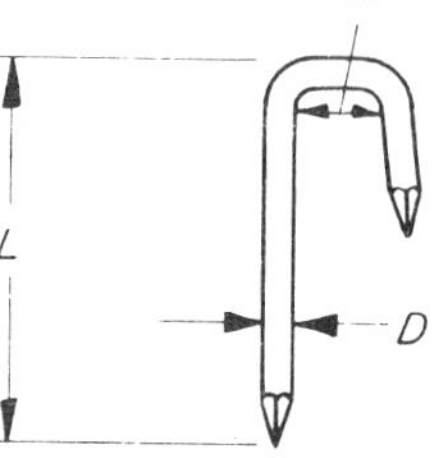

BS1202:Part 1:1974 Annular Ringed Shank Flat Head Nails

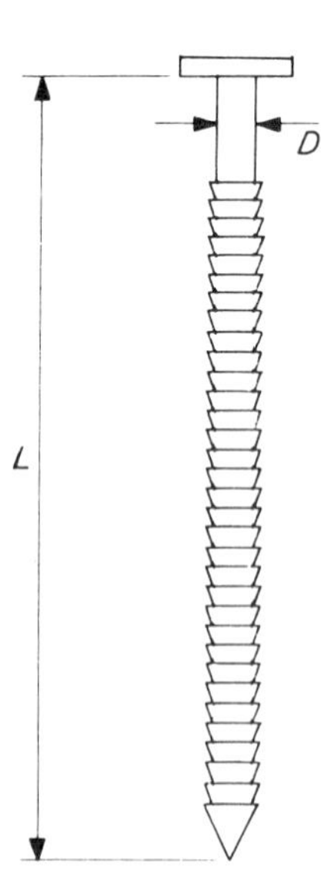

Length L	Shank Diameter D	swg Size	Approximate Number of Nails per kg
mm	mm		
20	2.00	14	1 900
25	2.00	14	1 430
30	2.00	14	1 170
30	2.36	13	840
40	2.36	13	750
40	2.65	12	575
45	2.36	13	640
45	2.65	12	510
50	2.65	12	440
50	3.00	11	340
50	3.35	10	290
60	2.65	12	385
60	3.00	11	310
60	3.35	10	255
65	3.00	11	275
65	3.35	10	230
65	3.75	9	175
75	3.35	10	194
75	3.75	9	154
75	4.00	8	121
90	3.35	10	152
90	3.75	9	123
90	4.00	8	106
100	5.00	6	66
115	5.00	6	57
125	5.60	5	35
150	6.00	4	29
180	6.70	3	22
200	8.00	5/16	13

BS1202:Part 1: 1974 Helical Threaded Shank Flat Head Nails

Length L	Shank Diameter D	swg Size	Approximate Number of Nails per kg
mm	mm		
40	2.65	12	575
45	2.65	12	510
50	2.65	12	440
50	3.00	11	340
50	3.35	10	290
60	2.65	12	385
60	3.00	11	310
60	3.35	10	255
65	3.00	11	275
65	3.35	10	230
65	3.75	9	175
75	3.35	10	194
75	3.75	9	154
75	4.00	8	121
90	3.35	10	152
90	3.75	9	123
90	4.00	8	106
100	5.00	6	66
115	5.00	6	57
125	5.60	5	35
150	6.00	4	29
180	6.70	3	22
200	8.00	5/16	13

Head diameters are the same as in Round Plain Head Nails

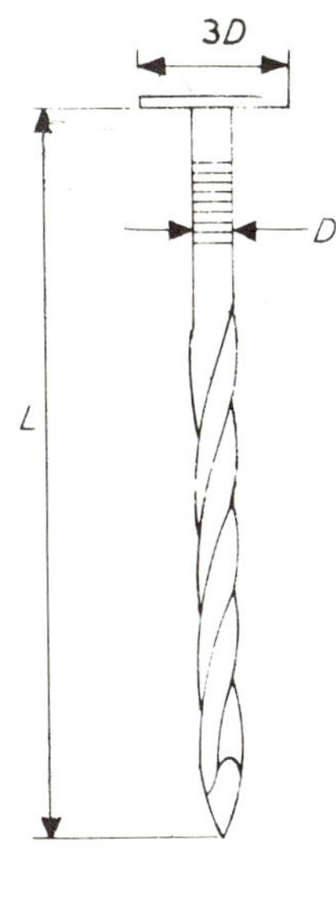

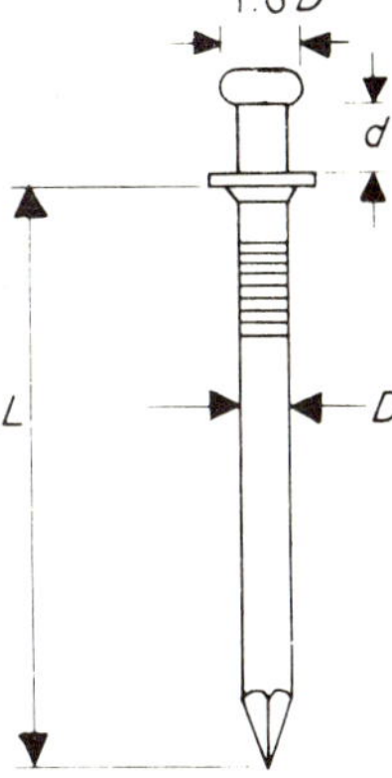

BS1202: Part 1: Duplex Head Nails

Length L	Shank Diameter D	swg Size	Lower Head Diameter	Approximate Number of Nails per kg	d
mm	mm		mm		mm
45	3.00	11	6.50	345	6.50
60	3.35	10	7.00	200	6.50
70	3.75	9	8.00	130	8.00
75	4.00	8	8.50	100	9.50
90	5.00	6	9.50	62	9.50
100	5.60	5	11.00	46	9.50

BS1202:Part 1:1974 Cut Clasp Nails

Length L	Shank Dimension D	swg Size	Approximate Number of Nails per kg
mm	mm		
25	1.60	16	1 384
30	1.80	15	858
40	2.00	14	616
50	2.65	12	286
60	2.65	12	202
65	3.00	11	171
75	3.35	10	103
90	3.75	9	66
100	4.00	8	48
125	5.00	6	30
150	5.60	5	19
175	5.60	5	13
200	6.00	4	11

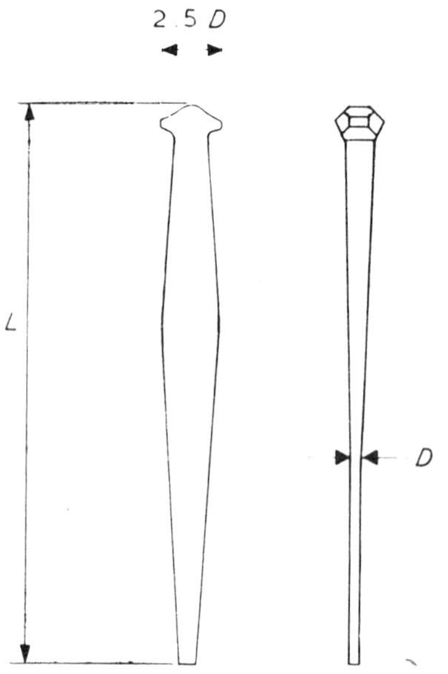

2.5 D

L

D

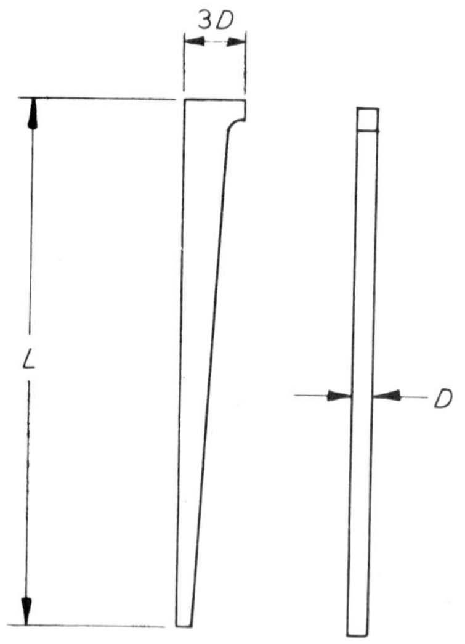

3D

L

D

BS1202:Part 1:1974 Cut Floor Brads

Length L	Shank Dimension D	swg Size	Approximate Number of Nails per kg
mm	mm		
40	2.36	13	396
45	2.36	13	330
50	2.65	12	264
60	3.00	11	198
65	3.35	10	154
75	3.35	10	100

Copper Nails

Copper nails to BS1202:Part 2 cover the following types:-

(i) Round lost head nails (40, 50 and 65mm long)

(ii) Clout and shingle nails (also known as slate, tile or felt nails), (20, 25, 30, 40, 45, 50 and 65mm long)

(iii) Flat head countersunk and rosehead square shank boat nails with a round or diamond point. (20 to 150mm length).

(iv) Clout cut nails, also known as cut slate nails. (25, 30, 40, 45 and 50mm long)

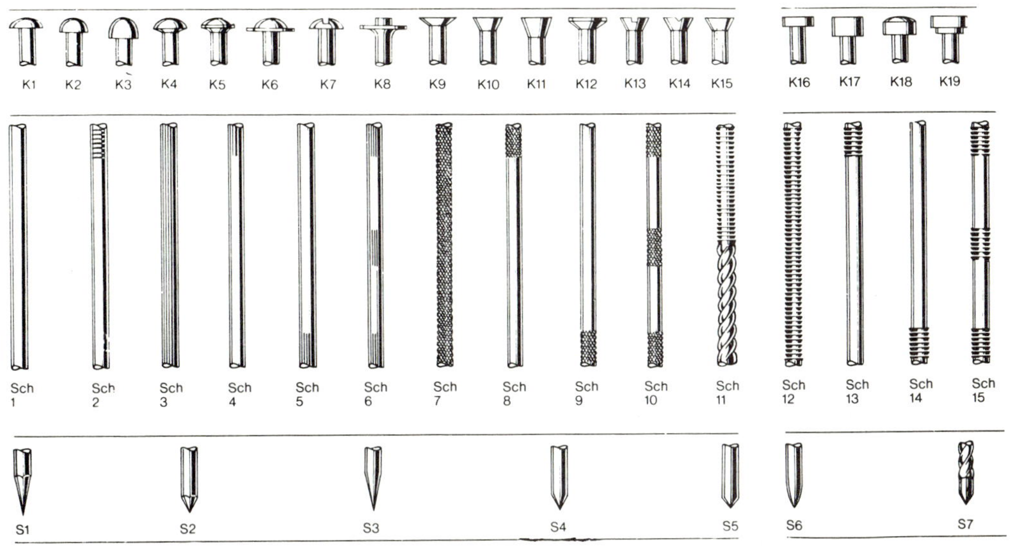

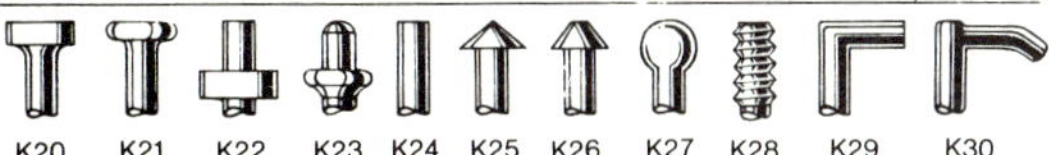

Nail heads, shanks and points (Bierbach).

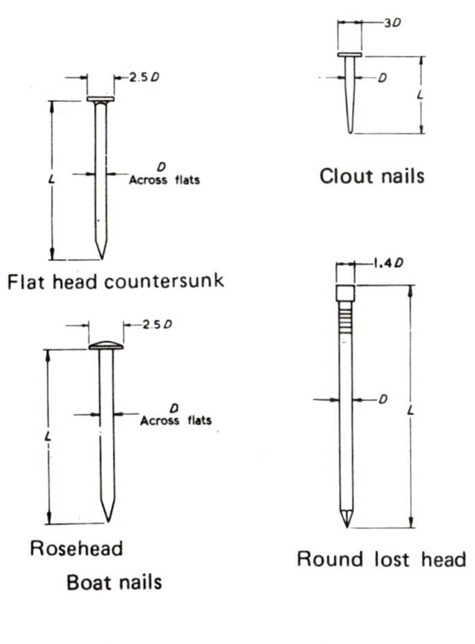

Flat head countersunk

Rosehead

Boat nails

Round lost head

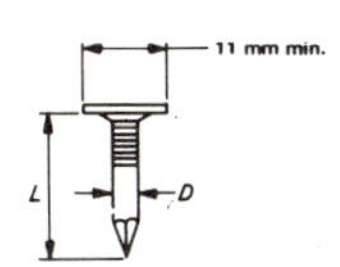

Clout nails

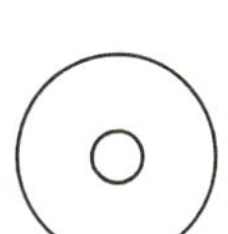

Extra large head felt nails

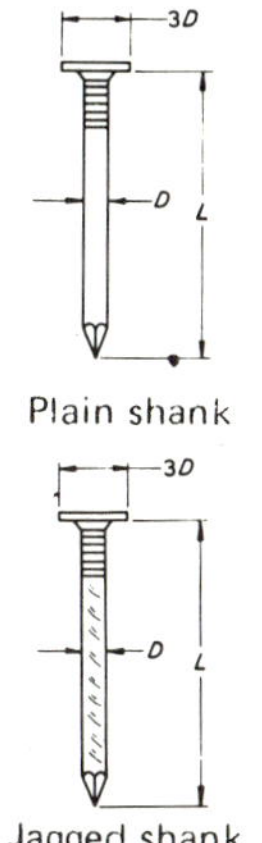

Plain shank

Jagged shank

Clout nails

Copper nails BS 1202 : Part 2.

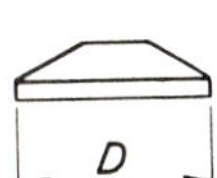

BS1202:Part 2:1974 Copper Roves

Suitable for Nails of Shank Diameter	swg Shank	Outside Diameter D
mm		mm
1.80	15	6.4
2.00	14	7.9
2.36	13	9.5
2.65	12	11.1
3.00	11	11.1
3.35	10	12.7
3.75	9	14.3
4.00	8	15.9
5.00	6	19.1

Aluminium Nails

Aluminium nails to BS1202:Part 3 include the following types:-

(i) Round plain head nails

(ii) Round lost head nails

(iii) Clout, slate and tile nails

(iv) Extra large head felt nails

(v) Nipple head roofing nails

(vi) Panel pins

(vii) Gimp pins

(viii) Tile pegs, sharp and round point

(ix) Annular ring shank flat head nails

(x) Helical threaded shank flat head nails

Aluminium round wire nails are made from HG 9 alloy (tensile strength not less than 280 N/mm^2) or HG 20 alloy (tensile strength not less than 370 N/mm^2). Nails thinner than 2mm diameter are made only from HG 20 alloy.

Aluminium nails by Bierbach.

BS1202: Part 3: 1974 Aluminium Round Lost Head Nails

Length L	Shank Diameter D	swg Size	Approximate Number of Nails per kg
mm	mm		
40	2.36	13	2 128
40	2.64	12	1 390
50	3.00	11	1 008
50	3.35	10	860
60	3.35	10	756
65	3.35	10	672
75	3.75	9	448

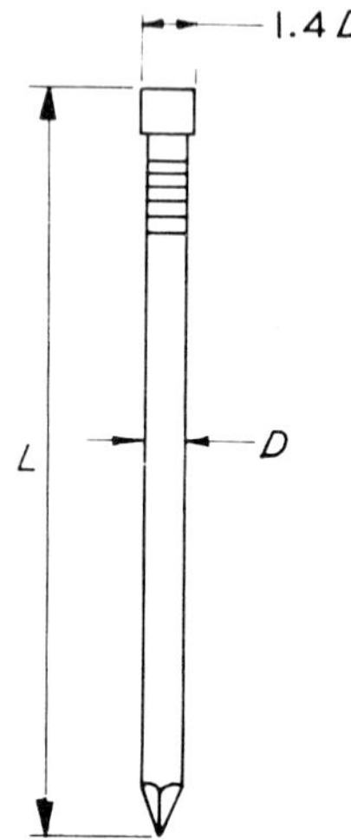

BS1202:Part 3:1974 Aluminium Round Plain Head Nails

Length L	Shank Diameter D	swg Size	Approximate Number of Nails per kg
mm	mm		
20	1.60	16	7 588
25	1.60	16	5 936
25	1.80	15	4 816
25	2.00	14	4 004
30	1.80	15	2 948
30	2.00	14	3 276
40	2.00	14	2 716
40	2.36	13	2 100
40	2.65	12	1 610
45	2.36	13	1 792
45	2.65	12	1 428
50	2.65	12	1 232
50	3.00	11	952
50	3.35	10	812
60	3.00	11	868
60	3.35	10	714
65	3.00	11	770
65	3.35	10	644
65	3.75	9	490
75	3.35	10	543
75	3.75	9	431
75	4.00	8	338
90	4.00	8	296
90	4.50	7	246
100	4.50	7	215
100	5.00	6	184
115	5.00	6	159

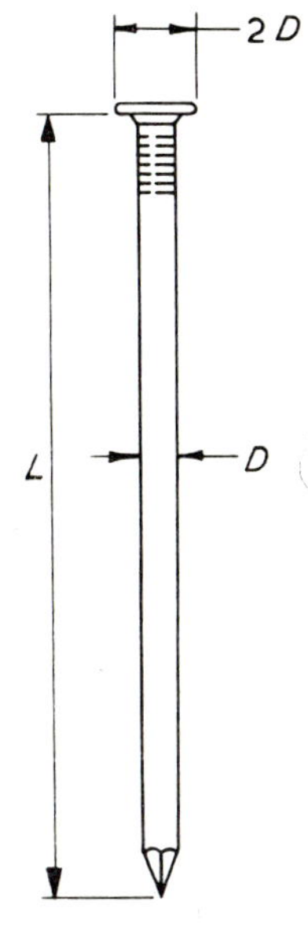

BS1202:Part 3:1974 Aluminium Clout,Slate and Tile Nails

Length L	Shank Diameter D	swg Size	Approximate Number of Nails per kg
mm	mm		
20	2.65	12	2 898
20	3.00	11	2 300
25	2.00	14	3 800
25	2.65	12	2 282
25	3.00	11	1 750
25	3.35	10	1 540
30	2.00	14	3 000
30	2.36	13	2 324
30	2.65	12	1 848
30	3.00	11	1 512
40	2.36	13	1 960
40	2.65	12	1 596
40	3.00	11	1 200
40	3.35	10	980
45	3.00	11	1 060
45	3.35	10	924
50	3.00	11	952
50	3.35	10	812
50	3.75	9	644
60	3.35	10	680
60	3.75	9	550
65	3.75	9	504

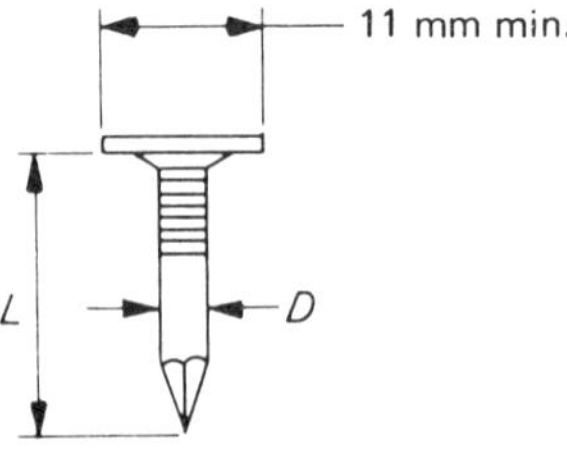

BS1202:Part 3:1974 Aluminium Extra Large Head Felt Nails

Length L	Shank Diameter D	swg Size	Approximate Number of Nails per kg
mm	mm		
15	3.00	11	2 283
15	3.35	10	1 840
20	3.00	11	2 130
20	3.35	10	1 848
25	3.00	11	1 636
25	3.35	10	1 296

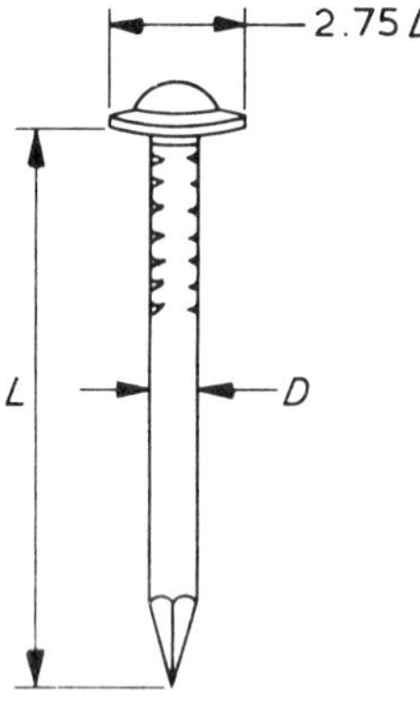

BS1202:Part 3:1974 Aluminium Nipple Head Roofing Nails

Length L	Shank Diameter D	swg Size	Approximate Number of Nails per kg
mm	mm		
25	3.35	10	1 410
25	3.75	9	1 200
45	3.75	9	682
50	3.75	9	594
50	4.50	7	390
65	5.00	6	316

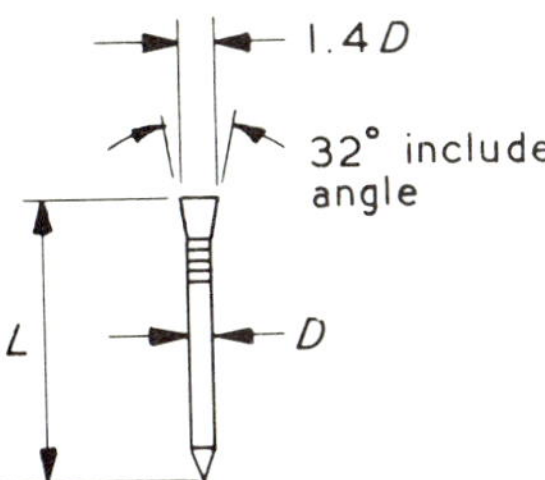

BS1202:Part 3:1974 Aluminium Panel Pins

Length L	Shank Diameter D	swg Size	Approximate Number of Nails per kg
mm	mm		
15	1.60	16	11 800
20	1.60	16	8 790
25	1.60	16	6 550

BS1202:Part 3:1974 Aluminium Gimp Pins

Length L	Shank Diameter D	swg Size	Approximate Number of Nails per kg
15	1.60	16	9 600
20	1.60	16	7 588
25	1.60	16	5 936

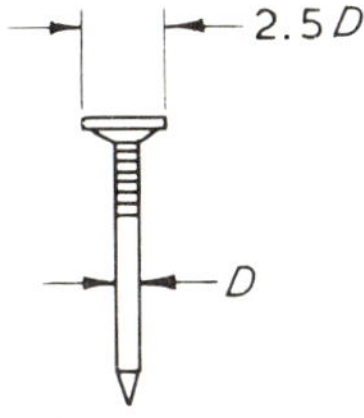

14.5 mm L D

Sharp point

14.5 mm L D

Round point

BS1202:Part 3:1974 Aluminium Tile Pegs

Length L	Shank Diameter D	swg Size	Approximate Number of Nails per kg
mm	mm		
30	4.50	7	600
30	5.00	6	545
40	4.50	7	490
40	5.00	6	450

BS1202:Part 3:1974 Aluminium Annular Ringed Shank Flat Head Nails

Length L	Shank Diameter D	swg Size	Approximate Number of Nails per kg
mm	mm		
25	2.65	12	2 280
40	3.00	11	1 200
40	3.35	10	980
50	3.35	10	812
65	4.00	8	440
75	4.00	8	345

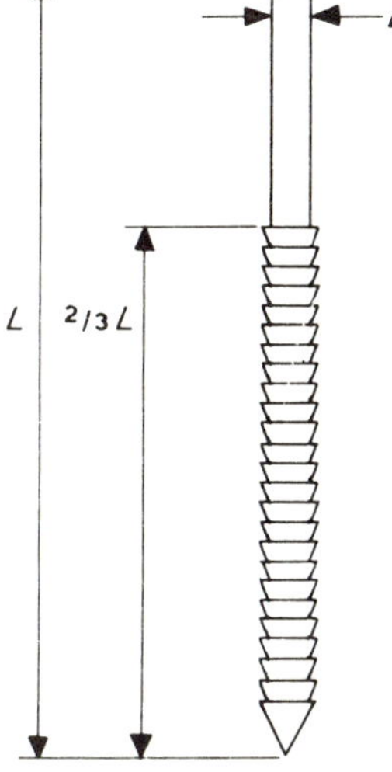

BS1202:Part 3:1974 Aluminium Helical Threaded Shank Flat Head Nails

Length L	Shank Diameter D	swg Size	Approximate Number of Nails per kg
mm	mm		
25	2.65	12	2 280
40	3.00	11	1 200
40	3.35	10	980
50	3.35	10	812
65	4.00	8	440
75	4.00	8	345

Note: This type of nail is also available with a countersunk head.

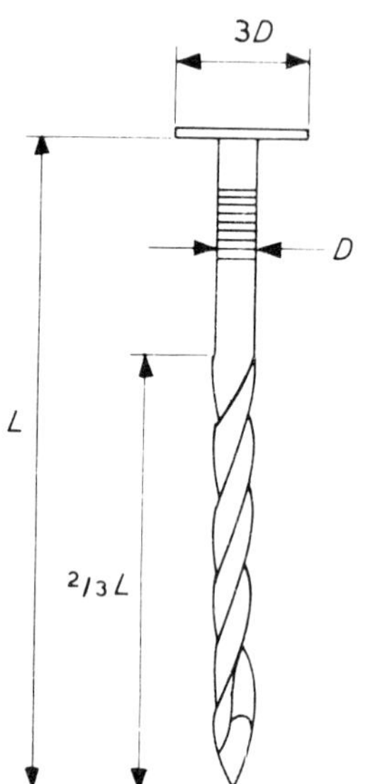

Other types of nail in common use are described in the following Table:-

Name	Description and Use	Typical Sizes
Sprig	Headless tack used for fixing glass in wood frames before puttying; also for fixing thick lino	½in and ¾in
*Corrugated ('Wiggle nail')	Tapered corrugated fastener for driving into corner joints in frames, etc.	7/8in – 1¼in long ¼in – 7/8in deep
Chisel Point Nail	Wedge-shaped point to drive into brickwork, etc – used for fastening iron rainwater goods.	1½in–4in
Double Nail	Headless wire nail with a point at each end for making hidden joints.	1½in–2in
Double Head Shutter Nail	Intended for temporary assemblies etc, the upper head making it easy to remove the nail.	1in–4in
Roofing Nail	Hammer-driven screw nail for fixing corrugated roof sheeting	2½in std ¾in head dia
Chair Nail	Nail with a large hollow cup head; used to cover tacks in upholstery work.	1/8in–½in dia

*There are other forms of corrugated nail-type fasteners – eg corrugated rings and toothed plates. The latter are also known as *timber connectors.*

Special Nails

Screwnails

Screwnails can be described as a cross between a wire nail and a coarse-threaded screw with a shallow thread terminating in a firmly pointed tip. They are intended to be hammer driven as an alternative fastening to a nail with about four times the holding power in typical materials. Screwnails should preferably be hardened when they can be used for fastening thin sheet metals, etc, to

TABLE I – GKN HARDBOARD SCREWNAILS

American Wire Gauge	All Head Types		Countersunk Head				Round Head			
	Major Diameter d Nom		Head Diameter D Nom		Head Depth k Nom		Head Diameter D Nom		Head Depth k Nom	
	in	mm	in	mm	in	mm	in	mm	in	mm
15	0.089	2.26	0.154	3.91	0.053	1.35	0.144	3.65	0.048	1.21
14	0.099	2.51	0.171	4.31	0.058	1.47	0.160	4.06	0.053	1.34
13	0.111	2.81	0.196	4.97	0.066	1.68	0.184	4.67	0.061	1.54
12	0.127	3.22	0.225	5.71	0.074	1.88	0.212	5.38	0.071	1.80
11	0.144	3.65	0.256	6.50	0.084	2.13	0.242	6.14	0.081	2.05
10	0.165	4.19	0.285	7.23	0.092	2.34	0.270	6.85	0.090	2.28

American Wire Gauge	Flat Head				Large Round Head			
	Head Diameter D Nom		Head Depth k Nom		Head Diameter D Nom		Head Depth k Nom	
	in	mm	in	mm	in	mm	in	mm
15	0.156	3.96	0.030	0.76				
14	0.188	4.77	0.034	0.86				
13	0.219	5.56	0.038	0.97	0.308	7.82	0.062	1.57
12	0.250	6.35	0.042	1.07				
11	0.263	6.68	0.046	1.17				
10	0.281	7.13	0.050	1.27				

Dimensions in inches

wood as well as wood to wood. They are produced in (hardened) steel and also in non-ferrous metals (eg silicon bronze for marine applications) — see Fig 1 and Table I.

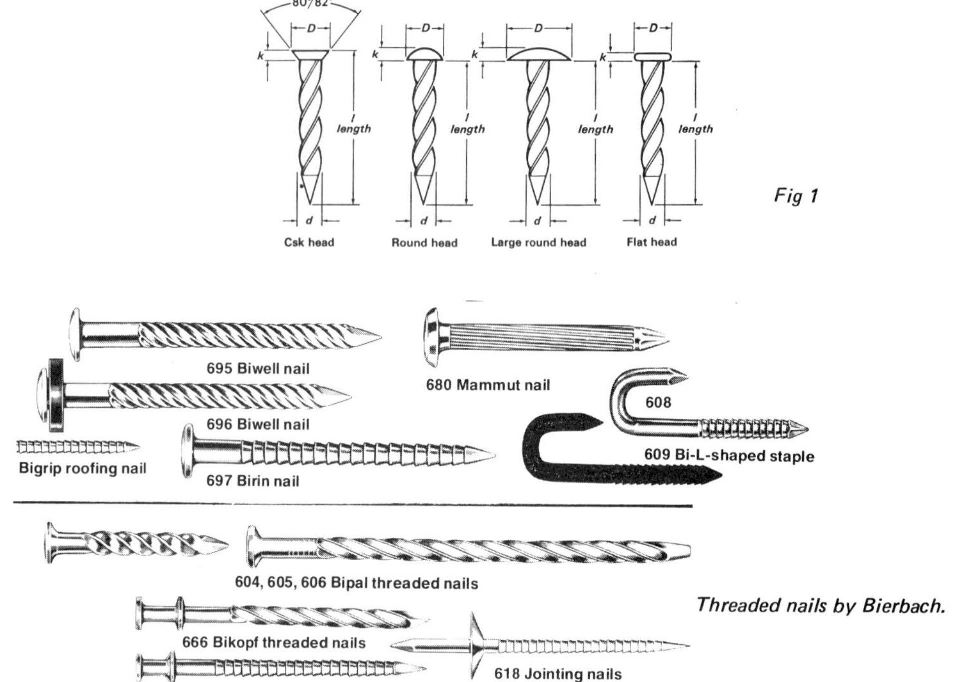

Threaded nails by Bierbach.

Masonry Nails

Masonry nails are hardened steel nails for driving into brickwork, masonry, concrete, etc. They may be similar to wire nails (ie with a plain shank), when they may also be called masonry pins, or have a fluted shank for enhanced driving properties and grip — Fig 2.

Hardened screw nail and types of head.

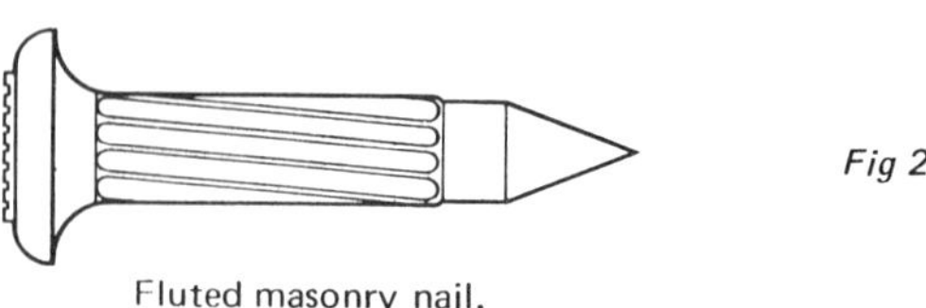

Fig 2

Fluted masonry nail.

TABLE II – 'TOWER' MASONRY NAILS

NAIL LENGTH mm	approx ins	AUSTEMPERED HIGH CARBON STEEL BRIGHT ZINC PLATED
16	5/8	DOME 2.5 mm diameter
19	3/4	
22	7/8	
25	1	
22	7/8	LIGHT 2.5 mm diameter
25	1	
30	1.1/8	
35	1.3/8	
40	1.5/8	
45	1.3/4	
50	2	
60	2.3/8	
22	7/8	MEDIUM 3.0 mm diameter
25	1	
30	1.1/8	
35	1.3/8	
40	1.5/8	
45	1.3/4	
50	2	
60	2.3/8	
70	2.3/4	
80	3.1/8	
25	1	HEAVY 3.5 mm diameter
30	1.1/8	
35	1.3/8	
40	1.5/8	
45	1.3/4	
50	2	
60	2.3/8	
70	2.3/4	
80	3.1/8	
90	3.1/2	
100	4	

Various head forms are used for masonry pins – eg countersunk, dome, butt and mushroom – depending on the individual manufacturer's preference – see Table II. Flat-headed pins – eg butt or mushroom – may be used with washers for increased bearing area. Matching washer sizes are normally a drive fit on the point of the pin.

Typical diameter sizes for masonry pins are 2.5mm, 3.0mm and 3.5mm. Lengths can be between 16mm (5/8in) and 60mm (2.3/8in) in 2.5mm diameter; 22mm (7/8in) and 80mm (3.1/8in) in 3.0mm diameter; and 25mm (1in) and 100mm (4in) in 3.5mm diameter.

Choice of length should allow 16–20mm (5/8in–3/4in) penetration of the pin into brickwork, beyond any facing plaster, etc. Deeper penetration may be required in softer materials and shallower penetration in harder materials. Attempting to drive a masonry pin too deep into hard materials will generally result in reduced efficiency of the fastening (see Tables III and IV).

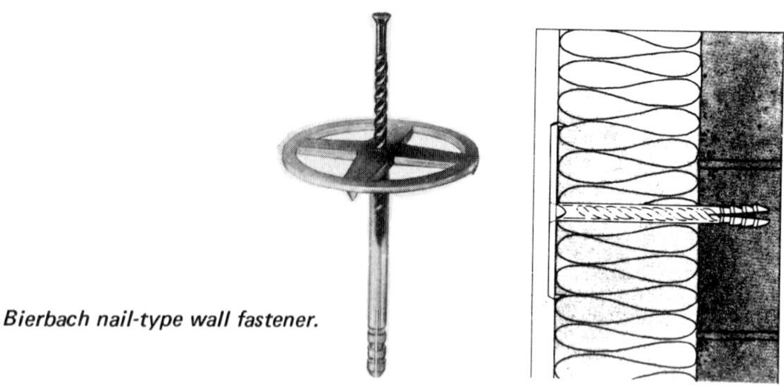
Bierbach nail-type wall fastener.

TABLE III – GKN HARDENED FIXING PINS (MASONRY PINS)

Butt Head	Dome Head	Mushroom Head
length x diameter	**length x diameter**	**length x diameter**
¾in x 2.5mm	1½in x 3.5mm	¾in x 2.5mm
1in	1¾in	1in
1¼in	2in	
1½in	2½in	
1¾in	3in	
2in	3½in	
2¼in	4in	
2½in		

TABLE IV – RECOMMENDATIONS FOR SELECTING LENGTH OF MASONRY PINS

Thickness of Part Being Attached		Recommended Length of Masonry Pin	
mm	in	mm	in
up to 3	up to 1/8	20	¾
6–9.5	1/4–3/16	25	1
12.5–16	1/2–5/8	30	1¼
19–22	3/4–7/8	35	1½
25–28	1–1.1/8	40	1¾
30–35	1¼–1.3/8	45	2
37.5–40	1½–1.5/8	50	2
45–48	1¾–1.7/8	60 or 70	2½
50–55	2–2.1/8	80	3
65–70	2½–2.7/8	90	3½
75–80	3–3.1/4	100	4

Threaded pins are another special form of nail for gun-driving into brickwork, concrete, etc (and some types are also suitable for driving into steel). These comprise hardened nails with elongated, studded heads. They are also produced with female threaded heads – Fig 3.

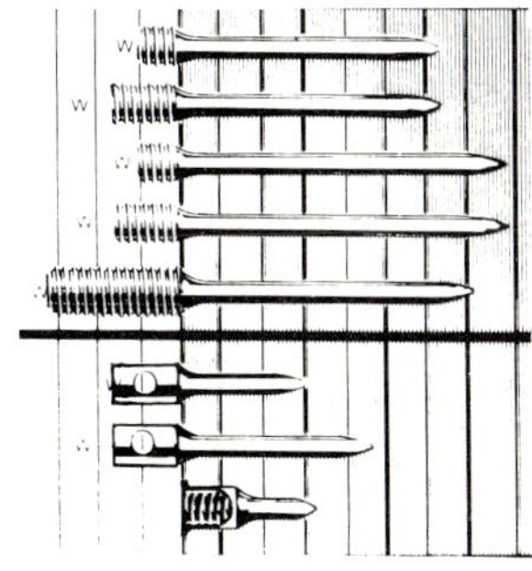

Fig 3 Examples of 'Spit' threaded pins.

Fig 4 Tower 'Gripfast' barbed nail.

Barbed Nails

Barbed nails or 'ring' nails have a barbed shank combining the straight driving characteristics of a nail with the holding power of a screw. The 'Gripfast' barbed ring nail – Fig 4 – has a parallel shank with buttress form and can be regarded as a direct alternative to a screw with the advantage of being hammer driven. 'Gripfast' nails are made in silicon bronze and are particularly suitable for marine use, and also for general work requiring a non-rusting, non-staining nail/screw type fastener.

Where 'Gripfast' nails are to take the place of the screws specified in a design, in general it is safe to choose a nail slightly longer than the screw (if the thickness of the wood permits) and rather thinner (see Table V). Nails may be driven at the same or slightly closer spacing than specified for screws. In the same way as screws they should be placed closer at the points of greatest strain.

TABLE V – SUGGESTED SIZES OF 'GRIPFAST' NAILS AS ALTERNATIVES TO SCREWS

Screw	'Gripfast' Nail
½in x 4g	5/8in x 16g
5/8in x 4g	¾in or 7/8in x 14g
¾in x 5 or 6g	7/8in x 12g
1in x 6g	1in or 1¼in x 12g
1¼in x 6g	1½in x 10g
1½in x 8g	1¾in x 10g
2in x 10g	2¼in x 8g
2in x 12g or 14g	2½in x 6g

Note: Nail gauge sizes are Standard Wire Gauge (swg) which differs from the number size for woodscrews.

TABLE VI – 'GRIPFAST' NAIL DATA FOR PLYWOOD SKINS

Plywood	Nail	Edge Spacing mm	in	Spacing Elsewhere mm	in
1/8in or 4mm	5/8in x 16g	40	1½	100	4
3/16in or 5mm	3/4in x 14g	50	2	100	4
1/4in or 6mm	3/4in x 14g	50	2	125	5
5/16in or 8mm	7/8in x 12g	65	2½	150	6
3/8in or 9mm	1in x 12g	75	3	150	6
1/2in or 12mm	1½in x 10g	90	3½	150	6
3/4in or 20mm	2¼in x 8g	100	4	150	6

The smaller sizes may be driven without a pilot hole of any sort. For larger sizes there may be a clearance hole in the top piece and a hole about half the nail diameter in the lower piece. The point at which drilling becomes necessary depends on the size and type of wood as well as the nail size.

See also chapter on *Building Construction Fasteners.*

Building Construction Fixings

ROOFING AND CLADDING FASTENERS

Standard fixings for roofing and cladding are specified in BS1494 and include the following types:-

(i) Bent Bolts (Hook Bolts) — (Table I)

(a) Standard hook bolts for securing roof sheets to tubular purlins.

(b) Square head hook bolts for securing stanchions, etc.

TABLE I – TYPES AND SIZES OF BENT BOLTS

Length mm	DIAMETER SIZE													
	Hook Bolts						Square Bend Hook		Crook Bolts		'J' Bolts		'U' Bolts	
	M6		M8		M10		M8		M8		M8		M8	
60	●	○	●	○										
70	●	○	●	○	●	○			●	○				
80	●	○	●	○	●	○			●	○				
100	●	○	●	○	●	○	●	○	●	○	●	○	●	○
120	●	○	●	○	●	○	●	○	●	○	●	○	●	○
140	●	○	●	○	●	○	●	○	●	○	●	○	●	○
160	●	○	●	○	●	○	●	○	●	○	●	○	●	○
180	●	○	●	○	●	○	●	○	●	○	●	○	●	○
200	●	○	●	○	●	○	●	○	●	○	●	○	●	○
220			●	○	●	○	●	○	●	○	●	○	●	○
240			●	○	●	○	●	○	●	○	●	○	●	○
260			●	○	●	○	●	○	●	○	●	○	●	○
280			●	○	●	○	●	○	●	○	●	○	●	○
300			●	○	●	○	●	○	●	○	●	○	●	○

● steel, bright zinc plated
○ aluminium

Steel bent bolts may be used with square or hexagon steel nuts; aluminium bent bolts are used with square aluminium nuts. All types may also be used with plastic sealing caps.

(c) Crook bolts for securing stanchions, etc.

(d) 'J' bolts for securing roof sheets to tubular purlins.

(e) 'U' bolts – round head for securing to tubular members where greater strength is required than that produced by a 'J' bolt; square head for securing to rectangular members where greater strength is required than that provided by a square head hook bolt.

There are also special forms of hook bolts which can be threaded through a hole drilled in the sheeting from the outside and hooked around a purlin before the nut is tightened up.

(ii) Roofing Bolts

These are an alternative to using hook bolts on metal frames, and may also be used with Oakley clips; they are also used for seam joining overlapping sheets. Standard sizes available are detailed in Table II. 'Snaprib' bolts have a square head for use with Alcan Snaprib roofing sheets.

Bolts may be used with metal washers, with or without sealing washers – see Table III.

TABLE II – SIZES OF ROOFING BOLTS

Length mm	DIAMETER SIZE							
	M5		M6		M8		M10	
8	●	○	●					
10	●	○	●					
12	●	○	●	○	●	○	●	
16	●	○	●	○	●	○	●	
20	●	○	●	○	●	○	●	
25	●	○	●	○	●	○	●	○
30	●	○	●	○	●	○	●	○
35	●	○	●	○	●	○	●	○
40	●	○	●	○	●	○	●	○
50	●	○	●	○	●	○	●	○
60	●	○	●	○	●	○	●	○
70	●	○	●	○	●	○	●	○
80			●	○	●	○	●	
100			●	○	●	○	●	
120			●		●	○	●	
140			●		●	○	●	
160			●		●	○	●	
180			●		●		●	
200			●		●		●	
220					●		●	
240					●			
260					●			
280					●			
300					●			

● steel, bright zinc plated

○ aluminium

Roofing bolts may be used with square or hexagon nuts or with Oakley clips (sizes M6 and M8). Bolt heads may be sealed with plastic screw caps or special covers.

TABLE III – TYPES AND SIZES OF WASHERS

Type	Material	Diameter Size M5	M6	M8	M10	Overall Size mm
Round flat	Steel, zinc plated Aluminium PVC Bitumen	x x	x x x x	x x x x	x x x	15–34 dia 15–34 dia 32 dia 18–24 dia
Round flat (sealed)	PVC coated steel		x	x		20–35 dia
Round curved	Steel, zinc plated Aluminium		x x	x x	x x	18–24 dia 18–24 dia
Diamond curved	Steel, zinc plated Aluminium Bitumen		x x x	x x x	x x x	35 x 35 35 x 35 38 x 38
Diamond cranked	Aluminium		x	x	x	35 x 35
Serrated conical	Lead		x	x	x	18–24 dia
Specials		*	*	*	*	*

*Specials include a wide variety of proprietary designs of sealing washers and/or covers to match standard sizes of screws; also individual designs of self-sealing screws, bolts, self-tapping screws, studs, etc.

Fig 1

(iii) **Drive Screws** – for fastening cladding, etc. – see Table IV.
(iv) **Drywall Screws** – special designs of screw for drywall fastening application.
(v) **Roofing Screws** – for fastening roofing sheets, etc, to timber.
(vi) **Gutter Bolts** – see Table V.
(vii) **Sheeting Clips** and **Proprietary Interlocking Systems.**
(viii) **Self-tapping Screws** – for fixing overlapping sheets together, or cladding to metal purlins, etc.

Jointing nails by Bierbach.

TABLE IV – SIZES OF DRIVE SCREWS

Length mm	Diameter Size Screw Gauge (mm)				
	14 (6.3)		18 (7.7)		20 (8.4)
35	●R	○P			
40	●R	○P			
50	●R	○P	●R	○P	
60	●R	○P	●R	○P	
70	●R	○P	●R	○P	
80	●R	○P	●R	○P	●R
100	●R	○P	●R	○P	●R
120	●R	○P	●R	○P	●R
140	●R		●R	○P	●R
160	●R		●R	○P	●R
180					●R
200					●R
220					●R
240					●R
260					●R

●R – roundhead, steel, bright and zinc plated
○P – pan head, aluminium

Drive screws can be used with a variety of metal and plastic washers, also with sealing covers.

TABLE V – SIZES OF GUTTER BOLTS

Length mm	DIAMETER SIZE				
	M5	M6		M8	
12	●	●	○	●	
16	●	●	○	●	
20	●	●	○	●	○
25	●	●	○	●	○
30	●	●	○	●	○
35	●	●		●	
40	●	●	○	●	○
50	●	●	○	●	○
60	●	●		●	
70	●	●		●	
80		●		●	
100		●		●	
120		●		●	
140		●			

● steel, bright zinc plated
○ aluminium

Gutter bolts may be used with square or hexagon nuts

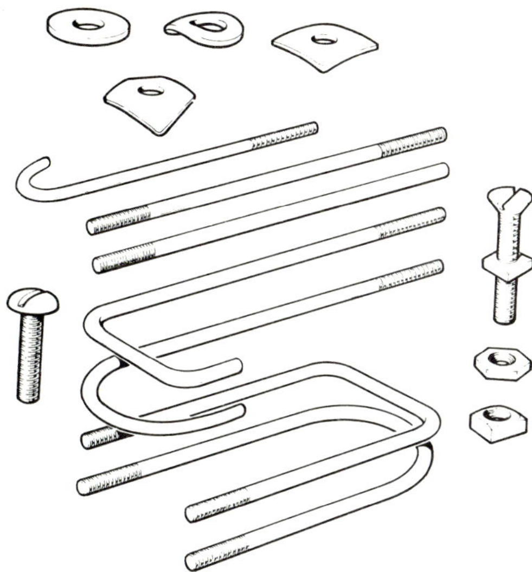

Fig 2 Examples of roof cladding fittings. (Baxters, (Bolts,Screws & Rivets) Ltd).

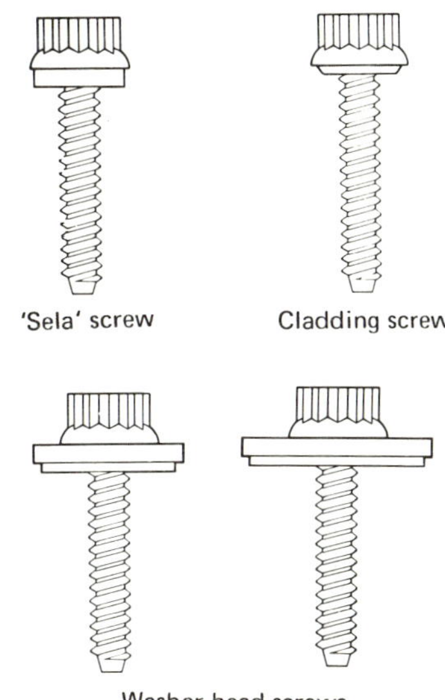

Fig 3 Examples of 'Sela' 12-point head cladding screws. (The British Screw Ltd).

Self-tapping screws embrace a range of both standard (hexagon head) and proprietary designs. Standard self-tapping screws may be used with sealing washers. Cap seals may also be used. Proprietary designs may incorporate sealing washers and modified head forms (eg 12-point with or without washer heads – see Fig 2). Plastic headed screws are also produced in colours to match coloured sheeting. Other items include self-tapping studs.

Self-piercing and self-drilling self-tapping screws offer the advantage of more rapid assembly and can be used on a variety of combinations of materials. Examples of a proprietary range are shown in Fig 3.

MASONRY FIXINGS

Ties and Cramps

Standard brick-to-brick and cavity ties are specified in BS1243. The three configurations are vertical twist, butterfly and double triangle. There are also numerous individual designs of proprietary ties.

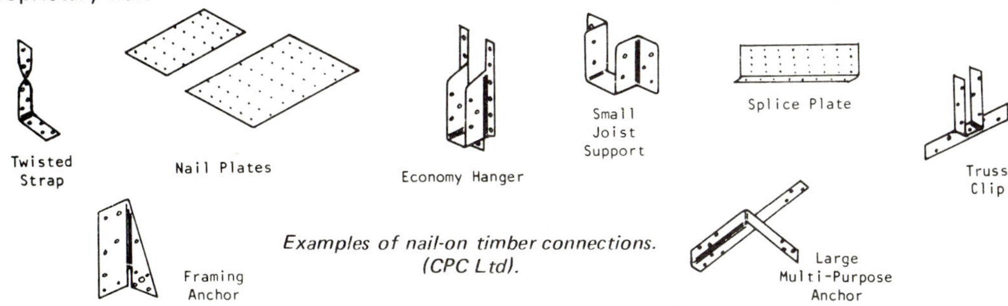

Examples of nail-on timber connections. (CPC Ltd).

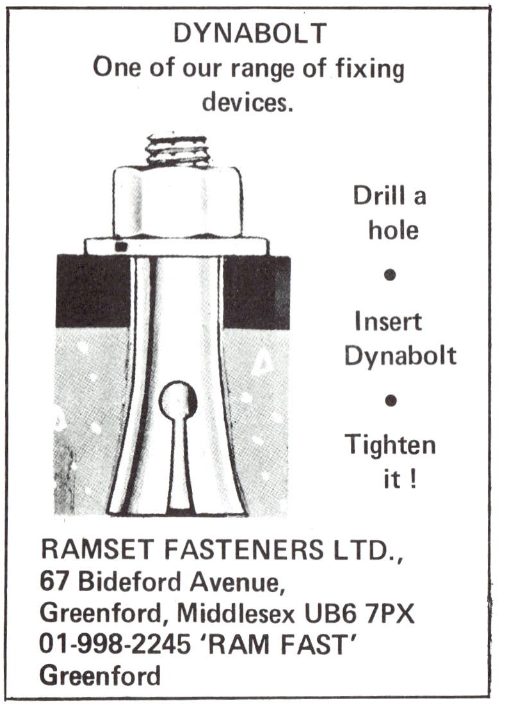

DYNABOLT
One of our range of fixing devices.
Drill a hole
Insert Dynabolt
Tighten it !
RAMSET FASTENERS LTD.,
67 Bideford Avenue,
Greenford, Middlesex UB6 7PX
01-998-2245 'RAM FAST'
Greenford

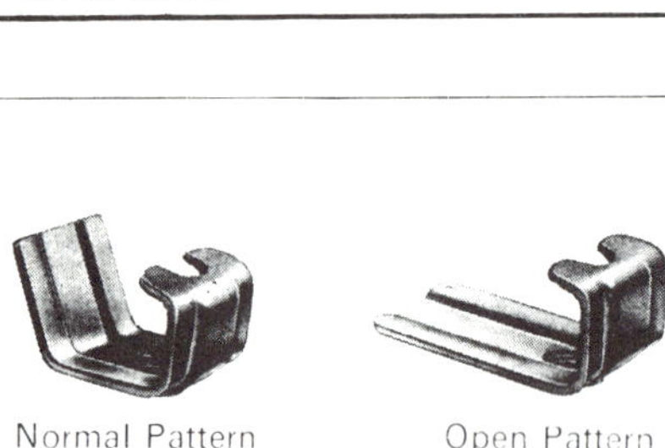

Oakley clips

Type S, bugle head for ½in–5/8in drywall to metal.

Type G bugle head for wallboard to wallboard.

Type W bugle head for wallboard to wood framing.

Type S pan head or hex washer head for metal studs to metal runners.

Type S, twin head for trim to steel studs.

Type S-12 bugle head for wallboards, etc to metal framing.

Type S-12 trim head for trim to metal framing.

Type S-12 pan head, for metal studs to runners.

Fig 4 'Teks' drywall screws (ITW Ltd).

Wall ties are made in zinc coated steel, copper, copper alloy and stainless steel, and also in plastic (although this is not usually approved by local authorities).

Cramps are usually made from strip material 1in to 1½in (25 to 37.5mm) wide, and 1/8in (3mm) to 3/16in (5mm) thick. Cramps may be specially designed for use with dowels, expanding bolts, etc, and may be produced in a wide variety of materials from galvanized mild steel to stainless steel and phosphor bronze, depending on the application requirements.

Cramps can be classified under two main types:-

(i) Those designed for restraint only, eg joining adjacent bricks on a course. The simplest form is a bar or rod with the ends bent up at right angles.

(ii) More elaborate shapes designed for both restraint and loadbearing in the same way as ties, but for use with dissimilar materials – eg stone to brick, brick to concrete. Whilst generally described as cramps, these are more correctly called *cavity anchors.*

TEKS/2 for sheet metal to metal.

TEKS/3 for attachment to medium gauge framing.

TEKS/4 for decking to bar posts and bars.

STITCH/TECK for lap jointing of unsupported metal sheets.

TEKS/5 for metal, siding, etc to heavy structural members.

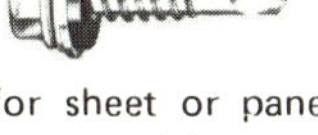

MB/TEKS for sheet or panel to purlin or girt.

MB/bit TEKS, for sheet to purlin or sheet to sheet.

PLYMETAL TEKS for wood or ply to metal.

HEADER TEKS, for 4 x 2 timber to metal.

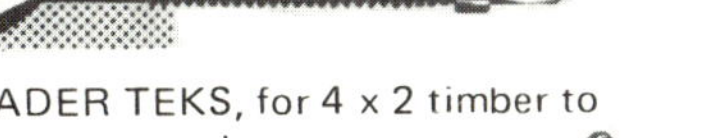

MB/W TEKS, for light gauge metal to wood.

STANDOFF TEKS for sheet or rigid insulation to purlins or girt.

Fig 5

Dowels are plain pins which may be used to locate and anchor cramps or may be grouted in on their own to form a connection – eg between a part of the structure and a supporting rib. Length, diameter and material are selected according to the application.

Corbels

A corbel is a principal load-bearing fixing in the form of a shaped plate or frame grouted into a pocket cast into the structure and carrying the component or cladding to be supported. It is described by its actual geometry, eg

Flat corbel – flat plate

Cranked corbel – plate with angle bend

Fishtail corbel – a flat or cranked plate with a splayed 'fishtail' at one end.

Toe corbel – a flat fishtail corbel with the other end turned up at right angles.

Frame corbel – U-shaped, for lighter loading (also known as a step corbel).

Z corbel – a plate bent to a Z section. Other sections may also be used as corbels.

Anchor Bolts

A summary of the types of anchor bolt used for heavy masonry fixings is given in Table VI. See also chapter on *Screw Anchors and Anchor Bolts.*

See also chapters on *Bolts, Self-Tapping Screws, Woodscrews, Special Screws, Nails, Special Nails, Blind Rivets.*

TABLE VI – TYPES OF ANCHOR BOLTS

Classification	Proprietary Types	Failure Strength kg of 1½in bolt*	Anchor Method	Loading Suitability			
				Dead	Variable	Vibration	Shock
Caulking anchor	Rawlbolt bolt anchor	3060	Profile and compression	√	√	√	√
		3560	Profile and compression	√	√		√
Expansive anchor (manually set)	Phillips Redhead	4480/3460†	Expansion against deformed shield	√	√	√	
	Phillips Redhead	2550	Expansion against deformed shield	√	√		
	Spitroc Anchor	3860	Expansion against deformed shield	√	√	√	
	Star Self-drill	4850	Expansion against deformed shield	√	√	√	
	Rapid Bighead Anchor	3270	Expansion against deformed shield	√	√	√	
	Rawlplug Sabretooth	4090	Expansion against deformed shield	√	√	√	
	Ucan Self-driving	4310	Expansion against deformed shield	√	√	√	
	Rotospit		Expansion against deformed shield	√	√		
	Rawlplug Silver Nugget	3860	Expansion against deformed shield	√	√	√	√
Expansive anchor (mechanically set)	Rawlbolt Loose Bolt	5700	Compression	√	√		√
	Sertbolt	3400	Compression	√	√	√	√
	Dynabolt	3400	Compression	√	√	√	√
	Rawlplug Stud Anchor	4490	Compression	√	√	√	√
	Liebig Bolt	6625	Compression	√	√	√	√
	Wejit (Armstrong)	3320	Compression	√	√	√	√
	Insit (Unifix)	5740	Compression	√	√	√	√
	Febolt	3400	Compression	√	√	√	√
	Forway	4170	Profile and expansion	√	√	√	√

† non-drill   * Architects Journal Information Library

Circlips and Retaining Rings

Retaining rings fall broadly into two categories – those which locate and seat in grooves, and self-locking types which rely purely on frictional grip to maintain their position and resist axial loading. Groove-fitting retaining rings, also known as circlips, are the more widely used and comprise some fifty functionally different types for axial assembly, radial assembly, endplate take-up, etc, although the majority are used for positioning components on shafts. Standard sizes range from miniature rings fitting 5mm (0.188in) shafts or smaller, up to about 250mm (10in).

The two basic types of groove-fitting ring are –

(i) internal and

(ii) external – Fig 1.

Internal rings are used in bores, housings, etc. External rings are fitted to shafts, bosses, studs, etc. The principal difference is in the shape and positioning of the lugs which permit the ring, once installed, to assume a circular form with constant pressure against the bottom of the locating groove. In the case of large rings installed on a rotating component, the ring plan form and lug shape may be specially proportioned to maintain balance under rotation.

Fig 1

Fig 2

Inverted retaining rings are, again, either internal or external, but with the lug positions 'inverted' – Fig 2. The particular advantage offered by this configuration is that they can be fitted with smaller clearance dimensions, and present a higher uniform shoulder when assembled. Their actual radial thrust capacity is, however, somewhat reduced.

Further variants on the basic forms are *bevelled* and *bowed* retaining rings – Fig 3. Both types are designed to take up end play. The main difference between the two is that with a bevelled ring end play is taken up in a rigid, positive manner and with a bowed ring by spring action. Both types are restricted to basic forms (ie they cannot be used in inverted configuration).

You haven't seen anything like the 'New'

CIRCLEX QUICK CLICK CLIP

'click'
AND
ONLY
4½p*
FOR THE
1x
SIZE

Fitting a worm drive hose clip was never a one handed operation. **It is now.** With the 'New' CIRCLEX QCC, a quick fit—quick release mild steel zinc plated worm drive hose clip with an extra deep perforated band, the quick release facility allows almost instant disconnect/connect of the clip from the assembly.

*More than competitively priced and subject to volume discounts and available in a wide range of sizes.

Send for your sample Quick-Click Clip QUICKLY!

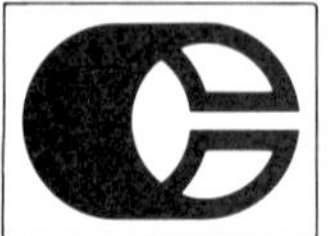

CIRCLEX MARKETING LIMITED
Circlex House, London Road,
Camberley, Surrey.
Telephone: Camberley 62461/2

* OVER 650 PAGES

* COMPLETELY REVISED

SECTION 1 Historical Notes; Properties of Air; Principles of Pneumatics; Pneumatic Circuits; Compressible Gas Flow; Compressed Air Safety; Compressed Air Economics; Air Hydraulics; Low Temperature Techniques; High Pressure Pneumatics; Noise Control; Fluidics; Mechanisation/Automation; Vacuum Techniques. **SECTION 2** Compressors; Compressor Selection; Compressor Installation; Compressor Controls; Pressure Vessels (General); Air Receivers and Pressure Vessels; Air Lines; Air Line Fittings; Pneumatic Valves; Heat Exchangers; Measurement and Instrumentation; Pressure Gauges; Seals and Packings; System and Component Maintenance; Air Cylinders; Air—Hydraulic Cylinders; Pneumatic Tools and Appliances; Workshop Tools; Air Starters; Air Motors; Bellows and Diaphragms; Bursting Discs; Blowers and Fans; Pneumatic Springs; Lifts, Hoists and Air Winches; Vacuum Pumps. **SECTION 3** Applications. **SECTION 4** Surveys of Air Motors; Cylinders; Compressors; Valves. **SECTION 5** Data. **SECTION 6** Manufacturers Buyers Guide.

**TRADE & TECHNICAL PRESS LTD.
CROWN HOUSE, MORDEN, SURREY**

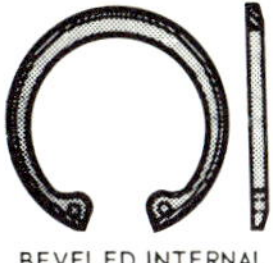

BEVELED INTERNAL RING   BEVELED EXTERNAL RING   BOWED INTERNAL RING   BOWED EXTERNAL RING   BOWED E-RING

Fig 3

Retaining rings may have a uniform section, when they are also known as *snap rings.* Snap rings do not have quite the same axial thrust capacity as retaining rings (circlips), but provide better clearance and a shoulder of uniform depth which enables them to be used as spacer or thrust washers as well as retainers. They are produced with a variety of gap profiles, in both internal and external types — see also *Retainer Clips*.

All types so far described are assembled in an axial direction, being either contracted (internal rings) or expanded (external rings) for manoeuvring into position or removal. Where it is necessary (or more convenient) to assemble or remove a retaining ring in a radial direction, then either an open or interlocking ring form can be used.

Basic forms of the open ring are the C- or crescent ring and the E-ring — Fig 4. The C-ring is normally produced with a tapered section and mainly chosen to provide security against involuntary movement, impact loading or vibration rather than to resist a constant thrust load, although its rating in this respect can be quite high. E-ring geometry can readily be varied to enlarge the bearing area, and the thickness increased to stiffen the section. The thrust capacity will, however, inevitably be less than that of a plain (basic) ring of the same thickness.

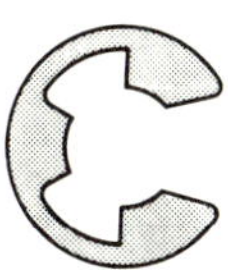
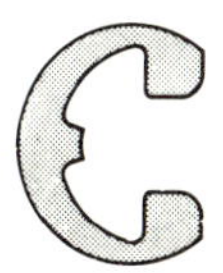

CRESCENT RING   E-RING   REINFORCED E-RING

Fig 4

INTERLOCKING RING

Fig 5

Both the C-ring and the E-ring are produced in flat and bowed forms, the latter providing for take up of end play in a resilient manner.

The interlocking ring is a split ring consisting of two identical halves which locate and lock together when assembled by means of prongs at the free end — Fig 5. In general it has about 75% of the thrust capacity of a plain (basic) ring, but has the advantage that it is statically balanced. It is thus well suited for use on high speed shafts, etc.

Some self-locking ring forms are shown in Fig 6. These are normally employed only for positioning or locking on assemblies not likely to be subject to appreciable axial thrusts, or on materials which are difficult to groove. Depending on their actual form (and material) they may or may not be re-usable. Apart from circular rings, other self-locking retainers or 'spring fasteners' may be used for such purposes, including both stamped sections and wire-formed or spiral-wound configurations.

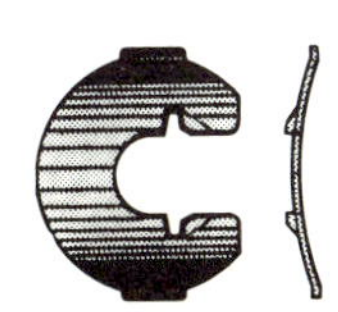

EXTERNAL CIRCULAR SELF-LOCKING RING   INTERNAL CIRCULAR SELF-LOCKING RING   LOCKING PRONG RING

Fig 6

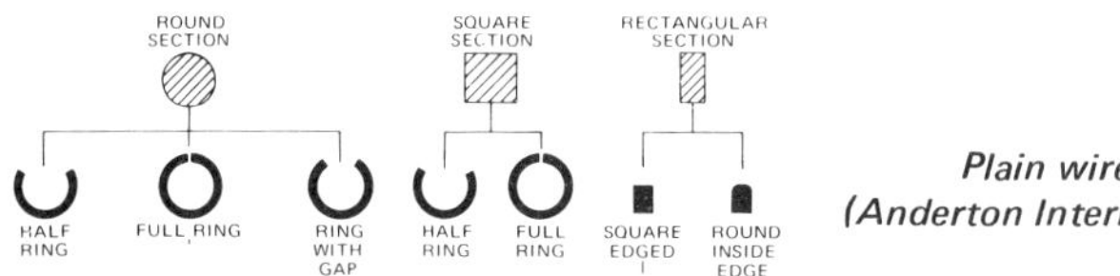

Plain wire rings. (Anderton International Ltd.)

Materials

Standard retaining rings are normally made from carbon steel of spring quality. Other materials used include aluminium, beryllium copper, phosphor bronze and stainless steel. Plated carbon steel rings are alternatives for use in corrosive ambiences.

Selection Factors

The basic ring forms (internal or external) provide the highest axial thrust capacity for a given ring diameter and thickness, and are thus normally a preferred choice for applications involving high axial loads. Where clearance for fitting is limited,, then the inverted ring form may be a better choice. Assembly (or disassembly) is by using retaining ring pliers to engage the eyes in the lugs of the ring which will expand or compress the ring as necessary. Tool clearance may also have to be considered in some applications.

End play can result from accumulated tolerances on component dimensions, ring thickness and groove width. Bevelled or bowed rings can take this up and eliminate the need for shims. For *rigid* take-up, a *bevelled* ring should be used, for *resilient* take-up, a *bowed* ring.

Where radial assembly of the ring is required, then either the C-ring or E-ring is a logical choice. Such rings can be assembled directly in the plane of the groove with a screwdriver or ring applicator

Groove Requirements

The groove depth together with the ring geometry will determine the load capacity of the ring. It is important that the groove be square cut and also that there be sufficient width of shoulder between the outer groove wall and the end of the shaft or housing (excluding any chamfer on these parts) to accommodate the full thrust load. Equally, any chamfer or radius on the abutting surface of a component will reduce the load capacity of a ring.If the chamfer cannot be eliminated, then a rigid square-cut backing washer can be used between the component and the ring.

Speed Limitations

In general, internal rings are not affected by rotation, and small sizes of external rings are generally light enough to be unaffected. Both types, however, may be sprung out of their grooves by adverse operating conditions. In the case of high speed shafting, and particularly where relatively large retaining rings are used on rotating shafts, due consideration may have to be given to the selection of a statically balanced ring form.

On shafts not exceeding 12.5mm (½in) diameter, none of the standard ring forms is likely to show any speed limitations, except possibly E-rings at speeds in excess of 20 000rev/min. With shaft diameters of 25mm (1in), most basic types of retaining ring can be considered likely to be trouble-free at speeds up to 20 000rev/min and E-rings up to 5 000rev/min. With larger shafts, or higher speeds, performance is best verified by practical tests. An empirical formula which can be used to establish speed limitations with standard ring types (except E-rings) is:-

$$\text{maximum safe rev/min} = \frac{20\,000}{\text{shaft diameter in inches}}$$

$$= \frac{500\,000}{\text{shaft diameter in millimetres}}$$

Standard Rings and Grooves

Individual manufacturers specify retaining rings by diameter size, referring to the matching shaft or housing diameter — eg 25mm (1in) external ring is designed to fit a 25mm (1in) diameter shaft. Matching groove geometry is then specified for each size of ring, based on correctly proportioned locking. Load figures are then quoted for permissible axial thrust, normally based on a safety factor of 4. In the case of bowed rings, the load required to straighten the ring may also be given. British Standard specifications for Spring Retaining Rings are given in BS3673 — see Tables I to IV.

Load Capacity Calculations

In an axially loaded condition, the ring, located in a rectangular groove, is subject to shear and the groove material to tensional loading. Optimum design would provide for simultaneous failure of ring and groove, this giving the maximum load capacity of the combination. However, in practice it may be desirable to design for the ring to fail first so that any damage on overload is restricted to the easily replaceable ring.

Failure load of the ring is given by

$$T = \pi\, t\, D\, S_s$$

where t = ring thickness

D = ring diameter

S_s = ultimate shear stress of ring material

Allowing a safety factor of 4, this reduces to

Maximum thrust load = $0.785\, t\, D\, S_s$ 1

Similarly, in the case of the groove and again allowing a safety factor of 4:

Maximum thrust load = $0.785 h\, D\, S_t$ 2

where D = shaft or housing diameter

S_t = ultimate tensile stress of shaft or housing material

h = groove depth

It follows from formulas (1) and (2) that maximum thrust load capacity is given when

$$t\, S_s \text{ (ring)} = h\, S_t \text{ (groove)}$$

Groove geometry can be adjusted as necessary to ensure that $t\, S_s$ is less than $h\, S_t$ for the ring to fail first.

Calculations on this basis do not account for the effect of dynamic or impact loads, or for that of ring deflection (or distortion). Neither is simple to evaluate. To account for the effect of possible dynamic loading, the following empirical factors are commonly applied:

Ring – use 0.5 t in formula (1)

Groove – use 0.5h in formula (2)

In critical applications it may be necessary to analyse the effect of ring deflection which will modify the stresses, or alternatively (and more simply), to determine the failure load by taking the average result of a number of empirical tests.

The straightforward method of calculation given by formulas (1) and (2) applies only to basic ring forms where the whole perimeter of the ring is in contact with the groove. The following empirical factors to maximum thrust load can be applied in the case of other ring forms:

type of ring	factor
inverted	0.66
E-rings	0.33
C-rings	0.50
interlocking	0.75

Note, however, that individual manufacturers' designs of C-rings, E-rings and interlocking rings may have higher (or lower) factors, depending on actual geometry. Also different factors may apply for ring and grooves (see Table V).

The nomogram of Fig 7 can be used for rapid assessment of the likely load capacity of standard retaining rings (using the aforementioned factors if appropriate). To determine the maximum thrust loading of the ring, connect ring thickness on left hand scale to maximum permissible shear stress of the ring material. From the point established on the reference scale, connect to shaft or housing diameter and read off maximum thrust on scale C.

To check the groove thrust rating, connect groove depth (on same scale as t) to maximum permissible tensile stress of groove material. From the point established on the reference line, connect to ring diameter and read off maximum thrust on scale C.

Plastic Retaining Rings

Plastic retaining rings are simple rings (O-rings) moulded in nylon or vinyl (usually) to locate in a groove on a shaft or spindle and provide a lightweight shoulder – Fig 8. The matching groove is normally proportioned so that more than 50% of the ring diameter is embedded. The elasticity of plastic retaining rings provides easy assembly by rolling the ring onto the shaft or spindle until it snaps into the groove.

Special-Purpose Retaining Rings

Special-purpose retaining rings are those with geometry designed specifically to meet an unusual requirement which cannot be covered by standard ring forms, such as special lug designs, modified or reinforced sections. They may be of stamped or wire-formed construction.

Individual manufacturers may include a range of special-purpose retaining rings as standard productions, with individual designs produced to special order.

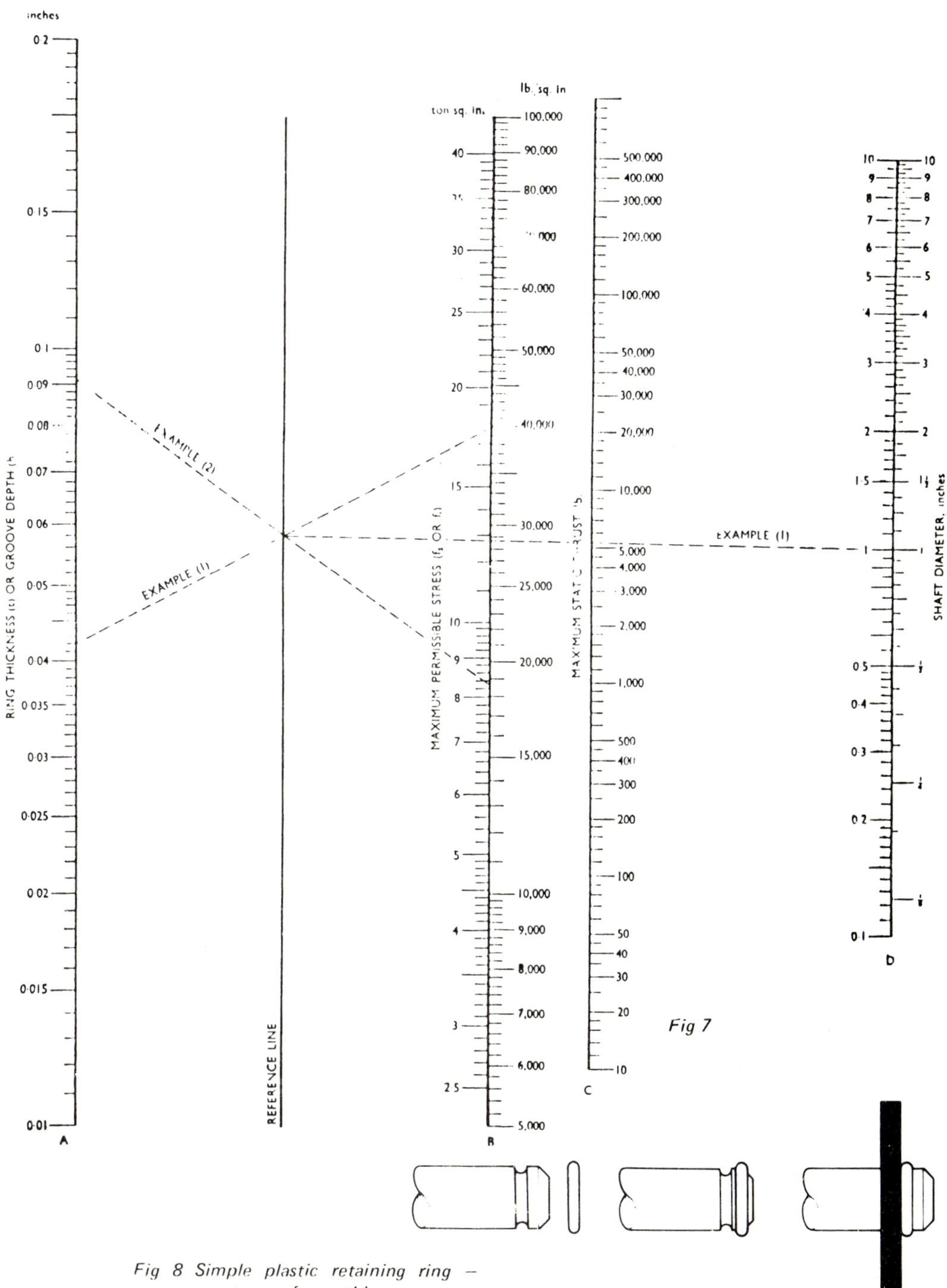

Fig 7

Fig 8 Simple plastic retaining ring – stages of assembly.

Snap Rings

Snap ring is the name given to wire-formed retaining rings. These may be bent from round or flat wire, with ends bent or stamped to form lugs. Some examples of snap ring geometry are shown in Fig 9.

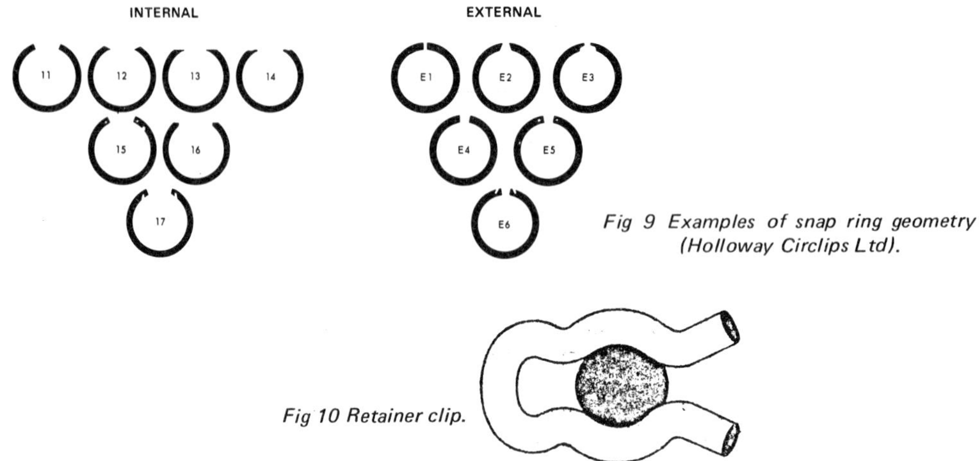

Fig 9 Examples of snap ring geometry (Holloway Circlips Ltd).

Fig 10 Retainer clip.

Retainer Clips

This description is given to open-ended spring wire clips designed to snap into a groove on a shaft or spindle – Fig 10. They may be formed from square or circular wire. Usually one size of retainer clip will fit several shaft diameters. They can be assembled and disassembled with pliers.

Retainer clips may be classified as light duty (formed from circular wire) and heavy duty (formed from square wire); 'duty' refers to the thrust rating.

Retainer clips are used on shafts and spindles from 1.5mm (1/16in) diameter up to about 25mm (1in). They rotate with the shaft and adequate clearance must be available to accommodate the specific circumference relative to the clip length.

Self-Locking Clips and Rings

This category defines push-on external rings of sprung and/or serrated form, the holding action being produced by spring pressure generated at the prongs or gripping surfaces. They produce suitable grip on plain shafts, but holding power is increased if they locate in a shallow groove. Fasteners of this type are described in the separate chapter *Spring Retainers.*

Examples of proprietary retaining rings and self-locking clips. (Salterfix Ltd).

TABLE I – CARBON STEEL EXTERNAL CIRCLIPS – BS3673 (ABRIDGED LIST);

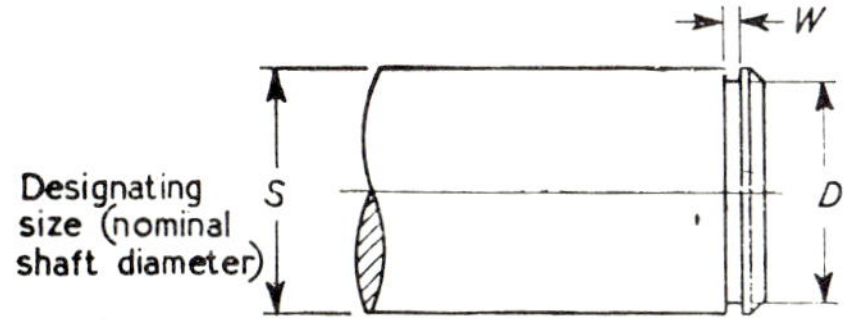

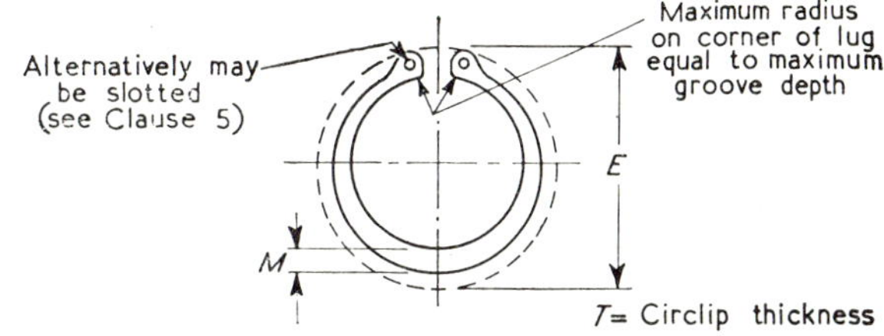

NOTE. Bottoms of grooves should have radii approximating to 10% of groove depth

Reference Number of Circlip	Designating Size (Nominal Shaft Diameter) S		Shaft Diameter	Thickness of Circlip T		Major Radial Depth M		External clearance diameter during assembly E	Lug-hole diameter (or slot width)	Groove Dimensions			
										Diameter D		Width W	
	inch	metric	max	min	max	min	max	min	min	min	max	min	max
	in	mm	in	in	in	in	in	in	in	in	in	in	in
S0015	0.156	–	0.158	0.0135	0.0155	0.028	0.034	0.272	0.024	0.145	0.148	0.018	0.021
S004M	–	4	0.159	0.0135	0.0155	0.028	0.034	0.274	0.024	0.147	0.150	0.018	0.021
S0018	0.188	–	0.190	0.0135	0.0155	0.030	0.037	0.304	0.024	0.176	0.179	0.018	0.021
S005M	–	5	0.199	0.0210	0.0230	0.031	0.038	0.369	0.024	0.184	0.187	0.025	0.028
S006M	–	6	0.238	0.0250	0.0270	0.035	0.043	0.439	0.024	0.222	0.225	0.029	0.032
S0025	0.250	–	0.252	0.0250	0.0270	0.036	0.044	0.452	0.039	0.234	0.238	0.029	0.032
S007M	–	7	0.278	0.0290	0.0320	0.038	0.047	0.516	0.039	0.259	0.262	0.035	0.038
S0031	0.312	–	0.314	0.0290	0.0320	0.042	0.051	0.552	0.039	0.293	0.297	0.035	0.038
S008M	–	8	0.317	0.0290	0.0320	0.042	0.051	0.555	0.039	0.296	0.300	0.035	0.038
S0037	0.375	–	0.377	0.037	0.0400	0.047	0.058	0.675	0.039	0.353	0.357	0.043	0.047
S010M	–	10	0.396	0.037	0.0400	0.049	0.060	0.694	0.039	0.371	0.374	0.043	0.047
S012M	–	12	0.475	0.037	0.0400	0.056	0.068	0.772	0.039	0.445	0.449	0.043	0.047
S0050	0.500	–	0.503	0.037	0.0400	0.059	0.071	0.800	0.045	0.472	0.476	0.043	0.047
S0056	0.562	–	0.565	0.037	0.0400	0.065	0.078	0.862	0.045	0.531	0.535	0.043	0.047
S015M	–	15	0.594	0.037	0.0400	0.067	0.081	0.890	0.045	0.558	0.562	0.043	0.047
S0062	0.625	–	0.628	0.037	0.0400	0.070	0.085	0.925	0.045	0.591	0.595	0.043	0.047
S016M	–	16	0.633	0.037	0.0400	0.071	0.085	0.930	0.045	0.595	0.599	0.043	0.047
S0068	0.688	–	0.691	0.037	0.0400	0.076	0.091	0.988	0.050	0.651	0.655	0.043	0.047
S018M	–	18	0.712	0.0445	0.0475	0.078	0.094	1.065	0.050	0.671	0.675	0.051	0.055
S0075	0.750	–	0.754	0.0445	0.0475	0.082	0.098	1.106	0.050	0.709	0.714	0.051	0.055
S020M	–	20	0.791	0.0445	0.0475	0.086	0.102	1.144	0.050	0.745	0.750	0.051	0.055
S0081	0.812	–	0.816	0.0445	0.0475	0.088	0.105	1.168	0.050	0.768	0.773	0.051	0.055
S022M	–	22	0.870	0.0445	0.0475	0.093	0.110	1.222	0.050	0.820	0.825	0.051	0.055
S0087	0.875	–	0.879	0.0445	0.0475	0.094	0.111	1.231	0.050	0.828	0.833	0.051	0.055
S024M	–	24	0.949	0.0445	0.0475	0.100	0.119	1.301	0.076	0.895	0.900	0.051	0.055
S025M	–	25	0.989	0.0445	0.0475	0.104	0.123	1.340	0.076	0.933	0.938	0.051	0.055
S0100	1.000	–	1.004	0.0445	0.0475	0.105	0.125	1.356	0.076	0.948	0.953	0.051	0.055

Note: Sizes extend up to 10in (250mm)

TABLE II – BRITISH STANDARD INTERNAL CIRCLIPS (BS3673) (ABRIDGED LIST)

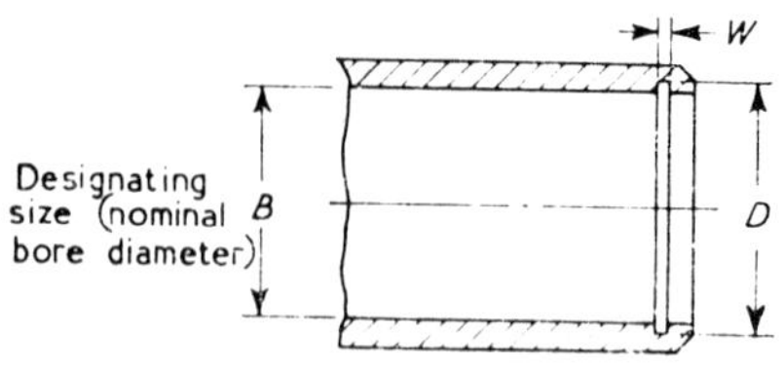

NOTE. Bottoms of grooves should have radii approximating to 10% of groove depth

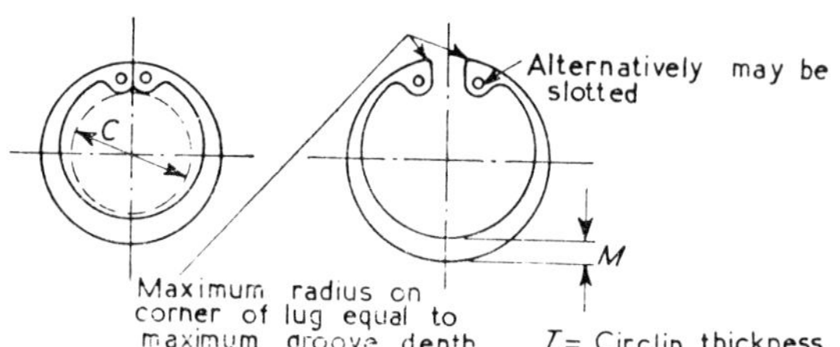

Maximum radius on corner of lug equal to maximum groove depth

T = Circlip thickness

Reference Number of Circlip	Designating Size (Nominal bore diameter) B		Bore Diameter	Thickness of Circlip T		Major Radial Depth M		Internal clearance diameter during assembly C	Lug-hole diameter	Groove Dimensions			
										Diameter D		Width	
	inch	metric	min	min	max	min	max	max	min	min	max	min	max
	in	mm	in	in	in	in	in	in	in	in	in	in	in
B008M	–	8	0.313	0.0290	0.0320	0.041	0.050	*	0.039	0.331	0.334	0.035	0.038
B009M	–	9	0.354	0.0290	0.0320	0.044	0.053	*	0.039	0.372	0.375	0.035	0.038
B0037	0.375	–	0.373	0.0370	0.0400	0.045	0.055	0.035	0.039	0.393	0.397	0.043	0.047
B010M	–	10	0.392	0.0370	0.0400	0.046	0.057	0.054	0.039	0.413	0.417	0.043	0.047
B0043	0.438	–	0.436	0.0370	0.0400	0.050	0.060	0.098	0.039	0.459	0.463	0.043	0.047
B012M	–	12	0.470	0.0370	0.0400	0.052	0.064	0.132	0.045	0.496	0.500	0.043	0.047
B0050	0.500	–	0.498	0.0370	0.0400	0.054	0.066	0.160	0.045	0.524	0.528	0.043	0.047
B014M	–	14	0.549	0.0370	0.0400	0.058	0.070	0.211	0.045	0.578	0.582	0.043	0.047
B0056	0.562	–	0.559	0.0370	0.0400	0.059	0.071	0.222	0.045	0.589	0.593	0.043	0.047
B0062	0.625	–	0.622	0.0370	0.0400	0.063	0.077	0.285	0.060	0.655	0.659	0.043	0.047
B016M	–	16	0.627	0.0370	0.0400	0.064	0.077	0.290	0.060	0.660	0.664	0.043	0.047
B0068	0.688	–	0.685	0.0370	0.0400	0.068	0.082	0.348	0.060	0.721	0.726	0.043	0.047
B018M	–	18	0.706	0.0370	0.0400	0.070	0.084	0.369	0.060	0.743	0.748	0.043	0.047
B0075	0.750	–	0.747	0.0370	0.0400	0.073	0.087	0.410	0.060	0.786	0.791	0.043	0.047
B020M	–	20	0.784	0.0370	0.0400	0.075	0.091	0.447	0.060	0.825	0.830	0.043	0.047
B0081	0.812	–	0.809	0.0370	0.0400	0.077	0.093	0.472	0.060	0.851	0.856	0.043	0.047
B021M	–	21	0.824	0.0370	0.0400	0.078	0.094	0.487	0.060	0.866	0.871	0.043	0.047
B022M	–	22	0.863	0.0370	0.0400	0.081	0.097	0.526	0.060	0.908	0.912	0.043	0.047
B0087	0.875	–	0.872	0.0370	0.0400	0.082	0.098	0.535	0.060	0.917	0.922	0.043	0.047
B0093	0.938	–	0.935	0.0445	0.0475	0.087	0.103	0.534	0.060	0.983	0.988	0.051	0.055
B024M	–	24	0.942	0.0445	0.0475	0.087	0.104	0.541	0.060	0.990	0.995	0.051	0.055
B025M	–	25	0.980	0.0445	0.0475	0.090	0.107	0.580	0.060	1.031	1.036	0.051	0.055
B0100	1.000	–	0.996	0.0445	0.0475	0.091	0.109	0.596	0.060	1.048	1.053	0.051	0.055

Note : Sizes extend up to 10in (250mm)

TABLE III – BRITISH STANDARD CARBON STEEL E CLIPS (BS3673:Part 2)

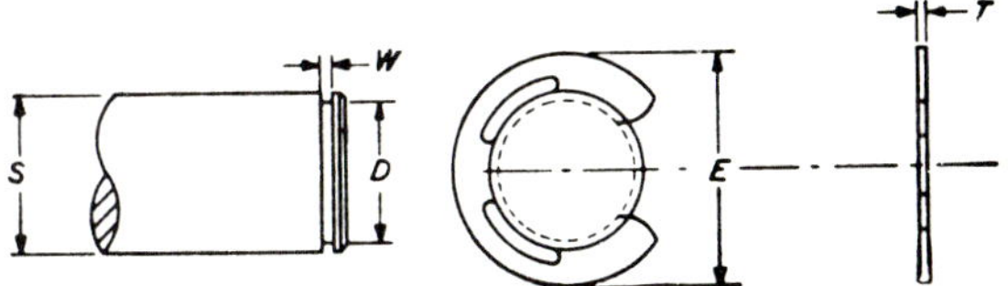

PREFERRED INCH SIZES – DIMENSIONS IN INCHES

Reference No	Recommended Groove Dimensions				Envelope dia fitted E	Clip Thickness		Shaft Diameter	
	Groove diameter D		Groove width W			T		S	
	min	max	min	max	max	min	max	min	max
028PS	0.0255	0.028	0.012	0.0145	0.090	0.0085	0.011	–	0.05
054PS	0.0515	0.054	0.012	0.0145	0.176	0.0085	0.011	0.06	0.09
076PS	0.0735	0.076	0.019	0.0215	0.193	0.013	0.017	0.09	0.13
097PS	0.0945	0.097	0.019	0.0215	0.213	0.013	0.017	0.12	0.17
104PS	0.1015	0.104	0.019	0.0215	0.230	0.013	0.017	0.14	0.17
118PS	0.1155	0.118	0.024	0.0265	0.300	0.018	0.022	0.15	0.20
129PS	0.1260	0.129	0.029	0.0315	0.330	0.023	0.027	0.17	0.22
149PS	0.1460	0.149	0.029	0.0315	0.350	0.023	0.027	0.18	0.25
190PS	0.1870	0.190	0.029	0.0315	0.450	0.023	0.027	0.22	0.32
212PS	0.2090	0.212	0.029	0.0315	0.540	0.023	0.027	0.25	0.38
252PS	0.2485	0.252	0.029	0.0315	0.540	0.023	0.027	0.31	0.38
306PS	0.3015	0.306	0.040	0.0440	0.680	0.033	0.037	0.37	0.47
346PS	0.3415	0.346	0.040	0.0440	0.710	0.033	0.037	0.43	0.47
399PS	0.3945	0.399	0.050	0.0540	0.820	0.040	0.044	0.50	0.63
488PS	0.4840	0.488	0.050	0.0540	0.960	0.040	0.044	0.62	0.75
583PS	0.5790	0.583	0.056	0.0600	1.140	0.048	0.052	0.75	0.82
678PS	0.6740	0.678	0.056	0.0600	1.320	0.048	0.052	0.87	1.00

METRIC SERIES – DIMENSIONS IN MILLIMETRES

Reference No	Recommended Groove Dimensions				Envelope dia fitted E	Clip Thickness		Shaft Diameter	
	Groove diameter D		Groove width W			T		S	
	min	max	min	max	max	min	max	min	max
008MS	0.74	0.8	0.24	0.26	2.0	0.18	0.22	1.0	1.4
012MS	1.14	1.2	0.34	0.36	3.0	0.28	0.32	1.4	2.0
015MS	1.44	1.5	0.44	0.46	4.0	0.38	0.42	2.0	2.5
019MS	1.84	1.9	0.54	0.56	4.5	0.48	0.52	2.5	3.0
023MS	2.24	2.3	0.64	0.66	6.0	0.58	0.62	3.0	4.0
032MS	3.13	3.2	0.64	0.66	7.0	0.58	0.62	4.0	5.0
040MS	3.93	4	0.74	0.76	9.0	0.68	0.72	5.0	7.0
050MS	4.93	5	0.74	0.76	11.0	0.68	0.72	6.0	8.0
060MS	5.93	6	0.74	0.76	12.0	0.68	0.72	7.0	9.0
070MS	6.89	7	0.94	0.96	14.0	0.88	0.92	8.0	11.0
080MS	7.91	8	1.05	1.11	16.0	0.97	1.03	9.0	12.0
090MS	8.91	9	1.15	1.21	18.5	1.07	1.13	10.0	14.0
100MS	9.89	10	1.25	1.31	20.0	1.17	1.23	11.0	15.0
120MS	11.89	12	1.35	1.41	23.0	1.27	1.33	13.0	18.0
150MS	14.89	15	1.55	1.61	29.0	1.47	1.53	16.0	24.0
190MS	18.87	19	1.85	1.91	37.0	1.72	1.78	20.0	25.0
240MS	23.87	24	2.05	2.11	44.0	1.97	2.03	25.0	38.0

TABLE IV – BRITISH STANDARD CARBON STEEL C-CLIPS (BS3673:Part 3)

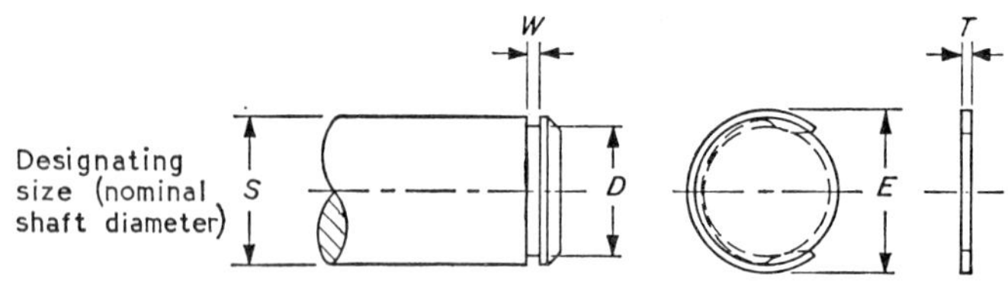

Reference No of Clip	Designating Size (nominal Shaft dia S	Recommended Groove Dimensions Groove diameter D		Groove width W	Clearance diameter fitted E	Clip Thickness T	
Inch Series – Dimensions in inches							
		minimum	maximum	minimum	maximum	minimum	maximum
C0012	0.125	0.1045	0.1075	0.018	0.180	0.013	0.017
C0015	0.156	0.1335	0.1365	0.018	0.220	0.013	0.017
C0018	0.187	0.1635	0.1665	0.018	0.250	0.013	0.017
C0021	0.219	0.191	0.195	0.029	0.290	0.023	0.027
C0023	0.234	0.206	0.210	0.029	0.310	0.023	0.027
C0025	0.250	0.218	0.222	0.029	0.330	0.023	0.027
C0026	0.265	0.228	0.232	0.029	0.330	0.023	0.027
C0028	0.281	0.245	0.249	0.029	0.360	0.023	0.027
C0031	0.312	0.274	0.278	0.029	0.390	0.023	0.027
C0037	0.375	0.333	0.337	0.029	0.470	0.023	0.027
C0040	0.406	0.362	0.366	0.029	0.500	0.023	0.027
C0043	0.437	0.391	0.395	0.029	0.530	0.023	0.027
C0046	0.469	0.419	0.423	0.029	0.570	0.023	0.027
C0050	0.500	0.447	0.453	0.039	0.600	0.033	0.037
C0056	0.562	0.504	0.510	0.039	0.670	0.033	0.037
C0062	0.625	0.560	0.566	0.039	0.740	0.033	0.037
C0068	0.687	0.616	0.622	0.046	0.800	0.040	0.044
C0075	0.750	0.673	0.679	0.046	0.870	0.040	0.044
C0081	0.812	0.729	0.735	0.046	0.940	0.040	0.044
C0087	0.875	0.786	0.792	0.046	0.010	0.040	0.044
C0093	0.937	0.840	0.846	0.046	1.080	0.040	0.044
C0100	1.000	0.897	0.903	0.046	1.150	0.040	0.044
Metric Series – Dimensions in millimetres							
C005M	5	3.925	4.0	0.64	6.5	0.58	0.62
C006M	6	4.925	5.0	0.74	8.5	0.68	0.72
C007M	7	5.910	6.0	0.85	9.2	0.78	0.82
C008M	8	6.910	7.0	0.85	9.9	0.78	0.82
C009M	9	7.910	8.0	1.10	11.5	0.97	1.03
C010M	10	8.910	9.0	1.10	12.3	0.97	1.03
C011M	11	9.890	10.0	1.10	13.5	0.97	1.03
C012M	12	10.790	10.9	1.10	14.5	0.97	1.03
C013M	13	11.690	11.8	1.10	15.5	0.97	1.03
C014M	14	12.590	12.7	1.10	17.0	0.97	1.03
C015M	15	13.490	13.6	1.10	18.0	0.97	1.03
C016M	16	14.390	14.5	1.10	19.0	0.97	1.03
C017M	17	15.290	15.4	1.10	20.0	0.97	1.03
C018M	18	16.190	16.3	1.30	21.0	1.17	1.23
C019M	19	17.090	17.2	1.30	22.0	1.17	1.23
C020M	20	17.890	18.1	1.30	23.0	1.17	1.23
C022M	22	19.690	19.9	1.30	25.5	1.17	1.23
C023M	23	20.590	20.8	1.30	26.3	1.17	1.23
C024M	24	21.490	21.7	1.30	27.6	1.17	1.23
C025M	25	22.390	22.6	1.30	29.2	1.17	1.23
C026M	26	23.290	23.5	1.30	30.0	1.17	1.23

TABLE V – TYPICAL LOAD CAPACITIES OF RETAINING RINGS

Type of Ring	Diameter in	Diameter mm	Thickness in	Thickness mm	Maximum Static Load lb	Maximum Static Load kg	Load Correction Factors Ring	Load Correction Factors Groove
Basic, external	1/8	3.0	.010	0.25	110	50	1.0	1.0
	1/4	6.0	.025	0.63	590	268		
	1/2	12.5	.035	0.89	1 650	750		
	1	25.0	.042	1.07	4 950	2 245		
	5	125 0	.110	2.79	64 200	29 100		
Basic, internal	1/4	6.0	.015	0.35	350	159	1.0	1.0
	1/2	12.5	.035	0.89	1 650	750		
	1	25.0	.042	1.07	4 950	2 245		
	5	125.0	.110	2.79	64 200	29 100		
Inverted, external	1/2	12.5	.035	0.89	1 100	900	0.66	0.33
	3/4	19.0	.042	1.07	3 300	1 500		
	1	25.0	.042	1.07	34 300	15 558		
	4	100.0	.110	2.79				
Inverted, internal	1/2	12.5	.035	0.89	110	50	0.66	0.33
	3/4	19.0	.042	1.07	3 300	1 500		
	1	25.0	.042	1.07	34 300	15 558		
	4	100.0	.110	2.79				
Bowed, external	1/4	6.0	.025	0.63	590	268	1.0	0.8
	1/2	12.5	.035	0.89	1 800	820		
	1	25.0	.042	1.07	4 950	2 245		
Bowed, internal	1/4	6.0	.025	0.63	330	150	1.0	0.8
	1/2	12.5	.035	0.89	1 650	750		
	1	25.0	.042	1.07	4 950	2 245		
Bevelled, external	1	25.0	.042	1.07	4 950	2 245	1.0	0.5
	2	50.0	.062	1.57	14 600	6 620		
	5	125.0	.110	2.79	64 200	29 100		
Bevelled, internal	1	25.0	.042	1.07	4 950	2 245	1.0	0.5
	2	50.0	.062	1.57	14 600	6 620		
	5	125.0	.110	2.79	64 200	29 100		
C-Ring	1/8	3.0	.015	0.38	70	32	0.5	0.5
	1/4	6.0	.025	0.63	290	132		
	1/2	12.5	.035	0.89	530	240		
	1	25.0	.042	1.07	2 480	1 125		
Bowed C-Ring	1/8	3.0	.015	0.38	70	32	0.5	0.5
	1/4	6.0	.025	0.63	290	132		
	1/2	12.5	.035	0.89	530	240		
	1	25.0	.042	1.07	2 480	1 125		
E-Ring	1/4	6.0	.025	0.63	255	116	0.33	0.66
	1/2	12.5	.042	1.07	1 110	504		
	1	25.0	.050	1.27	2 750	1 250		
Bowed E-Ring	1/4	6.0	.025	0.63	255	116	0.33	0.66
	1/2	12.5	.042	1.07	1 110	504		
	1	25.0	.050	1.27	2 750	1 250		

STANDARD EUROPEAN INTERNAL CIRCLIPS* (DIN1300) *Anderton International Ltd

Alternative lug sizes over 38mm

Most sizes over 165mm are without lugs

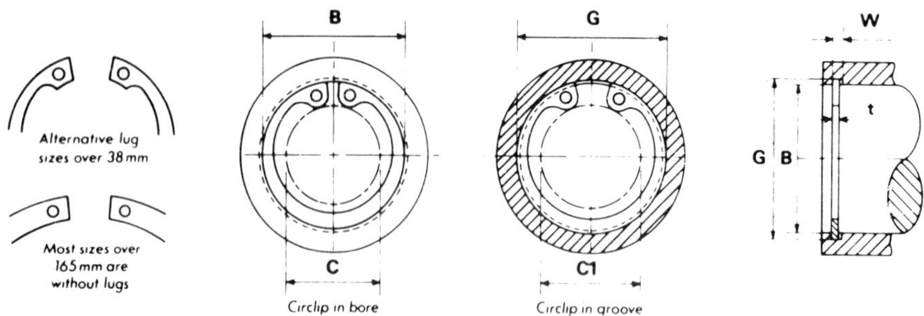

Circlip in bore
Circlip in groove

Range incorporates European Standard DIN 472 Blatt 1, (shown in heavy type column 1)

Bore Size (mm)	Groove				Circlip thickness (mm)		Clearance diameter (mm)		Application				Bore Size (mm)
	Diameter (mm)		Width (mm)						Allowable thrust load in Newtons (N)				
B	G	Tol.	W	Tol.	t	Tol.	C	C1	Circlip	Groove			B
8	8.4	+.090 −.000	0.9		0.8		2.8	3.6	6200	620			8
9	9.4		0.9		0.8		3.5	4.4	7000	690			9
9.5	9.9		1.1		1.0		3.5	3.9	9300	730			9.5
10	10.4		1.1		1.0		3.1	4.0	9700	770			10
11	11.4		1.1		1.0		3.9	4.8	10700	840			11
12	12.5		1.1		1.0		4.7	5.7	11700	1150			12
13	13.6		1.1		1.0		5.3	6.4	12700	1500			13
14	14.6	+.110 −.000	1.1		1.0		6.0	7.2	13600	1620			14
15	15.7		1.1		1.0		7.0	8.3	14600	2030			15
16	16.8		1.1		1.0		7.7	9.2	15600	2470			16
17	17.8		1.1		1.0		8.4	10.0	16600	2620			17
18	19.0		1.1		1.0		8.9	10.8	17500	3490			18
19	20.0		1.1		1.0		9.8	11.8	18500	3680			19
20	21.0		1.1		1.0		10.6	12.6	19500	3860			20
21	22.0		1.1		1.0		11.6	13.6	20400	4050			21
22	23.0		1.1		1.0		12.6	14.6	21400	4240			22
23	24.1	+.210 −.000	1.3		1.2		13.6	15.7	26900	4880			23
24	25.2		1.3		1.2		14.2	16.4	28000	5560			24
25	26.2		1.3		1.2		15.0	17.2	29200	5790			25
26	27.2		1.3	+.140 −.000	1.2	+.000 −.060	15.6	17.8	30400	6020			26
27	28.4		1.3		1.2		16.6	19.0	31600	7310			27
28	29.4		1.3		1.2		17.4	19.8	32700	7570			28
29	30.4		1.3		1.2		18.4	20.8	33900	7840			29
30	31.4		1.3		1.2		19.4	21.8	35100	8100			30
32	33.7		1.3		1.2		20.2	22.9	37400	10500			32
33	34.7		1.3		1.2		21.2	23.9	38600	10900			33
34	35.7		1.6		1.5		22.2	24.9	49700	11200			34
35	37.0		1.6		1.5		23.2	26.2	51100	13600			35
36	38.0		1.6		1.5		24.2	27.2	52600	14000			36
37	39.0		1.6		1.5		25.0	28.0	54100	14300			37
38	40.0	+.250 −.000	1.6		1.5		26.0	29.0	55500	14700			38
40	42.5		1.85		1.75		27.4	30.9	56600	19400			40
41	43.5		1.85		1.75		28.2	31.7	58100	19900			41
42	44.5		1.85		1.75		29.2	32.7	59500	20400			42
45	47.5		1.85		1.75		31.6	35.1	63700	21800			45
47	49.5		1.85		1.75		33.2	36.7	66600	22700			47
48	50.5		1.85		1.75		34.6	37.7	68000	23200			48
50	53.0	+.300 −.000	2.15		2.0		36.0	40.0	80900	29100			50
51	54.0		2.15		2.0		37.0	41.0	82500	29700			51
52	55.0		2.15		2.0		37.6	41.6	84100	30300			52

cont...

Bore Size (mm)	Groove Diameter (mm)		Groove Width (mm)		Circlip thickness (mm)		Clearance diameter (mm)		Application: Allowable thrust load in Newtons (N)				Bore Size (mm)
B	G	Tol.	W	Tol.	t	Tol.	C	C1	Circlip	Groove			B
55	58.0		2.15		2.0		40.4	44.4	89000	32000			55
56	59.0		2.15		2.0		41.4	45.4	90600	32500			56
57	60.0		2.15		2.0		42.4	46.4	92200	33100			57
58	61.0		2.15		2.0		43.2	47.2	93900	33700			58
60	63.0		2.15		2.0		44.4	48.4	97100	34800			60
62	65.0		2.15		2.0		46.4	50.4	100000	35900			62
63	66.0	+.300 −.000	2.15		2.0		47.4	51.4	102000	36800			63
65	68.0		2.65	+.140 −.000	2.5		48.8	52.8	131000	37600			65
67	70.0		2.65		2.5		50.6	54.6	136000	38700			67
68	71.0		2.65		2.5		51.4	55.4	138000	39300			68
70	73.0		2.65		2.5		53.4	57.4	142000	40400			70
72	75.0		2.65		2.5	+.000 −.060	55.4	59.4	146000	41600			72
75	78.0		2.65		2.5		58.4	62.4	152000	43300			75
76	79.0		2.65		2.5		59.4	63.4	154000	43800			76
78	81.0		2.65		2.5		60.0	64.0	158000	45000			78
80	83.5		2.65		2.5		62.0	66.5	162000	53900			80
82	85.5		2.65		2.5		64.0	68.5	166000	55300			82
85	88.5		3.15		3.0		66.8	71.3	206000	57200			85
88	91.5	+.350 −.000	3.15		3.0		69.8	74.3	214000	59200			88
90	93.5		3.15		3.0		71.8	76.3	218000	60500			90
92	95.5		3.15		3.0		73.6	78.1	223000	61900			92
95	98.5		3.15		3.0		76.4	80.9	231000	63800			95
98	101.5		3.15		3.0		79.0	83.5	238000	65800			98
100	103.5		3.15		3.0		81.0	85.5	243000	67100			100
102	106		4.15		4.0		82.6	87.6	330000	78400			102
105	109		4.15		4.0		85.6	90.6	340000	80700			105
108	112	+.540 −.000	4.15		4.0		88.0	93.0	350000	83000			108
110	114		4.15		4.0		88.2	93.2	356000	84500			110
112	116		4.15		4.0		90.0	95.0	362000	86000			112
115	119		4.15		4.0		93.0	98.0	372000	88200			115
120	124		4.15		4.0		97.0	102	388000	92000			120
125	129		4.15		4.0		102	107	405000	95800			125
127	131		4.15		4.0		104	109	411000	97300			127
130	134		4.15		4.0		107	112	421000	99500			130
135	139		4.15		4.0		112	116	437000	103000			135
140	144		4.15		4.0		117	121	453000	107000			140
145	149	+.630 −.000	4.15		4.0		122	126	469000	111000			145
150	155		4.15	+.180 −.000	4.0		125	131	485000	144000			150
155	160		4.15		4.0		130	136	502000	148000			155
160	165		4.15		4.0		133	139	518000	153000			160
165	170		4.15		4.0	+.000 −.075	138	144	534000	158000			165
170	175		4.15		4.0		145	150	550000	163000			170
175	180		4.15		4.0		149	155	566000	167000			175
180	185		4.15		4.0		153	158	583000	172000			180
185	190		4.15		4.0		157	162	599000	177000			185
190	195		4.15		4.0		162	167	615000	181000			190
195	200	+.720 −.000	4.15		4.0		167	172	631000	186000			195
200	205		4.15		4.0		171	177	647000	191000			200
210	216		5.15		5.0		181	188	739000	241000			210
220	226		5.15		5.0		191	198	774000	252000			220
230	236		5.15		5.0		201	208	809000	264000			230
240	246		5.15		5.0		211	218	845000	275000			240
250	256		5.15		5.0		221	228	880000	286000			250
260	268		5.15		5.0		227	236	915000	398000			260
270	278	+.810 −.000	5.15		5.0		237	246	950000	413000			270
280	288		5.15		5.0		247	256	985000	428000			280
290	298		5.15		5.0		257	266	1020000	443000			290
300	308		5.15		5.0		267	276	1060000	458000			300
310	320		6.2		6.0		271	282	1310000	594000			310
320	330		6.2		6.0		281	292	1350000	613000			320
330	340		6.2		6.0		291	302	1390000	632000			330
340	350		6.2	+.220 −.000	6.0		301	312	1440000	650000			340
350	360	+.890 −.000	6.2		6.0	+.000 −.180	311	322	1480000	669000			350
360	370		6.2		6.0		321	332	1520000	688000			360
370	380		6.2		6.0		331	342	1560000	707000			370
380	390		6.2		6.0		341	352	1610000	726000			380
390	400		6.2		6.0		351	362	1650000	745000			390
400	410		6.2		6.0		361	372	1690000	764000			400

cont...

STANDARD EUROPEAN EXTERNAL CIRCLIPS* (DIN1400) *Anderton International Ltd

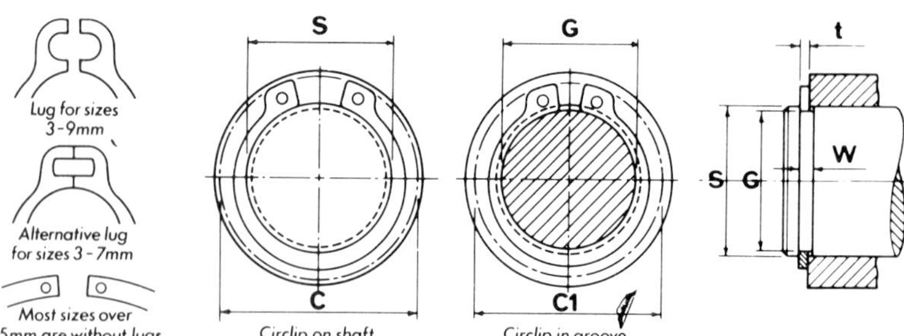

Range incorporates European Standard DIN 471 Blatt 1 (shown in heavy type column 1)

Shaft size (mm)	Groove				Circlip thickness (mm)		Clearance diameter (mm)		Application				Shaft size (mm)
	Diameter (mm)		Width (mm)						Allowable thrust load in Newtons (N)				
S	G	Tol.	W	Tol.	t	Tol.	C	C1	Circlip	Groove			S
3	2.8		0.5		0.4		7.2	6.6	1100	110			**3**
4	3.8	+0.00 −0.075	0.5		0.4		8.8	8.2	1600	150			**4**
5	4.8		0.7		0.6	+0.00 −0.40	10.7	9.8	2000	180			**5**
6	5.7		0.8		0.7		12.2	11.1	4100	330			**6**
7	6.7		0.9		0.8		13.8	12.9	5500	390			**7**
8	7.6	+0.00 −0.90	0.9		0.8		15.2	14.0	6200	590			**8**
9	8.6		1.1		1.0		16.4	15.2	8800	660			**9**
10	9.6		1.1		1.0		17.6	16.2	9700	740			**10**
11	10.5		1.1		1.0		18.6	17.1	10700	1010			**11**
12	11.5		1.1		1.0		19.6	18.1	11700	1110			**12**
13	12.4		1.1		1.0		20.8	19.2	12700	1440			**13**
14	13.4	+0.00 −0.11	1.1		1.0		22.0	20.4	13600	1550			**14**
15	14.3		1.1		1.0		23.2	21.5	14600	1930			**15**
16	15.2		1.1		1.0		24.4	22.6	15600	2350			**16**
17	16.2		1.1		1.0		25.6	23.8	16600	2500			**17**
18	17.0		1.3		1.2		26.8	24.8	21000	3300			**18**
19	18.0		1.3		1.2		27.8	25.8	22200	3490			**19**
20	19.0		1.3		1.2		29.0	27.0	23400	3680			**20**
21	20.0		1.3		1.2		30.2	28.2	24500	3860			**21**
22	21.0		1.3		1.2		31.4	29.4	25700	4050			**22**
23	22.0		1.3		1.2		32.6	30.6	26900	4240			23
24	22.9	+0.00 −0.21	1.3		1.2		33.8	31.7	28000	4860			**24**
25	23.9		1.3		1.2		34.8	32.7	29200	5070			**25**
26	24.9		1.3	+0.14 −0.00	1.2		36.0	33.9	30400	5280			**26**
27	25.6		1.3		1.2		37.2	34.8	31600	6940			27
28	26.6		1.6		1.5	+0.00 −0.60	38.4	36.0	40900	7200			**28**
29	27.6		1.6		1.5		39.6	37.2	42400	7470			**29**
30	28.6		1.6		1.5		41.0	38.6	43800	7730			**30**
32	30.3		1.6		1.5		43.4	40.7	46700	9980			**32**
33	31.3		1.6		1.5		44.4	41.7	48200	10300			33
34	32.3		1.6		1.5		45.8	43.1	49700	10600			**34**
35	33.0		1.6		1.5		47.2	44.2	51100	12800			**35**
36	34.0		1.85		1.75		48.2	45.2	61400	13200			**36**
38	36.0		1.85		1.75		50.6	47.6	64800	14000			**38**
40	37.5	+0.00 −0.25	1.85		1.75		53.0	49.5	56600	18300			**40**
42	39.5		1.85		1.75		56.0	52.5	59500	19200			**42**
45	42.5		1.85		1.75		59.4	55.9	63700	20600			**45**
46	43.5		1.85		1.75		60.4	56.9	65100	21100			46
47	44.5		1.85		1.75		61.6	58.1	66500	21600			47
48	45.5		1.85		1.75		62.8	59.3	67900	22000			**48**
50	47.0		2.15		2.0		64.8	60.8	80900	27400			**50**
52	49.0		2.15		2.0		67.0	63.0	84100	28600			**52**
54	51.0		2.15		2.0		69.2	65.2	87400	29700			54
55	52.0	+0.00 −0.30	2.15		2.0		70.4	66.4	89000	30300			**55**
56	53.0		2.15		2.0		71.6	67.6	90600	30800			**56**

cont...

Shaft size (mm)	Groove				Circlip thickness (mm)		Clearance diameter (mm)		Application				Shaft size (mm)
	Diameter (mm)		Width (mm)						Allowable thrust load in Newtons (N)				
S	G	Tol.	W	Tol.	t	Tol.	C	C1	Circlip	Groove†			S
58	55.0		2.15		2.0		73.6	69.6	93800	32000			58
60	57.0		2.15		2.0		75.8	71.8	97100	33100			60
62	59.0		2.15		2.0		78.0	74.0	100000	34200			62
63	60.0		2.15		2.0		79.2	75.2	102000	34800			63
65	62.0		2.65		2.5		81.6	77.6	131000	35900			65
67	64.0		2.65		2.5		83.8	79.8	135000	37000			67
68	65.0	+0.00 −0.30	2.65	+0.14 −0.00	2.5		85.0	81.0	138000	37600			68
70	67.0		2.65		2.5		87.2	83.2	142000	38700			70
72	69.0		2.65		2.5		89.4	85.4	146000	39900			72
75	72.0		2.65		2.5	+0.00 −0.60	92.8	88.8	152000	41600			75
77	74.0		2.65		2.5		95.0	91.0	156000	42700			77
78	75.0		2.65		2.5		96.2	92.2	158000	43300			78
80	76.5		2.65		2.5		98.2	93.7	162000	51600			80
82	78.5		2.65		2.5		101	95.9	166000	53000			82
85	81.5		3.15		3.0		104	98.9	206000	54900			85
88	84.5		3.15		3.0		107	102	214000	56900			88
90	86.5	+0.00 −0.35	3.15		3.0		109	104	218000	58200			90
95	91.5		3.15		3.0		115	110	231000	61500			95
98	94.5		3.15		3.0		119	113	238000	63500			98
100	96.5		3.15		3.0		121	116	243000	64800			100
102	98.0		4.15		4.0		123	117	330000	75400			102
105	101		4.15		4.0		126	121	340000	77700			105
108	104	+0.00 −0.54	4.15		4.0		130	124	349000	79900			108
110	106		4.15		4.0		132	126	356000	81400			110
115	111		4.15		4.0		138	132	372000	85200			115
120	116		4.15		4.0		143	138	388000	89000			120
125	121		4.15		4.0		149	144	404000	92800			125
130	126		4.15		4.0		155	149	421000	96500			130
135	131		4.15		4.0		160	155	437000	100000			135
140	136		4.15		4.0		165	160	453000	104000			140
145	141		4.15		4.0		171	165	470000	108000			145
150	145		4.15		4.0		177	171	485000	139000			150
155	150	+0.00 −0.63	4.15	+0.18 −0.00	4.0		182	176	501000	144000			155
160	155		4.15		4.0		188	182	518000	148000			160
165	160		4.15		4.0		193	187	534000	153000			165
170	165		4.15		4.0		197	191	550000	158000			170
175	170		4.15		4.0	+0.00 −0.75	202	196	566000	163000			175
180	175		4.15		4.0		208	202	582000	167000			180
185	180		4.15		4.0		213	207	598000	172000			185
190	185		4.15		4.0		219	213	615000	177000			190
195	190		4.15		4.0		224	218	631000	181000			195
200	195		4.15		4.0		229	223	647000	186000			200
205	199	+0.00 −0.72	5.15		5.0		234	227	721000	228000			205
210	204		5.15		5.0		239	232	739000	234000			210
220	214		5.15		5.0		249	242	775000	245000			220
230	224		5.15		5.0		259	252	809000	257000			230
240	234		5.15		5.0		269	262	844000	268000			240
250	244		5.15		5.0		279	272	880000	279000			250
260	252		5.15		5.0		293	284	915000	386000			260
270	262		5.15		5.0		303	294	950000	401000			270
280	272		5.15		5.0		313	304	985000	416200			280
290	282	+0.00 −0.81	5.15		5.0		323	314	1020000	431200			290
300	292		5.15		5.0		333	324	1056000	446300			300
310	**300**		6.2		6.0		349	340	1308920	575000			310
320	**310**		6.2		6.0		359	350	1351000	593800			320
330	**320**		6.2		6.0		369	360	1393000	612600			330
340	**330**		6.2		6.0		379	370	1436000	631500			340
350	**340**		6.2	+0.22 −0.00	6.0	+0.00 −0.18	389	380	1478000	650300			350
360	**350**	+0.00 −0.89	6.2		6.0		399	390	1520000	669000			360
370	**360**		6.2		6.0		409	400	1562000	688000			370
380	**370**		6.2		6.0		419	410	1604000	706900			380
390	**380**		6.2		6.0		429	420	1646000	725700			390
400	**390**		6.2		6.0		439	430	1689000	744600			400

STANDARD AMERICAN INTERNAL CIRCLIPS* (NAM1300)

*Anderton International Ltd

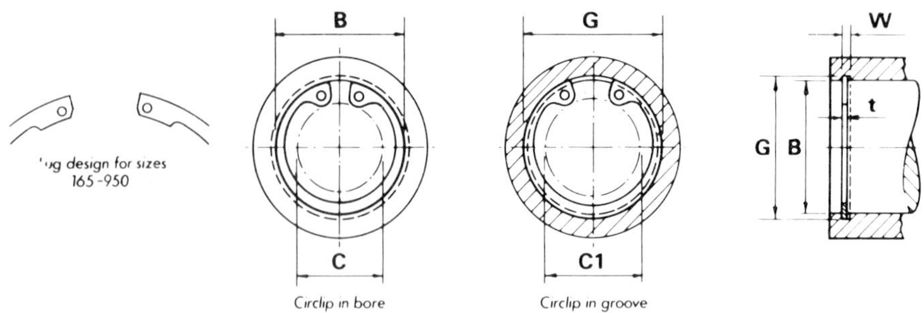

Circlip in bore

Circlip in groove

Size ref. No.	Bore Size (in)	Bore Size (in)	Groove Diameter (in)		Groove Width (in)		Circlip thickness (in)		Clearance diameter (in)		Application Thrust loads (lbs)				Size ref. No.
	B	B	G	Tol.	W	Tol.	t	Tol.	C	C1	Circlip	Groove			
25	1/4	.250	.268	±.001	.018	+ .002	.015		.11	.133	530	190			25
31	5/16	.312	.330		.018	− .000	.015		.17	.191	660	240			31
37	3/8	.375	.397		.029		.025		.20	.226	1320	350			37
43	7/16	.438	.461		.029		.025		.23	.254	1540	440			43
45	29/64	.453	.477		.029		.025		.25	.274	1600	460			45
50	1/2	.500	.530		.039		.035		.26	.29	2470	510			50
51	–	.512	.542	±.002	.039		.035		.27	.30	2530	520			51
56	9/16	.562	.596		.039		.035		.28	.305	2780	710			56
62	5/8	.625	.665		.039		.035		.34	.38	3090	1050			62
68	11/16	.688	.732		.039	+ .003	.035		.40	.44	3400	1280			68
75	3/4	.750	.796		.039	− .000	.035		.45	.49	3710	1460			75
77	–	.777	.825		.046		.042		.48	.52	4610	1580			77
81	13/16	.812	.862		.046		.042		.49	.54	4820	1710			81
86	–	.866	.920		.046		.042		.54	.59	5140	1980			86
87	7/8	.875	.931	±.003	.046		.042	±.002	.55	.60	5190	2080			87
90	–	.901	.959		.046		.042		.56	.62	5350	2200			90
93	15/16	.938	1.000		.046		.042		.61	.67	5570	2450			93
100	1	1.000	1.066		.046		.042		.66	.73	5940	2800			100
102	–	1.023	1.091		.046		.042		.69	.755	6070	3000			102
106	1.1/16	1.062	1.130		.056		.050		.69	.75	7500	3050			106
112	1.1/8	1.125	1.197		.056		.050		.74	.815	7950	3400			112
118	1.3/16	1.188	1.262		.056		.050		.80	.87	8400	3700			118
125	1.1/4	1.250	1.330		.056		.050		.87	.955	8850	4250			125
131	1.5/16	1.312	1.396	±.004	.056		.050		.93	1.01	9300	4700			131
137	1.3/8	1.375	1.461		.056		.050		.99	1.07	9700	5050			137
143	1.7/16	1.438	1.528		.056		.050		1.06	1.15	10200	5500			143
145	–	1.456	1.548		.056	+ .004	.050		1.08	1.17	10300	5700			145
150	1.1/2	1.500	1.594		.056	− .000	.050		1.12	1.21	10600	6000			150
156	1.9/16	1.562	1.658		.068		.062		1.14	1.23	10700	6350			156
162	1.5/8	1.625	1.725		.068		.062		1.15	1.25	11100	6900			162
165	–	1.653	1.755		.068		.062		1.17	1.27	11300	7200			165
168	1.11/16	1.688	1.792	±.005	.068		.062	±.003	1.21	1.31	11500	7450			168
175	1.3/4	1.750	1.858		.068		.062		1.26	1.36	11900	8050			175
181	1.13/16	1.812	1.922		.068		.062		1.32	1.43	12400	8450			181
185	–	1.850	1.962		.068		.062		1.36	1.47	12600	8750			185

cont...

Size ref. No.	Bore Size (in)	Bore Size (in)	Groove Diameter (in)		Groove Width (in)		Circlip thickness (in)		Clearance diameter (in)		Application Thrust loads (lbs)				Size ref. No.
	B	B	G	Tol.	W	Tol.	t	Tol.	C	C1	Circlip	Groove			
187	1.7/8	1.875	1.989		.068	+ .004 − .000	.062		1.39	1.50	12800	9050			187
193	1.15/16	1.938	2.056	±.005	.068		.062		1.45	1.56	13200	9700			193
200	2	2.000	2.122		.068		.062		1.50	1.62	13600	10300			200
206	2.1/16	2.062	2.186		.086		.078		1.54	1.66	17700	10850			206
212	2.1/8	2.125	2.251		.086		.078		1.58	1.70	18200	11350			212
218	2.3/16	2.188	2.318		.086		.078		1.64	1.77	18800	12050			218
225	2.1/4	2.250	2.382		.086		.078		1.69	1.82	19300	12600			225
231	2.5/16	2.312	2.450		.086		.078		1.75	1.88	19800	13550			231
237	2.3/8	2.375	2.517		.086		.078		1.81	1.95	20400	14300			237
244	2.7/16	2.438	2.584		.086		.078		1.86	2.00	20900	14900			244
250	2.1/2	2.500	2.648		.086		.078		1.91	2.05	21400	15650			250
	2.17/32	2.531	2.681		.086		.078		1.94	2.09	21700	15650			
256	2.9/16	2.562	2.714		.103		.093		1.95	2.10	26200	16500			256
262	2.5/8	2.625	2.781		.103		.093		2.02	2.17	26800	17350			262
268	2.11/16	2.688	2.848		.103		.093		2.06	2.22	27500	18250			268
275	2.3/4	2.750	2.914		.103		.093		2.12	2.28	28100	19200			275
281	2.13/16	2.812	2.980		.103		.093		2.18	2.34	28800	20050			281
287	2.7/8	2.875	3.051		.103		.093		2.22	2.39	29400	21500			287
300	3	3.000	3.182		.103		.093		2.35	2.53	30700	23150			300
306	3.1/16	3.062	3.248		.120		.109	±.003	2.41	2.59	36700	24100			306
312	3.1/8	3.125	3.315	±.006	.120	+ .005 − .000	.109		2.47	2.66	37500	25200			312
315	3.5/32	3.156	3.348		.120		.109		2.50	2.69	37800	25700			315
325	3.1/4	3.250	3.446		.120		.109		2.54	2.73	39000	27000			325
334	3.11/32	3.346	3.546		.120		.109		2.63	2.83	40100	28300			334
347	3.15/32	3.469	3.675		.120		.109		2.76	2.96	41600	30200			347
350	3.1/2	3.500	3.710		.120		.109		2.79	3.00	41900	31200			350
354	3.9/16	3.562	3.776		.120		.109		2.85	3.06	42700	31800			354
362	3.5/8	3.625	3.841		.120		.109		2.91	3.12	43400	33200			362
375	3.3/4	3.750	3.974		.120		.109		3.03	3.25	44900	35600			375
387	3.7/8	3.875	4.107		.120		.109		3.11	3.34	46400	38000			387
393	3.15/16	3.938	4.174		.120		.109		3.17	3.40	47200	39300			393
400	4	4.000	4.240		.120		.109		3.23	3.47	47900	40700			400
412	4.1/8	4.125	4.365		.120		.109		3.36	3.60	49400	42000			412
425	4.1/4	4.250	4.490		.120		.109		3.48	3.72	50900	43200			425
433	–	4.331	4.571		.120		.109		3.50	3.74	51900	44500			433
450	4.1/2	4.500	4.740		.120		.109		3.66	3.90	54000	45800			450
462	4.5/8	4.625	4.865		.120		.109		3.79	4.03	55400	47000			462
475	4.3/4	4.750	4.995		.120		.109		3.90	4.14	56900	49000			475
500	5	5.000	5.260		.120		.109		4.08	4.34	59900	55000			500
525	5.1/4	5.250	5.520		.139		.125		4.31	4.58	72200	60000			525
537	5.3/8	5.375	5.650		.139	+ .006 − .000	.125		4.41	4.68	73900	61500			537
550	5.1/2	5.500	5.770	±.007	.139		.125	±.004	4.53	4.80	75600	63300			550
575	5.3/4	5.750	6.020		.139		.125		4.78	5.05	79000	65900			575
600	6	6.000	6.270		.139		.125		5.03	5.30	82500	68600			600
625	6.1/4	6.250	6.530		.174		.156		5.24	5.52	107200	74100			625
650	6.1/2	6.500	6.790		.174		.156		5.49	5.78	111500	79900			650
662	6.5/8	6.625	6.925		.174		.156		5.60	5.90	113700	84200			662
675	6.3/4	6.750	7.055		.174		.156		5.65	5.95	115800	87000			675
700	7	7.000	7.315		.174		.156		5.88	6.19	120000	93100			700
725	7.1/4	7.250	7.575		.209		.187		6.08	6.40	138400	99600			725
750	7.1/2	7.500	7.840		.209		.187		6.33	6.67	143200	108100			750
775	7.3/4	7.750	8.100		.209	+ .008 − .000	.187		6.58	6.93	148000	115000			775
800	8	8.000	8.360	±.008	.209		.187	±.005	6.75	7.11	152700	122000			800
825	8.1/4	8.250	8.620		.209		.187		7.00	7.37	157600	129300			825
850	8.1/2	8.500	8.880		.209		.187		7.13	7.51	162300	136900			850
875	8.3/4	8.750	9.145		.209		.187		7.38	7.77	167000	145500			875
900	9	9.000	9.405		.209		.187		7.63	8.03	171800	154100			900
925	9.1/4	9.250	9.668		.209		.187		7.88	8.30	176600	163600			925
950	9.1/2	9.500	9.930		.209		.187		7.98	8.41	181400	173100			950

STANDARD AMERICAN EXTERNAL CIRCLIPS* (NAM1400) *Anderton International Ltd

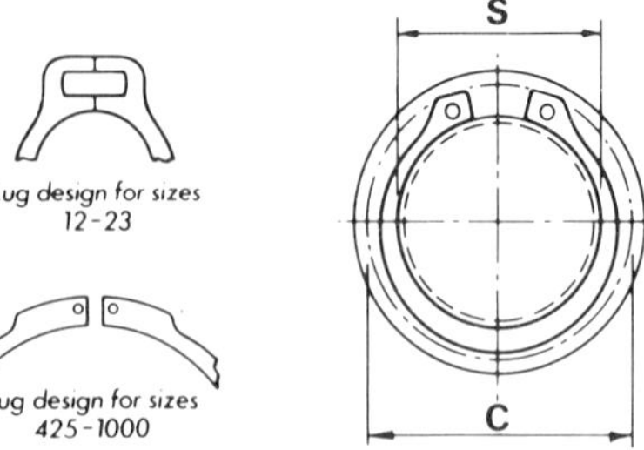

Circlip on shaft

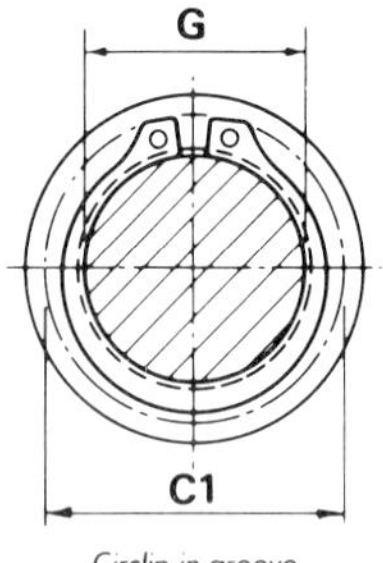

Circlip in groove

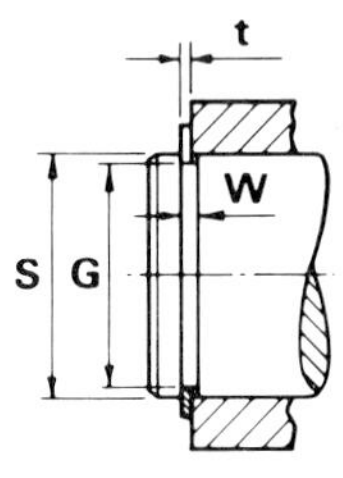

Standard materials:
Sizes 12-23 – Beryllium copper
Sizes 25 up – Carbon spring steel

Standard finish:
Sizes 12-23 – Self finish
Sizes 25 up – Phosphate

Size ref. No.	Shaft Size (in)	Shaft Size (in)	Groove Diameter (in)		Groove Width (in)		Circlip thickness (in)		Clearance diameter (in)		Application Thrust loads (lbs)				Size ref. No.
	S	S	G	Tol.	W	Tol.	t	Tol.	C	C1	Circlip	Groove			
12	1/8	.125	.117		.012		.010	±.001	.22	.214	110	35			12
15	5/32	.156	.146		.012		.010		.27	.260	130	55			15
18	3/16	.188	.175	±.0015	.018	+.002 −.000	.015		.30	.286	240	80			18
19	–	.197	.185		.018		.015		.32	.307	250	85			19
21	7/32	.219	.205		.018		.015		.34	.324	280	110			21
23	15/64	.236	.222		.018		.015		.36	.341	310	120			23
25	1/4	.250	.230		.029		.025		.45	.43	880	175			25
27	–	.276	.255		.029		.025		.48	.46	980	195			27
28	9/32	.281	.261		.029		.025		.49	.47	990	200			28
31	5/16	.312	.290		.029		.025		.54	.52	1100	240			31
34	11/32	.344	.321		.029		.025		.57	.55	1210	265			34
35	–	.354	.330		.029		.025		.59	.57	1250	300			35
37	3/8	.375	.352	±.002	.029		.025		.61	.59	1320	320			37
39	–	.394	.369		.029		.025		.62	.60	1390	335			39
40	13/32	.406	.382		.029		.025		.63	.61	1430	350			40
43	7/16	.438	.412		.029		.025		.66	.64	1550	400			43
46	15/32	.469	.443		.029		.025		.68	.66	1660	450			46
50	1/2	.500	.468		.039		.035		.77	.74	2470	550			50
55	–	.551	.519		.039	+.003 −.000	.035		.81	.78	2730	600			55
56	9/16	.562	.530		.039		.035		.82	.79	2780	650			56
59	19/32	.594	.559		.039		.035		.86	.83	2940	750			59
62	5/8	.625	.588		.039		.035	±.002	.90	.87	3090	800			62
66	43/64	.672	.631		.039		.035		.93	.89	3320	950			66
68	11/16	.688	.646		.046		.042		1.01	.97	4080	1000			68
75	3/4	.750	.704		.046		.042		1.09	1.05	4450	1200			75
78	25/32	.781	.733	±.003	.046		.042		1.12	1.08	4640	1300			78
81	13/16	.812	.762		.046		.042		1.15	1.10	4820	1450			81
87	7/8	.875	.821		.046		.042		1.21	1.16	5190	1650			87
93	15/16	.938	.882		.046		.042		1.34	1.29	5570	1850			93
98	63/64	.984	.926		.046		.042		1.39	1.34	5840	2000			98
100	1	1.000	.940		.046		.042		1.41	1.35	5940	2100			100
102	–	1.023	.961		.046		.042		1.43	1.37	6070	2250			102
106	1.1/16	1.062	.998		.056		.050		1.50	1.44	7500	2400			106
112	1.1/8	1.125	1.059	±.004	.056	+.004 −.000	.050		1.55	1.49	7900	2600			112
118	1.3/16	1.188	1.118		.056		.050		1.61	1.54	8400	2950			118

cont...

Size ref No.	Shaft Size (in)	Shaft Size (in)	Groove Diameter (in)		Groove Width (in)		Circlip thickness (in)		Clearance diameter (in)		Application Thrust loads (lbs)				Size ref. No.
	B	B	G	Tol.	W	Tol.	t	Tol.	C	C1	Circlip	Groove			
125	1.1/4	1.250	1.176	±.004	.056		.050		1.69	1.62	8800	3250			125
131	1.5/16	1.312	1.232		.056		.050		1.75	1.67	9300	3700			131
137	1.3/8	1.375	1.291		.056		.050	± .002	1.80	1.72	9700	4100			137
143	1.7/16	1.438	1.350		.056		.050		1.87	1.79	10200	4500			143
150	1.1/2	1.500	1.406		.056	+ .004	.050		1.99	1.90	10600	5000			150
156	1.9/16	1.562	1.468		.068	– .000	.062		2.10	2.01	10700	5200			156
162	1.5/8	1.625	1.529		.068		.062		2.17	2.08	11100	5500			162
168	1.11/16	1.688	1.589		.068		.062		2.24	2.15	11500	5850			168
175	1.3/4	1.750	1.650		.068		.062		2.31	2.21	11900	6200			175
177	–	1.772	1.669	±.005	.068		.062		2.33	2.23	12100	6400			177
181	1.13/16	1.812	1.708		.068		.062		2.38	2.28	12400	6650			181
187	1.7/8	1.875	1.769		.068		.062		2.44	2.34	12800	7000			187
196	–	1.968	1.857		.068		.062		2.54	2.43	13400	7800			196
200	2	2.000	1.886		.068		.062		2.55	2.44	13600	8050			200
206	2.1/16	2.062	1.946		.086		.078		2.68	2.57	17700	8450			206
212	2.1/8	2.125	2.003		.086		.078		2.75	2.63	18200	9150			212
215	2.5/32	2.156	2.032		.086		.078		2.78	2.66	18500	9450			215
225	2.1/4	2.250	2.120		.086		.078		2.87	2.74	19300	10350			225
231	2.5/16	2.312	2.178		.086		.078		2.94	2.81	19800	10950			231
237	2.3/8	2.375	2.239		.086		.078		3.01	2.88	20400	11400			237
243	2.7/16	2.438	2.299		.086		.078		3.07	2.94	20900	11900			243
250	2.1/2	2.500	2.360		.086		.078		3.12	2.98	21400	12350			250
255	–	2.559	2.419		.086		.078		3.18	3.04	21900	12650			255
262	2.5/8	2.625	2.481		.086		.078		3.25	3.11	22500	13350			262
268	2.11/16	2.688	2.541		.086		.078		3.32	3.18	23000	13850			268
275	2.3/4	2.750	2.602		.103		.093		3.45	3.31	28100	14400			275
287	2.7/8	2.875	2.721		.103		.093		3.57	3.42	29400	15650			287
293	2.15/16	2.938	2.779		.103		.093		3.64	3.49	30000	16400			293
300	3	3.000	2.838		.103		.093		3.69	3.53	30700	17200			300
306	3.1/16	3.062	2.898		.103		.093		3.74	3.58	31300	17750			306
312	3.1/8	3.125	2.957		.103	+ .005	.093		3.82	3.66	32000	18550			312
315	3.5/32	3.156	2.986	±.006	.103	– .000	.093	±.003	3.85	3.68	32300	18950			315
325	3.1/4	3.250	3.076		.103		.093		3.95	3.78	33200	20000			325
334	3.11/32	3.346	3.166		.103		.093		4.04	3.87	34200	21000			334
343	3.7/16	3.438	3.257		.103		.093		4.14	3.96	35200	21900			343
350	3.1/2	3.500	3.316		.120		.109		4.25	4.07	42000	22800			350
354	–	3.543	3.357		.120		.109		4.29	4.11	42500	23300			354
362	3.5/3	3.625	3.435		.120		.109		4.37	4.18	43400	24300			362
368	3.11/16	3.688	3.493		.120		.109		4.43	4.24	44200	25300			368
375	3.3/4	3.750	3.552		.120		.109		4.50	4.31	44900	26200			375
387	3.7/8	3.875	3.673		.120		.109		4.60	4.40	46400	27700			387
393	3.15/16	3.938	3.734		.120		.109		4.70	4.50	47200	28400			393
400	4	4.000	3.792		.120		.109		4.78	4.58	47900	29400			400
425	4.1/4	4.250	4.065		.120		.109		5.09	4.91	50900	27600			425
437	4.3/8	4.375	4.190		.120		.109		5.22	5.04	52400	28400			437
450	4.1/2	4.500	4.310		.120		.109		5.37	5.18	53900	30200			450
475	4.3/4	4.750	4.550		.120		.109		5.67	5.47	56900	33600			475
500	5	5.000	4.790		.120		.109		5.96	5.75	59900	37100			500
525	5.1/4	5.250	5.030		.139		.125		6.27	6.05	72200	40800			525
550	5.1/2	5.500	5.265	±.007	.139	+ .006	.125	±.004	6.57	6.34	75600	45500			550
575	5.3/4	5.750	5.505		.139	– .000	.125		6.86	6.62	79000	49600			575
600	6	6.000	5.745		.139		.125		7.16	6.91	82500	53800			600
625	6.1/4	6.250	5.985		.174		.156		7.46	7.20	107200	58300			625
650	6.1/2	6.500	6.225		.174		.156		7.87	7.60	111500	62900			650
675	6.3/4	6.750	6.465		.174		.156		8.06	7.78	115800	67700			675
700	7	7.000	6.705		.174		.156		8.36	8.07	120000	72700			700
750	7.1/2	7.500	7.180	±.008	.209	+ .008	.187	±.005	8.96	8.64	143200	84800			750
800	8	8.000	7.660		.209	– .000	.187		9.56	9.22	152700	96100			800
850	8.1/2	8.500	8.140		.209		.187		10.16	9.80	162300	108100			850
1000	10	10.000	9.575		.209		.187		11.94	11.52	190900	149800			1000

Machine Pins

The classification Machine Pins covers parallel and tapered pins (dowel pins) and grooved pins used primarily as machine element fastenings. There are also a variety of chamfered and/or slotted proprietary spring pins used as fastenings. Split pins (cotter pins) and clevis pins come in a separate category but may also be included under the general heading of machine pins.

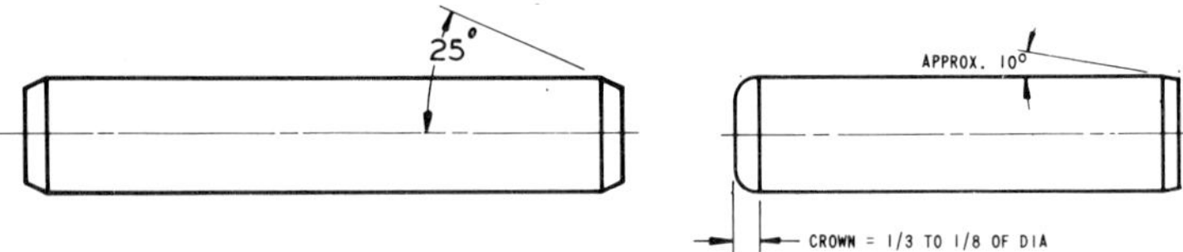

Fig 1 Parallel pins (dowel pins).

Standard sizes of plain (parallel) steel pins (dowel pins) are ground to finish. Ends may be square, chamfered both ends, or chamfered one end and crowned the other end — Fig 1. They may be hardened or unhardened, depending on the strength required. Sizes are expressed in terms of *overall length,* typically:

Inch sizes — inch fraction lengths, tolerance ± 0.012in

Metric sizes — millimetre lengths, tolerance ± 0.3mm

and *diameter* typically:-

nominal diameter — inch fraction or millimetre sizes

standard diameter — actual diameter size, slightly larger than a nominal diameter, expressed in decimal inches or millimetres. This corresponds to a 'tap' fit in a matching nominal diameter hole.

oversize diameter — actual diameter about 1% up on nominal diameter size. This corresponds to a 'press' fit in a matching nominal diameter size hole.

Tolerances on both standard and average diameters are very small — eg typically ± 0.001in or ±0.025mm. Diameter is normally expressed as the maximum (large end) diameter. For a standard 20mm to the metre (¼in to the foot) taper the smaller diameter can be determined by multiplying the length of the pin (in millimetres or inches) by 0.02083 and subtracting this figure from the large end diameter.

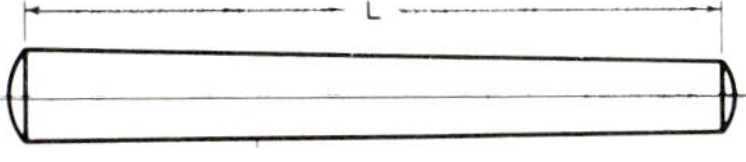

Fig 2 Taper pins.

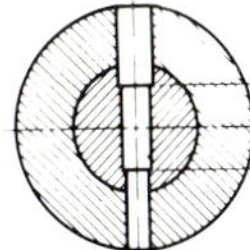

Fig 3 Step drilling for taper pins.

Taper pins are normally ground to a taper of 20mm per metre (¼in per foot) with close tolerances on length, diameter and linearity (in the case of precision taper pins) – Fig 2. The smaller size pins are a tap fit in a single matching size hole – eg up to about 6mm diameter (¼in). Larger pin diameters are intended to be a tap fit in a tapered hole produced by a fluted tapered reamer or a series of two or more stepped holes produced by consecutive drillings – Fig 3. In this case the first or through hole is the same diameter as the small end diameter of the taper pin – see Table II. Each step section should be the same length.

TABLE I – TYPICAL TAPER PIN SIZES

Number	7/0	6/0	5/0	4/0	3/0	2/0	0	1	
Large end diameter (inches)	0.0625	0.0780	0.0940	0.1090	0.1250	0.1410	0.1560	0.1720	
(mm)	1.60	2.00	2.40	2.77	3.175	3.58	3.96 (4.00)	4.37	
Number	2	3	4	5	6	7	8	9	10
Large end diameter (inches)	0.1930	0.2190	0.2500	0.2890	0.3410	0.4090	0.4920	0.5910	0.7060
(mm)	4.90	5.56	6.35	7.34	8.66	10.4	12.5	15	18

Grooved Pins

A variety of grooved pin types is shown in Fig 4. Grooving expands the nominal diameter of the pin so that when driven into a matching sized round hole the groove is closed and the pin is clamped in position. The pin invariably retains a certain plain length (pilot length) to facilitate assembly. There are also serrated, notched and knurled forms of pins. All these types are designed to be self-locking in situ.

Fig 4 Examples of grooved pins.

TABLE II – DRILL SIZES FOR TAPER PINS

Pin Size	1st Drill in	1st Drill mm	2nd Drill in	2nd Drill mm	3rd Drill in	3rd Drill mm	4th Drill in	4th Drill mm	5th Drill in	5th Drill mm
7/0	0.0469 (3/64)	1.15								
6/0	0.0469 (3/64)	1.20								
5/0	0.0625 (1/16)	1.55								
4/0	0.0781 (5/64)	2.00								
3/0	0.0938 (3/32)	2.35								
2/0	0.0938 (3/32)	2.40	0.1094 (7/64)	2.75						
0	0.0938 (3/32)	2.40	0.1250 (1/8)	3.10						
1	0.1094 (7/64)	2.75	0.1406 (9/64)	3.50						
2	0.1094 (7/64)	2.80	0.1406 (9/64)	3.60						
3	0.1406 (9/64)	3.50	0.1719 (11/64)	4.30						
4	0.1719 (11/64)	4.30	0.2031 (13/64)	5.10						
5	0.1875 (3/16)	4.70	0.2349 (15/64)	5.90						
6	0.2349 (15/64)	5.90	0.2656 (17/64)	6.70	0.2969 (19/64)	7.50				
7	0.2969 (19/64)	7.50	0.3281 (21/64)	8.33	0.3750 (3/8)	9.50				
8	0.3906 (25/64)	10.00	0.4219 (27/64)	10.70	0.4531 (29/64)	11.50				
9	0.4688 (15/32)	12.00	0.5000 (1/2)	12.60	0.5469 (35/64)	13.80				
10	0.5781 (37/64)	14.30	0.6094 (39/64)	15.25	0.6562 (21/32)	16.50				
11	0.6719 (43/64)	17.00	0.7188 (46/64)	18.25	0.7656 (49/64)	19.25	0.8125 (13/16)	20.50		
12	0.8438 (27/32)	21.50	0.8906 (57/64)	22.50	0.9375 (15/16)	24.00	0.9844 (63/64)	25.00		
13	1.0000 (1)	25.00	1.0469 (1.3/64)	26.50	1.0938 (1.3/32)	27.50	1.1406 (1.9/64)	28.50	1.1875 (1.3/16)	30.00
14	1.2500 (1¼)	31.70	1.2969 (1.19/64)	32.50	1.3594 (1.23/64)	34.50	1.4062 (1.13/32)	35.50	1.4688 (1.15/32)	37.00

Spring Pins

The (patented) Rollpin is a hollow (tubular) straight pin slotted along its length – Fig 5. When driven into a matching sized hole the pin is compressed diametrically, the slot dimension, pin/hole diameter and elastic limits of the material being designed to ensure positive locking action in holes

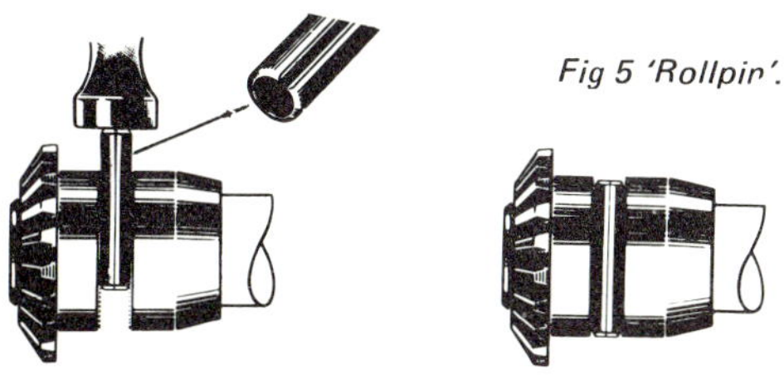

Fig 5 'Rollpin'.

drilled to normal production tolerances. Strength is comparable to or better than that of ground/dowel pins or grooved pins. The position of the slot is not normally significant, but where maximum shear strength is required the slot gap should be aligned with the direction of loading.

Due to their elastic nature, Rollpins are less affected by loose hole tolerances (oversize holes) than dowel pins and about equal to grooved pins. The Rollpin has an advantage over a grooved pin in that it is much easier to remove and it does not appreciably affect or deform the hole material when being inserted or removed. Rollpins may be used as a longitudinal key as well as for transverse pinning. They can also be used in combination (one pin inside another) when greater shear strength is required. Rollpin data are given in Tables IIIA and IIIB.

TABLE IIIA – 'ROLLPIN' SIZES AND DATA

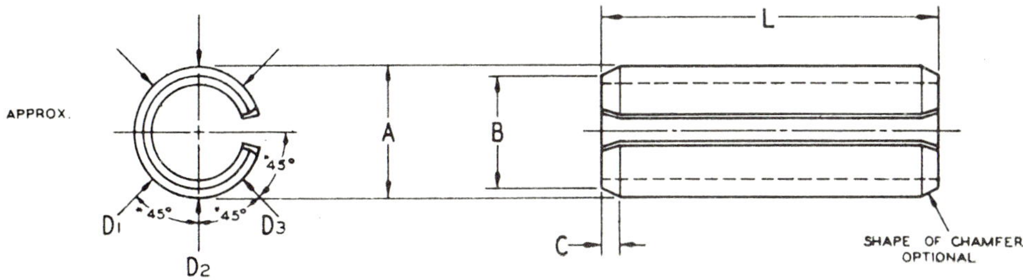

A			B	C	Stock	Rec Hole Size		Minimum Double Shear Strength, Pounds	
Nominal	Maximum (Go Ring Gauge)	Minimum 1/3 $(D_1+D_2+D_3)$	Max	Nom	Thickness	Min	Max	Carbon Steel and Stainless Steel	Equivalent Solid Cold Rolled Steel Pin
0.062	0.069	0.066	0.059	0.011	0.012	0.062	0.065	425	400
0.078	0.086	0.083	0.075	0.014	0.018	0.078	0.081	650	625
0.094	0.103	0.099	0.091	0.018	0.022	0.094	0.097	1 000	900
0.125	0.135	0.131	0.122	0.024	0.028	0.125	0.129	2 100	1 600
0.156	0.167	0.162	0.151	0.028	0.032	0.156	0.160	3 000	2 500
0.187	0.199	0.194	0.182	0.036	0.040	0.187	0.192	4 400	3 600
0.219	0.232	0.226	0.214	0.042	0.048	0.219	0.224	5 700	4 900
0.250	0.264	0.258	0.245	0.042	0.048	0.250	0.256	7 700	6 400
0.312	0.328	0.321	0.306	0.060	0.062	0.312	0.318	11 500	10 000
0.375	0.392	0.385	0.368	0.060	0.077	0.375	0.382	17 600	14 400
0.437	0.456	0.448	0.430	0.060	0.077	0.437	0.445	20 000	19 600
0.500	0.521	0.513	0.485	0.060	0.094	0.500	0.510	25 800	25 600

TABLE IIIB – AVERAGE MAXIMUM REMOVAL FORCE IN STEEL–POUNDS
Plain...Minimum Hole *Italic*...**Maximum Hole**

Nominal Diameter ins	Rollpin Size Number	Engaged Length – Inches							
		¼	½	¾	1	1½	2	2½	3
5/64	078	160	310	440	540				
		100	*210*	*300*	*370*				
3/32	094	250	470	620	720				
		120	*230*	*350*	*460*				
1/8	125	260	500	730	950	1 360			
		160	*300*	*460*	*620*	*880*			
5/32	156	220	420	610	800	1 150			
		150	*300*	*440*	*580*	*840*			
3/16	187	650	1 050	1 250	1 650	2 300			
		450	*800*	*1 050*	*1 250*	*1 700*			
7/32	219	350	700	1 000	1 350	1 900	2 400		
		200	*350*	*550*	*750*	*1 100*	*1 500*		
1/4	250	600	900	1 200	1 500	2 100	2 750		
		125	*250*	*350*	*450*	*660*	*850*		
5/16	312	500	800	900	1 100	1 450	1 800	2 100	2 500
		175	*300*	*450*	*550*	*800*	*1 050*	*1 350*	*1 600*
3/8	375	1 000	1 500	1 850	2 200	2 800	3 300	3 700	4 000
		200	*350*	*500*	*700*	*1 100*	*1 500*	*1 800*	*2 200*
7/16	437	850	1 350	1 750	2 200	3 000	3 600	4 100	4 400
		150	*300*	*400*	*550*	*800*	*1 100*	*1 350*	*1 600*
1/2	500	1 100	2 100	3 000	3 800	4 900	6 100	6 800	7 500
		450	*850*	*1 250*	*1 600*	*2 400*	*3 000*	*3 500*	*4 100*

The 'Spirol' spring pin is another patented design incorporating 2¼ to 2½ turn spiral construction – Fig 6. The multiple coil configuration permits the use of thin strip which reduces the stresses on the material while forming the pin and gives good flexibility and fatigue resistance. Choice of thickness enables the pins to be produced in light, standard and heavy duty ratings, ie offering a choice of radial tensions and flex strengths. Ends have a uniform swaged chamfer to facilitate fitting; and headed pins of this type are also produced for installations requiring positive location of the pin. Because of the spiral spring form, hole tolerance is more generous than for plain pins or slotted tubular pins – see Tables IVA and IVB.

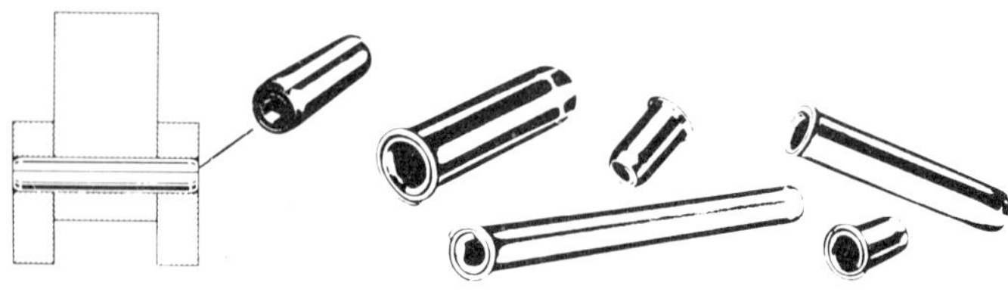

Fig 6 'Spirol' pins.

TABLE IVA – 'SPIROL' PIN SIZES AND DATA (see also Fig 6)

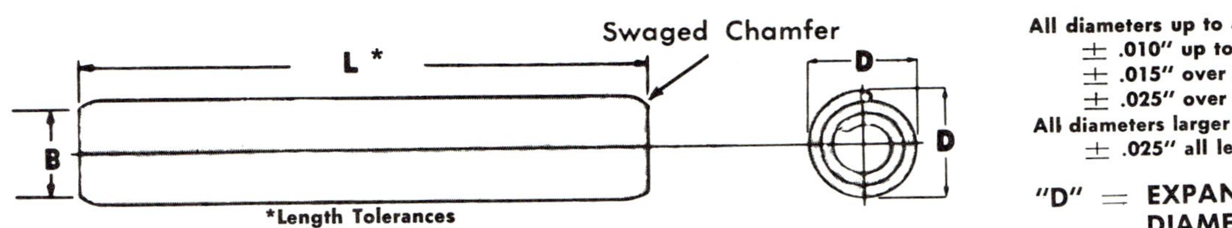

All diameters up to and including $\frac{3}{8}''$
± .010″ up to and including 2″ long
± .015″ over 2″ up to and including 3″ long
± .025″ over 3″ long
All diameters larger than $\frac{3}{8}''$
± .025″ all lengths

"D" = EXPANDED DIAMETER

Nominal Diameter		Recommended Drilled Hole Tolerance		Dimension "B"	Standard Duty (Formerly Medium Duty)				Heavy Duty				Light Duty			
					"D"		Min Double Shear in lbs		"D"		Min Double Shear in lbs		"D"		Min Double Shear in lbs	
		Plus	Minus	Max	Min	Max	302	Carbon Steel & 420	Min	Max	302	Carbon Steel & 420	Min	Max	302	Carbon Steel & 420
1/32	0.031	0.001	0.000	0.029	0.033	0.035	60	75								
0.039	0.039	0.001	0.000	0.037	0.041	0.044	100	120								
3/64	0.047	0.001	0.001	0.045	0.049	0.052	140	170								
0.052	0.051	0.001	0.001	0.050	0.054	0.057	190	230								
1/16	0.062	0.003	0.001	0.059	0.067	0.072	250	300	0.066	0.070	350	450	0.067	0.073	135	
5/64	0.078	0.003	0.001	0.075	0.083	0.088	400	475	0.082	0.086	550	700	0.083	0.089	225	
3/32	0.094	0.003	0.001	0.091	0.099	0.105	550	700	0.098	0.103	800	1 000	0.099	0.106	300	375
7/64	0.109	0.003	0.001	0.106	0.114	0.120	750	950	0.113	0.118	1125	1 400	0.114	0.121	425	525
1/8	0.125	0.004	0.001	0.121	0.131	0.138	1 000	1 250	0.130	0.136	1 700	2 100	0.131	0.139	550	675
5/32	0.156	0.004	0.001	0.152	0.163	0.171	1 550	1 925	0.161	0.168	2 400	3 000	0.163	0.172	875	1 100
3/16	0.187	0.005	0.002	0.182	0.196	0.205	2 250	2 800	0.194	0.202	3 500	4 400	0.196	0.207	1 200	1 500
7/32	0.219	0.005	0.002	0.214	0.228	0.238	3 000	3 800	0.226	0.235	4 600	5 700	0.228	0.240	1 700	2 100
1/4	0.250	0.006	0.003	0.243	0.260	0.271	4 000	5 000	0.258	0.268	6 200	7 700	0.260	0.273	2 200	2 700
5/16	0.312	0.007	0.004	0.304	0.324	0.337	6 200	7 700	0.322	0.334	9 200	11 500	0.324	0.339	3 500	4 400
3/8	0.375	0.008	0.005	0.366	0.388	0.403	9 000	11 200	0.386	0.400	14 000	17 600	0.388	0.405	5 000	
7/16	0.437	0.009	0.006	0.427	0.452	0.469	13 000	15 200	0.450	0.466	18 000	22 500	0.452	0.471	6 700	
1/2	0.500	0.010	0.007	0.488	0.516	0.535	16 000	20 000	0.514	0.532	24 000	30 000	0.516	0.537	8 800	
5/8	0.625	0.010	0.007	0.613	0.642	0.661	25 000	31 000	0.640	0.658	37 000	46 000				
3/4	0.750	0.010	0.007	0.738	0.768	0.787	36 000	45 000	0.766	0.784	53 000	66 000				

All dimensions in inches

TABLE IVB – METRIC SIZE 'SPIROL' PINS

NOMINAL DIAMETER	Recommended Drilled Hole Tolerance		Dimension "B"	STANDARD DUTY				HEAVY DUTY			
				'D' DIMENSION		MIN DOUBLE SHEAR IN KG		'D' DIMENSION		MIN DOUBLE SHEAR IN KG	
	Plus	Minus †	Max	Min	Max	302	1070 420	Min	Max	302	1070 420
1.0	0.03	0.00	0.94	1.04	1.12	44	55*				
1.5	0.07	0.00	1.40	1.62	1.72	110	135	1.60	1.68	160	200
2.0	0.07	0.03	1.90	2.11	2.24	170	215	2.08	2.18	250	320
2.5	0.09	0.03	2.35	2.65	2.80	250	320	2.60	2.70	360	450
3.0	0.09	0.03	2.85	3.15	3.30	340	430	3.12	3.25	500	640
4.0	0.10	0.03	3.80	4.14	4.35	700	875	4.10	4.27	1 100	1 350
5.0	0.12	0.05	4.80	5.25	5.50	1 000	1 250	5.15	5.35	1 600	2 000
6.0	0.12	0.05	5.80	6.25	6.50	1 400	1 750	6.20	6.45	2 100	2 600
8.0	0.15	0.07	7.70	8.23	8.56	2 800	3 500	8.18	8.49	4 100	5 200
10.0	0.15	0.07	9.60	10.45	10.80	4 000	5 000	10.40	10.80	6 400	8 000
12.0	0.18	0.10	11.50	12.50	12.85	5 500	7 000	12.45	12.80	8 000	10 000

† Use nominal drills - minus tolerance for permissible drill wear and shrinkage
* 420 only

Length Tolerances

+ 0.5 up to and including 10mm long
+ 1.0 over 10mm up to and including 50mm long
+ 1.5 over 50mm long

Split Pins (Cotter Pins)

Split pins are formed from a single length of substantially half-round wire with a loop head. The head loop may be symmetrical or asymmetrical. The legs are usually of unequal length and straight, or the top end turned down – Fig 7. (Other types of points are also used). Head sizes may vary, but normally conform to a specified minimum outside diameter. The length of the pin is specified to the bottom of the head. Pin diameter is the total (full circle) shank diameter. Recommended hole sizes are normally of the order of 125% of the shank diameter for small split pins, down to 105% of the shank diameter for large split pins (19mm or ¾in diameter). Typical production sizes are given in Table V.

Fig 7 Split pins (cotter pins).

British Standard split cotter pins are to the geometry shown in Fig 8, with dimensional data given in Table VIA and VIB.

Split pins are normally made from carbon steel but are also manufactured in stainless steel, brass and monel.

See also *Clevis Pins.*

TABLE V – SPLIT COTTER PINS: BS1574(INCH SIZES)

Diameter	Length 'L' Under Eye
1/32	1/4. 3/8. 7/8. 1½, 2
3/64	1/4, 3/8, 1/2, 7/8, 1 1.3/8
1/16	3/8, 1/2, 5/8, 3/4, 7/8, 1, 1.1/8, 1.3/8, 1½, 1.7/8, 2
5/64	3/8, 1/2, 7/8, 1, 1.1/8, 1¼, 1.3/8, 1½, 1.7/8, 2
3/32	3/8, 1/2, 5/8, 3/4, 7/8, 1, 1¼, 1½, 1¾, 2, 2½
7/64	3/4, 1, 1¼, 1½, 1¾, 2, 2¾
1/8	1/2, 5/8, 3/4, 7/8, 1, 1.1/8, 1¼, 1½, 1¾, 2, 2¼, 2½, 2¾, 3
9/64	1, 1¼, 1½, 1¾, 2, 2½
5/32	1, 1.1/8, 1¼, 1.3/8, 1½, 1.5/8, 1.7/8, 2, 2¼, 2½, 3
11/64	2
3/16	5/8, 1, 1.1/8, 1¼, 1.3/8, 1½, 1.5/8, 1¾, 2, 2¼, 2½, 3, 3½
7/32	1, 1½, 2, 2½, 3
1/4	1, 1¼, 1½, 1¾, 2, 2¼, 2½, 2¾, 3, 3½, 4, 4½
5/16	1½, 2, 2¼, 2½, 3, 3¼, 4¼, 6
3/8	2¼, 2¾, 3, 3¼, 3½, 4, 4¼, 5, 5¾
1/2	1½, 2, 2½, 3, 3½, 4, 5, 6

All dimensions in inches

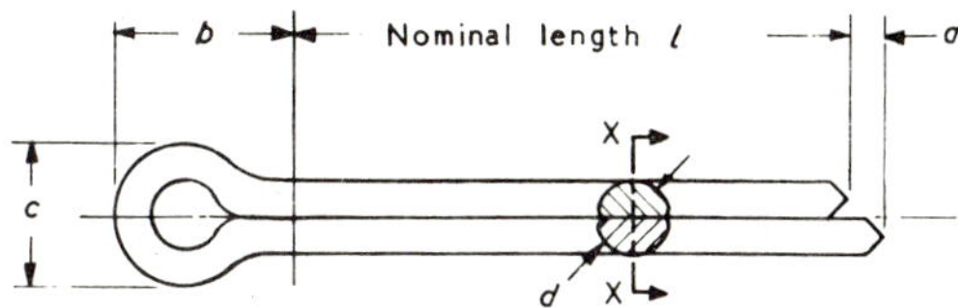

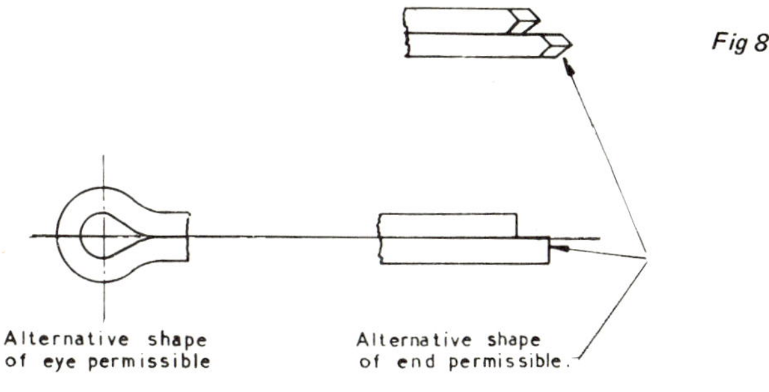

Fig 8

TABLE VIA – METRIC SERIES SPLIT COTTER PINS TO BS1574 : 1972

Nominal diameter of pin (hole diameter)	Shank diameter *d*		Outside diameter of eye *c*		Length of eye *b*	Radius on corner of wire *r*	Length of extended prong *a*	
	max.	min.	max.	min.	approx.	max.	max.	min.
mm	mm	mm	mm	mm	mm	mm	mm	mm
0.6	0.5	0.4	1.0	0.90	2.0	0.05	1.6	0.8
0.8	0.7	0.6	1.4	1.20	2.4	0.05	1.6	0.8
1	0.9	0.8	1.8	1.60	3.0	0.10	1.6	0.8
(1.2)	1.0	0.9	2.0	1.70	3.0	0.10	2.5	1.2
1.6	1.4	1.3	2.8	2.40	3.2	0.15	2.5	1.2
2	1.8	1.7	3.6	3.20	4.0	0.15	2.5	1.2
2.5	2.3	2.1	4.6	4.0	5.0	0.20	2.5	1.2
3.2	2.9	2.7	5.8	5.1	6.4	0.25	3.2	1.6
4	3.7	3.5	7.4	6.5	8.0	0.30	4.0	2.0
5	4.6	4.4	9.2	8.0	10.0	0.40	4.0	2.0
6.3	5.9	5.7	11.8	10.3	12.6	0.50	4.0	2.0
8	7.5	7.3	15.0	13.1	16.0	0.60	4.0	2.0
10	9.5	9.3	19.0	16.6	20.0	0.70	6.3	3.1
13	12.4	12.1	24.8	21.7	26.0	0.95	6.3	3.1
(16)	15.4	15.1	30.8	27.0	32.0	1.20	6.3	3.1
(20)	19.3	19.0	38.6	33.8	40.0	1.50	6.3	3.1

TABLE VIB – INCH SERIES SPLIT COTTER PINS TO BS1574 : 1972

Nominal diameter of pin (hole diameter)	Shank diameter *d*		Outside diameter of eye *c**	Length of eye *b*	Radius on corner of wire *r*	Length of extended prong *a*	
	max.	min.	max.	approx.	max.	max.	min.
in	in	in	in	in	in	in	in
1/32	0.031	0.024	0.062	0.09	0.002	0.06	0.03
3/64	0.046	0.035	0.092	0.11	0.004	0.10	0.05
1/16	0.062	0.051	0.124	0.12	0.006	0.10	0.05
5/64	0.077	0.067	0.154	0.16	0.006	0.10	0.05
3/32	0.093	0.083	0.186	0.19	0.008	0.10	0.05
(7/64)	0.108	0.101	0.216	0.22	0.008	0.10	0.05
1/8	0.124	0.106	0.248	0.25	0.010	0.10	0.05
(9/64)	0.140	0.132	0.280	0.28	0.011	0.16	0.08
5/32	0.155	0.138	0.310	0.31	0.012	0.16	0.08
(11/64)	0.171	0.161	0.342	0.34	0.013	0.16	0.08
3/16	0.186	0.173	0.372	0.38	0.016	0.16	0.08
(7/32)	0.218	0.208	0.436	0.44	0.016	0.16	0.08
1/4	0.249	0.224	0.498	0.50	0.020	0.16	0.08
(9/32)	0.280	0.270	0.560	0.56	0.021	0.16	0.08
5/16	0.311	0.287	0.622	0.62	0.024	0.16	0.08
3/8	0.373	0.358	0.746	0.75	0.028	0.25	0.12
(7/16)	0.436	0.421	0.872	0.87	0.033	0.25	0.12
1/2	0.498	0.476	0.996	1.00	0.038	0.25	0.12

NOTE. The sizes shown in brackets are non-preferred.

Clevis Pins

Clevis pins are shallow headed dowel pins, drilled with a hole at the other end to accommodate a split pin or cotter pin for locking – Fig 1. They are specified by diameter size, length (from underside of head), and head to centre of hole length. They are normally made of carbon steel, hardened steel or alloy steel, but may also be produced in stainless steel, brass and monel. Steel and brass pins may be surface coated or plated.

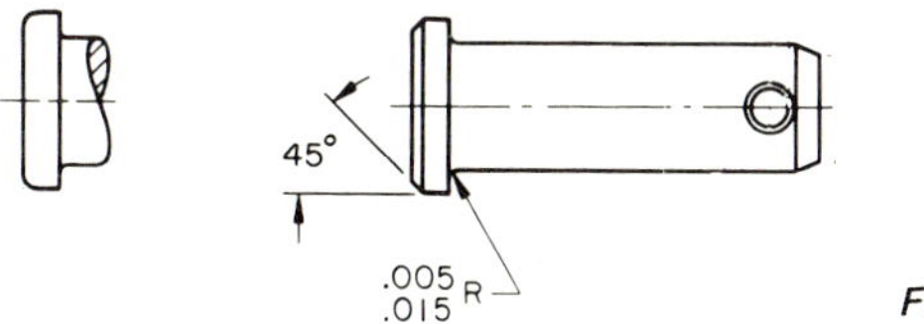

Fig 1

Clevis pins should be of good quality manufacture, free from burrs, loose scale, sharp edges and other defects. Size is generally matched to specific design requirements,but typical standard sizes are:-

TYPICAL STANDARD CLEVIS PIN SIZES

Nominal Pin Diameter		Minimum Length*		Head to Centre of Hole (min) *		Cotter Pin Size	
in	mm	in	mm	in	mm	in	mm
3/16	4.76	0.58	15	0.484	12.3	1/16	1.6
1/4	6.35	0.77	20	0.672	17.0	1/16	1.6
5/16	7.94	0.94	24	0.812	20.5	3/32	2.4
3/8	9.525	1.06	27	0.938	23.8	3/32	2.4
7/16	11.10	1.19	30	1.062	27.0	3/32	2.4
1/2	12.70	1.36	34.5	1.203	30.5	1/8	3.2
5/8	15.90	1.61	40	1.453	37.0	1/8	3.2
3/4	19.00	1.91	48.5	1.719	43.5	5/32	4.0
7/8	22.00	2.16	55	1.969	50.0	5/32	4.0
1	25.40	2.41	61	2.219	56.0	5/32	4.0

*Other lengths in standard increments

Keys and Splines

Keys and feathers are common forms of fastener used for mounting gearwheels, flywheels, etc, on shafts. Standard keys may be parallel or tapered. Parallel keys may be of square or rectangular section. Dimensions are normally arranged to give a small interference fit (equivalent to a push fit) at the sides and a clearance fit at the top and bottom. Taper keys have a standard taper of 100:1, providing an interference or driving fit all round. The nominal depth (thickness) of a taper key is measured at the thick end. Taper keys may be gib-headed — see Fig 2 — and the nominal thickness is then measured at the point where the radius of the gib-head finishes.

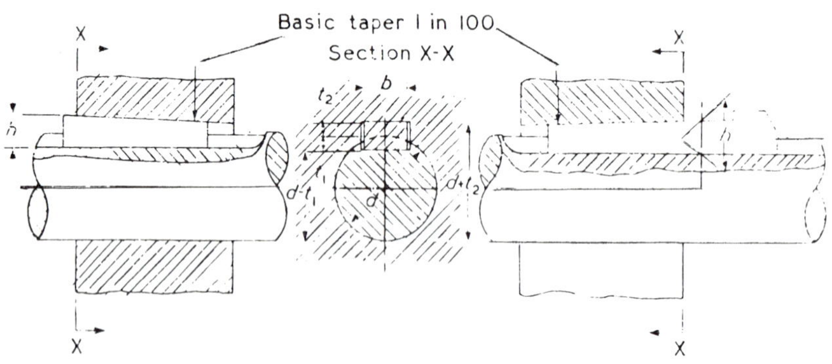

Fig 1 BS rectangular keys and keyways.

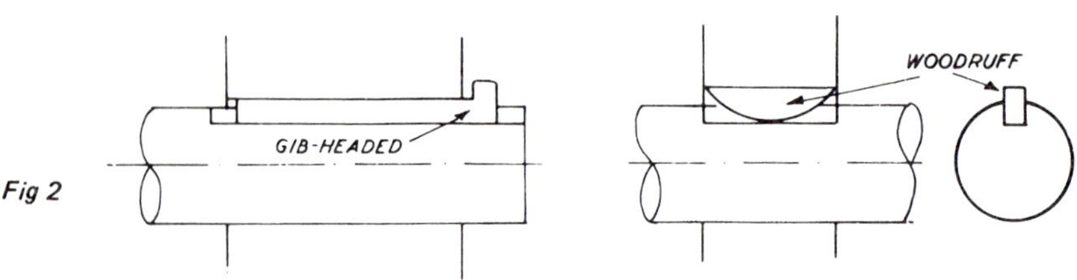

Fig 2

The Woodruff key (Fig 2) is segmental in shape. The corresponding keyway is usually cut with a milling cutter of the same radius and width as the key, but limiting the depth of cut to that which will leave the key sufficiently proud when fitted up. This type of key is particularly favoured for lighter duty installations.

Where hub and shaft have a relative sliding movement, as well as being locked together for torque transmission, the key needs to be self-locating so that it cannot be displaced with translational movement. Two basic forms of such a key are shown in Fig 3. They are of a type known as a feather. Feathers are normally square in cross section.

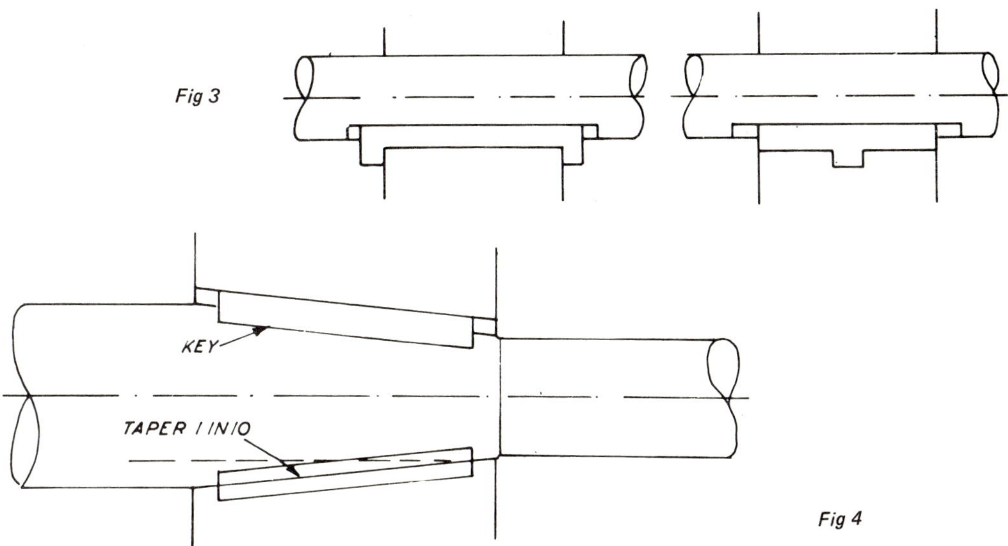

Fig 3

Fig 4

In the case of tapered shaft ends, rectangular or square parallel keys are employed with the key parallel to the side of the cone – Fig 4. The key size is based on a nominal diameter of shaft taken at the larger end of the cone. The standard taper is 1 in 10 on the diameter, but other tapers may be used for specific applications, eg:-

where a self-releasing taper is required: 1 in 3.428
locomotive piston rods: 1 in 4
fuel injection pumps (diesel engines): 1 in 5
automobile engineering (general): 1 in 8
marine shafting: 1 in 12 or 16

Materials, nominal sizes and tolerances for parallel keys, taper keys, gib-headed rectangular and square keys and Woodruff keys, and nominal sizes and dimensions for tangential keys, tapered shaft-end keys and marine tailshaft keys, are detailed in BS46: Part 1: 1958, with amendments May 1959, January 1961 and January 1964. British Standard 4325:Part 1: 1967 covers metric sizes of parallel and taper keyways.

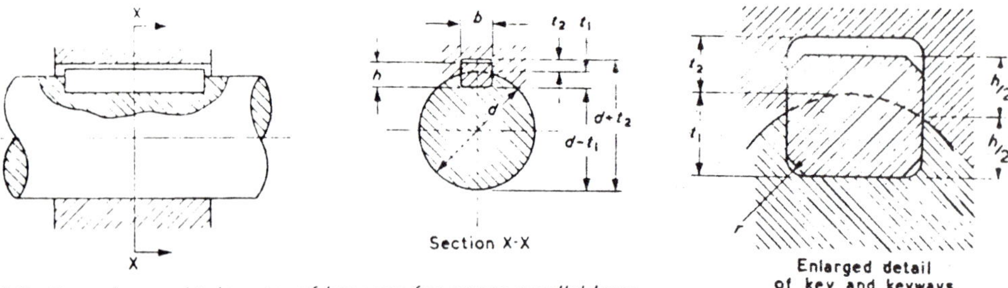

Fig 5 B.S. dimensions and tolerances of keyways for square parallel keys.

The basic dimension in the case of keyways is the depth of the key. In the case of parallel keys the depth of the keyway is normally proportioned so that one half the depth of the key is accommodated in this way – Fig 5. Nominally, the length of the key should not be less than the diameter of the shaft, or 1½ times the diameter of the shaft in the case of a tapered shaft end, although this may be influenced by the length of boss available.

Splines

Splines are a direct alternative to keys and keyways and involve no separate fastener as such. There are two basic forms of spline – *straight-sided splines* which may number 4, 6, 10 or up to 16 splines equally distributed around the circumference of a shaft, and *serrated splines* which are in the form of adjacent triangular teeth. The latter can further be described as 'fine' or 'coarse', according to the pitch of the teeth. Only 'coarse' splines are now admitted in British Standards.

Three types of fit are specified for splines –

(i) Clearance or sliding fit

(ii) Transition or light duty fit

(iii) Interference, requiring force fitting

British Standard straight-sided splines and serrations are specified in BS2059:1953, with amendment PD1843 issued April 1954. Part 1 details dimensions of internal and external straight-sided splines in the size range ½in to 6in inclusive, together with dimensions of the necessary Full Form GO gauges and the NOT GO feature gauges. Part II details dimensions of straight-sided 90 serrations over a size range of ¼in to 6in inclusive, together with dimensional details of gauges.

British Standard 3550:1963 specifies dimensions of involuted splined shafts and splined holes with a 30° pressure angle in the range–

Major diameter fit, flat roof 3/6 to 16/32 pitch 6 to 60 teeth

Side fit, flat root 2.5/5 to 32/64 pitch 6 to 60 teeth

Side fit, fillet root 2.5/5 to 48/96 pitch 6 to 60 teeth

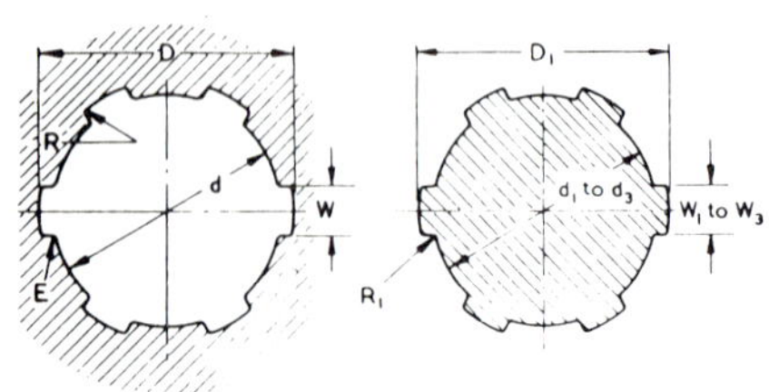

Fig 6

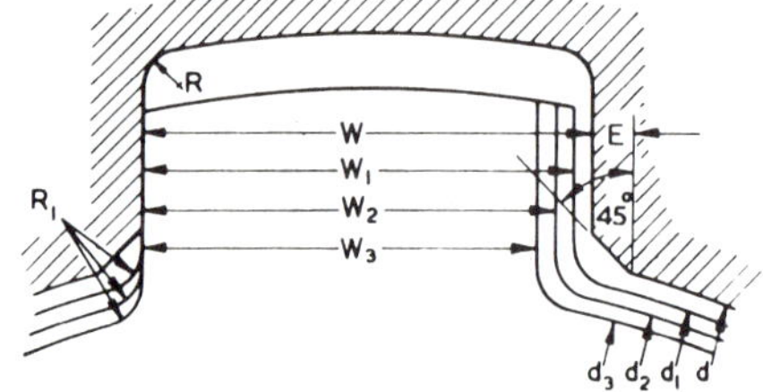

Fig 7 Spline section geometry

Nominal geometry for straight-sided splines is shown in Fig 6. The depth of spline (h) is dependent both on the number of splines and the type of fit. For typical automobile practice, values would be of the following order for a 6-spline fitting:-

Sliding Fit	**Transition Fit**	**Interference Fit**
h = 0.1D	h = 0.075D	h = 0.05D

spline width W = 0.25D

The torque capacity of straight-sided splines can be determined directly from the total load bearing area and pressure on the spline sides, thus

Torque capacity (per unit length of spline) = n x P x mean radius x h

$$= n \times P \times \frac{D - d}{2} \times h$$

or

$$= n \times P \times \frac{d + h}{2} \times h$$

where n = number of splines
P = pressure on spline sides

TABLE I – BS SQUARE PARALLEL KEYS (METRIC)
(All dimensions in millimetres)

Shaft		Key	Keyway											
nominal diameter d		section $b \times h$ width x thickness	width b						depth				radius r	
				tolerance for class of fit					shaft t_1		hub t_2			
				free		normal		close						
over	incl		nom	shaft (H9)	hub (D10)	shaft (N9)	hub (J_s9)	shaft and hub (P9)	nom	tol	nom	tol	max	min
6	8	2 x 2	2	+ 0.025	+ 0.060	- 0.004	+ 0.012	– 0.006	1.2		1		0.16	0.08
8	10	3 x 3	3	0	+ 0.020	–0.029	–0.012	– 0.031	1.8		1.4		0.16	0.08
10	12	4 x 4	4						2.5	+ 0.1 0	1.8	+ 0.1 0	0.16	0.08
				+0.030	0.078	0	+ 0.015	– 0.012						
12	17	5 x 5	5						3		2.3		0.25	0.16
				0	+ 0.030	–0.030	– 0.015	– 0.042						
17	22	6 x 6	6						3.5		2.8		0.25	0.16

TABLE II – BS STRAIGHT-SIDED SPLINES (SHALLOW) (See also Figs 5 and 6)

Nom. Size	HOLE					SHAFT								
	D	d	W	R	E Chamfer at 45°	D_1	Fit 1		Fit 2		Fit 3		RR_1
							d_1	W_1	d_2	W_2	d_3	W_3	
1/2	0.520 0.530	0.4500 0.4510	0.1250 0.1265	0.010 0.005	0.007 0.010	0.4990 0.4970	0.4500 0.4485	0.1240 0.1225	0.4485 0.4465	0.1232 0.1215	0.4470 0.4445	0.1224 0.1205	0.006 0.003
5/8	0.645 0.655	0.5630 0.5640	0.1560 0.1575	0.010 0.005	0.007 0.010	0.6240 0.6210	0.5630 0.5615	0.1550 0.1535	0.5615 0.5595	0.1542 0.1525	0.5600 0.5575	0.1534 0.1515	0.006 0.003
3/4	0.770 0.780	0.6750 0.6760	0.1870 0.1885	0.010 0.005	0.008 0.011	0.7490 0.7460	0.6750 0.6735	0.1860 0.1845	0.6735 0.6715	0.1852 0.1835	0.6720 0.6695	0.1844 0.1825	0.007 0.004
7/8	0.905 0.915	0.7880 0.7890	0.2180 0.2195	0.015 0.010	0.009 0.012	0.8740 0.8710	0.7880 0.7865	0.2170 0.2155	0.7865 0.7845	0.2162 0.2145	0.7850 0.7825	0.2154 0.2135	0.008 0.005
1	1.030 1.040	0.9000 0.9010	0.2490 0.2510	0.015 0.010	0.010 0.013	0.9990 0.9950	0.9000 0.8985	0.2480 0.2460	0.8985 0.8965	0.2472 0.2450	0.8970 0.8945	0.2464 0.2440	0.009 0.005
1 1/8	1.155 1.165	1.0130 1.0140	0.2800 0.2820	0.015 0.010	0.012 0.015	1.1200 1.1190	1.0130 1.0115	0.2790 0.2770	1.0115 1.0090	0.2782 0.2757	1.0100 1.0065	0.2774 0.2744	0.011 0.005
1 1/4	1.280 1.290	1.1250 1.1260	0.3120 0.3140	0.015 0.010	0.013 0.017	1.2490 1.2450	1.1250 1.1235	0.3110 0.3090	1.1235 1.1210	0.3102 0.3077	1.1220 1.1185	0.3094 0.3064	0.012 0.005
1 3/8	1.405 1.415	1.2380 1.2390	0.3430 0.3450	0.015 0.010	0.014 0.018	1.3740 1.3700	1.2380 1.2365	0.3420 0.3400	1.2365 1.2340	0.3412 0.3387	1.2350 1.2315	0.3404 0.3374	0.013 0.005
1 1/2	1.560 1.570	1.3500 1.3510	0.3740 0.3765	0.030 0.015	0.015 0.019	1.4990 1.4940	1.3500 1.3485	0.3730 0.3705	1.3485 1.3460	0.3722 0.3692	1.3470 1.3435	0.3714 0.3679	0.014 0.005
1 5/8	1.685 1.695	1.4630 1.4645	0.4050 0.4075	0.030 0.015	0.017 0.021	1.6240 1.6190	1.4630 1.4615	0.4035 0.4010	1.4615 1.4585	0.4027 0.3995	1.4600 1.4555	0.4019 0.3980	0.015 0.005
1 3/4	1.810 1.820	1.5750 1.5765	0.4370 0.4395	0.030 0.015	0.018 0.022	1.7490 1.7440	1.5750 1.5735	0.4355 0.4330	1.5735 1.5705	0.4347 0.4315	1.5720 1.5675	0.4339 0.4300	0.016 0.005
1 7/8	1.935 1.945	1.6880 1.6895	0.4680 0.4705	0.030 0.015	0.019 0.023	1.8740 1.8690	1.6880 1.6865	0.4665 0.4640	1.6865 1.6835	0.4657 0.4625	1.6850 1.6805	0.4649 0.4610	0.017 0.005
2	2.060 2.070	1.8000 1.8015	0.4980 0.5010	0.030 0.015	0.021 0.026	1.9990 1.9940	1.8000 1.7985	0.4965 0.4934	1.7985 1.7955	0.4957 0.4920	1.7970 1.7925	0.4949 0.4905	0.019 0.005

$2\frac{1}{4}$	2.330 2.340	2.0250 2.0265	0.5610 0.5640	0.040 0.020	0.023 0.028	2.2490 2.2440	2.0250 2.0235	0.5595 0.5565	2.0235 2.0200	0.5587 0.5547	2.0220 2.0165	0.5579 0.5529	0.021 0.005
$2\frac{1}{2}$	2.580 2.590	2.2500 2.2515	0.6230 0.6260	0.040 0.020	0.025 0.030	2.4990 2.4940	2.2500 2.2485	0.6215 0.6185	2.2485 2.2450	0.6207 0.6167	2.2470 2.2415	0.6199 0.6149	0.023 0.005
$2\frac{3}{4}$	2.830 2.840	2.4750 2.4765	0.6860 0.6895	0.040 0.020	0.028 0.033	2.7490 2.7440	2.4750 2.4735	0.6845 0.6810	2.4735 2.4700	0.6837 0.6792	2.4720 2.4665	0.6829 0.6774	0.026 0.005
3	3.080 3.090	2.7000 2.7015	0.7480 0.7515	0.040 0.020	0.031 0.036	2.9990 2.9930	2.7000 2.6985	0.7465 0.7430	2.6985 2.6950	0.7457 0.7412	2.6970 2.6915	0.7449 0.7394	0.029 0.010
$3\frac{1}{4}$	3.330 3.340	2.9250 2.9265	0.8110 0.8145	0.040 0.020	0.033 0.038	3.2490 3.2430	2.9250 2.9230	0.8090 0.8055	2.9230 2.9190	0.8080 0.8035	2.9210 2.9150	0.8070 0.8015	0.030 0.010
$3\frac{1}{2}$	3.580 3.590	3.1500 3.1515	0.8730 0.8765	0.040 0.020	0.036 0.042	3.4990 3.4930	3.1500 3.1480	0.8710 0.8675	3.1480 3.1440	0.8700 0.8655	3.1460 3.1400	0.8690 0.8635	0.033 0.010
$3\frac{3}{4}$	3.830 3.840	3.3750 3.3765	0.9360 0.9400	0.040 0.020	0.039 0.045	3.7490 3.7420	3.3750 3.3730	0.9340 0.9300	3.3730 3.3685	0.9330 0.9277	3.3710 3.3640	0.9320 0.9254	0.036 0.010
4	4.080 4.090	3.6000 3.6015	0.9980 1.0020	0.040 0.020	0.041 0.047	3.9990 3.9920	3.6000 3.5980	0.9960 0.9920	3.5980 3.5935	0.9950 0.9897	3.5960 3.5890	0.9940 0.9874	0.038 0.010
$4\frac{1}{4}$	4.330 4.340	3.8250 3.8270	1.0600 1.0640	0.040 0.020	0.043 0.049	4.2490 4.2420	3.8250 3.8230	1.0580 1.0540	3.8230 3.8185	1.0570 1.0517	3.8210 3.8140	1.0560 1.0494	0.040 0.010
$4\frac{1}{2}$	4.580 4.590	4.0500 4.0520	1.1220 1.1260	0.040 0.020	0.045 0.051	4.4990 4.4910	4.0500 4.0480	1.1200 1.1160	4.0480 4.0435	1.1190 1.1137	4.0460 4.0390	1.1180 1.1114	0.042 0.010
$4\frac{3}{4}$	4.830 4.840	4.2750 4.2770	1.1850 1.1895	0.040 0.020	0.048 0.054	4.7490 4.7410	4.2750 4.2730	1.1825 1.1780	4.2730 4.2680	1.1815 1.1765	4.2710 4.2630	1.1805 1.1730	0.044 0.010
5	5.080 5.090	4.5000 4.5020	1.2470 1.2515	0.040 0.020	0.051 0.057	4.9990 4.9910	4.5000 4.4980	1.2445 1.2400	4.4980 4.4930	1.2435 1.2375	4.4960 4.4880	1.2425 1.2350	0.047 0.010
$5\frac{1}{4}$	5.330 5.340	4.7250 4.7270	1.3100 1.3145	0.040 0.020	0.053 0.060	5.2490 5.2410	4.7250 4.7230	1.3075 1.3030	4.7230 4.7180	1.3065 1.3005	4.7210 4.7130	1.3055 1.2980	0.049 0.010
$5\frac{1}{2}$	5.580 5.590	4.9500 4.9520	1.3720 1.3765	0.040 0.020	0.055 0.062	5.4990 5.4900	4.9500 4.9480	1.3695 1.3650	4.9480 4.9430	1.3685 1.3625	4.9460 4.9380	1.3675 1.3600	0.051 0.010
$5\frac{3}{4}$	5.830 5.840	5.1750 5.1770	1.4350 1.4395	0.040 0.020	0.060 0.067	5.7490 5.7400	5.1750 5.1730	1.4325 1.4280	5.1730 5.1680	1.4315 1.4255	5.1710 5.1630	1.4305 1.4230	0.056 0.015
6	6.080 6.090	5.4000 5.4020	1.4970 1.5015	0.040 0.020	0.061 0.068	5.9990 5.9900	5.4000 5.3980	1.4945 1.4900	5.3980 5.3930	1.4935 1.4875	5.3960 5.3880	1.4925 1.4850	0.057 0.015

TABLE III — BS STRAIGHT-SIDED SPLINES (DEEP) (See also Figs 5 and 6)

Nom. Size	HOLE					SHAFT							
							Fit 1		Fit 2		Fit 3		
	D	d	W	R	E Chamfer at 45°	D_1	d_1	W_1	d_2	W_2	d_3	W_3	R_1
$\frac{1}{2}$	0.520 0.530	0.4250 0.4260	0.1250 0.1265	0,010 0.005	0.010 0.013	0.4990 0.4970	0.4250 0.4235	0.1240 0.1225	0.4235 0.4215	0.1232 0.1215	0.4220 0.4195	0.1224 0.1205	0.009 0.005
$\frac{5}{8}$	0.645 0.655	0.5320 0.5330	0.1560 0.1575	0.010 0.005	0.010 0.013	0.6240 0.6210	0.5320 0.5305	0.1550 0.1535	0.5305 0.5285	0.1542 0.1525	0.5290 0.5265	0.1534 0.1515	0.009 0.005
$\frac{3}{4}$	0.770 0.780	0.6380 0.6390	0.1870 0.1885	0.010 0.005	0.012 0.015	0.7490 0.7460	0.6380 0.6365	0.1860 0.1845	0.6365 0.6345	0.1852 0.1835	0.6350 0.6325	0.1844 0.1825	0.010 0.005
$\frac{7}{8}$	0.905 0.915	0.7440 0.7450	0.2180 0.2195	0.015 0.010	0.014 0.018	0.8740 0.8710	0.7440 0.7425	0.2170 0.2155	0.7425 0.7405	0.2162 0.2145	0.7410 0.7385	0.2154 0.2135	0.012 0.005
1	1.030 1.040	0.8500 0.8510	0.2490 0.2510	0.015 0.010	0.016 0.020	0.9990 0.9950	0.8500 0.8485	0.2480 0.2460	0.8485 0.8465	0.2472 0.2450	0.8470 0.8445	0.2464 0.2440	0.014 0.005
$1\frac{1}{8}$	1.155 1.165	0.9560 0.9570	0.2800 0.2820	0.015 0.010	0.018 0.022	1.1240 1.1190	0.9560 0.9545	0.2790 0.2770	0.9545 0.9520	0.2782 0.2757	0.9530 0.9495	0.2774 0.2744	0.016 0.005
$1\frac{1}{4}$	1.280 1.290	1.0630 1.0640	0.3120 0.3140	0.015 0.010	0.020 0.025	1.2490 1.2450	1.0630 1.0615	0.3110 0.3090	1.0615 1.0590	0.3102 0.3077	1.0600 1.0565	0.3094 0.3064	0.018 0.005
$1\frac{3}{8}$	1.405 1.415	1.1690 1.1700	0.3430 0.3450	0.015 0.010	0.020 0.025	1.3740 1.3700	1.1690 1.1675	0.3420 0.3400	1.1675 1.1650	0.3412 0.3387	1.1660 1.1625	0.3404 0.3374	0.018 0.005
$1\frac{1}{2}$	1.560 1.570	1.2750 1.2760	0.3740 0.3765	0.030 0.015	0.023 0.028	1.4990 1.4940	1.2750 1.2735	0.3730 0.3705	1.2735 1.2710	0.3722 0.3692	1.2720 1.2685	0.3714 0.3679	0.021 0.005
$1\frac{5}{8}$	1.685 1.695	1.3810 1.3825	0.4050 0.4075	0.030 0.015	0.025 0.030	1.6240 1.6190	1.3810 1.3795	0.4035 0.4010	1.3795 1.3765	0.4027 0.3995	1.3780 1.3735	0.4019 0.3980	0.023 0.005
$1\frac{3}{4}$	1.810 1.820	1.4880 1.4895	0.4370 0.4395	0.030 0.015	0.027 0.032	1.7490 1.7440	1.4880 1.4865	0.4355 0.4330	1.4865 1.4835	0.4347 0.4315	1.4850 1.4805	0.4339 0.4300	0.024 0.005
$1\frac{7}{8}$	1.935 1.945	1.5940 1.5955	0.4680 0.4705	0.030 0.015	0.030 0.035	1.8740 1.8690	1.5940 1.5925	0.4665 0.4640	1.5925 1.5895	0.4657 0.4625	1.5910 1.5865	0.4649 0.4610	0.027 0.010
2	2.060 2.070	1.7000 1.7015	0.4980 0.5010	0.030 0.015	0.030 0.035	1.9990 1.9940	1.7000 1.6985	0.4965 0.4935	1.6985 1.6955	0.4957 0.4920	1.6970 1.6925	0.4949 0.4905	0.027 0.010

$2\frac{1}{4}$	2.330 2.340	1.9130 1.9145	0.5610 0.5640	0.040 0.020	0.034 0.040	2.2490 2.2440	1.9130 1.9115	0.5595 0.5565	1.9115 1.9080	0.5587 0.5547	1.9100 1.9045	0.5579 0.5529	0.031 0.010
$2\frac{1}{2}$	2.580 2.590	2.1250 2.1265	0.6230 0.6260	0.040 0.020	0.038 0.044	2.4990 2.4940	2.1250 2.1235	0.6215 0.6185	2.1235 2.1200	0.6207 0.6167	2.1220 2.1165	0.6199 0.6149	0.034 0.010
$2\frac{3}{4}$	2.830 2.840	2.3380 2.3395	0.6860 0.6895	0.040 0.020	0.042 0.048	2.7490 2.7440	2.3380 2.3365	0.6845 0.6810	2.3365 2.3330	0.6837 0.6792	2.3350 2.3295	0.6829 0.6774	0.038 0.010
3	3.080 3.090	2.5500 2.5515	0.7480 0.7515	0.040 0.020	0.046 0.052	2.9990 2.9930	2.5500 2.5485	0.7465 0.7430	2.5485 2.5450	0.7457 0.7412	2.5470 2.5415	0.7449 0.7394	0.042 0.010
$3\frac{1}{4}$	3.330 3.340	2.7630 2.7645	0.8110 0.8145	0.040 0.020	0.050 0.056	3.2490 3.2430	2.7630 2.7610	0.8090 0.8055	2.7610 2.7570	0.8080 0.8035	2.7590 2.7530	0.8070 0.8015	0.046 0.010
$3\frac{1}{2}$	3.580 3.590	2.9750 2.9765	0.8730 0.8765	0.040 0.020	0.053 0.060	3.4990 3.4930	2.9750 2.9730	0.8710 0.8675	2.9730 2.9690	0.8700 0.8655	2.9710 2.9650	0.8690 0.8635	0.048 0.010
$3\frac{3}{4}$	3.830 3.840	3.1880 3.1895	0.9360 0.9400	0.040 0.020	0.056 0.063	3.7490 3.7420	3.1880 3.1860	0.9340 0.9300	3.1860 3.1815	0.9330 0.9277	3.1840 3.1770	0.9320 0.9254	0.051 0.010
4	4.080 4.090	3.4000 3.4015	0.9980 1.0020	0.040 0.020	0.061 0.068	3.9990 3.9920	3.4000 3.3980	0.9960 0.9920	3.3980 3.3935	0.9950 0.9897	3.3960 3.3890	0.9940 0.9874	0.055 0.015
$4\frac{1}{4}$	4.330 4.340	3.6130 3.6150	1.0600 1.0640	0.040 0.020	0.064 0.072	4.2490 4.2420	3.6130 3.6110	1.0580 1.0540	3.6110 3.6065	1.0570 1.0517	3.6090 3.6020	1.0560 1.0494	0.058 0.015
$4\frac{1}{2}$	4.580 4.590	3.8250 3.8270	1.1220 1.1260	0.040 0.020	0.068 0.076	4.4990 4.4910	3.8250 3.8230	1.1200 1.1160	3.8230 3.8185	1.1190 1.1137	3.8210 3.8140	1.1180 1.1114	0.062 0.015
$4\frac{3}{4}$	4.830 4.840	4.0380 4.0400	1.1850 1.1895	0.040 0.020	0.072 0.080	4.7490 4.7410	4.0380 4.0360	1.1825 1.1780	4.0360 4.0310	1.1815 1.1755	4.0340 4.0260	1.1805 1.1730	0.066 0.015
5	5.080 5.090	4.2500 4.2520	1.2470 1.2515	0.040 0.020	0.076 0.084	4.9990 4.9910	4.2500 4.2480	1.2445 1.2400	4.2480 4.2430	1.2435 1.2375	4.2460 4.2380	1.2425 1.2350	0.069 0.020
$5\frac{1}{4}$	5.330 5.340	4.4630 4.4650	1.3100 1.3145	0.040 0.020	0.080 0.088	5.2490 5.2410	4.4630 4.4610	1.3075 1.3030	4.4610 4.4560	1.3065 1.3005	4.4590 4.4510	1.3055 1.2980	0.073 0.020
$5\frac{1}{2}$	5.580 5.590	4.6750 4.6770	1.3720 1.3765	0.040 0.020	0.083 0.091	5.4990 5.4900	4.6750 4.6730	1.3695 1.3650	4.6730 4.6680	1.3685 1.3625	4.6710 4.6630	1.3675 1.3600	0.076 0.020
$5\frac{3}{4}$	5.830 5.840	4.8880 4.8900	1.4350 1.4395	0.040 0.020	0.087 0.095	5.7490 5.7400	4.8880 4.8860	1.4325 1.4280	4.8860 4.8810	1.4315 1.4255	4.8840 4.8760	1.4305 1.4230	0.080 0.020
6	6.080 6.090	5.1000 5.1020	1.4970 1.5015	0.040 0.020	0.090 0.100	5.9990 5.9900	5.1000 5.0980	1.4945 1.4900	5.0980 5.0930	1.4935 1.4875	5.0960 5.0880	1.4925 1.4850	0.082 0.020

TABLE IV – BS RECTANGULAR TAPER KEYS (METRIC)
(All dimensions in millimetres)

Shaft		Key †	Keyway							
nominal diameter † d		section $b \times h$ width $\times$ thickness	width b shaft and hub		depth				radius r	
					shaft t_1		hub t_2			
over	incl		nom	tol (D10)	nom	tol	nom	tol	max	min
22	30	8 × 7	8	+ 0.098	4		2.4		0.25	0.16
30	38	10 × 8	10	+ 0.040	5		2.4		0.40	0.25
38	44	12 × 8	12		5		2.4		0.40	0.25
44	50	14 × 9	14	+ 0.120	5.5		2.9		0.40	0.25
50	58	16 × 10	16	+ 0.050	6		3.4		0.40	0.25
58	65	18 × 11	18		7	+ 0.2 0	3.4	+ 0.2 0	0.40	0.25
65	75	20 × 12	20		7.5		3.9		0.60	0.40
75	85	22 × 14	22	+ 0.149	9		4.4		0.60	0.40
85	95	25 × 14	25	+ 0.065	9		4.4		0.60	0.40
95	110	28 × 16	28		10		5.4		0.60	0.40
110	130	32 × 18	32		11		6.4		0.60	0.40
130	150	36 × 20	36	+ 0.180	12		7.1		1.00	0.70
150	170	40 × 22	40	+ 0.080	13		8.1		1.00	0.70
170	200	45 × 25	45		15		9.1		1.00	0.70
200	230	50 × 28	50		17		10.1		1.00	0.70
230	260	56 × 32	56		20	+ 0.3	11.1	+ 0.3	1.60	1.20
260	290	63 × 32	63	+ 0.220	20	0	11.1	0	1.60	1.20
290	330	70 × 36	70	+ 0.120	22		13.1		1.60	1.20
330	380	80 × 40	80		25		14.1		2.50	2.00
380	440	90 × 45	90	+ 0.260	28		16.1		2.50	2.00
440	500	100 × 50	100	+ 0.120	31		18.1		2.50	2.00

The relations between shaft diameter and key section given above are for general applications. The use of smaller key sections is permitted if suitable for the torque transmitted.

In cases such as stepped shafts when larger diameters are required, for example, to resist bending, and when fans, gears and impellers are fitted with a smaller key than normal, an unequal disposition of key in shaft with relation to the hub results. Therefore, dimensions $d - t_1$ and $d - t_2$ should be recalculated to maintain the h/2 relationship.

The use of larger key sections is not permitted.

Spring Nuts

The basic form of spring nut is a pressing in spring steel having an arched base and arched prongs. Each prong engages a single turn of a screw thread so that when tightened the base is straightened and the prongs are drawn inwards – Fig 1. The straightened base provides a self-energised spring lock, and the prongs a compensating thread lock – ie the fastener is self-locking. Such spring nuts may be designed for use with self-tapping screws or machine screws – see Table I. In the latter case locking performance is generally reduced.

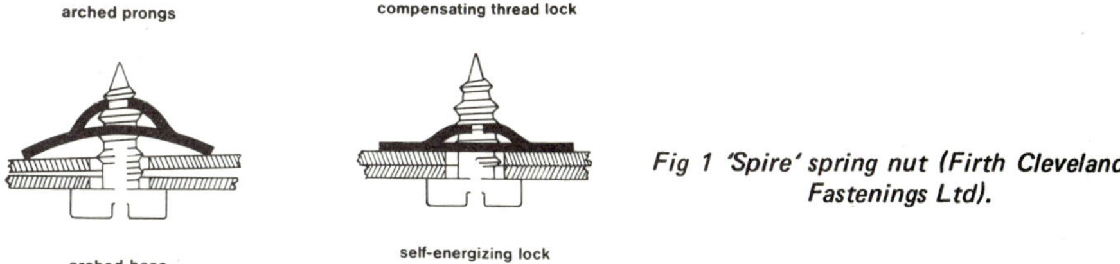

Fig 1 'Spire' spring nut (Firth Cleveland Fastenings Ltd).

Unlike threaded nuts, spring nuts do not have to be tightened with any great degree of torque. Their holding power and vibration resistance are dependent solely upon spring tension. They must be tightened sufficiently to produce the thread and spring lock, but any torque applied beyond this point is excessive, and very high torque might even distort the arch of the prongs and affect their spring tension resilience.

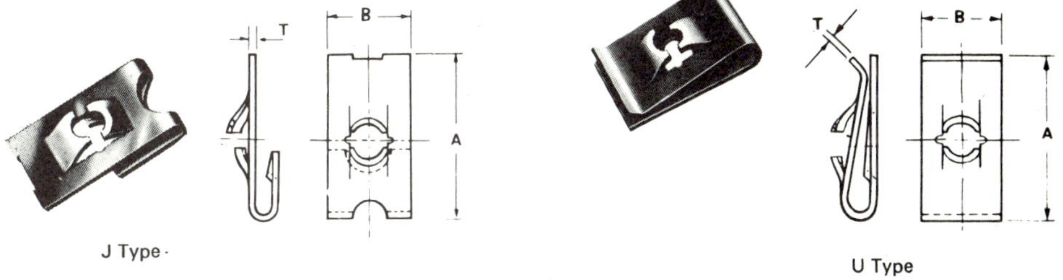

Fig 2 Type 'J' and 'V' captive spring nuts.

Spring nuts may also be designed in a modified form with self-retaining characteristics (eg edge fitting captive nuts – Fig 2 – including prevailing torque nuts) or as weld nuts and wood anchors (Fig 3).

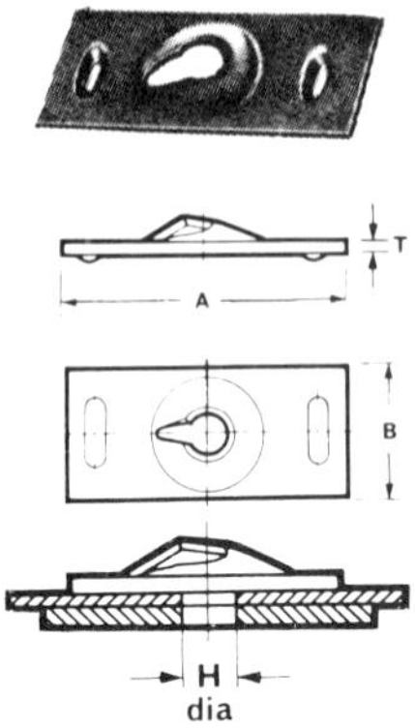

T
A
B
H
dia

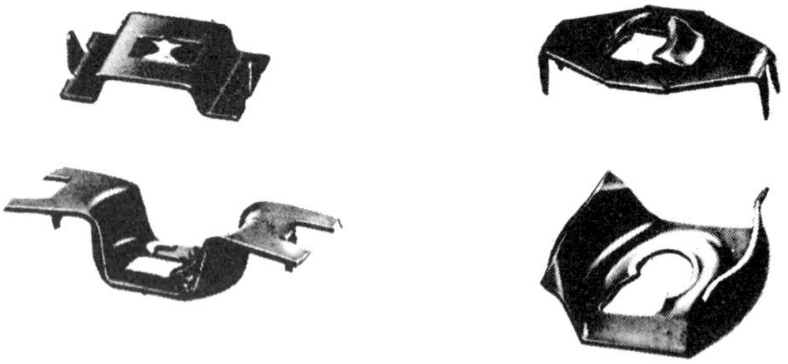

Fig 3 'Spire' weld nut (left) and wood anchor nuts (right).

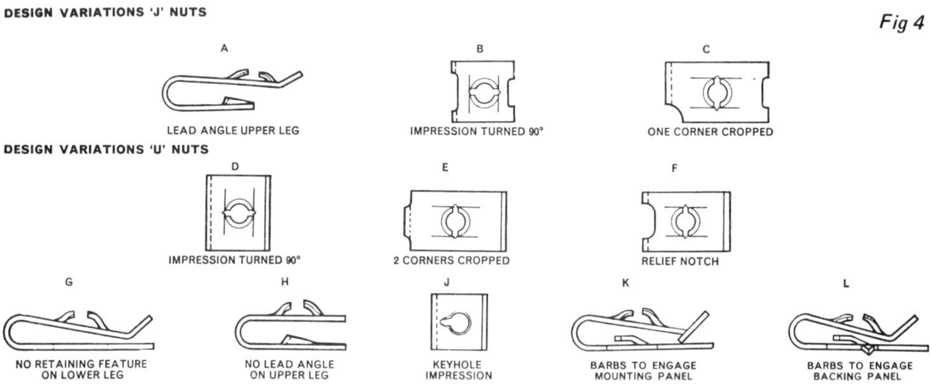

DESIGN VARIATIONS 'J' NUTS

Fig 4

A
LEAD ANGLE UPPER LEG

B
IMPRESSION TURNED 90°

C
ONE CORNER CROPPED

DESIGN VARIATIONS 'U' NUTS

D
IMPRESSION TURNED 90°

E
2 CORNERS CROPPED

F
RELIEF NOTCH

G
NO RETAINING FEATURE ON LOWER LEG

H
NO LEAD ANGLE ON UPPER LEG

J
KEYHOLE IMPRESSION

K
BARBS TO ENGAGE MOUNTING PANEL

L
BARBS TO ENGAGE BACKING PANEL

T
D
A
B
T
D
A
B
EDGE BUILD UP OF VITREOUS ENAMEL
E max
C
ASSEMBLY DETAILS

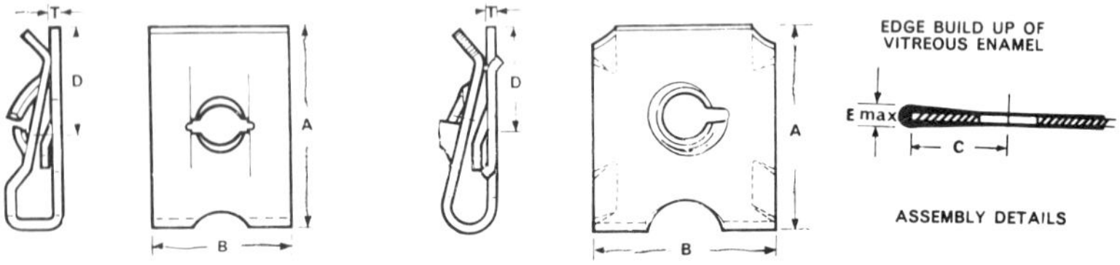

Fig 5 SNU type captive nuts.

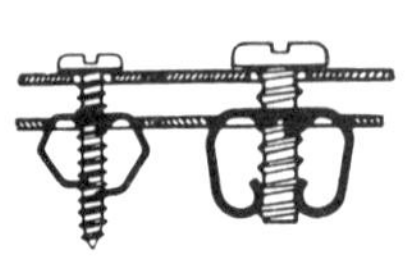

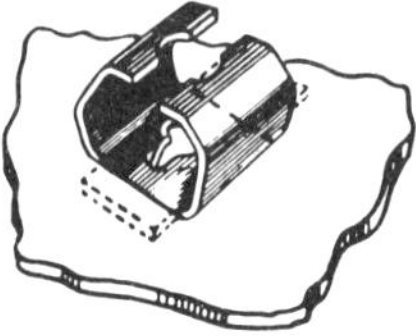

Fig 6 Captive expansion type spring nuts for blind assemblies.

TABLE I – 'SPIRE' FLAT SPRING NUTS – SHEET METAL SCREW TYPES

Thread Size	Type	A		B		H		P		T	
		mm	in	mm	in	mm	in	mm	in	mm	in
No 4 AB SMS	SNR	14.2	0.56	–	–	2.39	0.094	9.50	0.374	0.51	0.020
No 6 AB SMS	SNP	13.5	0.53	8.6	0.34	–	–	–	–	0.56	0.022
No 6 AB SMS	SNR	14.2	0.56	–	–	2.39	0.094	9.50	0.374	0.56	0.022
No 8 AB SMS	SNP	15.2	0.60	9.7	0.38	–	–	–	–	0.61	0.024
No 8 AB SMS	SNR	18.8	0.74	–	–	3.18	0.125	12.70	0.500	0.61	0.024
No 10 AB SMS	SNP	16.8	0.66	10.4	0.41	–	–	–	–	0.71	0.028
No 10 AB SMS	SNR	20.6	0.81	–	–	3.18	0.125	14.27	0.562	0.71	0.028

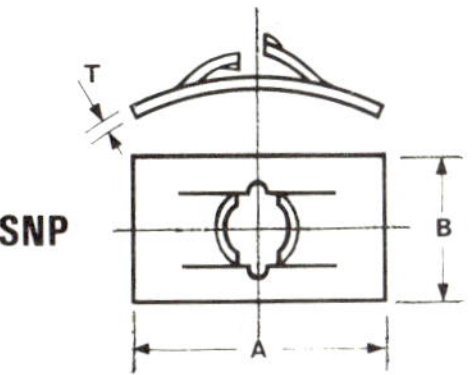

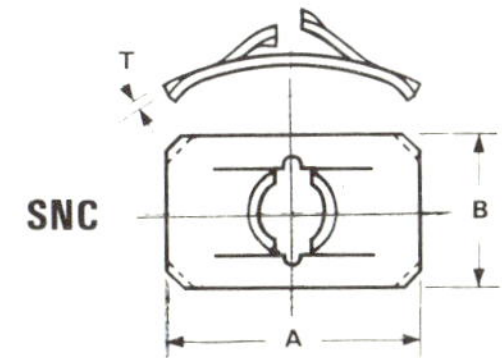

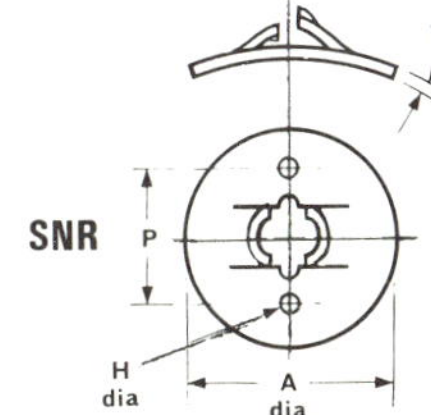

'SPIRE' FLAT SPRING NUTS – MACHINE SCREW TYPES

Thread Size	Type	A		B		H		P		T	
		mm	in	mm	in	mm	in	mm	in	mm	in
6 BA	SNP	8.6	0.34	5.8	0.23	–	–	–	–	0.28	0.011
4 BA	SNP	11.2	0.44	7.1	0.28	–	–	–	–	0.35	0.014
M4 x 0.7	SNP	14.2	0.56	8.6	0.34	–	–	–	–	0.41	0.016
No 10 UNF/2 BA	SNP	15.7	0.62	9.6	0.38	–	–	–	–	0.41	0.016
No 10 UNC/ 3/16 BSW	SNC	15.7	0.62	9.6	0.38	–	–	–	–	0.56	0.022
No 10 UNC 3/16 BSW	SNR	20.8	0.82	–	–	3.18	0.125	14.27	0.562	0.56	0.022
No 10 UNC 3/16 BSW	SNP	15.7	0.62	9.6	0.38	–	–	–	–	0.56	0.022
¼ UNC ¼ BSW	SNP	19.0	0.75	12.7	0.50	–	–	–	–	0.61	0.024
¼ UNC ¼ BSW	SNC	19.0	0.75	12.7	0.50	–	–	–	–	0.61	0.024
¼-12 ACME/ No 14 AB SMS	SNP	24.9	0.98	16.0	0.63	–	–	–	–	0.91	0.036

Installation torque values for *Spire* spring nuts are given in Table II.
See also chapters on *Spring Fasteners* and *Blind Fasteners*, etc.

TABLE II – INSTALLATION TORQUE FOR 'SPIRE' FLAT NUTS

Type	Thread Size	Installation Torque*	
		kg-Metre	lb-in
SNP	10 UNF/2BA	0.081	7
	10UNC/3/16 BSW	0.127	11
	¼UNC/¼ BSW	0.231	20
	No 8 AB SMS	0.208	18
	No 10 AB SMS	0.277	24
SNC	10 UNC/3/16 BSW	0.127	11
	¼ UNC/¼ BSW	0.231	20
SNR	10 UNC/3/16 BSW	0.127	11
	No 8 AB SMS	0.208	18
	No 10 AB SMS	0.277	24

**to provide maximum tensile strength*

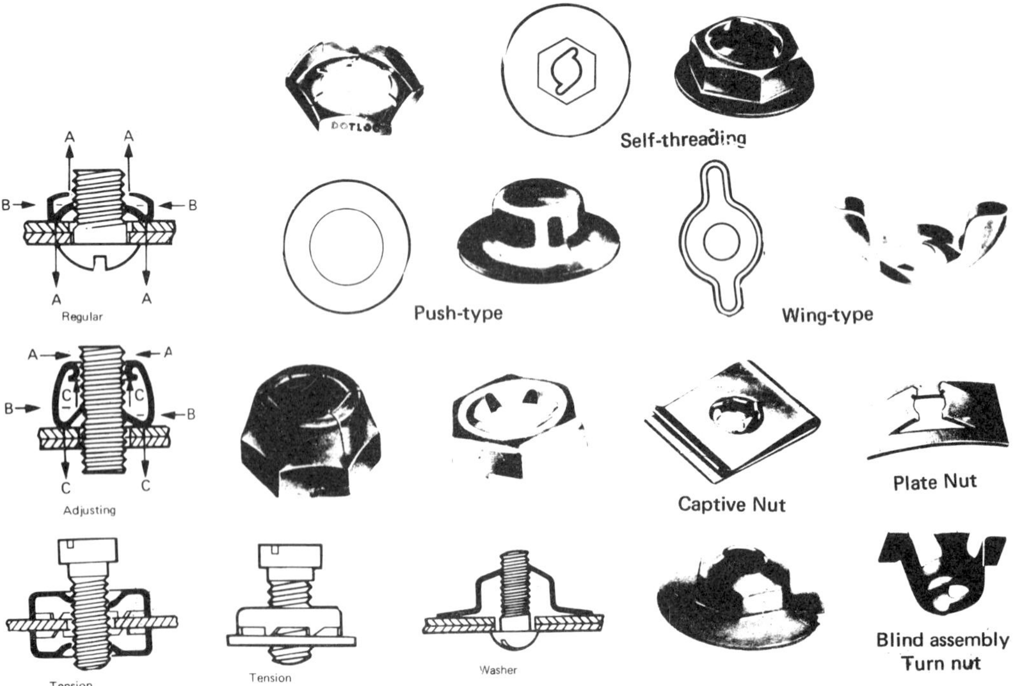

Types of 'Dotloc' single-thread lock nuts. (Carr Fastener Co.Ltd).

Spring Fasteners

Examples of spring fasteners (Firth Cleveland).

Some thirty five years ago a start was made in the development of spring fasteners. These were made from spring steel and designed to replace conventional fasteners for a large variety of applications. The spring fastener can simplify assembly and cut costs, a single fastener often being capable of replacing three or four parts required with conventional fasteners. Since that time the development and application of spring fasteners has continued to expand and the number of types now available are too numerous to attempt to classify or cover. Further, the uses of these fasteners

Ratchet plates (unthreaded stud fasteners).

Edge clips

Body trim clips

Examples of spring fasteners. (Carr Fastener Co Ltd).

are limited only by the imagination of designers, draughtsmen and engineers. New applications are continually emerging, and the availability of so many different types has inspired many new techniques in design and assembly.

Spring nuts are described in a separate chapter. Closely allied to these are push-on spring fixes designed to lock onto an unthreaded fastener, stud, tubing, wire, etc. The original rectangular or circular push-on fix for circular studs (Fig 1) has led to many further variations in form to suit both circular and rectangular studs – see Figs 2 and 3. There are also multi-tooth forms for applications where rotational conditions preclude the use of a conventional push-on fix (Fig 4). There are also tubular-type push-on fixes.

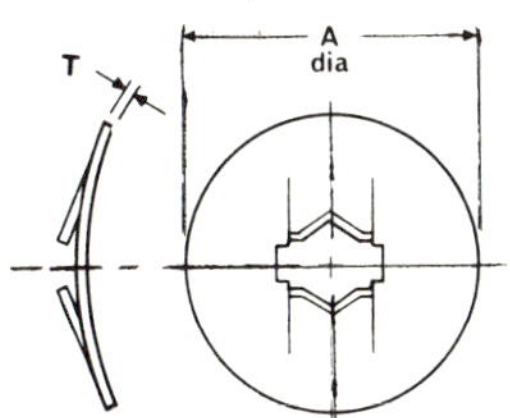

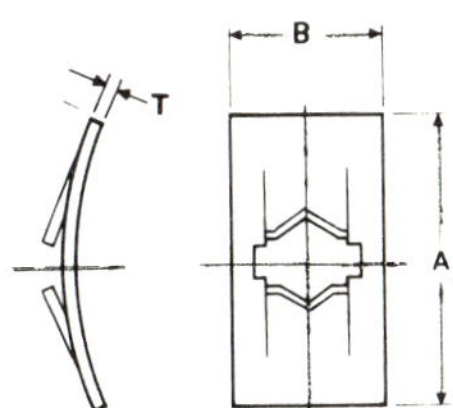

Fig 1 Spire push-on spring fastener fixes.

IMPRESSIONS

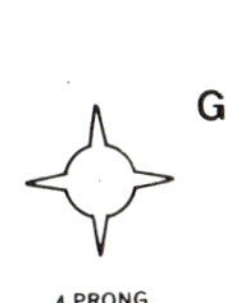

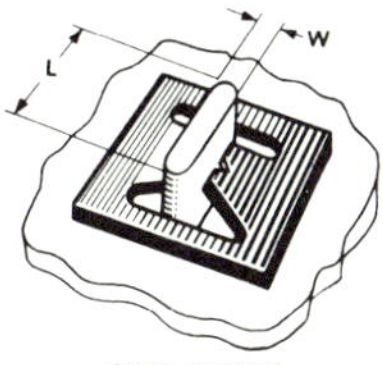

Fig 2 Design variations of Spire push-on fixes.

PROFILES

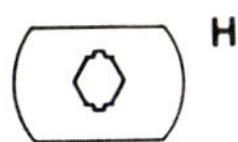

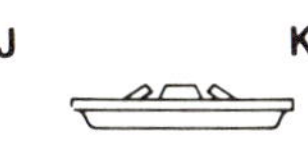

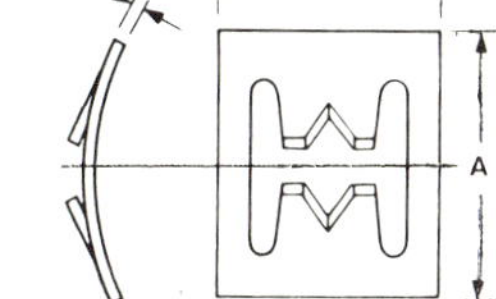

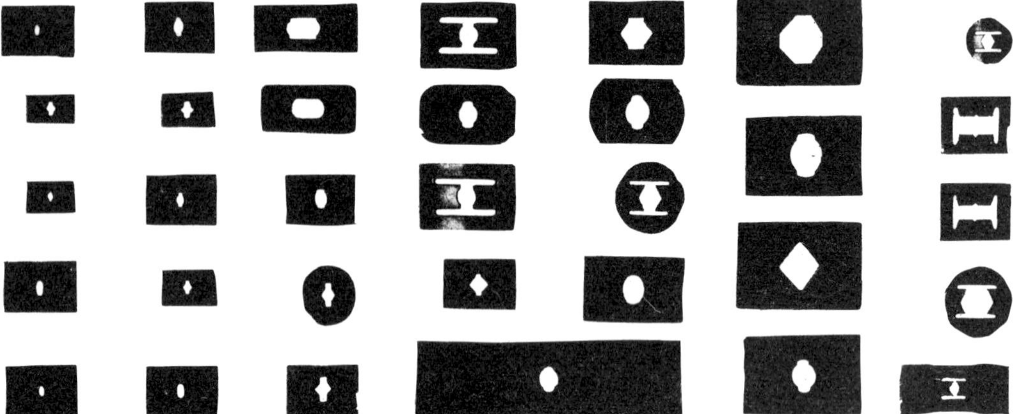

Fig 3 Push-on fixes (A. Raymond).

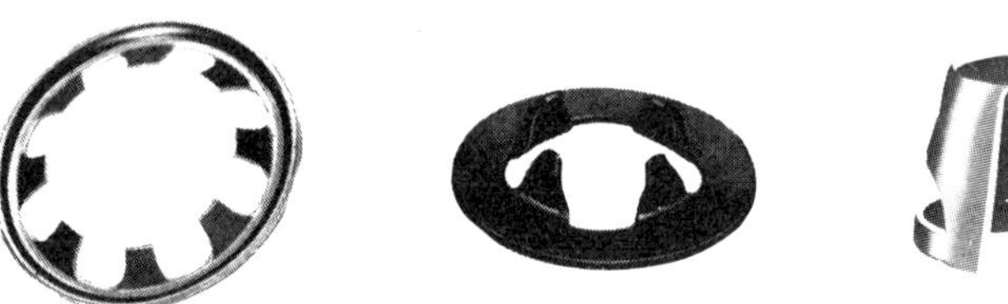

Fig 4 Spire SFO push-on fix (left) and SCO retainer clip (centre). The latter replaces a washer and split pin. (Right) tubular-type push-on for unthreaded studs or nuts. Both locking and removable types are produced.

Fig 5 Single-U edge-fixing clip-basic patterns.

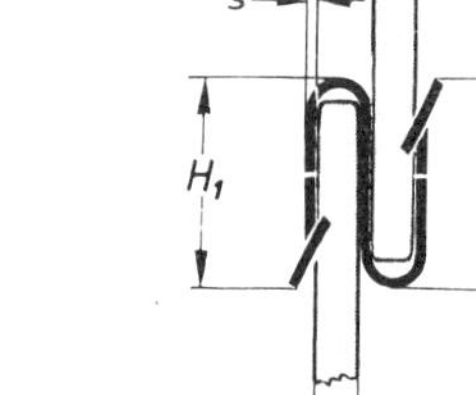

Fig 6 Double-U edge-fixing clips. (A. Raymond).

Fig 7 Examples of edge-fixing clips. (A. Raymond).

Another main group comprises edge-fixing clips, or variations on single-U or double-U spring clips for securing assemblies, trim, etc – see Figs 5, 6 and 7.

Other types of spring fastener fall under the headings of panel clips, knob clips, pipe clips, trim fasteners and spring nails, linkage clips, moulding fasteners, tubular clips, blind fasteners and retainers, spring latches, etc, as well as a widening variety of special purpose clips and fasteners. See also chapter on *Circlips and Retaining Rings*.

Spring nails may be designed as permanent or removable fasteners for joining two panels of relatively thin material. They are blind fasteners which are simply pressed in place through pre-drilled holes, locking in place by the spring action of the legs – Fig 8.

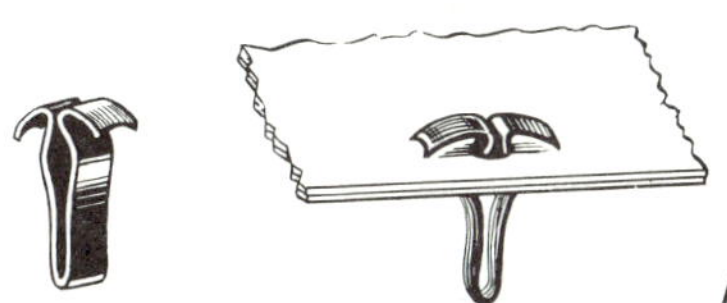

Fig 8 Spring nail (Carr Fastener Co Ltd).

Similar geometry may be used with a button head; or similar locking action in all-plastic drive fasteners, button fasteners and trim pad clips. In the all-plastic type of fastener the shank is oversize relative to the matching hole, flexing on entry to expand after passing through the hole and providing a wedging action to remain secure – see Fig 9 and table. See also chapters on *Blind Fasteners* and *Miscellaneous Clips*

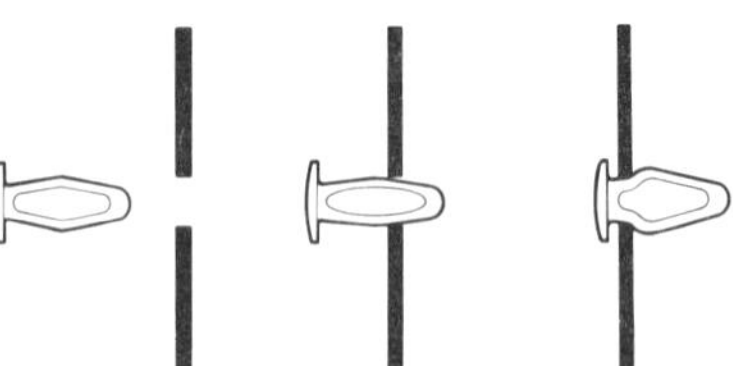

Fig 9 Fastex nylon drive fastener or plastic spring nail.

FASTEX NYLON DRIVE FASTENER DATA

Nominal Diameter		Driving Force		Pull-Out Load		Single Shear Load	
mm	in	kg	lb	kg	lb	kg	lb
3.45	0.136	5.7	12.5	18.0	40	11.0	25
7.14	0.281	5.4	12.0	41.0	90	27.0	60
5.54	0.218	5.7	12.5	32.0	70	9.0	20
2.69	0.106	6.3	14.0	7.7	17	6.8	15
8.15	0.321	9.5	21.0	16.0	35	18.0	40
5.54	0.218	3.2	7.0	20.0	45	22.6	50
6.35	0.250	6.1	13.5	9.0	20	20.0	45
10.00	0.394	3.6	8.0	25.0	55	11.0	25
4.75	0.187	6.8	15.0	8.6	19	13.6	30

Threaded Inserts

Threaded inserts are captive devices incorporating a tapped hole which perform the function of nuts. They are used particularly to provide strong threads in a relatively soft material (eg plastics) or sheet material too thin or weak for tapping, or where additional strength is required (eg in die castings). They are produced in a variety of types and designs to meet almost every possible requirement and method of fitting: expansion fit, press-in, screwed fit, self-tapping, self-clenching, interference fit, moulded-in, rivet-type, sandwich panel inserts, etc. Individual designs may offer thread locking systems as well as free running threads.

The classic example of the threaded insert is the 'Heli-Coil' which consists of a helical coil of diamond section stainless steel wire which can be wound into a tapped hole – Fig 1. The insert is self adjusting and compensates for lead and angle errors between mating members and generally eliminates thread failure due to stripping, vibration corrosion and seizing as well as increasing thread strength. The insert is self anchoring by virtue of the radial pressure generated when the bolt is inserted, although the insert itself is free running as regards fitting. There is also a screw-locking Heli-Coil insert where one or more of the intermediate coils are polygroove in form – Fig 2. When the bolt is screwed in these generate a strong elastic pressure on the thread flanks to provide a locking action. See Table I for Heli-Coil installation data.

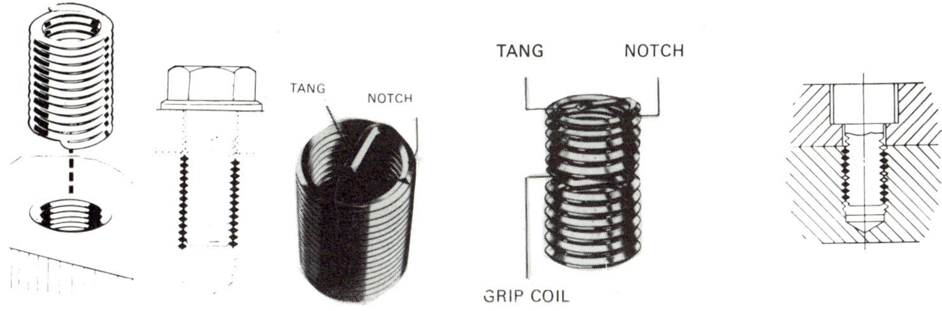

Fig 1 Heli-coil threaded insert.

Fig 2 Heli-coil screw lock insert.

TABLE I — HELI-COIL DATA

Thread Series	Standards	Size (diameter)	Fits
UNF	to BS 1580	Up to 1½in diameter	3B 2B
UNC	to BS 1580	Up to 1½in diameter	3B 2B
BSF	to BS 84	Up to 1½in diameter	Close and medium
BSW	to BS 84	Up to 1½in diameter	Close and medium
BSPF	to BS 2779	Up to 1½in diameter	Close and medium
BA	to BS 93	Up to 0.BA	Normal
Spark Plug:	Aviation and Automotive	10 x 1.00mm 14 x 1.25mm 18 x 1.50mm	
Metric	ISO	Sizes on request	

Lengths: Each size insert is normally stocked in 5 standard lengths, ie 1—1½, 2—2½ and 3 times the nominal diameter

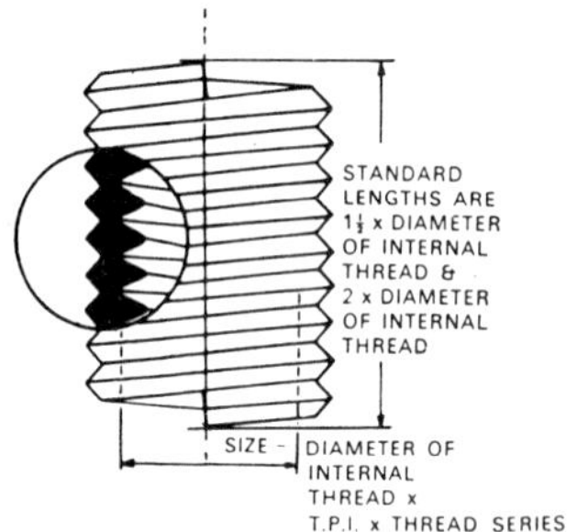

Heli-coil push-type insert for plastics can be pushed directly into blind or through holes in moulded plastics.

Fig 3 Heli-coil

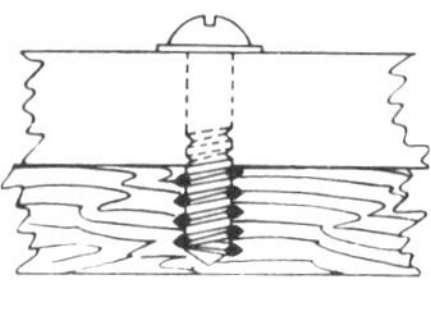

'Woodsert' (self-tapping insert).

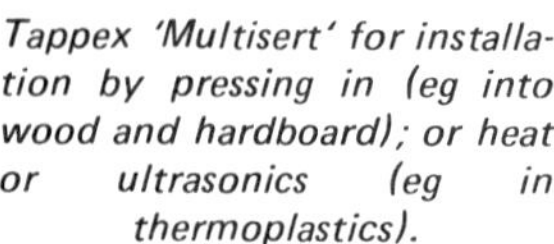

Tappex 'Multisert' for installation by pressing in (eg into wood and hardboard); or heat or ultrasonics (eg in thermoplastics).

Barbed rings aid installation and provide high pullout resistance.

Pilot provides easy location.

Knurl gives high rotational resistance.

The self-tapping form of the 'Heli-Coil' is shown in Fig 3; it is again wound from diamond section wire but this time in carbon steel. It is designed to provide threaded fastenings in wood, particle boards, and other similar fibrous materials. It is entered in a drilled hole of matching diameter and depth and screwed into place with an inserting mandrel – see Fig 4 and Table II. This form of Heli-Coil is also known as a 'Woodsert'. As a general figure, the pull out load of a 'Woodsert' and matching screw is about twice that of a similar size of wood screw used in the same material.

TABLE II – DRILL SIZES FOR HELI-COIL 'WOODSERTS'

Thread Size	Insert Length in	Drill Size in
10–12	3/8 1/2 5/8	3/16
	3/4 7/8 1 1.1/4 1.1/2	1/4
1/4in–10	3/8 1/2 5/8	1/4
	3/4 7/8 1 1.1/4 1.1/2	9/32
5/16in–9	3/8 1/2 5/8	5/16
	3/4 7/8 1 1.1/4 1.1/2	11/32

Fig 4 Inserting mandrel for 'Woodsert'.

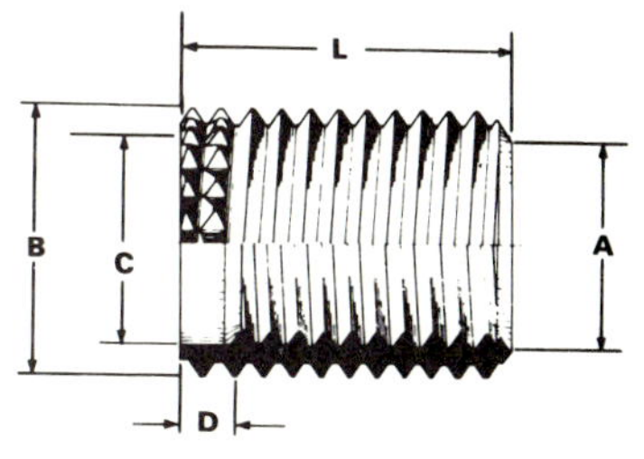

Fig 5 'Slim-sert'. (Instrument Screw Co).

Simple screw-in inserts can be used to provide thread protection in relatively weak materials – eg light alloy castings. They are basically bushings threaded internally and externally, screwing into tapped holes. The external thread form can incorporate some form of self-locking device to provide a self-anchoring feature – eg swage-locking provided by the 'Slim-Sert' illustrated in Fig 5; see also Table III.

TABLE III 'SLIM-SERT' DIMENSIONAL DATA

A	B	C ±0.1	D ±0.2	LENGTH Type A	LENGTH B	HOLE PREPARATION TAP DRILL	HOLE PREPARATION TAP FINE ISO
M2.5	M5 × 8	3.17	1.50	3.00	4.62	4.1	M5 × 8
M3	M5 × 8	3.17	1.50	3.75	5.35	4.1	M5 × 8
M4	M6 × 1	4.21	1.50	4.50	6.50	5.0	M6 × 1
M5	M7 × 1	5.23	2.00	5.75	8.25	6.0	M7 × 1
M6	M8 × 1	6.28	2.00	6.50	9.60	7.1	M8 × 1
M8	M10 × 1	8.32	2.50	8.50	12.50	9.1	M10 × 1
M10	M12 × 1	10.37	3.00	10.50	15.50	11.2	M12 × 1
M12	M14 × 1	12.37	3.00	12.00	18.00	13.2	M14 × 1

Self-tapping inserts for use in plastics, etc are basically metal bushings threaded both internally and externally. The external thread may be either thread cutting or thread forming. The internal thread matches standard screw thread sizes and may be of free-running or self-locking form. Thread-cutting inserts are used on more brittle plastic materials. Thread-forming inserts are suitable for softer and more ductile plastics and laminate materials. Both produce an interference fit with good resistance to loosening under conditions of vibration or shock.

Self-tapping inserts are screwed or driven into drilled or cored holes of matching size (see Figs 6 and 7). Pull out figures and torque strength of such an insert will normally exceed the screw strength in all but weak plastics.

Expansion and interference type inserts are pressed or driven into drilled or moulded holes. In the case of an expansion type insert, the self-anchoring force is generated by the radial expansion of grooved or knurled surfaces to provide the equivalent of an interference fit without subjecting the wall holes to an axial force during assembly. They are expanded in position with special tools. In the case of interference fit inserts, grooved or knurled surfaces again provide the self-anchoring grip against pull out torque, but do interfere with the wall surface during assembly. Only simple

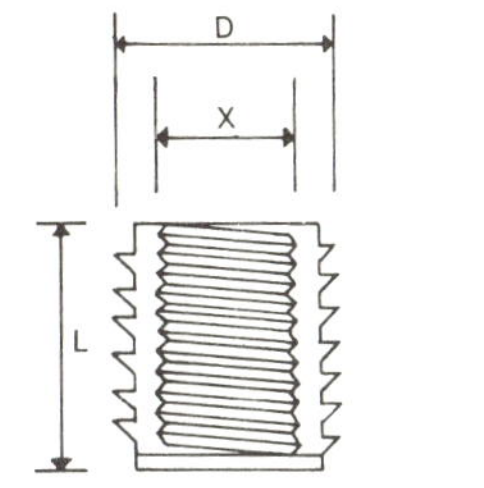

Fig 6 Spirol self-tapping insert.

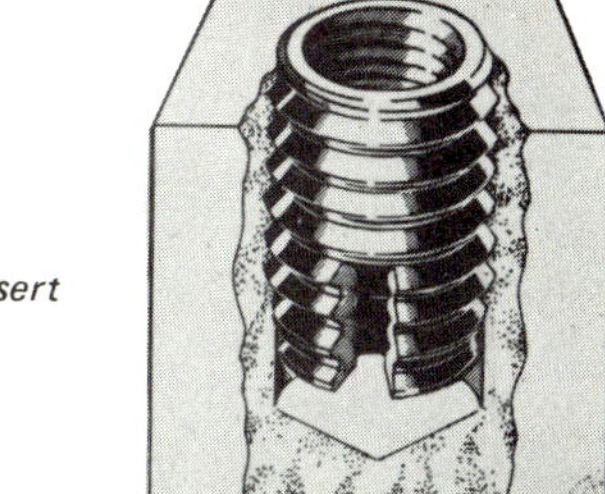

Tappex insert

Fig 7

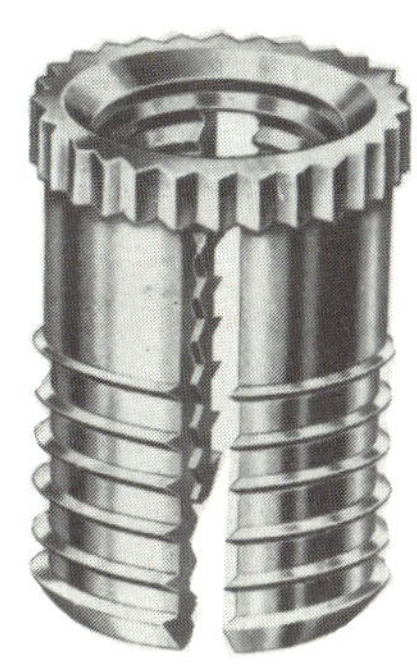

The PSM multivane self-locking insert for thermoplastics

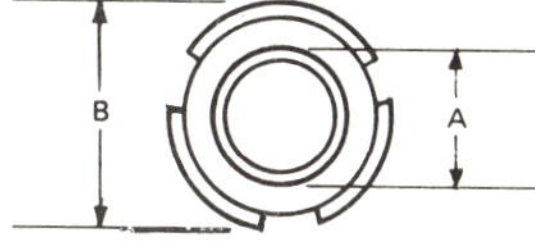

Tappex tri-sert

pressure is needed to fit these inserts. There is a further class of expansion-interference fit insert consisting basically of an externally grooved or knurled sleeve and an internally threaded expander nut which is pulled into the sleeve by a special tool, expanding the grooved surface of the sleeve into the walls of the hole. The locking action of the internal threads is normally provided by an out-of-round section at the bottom of the expander nut.

Both expansion and interference types of insert tend to have low axial load capacity and generate rather higher radial stresses in the surrounding material than self-tapping inserts.

'Armasert' threaded inserts for thermoplastics, inserted by a heater tool after remoulding.

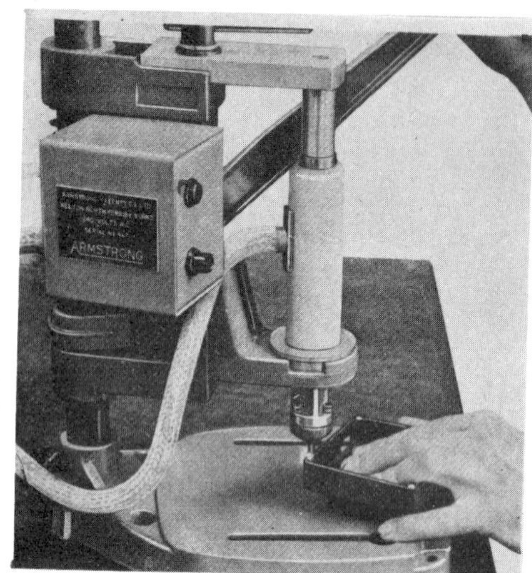

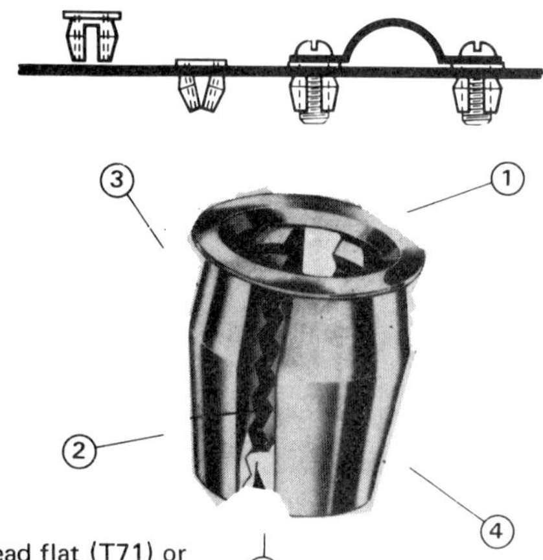

1—head flat (T71) or countersunk (T81)
2—thread
3—upper taper
4—lower taper
5—slot

Banc-Lok self-locking threaded insert.

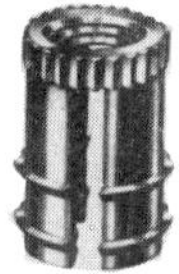

Banc-Lok inserts.

The 'Multivane' type for use in thermoplastics.
The 'Knurled' type for use in the harder thermo-setting plastics.
The 'Headed' type for use in laminated plastic sheet or plastic walls of mouldings.
The 'Fin' type for use in wood and other fibrous materials.

TABLE IV SPIROL SELF-TAPPING INSERT DATA

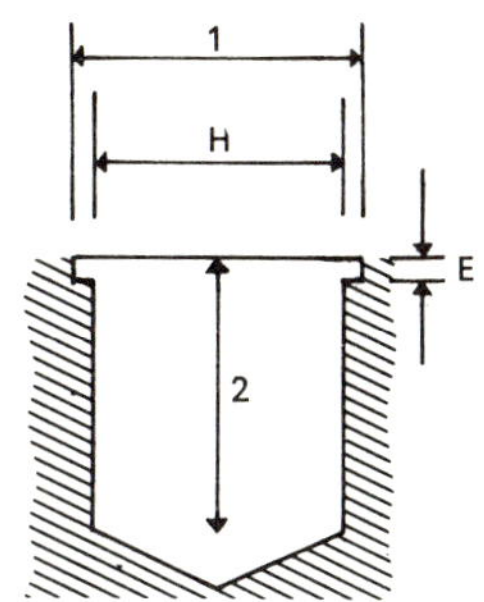

1 — Counter bore diameter = 'D'
2 — Blind hole depth 1.2 x length of insert
L — Insert length
D — External thread
X — Internal thread size
H — Recommended hole dimension

INSERT NUMBER	INTERNAL THREAD SIZE "X"				EXTERNAL THREAD DIA "D"	LENGTH "L" ± .010	RECOMMENDED HOLES ✱	
	BA	BSW	UNIFIED	ISO			"E"	"H"
INS/100	6		4–40		11/64	.187	.040	.152/.156
INS/101	4		6–32	3	7/32	.219	.050	.198/.202
INS/102			8–32	4	1/4	.250	.050	.228/.232
INS/103	2	3/16	10–32 10–24	5	9/32	.281	.055	.257/.262
INS/104		1/4	1/4–20	6	11/32	.344	.060	.316/.321

* These recommended holes may vary under specific conditions. The best hole size for ease of installation and optimum performance depends on the physical characteristics of the material into which the insert is to be assembled and should be established by test or consultation with the manufacturer.

TAB!.E V TAPPEX SLOTTED THREAD INSERT

THREAD SIZES				External Thread Diameter	LENGTHS		SUGGESTED HOLE SIZE		
ISO Metric	BSW BSF	UNC UNF	BA		Regular	Short	Material Class I	Material Class II	Material Class III
M2.5	–	4	8	.172	.234	.156	.157	.156	.152
M3	–	–	6	.187	.234	.156	.177	.177	.173
M3.5	–	6	–	.218	.281	.187	.191	.191	.187
M4	–	8	4	.250	.281	.187	.234	.228	.221
M5	3/16	10	2	.312	.375	.250	.295	.290	.281
M6	¼	¼	0	.375	.484	.312	.354	.348	.343
M8	5/16	5/16	–	.500	.562	.375	.472	.468	.453
–	3/8	3/8	–	.562	.687	.437	.531	.522	.515
M10	7/16	7/16	–	.625	.781	.500	.594	.590	.578
M12	½	½	–	.750	.906	.562	.719	.703	.687
M14/16	5/8	5/8	–	.875	1.125	.687	.844	.828	.812
–	¾	¾	–	1.125	1.375	.812	1.078	1.062	1.045

CLASS I For light alloys of poor machineability cast iron, malleable iron and mild steel
CLASS II For light alloys of average machineability, thermosetting and laminated plastics
CLASS III For light alloys of good machineability, thermo-plastics and thermosetting plastics.

The application of moulded-in inserts in plastic productions is obvious and needs little description. Inserts used may range from standard nuts or threaded bushings to special designs with knurled peripheries to improve torque strength. Disadvantages which may be associated with moulded-in inserts are the possibility of introducing high stresses in the plastic in the region of the insert and (usually) the necessity for a secondary operation to remove flash from internal threads. The increased moulding cost is generally offset by the elimination of the need for matching hole forming and separate insert installation needed with other types of insert.

Rivet-type inserts take the form of internally threaded blind nuts, of which there are a variety of proprietary types. They are inserted in matching holes and locked in place by a pull-up force expanding or clenching the shank. Alternatively, the rivet may incorporate a blind nut which is either drawn up into the shank to expand it when the bolt is screwed up tight, or the nut section is drawn up with a special tool to form a clenched assembly.

Sandwich panel inserts may be through-hole, blind or moulded-in. Load-bearing inserts must also be designed to prevent panel collapse. Inserts may be one-piece or two-piece, with the majority designed for flush fitting on one or both sides of the panel. Through-hole inserts normally provide the greatest resistance to pull out, but moulded-in inserts may offer superior resistance to compression and shear loads, depending on the individual designs.

Dodge Expansion Inserts

Dodge expansion inserts are designed to provide strong brass threads in plastic parts of all types after moulding. They are assembled in a plain moulded or drilled hole in one operation by placing the insert in the hole and pushing a spreader plate down to the bottom of the insert. The spreader expands the knurled section at the bottom of the insert to give a tight grip.

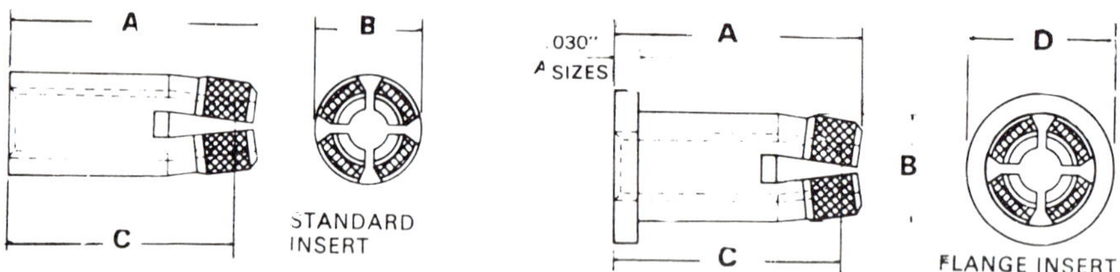

Dodge expansion insert.

DODGE FLANGED INSERTS

Thread Series	Size	A Insert Length	B Body Dia	C Thread Length	D Flange Dia
		in	in	in	in
UNC	4–40	0.219	0.156	0.156	0.219
	6–32	0.281	0.188	0.219	0.250
	8–32	0.281	0.219	0.219	0.281
	10–24	0.313	0.250	0.250	0.313
	¼–20	0.438	0.313	0.375	0.406
UNF	10–32	0.313	0.250	0.250	0.313

DODGE STANDARD INSERTS

Thread Series	Size	A Insert Length	B Body Dia	C Thread Length
		in	in	in
BA	6	0.250	0.156	0.188
	5	0.313	0.188	0.250
	4	0.313	0.188	0.250
	4	0.375	0.188	0.313
	3	0.313	0.219	0.250
	3	0.375	0.219	0.313
	2	0.313	0.250	0.250
	2	0.375	0.250	0.313
	2	0.438	0.250	0.375
	0	0.500	0.313	0.438
UNC	4–40	0.250	0.156	0.188
	5–40	0.313	0.188	0.250
	6–32	0.313	0.188	0.250
	6–32	0.375	0.188	0.313
	8–32	0.313	0.219	0.250
	8–32	0.375	0.219	0.313
	10–24	0.375	0.250	0.313
	10–24	0.438	0.250	0.375
	¼ –20	0.500	0.313	0.438
UNF	10–32	0.313	0.250	0.250
	10–32	0.375	0.250	0.313
	10–32	0.438	0.250	0.375
		mm	mm	mm
ISO METRIC COARSE	3mm x 0.5	6.35	3.91	4.78
	3.5mmx0.6	7.95	4.70	6.38
	4mm x 0.7	9.50	5.51	7.95
		7.95	5.51	6.38
	5mm x 0.8	9.50	6.30	7.95
	6mm x 1	12.70	7.87	11.13

Wallplugs and Through Fixes

Simple wallplugs for accommodating screws may be made of fibre or plastic. Fibre Rawlplugs are rigid tubes of high density fibre impregnated with a waterproof binding agent. The tube is sufficiently elastic to allow the screw to form its own thread in the material and effect expansion of the plug without damage to the fibre structure. It offers a grip as strong as the surrounding masonry, provided the plug, drill and screw are correctly matched in size – see Tables I and II.

The correct method of fixing with fibre Rawlplugs is:-

Turn the screw into the Rawlplug for one or two threads. Using the screw, push the Rawlplug right into the hole until it is just below the surface. (If making fixings in glazed tile, the Rawlplug must go deeper).

TABLE I – RAWLPLUG DATA

Reference* No.	Drill diameter mm	in	Lengths available ins mm equivalent in bracket	Coach screw sizes in (mm)
3	3.0	1/8		–
6	4.0	5/32	1/2 (12.5), 3/4 (20),	–
8	5.0	3/16	1 (25), 1¼ (32),	–
10	5.5	7/32	1½ (38) and 2 (51)	–
12	6.0	1/4		–
14	7.0	9/32		–
16	8.0	5/16	1½ (38), 2 (51),	–
18	8.5	11/32	and 2½ (63.5)	–
20	9.5	3/8		–
22	11.0	7/16	1½ (38), 2 (51),	1/4 (6)
24	12.5	1/2	2½ (63.5), 3 (76)	1/4 (6)
26	16.0	5/8	3½ (89) and 4 (102)	5/16 (8)

**Matching wood screw size is same as reference number.*

TABLE II - LOADS SUPPORTED BY FIBRE RAWLPLUGS*

Size of Rawlplug x Length in	mm	Ultimate Load lb	kg	Direct Pull — Recommended Loads: Safe Steady Load lb	kg	Safe Shock Load lb	kg
No. 8 × $\frac{3}{4}$	20	350	159	70	32	35	16
,, × 1	25	600	272	120	54	60	27
,, × 1$\frac{1}{2}$	38	750	340	150	68	75	34
,, × 2	51	900	408	180	82	90	41
No. 10 × $\frac{3}{4}$	20	400	181	80	36	40	18
,, × 1	25	650	295	130	59	65	29
,, × 1$\frac{1}{2}$	38	800	363	160	73	80	36
,, × 2	51	1200	544	240	109	120	54
No. 12 × 1	25	550	249	110	50	55	25
,, × 1$\frac{1}{2}$	38	1000	453	200	91	100	45
,, × 2	51	1300	589	260	118	130	59
No. 14 × 1	25	750	340	150	68	75	34
,, × 1$\frac{1}{2}$	38	1120	508	224	102	112	51
,, × 2	51	1500	680	300	136	150	68
No. 16 × 1$\frac{1}{2}$	38	1180	535	236	107	118	54
,, × 2	51	1650	748	330	150	165	75
No. 18 × 2	51	2000	907	400	181	200	91
,, × 2$\frac{1}{2}$	64	3000	1360	600	272	300	136
No. 20 × 2	51	2500	1134	500	227	250	113
,, × 2$\frac{1}{2}$	64	3200	1451	640	290	320	145

Recommended loads given above are for Fibre Rawlplugs with the correct size screws fitted into them.

Plastic wallplugs are of split, tapered tube form — Figs 1 and 2. The split enables the plug to expand whilst the taper automatically centres the screw. Ribs along the length of the plug prevent rotation in the hole. The most common plastic material used is polypropylene. Plastic plugs are normally moulded with a small lip to help prevent the plug sinking below the surface in over-deep holes although it is small enough and pliable enough to enable the plug to be sunk below the surface if required — eg in materials where the expansive force might cause surface cracking.

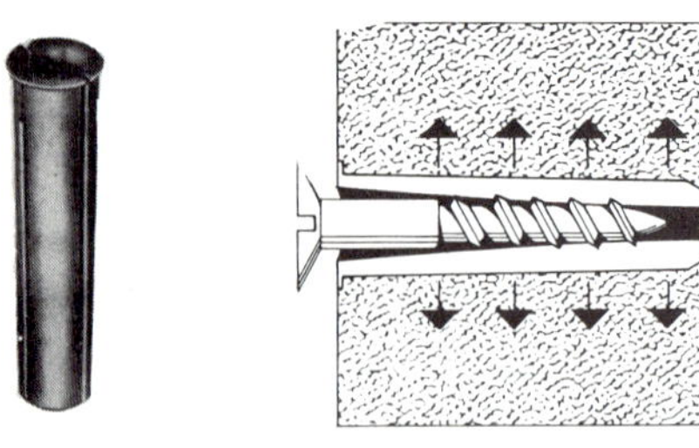

Fig 1 Plastic Rawlplug.

Fig 2 Tower plastic tri-plug.

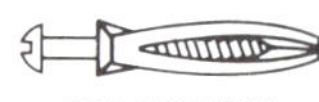

TABLE III – PLASTIC RAWLPLUG DATA

Plug Colour	Length		Matching Screw Sizes	Drill Size	
	mm	in		No	mm
Green	20	3/4	4, 6, 8	8	5.0
Pink	25	1	6, 8, 10	10	5.5
Orange	25	1	6, 8, 10	12	6.5
Grey	35	1.3/8	6, 8, 10	10	5.5
White	35	1.3/8	8, 10, 12	12	6.5
Blue	35	1.3/8	10, 12, 14	16	8.0
Yellow	50	2	16, 18, 20	22	10.5

TABLE IV – TOWER TRI-PLUGS

Plug Colour	Length		Matching Screw Sizes	Drill Size	
	mm	in		No	mm
Yellow	25	1	4, 6 8, 10	8 10	5.0 5.5
Red	35	1.3/8	6, 8, 10	10 14	5.5 7.0
Brown	40	1.9/16	10, 12 14	16	8.0
Blue	45	1.3/4	14 16	18 20	9.0 9.5

**Note: '14' screw gauge size is suitable for use with 1/4in coachscrew*
'16' screw gauge size is suitable for use with 5/16in coachscrew

The greater elasticity of plastic plugs enables a reduced range of sizes to accommodate the normal range of woodscrews (and coach-screws) with which they are normally used. Plastic plugs are normally self-coloured, the colour identifying the plug size and matching screw sizes. Manufacturers do not, however, adopt a common colour code – see Tables III and IV.

Combination Plugs

The term 'combination plug' describes a plug type fastening combined with a special nail. After the plug is inserted into a predrilled hole the fastener is secured by hammering the head of the nail. This expands the plug to secure the fastening.

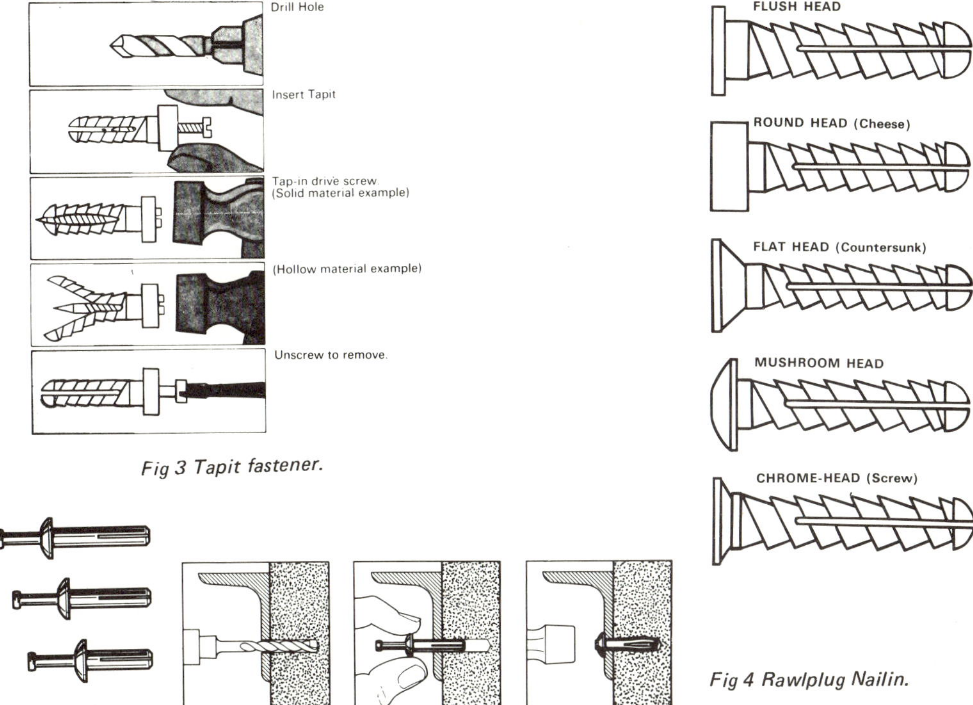

Fig 3 Tapit fastener.

Fig 4 Rawlplug Nailin.

Two proprietary examples are illustrated in Figs 3 and 4. The 'Tapit' consists of a nylon shell with helical serrations and a drive screw. After the fastener is inserted into a hole the screw is tapped home to swage and compress the nylon shell in solid material, or flare the legs in hollow applications. In each case the fastener is removable by unscrewing.

The 'Tapit' has the profile of a No 8–10 woodscrew in the 5mm (3/16in) diameter size; and No 12–14 woodscrew in the 6mm (¼in) diameter size. It is produced in a variety of head forms. Average holding power in concrete is 90.7kg (200lb) tensile and 363 kg (800lb) shear.

The Rawlplug 'Nailin' is an all-metal fastener consisting of a flanged expander sleeve in corrosion resistant zinc alloy containing a nail type expander. It is fixed by driving the nail head home after the sleeve has been inserted in its hole. Average holding power in concrete is 225–520kg (500–1 150lb), depending on diameter size.

Toggles

Spring and gravity toggles (Fig 5) provide all-metal through fastening systems for screws and bolts as an alternative to blind anchors, etc. The choice of which type to use depends primarily on the surface material involved – see Table V. In general, gravity toggles are used in cavity walls faced with building boards. Spring toggles can be used in cavities of less depth than that required for gravity toggles and are particularly useful for lath and plaster construction. In both cases, holes in all types of cavity material should be drilled and not jumped.

See also chapters on *Screw Anchors, Bolt Anchors,* and *Building Construction Fasteners.*

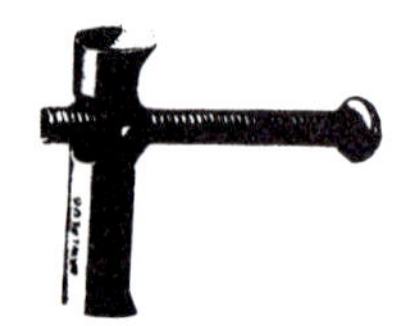

Fig 5 Gravity (left) and spring toggles (right)

TABLE V – SUITABILITY OF TOGGLES

Wall Material	Spring Toggle	Gravity Toggle
Hollow clay pots	Yes*	Yes, walls only
Plaster board	Yes	Yes, walls only
Lath and plaster	Yes	No
Sheet metal partitions	No	No
Insulation board	Yes	Yes, walls only
Hardboard	Yes	Yes, walls only
Hollow cement blocks	Yes*	No
Acoustic tiles	Yes	Yes, walls only
Glass	No	No
Cement asbestos sheet	Yes	Yes, walls only
Plywood	Yes	Yes, walls only
Holoplast	No	No

**If cavity has enough depth*

TYPICAL PULL-OUT RESISTANCE OF WALL PLUGS IN CONCRETE (kgf)

Type of Plug	Type of Concrete			
	Dense Aggregate	Slag Aggregate	Ash Aggregate	Expanded Clay Aggregate
Fibre	120	110	90	90
White metal	100	90	50	50–55
Plastic	90–110	110	80–90	80–90
Rubber	130–140	150–155	90	70–75

EXAMPLES OF WALLPLUG TYPES

Type of Plug	Remarks	Proprietary Names	Manufacturer
Asbestos – fibre compound	–	Rawlplastic	Rawlplug Co Ltd
Fibre	–	Rawlplug	Rawlplug Co Ltd
	–	Kuli	Ucan Products Ltd
Plastic	–	Rawlplug	Rawlplug Co Ltd
	–	Tri-Plugs	Tower Manufacturing
	–	Wallplug Type S	Fischer Ltd
	–	Thunderplug	Thunder Screw Anchors Ltd
	–	Screw Anchor	US Expansion Bolt Co Ltd
	for nails	Tapit Plug	US Expansion Bolt Co Ltd
Rubber	expansion plug	Rawlnut	Rawlplug Co Ltd
Expansion plugs	nylon	Wallplug Type P	Fischer Ltd
	–	Wallplug Type 512D	Fischer Ltd
	–	WallplugType M–5	Fischer Ltd
	–	Rawlset	Rawlplug Co Ltd
	brass	Expansion plug	Ucan Products Ltd
Plastic	slotted form	Tor-Plug	Tornado Fixings Ltd
	PVC	Philplug	Whitehouse Industries Ltd
	polythene	Ucan plastic anchor	Ucan Products Ltd
Expansible (umbrella)	–	Rawlanchor	Rawlplug Co Ltd
Cavity fixing	–	Fast Brolly	DOM Products Ltd
Toggles	–	Spring Toggle	Rawlplug Co Ltd
	–	Gravity Toggle	Rawlplug Co Ltd
	nylon cord-retention toggle	Toggle Type K 'H' Type (gravity)	Riston Ltd Rawlplug Co Ltd
	plastic	Toggle plug	Thunder Screw Anchors Ltd

Metal Plugs

Metal plugs are available for use in conditions which are unfavourable or not suitable for fibrous, plastic or rubber types. They must be used with compatible screw materials.

Whitemetal (zinc alloy) plugs are used for very 'wet' conditions, or where high temperatures are involved. Pullout resistance is about 60% of a fibre plug. Whitemetal plugs should only be used with cadmium plated or stainless steel screws.

Aluminium plugs may be used under corrosive or 'wet' conditions. They can be used with cadmium plated, stainless steel or aluminium screws. Proprietary plugs in this metal are usually pressings. Pullout resistance is about 30–50% of a fibre plug.

Lead plugs (usually antimonial lead or lead alloy) are normally only used where the conditions are excessively acid. Pullout resistance is about 50% of a fibre plug. Lead plugs should only be used with cadmium plated or stainless steel screws. These are available for screw sizes from 6 to 20 gauge in white bronze, but generally in a more limited range of sizes in aluminium (12 and 20 gauge) and lead (8–14 gauge).

Rubber plugs

Rubber wallplugs were originally intended for use in cavity walls, but have subsequently proved suitable for use in solid masonry, etc, particularly where the waterproof and vibration resistant properties of rubber can offer advantages. They are designed for use with machine screws, the rubber plug incorporating a nut which, when the screw is tightened, expands the plug section.

Mouldable Plugs

Filler materials based on a fibrous reinforcement in a hard-setting filler may be used in place of wallplugs where it is impossible to drill matching size holes for plugs, or the drilled hole is too distorted or enlarged to accommodate a plug. The filler material is usually dry and mixed with a little water before use to work into a puttylike consistency. It is then packed tightly in the hole and a pilot hole for the screw pierced in it before the filler hardens. The fixing screw should not be driven in until the filler has hardened.

Blind Fasteners

By definition a *blind fastener* is one which can be assembled and closed from one side without the necessity of having access to the other. It does not follow that they are used only in such circumstances. They may be a preferred choice – eg because of simpler or speedier assembly – where access to both sides of an assembly is readily available.

As a specific example of the latter, the following comparative costs and performance figures between using a screw, nut and washer and a self-tapping screw are based on data originated by GKN.

Fastener	**Installation rate per 8 hour day**
Screw, nut and washer	Manual – 180 per hour
	Power – 270 per hour
Self tapping screw	Manual – 270 per hour
	Power – 450 per hour

Cost per thousand holes:-

	Screw, nut, washer				Self tapping screw			
	Hand	%	Power	%	Hand	%	Power	%
Parts	1.60	23.0	1.60	26.5	1.20	23.5	1.20	28.4
Prepared hole	1.00	14.4	1.00	16.6	1.00	19.6	1.00	23.6
Applicator	0.04	0.5	0.36	6.0	0.04	0.8	0.12	2.8
Bit	0.00	0.0	0.15	2.5	0.00	0.0	0.15	3.5
Handling	0.06	0.9	0.06	1.0	0.04	0.8	0.04	0.9
Storage	0.04	0.5	0.04	0.7	0.03	0.6	0.03	0.7
Air	0.00	0.0	0.01	0.2	0.00	0.0	0.01	0.2
Total;	2.74	39.3	3.22	53.5	2.31	45.3	2.55	60.1
Total in-place cost	6.94	100	6.02	100	5.11	100	4.23	100

A significant fact to emerge from this analysis is that on a mass-production basis, operator costs can range between 30% and 70% of the total in-place cost, and hole preparation and fastener parts between 30% and 60% of the total in-place cost.

Much also depends on the strength required of the fastening. Whilst a screw and nut, a rivet, or a self-tapping screw may offer similar mechanical performance for joining sheet metals, etc, a spring clip which can be snapped home by finger pressure is much quicker and cheaper than any of these if only low mechanical strength is required (eg securing a trim panel). About the only limitation in this particular field is the ingenuity of designers to produce simple-to-fit fasteners capable of providing the necessary mechanical performance to meet general and specific applications. Many blind fasteners in this category, in fact, are initially designed with specific applications in mind, and subsequently adopted for other work, or further modified or developed for general application.

It is impossible to describe and treat *blind fasteners* under one general heading. The types and varieties, and even the materials used, are too numerous, and are more realistically included under separate headings. The main types of blind fasteners can be categorised as follows:-

Self Tapping Screws — see under this chapter heading
Blind Rivets — see under this chapter heading
Drive Rivets — see chapter on *Blind Rivets.*
Spring Nuts — see under this chapter heading
Spring Clips — see chapter on *Spring Fasteners* also *Miscellaneous Clips*
Push Nuts — see chapter on *Threaded Inserts and Captive Nuts*
Friction Bushes — see chapter on *Grommets and Bushes*
Threaded Washers — see under this chapter heading
Through Fixes — see chapter on *Wallplugs and Through Fixes*
Plastic Fasteners designed for blind assembly — see chapter on *Plastic Fasteners.*
Miscellaneous Types described under specific descriptions as: Anchor Clips, Push Nuts, Push Fixes, Spring Nails, etc.
Latch-Type Fasteners — see chapter on *Quick-Operating Fasteners.*

There are also numerous other types of fastener which could be described marginally as blind fasteners, eg: Staples — used in unclenched form, Captive Nuts, Screw Anchors, Bolt Anchors,Nails,Building Construction Fasteners; also Bifurcated Rivets in some applications.

There are also numerous stud, spring or clip type fasteners designed for fastening specific combinations of materials. These are described in the chapters most appropriate to their geometric/mechanical properties — eg spring-type fasteners under *Spring Fasteners,* clip-type fasteners under *Pipe Clips, Hose Clips* and *Miscellaneous Clips.*

Finally, engineering adhesives must also be considered as possible alternatives for performing blind-fastening operations — see *Adhesives Section.*

Avdel Monobolt System

The Avdel Monobolt is a new all-steel blind fastener designed to provide a vibration-proof fastener with a wide grip range (2.03 to 8.13mm) at low installed-cost. Recommended hole size is 0.260in–0.275in (6.6mm–7.0mm). The fastener is installed with a special pneumatically operated placing tool, each placing taking six seconds. The fastener itself is of steel, zinc plated and supplied lubricated (it must not be degreased). Minimum clearance required to behind sheet to any obstruction is 0.475in (12.07mm).

Avdel Monobolt placing sequence.

1 2 3 4a

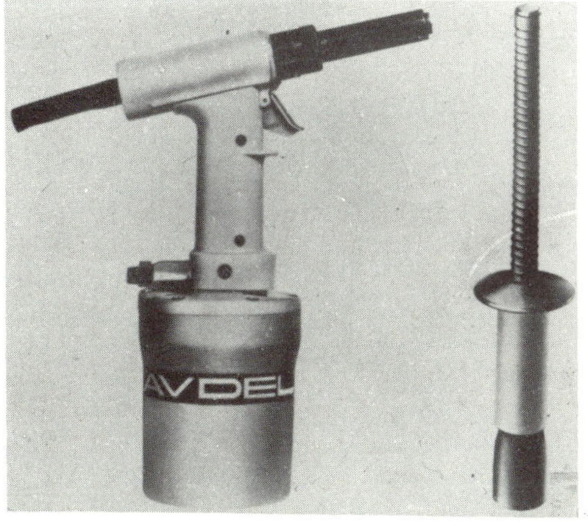

The Avdel Monobolt bolt type fastener and its complementary high speed placing tool.

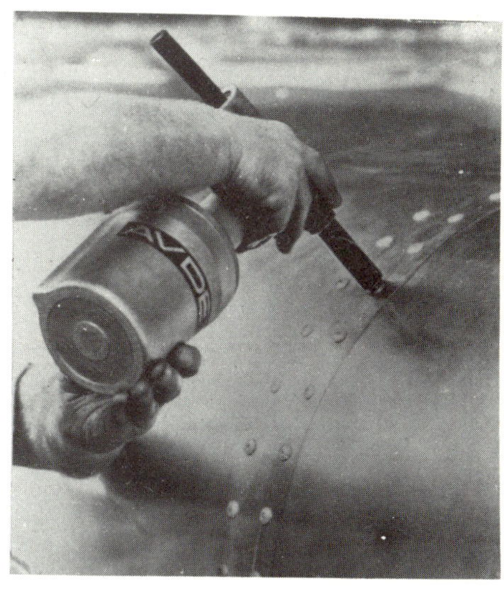

The Avdel Monobolt system securing prefabricated assemblies to the roof structure on a Leyland double-decker bus being manufactured at Park Royal Vehicles Ltd.

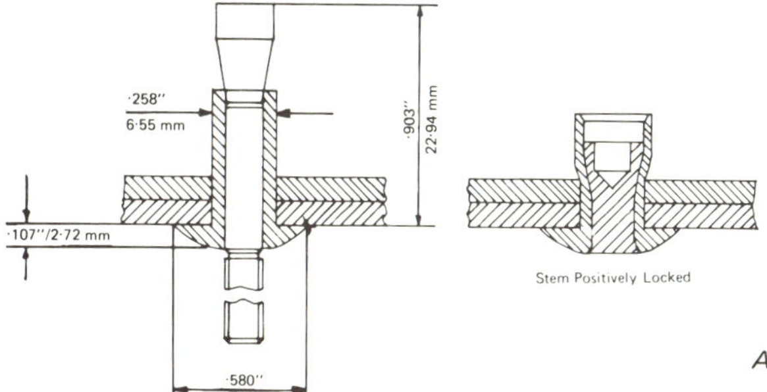

BEFORE PLACING AFTER PLACING

Avdel Monobolt

Quick-Operating Fasteners

Quick-operating fasteners are used to facilitate removal or opening of panels, doors, etc which are normally positively secured but require intermittent disassembly.

Quick-release pins or drawbolts are the basic types, although modern quick-operating fasteners are invariably of highly developed specialised designs. It is difficult to classify them except by the action involved in the fastening/unfastening operation, eg —

(i) partial turn
(ii) rotary locking
(iii) slide action
(iv) lever action
(v) latch action
(vi) cam action
(vii) press-pull.

Positive fastening can readily be provided by turn-operated stud fasteners, the principle involved being that the stud is engaged in a matching retainer and then turned to complete engagement, compression and hold-down. Fasteners of this type may have a screw thread, projecting lug or pawl on the end of the shaft to produce the draw-down action. The retaining device in the case of a lug or pawl end usually incorporates a detent or stop to lock the fastener closed after a certain amount of rotation (commonly one-quarter of a turn). Screw-type turn-operated fasteners may have a fast-lead thread or multiple thread and unrestricted rotation. For example, one-quarter or one-half turn may close the fastener, but further turning may be possible to increase the compression.

Fasteners of this type may be designed with 'grip' ends for manual operation, or with slotted ends for screwdriver (or tool) operation. Most are easy to install, have good tolerance as regards alignment, high strength and high resistance to impact and vibration. They are capable of providing higher closure pressure (compression) than most other types of quick-acting fastener, except certain lever-action types.

The original DZUS quarter-turn fastener was invented in 1932 and was the first engineering-type quick-release fastener, initially used almost exclusively on aircraft for fastening panels over areas requiring regular servicing. It remains a standard and widely-used design, considerably developed during recent years to reduce the in-place cost of the fastener assembly. It consists of a rotatable stud with spiral cam which engages on to a spring or a receptacle mounted permanently on the equipment frame or support – Fig 1. The spring or receptacle may be supplied in a form suitable for riveting, screwing, bolting, welding or simple fitting by hand. The stud is available in five diameter sizes and a number of different lengths. Head styles are available for hand operation (knob, ring or wing), screwdriver or special key.

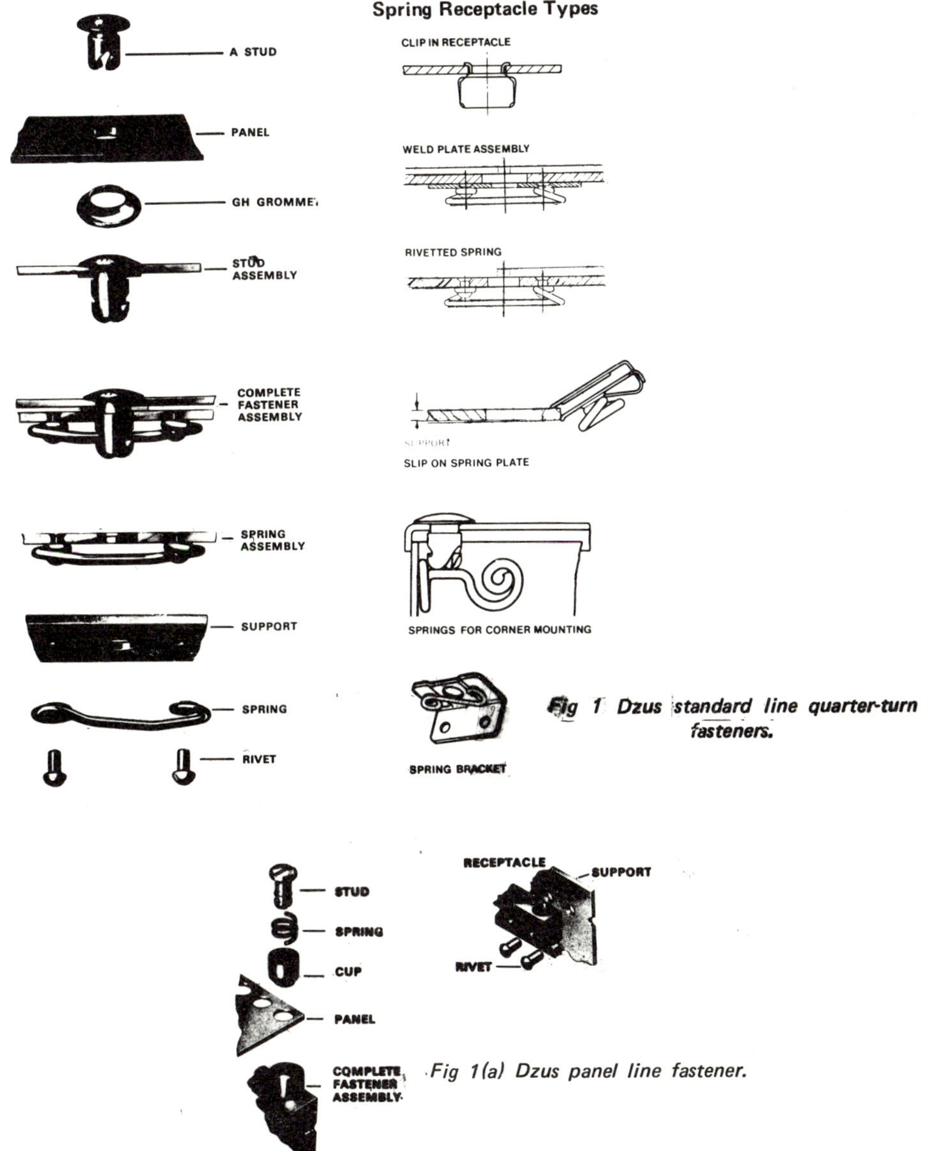

Fig 1 Dzus standard line quarter-turn fasteners.

Fig 1(a) Dzus panel line fastener.

TABLE I – DZUS STANDARD LINE FASTENERS

Application	Types									
Thin panels	Short undercut	A	AW	BR	F	SW	ALW	ASH	MA	K
General usage	Long undercut	AJ	AJW	BJR	FJ	SWJ	AJLW	ASHJ	MAJ	KJ

Note: Letters designate head style.

	Size 3	Size 4	Size 5	Size 6	Size 7
	Ø 5"/16 (8); L; Ø 3"/16 (5.96)	Ø 7"/16 (11); L; Ø 1"/4 (6.35)	Ø 9"/16 (14); L; Ø 5"/16 (7.9)	Ø 5"/8 (16); L; Ø 3"/8 (9.45)	Ø 3"/4 (19); L; Ø 7"/16 (11.05)
Normal holding tension locked	9kg/20lb	12kg/30lb	16kg/45lb	25kg/55lb	30kg/65lb
Max safe load without distortion	20kg/45lb	28kg/60lb	40kg/85lb	50kg/110lb	60kg/125lb
Breaking tension of stud	150kg/350lb	300kg/700lb	450kg/1000lb	560kg/1250lb	750kg/1800lb
Rivet diameter (Metric rivet)	3/32in (2.5mm)	3/32in (2.5mm)	1/8in (3.0mm)	1/8in (3.0mm)	1/8in (3.0mm)
Minimum overlap (2½ times body dia of fastener)	12mm/ 15/32in	16mm/ 5/8in	20mm/ 25/32in	24mm/ 15/16in	28mm/ 1.3/32in
Rivet spacing	16mm/ 5/8in	19mm/ 3/4in	25mm/ 1in	35mm/ 1.3/8in	36mm/ 1.7/16in

The CAMLOC quarter-turn fastener operates on a similar principle the receptacle being designed to provide a spring-loaded over-centre lock on the stud. It normally comprises three components, a stud, retaining ring and receptacle – Fig 2. Different types of receptacle design are available to suit different applications (see Table II). Studs have various head styles.

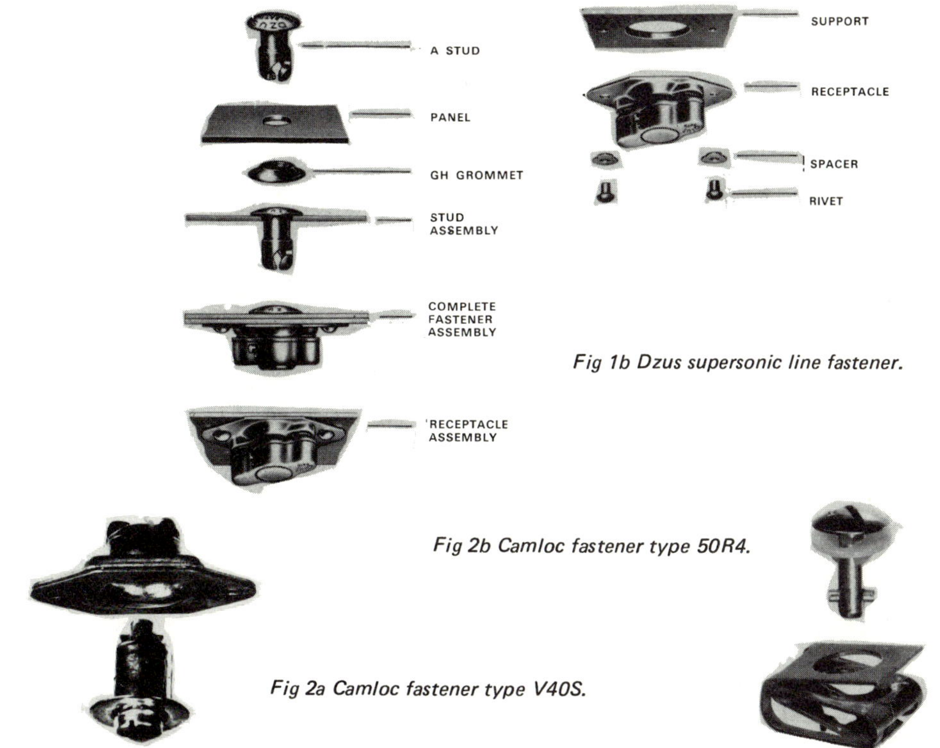

Fig 1b Dzus supersonic line fastener.

Fig 2b Camloc fastener type 50R4.

Fig 2a Camloc fastener type V40S.

TABLE II — CAMLOC QUARTER-TURN FASTENERS

Series	Applications	Ultimate Tensile Strength kg	Available Grip Range mm	Grip Range per Stud Length mm	Stud Headstyles	Stud Mounting	Receptacle Mounting
5F	Light-duty,low profile fastener for use where space is limited. Variety of head styles. Widely used in electronics industry.	68	0.51–10.02	0.37	slot,Phillips, Phillips (flush) knurled, wing.	drill sheet,insert, stud and fit retaining ring. c/s for flush style.	drill sheet, c/s and rivet
12F	Light-duty fastener for use where limited depth behind panel is available for receptacle. Widely used in electronics industry.	68	0.89–19.92	0.37	slot	drill sheet, insert stud and fit retaining ring	drill sheet, c/s and rivet.

cont...

TABLE II – CAMLOC QUARTER-TURN FASTENERS (contd.)

Series	Applications	Ultimate Tensile Strength kg	Available Grip Range mm	Grip Range per Stud Length mm	Stud Headstyles	Stud Mounting	Receptacle Mounting
2600/ 2700F	Med duty,low cost fastener. This is the most commonly used Camloc series and is available in a wide range of styles and finishes	135	0.50–48.0	0.76	slot,Phillips, slot (flush), Phillips(flush) wing knurled, bail handle, plastic knob	drill sheet, insert stud and fit re-taining ring	drill sheet, c/s and rivet
34F	Ultra heavy-duty fastener with adjustment screw at stud base for varying grip after installation	3650	6.35–25.35	6.35	hexagonal	drill sheet, insert stud and fit cross-pin with special Camloc tool	drill, c/s and rivet or drill and weld
4002F	Designed for use with GRP mater-ials.Wide self-compensating grip to take up moulding toler-ances.Broad rim-med stud bushing and receptacle base to spread load	300	5.9–16.7	3.0	slot	drill sheet, insert stud and secure by retaining ring	drill sheet, c/s and rivet
50F	Heavy-duty fast-ener with low-profile receptacle for a wide range of low cost appli-cations.Clip-in re-ceptacle available as alternative to weld/rivet styles	360	0.50–12.30	0.57	slot, wing	drill sheet, insert stud and secure by retaining ring	drill sheet, c/s and rivet or drill sheet and weld
91F	Heavy-duty fast-ener ideal for transport equip-ment, heavy in-dustrial plant and rolling-stock	1000	0.89–38.97	0.75	slot, wing, plastic knob	drill sheet, insert stud and secure by retaining ring	drill sheet, c/s or dimple and rivet. Or drill sheet and weld
99F	Light fastener needing neither rivets nor weld-ing.Wide grip range for each stud length	70	0.50–10.50	1.50	slot	drill sheet, insert stud and secure by retaining ring	drill sheet, c/s or dimple and secure receptacle by retaining ring

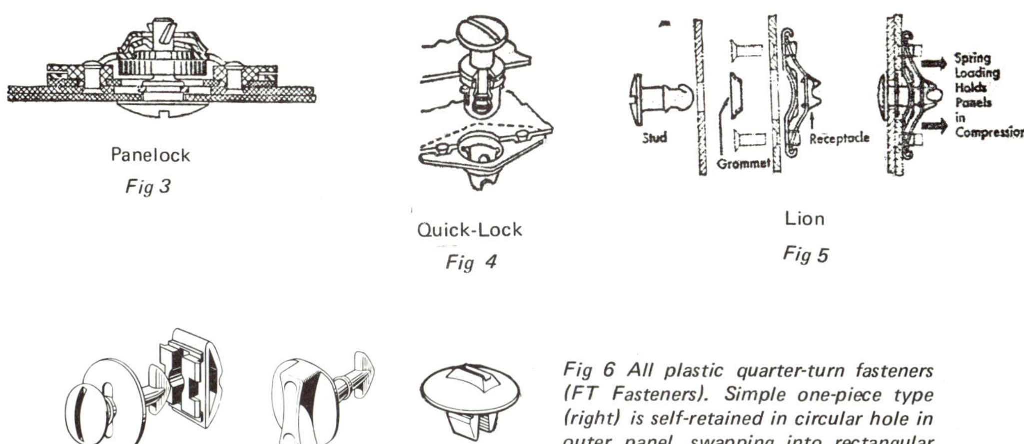

Panelock

Fig 3

Quick-Lock

Fig 4

Lion

Fig 5

Fig 6 All plastic quarter-turn fasteners (FT Fasteners). Simple one-piece type (right) is self-retained in circular hole in outer panel, swapping into rectangular hole in inner panel.

Other examples of proprietary partial-turn quick-operating fasteners are shown in Figs 3, 4, 5 and 6. The last is a non-metallic type, the stud and cam being moulded in acetal with the washer in nylon. Maximum service temperature for this particular type of fastener is 100°C.

Pawl latches are another type of partial-turn fastener, normally used on hinged panels or doors. A pawl latch consists, basically, of a shaft with a knob or lever on one end and a pawl on the other, together with a suitable retaining device for the assembly – Fig 7. The pawl catches against the frame or a separate keeper when the shaft is turned. The pawl may or may not be spring loaded. Other types may be key operated, employing a cylindrical mechanism with tumblers, a pawl latch and strike.

Fig 7 Pawl latch (Dzus).

Slide action fasteners may be based on a sliding pawl or bolt engaging a keeper, or an open-ended slide having a snap-action engagement on a stud. Alternatively the bolt movement may be spring-loaded. None of the basic types is capable of providing good compression, except for the sliding latch with tightening knob. All three types are designed to accommodate shear loads only, but all should have good resistance to vibration.

Simple lever-action fasteners are particularly suitable for edge-to-edge applications, with high leverage available for compression. Mechanically, however, they have moderate to low tensile strength and poor shear strength, as well as being limited in resistance to vibration and shock. The last two properties can be improved by incorporating springs in the lever. Speed of operation is a favourable factor, but tolerance to misalignment is relatively poor.

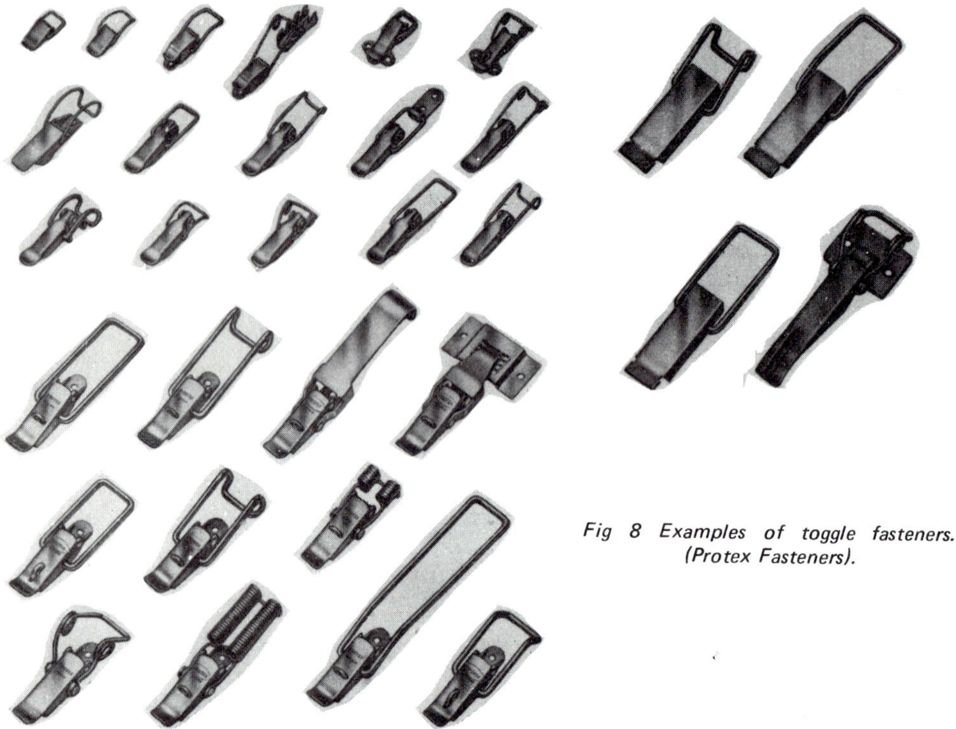

Fig 8 Examples of toggle fasteners. (Protex Fasteners).

More sophisticated designs of lever action fastener are generally called *toggles* or *toggle catches* and may be designed for light, medium or heavy duties, in fixed, sprung and adjustable types. They are associated with a matching shape of catchplate. Some examples of design geometry are shown in Fig 8.

Cam-action fasteners are based on a lever and cam arrangement, lever and cam being either separate or incorporated in a single lever component. The principle of operation is that the cam is notched or otherwise shaped to engage, pull in and lock on a pin mounted in a separate keeper.

Londesborough toggle latches in A100 maranyl nylon (also made in poly-carbonate).

Cam-action fasteners, in general, are quick and easy to operate, provide positive locking with high compression and excellent resistance to vibration. Tensile strength and fatigue strength are both high, but shear strength is only fair. Other disadvantages are that they need accurate alignment and are generally relatively bulky and heavy compared with other types of quick-acting fastener.

Simple snap-action fasteners are generally described as *latches* comprising a latching member which is a snap fit into a stud or matching hole – Fig 9. There is a wide variety of individual designs, all characterised by self-latching operation. Some types may require unlatching with a key, pushbutton, or lever.

Fig 9 Simple snap-action catch (Fastex)

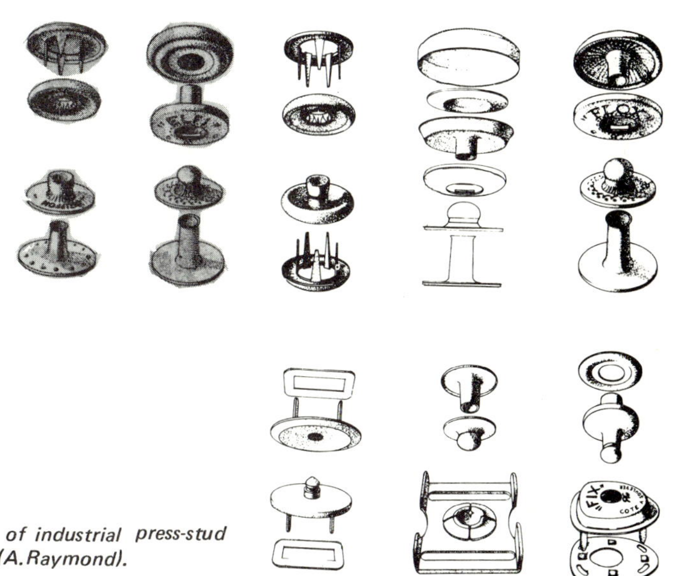

Fig 10 Examples of industrial press-stud fasteners (A.Raymond).

HERE'S THE DIFFERENCE
Speedy INSTALLATION
Rattle-proof ASSEMBLY
Fastener—*fast'*-en-er. Thanks to Quick Release Vibrex® the accent's on *fast! Snaps to* in *one* second! An exclusive rubber compound soaks shock, negates noise. Seats from one side. The proven answer to hundreds of design and production problems, it can solve yours too. Write *today* for comprehensive brochure.
UNLOCKED
LOCKED
QUICK-RELEASE
VIBREX
FASTENERS
Metric Screws
(Fastenings) Ltd
Advanced Fastening Techniques
Pease Pottage, Crawley, Sussex.
Phone: 0293-25811 (4lines) Telex: 87206. Sole U.K. Concessionaires

THE FIRST!
Over 600 pages containing thousands of diagrams, tables, charts, illustrations, etc., stiff board bound and gold blocked.
HANDBOOK OF POWER DRIVES
HANDBOOK OF POWER DRIVES
TRADE AND TECHNICAL PRESS LTD.
TRADE & TECHNICAL PRESS LTD.
CROWN HOUSE, MORDEN, SURREY. SM4 5EW

The basic press-stud is another type of snap-action fastener, developed in larger and more robust form for industrial fasteners. Locking action is provided by a simple wire spring in the female component engaging under the headed stud when the fastener is closed. The chief application of this type of fastener is for securing a flexible sheet or canopy to a rigid structure. It is not suitable for fastening two rigid components because it is difficult to release in the absence of flexibility to spring the fastener apart. Equally, it is lacking in compression and rigid hold-down capability, and generally lacks both tensile and shear strength.

Industrial stud-type or push-pull fasteners are therefore normally designed with more positive location and locking. In one basic configuration the stud has a grooved end which, when pushed into the mating member, expands spring fingers which clip into the groove to lock the stud in position. The actual geometry is so designed that further pressure on the stud engages a collar behind the groove to expand the spring fingers, allowing the stud to retract and be removed once pressure is released again. This type of fastener is known as a finger latch.

More positive locking with a tighter hold-down is provided by the expandable latch. The complete fastener, comprising a headed stud in a housing fitted with spring prongs at the lower end is simply inserted in a hole in the mating panel to be retained. Pressing the stud downwards expands the prongs on the other side of the hole drawing the two panels together under compression. Pulling the stud outwards allows the prongs to close so that the two panels can be separated, the stud being retained in its housing so that it cannot be withdrawn completely. This is a quick-acting fastener particularly suitable for light to medium duties where good resistance to vibration or fatigue is required. It is relatively poor in both tension and shear strength, however, and has very little tolerance to misalignment. Its particular advantages are speed of operation, lightness of weight and the fact that only a single component fitted to one panel is involved.

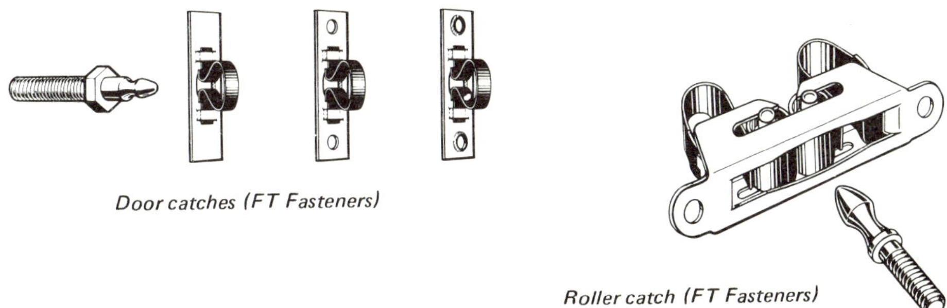

Door catches (FT Fasteners)

Roller catch (FT Fasteners)

Fastener Selection Guide

The following parameters are relevant when selecting a quick-acting fastener for a particular application.

Cost – This is concerned with the cost of fitting as well as that of the fastener itself. Fasteners with a low tolerance to misalignment are generally to be avoided on low-cost productions where generous tolerances are normally involved. Fasteners with good tolerance to misalignment are:-

(i) Most turn-operated fasteners, except multi-thread screws.

(ii) Hook-type lever latches.

(iii) Spring-lever latches.

(iv) Sliding latches.

Ease of Operation. Ease of operation may be of primary importance, in which case push-pull, lever-action or slide-action fasteners are to be preferred. Lever action and push-pull fasteners are again the first choice for speed of operation, although some types of slide-action and turn-operated fasteners may be just as quickly operated. This is not normally a vital parameter.

Loads to be met. The suitability of different types of quick-acting fastener to meet particular classes of loading is basically as follows:

(i) *Tensile strength* – turn-operated fast-lead screw fasteners are generally best, with cam-lever fasteners almost as good.

(ii) *Shear strength* – turn-operated fast-lead screw fasteners are generally superior to other basic types, although most slide-action fasteners approximate to them.

(iii) *Resistance to vibration* – most lever-operated fasteners and cam- or pawl- quarter-turn fasteners are generally superior in this respect. Most push-pull fasteners have good resistance to vibration.

(iv) *Resistance to impact loading* – superior types are (some, but not all) lever-action fasteners and stud- or cam- quarter-turn fasteners. Sliding pawl fasteners are least resistant to impact loading.

Compression. Most lever- and turn-operated fasteners are capable of producing high compression if required. Push-pull fasteners and slide-action fasteners are normally poor (except for the sliding latch with tightening knob).

Tolerance to deformed surfaces. The following types can be regarded as self-compensating when sprung or deformed panels are to be closed.

(i) Lever-action latch with hook engagement.

(ii) Integral lever-cam latch.

(iii) Most types of turn-operated fastener (with the exception of quarter-turn fasteners).

(iv) Sliding-bolt fasteners (some).

Push-pull fasteners are generally unsuitable when deformed panels are to be fastened.

Weight. In general, push-pull and turn-operated fasteners are the lightest types (with the exception of certain types of adjustable pawl fasteners). Cam-lever fasteners are usually the heaviest.

Bulk Projection. Push-pull fasteners and turn-operated fasteners usually have the least bulk or projection on the outside of the assembly. Lever-cam fasteners also have only the lever projecting on the outside. Internally, the least bulk projection is provided by lever-action fasteners (not cam-action type) and slide-action fasteners, and some types of turn-operated fastener.

Single Component Fasteners. Only the following types of fastener are single component types – ie do not require a keeper, striker or matching component mounted on the second surface to be fastened.

(i) expanding prong push-pull fasteners

(ii) some turn-operated screw fasteners

(iii) some turn-operated pawl fasteners

It should be noted that all the above comments refer to the basic types of quick-acting fastener described under their separate headings. Proprietary designs offer variations on these types, and there are other patented forms of fastener which may modify the characteristics of the basic type.

'Oddie' Fasteners

The 'Oddie' quick-operating fastener consists of three parts – a stud, spring clip and washer. The washer provides resilience in the fastened assembly to resist vibration. For high temperature applications it may be replaced by a spring.

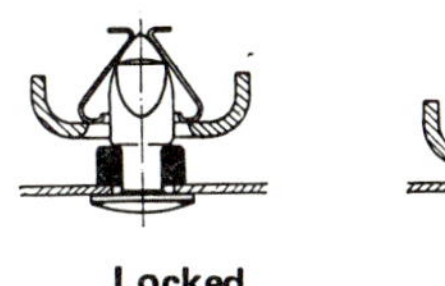

Locked

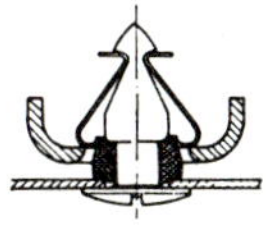

Opened

The diagrams above show the Oddie fastener in its locked and opened positions. The former shows how the positive lock is obtained, and the latter how the stud is ejected automatically by the pressure of the clip on the streamlined nose of the stud.

When unlocking, if the head of the stud is turned through 90° and back again to its original position, this leaves the fastener ready for locking on reassembly. Finger pressure on the head of the stud should engage the fastener with an audible 'click'.

When the fastener is locked, suction load on the panel tends to lock the stud more positively against the clip.

'Oddie' fasteners are made in three sizes – standard, midget and atom. The clip is also available as a 'clip-on' type for pressing into a rectangular hole.

A locking principle similar to that of the 'Oddie' fastener is employed in the 'Oddie' quick-release pin.

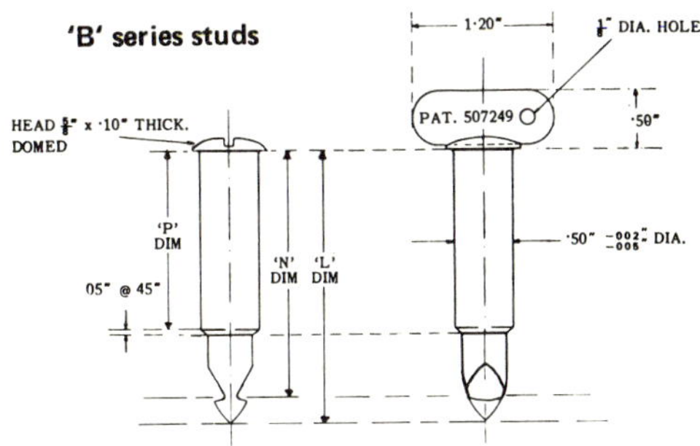

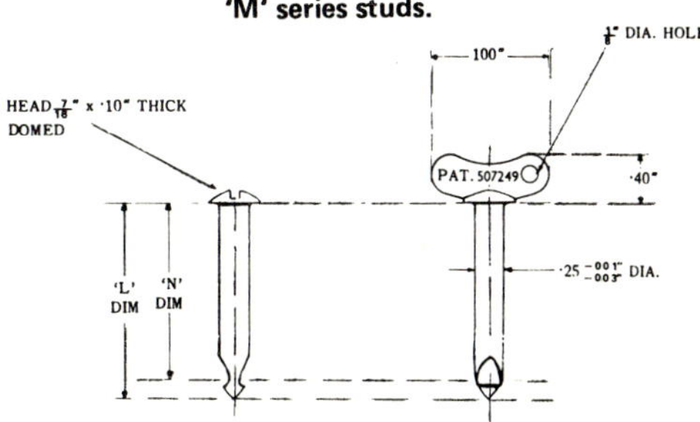

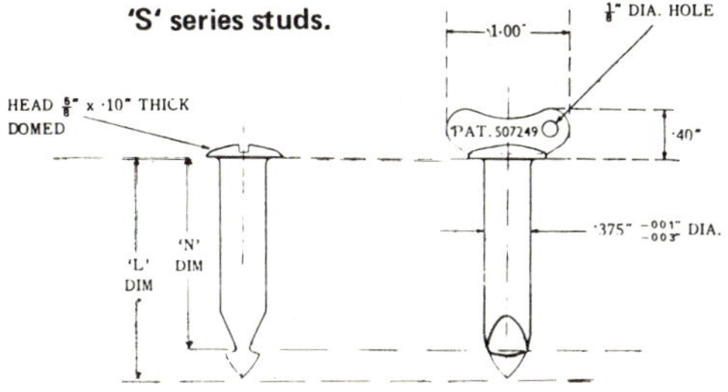

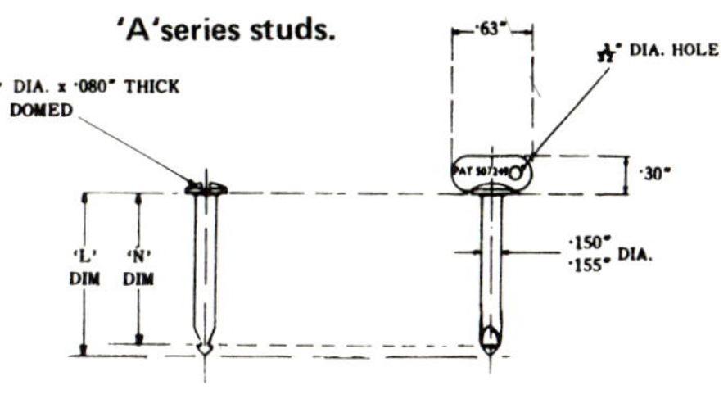

'Oddie' quick-release pins.

Plastic Fasteners

Plastic fasteners may be categorised under:-

(i) Screws, nuts and washers, etc, moulded in plastic in both standard and special sizes and forms (Fig 1).

(ii) Plastic-headed screws – ie metallic screws with heads moulded in various forms (Fig 3).

(iii) Special nuts

(iv) Plastic-coated fasteners – ie standard forms of fastener such as bolts, screws, studs and nuts having a steel core with a plastic shell for protection in corrosive surroundings. The usual form of plastic coating used is glass-filled nylon.

(v) Plastic rivets.

(vi) A wide and increasing range of special and individual designs of fastener in all plastic or plastic/metal combinations as alternatives to conventional fasteners where plastic fasteners offer specific advantages, eg

(a) low cost

(b) ease of assembly

(c) resistance to corrosion, rusting or staining

(d) good electrical insulation

(e) attractiveness of appearance

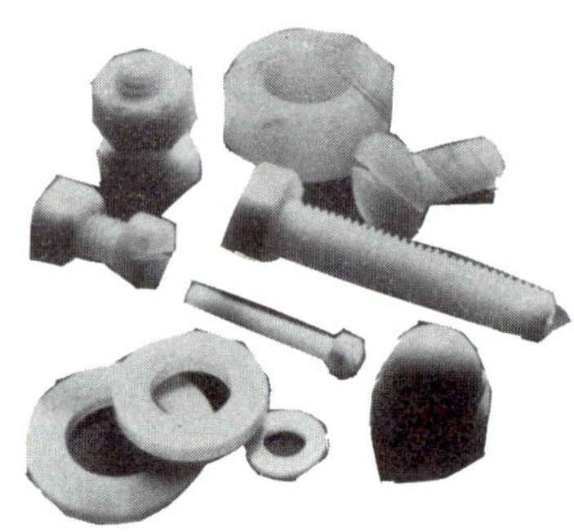

Fig 1 Nylon metric screws, nuts and washers from Plastic Screws Ltd.

Time is the most expensive component

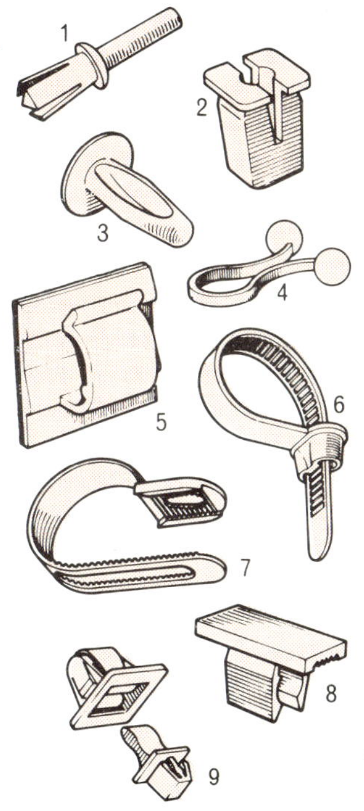

FASTEX helps you bring assembly costs down.

We make fasteners at Fastex, thousands of different fasteners, for putting together thousands of different products.
But anyone can do that.
On the other hand, not everyone can say they developed Rokut Rivets, or Lokut Nuts, or Wire Ties or . . .
There's a lot more too – but that's not the point here. The point is that these products were designed to bring assembly costs down – and that's exactly what they are doing.
Every product we make is the result of close collaborations between Fastex and the customers and there are many examples of how Fastex, working jointly with manufacturers, has contributed to bringing assembly costs down.
Virtually every industry imaginable has profited from Fastex knowledge and experience.
We would welcome the opportunity to work with you in support of your new design and development effort.

1. Rokut Rivet
2. Reverse Lokut Nut
3. Drive Fastener
4. Wire Tie
5. Cord Clip
6. Ratchet Cable Strap
7. 'P' Clip
8. Neon Lens Clip
9. Latch and Strike

FASTEX
A DIVISION OF ITW LTD.
Viables Estate, Basingstoke, Hampshire RG22 4BW
Telephone Basingstoke (0256) 61151, Telex 847872.

FASTEX
DIVISION DER ITW-ATECO GMBH.
2 Norderstedt 1, Stormarnstrasse 43-49, Hamburg, West Germany.
Telefon 040 5252001, Telex 02 12912.

"Remember, you're never more than a few feet away from a product of ITW"

A selection of set screws, nuts and bolts in aluminium alloy HE30-WP, a medium strength heat treated alloy with excellent resistance to corrosion.

NYLON & ALLOYS LTD.

Specialists in Plastic Fastenings and Materials

NUTS, BOLTS, SCREWS, TURNED AND MOULDED PARTS

Hymax Works, Half Acre Road, Hanwell, London W7 3JJ.
Tel: 01-579 5166

Specialist manufacturers and designers of metal and plastic clips, fasteners and general affixing principles for the automotive light industrial and general domestic appliance industries.

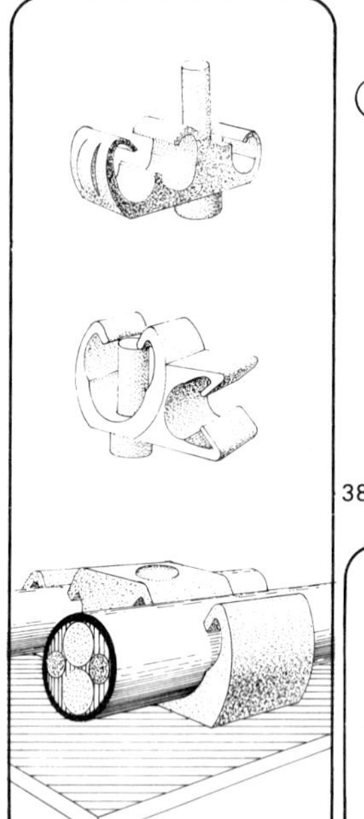

A. Raymond

IVOR SPRY & CO. LTD.
Marwood House, 33 Colville Road,
LONDON W3 8BJ
Tel: 01-992 6533/5
Telex 933625

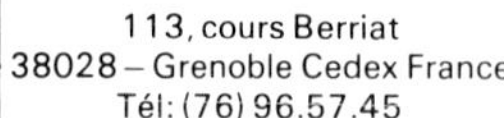

113, cours Berriat
38028 – Grenoble Cedex France
Tél: (76) 96.57.45

7850 Lorrach/Baden
Postfach 42. W. Germany
Tél: (07621) 88081

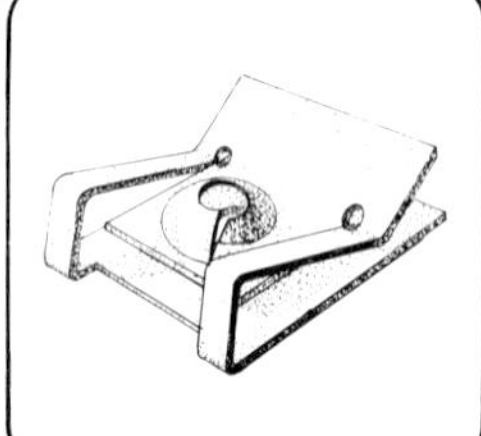

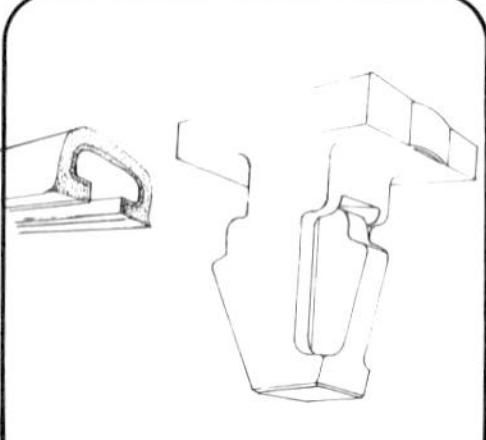

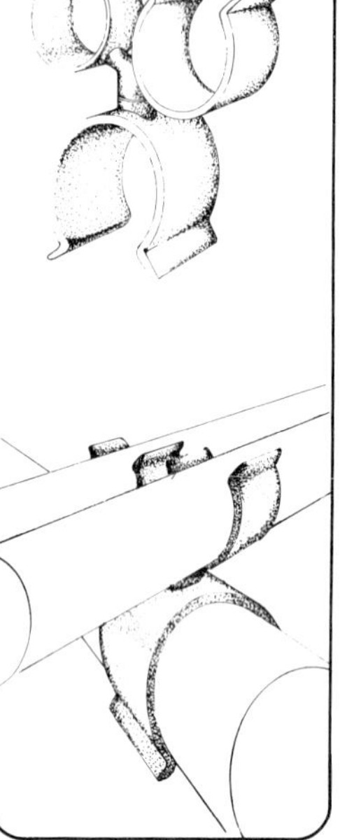

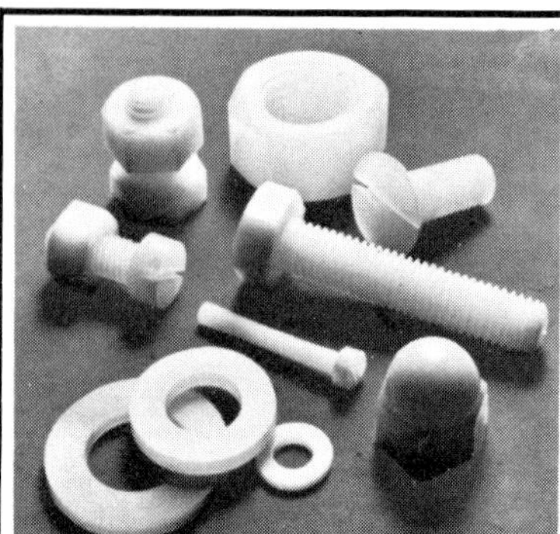

A range of metric screws, nuts and washers in Nylon

Made in Nylon to metric sizes, plastic screws make ideal insulators, are tough, non-magnetic, six times lighter than steel and won't rust or rattle.
Catalogue and free samples on request.

Plastic Screw Labpacks available too!

Uddens Trading Estate,
Nr. Wimborne, Dorset BH21 7NL
Tel. Ferndown (STD 02017) 71411/2/3.
Telex 41408

Plastic Screws, Nuts, etc

The material most widely used for non-metallic screws, nuts, washers, etc, is nylon. Grade 66 provides good mechanical properties, which may be enhanced by glass filling (33%). This also improves the dimensional stability of the nylon. Rigid grades of nylon (eg 11 and 12) may also be used. Small screws are normally made from unfilled nylon. Woodscrews are also included in this category, although nylon woodscrews are far less widely used than nylon machine screws.

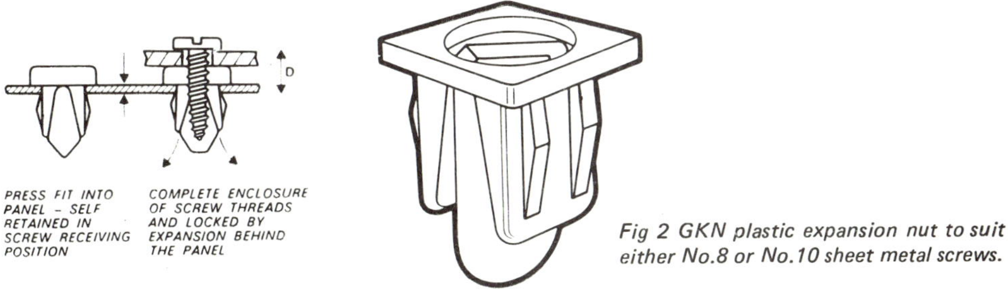
D
PRESS FIT INTO PANEL - SELF RETAINED IN SCREW RECEIVING POSITION
COMPLETE ENCLOSURE OF SCREW THREADS AND LOCKED BY EXPANSION BEHIND THE PANEL

Fig 2 GKN plastic expansion nut to suit either No.8 or No.10 sheet metal screws.

Special Plastic Nuts

Special nuts produced in plastic fall under the two main headings of *expansion nuts* and *self-threading nuts*. Both types are normally designed for use with self-tapping screws. Expansion nuts for blind fastening applications are designed for press-in or snap fitting in a preformed hole (usually square) – Figs 2 and 4. They offer the specific advantage that they can be used with all types of metal finishes without scratching or chipping the surface. Self-locking action is provided by the entering screw expanding the prongs or shank of the nut.

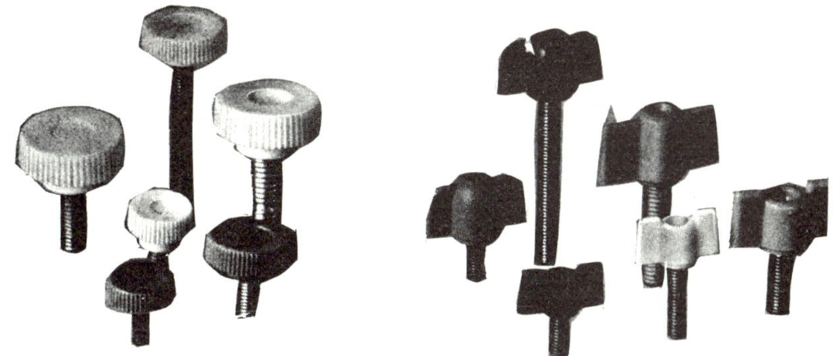

Fig 3 Examples of 'Unitec' plastic headed screws.

Fig 4 Examples of nylon expansion nuts.

Examples of plastic self-threading nuts are given in Fig 5. These offer a nut which is both corrosion resistant and non-scratch for decorative applications, etc. Thin-bodied plastic self-threading nuts with a blind end may also be designed for use as thread seals, ie applied over the thread protruding beyond a conventional nut.

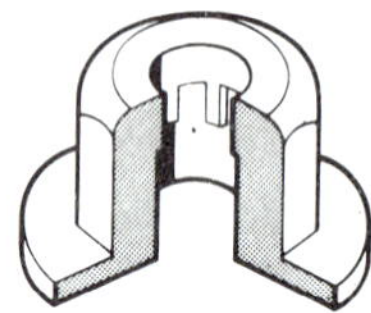
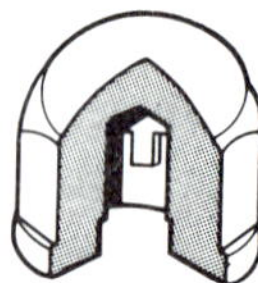

Fig 5 Self-threading nylon nuts (GKN).

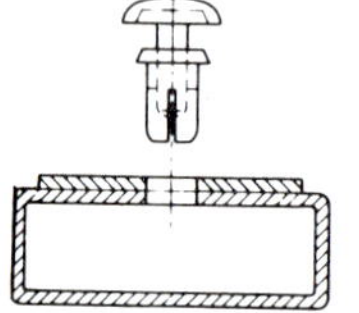
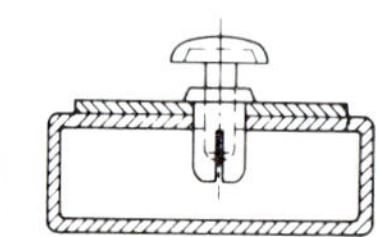
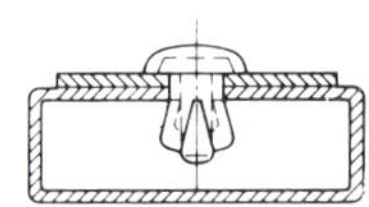

Fig 6 Clic-Rivet – plastic blind rivet, sequence of assembly. (Instrument Screw Co).

Plastic Rivets

Snap-fitting plastic rivet designs may be based on a split shank form locking in place by wedging action, or as two-part components which are self-setting when assembled – see Fig 7.

There are also other types which need a special tool for setting. An example of this type is shown in Fig 8. This rivet is set by a pin driven through the hollow shank, either with a manual or an air-operated tool. Set rivets are normally stronger fasteners than snap-in plastic rivets, some typical performance data being summarised in the Table.

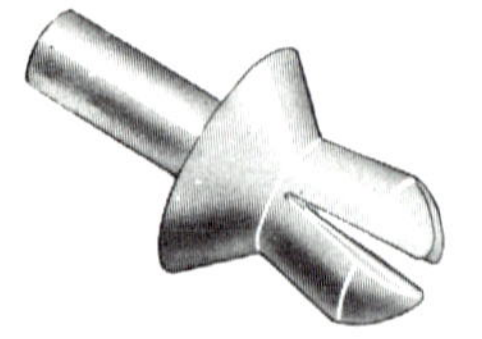
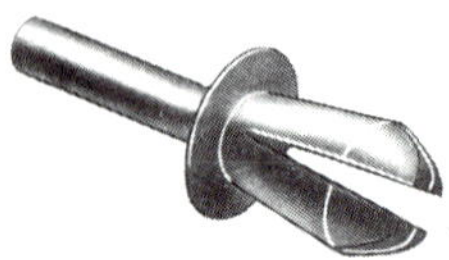
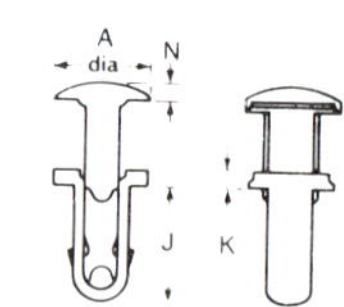

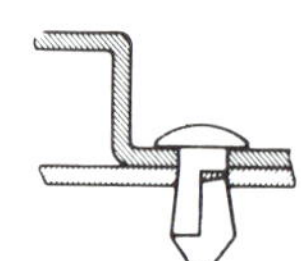
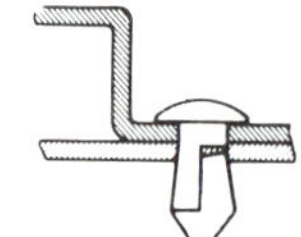
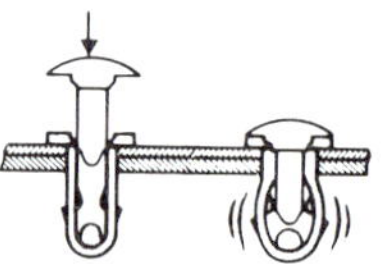

Fig 7 Snap-fitting plastic rivets – Carr Fastener (left), and GKN (right).

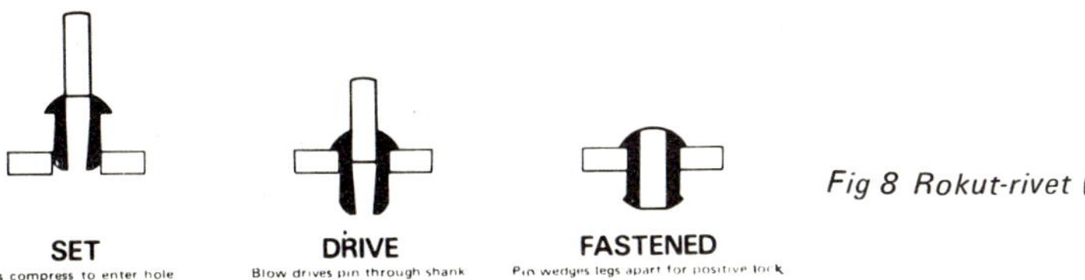
SET
Legs compress to enter hole
DRIVE
Blow drives pin through shank
FASTENED
Pin wedges legs apart for positive lock

Fig 8 Rokut-rivet (Fastex).

ROKUT – RIVETS – Dimensional and Performance Data

Shank Dia.		Ultimate Tension		Single Shear		Double Shear		Recommended Maximum Load				Recommended Maximum Intermittent Load			
								Tension		Shear		Tension		Shear	
Ins.	mm	Lb.	Kg.	Lb.	Kg.	Lb.	Kg.	Lb.	Kg.	Lb.	Kg.	Lb.	Kg.	Lb.	Kg.
.125	3.17	50	22.7	55	24.9	100	45.4	10	4.5	15	6.8	20	9.1	35	15.9
.156	3.96	75	34.0	100	45.4	180	81.6	15	6.8	25	11.3	40	18.1	60	27.2
.187	4.75	100	45.4	175	79.4	300	136	25	11.3	40	18.1	65	29.5	60	27.2
.218	5.54	150	68.0	230	104	440	200	35	15.9	55	24.9	95	43.1	150	68.0
.250	6.35	200	90.7	300	136	550	249	50	22.7	75	34.0	125	56.7	175	79.4
.312	7.92	280	127	450	204	800	363	70	31.8	120	54.4	160	72.6	240	109
.375	9.53	350	159	600	272	1200	544	80	36.3	160	72.6	180	81.6	360	163
.437	11.1	400	181	700	317	1400	635	90	40.8	180	81.6	200	90.7	400	181
.500	12.7	500	227	800	363	1600	726	100	45.4	200	90.7	250	113	500	227

Special Fasteners

Special fasteners can be categorised in a number of ways:—

(i) Types of fastener produced or developed for specific industries, but largely extensions or modifications of standard fasteners — eg fasteners for Building Construction.

(ii) Special designs of fastener produced for specific industries and largely exclusive to these industries — eg special and patterned high tensile bolts, special forms of bolts and screws etc for the Automobile Industry, Domestic Appliance Industries, Radio and Electronic Industries, etc.

(iii) Individual designs of fastener intended to be competitive alternatives to existing fasteners, incorporating original features which make them difficult to classify under any specific type.

(iv) 'One-off' or purpose designs of fastener produced for a specific individual requirement. These are usually modifications of a specific type of fastener (eg spring-clip) which would have little or no application outside the purpose for which they were designed. Original designs of fasteners in this category, however, may have applications considerably broader in scope than their original purpose. An example here is metal-to-rubber bonding inserts originally developed for the automobile industry but subsequently finding extensive application in the domestic appliance industries and other fields.

(v) Original and/or purpose designed fasteners which have a broad application or potential application, but are essentially unique. A classic example of this category of fastener is 'Velcro'.

VELCRO

'Velcro' is a patented all textile 'touch and close' fastener invented in Switzerland. A Velcro fastener consists of two nylon strips, one with thousands of tiny hooks and the other with tiny loops. When pressed together the hooks grip the loops to give a secure, tight closure — Fig 1. The two strips can be separated by pulling apart, the life of such a closure being virtually infinite. It can also be used with slide rings (D-rings) where it is necessary to increase tension — Fig 2.

Velcro is made in Nylon 66 in 15, 20, 30 and 50mm widths in 20 colours. 10mm wide Velcro is also made in white, black, nutmeg and chemical grey. It is supplied in 10 or 25 metre flanged rolls.

Sole licencee for the manufacture of Velcro in the UK and Europe is Selectus Ltd.

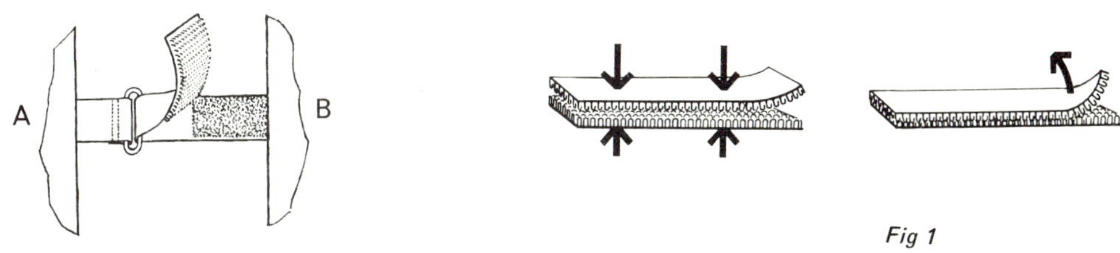

Fig 1

Fig 2 Purely diagrammatic.
For pulling one component A
up to another B.

Hedlock Fasteners

The Hedlock fastener is a recently introduced form of non-flexible 'touch and close' fastener based on interlocking spherical nylon or pear shaped acetal heads on stems – Fig 3. Hedlock 1 fasteners will engage only when the stems are parallel, ie in six different positions at $60^{0} \pm 8^{0}$ to each other. Hedlock II fasteners are fully rotational and the heads will engage at all angles. The two forms are not intended to be mated together.

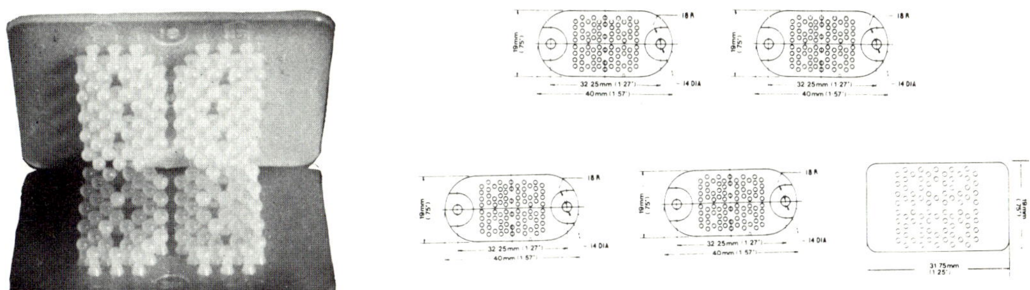

Fig 3 Hedlock fastener (3M).

Different engagement strengths are obtained by using different pairs (sizes) of Hedlock fasteners. Strengths available range from 1-2kg (2-4lb) to 18-23kg (40-50lb). Attachment method varies according to the surface on which the fastener is to be mounted, viz:–

woodscrews – for wood and plasterboard
self-tapping screws – for sheet metal
rivets – for rigid plastic and sheet metal
staples – for fabrics, fibreboard, wood, rigid plastics and plaster board
adhesive – for fabrics, wood, rigid plastics and flexible plastics
snap-in or lock rivets – for rigid plastics and sheet metal.

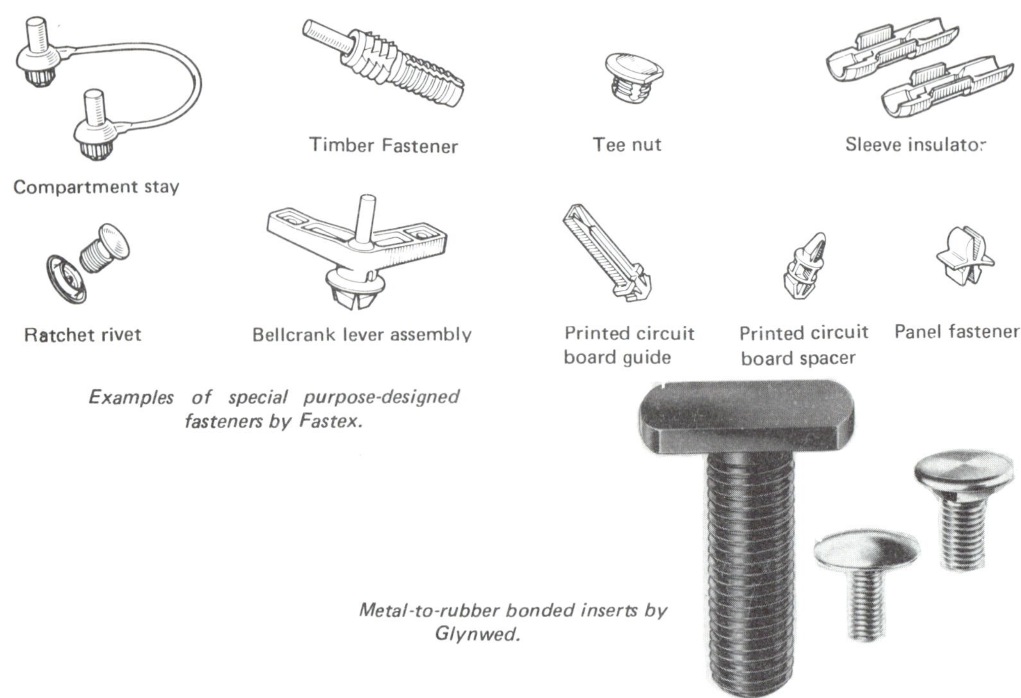

Examples of special purpose-designed fasteners by Fastex.

Metal-to-rubber bonded inserts by Glynwed.

BIGHEADS

The 'Bighead' is an original design of load-spreading bonded fastener for the 'secret' fixing of glass reinforced plastic, rubber, fibreboard and other mouldable materials, as well as ply, etc. A blind or sighted flat or curved disc provides the bonded anchor point, which can accommodate for instance a spring clip or a welded-on plain or threaded (male or female) stud. Possible combinations are numerous, covering a wide variety of fastening possibilities.

Some examples are shown in Fig 4.

When Bighead fasteners are bonded into glass reinforced plastic the perforations – visible here through the plastic – provide a strong key.

A selection of Bighead fasteners attached to different thicknesses of plywood by staples, Bigbond adhesive, screws and nails. A barbed fastener is shown (left) on the thinnest ply.

Fig 4

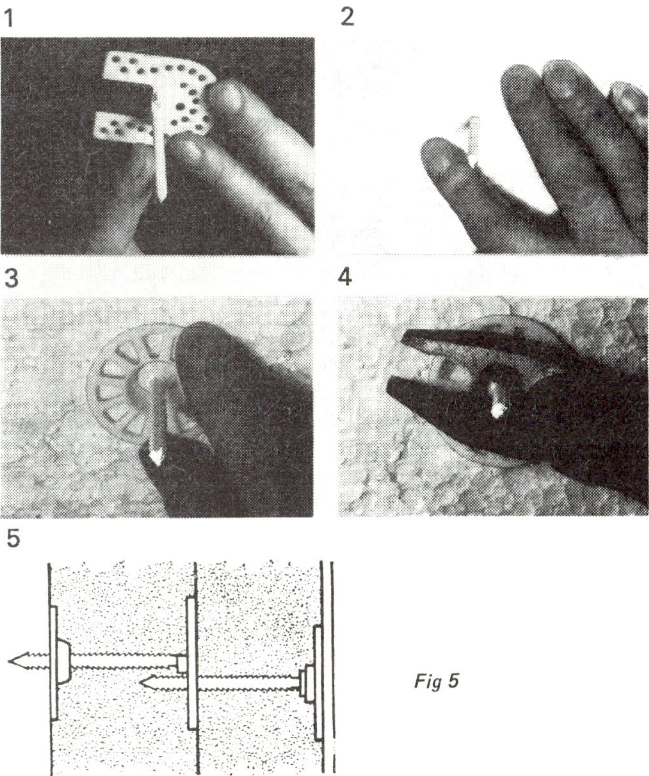

Fig 5

Clipfas Nylon Hangers

The Foster Clipfas nylon hanger is a non-corroding fastener designed for fastening insulation, etc. It is moulded in one piece with a round perforated base and a serrated nylon spindle hinged to lie flat until required. Then the spindle is quickly pulled up on its hinge, see Fig 5, snapping into place and locking erect, ready for impaling the insulating material (2). A perforated nylon washer is supplied for pressing down over the serrated spindle (3), recessed centre down, locking against the insulation. Any excess spindle length is snipped off in the washer recess with wire nippers (4), leaving a flush surface. If multiple courses of insulation are installed (more than 2ins thick), another hanger is snapped over the first to act as the washer for each intermediate course (5) thus providing additional spindle length, in piggy back fashion.

Clipfas neoprene adhesive 13-29 is used to cement the perforated base of the Clipfas hanger at 30cm centres on smooth metal surfaces.

Foster Flexfas adhesive 82-10 should be used when affixing hangers to uneven or rough concrete or brick surfaces. The base of the hanger is embedded in a spot of the 82-10 adhesive.

Rivets

Rivets may be divided into two broad classifications — 'light' rivets for general assembly work, etc, and 'heavy' rivets for larger engineering and structural applications. 'Heavy' rivets are usually of solid type. 'Light' rivets may be divided into five main categories (see also Table I).

(i) Solid rivets (ie rivets with solid shanks)

(ii) Tubular rivets

(iii) Semi-tubular rivets

(iv) Bifurcated rivets

(v) Drive rivets

Light rivets are produced in a variety of ferrous and non-ferrous metals; mainly steel and stainless steel, and aluminium, brass and copper. All types are available with a variety of head, both standard and 'specials', and with a variety of finishes. Typical finishes available on standard ranges of light rivets are given in Table II.

Modern Riveting

The modern light rivet is an extremely versatile low cost fastener which can be fed to the job automatically by suitable machines. On power operated machines a speed of setting of over 100 rivets per minute can be achieved, making riveting highly competitive with all other methods of fastening. Rivet setting machines can also incorporate pressure control to allow for slight variations in component thickness to achieve consistent setting and also avoid damage to brittle materials such as glass, plastic, fibre, etc.

Modern riveting can therefore be classified as a high speed, low cost precision fastening technique. It can also be used without difficulty by unskilled labour. Examples of typical applications in a variety of industries are given in Table III.

Semi-Tubular Rivets

When correctly set these rivets expand to fill the hole in the components, thereby providing a solid joint. Standard head forms for semi-tubular rivets are shown in Fig 1. The shank diameter is controlled by the nominal diameter of the hole in the component, the specified clearance allowing for the correct amount of shank expansion for a sound mechanical joint when the rivet is set. Standard dimensions are summarised in Tables IVA and IVB.

The Rivet Makers

Your guide to the largest manufacturers of cold formed rivets in the United Kingdom – The Bifurcated Engineering Group.

Bifurcated & Tubular Rivet Co. Ltd
P.O. Box 2, Mandeville Road, Aylesbury, Bucks, HP21 8AB
Tel: Aylesbury (0296) 5911. Telex: 83210
Makers of Aylesbury Rivets and other special Cold Formed Parts – we produce over 20 million a day – and Aylesbury Rivet Setting Machines. We offer a complete riveting system, with free advice on any fastening problem from our experienced technical sales staff. Contact us at the design stage, and save time and money.

Black & Luff Ltd
Birmingham Factory Centre, Kings Norton, Birmingham B30 3HQ
Tel: 021-459 2281. Telex: 338798
Makers of solid, tubular and semi-tubular rivets in all standard materials, cold formed clevis pins and special threaded parts. We offer a specialist service in custom-made cold forgings for all industries.

Clevedon Rivets & Tools Ltd
Reddicap Trading Estate, Sutton Coldfield, West Midlands B75 7DG
Tel: 021-354 5238. Telex: 339294
Makers of rivets in all sizes and types – solid, tubular, semi-tubular – in non-ferrous metals, light alloys, mild and stainless steel. We can supply the rivets you want in the quantities you want – small, medium, large.

Jesse Haywood & Co. Ltd
Foundry Lane, Smethwick, Warley, West Midlands B66 2LW
Tel: 021-558 3027
Makers of solid rivets – ferrous and non-ferrous – and special cold forged parts. We offer a highly specialised second operation service for all industries.

Alec Pine Fasteners Ltd
5 Glebe Road, Letchworth, Herts. SG6 1DS
Tel: Letchworth (046 26) 71840
Suppliers of a wide range of industrial fasteners.

TABLE I – BASIC TYPES OF LIGHT RIVETS

	Type	Remarks	Relevant BS
	Solid	Strongest type. Requires high clinching force	BS641
	Tubular	Designed for lighter clinching forces; also capable of being punched through a variety of materials. Can be clenched by automatic feed rivet setting machines or with a spinning or squeezing tool.	
(i) (ii) (iii)	Semi-Tubular	Preserves shear strength of solid rivet with easier clinching. Particularly suitable for use with automatic feed riveting tools. For maximum strength may be used with rivet caps, dimensions and construction of which may vary widely with different manufacturers. (i) Short hole, parallel taper, used where components vary in thickness (ii) Double taper, for joining brittle materials (shank expansion is minimised) (iii) Bullnose, for maximum strength in pre-pierced holes	BS1855
	Bifurcated	Can be used with a single pierce/clinch tool for fastening thin sheet metal, etc. May be used with rivet caps, dimensions, construction and finish of which may vary widely with different manufacturers.	Tool: BS1855 Part 1

TABLE II – PLATING AND OTHER FINISHES FOR RIVETS*

Finish	Shade	Properties and Use
BK	Matt black	Chemical finish on steel, mainly decorative, some corrosion resistance if oiled.
Brass	Brass	Mainly decorative but some corrosion resistance if lacquered.
Burnish	Bright	Barrel finish for all metals. Decorative only.
Chromium	Bright blue-white	Decorative finish always applied on nickel. Some corrosion resistance given by nickel.
Cadmium	Blue-grey	Good corrosion resistance. Improved by chromate passivation.
Copper	Copper	Decorative and good conductivity.
Ebonol	Matt black	Chemical finish on non-ferrous metals. Decorative only.
Lacquer	Clear	Maintains bright finish of brass or copper. Moderate corrosion resistance.
Nickel	Bright yellow-white	Decorative only with standard deposit. Heavier deposit gives good corrosion resistance.
Passivation	Yellow-brown	Chromatic dip on zinc or cadmium plate gives improved corrosion resistance.
Silver	Dull white	Good conductivity and good soldering surface.
Stove-enamel	Black and colours	Good paint surface to match coloured components.
Tin	Dull grey-white	Good corrosion resistance and soldering surface.
Tin-lead	Dull dark grey	High corrosion resistance. Fair soldering properties.
Zinc	Bright or dull silver-grey	Good corrosion resistance. Improved by chromate passivation.

*The Bifurcated & Tubular Rivet Co Ltd.

TABLE III – TYPICAL APPLICATIONS OF LIGHT RIVETS

Industry	Application	Type(s) Used	Material
Automotive	Joining friction linings to metal brake and clutch components	Solid, semi-tubular tubular	Brass, copper and steel
	External assemblies, and screen wipers, etc.		Stainless steel for rust-free fastening
	Assembly/mounting of electrical components	Solid, tubular, semi-tubular and specials	Steel and non-ferrous metals
	Plaster to metal/metal to metal multi-component assemblies	Solid, semi-tubular and tubular	Steel or brass
	Minor assemblies	Solid, semi-tubular, tubular and specials	Mainly steel
Domestic Appliance Manufacture	Washing machines	All types	Stainless steel for corrosion resistance
	Vacuum cleaners (assembly)	All types	Mainly steel
	Electric irons, etc	Mainly semi-tubular	Steel or brass
	Miscellaneous articles	All types: standard rivets may also be used as hinges or pivots; to secure handles, etc	Stainless steel for corrosion resistance; aluminium, steel or brass with suitable finish
Leather goods,etc	Case and briefcase assembly	Mainly semi-tubular, some bifurcated. Special rivets used in assembly of locks	Aluminium, steel or brass with suitable finish.
	Handbags	Semi-tubular rivets for framé hinge pins	
	Belts, etc	Bifurcated rivets with a variety of head designs for both functional and decorative use	Variety of plated and decorative finishes
	Feet (for cases, trunks, handbags, etc)	Bell head rivets	Plated or decorative finish
	Boot and shoe studs	Special heads	
Miscellaneous	Fibre container assembly	Bifurcated with or without rivet caps	
	Hinges and linkages	Solid, semi-tubular and shouldered rivets	Mainly steel
	Can handles	Shouldered rivets to secure metal wire or strip handles	Usually steel
	Umbrellas (assembly)	Small double-taper semi-tubular	
Electrical	Insulated component assembly	Special hollow rivets with insulating washer	Aluminium, steel and brass with nylon sheath and insulating washer
	Conductive assembly (printed circuits)	Special hollow rivets	Copper

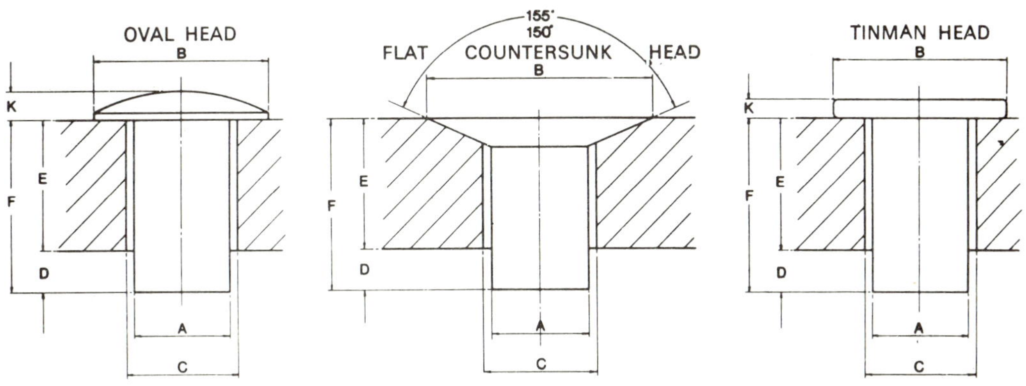

A = Diameter of rivet
B = Diameter of head
C = Diameter of hole in component
D = Normal riveting allowance
E = Component thickness
F = Length of rivet (measure under head for oval and tinman head, overall for countersunk)
Length F = Component thickness + riveting allowance
K = Head thickness

Fig 1 British standard semi-tubular rivets.

TABLE IVA – BRITISH STANDARD METRIC SEMI-TUBULAR RIVETS* (BS4895) – See also Fig 1

C	A	B	D for Short Hole Tubular	D for Bull Nose and Double Taper	K
1·2	1·16/1·08	2·26/ 2·06	0·65	0·65	0·45/0·35
1·6	1·53/1·45	2·98/ 2·78	0·65	0·80	0·58/0·48
2·0	1·92/1·82	3·72/ 3·48	1·20	1·20	0·74/0·58
2·5	2·39/2·29	4·62/ 4·38	1·20	1·60	0·91/0·75
3·0	2·87/2·77	5·52/ 5·28	1·60	2·00	1·08/0·92
3·5	3·38/3·28	6·42/ 6·18	2·00	2·40	1·24/1·08
4·0	3·83/3·71	7·34/ 7·06	2·00	2·80	1·43/1·23
5·0	4·77/4·65	9·14/ 8·86	2·40	3·60	1·76/1·56
6·0	5·76/5·62	10·98/10·62	3·20	4·00	2·10/1·90
7·0	6·76/6·62	12·78/12·42	4·00	4·50	2·76/2·56
8·0	7·74/7·58	14·60/14·20	4·00	4·50	2·78/2·54
10·0	9·72/9·56	18·20/17·80	4·50	5·00	3·45/3·21

All dimensions in millimetres

*The Bifurcated & Tubular Rivet Co Ltd

TABLE IVB – STANDARD PRODUCTION SEMI-TUBULAR RIVETS* (See also Fig 1)

RIVET GAUGE	'A'		'B'		'C'		'C' Nearest Drill Size		'D' for Bull-Nose and Double Taper		'D' for P T.	
	inch	mm	inch	mm	inch	mm	inch	mm	inch	mm	inch	mm
18	·048/·046	1·22/1·17	·103/·098	2·62/2·49	·052/·050	1·32/1·27	·052	1·3	·025	·64	·025	·64
17	·056/·054	1·42/1·37	·132/·127	3·35/3·23	·060/·058	1·52/1·47	·0595	1·5	·031	·79	·025	·64
$\frac{1}{16}$	·064/·061	1·63/1·55	·132/·127	3·35/3·23	·068/·066	1·73/1·68	·067	1·7	·047	1·19	·047	1·19
16	·068/·065	1·73/1·65	·132/·127	3·35/3·23	·072/·070	1·83/1·78	·070	1·8	·047	1·19	·047	1·19
15	·075/·072	1·91/1·83	·159/·154	4·04/3·91	·080/·078	2·03/1·98	·0785	2·0	·047	1·19	·047	1·19
14½	·081/·078	2·06/1·98	·159/·154	4·04/3·91	·087/·085	2·21/2·16	·086	2·2	·062	1·58	·047	1·19
14	·089/·085	2·26/2·16	·159/·154	4·04/3·91	·095/·093	2·41/2·36	$\frac{3}{32}$	2·4	·062	1·58	·047	1·19
13	·094/·090	2·39/2·29	·190/·185	4·83/4·70	·100/·098	2·54/2·49	·098	2·5	·062	1·58	·047	1·19
12½	·101/·097	2·56/2·46	·190/·185	4·83/4·70	·107/·105	2·72/2·67	·1065	2·7	·078	1·99	·062	1·58
12	·108/·104	2·74/2·64	·221/·216	5·62/5·48	·114/·112	2·90/2·84	·113	2·85	·078	1·99	·062	1·58
11	·121/·117	3·07/2·97	·221/·216	5·62/5·48	·127/·125	3·23/3·17	$\frac{1}{8}$	3·2	·086	2·19	·062	1·58
$\frac{1}{8}$	·126/·122	3·20/3·10	·221/·216	5·62/5·48	·132/·130	3·35/3·30	·1285	3·3	·093	2·38	·078	1·99
10	·130/·126	3·30/3·20	·221/·216	5·62/5·48	·136/·134	3·45/3·40	·136	3·4	·093	2·38	·078	1·99
9½	·144/·140	3·66/3·56	·316/·310	8·02/7·87	·152/·150	3·86/3·81	·1495	3·8	·109	2·78	·078	1·99
9	·155/·150	3·94/3·81	·316/·310	8·02/7·87	·163/·161	4·14/4·08	·161	4·1	·109	2·78	·078	1·99
8	·165/·160	4·19/4·06	·316/·310	8·02/7·87	·173/·171	4·40/4·34	$\frac{11}{64}$	4·4	·125	3·18	·078	1·99
7	·176/·171	4·47/4·34	·349/·342	8·87/8·69	·184/·182	4·67/4·62	·182	4·6	·125	3·18	·093	2·38
$\frac{3}{16}$	·188/·183	4·77/4·65	·380/·373	9·65/9·48	·198/·196	5·03/4·98	·196	5·0	·140	3·57	·093	2·38
6	·200/·195	5·08/4·96	·380/·373	9·65/9·48	·208/·205	5·28/5·21	·2055	5·2	·140	3·57	·125	3·18
5	·215/·210	5·46/5·33	·442/·434	11·23/11·02	·226/·223	5·74/5·67	·228	5·7	·156	3·95	·125	3·18
4	·237/·232	6·02/5·89	·505/·497	12·83/12·63	·248/·245	6·30/6·23	·246	6·3	·171	4·36	·156	3·95
3	·250/·245	6·35/6·23	·505/·497	12·83/12·63	·261/·258	6·63/6·55	·261	6·6	·171	4·36	·156	3·95

***The Bifurcated & Tubular Rivet Co Ltd**

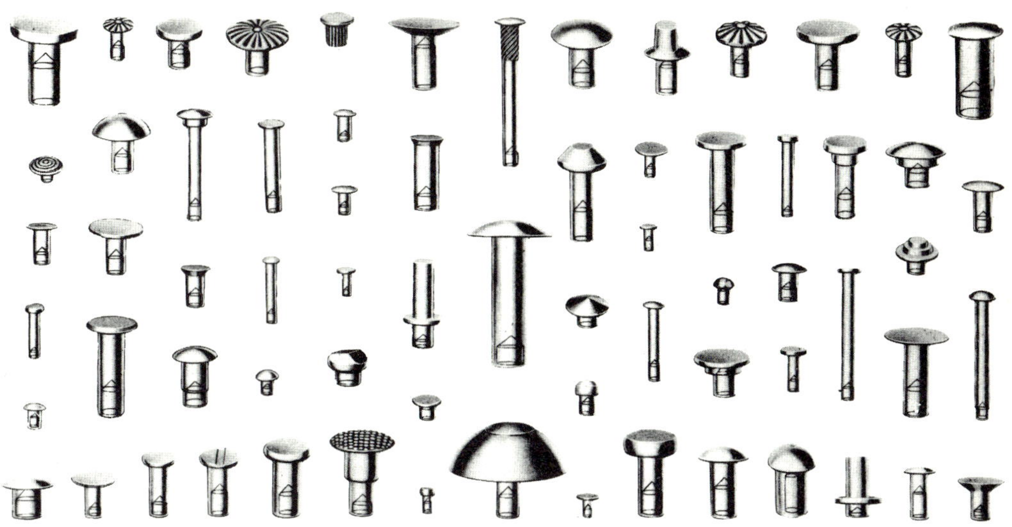

Examples of semi-tubular rivets. (Stimpson, USA).

The hole in the rivet shank is normally tapered to help the expansion in the setting operation But some schools of thought prefer a hole with parallel walls. This is particularly relevant when assembling components having a variation in total thickness. The hole depth in both cases is usually between 80%—100% of the rivet shank diameter.

Specific types of tubular rivets are also produced for attaching clutch facings and brake linings, notably:-

British Standard (BS3575) – Fig 2 and Table VA : brass or copper rivets

European Standard (DIN7338) – Fig 3 and Table VB : steel, brass, copper, aluminium and light alloy.

US Standard – Fig 4 and Table VC : brass rivets

Standard semi-tubular rivets are sometimes used.

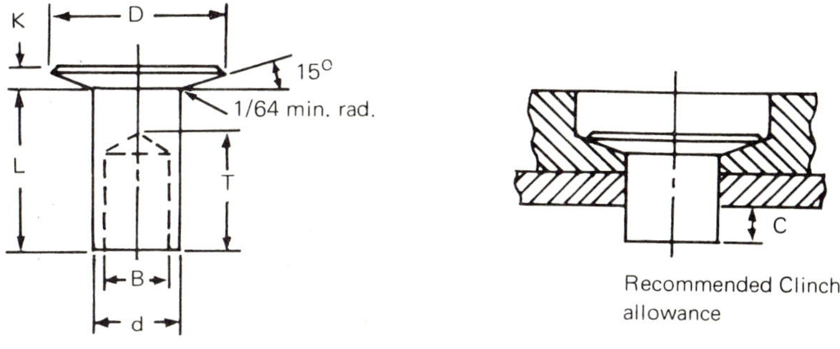

Friction lining rivets. (Tower Manufacturing).

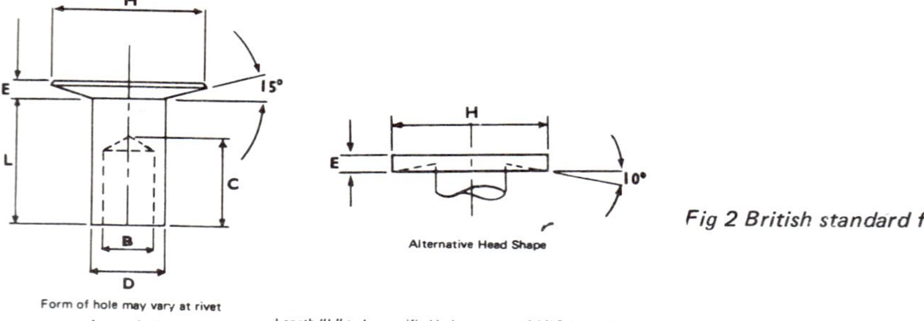

Fig 2 British standard friction lining rivets.

Length "L" to be specified in increments of 1/16, hence G.10 indicates a rivet size G of length 10/16.

Form of hole may vary at rivet manufacturer's discretion

TABLE VA – BRITISH STANDARD RIVETS FOR FRICTION LININGS ✶
(BS3575) – see also Fig 1

Rivet Size	Shank Diameter D			Head Diameter H			Head Thickness E		Depth of Hole C	Diameter B	Min. Hand Clenching Length R	Min. Length L
	Nom.	Max.	Min.	Nom.	Max.	Min.	Max.	Min.				
D	3/32	.094	.089	5/32	.156	.146	.030	.025	Solid	Solid	1/16 to 1/32	3/16
E	1/8	.125	.120	7/32	.218	.208	.030	.025	5/32*	.082	3/32	3/16
F	9/64	.144	.140	19/64	.300	.290	.035	.030	5/32*	.099	3/32	3/16
G	5/32	.156	.151	5/16	.317	.307	.040	.035	3/16†	.106	1/8	3/16
H	11/64	.176	.171	5/16	.317	.307	.040	.035	3/16†	.128	1/8	1/4
J	3/16	.188	.183	3/8	.364	.354	.045	.040	7/32**	.136	1/8	1/4
K	7/32	.215	.210	27/64	.430	.420	.060	.050	7/32**	.154	1/8	5/16
L	1/4	.250	.245	15/32	.478	.468	.060	.050	7/32	.180	5/32	3/8
M	5/16	.312	.307	9/16	.562	.547	.070	.060	1/4	.234	3/16	7/16
N	3/8	.375	.370	5/8	.625	.610	.080	.070	9/32	.281	7/32	1/2

*Rivets under 5/16 long C=1/8
†Rivets under 5/16 long C=5/32
**Rivets under 3/8 long C=5/32
✶ The Bifurcated & Tubular Rivet Co Ltd

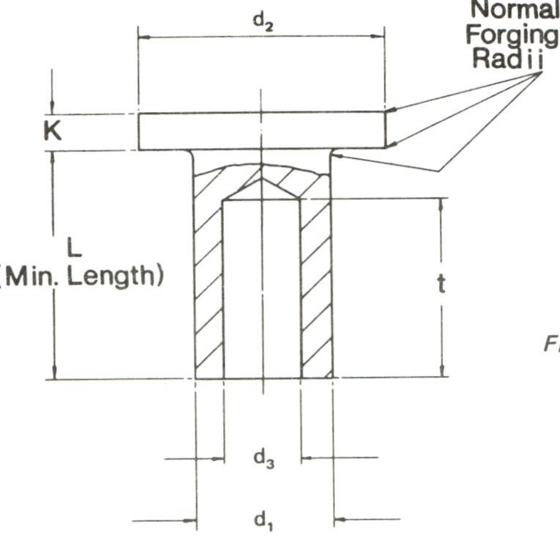

Fig 3 DIN standard friction lining rivets.

TABLE VB – EUROPEAN STANDARD RIVETS FOR FRICTION LININGS (DIN7338) – See also Fig 3 *

d_1	d_2	d_3	+0.0 K–0.2	+0.5 t–0.0	L
3	5.5	1.7	0.8	4.0*	5.0
4	7.5	2.7	1.0	5.0**	6.0
5	9.5	3.5	1.0	6.0***	6.0
6	11.5	4.2	1.2	8.0†	8.0
8	15.5	6.0	1.2	10.0††	10.0

All dimensions in millimetres

*The Bifurcated & Tubular Rivet Co Ltd

Tolerances are as shown in Standard DIN 7338

Note:–Hole depth for short rivets
* Where L is less than 6mm t=3.5mm
** Where L is less than 7mm t=4.0mm
***Where L is less than 8mm t=4.0mm
† Where L is less than 10mm t=6.0mm
†† Where L is less than 12mm t=8.0mm

Material Specification Abbreviations
Steel–ST
Brass–MS
Copper–C-CU
Aluminium–AL
Light Alloy–AL-LEG

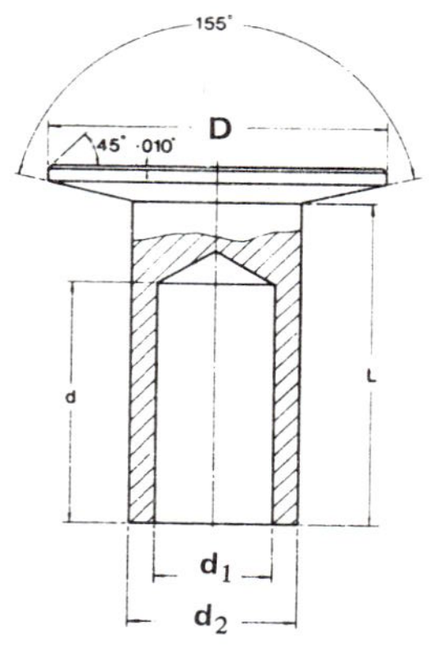

Fig 4 US standard friction lining rivets.

TABLE VC – US STANDARD RIVETS FOR FRICTION LININGS
(See also Fig 4)

Ref	L	d	D	d_1	d_2	Ref	L	d	D	d_1	d_2
3-3	3/16	5/32				7-4	4/16	3/16			
3-4	4/16	3/16				7-5	5/16	3/16			
3-5	5/16	1/4				7-6	6/16	3/16			
3-6	6/16	1/4	0.307		0.144	7-7	7/16	3/16	0.370		0.188
3-7	7/16	1/4	0.302	0.0995	0.140	7-8	8/16	3/16	0.365	0.136	0.183
3-8	8/16	1/4				7-10	10/16	3/16			
3-10	10/16	1/4				7-12	12/16	3/16			
4-4	4/16	3/16				8-6	6/16	3/16			
4-5	5/16	3/16				8-8	8/16	3/16			
4-6	6/16	3/16	0.307		0.144	8-10	10/16	3/16	0.495		0.188
4-7	7/16	3/16	0.302	0.0995	0.140	8-12	12/16	3/16	0.490	0.136	0.183
4-8	8/16	3/16				8-14	14/16	3/16			
						8-16	16/16	3/16			
5-3	3/16	3/16				10-6	6/16	5/16			
5-4	4/16	3/16				10-8	8/16	3/8			
5-5	5/16	3/16	0.370		0.144	10-10	10/16	3/8	0.495	0.180	0.250
5-6	6/16	3/16	0.365	0.0995	0.140	10-12	12/16	3/8	0.490	0.180	0.245
5-8	8/16	3/16				10-16	16/16	I			
5-10	10/16	3/16									
5-12	12/16	3/16									

All dimensions in inches

Tubular Rivets

The setting operation for tubular rivets simply involves clenching all or part of the hollow portion of the shank. The pressure from the setting machine does not affect the shank but can be adjusted to provide either a tight or loose joint. These rivets are often used as axis pins and do not normally need any support.

The holes in these rivets have parallel walls, their size will depend on the application involved.

Certain tubular rivets can now be punched through steel sheets up to 3/16in total thickness without difficulty. This method of fastening provides a watertight joint and speeds up the production cycle considerably, eliminating the necessity for pre-drilling or pre-piercing holes in the components.

Examples of bifurcated rivets. (Stimpson, USA).

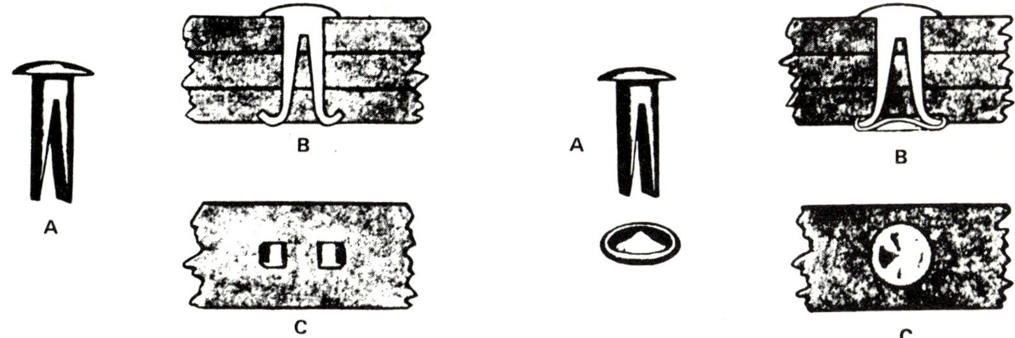

A. A bifurcated rivet before setting.
B. Cross-section of material and clinched rivet.
C. Plan view of the clinch.

A. A bifurcated rivet and cap before setting.
B. Cross-section of material and rivet with the prongs clinched in a cap.
C. Plan view of the cap in position.

Fig 5 Standard bifurcated rivets. (The Bifurcated & Tubular Rivet Co Ltd).

Bifurcated Rivets

Bifurcated rivets (also sometimes known as split rivets) are widely used on softer materials, such as leather, plywood, fibreboard, etc, as self-piercing rivets. They can also be driven (without setting) directly into chipboard, softwoods, etc, although in such applications a special point can be put on to produce a self-setting action as the rivet is driven into the material. Bifurcated rivets can also be driven through terneplate to fasten it onto wooden boxes.

When bifurcated rivets are used on soft materials they have the advantage that the legs can be forced back into the surface of the material when set, leaving a relatively smooth surface. Bifurcated rivets may also be used with caps which provide both a decorative appearance and enhanced holding power – Fig 5. They can be used in automatic feed rivet setting machines. Dimensions of standard steel bifurcated rivets are given in Table VI, and alternative head forms are shown in Fig 6. Some special purpose bifurcated rivets are shown in Fig 7.

TABLE VI – STANDARD STEEL BIFURCATED RIVET DIMENSIONS*

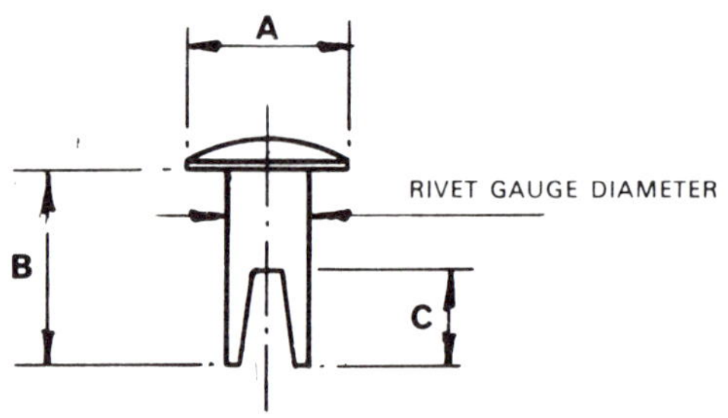

Rivet Length 'B'	RIVET GAUGE DIAMETER													
	16 ·068/·065		**14** ·089/·085		**12** ·108/·104		**11** ·121/·117		**9** ·155/·150		**6** ·200/·195		**3** ·250/·245	
	A	C	A	C	A	C	A	C	A	C	A	C	A	C
1/8	1/8	7/64	5/32	3/32	7/32	3/32	—	—	—	—	—	—	—	—
3/16	1/8	5/32	5/32	5/32	7/32	5/32	7/32	5/32	5/16	1/8	—	—	—	—
1/4	1/8	3/16	5/32	7/32	7/32	7/32	7/32	7/32	5/16	3/16	—	—	—	—
5/16	1/8	3/16	5/32	7/32	7/32	1/4	7/32	1/4	5/16	1/4	3/8	7/32	—	—
3/8	1/8	3/16	5/32	9/32	7/32	5/16	7/32	5/16	5/16	5/16	3/8	9/32	1/2	5/16
7/16	1/8	3/16	5/32	9/32	7/32	5/16	7/32	5/16	5/16	3/8	3/8	11/32	1/2	3/8
1/2	1/8	3/16	5/32	9/32	7/32	5/16	7/32	5/16	5/16	3/8	3/8	13/32	1/2	7/16
9/16	1/8	3/16	5/32	9/32	7/32	5/16	7/32	5/16	5/16	1/2	3/8	15/32	1/2	1/2
5/8	1/8	3/16	5/32	9/32	7/32	5/16	7/32	5/16	5/16	1/2	3/8	1/2	1/2	1/2
11/16	1/8	3/16	5/32	9/32	7/32	5/16	7/32	1/2	5/16	1/2	3/8	1/2	1/2	9/16
3/4	1/8	3/16	5/32	9/32	7/32	5/16	7/32	1/2	5/16	1/2	3/8	1/2	1/2	5/8
13/16	1/8	3/16	5/32	9/32	7/32	5/16	7/32	1/2	5/16	1/2	3/8	1/2	1/2	11/16
7/8	1/8	3/16	5/32	9/32	7/32	5/16	7/32	1/2	5/16	1/2	3/8	1/2	1/2	3/4
15/16	1/8	3/16	5/32	9/32	7/32	5/16	7/32	1/2	5/16	1/2	3/8	1/2	1/2	3/4
1	1/8	3/16	5/32	9/32	7/32	5/16	7/32	1/2	5/16	1/2	3/8	1/2	1/2	3/4

All dimensions in inches

* The Bifurcated & Tubular Rivet Co Ltd.

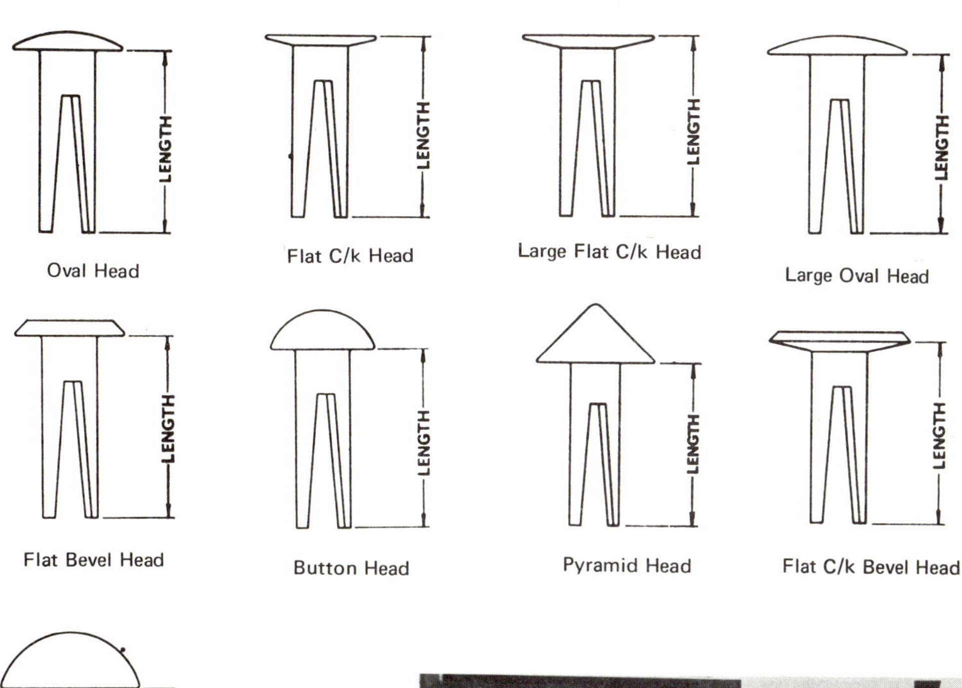

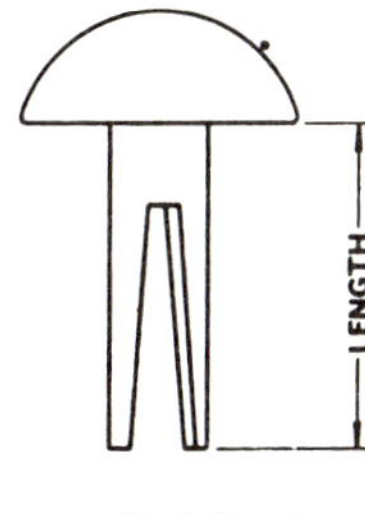

Fig 6

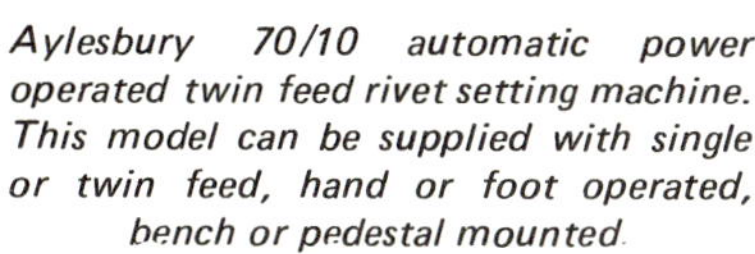

Aylesbury 70/10 automatic power operated twin feed rivet setting machine. This model can be supplied with single or twin feed, hand or foot operated, bench or pedestal mounted.

(The Bifurcated & Tubular Rivet Co Ltd)

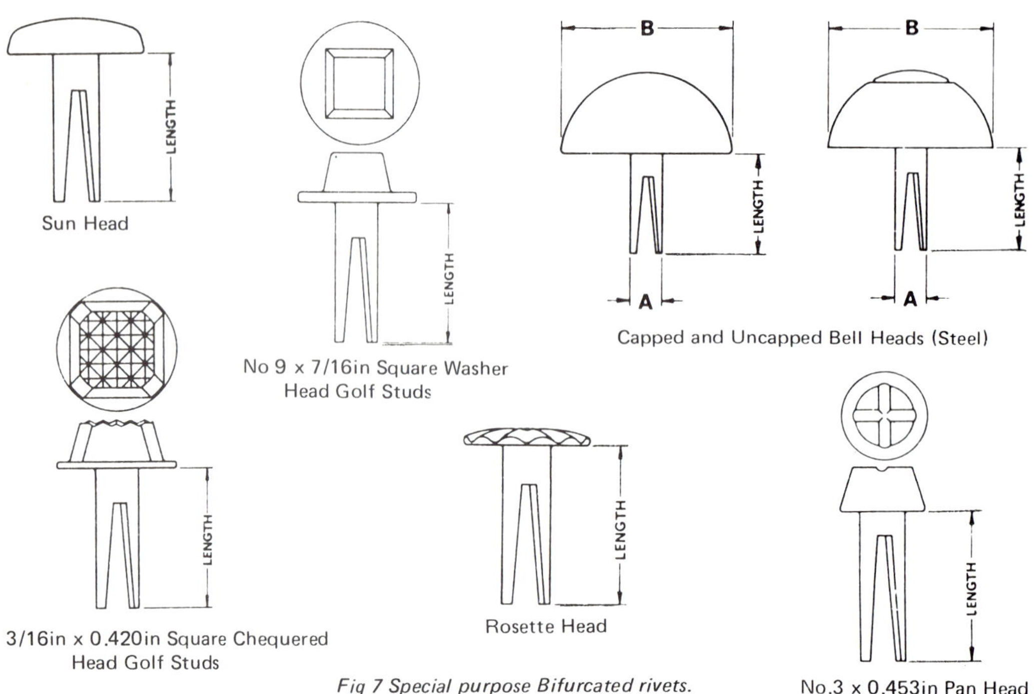

Fig 7 Special purpose Bifurcated rivets. (The Bifurcated & Tubular Rivet Co Ltd)

Drive Rivets

Drive rivets may be solid, tubular (or semi-tubular) or split. They have a screw or spline-form shank to provide grip when driven into matching size holes. Some examples of drive rivet forms are shown in Fig 8. Strictly speaking they cannot be classified as true rivets since they are not clenched over to complete the fastening (this same distinction may apply in certain types of bifurcated rivet which may be driven 'blind' into a material thicker than the length of the rivet).

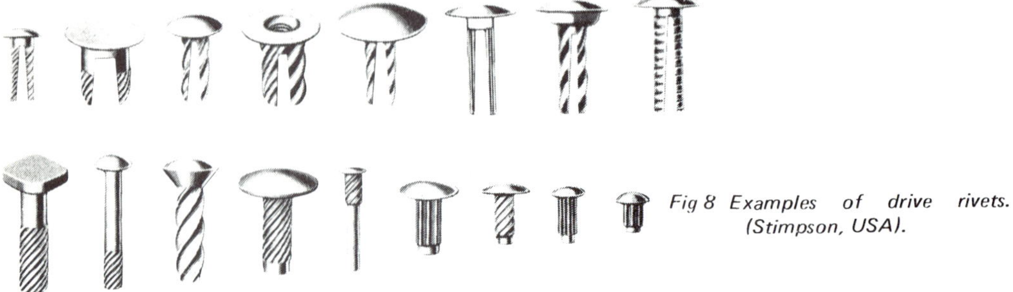

Fig 8 Examples of drive rivets. (Stimpson, USA).

'Tube' Rivets

Fully hollow tubular rivets may be used in the construction of some aircraft and are normally clenched with a spinning or squeezing tool. Their purpose is to span open or unsupported areas. Where the unsupported length of th. rivet is relatively large the rivet may be supported by a tubular distance piece – Fig 9.

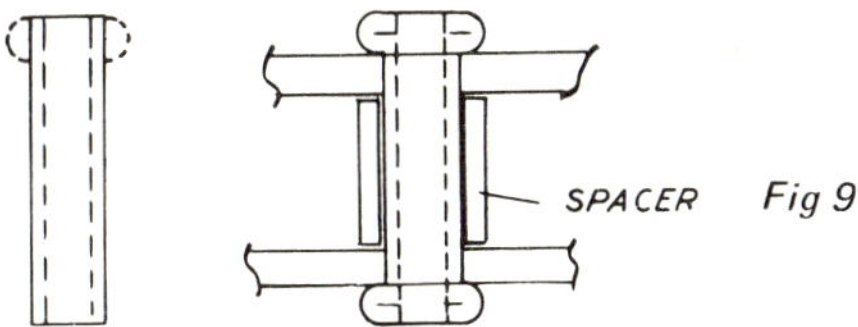

Fig 9

Cup Rivets

Cup rivets are light duty tubular rivets easily clenched by power tools, or manually with a hammer with or without a dolly. They are available with both plain (open) and capped heads. Head diameter is normally 1.3 times shank diameter, the formed head being of a similar diameter after clenching – Fig 10. Heading allowance is rather less than solid raised head rivets.

Cup rivets are only suitable for riveting plates with a total thickness of up to the rivet shank diameter. See also chapters on *Blind Rivets, Eyelets and Tools.*

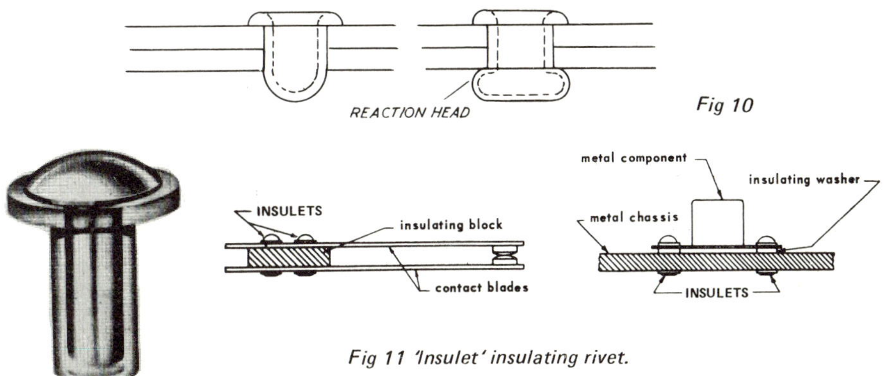

Fig 10

Fig 11 'Insulet' insulating rivet.

Electrical Rivets

A variety of special rivets are produced for electrical assemblies. Non-conducting or insulating rivets are used for joining electrical components which need to be insulated from each other. A proprietary example and two typical applications are shown in Fig 11. The rivet in this case has a metal body sheathed in nylon, expanding into a washer section under the head.

Various other rivets are designed specifically for use on printed circuit boards, eg 'terminal rivets'. These are usually fully tubular rivets in copper, shank length and diameter being designed for use with standard thicknesses of board drilled with holes of specific diameter. See also chapter on *Electrical Fasteners.*

Solid Rivets

Solid rivets in 'light' sizes are mainly used for general engineering purposes.

Rivet nomenclature is given in Fig 12 and typical manufactured head forms and proportions in Fig 13. Formed heads can vary in shape according to different industrial practice. For general engineering, and particularly with larger rivets, the formed head (after driving) is generally of similar form to the manufactured head. However the formed head may also be button shaped. Countersunk rivets are commonly formed with a button shaped driven head, but may also be formed with a button head.

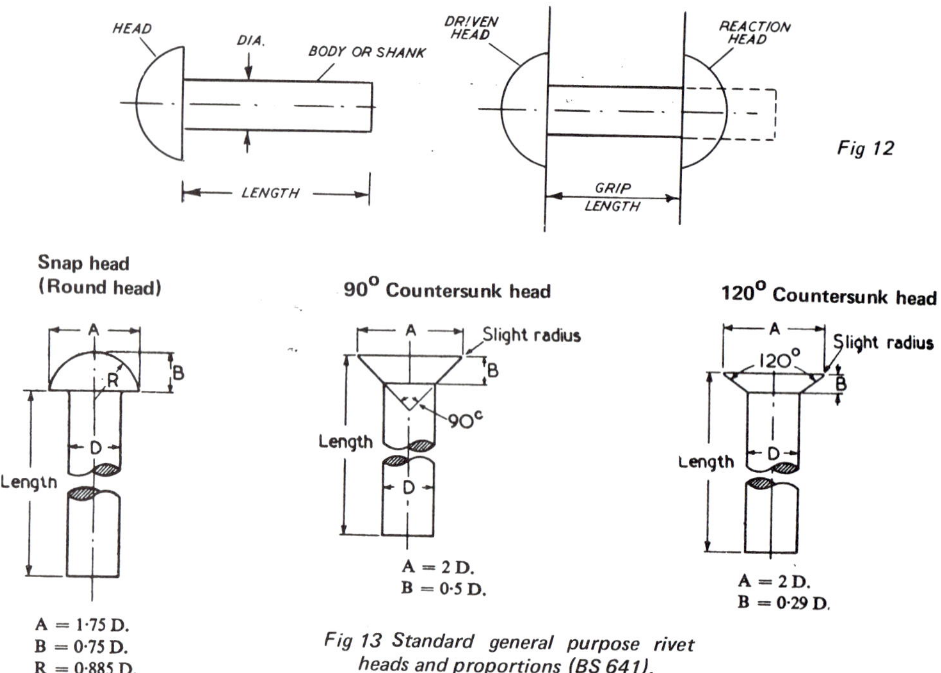

Fig 12

Fig 13 Standard general purpose rivet heads and proportions (BS 641).

Heading allowance, which determines the size of the formed head and also the grip when using a contoured dolly, is generally similar for all raised head rivets. A similar allowance may also be specified for countersunk rivets, but can be as much as 66% less. Typical heading allowances are summarised in Table VII.

In British practice, rivet diameter sizes are specified either in fractions of an inch or for intermediate sizes in equivalent swg numbers. The sizes available in the different types are shown in Table VIII. (British Standard metric size rivets are listed in Table X). Aircraft rivets are now specified in dash sizes – see Table IX.

The most economic size of rivet is approximately three times the thickness of the material through which it passes, although this general rule may not always be practical. Nominally a rivet diameter of 1.2 x the plate thickness is used. The length of rivet required is equal to the heading allowance plus the combined thickness of the materials to be formed. In the case of small rivets, lengths are usually to the nearest 1/32in or 1mm.

If the rivet is too long, excessive bearing is needed to form the driven head, which may crack the rivet and/or damage the surrounding material. If the rivet is formed with a snap action, an over-length rivet can lead to a flash or surplus metal being generated under the manufactured head. If the rivet is too short the driven head will be too small, and thus weak.

Other riveting faults which can occur are shown in Fig 14. If the dolly is not held square the rivet will be deformed and not properly clenched, and the surrounding material may also be damaged. If the plates are not held together securely during driving, then there is the possibility of the shank swelling and spreading between the plates.

TABLE VII – HEADING ALLOWANCE FOR SMALL RIVETS

Rivet Diameter		All Raised Heads		Heading Allowance Countersunk Heads (min)	
in	mm	in	mm	in	mm
1/16	1.6	0.09	2.3	0.03	0.75
3/32	2.4	0.12	3.0	0.05	1.27
1/8	3.2	0.16	4.0	0.06	1.5
5/32	4.0	0.19	4.8	0.08	2.0
3/16	4.75	0.23	5.8	0.09	2.3
1/4	6.4	0.31	7.9	0.13	3.3
5/16	8.0	0.39	9.9	0.16	4.1
3/8	9.5	0.47	11.9	0.19	4.8

TABLE VIII – SMALL GENERAL PURPOSE RIVETS TO BS641
(Mild Steel and Non-Ferrous)

Nominal Diameter (in) / Head Type	1/16	3/32	1/8	5/32	3/16	1/4	5/16	3/8	7/16	1/2
Snap Head	o	o	o	o	o	o	o	o	o	
Pan Head					o	o	o	o	o	
Mushroom Head	o	o	o	o	o	o	o	o	o	
Flat Head	o	o	o	o	o	o	o	o	o	o
60° Countersunk					o	o	o	o	o	
90° Countersunk	o	o	o	o	o	o	o	o	o	
120° Countersunk			o		o	o	o	o		

TABLE IX – 100° COUNTERSUNK TRUNCATED RADIUSED HEAD RIVETS
AEROSPACE SERIES (BS2SP142 and SP143)

Dash Number	Nominal Diameter mm	Nominal Diameter in
–24	2.4	0.094
–32	3.2	0.126
–40	4.0	0.157
–48	4.8	0.189
–56	5.6	0.220
–64	6.4	0.250
–80	8.0	0.315
–96	9.6	0.378

Materials:

SP142 L86, anodised, colour violet
SP143 L37, plain

Identification marks:

SP142 o indented
SP143 –o– embossed

TABLE X – BRITISH STANDARD METRIC SIZE RIVETS FOR GENERAL ENGINEERING PURPOSES (BS4620)

Nom Shank Dia (mm)	COLD FORGED				HOT FORGED		
	Snap Head	Universal Head	Flat Head	90° Countersunk	Snap Head	Universal Head	60° Countersunk
1	○	○	○	○			
1.2	○	○	○	○			
1.6	○	○	○	○			
1	○	○	○	○			
2.5	○	○	○	○			
3	○	○	○	○			
(3.5)	○	○	○	○			
4	○	○	○	○			
5	○	○	○	○			
6	○	○	○	○			
(7)	○	○	○	○			
8	○	○	○	○			
10	○	○	○	○			
12	○	○		○			
(14)	○	○			○	○	○
16	○	○			○	○	○
(18)					○	○	○
20					○	○	○
(22)					○	○	○
24					○	○	○
(27)					○	○	○
30					○	○	○
(33)					○	○	○
36					○	○	○
(39)					○	○	○

* Non-preferred sizes in brackets

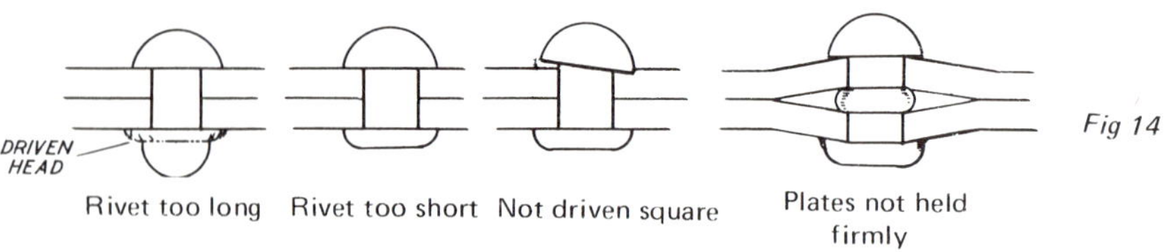

Fig 14

In the case of countersunk rivets, an oversize hole will result in a weak head, as will an undersize heading allowance. The heading allowance should be sufficient to leave enough material to form the head after the rivet shank has been expanded to fill the rivet hole. Weak rivets will also result if the countersink is too deep or too shallow, or not square with the surface of the plate.

Rivets may be loaded in tension or shear, or a combination of these loads, depending on the assembly involved. The most common application is for loading in shear when ultimate failure can occur in any one of the following ways:-

(i) Shearing of the rivet
(ii) Deformation of the rivet holes in the plate
(iii) Tearing of the plate at right angles to the axis of pull
(iv) Tearing of the plate at its edge at right angles to the axis of pull
(v) Crushing of the rivets

In a thin plate, the plate itself may fail in bearing before the rivet fails in shear, in which case the ultimate bearing pressure of the plate is the failure criterion. In thicker plates, the shear strength of the rivets will be the failure criterion, the rivet being either in single shear or double shear – see Fig 15.

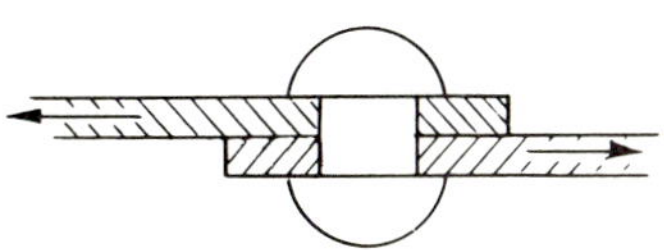

SINGLE SHEAR

Fig 15

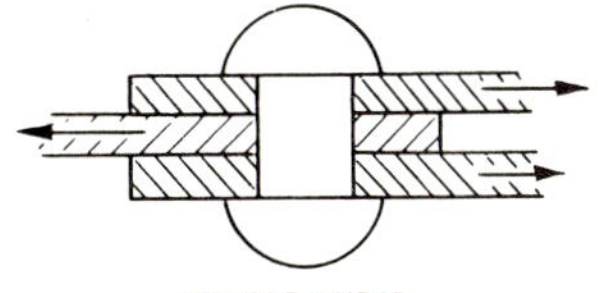

DOUBLE SHEAR

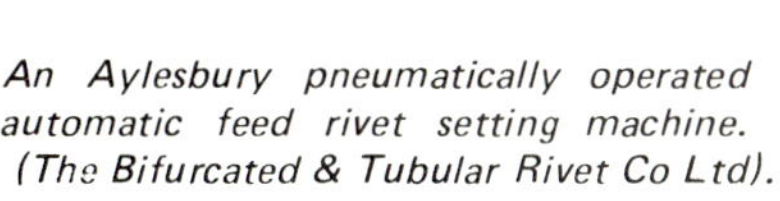

An Aylesbury pneumatically operated automatic feed rivet setting machine. (The Bifurcated & Tubular Rivet Co Ltd).

Basic formulas are:-

Rivet : single shear $L_s = 0.7854\, d^2\, f_s$

double shear $L_d = 0.15708\, d^2\, f_s$

Plate bearing: $L_b = d\, t\, f_b$

when d = diameter of rivet

t = thickness of plate

f_s = ultimate shear strength of rivet material

f_b = ultimate bearing pressure of plate material

Where a riveted joint is subject to both shear and tensional loads the resulting stresses can be determined as follows:-

$$\text{maximum shear stress} = \sqrt{\frac{f_t^2}{4} + f_s^2}$$

$$\text{maximum tensile stress} = \frac{f_t}{2} + \sqrt{\frac{f_t^2}{4} + f_s^2}$$

where f_t = tensile stress caused by tensile force
f_s = shear stress caused by shear force

For practical calculations maximum permissible material stresses would be substituted for ultimate strengths, and in cases of doubt both the resulting rivet load (L_s or L_d) and plate bearing load (L_b) calculated and compared. The lower value will be the true calculated strength.

In practice, the ultimate bearing stress for the plate will depend on the nature of the applied load. If the load acts in a single direction and there is little vibration, the practical value of f_b is given by

$$f_b \text{ (ultimate)} = \text{ft } (4.45 \frac{t}{d} + 1.22) \text{ for steel plates.}$$

The proof bearing pressure f_b (proof) is taken as equal to 1.4 times the ultimate tensile strength of the plate material (ft).

If the load is applied in alternating directions, or there is heavy vibration present, then the proof bearing pressure f_b (proof) should be taken as equal to the ultimate tensile strength.

The corresponding formula for duralumin plate is

$$f_b \text{ (ultimate)} = \text{ft } (1.9 \frac{t}{d} + 1.63)$$

proof bearing pressure f_b = ft for uni-directional loads and low vibration
= 0.8 ft for multi-directional loading and/or heavy vibration.

The actual load on a riveted assembly is normally shared by a number of rivets. Thus rivet and/or bearing strength can be calculated on the basis that the load is equally shared by the rivets – ie the riveting is 100% efficient. In practice this is unlikely to be realised, and so the likely rivet efficiency must also be estimated. Typical efficiencies for general work are:-

lap joints – single row of rivets – 55%
double row of rivets – 70%
triple row of rivets – 75–80%
butt joints – single row of rivets – 65%
double row of rivets – 80%
triple row of rivets – 85%

For high class riveting (eg to aircraft standards of production) a nominal rivet efficiency of 80% can be assumed. Efficiency figures can also be calculated from approximate formulas.

Rivet pitch and spacing depend on the particular application involved. Normally a minimum pitch of about 3d would be used, with a minimum spacing between rows of 2d or 2.5d.

Fig 16

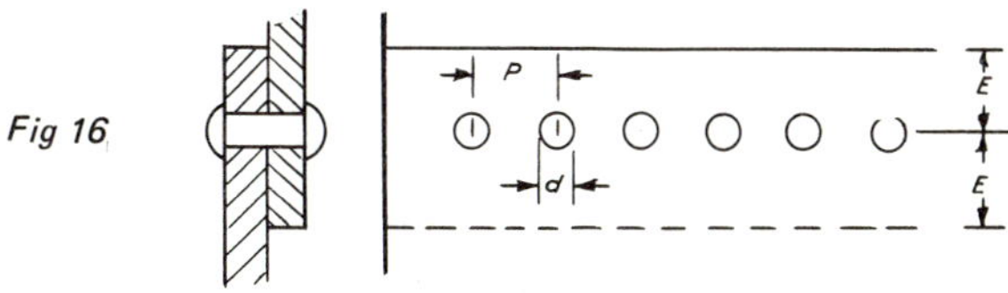

Lap Joints (Fig 16)

In the case of lap joints the following formulas can be applied for heavy engineering duties.

Nomenclature:

- t = thickness of plate
- d = diameter of rivet
- a_1 = sum of sectional areas of rivets in single shear (= area of single rivet x number of rivets)
- a_2 = sum of sectional areas of rivets in double shear
- S_r = 0.85 x tensile strength of rivet material (ton f/in^2)
- S_p = min tensile strength of plate material
- P = pitch of rivets
- V = spacing between lines of rivets

Single Riveted:

maximum pitch (P) = 1.31 t + 1.625 (inches)

minimum edge distance (E) = 1.5 d

$$a_1 = 0.78\, d^2$$

$$\text{Rivet efficiency (\%)} = \frac{100 \times S_r \times a_1}{S_p \times P \times t}$$

$$\text{Plate efficiency (\%)} = \frac{100\,(P - d)}{P}$$

Double Riveted:

P = 2.62 t + 1.625

E = 1.5 d

$a_1 = 1.57\, d^2$

V = 2 d

$$\text{Rivet efficiency (\%)} = \frac{100 \times S_r \times a_1}{S_p \times P \times t}$$

$$\text{Plate efficiency (\%)} = \frac{100\,(P - d)}{P}$$

Triple Riveted:

$P = 3.47\ t + 1.625$
$E = 1.5\ d$
$a_1 = 2.36\ d^2$
$V = 2\ d$

$$\text{Rivet efficiency (\%)} = \frac{100 \times S_r \times a_1}{S_p \times P \times t}$$

$$\text{Plate efficiency (\%)} = \frac{100\ (P - d)}{P}$$

Butt Joints (see Fig 17)

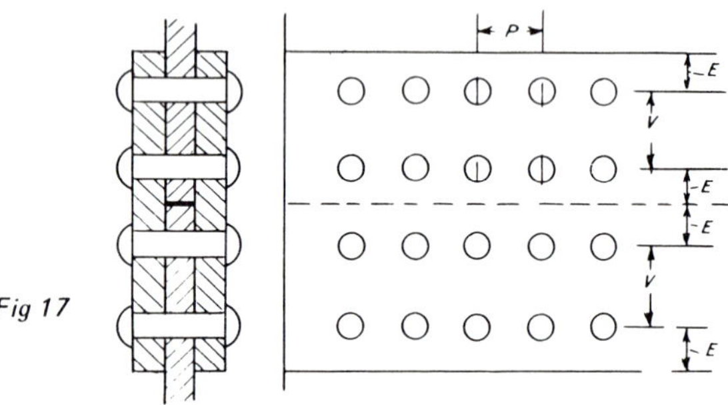

Fig 17

Double Riveted, Double Straps:

$P = 3.5\ t + 1.625$
$E = 1.5\ d$
$a_2 = 1.57\ d^2$
$V = 2\ d$

$$\text{Rivet efficiency (\%)} = \frac{100 \times S_r \times 1.875\ a_2}{S_p \times P \times t}$$

$$\text{Plate efficiency (\%)} = \frac{100\ (P - d)}{P}$$

Triple Riveted, Double Straps:

$P = 6\ t + 1.625$
$E = 1.5\ d$
$V = 0.165P + 0.67\ d$

$$\text{Rivet efficiency (\%)} = \frac{100 \times S_r \times 1.875\ a_2}{S_p \times P \times t}$$

$$\text{Plate efficiency (\%)} = \frac{100\ (P - d)}{P}$$

Blind Rivets

The general description 'blind rivet' applies to all types of hollow rivet designed so that they can be fitted and clenched from one side of the work only. Their application is not necessarily confined to blind assemblies. They can be classified according to the method by which the blind end of the rivet is expanded to set the rivet, viz:-

(i) *Pull-through mandrel* – where the basic components are a hollow rivet body and a mandrel. The mandrel is pulled right through the body during setting to upset the blind end of the body forming the head and leaving a hollow rivet.

(ii) *Pop rivets* – where the mandrel is pulled through the hollow rivet body to form the head. Further axial pull then causes the mandrel to break. This may leave the head of the mandrel loose in the rivet (break-stem mandrel) or break its head off against the formed rivet head (break-head mandrel). In the latter case the mandrel head is loose and can fall or be pushed out to leave a hollow rivet.

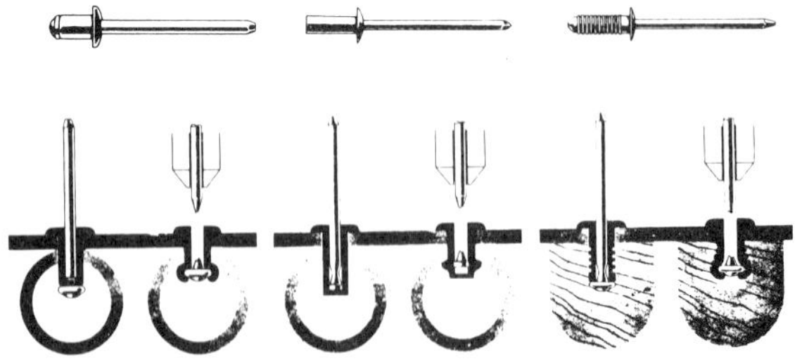

Standard sealed and grooved types of 'pop' rivets (Tucker Fasteners Ltd).

Short break (left) and long break (right) 'pop' rivets.

Break stem (left) and break head (right) 'pop' rivets.

(iii) *Non-break mandrel* – where the mandrel is pulled into or against the rivet body to form the head and then remains in this position unbroken. It is then necessary to dress the mandrel against the manufactured head.

(iv) *Drive mandrel* (drive rivets) – where the mandrel projects from the manufactured head and is pressed or driven into the body of the rivet to form the head on the blind side by a flaring operation. The mandrel ends up flush with the manufactured head and requires no dressing.

(v) *Threaded mandrel* – where the rivet body is internally threaded and the mandrel is externally threaded. Torquing or pulling the mandrel causes the rivet body to expand and form a head on the blind side of the assembly.

(vi) *Explosive rivet* – a hollow rivet with the blind end filled with an explosive chemical. After positioning, heat or electric current is applied to the rivet head to detonate the small explosive charge, expanding the walls to form a blind head.

See also *Plastic Rivets* in chapter on *Plastic Fasteners*

Rivet Selection (Tucker 'Pop' Rivets)

Rivet Diameter – whilst in a few cases this may be decided by the size of an existing hole in new designs the main factor is usually the shear or tensile strength required from the riveted joint. The rivet manufacturer's catalogue should therefore be consulted to obtain the requisite figures, bearing in mind that in general rivet price increases proportionately to size.

Rivet Length – the manufacturer's catalogue also publishes the correct rivet length for each thickness of material. Some users may prefer to standardise on one rivet for all thicknesses of material. This is perfectly possible as any rivet which is longer than that recommended for a particular thickness will always set satisfactorily in thinner material. However, the user must balance the gains obtained from such standardisation against the higher purchase price which must be paid for the longer rivet.

Rivet Material – choice of rivet material will probably be related to the strength required but other considerations such as cost, corrosion resistance, weight, performance under elevated temperatures, may also affect the decision.

Rivet Cost – broadly speaking, the cost of the rivets reviewed in the previous section is cheapest in the case of aluminium alloy, followed by copper, steel and monel in that order. Rivet prices naturally increase proportionately to diameter and length.

Rivet Head Style – the reasons for selection of domed or countersunk head rivets are self-evident.

Mandrel Type – the break stem mandrel is usually chosen for the 'Pop' rivet, as it acts as a weatherproof plug in the set rivet. Also it may be inconvenient to eject a mandrel head on the 'blind' side of an enclosed structure. The break stem mandrel can also form a suitable 'key' for any filling compound or solder which the user wishes to employ. The break head mandrel is usually specified where weight of the finished product is an important consideration, or where a completely clear hole is required through the set rivet.

'Pop' Sealed Rivets – many of the foregoing considerations apply in selection of the correct size of sealed rivets. However, in addition, this type of rivet should be chosen where a completely sealed and pressure-tight joint is necessary. As its high rate of radial expansion in setting completely fills the hole, it has a high proof-shear strength which also makes it particularly suitable for joints subject to vibratory conditions.

'Pop' Sealed Mandrel Types – the short break mandrel is most commonly employed, but where a completely flush finish is required to the head of the rivet, or where the added shear strength of the steel mandrel core is desired, the correct choice lies with the long break mandrel.

Recommended Riveting Practice (Tucker 'Pop' Rivets)

Rivet Diameter – in load-bearing joints the diameter of the rivet should be at least equal to the thickness of the thickest sheet, but should not be greater than three times the thickness of the sheet immediately under the rivet head.

Edge Distance – In lap or butt joints which are likely to be subjected to shear or tensile loads, rivet holes should not be drilled within a distance equal to two rivet diameters of the edge of the sheet, nor should this edge distance exceed 24 rivet diameters.

Pitch of Rivets – in joints carrying appreciable loads, the distance between rivets in the same row should not exceed six rivet diameters. In butt joints it may be desirable to consider the inclusion of a stiffening cover strip fastened to the underlying sheets by staggered rivets. Even in joints of non-load bearing character, rivet pitch should not exceed 24 times the thickness of the thinnest sheet being joined.

Rivet Material – as mentioned in the previous section, the most important consideration in the choice of rivet material will probably be strength. However, where this involves selecting a rivet of a different metal from that of the sheets being joined, it is advisable to guard against the possibility of electrolytic corrosion occurring in the riveted structure. As a general guide, it is inadvisable for the P.D. between any two metals in contact to exceed 0.5 volts or, where the joint is exposed to the weather, 0.25 volts. Anodising or plating of rivets inhibit electrolytic corrosion to some extent but where the P.D. between the rivet material and the surrounding sheets is great, the plating will necessarily be of a sacrificial nature and it may be necessary to consider some additional protection, such as painting of the hole prior to riveting, if the joint is likely to be exposed to conditions particulary conducive to corrosion.

Tool Clearance – whilst blind rivets can be set in positions often inaccessible to other fasteners, it is advisable to check with the manufacturer's catalogue on the minimum distance from a flange or corner to the centre-line of the rivet hole which is permissible. 'Tucker' blind riveting tools can approach to within 3/16in (4.8mm) or in some cases 3/32in (4.0mm) of such a projection. Depth of a channel or cylindrical section can likewise limit the satisfactory setting of a rivet although, again, the use of special heads or extended nosepieces can overcome most of the problems involved. Where possible, it is of course desirable to design a joint with unlimited access in order to obtain maximum riveting speeds.

Differing Sheet Thicknesses and Compressible Materials – where sheets of differing thickness or strength are being joined it is desirable that the rivet should set against the stronger member. Thus if plastic sheet is being fastened to metal, the former should lie immediately under the rivet head.

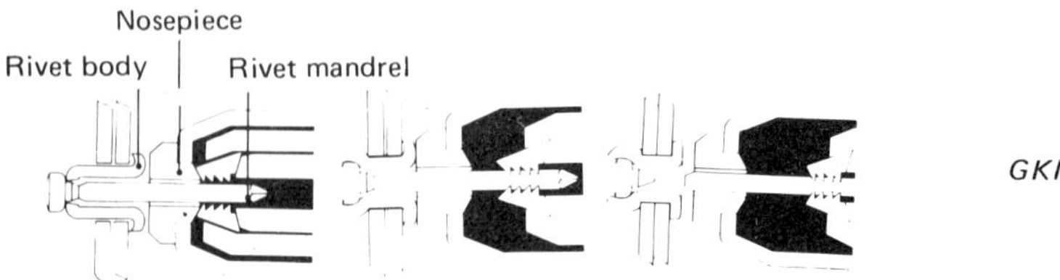

GKN blind rivet.

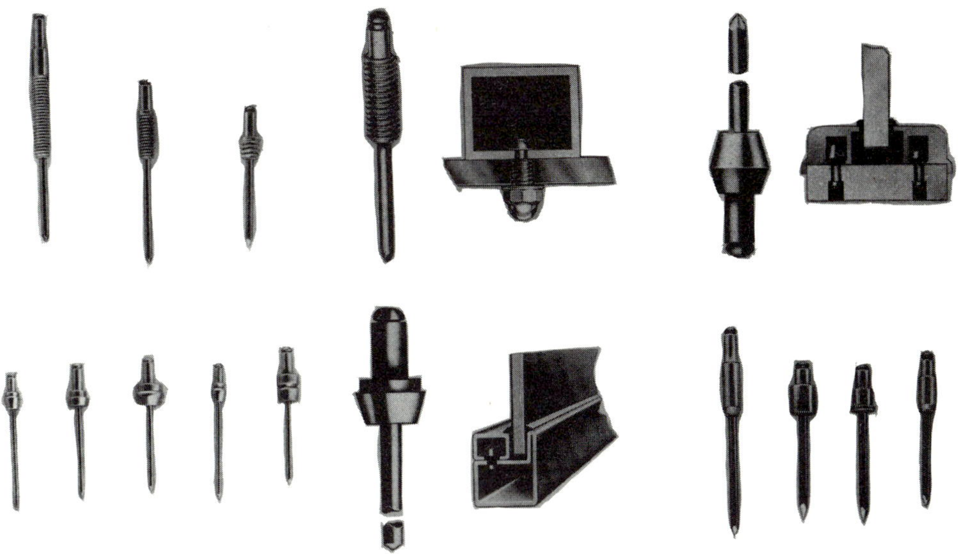

Special 'Gesipa' blind rivets.

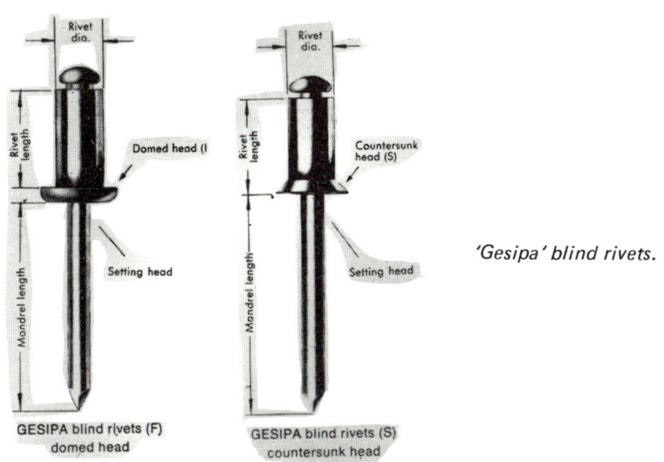

'Gesipa' blind rivets.

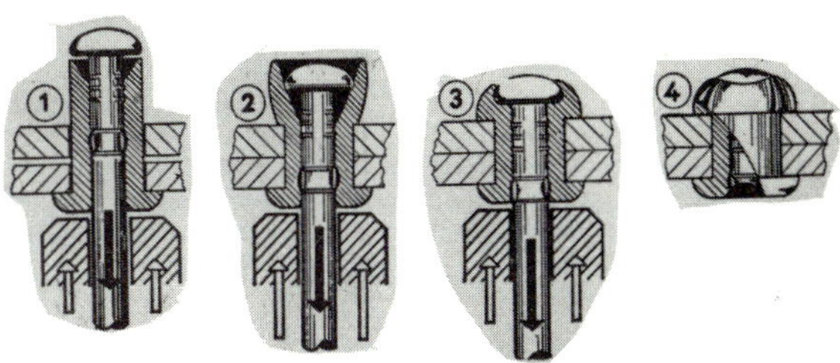

Stages in setting 'Gesipa' blind rivet.

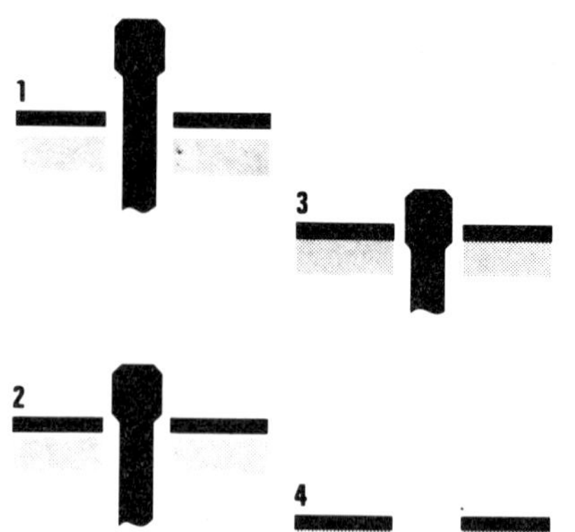

1. The pre-drilled holes of the component are placed over the rivet.
2. Actuation, draws the mandrel from the tail end of the rivet, compressing the rivet material towards its head.
3. The mandrel continues to be pulled through the rivet, symetrically expanding the shank to fill the hole. This action continues until the material is clamped tight.
4. The rivet is finally installed.

Briv high clench rivet (Avdel).

Hole Size – drill sizes for each diameter of rivet are usually specified in the manufacturer' catalogue. These are usually designed to give a clearance of 0.002–0.005ins (0.05–0.125mm between the rivet and the hole and in the case of rivets capable of only limited radial expansion these recommendations should be observed if the best performance from the rivet is to b obtained. The sealed rivet is not so susceptible to hole clearance as the majority of other blin rivets.

When set in holes of progressively increasing clearances up to 0.019ins (0.05mm) over nomin rivet diameter, its ultimate shear strength increases correspondingly.

MATERIAL SELECTION GUIDE FOR BLIND RIVETS (GKN)

SHEET MATERIAL:		ALUMINIUM				MILD STEEL			STAINLESS STEEL
HUMIDITY:		DRY†		WET		DRY†		WET	
CHOICE:		1st	2nd	1st	2nd	1st	2nd	1st	
SHEET FINISH	Unprotected	A	S	A	M	S	A	S	The use of Stainless Steel implies a corrosive environment, use: *Monel* Rivets.
	Prepainted or Galvanised	A	S	A	M	S	A	M	
	Painted after assembly	A	S	A	M	S	A	S	

†DRY refers to atmospheres in which the relative humidity is less than 75%.
A—Aluminium S—Steel M—Monel

DIAMETER SELECTION GUIDE FOR BLIND RIVETS

FAILURE LOAD:	IN SINGLE SHEAR					
RIVET DIAMETER:	1/8″		5/32″		3/16″	
	lbs	kgs	lbs	kgs	lbs	kgs
RIVET MATERIAL: Aluminium	140	63	210	95	320	145
Steel	260	118	370	168	540	245

GESIPA
We sell togetherness
GESIPA are experts in bringing any two surfaces together, speedily and permanently. Our GESIPA RIVETS (in an extensive range of designs, sizes and materials) are used in a wide variety of applications and the hand and pneumatic/hydraulic tools that set them ensure efficient fastening.
GESIPA
till death us do part
Gesipa Fasteners Ltd, Dalton Lane, Keighley, Yorks.
Tel: (05352) 7844 Telex: 51429 (Gesipa Keighley)

THE FIRST!
Over 600 pages containing thousands of diagrams, tables, charts, illustrations, etc., stiff board bound and gold blocked.
HANDBOOK OF POWER DRIVES
HANDBOOK of POWER DRIVES
TRADE AND TECHNICAL PRESS LTD.
Trade and Technical Press Ltd.
TRADE & TECHNICAL PRESS LTD.
CROWN HOUSE, MORDEN, SURREY. SM4 5EW

TABLE I – NEW 'POP' STANDARD OPEN TYPE RIVET RANGE

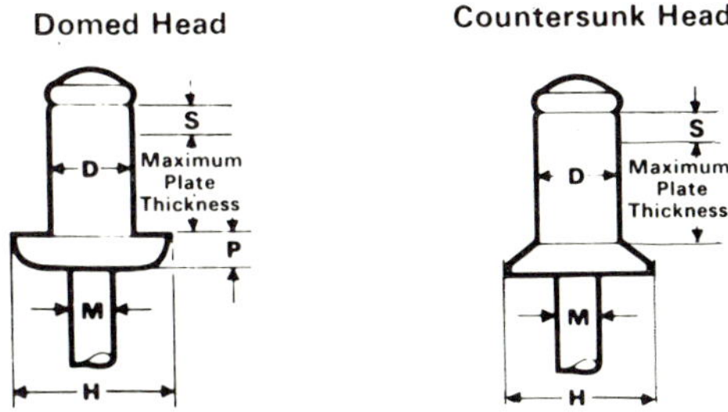

Rivet Diameter ins.	Hole Size ins.	Maximum Plate Thickness ins.	New⊕ Rivet Code No.	A Nearest Equivalent Size Old Code No.	D ins.	H Domed and Counter-sunk ins.	P Domed ins.	M 001 ins.	S 015″ — 005″ ins.	Minimum failing loads in Single Shear
		·062	**42**	420					·108	
		·125	**44**	423					·116	
125	130	·187	**46**	429	·125	·250	·036	·072	·123	*180 lb.
		·250	**48**	435		·007	·003		·132	
		·312	**410**	440					·139	
		·375	**412**	450					·147	
		·437	**414**						·155	
		·500	**416**	460					·163	
		·562	**418**	466					·170	
		·062	**52**	518					·132	
		·125	**54**	523					·141	
156	·161	·187	**56**	529	156	·312	·045	·090	·148	*300 lb.
		·250	**58**	537		·007	·003		·156	
		·312	**510**	545					·164	
		·375	**512**	550					·172	
		·437	**514**	560					·180	
		·500	**516**	565					·187	
		·094	**63**	625					·161	
		·125	**64**	629					·165	
·187	·193	·187	**66**	635	·187	·375	·054	·104	·174	*455 lb.
		·250	**68**	640		·010	·004		·181	
		·312	**610**	649					·189	
		·375	**612**	657					·197	
		·437	**614**						·204	
		·500	**616**	665					·212	
		·532	**617**	675					·216	
·250	·2559	·250	**88**	850	·250	·500	·075	·144	·250	*720 lb.
		·500	**816**	875		–·015	·005		·266	
		·750	**824**	8100		—·010			·282	

* Tensile strengths are approximately 40% above quoted Shear Strengths.

⊕ Tucker Fasteners Ltd

Bulb-Tite Blind Rivet (Linread Ltd)

This patented blind rivet has an unsetting action which forces the outer sleeve to form three 'bulbs' with a generous bearing area, preventing crazing on brittle materials and giving high resistance to pulling out in thin or soft materials. It is also available in self-sealing form with a pre-assembled neoprene sealing washer.

Nominal diameter sizes are 1/4in, 5/32in, 3/16in, and 9/32in. Standard material is aluminium, with raised or countersunk head, but standard sizes are also available in monel, steel and silicon bronze and in other head styles. Plastic snap-on caps are also available to fit certain head styles.

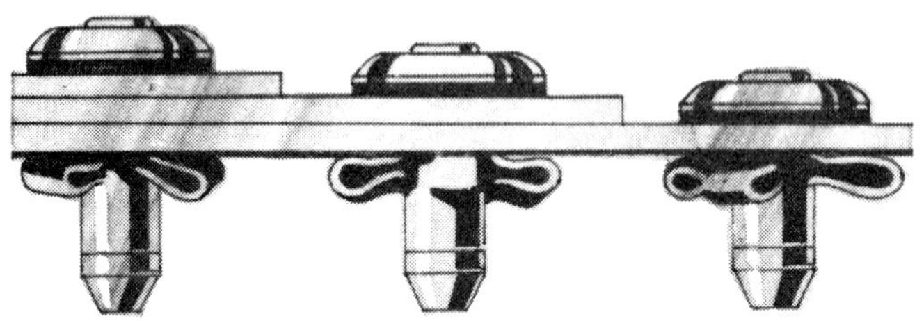

BULB-TITE PERFORMANCE

Material	Nom Dia	TYPICAL ULTIMATE VALUES — LBS		
		Shear	Tension	Clamp up
Aluminium	5/32″			
Aluminium	3/16″	740	460	130
Aluminium	1/4″	1100	675	250
Aluminium	9/32″	1500	1100	375
Monel	3/16″	1350	730	350
Steel	1/4″	1650	875	500
Silicon Bronze	9/32″	2250	1700	375

Eyelets

Eyelets are low cost hollow rivet type fasteners which may be manufactured in a variety of configurations.

In addition to fastening (riveting), eyelets may also be used for the following functions:-

(a) Hole reinforcement – eg in plastic sheet materials, canvas, etc.

(b) Light duty bushings or bearings.

(c) Connectors – eg in electronic assemblies.

(d) Pivots.

(e) Ventilators – eg breather holes in boots, clothing, mattresses, etc.

(f) Spacers – eg in mechanical sub-assemblies.

The various types of eyelet are round (type A), flat (Type B) and funnel (Type C) – see Fig 1. Other standard shapes are oval, sail and grommet eyelets, the last two types being associated with matching rings or washers – Fig 2.

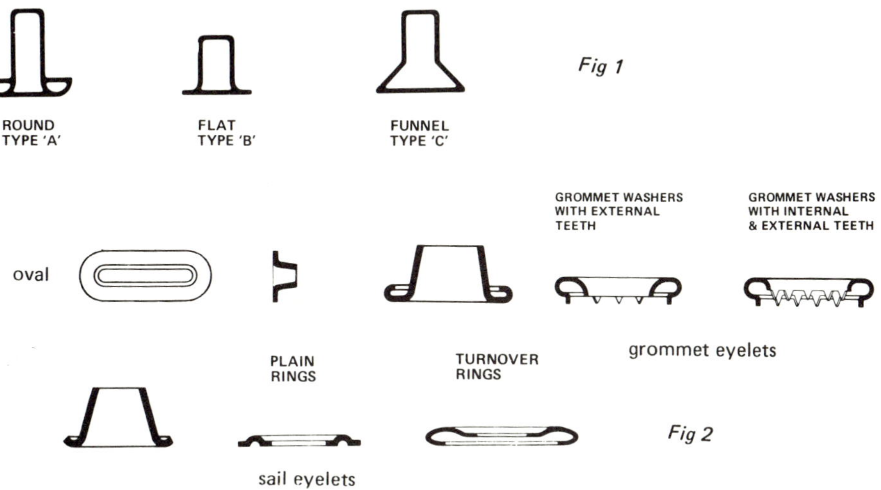

Fig 1

Fig 2

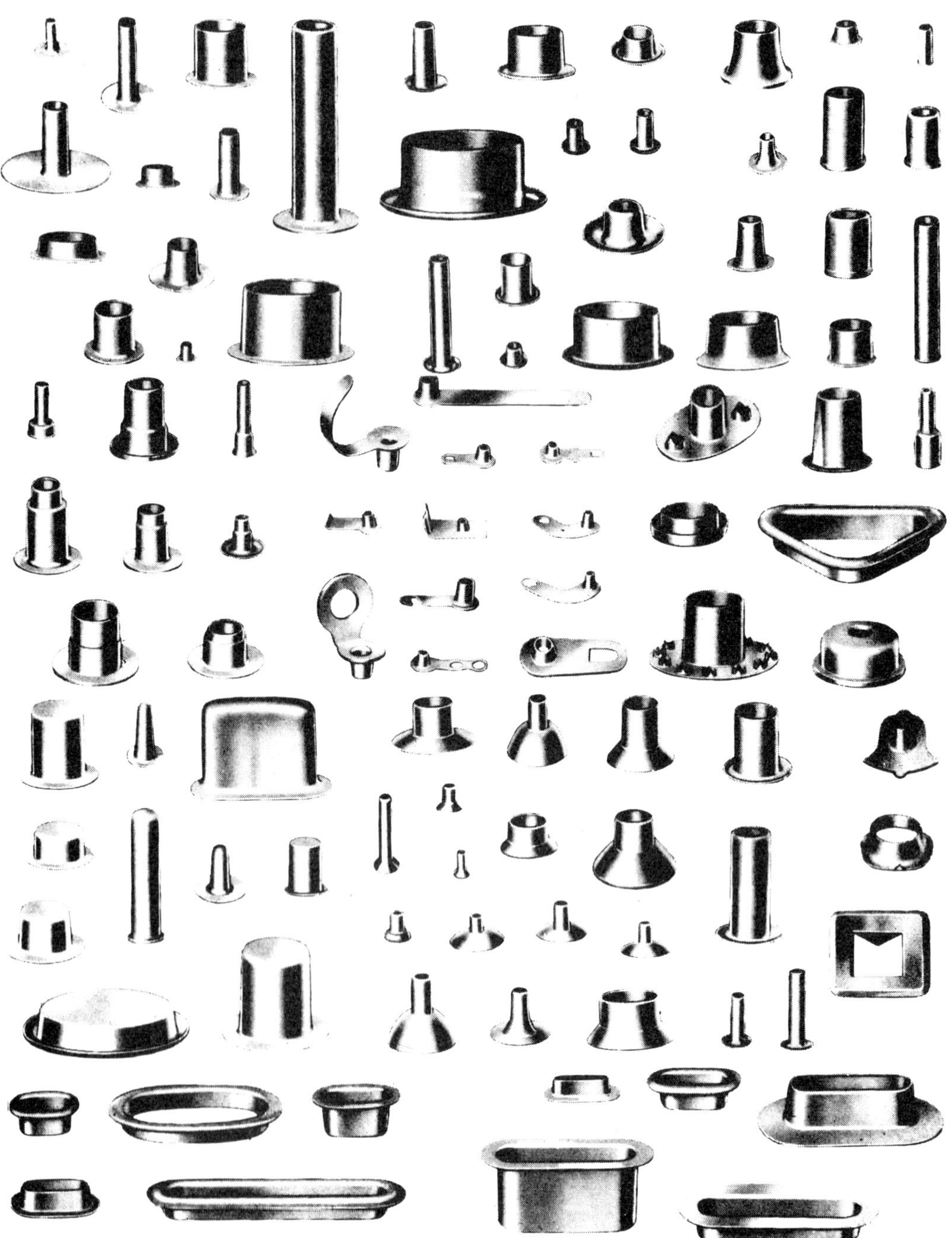

Examples of grommets and washers.
(Simpson USA)

TABLE I – GENERAL PURPOSE EYELETS

Barrel Diameter		Length Under Flange		Overall Length		Flange Diameter		Material	Flange Form	Weight per Million		Gauge of Material	
mm	in	mm	in	mm	in	mm	in			kg	lb	mm	in
1.27	0.050	2.06	0.081	2.54	0.100	2.16	0.085	Brass	A	16.78	37	0.008	0.203
1.68	0.066	3.60	0.142	4.06	0.160	2.79	0.110	Brass	A	37.65	83	0.330	0.013
1.68	0.066	4.95	0.195	5.41	0.213	2.79	0.110	Brass	A	47.63	105	0.305	0.012
1.98	0.078			2.59	0.102	3.07	0.121	Brass	A	38.10	84	0.330	0.013
1.98	0.078	5.99	0.236	6.40	0.252	3.20	0.126	Brass	A	80.29	177	0.330	0.013
1.98	0.078	7.97	0.314	8.45	0.333	3.17	0.125	Brass	A	92.99	205	0.203	0.008
2.18	0.086	3.55	0.140	4.46	0.176	5.58	0.220	Brass	A	109.32	241	0.381	0.015
2.28	0.090			6.35	0.250	3.81	0.150	Brass	B	99.79	220	0.381	0.015
2.28	0.090	5.96	0.235	6.47	0.255	5.58	0.220	Brass	A	149.69	330	0.381	0.015
2.29	0.090	12.70	0.500			3.81	0.150	Brass	A	176.90	390	0.203	0.008
2.54	0.100	2.08	0.082			5.99	0.236	Brass	A	68.49	151	0.254	0.010
2.79	0.110			2.54	0.100	6.73	0.265	Brass	B	85.28	188	0.254	0.010
2.84	0.112	2.66	0.105	3.05	0.120	5.58	0.220	Brass	B	72.12	159	0.203	0.008
2.94	0.116			6.73	0.265	4.31	0.170	Steel	B	132.45	292	0.279	0.011
3.76	0.148	8.50	0.335	9.06	0.357	5.20	0.205	Brass	A	238.59	526	0.254	0.010
3.81	0.150	3.65	0.144	4.57	0.180	7.92	0.312	Brass	A	276.24	609	0.305	0.012
4.19	0.165	4.44	0.175	5.20	0.205	6.14	0.242	Brass	B	161.93	357	0.254	0.010
4.19	0.165	5.55	0.218	6.35	0.250	6.35	0.250	Brass	C	244.94	540	0.305	0.012
4.29	0.169			3.56	0.140	6.10	0.240	Brass	A	71.21	157	0.203	0.008
4.44	0.175			5.97	0.235	7.75	0.305	Brass	A	186.43	411	0.203	0.008
4.62	0.182			9.91	0.390	6.35	0.250	Brass	B	320.24	706	0.254	0.010
4.72	0.186			3.76	0.148	7.11	0.280	Aluminium	A	50.88	112	0.229	0.009
4.82	0.190			5.08	0.200	9.01	0.355	Brass	A	207.30	457	0.203	0.008
4.82	0.190	5.60	0.220	6.35	0.250	9.01	0.355	Brass	A	223.17	492	0.203	0.008
4.90	0.193			3.76	0.148	7.11	0.280	Zinc	B	134.27	296	0.229	0.009
4.95	0.195			4.19	0.165	7.24	0.285	Zinc	B	130.63	288	0.229	0.009
5.08	0.200	3.18	0.125	3.94	0.155	7.11	0.280	Brass	C	122.47	270	0.203	0.008
5.08	0.200			7.62	0.300	7.62	0.300	Brass	A	345.18	761	0.305	0.012
5.20	0.205			1.98	0.078	8.00	0.315	Copper	B	97.52	215	0.229	0.009
5.33	0.210	7.18	0.283			10.03	0.395	Steel	A	344.74	760	0.229	0.009
5.38	0.212			5.64	0.222	9.96	0.392	Brass	A	340.19	750	0.203	0.008
5.46	0.215	3.96	0.156	4.44	0.175	11.09	0.437	Brass	B	445.89	983	0.381	0.015
5.59	0.220			11.89	0.468	9.52	0.375	Steel	B	766.57	1690	0.381	0.015
5.71	0.225			7.62	0.300	9.14	0.360	Steel	B	537.52	1185	0.381	0.015
5.92	0.233			7.75	0.305	8.00	0.315	Brass	A	333.85	736	0.229	0.009
5.94	0.234	16.62	0.655	17.26	0.680	7.92	0.312	Brass	A	876.81	1933	0.305	0.012
6.35	0.250	4.57	0.180	5.33	0.210	8.63	0.340	Brass	A	339.29	748	0.381	0.015
6.60	0.260			3.55	0.140	9.14	0.360	Zinc	A	163.75	361	0.229	0.009
7.92	0.312			1.57	0.062	10.31	0.406	Aluminium	B	317.52	700	0.406	0.016
8.12	0.320	3.17	0.125	3.68	0.145	11.04	0.435	Brass	A	301.64	665	0.305	0.012
10.15	0.400	4.22	0.166	5.20	0.250	13.58	0.535	Brass	A	612.36	1350	0.381	0.015
12.83	0.505			3.30	0.130	15.47	0.609	Brass	B	408.70	901	0.305	0.012
15.15	0.596	2.99	0.118	3.30	0.130	16.89	0.665	Brass	B	404.16	891	0.305	0.012
	0.120				0.288		0.340	Steel	A			0.203	0.008
		5.16	0.203	5.94	0.234	15.88	0.625	Brass	B	807.85	1781	0.330	0.013
				3.18	0.125	7.24	0.285	Brass	Snare	247.66	546	0.381	0.015

(Tucker Fasteners Ltd)

Eyelets may also be specified by purpose, eg

(i) *General purpose* – Types A, B and C in brass, aluminium, steel, copper and zinc: also plated, cellulose coated or plastic coated for decorative applications (eg clothing eyelets). See Tables 1 and 1A.

TABLE IA – METRIC EYELETS (TYPE A ONLY)

Barrel Diameter		Length Under Flange		Flange Diameter		Material	Weight per Million		Gauge of Material	
mm	in	mm	in	mm	in		kg	lb	mm	in
2.0	0.078	3.0	0.118	3.2	0.126	Brass	39.01	86	0.16	
2.0	0.078	4.5	0.177	3.2	0.126	Brass	58.97	130	0.16	
2.5	0.098	3.0	0.118	4.0	0.157	Brass	54.89	121	0.16	
2.5	0.098	4.5	0.177	4.0	0.157	Brass	77.11	170	0.16	
3.0	0.118	3.0	0.118	4.5	0.177	Brass	59.88	132	0.16	
3.0	0.118	4.5	0.177	4.5	0.177	Brass	86.18	190	0.16	
3.0	0.118	6.0	0.236	4.5	0.177	Brass	100.25	221	0.16	
3.5	0.137	3.0	0.118	5.5	0.216	Brass	92.99	205	0.20	
3.5	0.137	4.5	0.177	5.5	0.216	Brass	122.93	271	0.20	
3.5	0.137	6.0	0.236	5.5	0.216	Brass	154.22	340	0.20	
4.0	0.157	3.0	0.118	6.5	0.256	Brass	117.48	259	0.20	
4.0	0.157	4.5	0.177	6.5	0.256	Brass	148.78	328	0.20	
4.0	0.157	6.0	0.236	6.5	0.256	Brass	188.24	415	0.20	
4.5	0.177	3.0	0.118	7.0	0.275	Brass	168.29	371	0.25	
4.5	0.177	4.5	0.177	7.0	0.275	Brass	215.01	474	0.25	
4.5	0.177	6.0	0.236	7.0	0.275	Brass	237.23	523	0.25	
5.0	0.196	3.0	0.118	8.0	0.314	Brass	183.25	404	0.25	
5.0	0.196	4.5	0.177	8.0	0.314	Brass	237.69	524	0.25	
5.0	0.196	6.0	0.236	8.0	0.314	Brass	303.91	670	0.25	
6.0	0.236	4.5	0.177	9.5	0.374	Brass	307.09	677	0.25	
6.0	0.236	6.0	0.236	9.5	0.374	Brass	368.78	813	0.25	

(Tucker Fasteners Ltd)

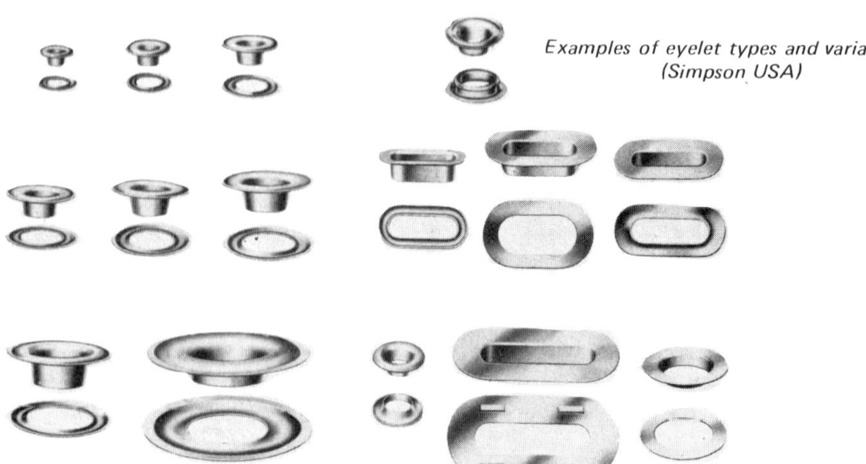

Examples of eyelet types and variations. (Simpson USA)

(ii) *Stationery eyelets* – Types A and B, normally in brass only but some sizes available in aluminium or steel (see Table II).

(iii) *Belt and leather eyelets* – Types A and B, usually in brass or steel (see Table III).

(iv) *Sail eyelets* – Special form with plain or turnover rings, manufactured in brass only (see Table IV).

(v) *Electrical and radio eyelets* – Types A, B and C in copper and brass with self-colour, solder-tinned, nickel plated and tinned, or other specified finishes – see chapter on *Electrical Fasteners.*

TABLE II – STATIONERY EYELETS

Barrel Diameter		Length Under Flange		Overall Length		Flange Diameter		Material	Flange Form	Weight per Million		Gauge of Material	
mm	in	mm	in	mm	in	mm	in			kg	lb	mm	in
2.54	0.100			2.54	0.100	3.30	0.130	Brass		27.22	60	0.203	0.008
3.68	0.145			2.28	0.090	5.08	0.200	Brass	B	40.82	90	0.152	0.006
3.68	0.145			3.17	0.125	5.08	0.200	Brass	B	48.99	108	0.152	0.006
4.01	0.158	2.40	0.095	3.04	0.120	5.84	0.230	Brass	A	60.78	134	0.152	0.006
4.06	0.160			4.45	0.175	6.10	0.240	Brass	B	93.98	207	0.178	0.007
4.57	0.180			3.42	0.135	6.74	0.255	Brass	B	93.90	207	0.178	0.007
4.70	0.185	2.54	0.100	3.30	0.130	7.23	0.285	Brass	A	109.77	242	0.254	0.010
4.70	0.185	3.20	0.126	3.55	0.140	6.98	0.275	Brass	A	92.99	205	0.178	0.007
4.70	0.185	4.19	0.165	4.57	0.180	6.98	0.275	Brass	A	125.65	277	0.178	0.007
4.70	0.185	5.47	0.215	5.84	0.230	6.98	0.275	Brass	A	165.56	365	0.203	0.008
4.70	0.185	7.00	0.275	7.49	0.295	6.98	0.275	Brass	A	343.83	758	0.330	0.013
4.70	0.185	8.75	0.344	9.39	0.370	6.98	0.275	Brass	A	394.63	870	0.305	0.012
4.75	0.187			2.92	0.115	6.61	0.260	Brass	B	75.75	167	0.178	0.007
4.75	0.187			2.72	0.107	6.61	0.260	Brass	B	71.21	157	0.178	0.007
4.75	0.187			2.92	0.115	6.61	0.260	Brass	B	75.75	167	0.178	0.007
4.82	0.190			5.08	0.200	7.23	0.285	Brass	A	213.19	470	0.254	0.010
4.90	0.193	3.25	0.128	3.76	0.148	7.11	0.280	Brass	A	145.15	320	0.229	0.009
5.84	0.230			3.43	0.135	8.00	0.315	Brass	B	122.47	270	0.178	0.007
5.97	0.235			2.79	0.110	7.88	0.310	Brass	B	97.98	216	0.178	0.007
6.47	0.255	2.80	0.110	3.04	0.120	9.65	0.380	Brass	B	157.40	347	0.254	0.010
6.98	0.275	2.77	0.110	3.30	0.130	9.70	0.382	Brass	B	283.50	625	0.305	0.012
7.87	0.310	4.19	0.165	4.44	0.175	10.54	0.415	Brass	B	196.41	433	0.178	0.007
9.40	0.370			2.79	0.110	11.68	0.460	Brass	B	254.92	562	0.254	0.010
25.4	1.00			7.62	0.300	29.59	1.16	Brass	A	1995.84	4400	0.305	0.012
26.1	1.02			4.96	0.195	30.20	1.19	Brass	A	1388.90	3062	0.303	0.012

(Tucker Fasteners Ltd)

TABLE III – BELT AND LEATHER EYELETS

Barrel Diameter		Length Under Flange		Overall Length		Flange Diameter		Material	Flange Form	Weight per Million		Gauge of Material	
mm	in	mm	in	mm	in	mm	in			kg	lb	mm	in
3.55	0.140	3.67	0.145	4.32	0.170	6.22	0.245	Brass	A	101.61	224	0.178	0.007
3.68	0.145			2.77	0.109	6.10	0.240	Brass	B	89.81	198	0.229	0.009
3.71	0.146	2.66	0.105	2.97	0.117	5.96	0.235	Brass	B	82.56	182	0.229	0.009
3.71	0.146	2.66	0.105	2.97	0.117	5.96	0.235	Steel	B	83.01	183	0.203	0.008
4.16	0.164	3.30	0.130	3.68	0.145	7.61	0.300	Brass	A	124.75	275	0.178	0.007
4.06	0.160	3.63	0.143	3.93	0.155	5.84	0.230	Brass	A	84.82	187	0.178	0.007
4.19	0.165	3.25	0.128	4.14	0.163	7.36	0.290	Brass	A	125.65	277	0.178	0.007
4.19	0.165	3.25	0.128	4.11	0.162	7.36	0.290	Alum	A	52.16	115	0.203	0.008
4.19	0.165	4.02	0.171	4.65	0.183	7.36	0.290	Brass	A	138.35	305	0.178	0.007
4.32	0.170	4.07	0.160	4.57	0.180	6.35	0.250	Brass	A	105.69	233	0.178	0.007
4.44	0.175	3.45	0.136	4.06	0.160	6.85	0.270	Brass	A	156.95	396	0.254	0.010
4.44	0.175	5.20	0.205	5.91	0.233	6.85	0.270	Brass	A	208.20	459	0.229	0.009
4.67	0.184	4.17	0.164	4.57	0.180	7.11	0.280	Brass	A	121.56	268	0.178	0.007
4.75	0.187	3.30	0.130	4.07	0.160	7.75	0.305	Brass	A	132.90	293	0.178	0.007
4.75	0.187	4.57	0.180	5.07	0.200	7.75	0.305	Brass	A	141.52	312	0.178	0.007
4.82	0.190	4.25	0.167	5.08	0.200	9.01	0.355	Brass	A	185.98	410	0.178	0.007
4.82	0.190	5.07	0.200	5.84	0.230	9.01	0.355	Brass	A	209.56	462	0.203	0.008
4.95	0.195			5.72	0.225	8.00	0.315	Brass	A	217.72	480	0.229	0.009
5.38	0.212	4.62	0.182	5.46	0.215	9.95	0.392	Brass	A	257.19	567	0.203	0.008
5.38	0.212	3.85	0.152	4.70	0.185	9.90	0.390	Brass	A	219.09	483	0.203	0.008
5.38	0.212	4.57	0.180	5.33	0.210	9.90	0.390	Brass	A	264.90	584	0.203	0.008
5.38	0.212	5.20	0.205	6.09	0.240	9.90	0.390	Brass	A	262.63	579	0.203	0.008
6.09	0.240	4.07	0.160	4.70	0.185	9.77	0.385	Brass	A	229.97	507	0.203	0.008
6.73	0.265	3.92	0.155	5.08	0.200	11.80	0.465	Brass	A	292.57	645	0.203	0.008
6.73	0.265	5.85	0.230	6.85	0.270	11.80	0.465	Brass	A	397.35	876	0.229	0.009
6.73	0.265	3.67	0.145	4.70	0.185	11.80	0.465	Brass	A	278.06	613	0.203	0.008
				1.90	0.075	8.76	0.345	Steel	Cap	261.72	577	0.305	0.012
				1.90	0.075	9.27	0.365	Steel	Cap	301.64	665	0.305	0.012
				0.51	0.020	8.38	0.330	Brass	Washer	56.70	125	0.178	0.007
				0.40	0.016	9.78	0.385	Brass	Washer	76.66	169	0.178	0.007
				0.64	0.025	11.68	0.460	Brass	Washer	141.07	311	0.254	0.010

TABLE IV – SAIL EYELETS

Eyelet						Matching Plain Ring				Matching Turnover Ring			
Nominal Flange Diameter		Nominal Overall Length		Minimum Interval Diameter After Setting		Nominal Outside Diameter		Hole Diameter		Nominal Outside Diameter		Hole Diameter	
in	mm	in	mm	in	mm	in	mm	in	mm	in	mm	in	mm
3/8	9.53	11/64	4.37	13/64	5.16	3/8	9.53	1/4	6.35				
13/32	10.32	3/16	4.76	15/64	5.95	13/32	10.32	9/32	7.14				
7/16	11.11	13/64	5.16	17/64	6.75	7/16	11.11	5/16	7.94	15/32	11.91	19/64	7.54
1/2	12.70	7/32	5.56	9/32	7.14	1/2	12.70	11/32	8.73	1/2	12.70	21/64	8.33
9/16	14.29	15/64	5.95	11/32	8.73	9/16	14.29	3/8	9.53	9/16	14.29	23/64	9.13
5/8	15.88	15/64	5.95	11/32	8.73	5/8	15.88	13/32	10.32	5/8	15.88	25/64	9.92
11/16	17.46	1/4	6.35	3/8	9.53	11/16	17.46	7/16	11.11	11/16	17.46	27/64	10.72
3/4	19.05	17/64	6.75	13/32	10.32	3/4	19.05	15/	11.91	3/4	19.05	29/64	11.51
13/16	20.64	9/32	7.14	7/16	11.11	13/16	20.64	1/2	12.70	13/16	20.64	31/64	12.30
7/8	22.23	19/64	7.54	15/32	11.91	7/8	22.23	17/32	13.49	7/8	22.23	33/64	13.10
15/16	23.81	5/16	7.94	1/2	12.70	15/16	23.81	18/32	15.08	15/16	23.81	37/64	14.68
1	24.5	21/64	8.33	9/16	14.29					1	25.4	39/64	15.48
1.1/8	28.58	11/32	8.73	5/8	15.88					1.1/8	28.58	43/64	17.07
1.1/4	31.75	3/8	9.53	11/16	17.46					1.1/4	31.75	47/64	18.65
1.3/8	39.93	13/32	10.32	3/4	19.05					1.3/8	34.93	53/64	21.03
1.1/2	38.10	27/64	10.72	13/16	20.64					1.1/2	38.10	7/8	22.23
1.5/8	41.28	15/32	11.91	15/16	23.81					1.5/8	41.28	31/32	24.61
1.3/4	44.45	1/2	12.70	1	25.4					1.3/4	44.45	1 5/64	27.388
1.7/8	47.63	9/16	14.29	1.1/8	28.58					1.7/8	47.63	1.11/64	29.77
2	50.80	39/64	15.48	1.1/4	31.75					2	50.80	1.5/16	33.34

(Tucker Fasteners Ltd)

TABLE V – OVAL EYELETS

Nominal Barrel Diameter		Nominal Flange Diameter		Nominal Overall Length		Approximate Size of Slot After Setting	
in	mm	in	mm	in	mm	in	mm
0.276 x 0.118	7.01 x 3.00	0.406 x 0.295	10.31 x 6.22	0.190	4.83	0.250 x 0.093	6.35 x 2.36
0.365 x 0.175	9.27 x 4.46	0.510 x 0.325	12.96 x 8.26	0.206	5.23	0.343 x 0.156	8.71 x 3.96
0.380 x 0.175	9.65 x 4.45	0.507 x 0.312	12.88 x 7.93	0.170	4.32	0.343 x 0.156	8.71 x 3.96
0.580 x 0.175	14.73 x 4.45	0.736 x 0.329	18.70 x 8.36	0.218	5.54	0.546 x 0.140	13.87 x 3.56
0.690 x 0.172	17.53 x 4.37	0.835 x 0.343	21.21 x 8.71	0.250	6.35	0.656 x 0.140	16.66 x 3.56
0.720 x 0.190	18.29 x 4.83	0.935 x 0.410	23.75 x 10.42	0.222	5.64	0.703 x 0.171	17.86 x 4.34
0.720 x 0.190	18.29 x 4.83	0.935 x 0.410	23.75 x 10.42	0.285	7.24	0.703 x 0.171	17.86 x 4.34
0.790 x 0.240	20.07 x 6.10	0.955 x 0.425	24.26 x 10.80	0.390	9.91	0.734 x 0.187	18.65 x 4.75
0.796 x 0.296	20.22 x 7.52	1.022 x 0.522	25.96 x 13.26	0.140	3.56	0.765 x 0.281	19.43 x 7.14
0.800 x 0.360	20.32 x 9.15	1.000 x 0.562	25.40 x 14.28	0.265	6.73	0.750 x 0.343	19.05 x 8.71
0.830 x 0.240	21.08 x 6.10	0.950 x 0.360	24.13 x 9.15	0.256	6.50	0.805 x 0.218	20.45 x 5.54
0.845 x 0.235	21.47 x 5.97	1.115 x 0.490	28.32 x 12.45	0.200	5.08	0.812 x 0.203	20.63 x 5.16
1.125 x 0.344	28.58 x 8.73	1.453 x 0.641	36.91 x 17.07	0.266	6.75	1.062 x 0.281	26.99 x 7.14

(Tucker Fasteners Ltd)

TABLE VI – GROMMET EYELETS

Grommet						Matching Washer External Teeth				Matching Washer Internal/External Teeth			
Nominal Flange Diameter		Nominal Overall Length		Min Internal Diameter After Setting		Nominal Outside Diameter		Hole Diameter		Nominal Outside Diameter		Hole Diameter	
in	mm	in	mm	in	mm	in	mm	in	mm	in	mm	in	mm
13/32	10.32	3/16	4.76	13/64	5.16	13/32	10.32	15/64	5.95				
						13/32	10.32	0.265	6.73				
9/16	14.29	3/32	5.56	1/4	6.35	9/16	14.29	5/16	7.94				
43/64	17.07	1/4	6.35	21/64	8.33	43/64	17.07	3/8	9.53				
13/16	20.64	19/64	7.54	3/8	9.53	13/16	20.64	29/64	11.51				
15/16	23.81	11/32	8.73	15/32	11.91	15/16	23.81	17/32	13.49				
1	25.4	23/64	9.13	1/2	12.70	1	25.4	9/16	14.29				
1.1/8	28.58	13/32	10.32	37/64	14.68	1.1/8	28.58	21/32	16.67				
1.11/32	34.13	15/32	11.91	43/64	17.07					1.11/32	34.13	3/4	19.05
1.9/16	39.69	13/32	13.49	25/32	19.84					1.9/16	39.69	59/64	23.42
1.3/4	44.45	17/32	13.49	29/32	23.02					1.3/8	44.45	1.1/16	26.99
1.7/8	47.63	17/32	13.49	1.1/16	26.99					1.7/8	47.63	1.7/32	30.96

(Tucker Fasteners Ltd)

Grommets and Bushes

Standard or modified forms of metal eyelets may be used as grommets or bushes (see chapter on *Eyelets).* A wide variety of push-fit grommets is also produced in plastic for use as bearings, bushes, etc. Designs are normally self-locating, either by snap-action or frictional grip, and suitable for blind assemblies. Some examples are shown in Fig 1.

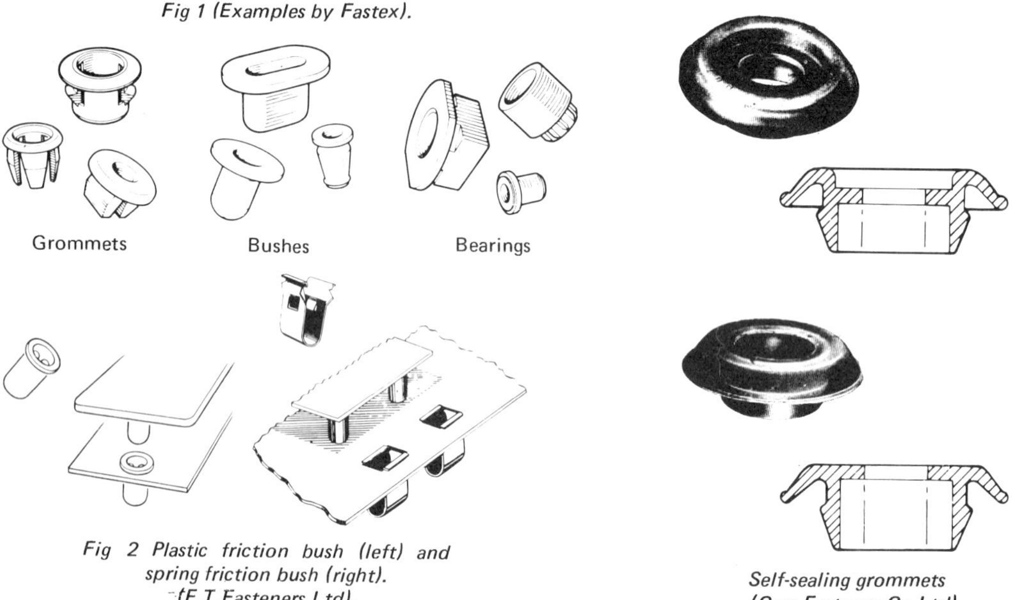

Fig 1 (Examples by Fastex).

Fig 2 Plastic friction bush (left) and spring friction bush (right). (F.T.Fasteners Ltd).

Self-sealing grommets (Carr Fastener Co.Ltd).

For bearing applications, mouldings are normally in nylon. Bushes may be in nylon, polythene or polypropylene. Grommets are moulded in a wide variety of thermoplastic materials. Stud anchors (blind bushes) are most commonly made in polythene and polypropylene.

Friction bushes are bushings designed as retainers for studs or non-threaded spigots, etc. They may be all-plastic or in the form of spring clips – see Fig 2.

Decorative Caps

Decorative caps to eliminate the poor appearance of exposed screw heads may have a threaded spigot for screwing into a tapped screw head or be press-fitted. The latter provides a simpler, and speedier fitting, suitable designs being developed by individual manufacturers.

Snap-Caps (Snap-Caps Ltd)

Snap-caps comprise a moulded plastic washer and cap manufactured in four sizes as follows:-

3/3 : 3 M/M to take 3/32in and 7/64in rivet

5/5 : 4 M/M to take 1/8in rivet

8/8 : 5 M/M to take 5/32in rivet

12/12 : 6.00M/M to take 3/16in and 1/4in rivet

The washer is secured under the head of the fastener, then the cap can be snapped in place. Snap-caps are normally moulded in UV polypropylene and are available in a range of thirty-five colours, including plated finishing. They are suitable for use with a wide variety of screw, rivet, bolt and double rivet heads and have a waterproof connection (Fig 1).

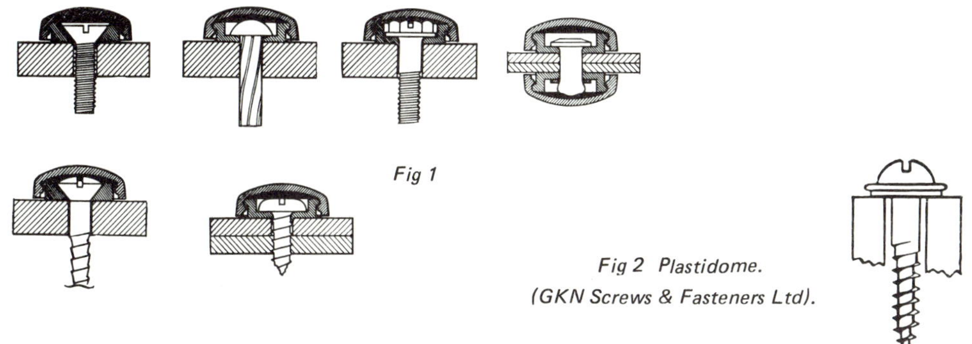

Fig 1

Fig 2 Plastidome. (GKN Screws & Fasteners Ltd).

Plastidome (GKN Screws & Fasteners Ltd)

The 'Plastidome' is a snap-on cover locking onto a collar assembled under the screw head. It can be used with 6 or 8 gauge woodscrews, machine screws and self-tapping screws, 4 or 6 UN, 3 or 4 BA, 1/8in or 5/32in BSW and M3 or M4. (Fig 2).

Pozitop (GKN Screws & Fasteners Ltd)

'Pozitop' is a simple plastic cap designed to press-fit and lock into a Pozidriv recess. Two sizes of Pozitop fit eight sizes of Pozidriv recess head woodscrews from 5 to 15 gauge and similar size heads in machine screws and self-tapping screws – Fig 3.

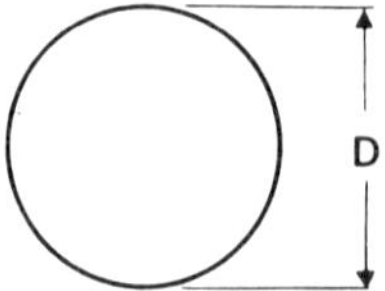

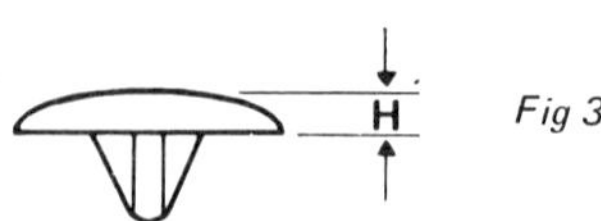

Fig 3

Size	Nominal Diameter D	Nominal Height of Cap Installed H
No 2	0.450in	0.065in
No 3	0.600in	0.105in

Wellburgh Limited

The Korrex is specially manufactured in nylon for bolt application (Fig 4).

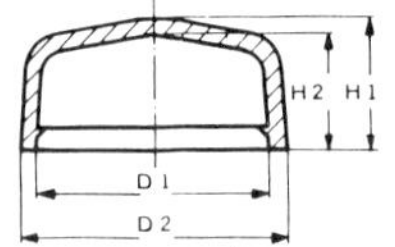

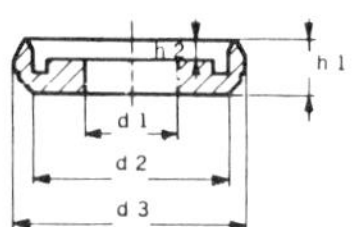

Fig 4

Size		Nr.	d1	d2	d3	h1	h2		D1	D2	H1	H2
M 6	1/4"	1406	6,5	13,6	16,2	3,7	1,2		16,5	18,9	9,0	7,7

dimensions in millimetres

Aluminium screw cap and matching screw cap retainers. (Carr Fastener Co Ltd).

Button Fasteners and Plugs

Button-type fasteners, used for securing trim panels, to metal panels, etc, may be button-headed spring nails, but are more usually of plastic, in nylon, acetal, polythene, etc. A large variety of individual designs has been produced by different manufacturers and these are normally described as trim clips, trim pad clips, trim fixes, trim studs, etc. Stud-locking button-headed fasteners are used as snap-in carpet fasteners.

Button-shaped self-locking fasteners are also designed for use as hole plugs.

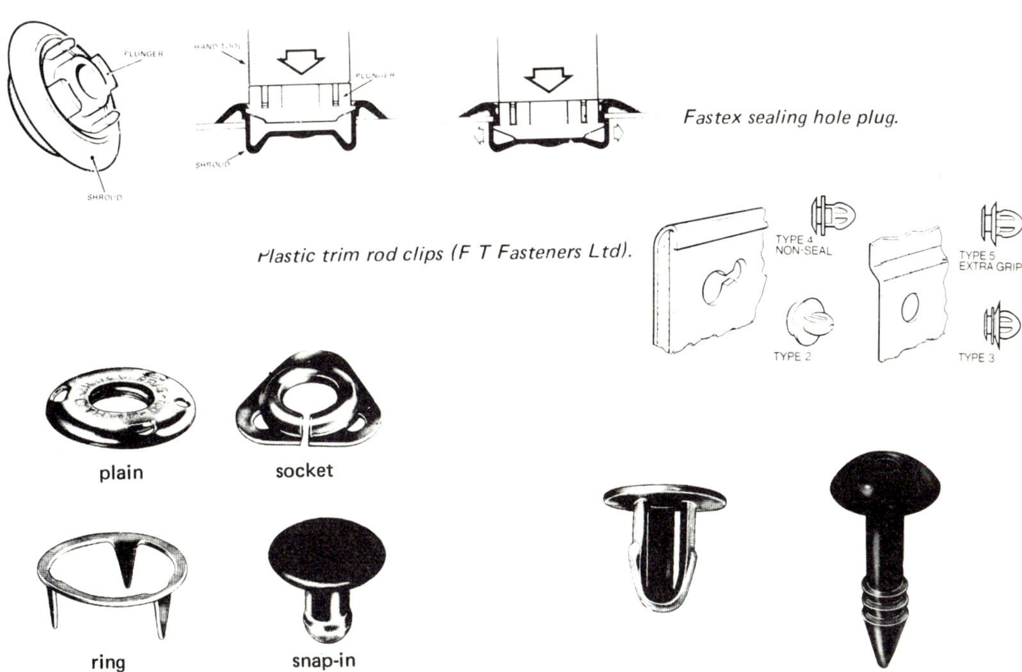

Fastex sealing hole plug.

Plastic trim rod clips (F T Fasteners Ltd).

Clinch type carpet fasteners. (Carr Fastener Co Ltd).

Trim clip (left) and trim stud (right). (Carr Fastener Co Ltd).

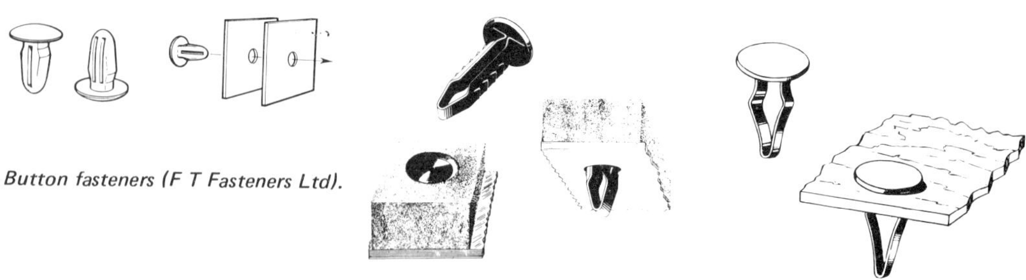

Button fasteners (F T Fasteners Ltd).

Snap-in carpet fasteners. (Carr Fastener Co Ltd).

Snap-in button (Carr Fastener Co Ltd).

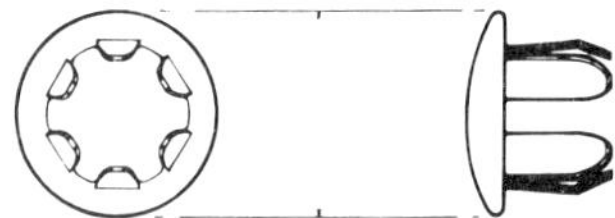

Plug buttons (F T Fasteners Ltd).

Electrical Fasteners

Standard types of electrical terminals and connectors are summarised in Table I (see also Fig 1). Individual manufacturers produce a wide range of variations on these basic shapes and forms – Fig 2.

TABLE I – ELECTRICAL TERMINALS AND CONNECTORS – BASIC TYPES

Type	Application	Remarks
Ring or eye	Terminal tongue	Basic type; also lends itself to staggered stacking in multiple units (manufactured or individually assembled).
Spade	Terminal tongue	Open ended for simplified assembly; can pull loose.
Flanged spade	Terminal tongue	Spade with turned up tips; self-locking in position.
Hook	Terminal tongue	Open eye for simplified assembly; less liable to pull loose than a spade terminal.
Rectangular eye	Terminal tongue	Parallel sides and ends can provide positive alignment on matching terminal blocks.
Flag	Terminal tongue	Ring tongue (usually with straight sides) with wire retainer at right angles to reduce overall length.
Quick disconnect	Two-piece connector	Matching tongue and receptable slide together with with spring locking action; can be connected and disconnected repeatedly.
Tubular	Connector	Circular sleeve accommodating two wires secured by a single crimp.
Sleeve	Connector	Oval sleeve accommodating two solid wires secured by twisting sleeve.
Butt	Connector	Extended tubular connector accommodating one wire at each end, each secured by separate crimp.
Snap plug	Connector	Locking type connector for single wire for pushing into a matching spring loaded female counterpart.

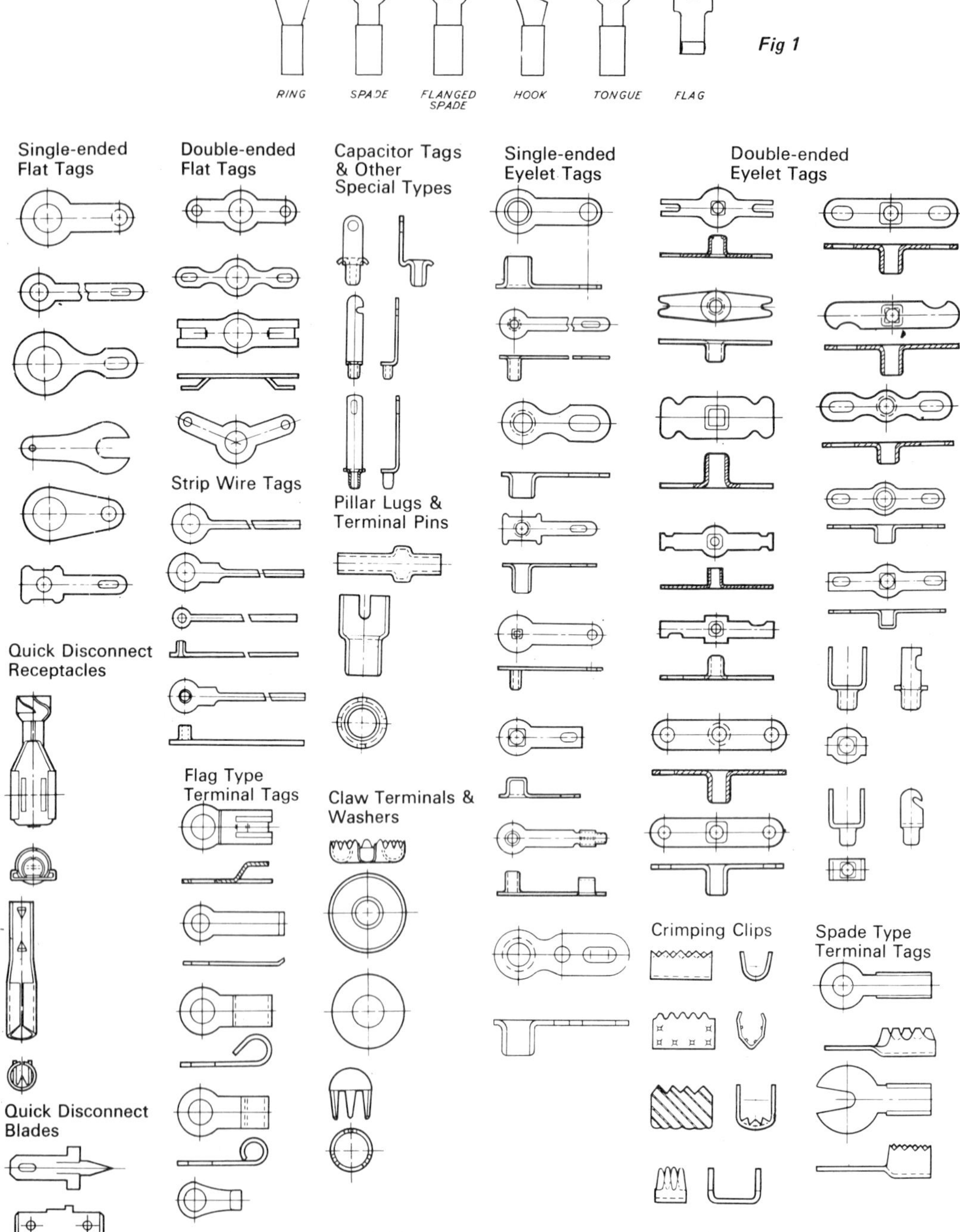

Fig 2 Selection of proprietary connections. (Tucker Fasteners).

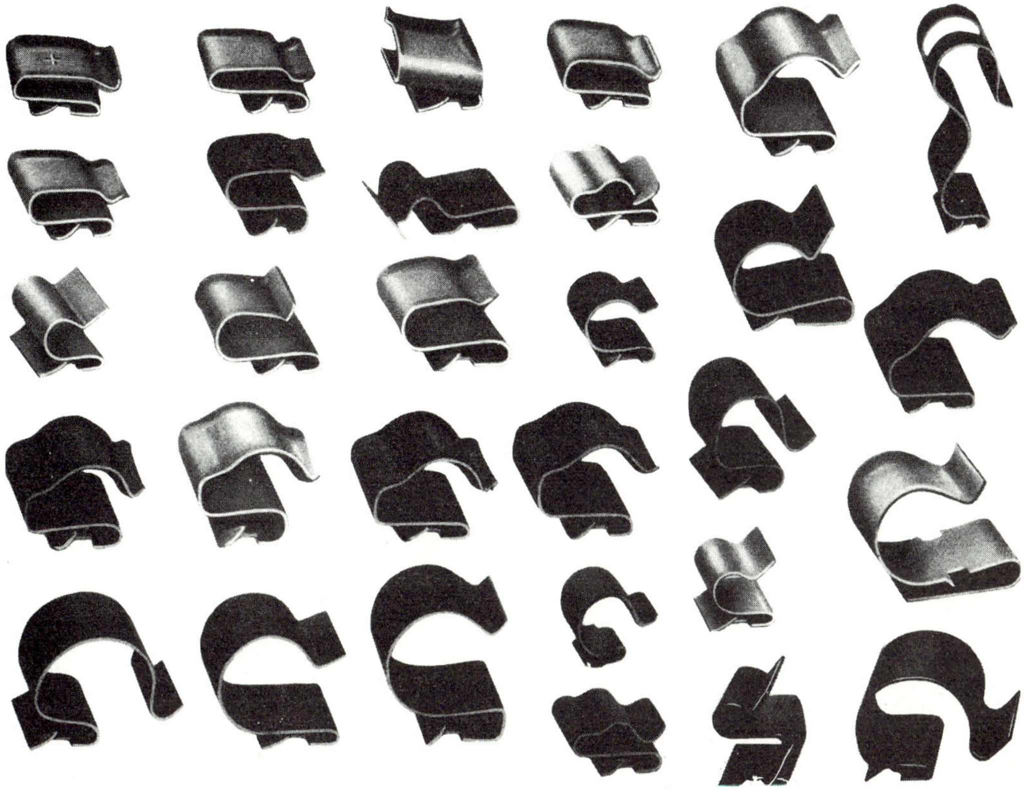

Spring cable clips (A.Raymond).

Wiring Clips, etc.

The range of wiring clips, mounting clips, etc, is equally vast and varied. These fall into the following main categories:-

Cable Clips –

(i) External wiring fasteners in the form of either insulated staples or moulded plastic clips secured with a masonry nail (Fig 3).

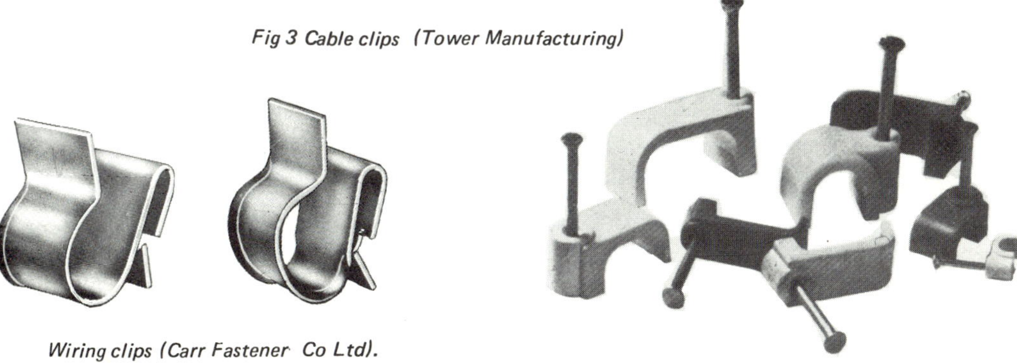

Fig 3 Cable clips (Tower Manufacturing)

Wiring clips (Carr Fastener Co Ltd).

ELECTRICAL FASTENERS

(ii) Spring-type cable clips for securing single wires, cables or multiple wires. These fall into three main types:-
 (a) Edge-type
 (b) Panel-type — self-retaining in a hole in the panel
 (c) Plastic clips of special design.

Wiring Straps — Strap-type clips secured by one or two screws, or self-anchoring in drilled holes (Fig 4).

Cable Fasteners — heavier designs of fasteners, straps or clamps to mount heavier electrical cabling or tubing (Fig 5).

Illustration of typical assembly

Clip used with minimum diameter cables

Clip when used with maximum diameter cable

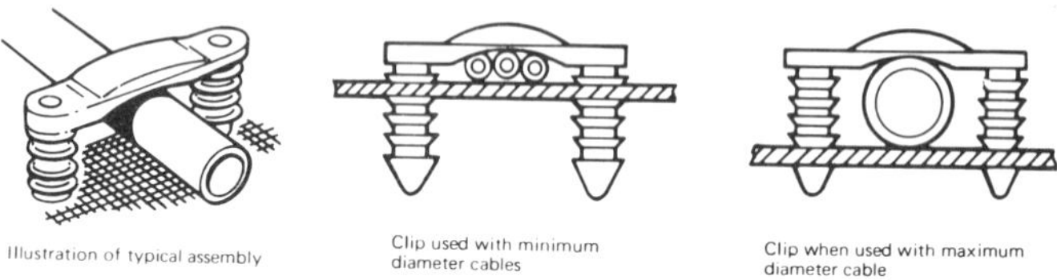

Fig 4 Wiring strap. (Carr Fastener Co).

Fig 5 'Sopra' cable suspension bracket.

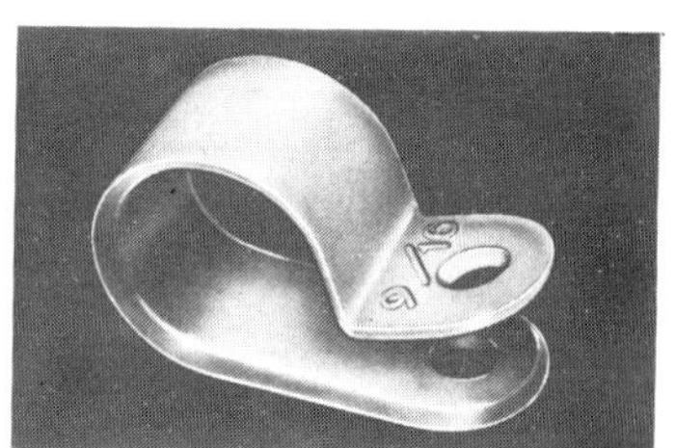

Heyco nylon cable clamp.

Heyco Nylon bushings do the lot for cables

ANCHOR — STRAIN RELIEFS — Completely absorb 'push', 'pull' and 'twist'.

INSULATE — SNAP BUSHINGS — Convert sharp edges to smooth, neat holes.

PROTECT — UNIVERSALS — Cushion and position one or more cables.

CONNECT — TERMINALS — Quick connection/disconnection thru' panel.

REPLACE — HOLE PLUGS — Neat, low cost seals for unwanted openings.

Free samples and catalogue on request.

HEYCO MANUFACTURING COMPANY LIMITED Uddens Trading Estate, Nr. Wimborne, Dorset BH21 7NL. Tel: Ferndown (STD: 02017) 71411/2/3. Telex: 41408

Component Clips – these are designed specifically to hold individual components captive on the basic panel, eg

(a) printed circuit clips for holding PC panels

(b) control cable clips to locate and anchor outer casing of flexible shaft controls.

(c) can clips for mounting IF transformers, etc.

(d) capacitor clips for mounting electrolytic capacitors

(e) angle nuts for cabinet and/or panel assemblies

(f) fixing clips for securing card back to cabinet

Some examples of these main types of fastenings are given in Figs 6 and 7.

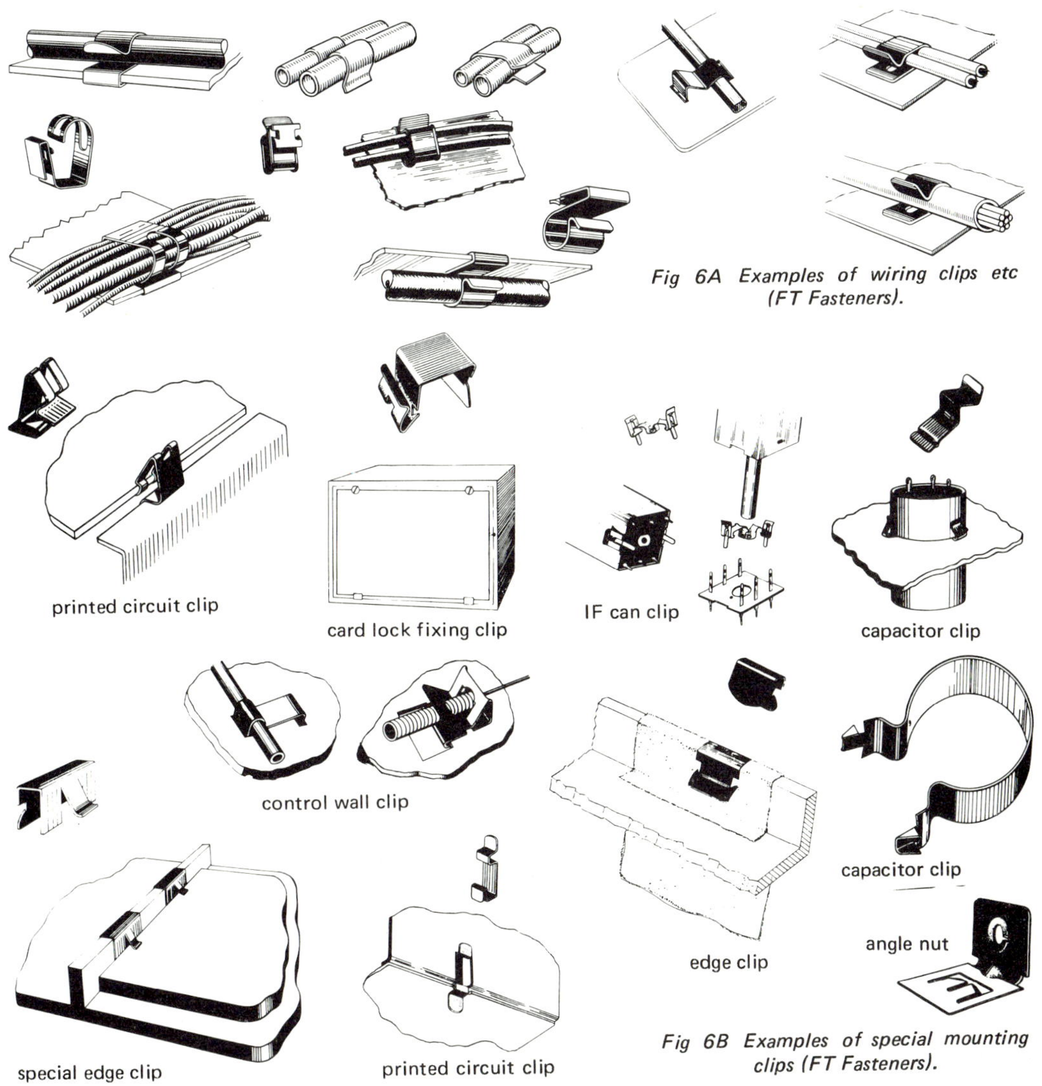

Fig 6A Examples of wiring clips etc (FT Fasteners).

Fig 6B Examples of special mounting clips (FT Fasteners).

NEON-LENS CLIP – STANDARD TYPE TO SUIT A RECTANGULAR HOLE

LENS CLIP – WITH INTEGRAL LATCH LEGS TO SUIT A ROUND HOLE

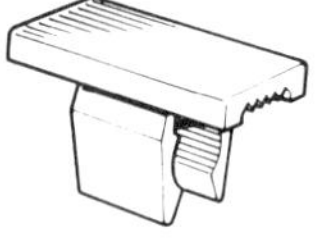
NEON-LENS CLIP – WITH LIGHT REFLECTING GROOVES

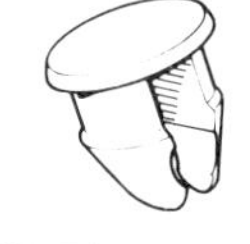
NEON-LENS CLIP – STANDARD TYPE TO SUIT A ROUND HOLE

NEON-LENS CLIP – WITH LOCKING LATCH LEG

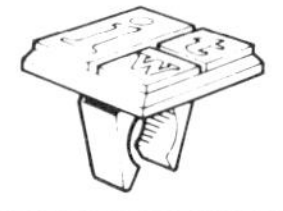
NEON-LENS CLIP SHOWING A TYPICAL HOUSE EMBLEM

Fig 7 Neon-lens clips (Fastex).

Electrical Eyelets

Electrical eyelets are normally manufactured in both copper and brass in standard eyelet types (see chapter on *Eyelets),* and special types for through-connecting. Eyelets provide stronger component fixing in printed circuit boards, etc.

Copper eyelets can be supplied cleaned (for subsequent plating by the user), solder-tinned, nickel-plated and solder-tinned or nickel-plated and electro-tinned. A typical range of sizes is given in Table II.

TABLE II – ELECTRICAL AND RADIO THROUGH CONNECTION EYELETS (FUNNEL TYPE)

Barrel Diameter		Length Under Flange		Flange Diameter		Suitable for Circuit Thickness		Recommended Hole Size		Material	Gauge	
±0.05 mm	±0.002 ins	±0.05 mm	±0.002 in	±0.05 mm	±0.002 in	mm	in	mm	in		mm	in
1.19	0.047	0.69	0.027	1.65	0.065	‡	‡	1.40	0.055	Copper	0.18	0.007
1.19	0.047	1.60	0.063	1.65	0.065	1.20	3/64	1.40	0.055	Copper	0.18	0.007
1.19	0.047	2.11	0.083	1.65	0.065	1.60	1/16	1.40	0.055	Copper	0.18	0.007
1.19	0.047	2.38	0.093	2.03	0.080	1.60	1/16	1.40	0.055	Copper	0.18	0.007
1.19	0.047	3.15	0.124	1.65	0.065	2.40	3/32	1.40	0.055	Brass	0.23	0.009
1,27	0.050	2.38	0.093	2.41	0.095	1.60	1/16	1.45	0.058	Copper	0.23	0.009
1.50	0.059	2.24	0.088	2.41	0.095	1.60	1/16	1.70	0.067	Copper	0.23	0.009
1.50	0.059	2.72	0.107	2.41	0.095	2.00	5/64	1.70	0.067	Copper	0.23	0.009
1.50	0.059	3.17	0.125	2.41	0.095	2.40	3/32	1.70	0.067	Copper	0.23	0.009
1.50	0.059	3.97	0.156	2.41	0.095	3.17	1/8	1.70	0.067	Copper	0.23	0.009
1.60	0.063	1.40	0.055	2.03	0.080	0.80	1/32	1.80	0.071	Copper	0.23	0.009
1.60	0.063	1.70	0.067	2.03	0.090	1.15	3/64	1.80	0.071	Copper	0.23	0.009
1.60	0.063	2.92	0.115	2.03	0.090	2.40	3/32	1.80	0.071	Copper	0.23	0.009
1.94	0.076	1.68	0.066	3.07	0.121	0.80	1/32	1.70	0.084	Copper	0.23	0.009
1.98	0.078	1.78	0.070	2.41	0.105	1.15	3/64	2.18	0.086	Copper	0.23	0.009
2.26	0.089	2.24	0.083	2.77	0.109	1.60	1/16	2.47	0.097	Copper	0.23	0.009
2.26	0.089	2.38	0.093	3.17	0.125	1.60	1/16	2.47	0.097	Copper	0.23	0.009
2.26	0.089	3.17	0.125	3.17	0.125	2.40	3/32	2.47	0.097	Copper	0.23	0.009
2.26	0.089	3.43	0.135	3.17	0.125	2.40	3/32	2.47	0.097	Copper	0.23	0.009
2.26	0.089	3.97	0.156	3.17	0.125	3.17	1/8	2.47	0.097	Copper	0.23	0.009
3.78	0.149	4.45	0.175	5.46	0.215	3.57	9/64	3.99	0.157	Brass	0.31	0.012
4.19	0.165	2.31	0.091	5.56	0.218	1.60	1/16	4.39	0.173	Copper	0.23	0.009
4.19	0.165	3.17	0.125	5.56	0.218	2.40	3/32	4.39	0.173	Copper	0.23	0.009

(Tucker Fasteners Ltd)

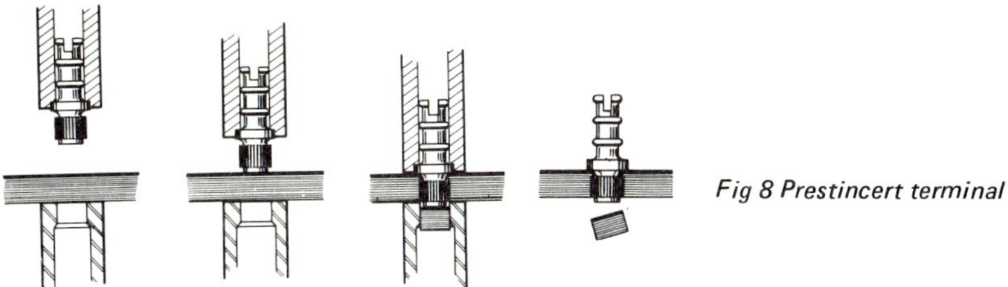

Fig 8 Prestincert terminal

Bushes

Metallic bushes for mounting in laminates and aluminium sheet are usually either of rivet bush type for clenching in preformed holes, or designed to be pushed directly into sheet material – Fig 8. They are normally made in brass. Typical sizes are:-

BA – 2, 4 and 6BA threads.

ISO Metric – M2.5, M2.6, M3, M4, M5 and M6.

Unified – 4-40UNC, 6-32UNC, 10-32UNF.

see also *Rivet Bushes and Tank Bushes*.

Snap bushings are used at chassis outlets as an alternative to rubber grommets to prevent chafing. Special bushings may also be designed to provide strain relief – ie absorb 'pull', 'push' or 'twist' forces on the wiring or – to absorb vibration.

Terminal bushings offer a quick-connect/disconnect facility at chassis or panel entry points. Some examples of special electrical bushings are shown in Fig 9.

Heyco nylon snap bushing | Heyco nylon universal bushing | Heyco strain relief bushing | Heyco nylon terminal bushing

Fig 9

Terminal Posts, etc

Terminal posts may be designed for press fitting in plain holes, screw assembly, rivet or clench assembly in matching size holes or self-clenching in preformed holes or when pushed directly into sheet material. Head-through posts are commonly designed for insertion into polythene or PTFE bushes to give a non-rotating, insulated lead-through terminal.

Knurled Heads and Nuts

Knurling is commonly adopted on screws, terminals and nuts to facilitate finger tightening or removal. Typical examples of knurling are shown in Fig 10.

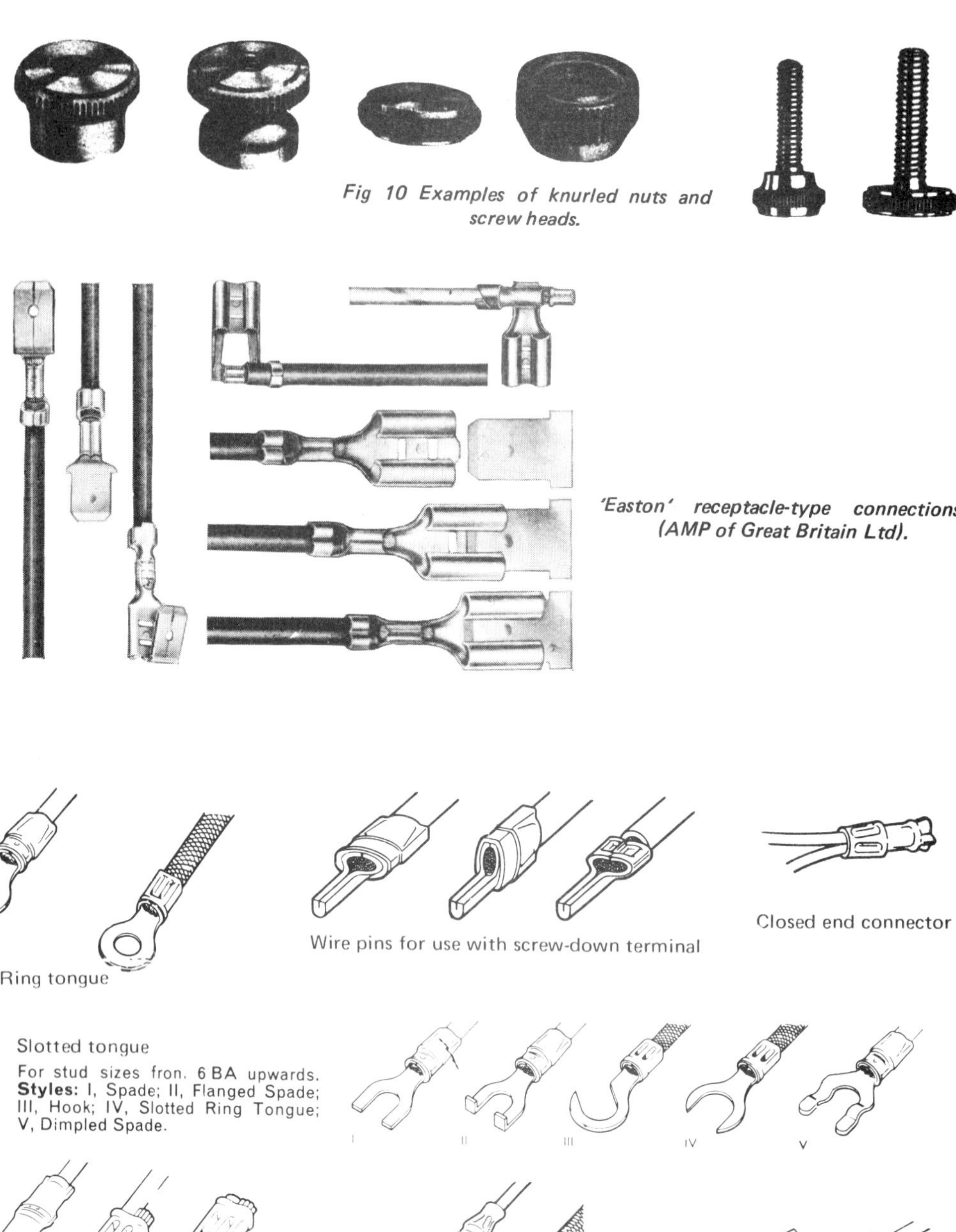

Fig 10 Examples of knurled nuts and screw heads.

'Easton' receptacle-type connections, (AMP of Great Britain Ltd).

Examples of proprietary terminals and splices (AMP of Great Britain Ltd).

Fig 11 Terminal block.

Terminal Blocks

Standard terminal blocks are moulded in rigid and flexible PVC, polythene and to a lesser extent thermoset plastics with brass inserts and screws – Fig 11. Thermoplastic blocks can easily be sub-divided with a knife. Typical single block dimensions are:-

Length		Width		Height		Fixing Holes		Fixing Centres		Cable Entry	
in	mm	in	mm	in	mm	in	mm	in	mm	in	mm
4.50	114.30	0.812	20.62	0.625	15.87	0.110	2.79	0.375	9.52	0.125	3.17

Insulation resistance is usually of the order of 10^{12} ohms (depending on the plastic used), with a continuous current rating of 5-10amps. Typical breakdown voltages for a flexible PVC terminal block are:-

Between terminals – 9kV }
Terminals to earth – 5kV } 50 cycles ac

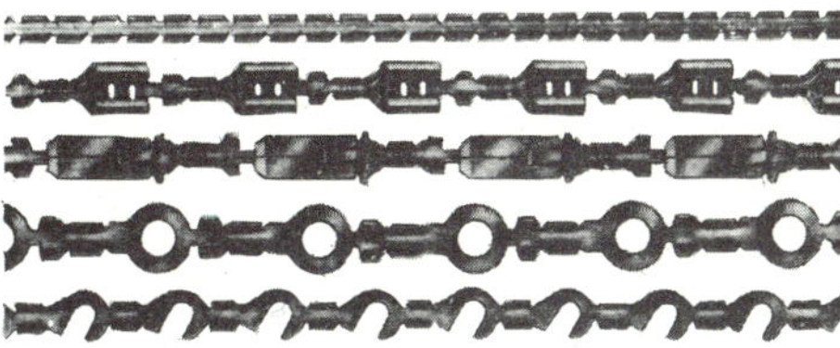

Examples of terminals and splices for automatic assembly by machine. (AMP of Great Britain Ltd).

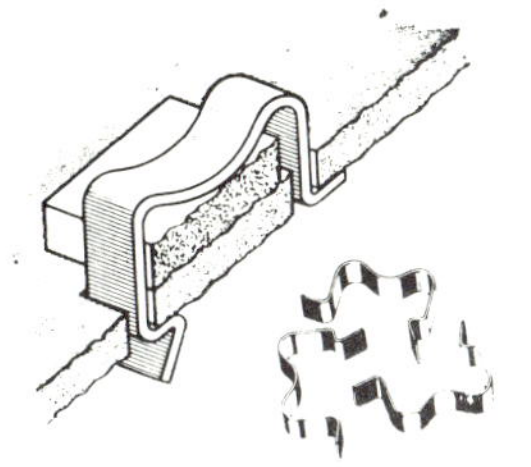

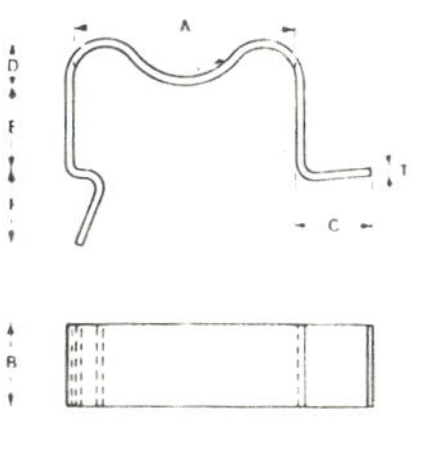

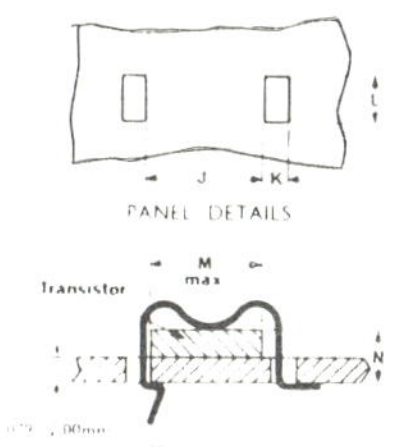

Transistor clip (GKN)

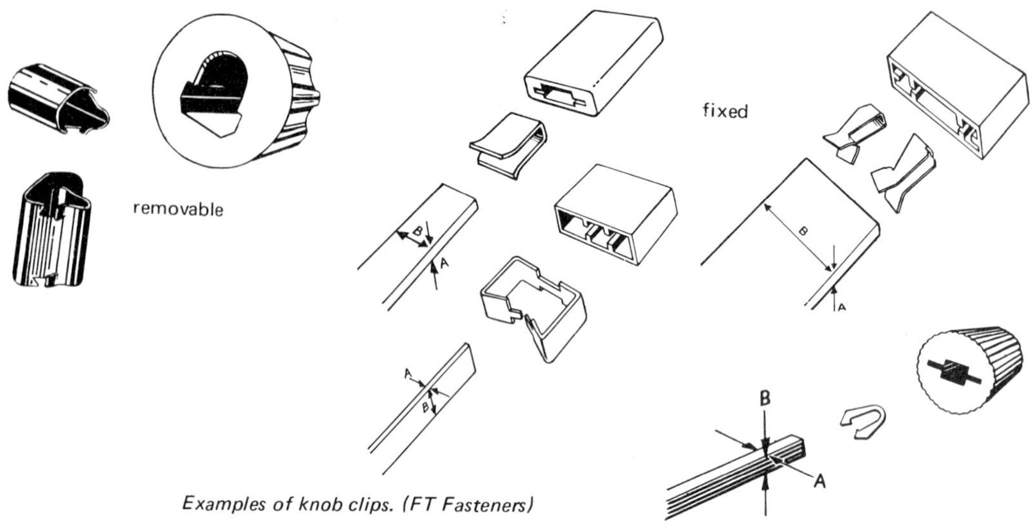

Examples of knob clips. (FT Fasteners)

Spring Friction Fasteners

Spring nut forms are also widely used for locking unthreaded studs, etc. Their bored form normally accommodates two identical prongs which grip tightly when the retainer is pushed into a position which produces some straightening of the arc. Such types are best described as spring friction fasteners (basically 'flat' shapes) or spring friction bushes, depending on their geometry. Types produced for special purposes may be known by their application – eg knob clips.

There is a wide range of designs based on circular, square and rectangular planforms. See also chapters on *Spring Nuts* and *Spring Fasteners.*

Metal Wire Stitching

Metal-stitching is a low cost, rapid fastening technique which can be used as an alternative to riveting, etc, in many fields of assembly. Metal stitching machines are capable of providing approximately five stitches a second from coils of high tensile steel wire, forming their own holes in the materials being stitched. The operation is similar to stapling except that stitchers cut their own lengths of wire from a coil (instead of using prepared staples) and form each fastener as the stitch is made.

Stitchable materials include the following:-

Metal — aluminium — brass — chrome plated steel — copper — manganese — monel — stainless steel — steel.

Non-Metals — asbestos — building boards — canvas — chipboards — cork — corrugated paper — felt — flexible plastics — gypsum — hardboards — laminated plastics — leather — padding — paper-board — rigid plastics — rubber — woods.

Main limitations as regards the suitability of a material for stitching are thickness and hardness, eg stitching heavier gauge materials, or substantial thicknesses of hard materials, is usually impractical because of the tendency of the legs of the stitch to wander when being driven through.

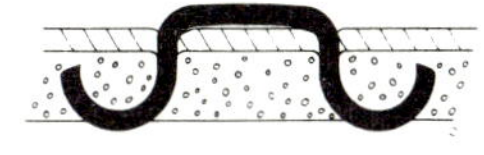

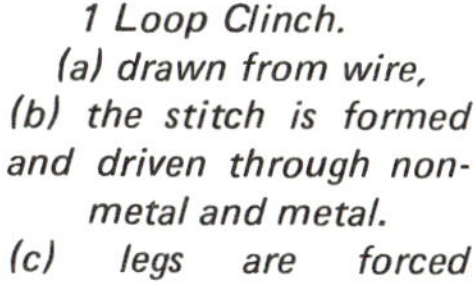

1 Loop Clinch. (a) drawn from wire, (b) the stitch is formed and driven through non-metal and metal. (c) legs are forced upward in a loop and (d) are clinched against metal.

2 Bypass Clinch (a) drawn from wire, (b) stitch is formed and driven through metal past rod. (c) legs overlap each other to encircle rod and (d) are clinched against metal.

3 Outwood Loop Clinch. (a) drawn from wire, (b) stitch is formed and driven through metal to non-metal, slug being punched out by stitch. (c) legs are clinched outward and (d) buried in non-metal

4 Flat Clinch. (a) drawn from wire, (b) stitch is formed and driven through two sections of metal, slug being punched out by stitch. (c) legs are bent sharply and (d) clinched flat against metal.

Fig 1 Types of Stitches (Acme Steel Co.)

The two basic types of stitch used are the flat clinch and loop clinch (inside, outside or bypass) – Fig 1. Loop stitches are formed with stationary solid-die clinchers. The flat stitch is formed by folding the legs flat against the bottom material with an upward-moving clincher, after it has been driven through the material. Loop clinchers are used for non-structural applications (approximately 90% of all wire stitching requirements). Flat stitches should always be used where maximum joint strength is required. Stitchers normally have the following crown widths (but they may vary for different applications):-

inside loop clinch – 6.5mm (¼in) or 11mm (7/16in)

bypass loop clinch – 6.5mm (¼in)

outside loop clinch – 6.5mm (¼in) or 11mm (7/16in)

flat-clinched stitch – 6.5mm (¼in) or 11mm (7/16in)

Smaller stitches have a neater appearance. If necessary 3mm (1/8in) stitches can be used. At the other end of the scale, stitches with a crown width of 25mm (1in) may be used for particular assemblies. In addition, other stitch profiles may be used for specific applications, depending on the fastening requirements. Most stitching machines are adaptable to take a range of clincher profiles and dies for accommodating a variety of stitch sizes and types.

Stitches can be applied perpendicularly, parallel, or diagonally to the line of pull. Perpendicular and diagonal stitches have the higher shear strength – see Fig 2. Diagonal stitches offer high strength in the case of several lines of pull being present. Loop clinch stitches have about 75% of the strength of flat clinch stitches in the same configuration. Stitches should normally be used for loading in shear. Stitches loaded in tension have substantially reduced strength – see Fig 3.

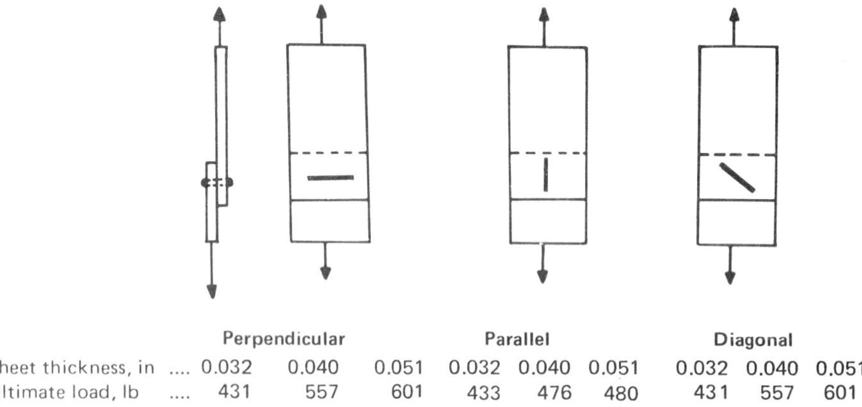

	Perpendicular			Parallel			Diagonal		
Sheet thickness, in	0.032	0.040	0.051	0.032	0.040	0.051	0.032	0.040	0.051
Ultimate load, lb	431	557	601	433	476	480	431	557	601

Fig 2 Shear strength of metal stitches 0.051in grade 290 wire in 24ST aluminium sheet.

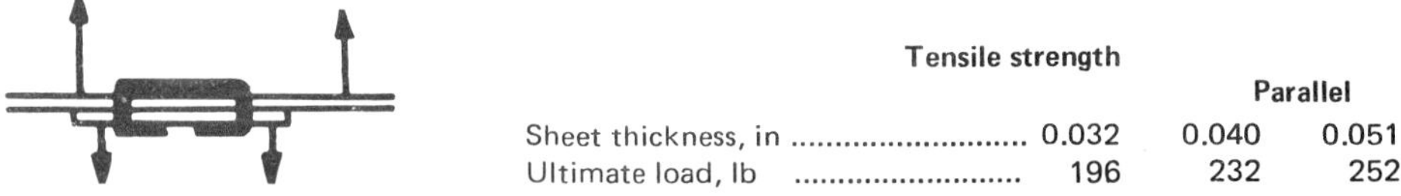

Tensile strength		Parallel	
Sheet thickness, in	0.032	0.040	0.051
Ultimate load, lb	196	232	252

Fig 3 Strength of typical flat clutch in tension – 0.051in grade 290 wire in 24ST aluminium sheet.

Parallel rows of flat clinched stitches should have a minimum clearance of 6.5mm (¼in) between rows, and a minimum spacing between stitches of 22mm (0.88in). Diagonal stitches should have a minimum spacing of 12.5mm (½in). With all types of stitch, minimum distance between any stitch and any edge should be 5mm (0.2in).

For fastening metal to metal with an overlap joint, minimum overlap should be 10mm (0.4in). When flexible materials are stitched to metals with an overlap joint, minimum overlap should be 12.5mm (0.5in), with stitches located in the centre of the overlap. In stitching brittle plastics or similar materials, it is generally recommended that a metal backing sheet be located under the stitch crown.

Stitching wires used are mainly low carbon steel, high carbon steel and stainless steel, although phosphor bronze is an alternative material when a non-rusting stitch is required – see Tables I and II.

TABLE I – COMMON SIZES AND FINISHES – FLAT STITCHING WIRES

Size inch	Size mm	Steel Material	Tensile lb/in^2	Lacquer	Finish Galvanised	Tinned	Other
0.103 x 0 028	2.6 x 0.71	Low carbon	105 000		X		
0.103 x 0.023	2.6 x 0.58	Low carbon	105 000		X		
0.103 x 0.020	2.6 x 0.5	Low carbon	105 000		X		
0.103 x 0.020	2.6 x 0.5	302 stainless					Bright
0.103 x 0.017	2.6 x 0.43	Low carbon	105 000		X		
0.103 x 0.014	2.6 x 0.35	Low carbon	105 000		X		
0.060 x 0.024	1.5 x 0.6	Low carbon	105 000		X		
0.060 x 0.024	1.5 x 0.6	302 stainless					Bright
0.060 x 0.205	1.5 x 0.5	Low carbon	105 000		X		
0.060 x 0.205	1.5 x 0.5	302 stainless					Bright
0.078 x 0.022	2.0 x 0.55	Low carbon	105 000		X		
0.0475 x 0.0348	1.2 x 0.88	Low carbon	125 000	X	X	X	
0.0475 x 0.0348	1.2 x 0.88	Hi carbon	230 000			X	
0.0410 x 0.030	1.0 x 0.76	Low carbon	125 000	X	X	X	
0.0348 x 0.0258	0.9 x 0.66	Low carbon	125 000	X	X	X	
0.0348 x 0.023	0.9 x 0.58	Low carbon	125 000	X	X	X	
0.0348 x 0.0204	0.9 x 0.5	Low carbon	125 000	X	X	X	
0.0348 x 0.0204	0.9 x 0.5	302 stainless					Bright
0.03175 x 0.0204	0.8 x 0.5	Low carbon	125 000	X	X	X	
0.03175 x 0.0162	0.8 x 0.4	Low carbon	125 000	X	X	X	
0.0286 x 0.0181	0.7 x 0.45	Low carbon	125 000	X	X	X	
0.0258 x 0.140	0.65 x 0.35	Low carbon	125 000	X	X	X	

Stitch penetration is limited by the shear resistance of the material being stitched. Basically this limits the thickness of material suitable for stitching – see Table III.

The formation of flat-clinched stitches is a punch and die operation in which the wire punches out small slugs of the same diameter as it is forced through the materials being joined. Usually an air blast blows the slugs out of the clincher. The full support of the stitch and the speed in which it is driven permits penetration through thicknesses up to four times the wire diameter.

Since penetration is limited mainly by the shear resistance of the work, it is usually easier to punch through a number of thin sheets than through one heavy sheet of the same total thickness. The reason for this is that the wire punches the slug from each sheet shortly before experiencing the full strain of punching the next sheet.

TABLE II – COMMON SIZES AND FINISHES – ROUND STITCHING WIRES

Diameter swg*	inch	mm	Material		Tensile Range lb/in^2	Lacquer	Finish Galvanised	Tinned	Other
16	0.0625	1.6	Hi-carbon	260	250 000 to 280 000				Bright
			Low-carbon		120 000 to 150 000		X		
17	0.054	1.4	Low-carbon		120 000 to 150 000	X	X	X	
18	0.0475	1.2	Hi-carbon		70 000 to 85 000	X	X		
			Soft-core	200	190 000 to 219 000	X	X	X	
			Hi-carbon	230	222 000 to 249 000	X	X	X	
			Hi-carbon	260	250 000 to 289 000	X	X	X	
			Hi-carbon	290	290 000 to 319 000	X	X	X	Black
			Hi-carbon	330	320 000 to 360 000	X	X	X	
			Music wire	330	330 000 MIM			X	
			Low-carbon		120 000 to 150 000	X	X		
			Soft-care		70 000 to 85 000	X	X		
			CMN stainless	260	240 000 to 265 000				Bright
			302 stainless	260	240 000 to 265 000				214
			Phos.bronze S-54	150	150 000				
18	0.051	1.3	Aircraft wire	290	290 000 to 319 000		X		
20	0.0348	0.61	Hi-carbon	260	250 000 to 289 000	X	X	X	
			Hi-carbon	290	290 000 to 319 000	X	X	X	
			Low-carbon		120 000 to 150 000	X	X	X	
21	0.0317	0.8	Low-carbon		120 000 to 150 000	X	X	X	
22	0.0286	0.7	Low-carbon		120 000 to 150 000	X	X	X	
23	0.0258	0.65	Low-carbon		120 000 to 150 000	X	X	X	
24	0.0230	0.6	Low-carbon		120 000 to 150 000	X	X	X	
25	0.0204	0.5	Low-carbon		120 000 to 150 000	X	X	X	
26	0.0181	0.46	Low-carbon		120 000 to 150 000	X	X	X	
26	0.0173	0.45	Low-carbon		120 000 to 150 000	X	X	X	
27	0.0162	0.4	Low-carbon		120 000 to 150 000	X	X	X	
29	0.0140	0.35	Low-carbon		120 000 to 150 000	X	X	X	

*Nearest equivalent to American specified gauge

The majority of applications, and those on which the largest time and material economies are effected, are the ones in which non-metallic materials are joined to metal sections. The economy is apparent when assembly of materials by riveting and by stitching is compared. Drilling of holes through materials such as woven asbestos, sponge rubber, sheet cork and leather together with the subsequent location and driving of rivets is time-consuming. In contrast, the legs of a metal stitch form their own holes.

The most frequently stitched metals are: aluminium, clad aluminium, aluminium extrusions, cold-rolled steel, galvanised sheet, stainless steel (full hard, ½ hard, ¼ hard and annealed), soft sheet brass and sheet copper.

The most frequently stitched non-metallic materials are: sheet cork, leather, sheet asbestos, fibreboard, standard and tempered Masonite, sponge and solid rubber, the phenolics, the plastics, solid wood and plywood.

TABLE III – RECOMMENDED MAXIMUM THICKNESSES FOR WIRE STITCHING

Metals	Metal-to-Metal†				Metal-to-Non-Metal*				Non-Metal Material Thicknesses
	Loop Clinch		Flat Clinch		Loop Clinch		Flat Clinch		
	in	mm	in	mm	in	mm	in	mm	
Aluminium (¼ hard)	0.093	2.4	0.093	2.4	0.125	3.0	0.125	3.0	
Aluminium (½ hard)	0.064	1.6	0.064	1.6	0.080	2.0	0.080	2.0	½in (12.5mm) sheet cork
Aluminium hard and dural	0.040	1.0	0.040	1.0	0.064	1.6	0.064	1.6	3/8in (9mm) leather
Aluminium extrusion	0.062	1.55	0.062	1.55	0.093	2.4	0.093	2.4	¼in (6mm) sheet asbestos
Cold-rolled steel	0.050	1.25	0.040	1.0	0.080	2.0	0.040	1.0	½in (12.5mm) fibre
Hot rolled steel	0.050	1.25	0.035	0.9	0.0625	1.95	0.035	0.9	½in (12.5mm) sponge rubber
Galvanised sheet	0.037	1.0	0.032	0.8	0.0475	1.2	0.032	0.8	¼in (6mm) solid rubber
Stainless (type 302) full hard	0.010	0.25	0.010	0.25	0.020	0.5	0.010	0.25	1/8in (3mm) phenolics ‡
Stainless, ½ hard	0.012	0.30	0.012	0.30	0.025	0.65	0.012	0.30	3/16in (4mm) plastic ‡
Stainless, ¼ hard	0.015	0.40	0.015	0.40	0.030	0.8	0.015	0.40	3/8in (9mm) standard masonite
Stainless, annealed	0.020	0.50	0.020	0.50	0.040	1.0	0.020	0.50	¼in (6mm) tempered masonite
Sheet brass, soft	0.030	0.80	0.030	0.80	0.050	1.25	0.040	1.0	3/8in (9mm) wood
Sheet copper	0.035	0.90	0.035	0.90	0.064	1.6	0.045	1.1	

† 1 piece to 1 piece of specified thickness

* Metal thickness

‡ Must be soft enough to penetrate without cracking

Frequently the maxima shown in Table III can be exceeded. However, where the metal has been work-hardened through forming, the recommended maximum thickness it may be impossible to stitch. Another factor affecting stitchable thicknesses is the condition of the machine. Most of the thicknesses tabulated are for parts stitched with 18 gauge, grade 330 wire. Wherever possible the lowest satisfactory tensile-strength wire should be used in order to keep machine maintenance at a minimum.

Metal stitches have high shear and tensile strengths and are resistant to fatigue and vibration. Loading strengths of flat-clinched stitches are at least equal to those of rivets. The total cross-sectional area of the two legs of a single 18 gauge flat-clinched stitch and the total shear load this stitch can withstand approximately equal one-third the total area and the shear load values of an 1/8in full-hard aluminium rivet.

Very little power is consumed by the stitching operation. Motors of 1/3 and ½ hp are more than sufficient for driving stitches through the maximum recommended material thicknesses. Dissimilar metals that are not readily welded, such as steel and aluminium, are very easily fastened by this method. Metal stitching does not remove galvanised coatings as do the attachment methods utilizing heat. Also, materials need not be cleaned prior to joining. Stitches applied to parts that have been painted do not disturb the finished surfaces.

Machine speeds range from 280 to 325 stitches per minute. Operating speeds are considerably lower than machine speeds because of the time delay of setting up and positioning the work. On most production applications operating speeds of from 80 to 100 stitches per minute can be attained. Other factors affecting operating speeds are the shape of the parts being joined, the number of stitches required per assembly, the speed of the machine being used and the mobility of the operator. As the time for set-up and positioning is diminished, operator speed approaches the machine speed.

Metal stitching is currently used for joining non-metallic materials such as canvas, rubber, plywood, leather, cardboard, asbestos, twisted paper and felt to a full range of metals. This includes aluminium, steel, brass, copper and stainless steel. The technique is being used in the manufacture of automobiles, buses, trucks, refrigerators, electric cookers, furnaces, toys, novelties, aircraft, metal furniture and other items.

Staples

Staples are semi-shaped wire stitches which can be loaded in a suitable tool, the operation of which will press the staple legs through the materials to be fastened and clinch the staple to produce a fastening. The legs are conventionally clinched inwards to form a loop with the tips of the staple legs pressing against the workpiece – Fig 1. Alternatively, an outward clinched loop may be formed if this is more suitable (usually by reversal of the stapler anvil, or clincher). Staples may also be driven directly into the work without clinching for tacking, etc (eg fastening thin material to a thicker base material). For this type of work, special staples are available with divergent legs which will spread apart under the work to improve the holding capacity.

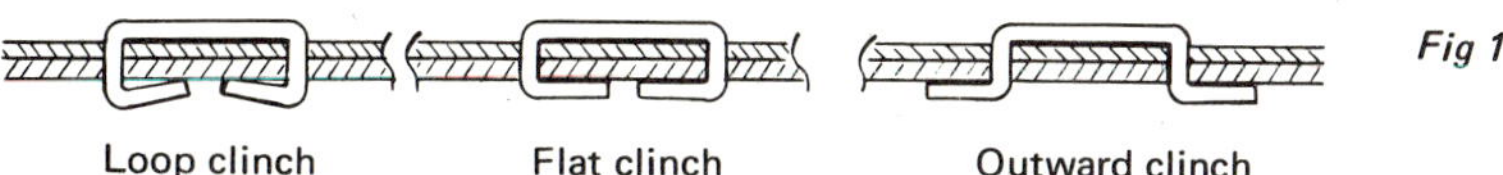

Fig 1

Staple crown widths (and wire sizes) vary according to the duty required – from miniature staples in round wire with crown widths of the order of 1.5mm (0.060in) up to flat wire staples of 25mm (1in) crown width or more for stitching packages. Staples are also produced in different (preformed) crown shapes – eg *round crown* to fasten over tubular workpieces, *shaped crown with shoulders* and *contour crown* – see Fig 2. Shaped crown and shouldered staples tend to drive better in soft materials and clinch better than round crown staples in hard materials. Crown contouring can also be produced by altering the shape of the driver and clincher.

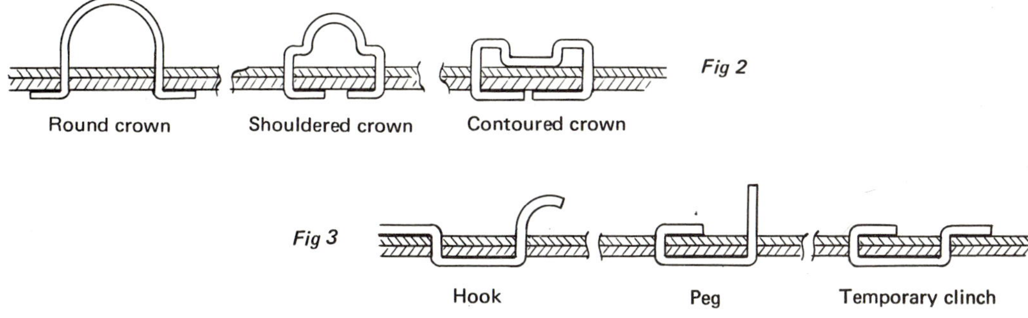

Fig 2

Fig 3

Further clench forms are shown in Fig 3. Where only one leg is clinched, the other can be left straight or slightly bent, to form a peg or loop. Alternatively, both legs can be turned the same way to form a temporary clinch. This will hold two workpieces together, but the top layer can be removed by a sidelong pull and 'peeling' action without disturbing the other layer.

A flat clinch can be produced when greater holding power is required. A movable clencher is used which wipes the staple legs into a flat shape with the whole clinched length of each leg pressing against the workpiece. Other clinch forms employed include circular or elliptic loops, with ends either clinched to contact and grip without piercing the workpiece or turned in to prevent the staple from sliding off. Again the actual shape and form of the loop are controlled by the type of clincher. – Fig 4.

Other special types of staple include those developed for electrical work, notably the *terminal staple* and *printed circuit staple* – Fig 5. Terminal staples (usually in tinned copper) provide easily fitted soldering terminals for electrical circuits and electronic components. Printed circuit staples are designed specifically for stapling to printed circuit boards to form jumpers or other connections which cannot be accommodated on the copper lands. After stapling in place, they are normally soldered to the corresponding copper lands.

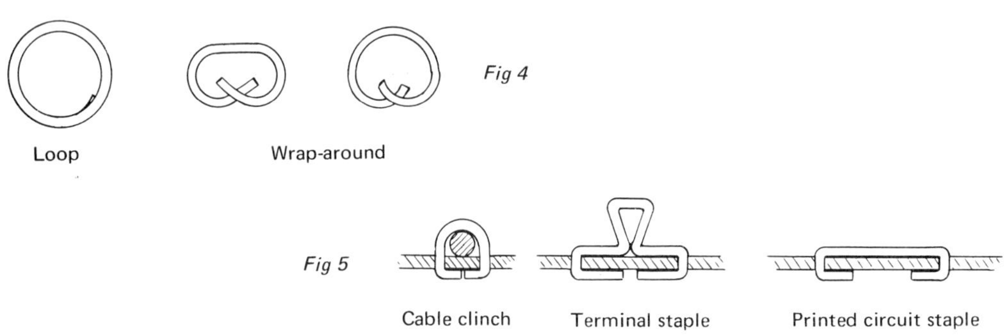

Fig 4

Fig 5

Stapling machines range from tackers (from miniature and standard office types upward), self-feed hammers for general industrial use, to top stitchers and bottom stitchers, foot or motor operated, for packaging, etc, and motor-driven machines for continuous production line work.

Some typical staple sizes are:-

Office – 0.5mm x 0.4mm; 6mm crown 3mm legs
(0.019in x 0.0175in; ¼in crown 1/8in legs)
0.7mm x 0.4mm; 12.5mm crown, 6mm legs
(0.027in x 0.0175in; ½in crown, ¼in legs)

Light Industrial		
	0.5mm round (0.19in round)	6, 9 and 12mm legs (¼in, 3/8in and ½in legs)
	0.7mm x 0.5mm (0.026in x 0.019in)	6mm, 9mm and 12.5mm crown widths (¼in, 3/8in and ½in crown widths)
	2.3mm x 0.5mm 0.090in x 0.019in)	

Automobile trim	0.019in (0.7mm) round 0.026in (0.7mm) x 0.019in (0.8mm) 0,090in (2.3mm) x 0.019in (0.8mm)	Crown widths ¼in (6mm) ¾in (9mm)

Carton stapling	1.3mm x 0.5mm (0.050in x 0.019in) 1.3mm x 0.7mm (0.050in x 0.025in) 2mm x 1mm (0.075in x 0.037in) 2.6mm x 0.5mm (0.103in x 0.020in)	6, 9, 12 and 14mm legs (¼in, 3/8in, ½in and 9/16in legs) Crown widths up to 50mm (Crown widths up to 2in)

See also the chapter *Metal Wire Stitching* for further information on the applications, characteristics and performance of clinched wire stitches.

Pipe Clips,Clamps and Fasteners

Permanent installations in rigid (metallic) piping need adequate support, both to prevent sagging under their own weight and also to eliminate strain on fittings or components to which they are attached.

As a general rule, pipe clips, hangers or supports should be spaced close enough together to support the weight of straight horizontal runs without sagging. Wider spacing can be used on vertical runs, but with all runs the possibility of an unsupported length of pipe being accidentally snagged and pulled out of position should be considered. Clips should also be placed close to fittings which are not themselves mounted so that the pipe run does not carry the weight of the fitting. Pipes should approach end fittings in as straight a line as possible, the radius of any adjacent bends being adjusted as necessary.

TABLE I – RECOMMENDED PIPE SUPPORT SPACING FOR HIGH PRESSURE LINES

Tube o.d. mm	Maximum Distance Between Clips* mm	Tube o.d. inches	Maximum Distance Between Clips* inches
3	250	1/8	10
4	275	3/16	12
5	300	1/4	15
6	350	5/16	18
7	400	3/8	18
8	450	7/16	24
9	500	1/2	24
10	550	9/16	24
15	600	5/8	24
20	700	3/4	30
25	750	7/8	30
over 25	30 x diameter	1	30
		over 1	30 x diameter

*up to twice this spacing may be used on vertical runs

Where pipes carry pressurised fluids, adequate support should be provided before and after bends as any sudden interruption in flow will tend to produce a straightening force in the bends, with the possibility of 'whipping' if the pipe is not properly supported. Recommended spacing for pipe clips on high pressure lines (eg hydraulic lines) is given in Table I.

Pipe clips should provide resilient rather than rigid support to allow sufficient movement to accommodate thermal expansion and contraction. Various proprietary pipe clips are available, most of which have an elastic or semi-elastic lining bonded to the inside to grip the pipe without locking it rigidly in place. Others take the form of simple clips which can accommodate a resilient sleeve fitted round the pipe at the point of support.

Besides providing support for pipes (and particularly pressure lines), pipe clips also assist in damping vibrations which may be set up in the pipe run. In certain systems all metallic piping is required to be electrically earthed and provision for this is often made in the design of the pipe clip – eg it has a small tongue or similar projection that will contact the tube wall and effectively earth the pipe through the clip. With a plain type of clip earthing can be introduced by inserting a layer of metal gauze between the pipe and clip, and connecting the gauze to an earth point such as the fixing screws or bolts of the clip.

Clips providing resilient mounting and vibration damping may only be effective in reducing noise if the complete system is isolated or decoupled from the original source of vibration. In other words, isolation treatment may have to be applied at through-ways and end fittings, as the efficiency of isolation depends largely on the mass to which the isolation clips themselves are attached. The more solid and massive the fittings or structure to which they are secured, the more effective they will be.

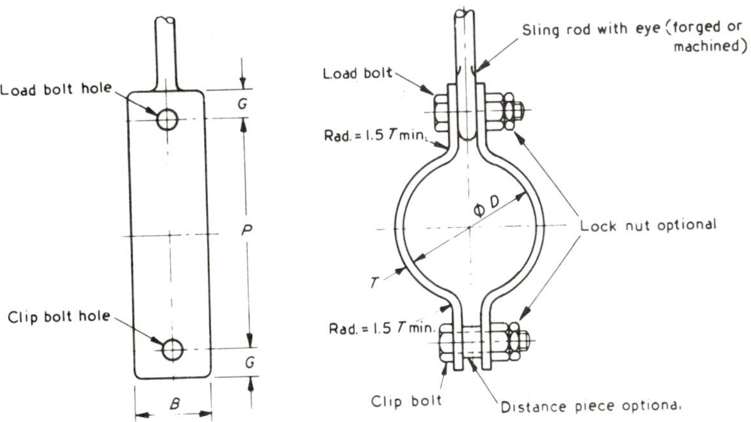

British Standard pipe clip for steel and cast iron pipes (BS3974)

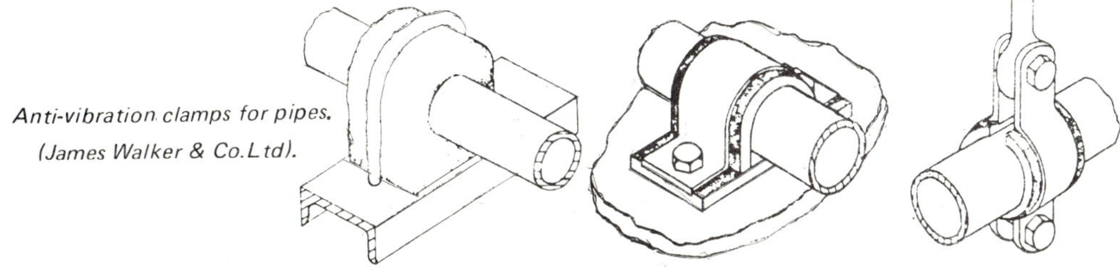

Anti-vibration clamps for pipes. (James Walker & Co. Ltd).

Range A for $-20^{o}C$ to $100^{o}C$
Range B for $-20^{o}C$ to $400^{o}C$
Range C above $400^{o}C$ to $470^{o}C$

TABLE II – DIMENSIONS OF BRITISH STANDARD PIPE CLIPS – RANGE A

Pipe Size		Light Series								Heavy Series							
Nom Size	o.d. dia	Sling Rod dia	D dia	Clip dimensions B x T	P	Clip and load bolts Bolt	Clip and load bolts Hole dia	G min	Safe Working load	Sling Rod dia	D dia	Clip dimensions B x T	P	Clip and load bolts Bolt	Clip and load bolts Hole dia	G min	Safe Working load
									kgf								kgf
15	21.3	USE HEAVY SERIES								10	23	35 x 5	65	M10	12	15	280
20	26.9									10	28	35 x 5	70	M10	12	15	280
25	33.7									10	36	35 x 5	75	M10	12	15	280
32	42.4									12	44	35 x 5	90	M12	15	18	280
40	48.3									12	50	35 x 5	95	M12	15	18	280
50	60.3									12	62	35 x 5	105	M12	15	18	280
65	76.1	12	80	35 x 5	125	M12	15	18	165	16	80	35 x 8	155	M16	19	24	450
80	88.9	12	92	35 x 5	135	M12	15	18	165	16	92	35 x 8	165	M16	19	24	450
100	114.3	12	118	35 x 5	170	M12	15	18	165	16	118	35 x 8	190	M16	19	24	450
125	139.7	16	144	35 x 5	195	M16	19	24	280	16	144	35 x 8	215	M16	19	24	450
150	168.3	16	172	35 x 5	225	M16	19	24	280	16	172	35 x 8	245	M16	19	24	450
175	193.7	16	198	35 x 8	270	M16	19	24	450	16	198	45 x 10	280	M16	19	24	900
200	219.1	16	224	35 x 8	295	M16	19	24	450	16	224	45 x 10	305	M16	19	24	900
225	244.5	16	248	35 x 8	320	M16	19	24	450	20	248	60 x 10	340	M20	24	30	1 350
250	273.0	16	278	35 x 8	350	M16	19	24	450	20	278	60 x 10	365	M20	24	30	1 350
300	323.9	20	330	45 x 10	420	M20	24	30	900	24	330	65 x 15	455	M24	28	36	1 800
350	355.6	24	362	55 x 10	460	M24	28	36	900	30	362	65 x 15	500	M30	35	45	2 250
400	406.4	24	412	60 x 15	535	M24	28	36	1 350	30	412	65 x 20	575	M30	35	45	2 700
450	475.0	30	464	65 x 20	625	M30	35	45	2 250	36	464	80 x 20	635	M36	42	54	3 600
500	508.0	30	516	65 x 20	675	M30	35	45	2 250	36	516	90 x 25	715	M36	42	54	4 500
550	559.0	30	566	65 x 20	725	M30	35	45	2 250	36	566	90 x 25	765	M36	42	54	4 500
600	610.0	30	618	80 x 20	780	M30	35	45	2 700	42	618	110 x 25	830	M42	48	63	5 900

All dimensions in millimetres

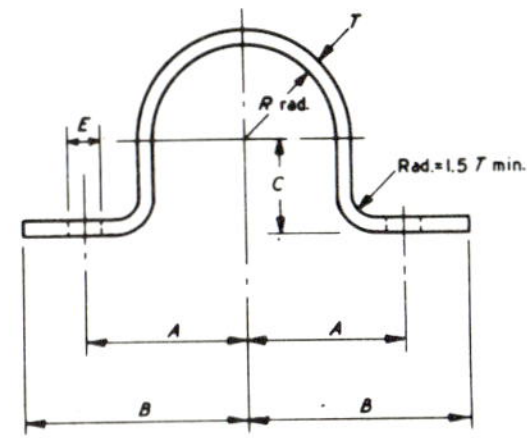

TABLE III – OVERSTRAPS FOR PIPES (BS3974)

Nominal pipe size	Pipe o.d.	A	B	Steel Size W x T	C	R	Hole E	Bolt dia
15	21.3	53	91	35 x 5	10	11.5	12	10
20	26.9	55	93	35 x 5	13	14	12	10
25	33.7	57	95	35 x 5	16	18	12	10
32	42.4	64	102	35 x 8	20	22	15	12
40	48.3	79	117	35 x 8	23	25	15	12
50	60.3	81	119	35 x 8	29	31	15	12
65	76.1	89	127	45 x 10	36	40	19	16
80	88.9	99	137	45 x 10	43	46	19	16
100	114.3	108	146	45 x 10	55	59	19	16
125	139.7	119	160	60 x 10	68	72	24	20
150	168.3	136	174	60 x 10	82	86	24	20
175	193.7	150	188	75 x 15	95	99	24	20
200	219.1	162	200	75 x 15	107	112	24	20

All dimensions in millimetres

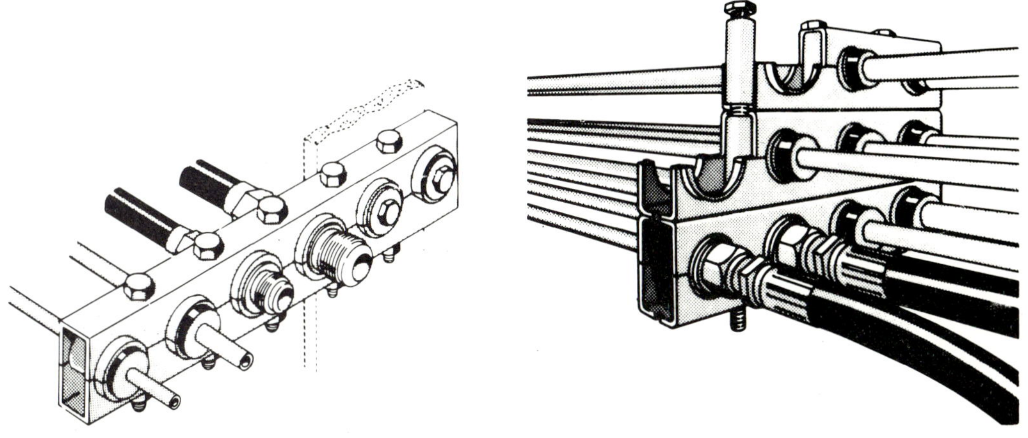

UCC International pipe clamps

Example of clamped pipework (UCC International).

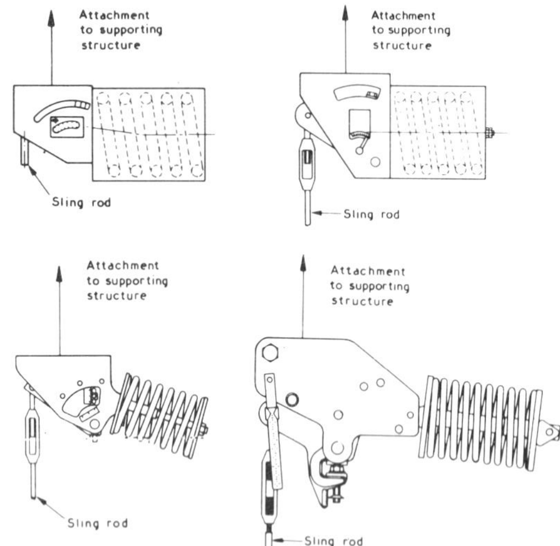

Examples of constant load spring hangers (BS 3974).

Metal spring pipe clip. (Carr Fastener Co.)

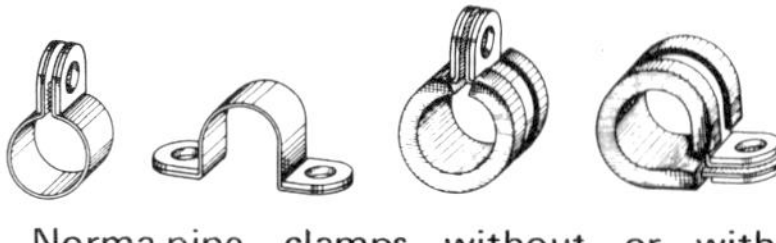

Norma-pipe clamps without or with rubber cushions.

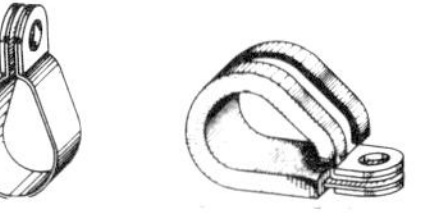

Norma-pipe clamps of special design.

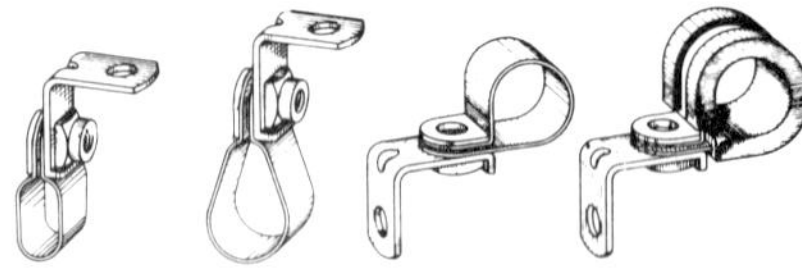

Norma-pipe clamps with angle brackets.

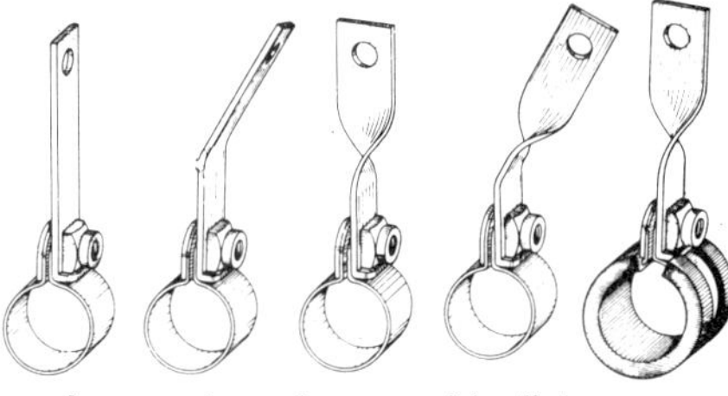

Norma-pipe clamps with fixing strap without or with rubber cushions.

Examples of proprietary pipe clips.

Self-locating plastic pipe clips. (Carr Fastener Co.)

Hose Clips

Two classes of hose clip are covered by British Standards:-

(i) Worm drive clips for general purpose industrial and domestic use, in a range of sizes from 12.5mm (½in) to 150mm (6¼in) maximum outside diameter (BS3628) – Fig 1.

TABLE I – BS WORM DRIVE HOSE CLIP DIMENSIONS (BS3678)

Clip Designating Size	Working* diameter		Housing Length	Length	Clip Designating Size	Working* diameter		Housing Length	Length
	min	max	max	max		min*	max*	max	max
050	0.375	0.500	0.550	1.00					
062	0.500	0.625	0.550	1.00	237	1.750	2.375	1.095	1.30
075	0.500	0.750	0.550	1.00	275	2.000	2.750	1.095	1.30
087	0.625	0.875	0.785	1.00	312	2.375	3.125	1.095	1.30
100	0.750	1.000	0.785	1.00	350	2.750	3.500	1.095	1.30
112	0.875	1.125	0.950	1.16	400	3.250	4.000	1.095	1.30
137	1.000	1.375	0.950	1.16	450	3.750	4.500	1.095	1.30
162	1.125	1.625	0.950	1.16	500	4.125	5.000	1.095	1.30
187	1.250	1.875	1.095	1.30	575	5.000	5.750	1.095	1.30
212	1.500	2.125	1.095	1.30	625	5.500	6.250	1.095	1.30

All dimensions in inches

*Size range for matching hose o.d.

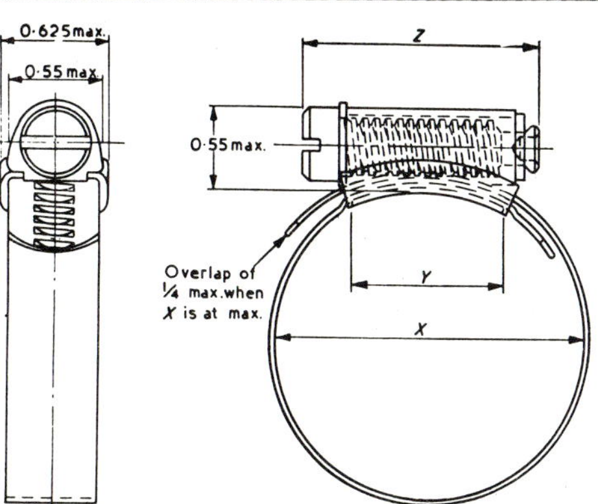

Fig 1 BS worm drive hose clip (BS3628)

(ii) Adjustable clips of channelled ring design for securing up to 50mm (2in) diameter hose under internal pressures not exceeding 1.5bar (20lb/in^2) (BS4055) – Fig 2.

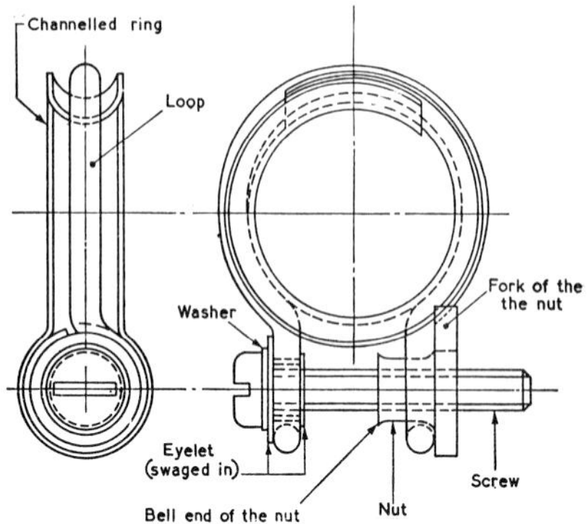

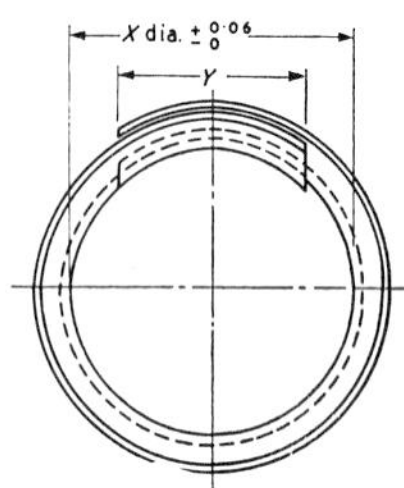

Fig 2 BS channelled ring hose clips (BS 4055).

TABLE II – BS CHANNELLED RING HOSE CLIPS (BS4055)

Size Designating Symbol	X Diameter	Y Lengths	Size Designating Symbol	X Diameter	Y Lengths
A	0.31	0.2	N	1.13	0.6
B	0.38	0.2	P	1.25	0.6
C	0.44	0.2	Q	1.38	0.7
D	0.50	0.2	R	1.50	0.7
E	0.56	0.4	S	1.63	0.8
F	0.63	0.4	T	1.75	0.8
G	0.69	0.4	U	1.88	0.9
H	0.75	0.5	V	2.00	0.9
J	0.81	0.5			
K	0.88	0.5			
L	0.94	0.5			
M	1.00	0.6			

All dimensions in inches

Metric sizes for worm drive hose clips are summarised in Table III.

There are numerous types of individual proprietary designs of hose clip, some examples of which are given under:-

Unix Hose Clip

This is a double loop clip providing a completely uniform turn contacting the periphery of the hose, capable of withstanding pressures up to 140bar (2 000lb/in^2). The clip will also accommodate elliptic and other profiles. It is produced in two types – type NH with a hex head for hand adjustment and type N with a slotted screw for screwdriver adjustment – Fig 3.

The standard range is produced in stainless steel, but all sizes are also available in phosphor bronze. Standard sizes are detailed in Table IV.

TABLE III – METRIC WORM DRIVE HOSE CLIPS (DIN3017)

Size Range mm	Size Range inch	Band Width* mm
12–20	½–¾	9
16–25	5/8–1	9
20–32	¾–1¼	9 and 14
23–35	7/8–1.3/8	9 and 14
25–40	1–1.5/8	9 and 14
32–50	1¼–2	9 and 14
40–60	1.5/8–2.3/8	9 and 14
50–70	2–2¾	9 and 14
60–80	2.3/8–3.1/8	9 and 14
70–90	2¾–3½	9 and 14
80–100	3.1/8–4	9 and 14
90–110	3½–4.3/8	9 and 14
110–140	4.3/8–5½	14
125–160	4.7/8–6¼	14

*Band widths used include 9mm, 12mm, 15mm, 20mm and 25mm
Broad band clips encompass band widths of 20mm, 25mm, 30mm, 40mm, 48mm and 60mm.

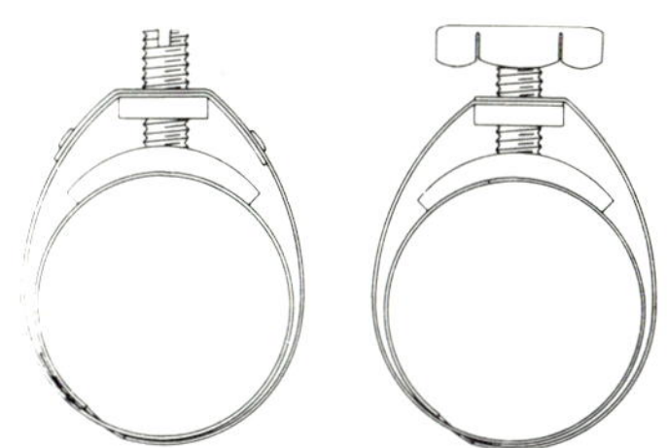

Fig 3 'Unix' stainless steel hose clip. Type N (left), type NH (right). (Deltight Industries).

TABLE IV – COMMERCIAL RANGE OF 'JUBILEE' CLIPS

Size No.	BS 3628	Inches	m.m.	SP91 Equiv.
000	050	⅜"- ½"	9- 13	—
MOO	062	½"- ⅝"	13- 16	A
OO	075	½"- ¾"	13- 19	B
O	087	⅝"- ⅞"	16- 22	C
OX	100	¾"-1"	19- 25	D
1A	112	⅞"-1⅛"	22- 28	E
1	137	1" -1⅜"	25- 35	F
1X	162	1⅛"-1⅝"	28- 41	G
2A	187	1¼"-1⅞"	32- 48	H
2	212	1½"-2⅛"	38- 54	J
2X	237	1¾"-2⅜"	44- 60	K
3	275	2" -2¾"	50- 70	L
3X	312	2⅜"-3⅛"	60- 80	M
4	350	2¾"-3½"	70- 90	N
4X	400	3¼"-4"	82-100	P
5	450	3¾"-4½"	95-115	Q
6	500	4⅛"-5"	105-125	R
6X	575	5" -5¾"	127-146	—
7	625	5¼"-6¼"	133-158	—

Large sizes from 6½" (165 m.m.) to 18" (457.2 m.m.) in diameter.

Delclip

The Delclip (Fig 4) is a fully openable clip for fitting in situations where the hose cannot be dismantled. The design accommodates a large contraction ratio and also elliptic as well as circular profiles. Standard production is in stainless steel.

Examples of lighter spring hose clips are also illustrated.

Fig 4 'Delclip' stainless steel hose clip.

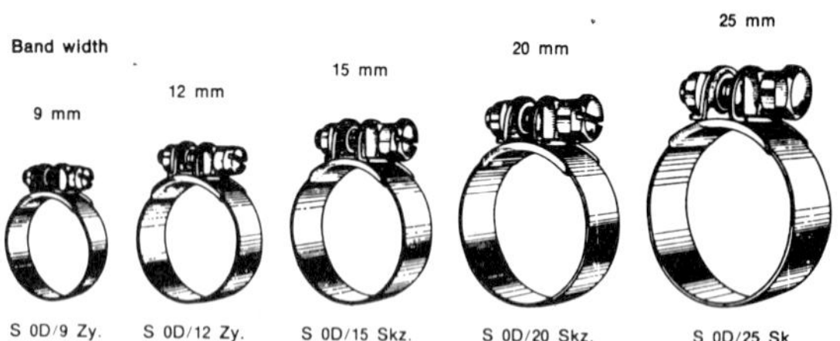

'Norma' metric size hose clips.

Examples of clamping clips. (Norma Schellen).

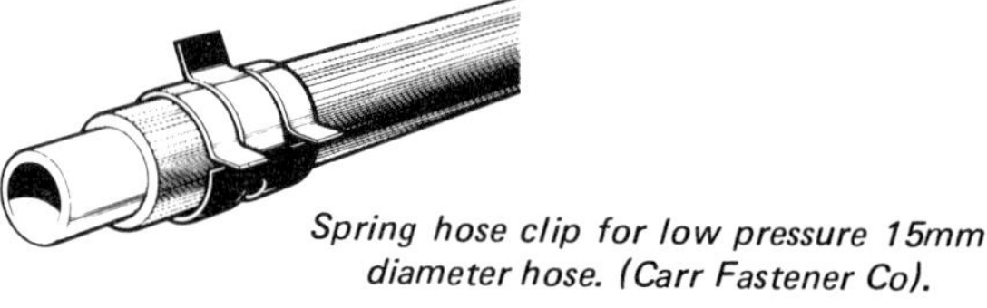

Spring hose clip for low pressure 15mm diameter hose. (Carr Fastener Co).

'Jubilee' hose clips.

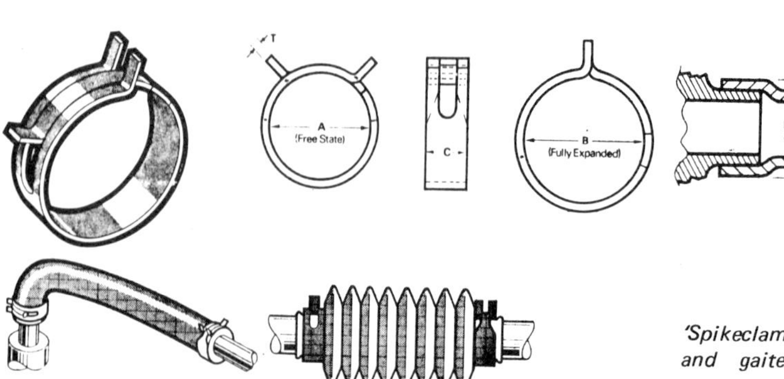

'Spikeclamp' for plastic and rubber hose and gaiters up to 30mm (1¼inch) diameter.

Miscellaneous Clips

The immense variety of metal and plastic clip-type fasteners and variations in individual designs precludes specific classification, but the following tables and illustrations offer a general guide as to types and applications. See also chapters on *Hose Clips, Pipe Clips, Spring Fasteners,* etc.

TYPES OF CLIP IN PLASTIC, METAL OR METAL/PLASTIC COMBINATION

Types of Clip	Description
Linkage clips	All-plastic, metal and plastic or all-metal (Note: all-metal linkage clips are to be avoided where electrical noise is a consideration)
Beading clips	For rattle-proof securing of beading. Both plastic and all-metal types are used
Tube clips	For securing fabrics, rubbers, etc to metal tube. Usually metal spring clips.
Can lid clip	Security fastening for lever lid cans. All-metal spring clips.

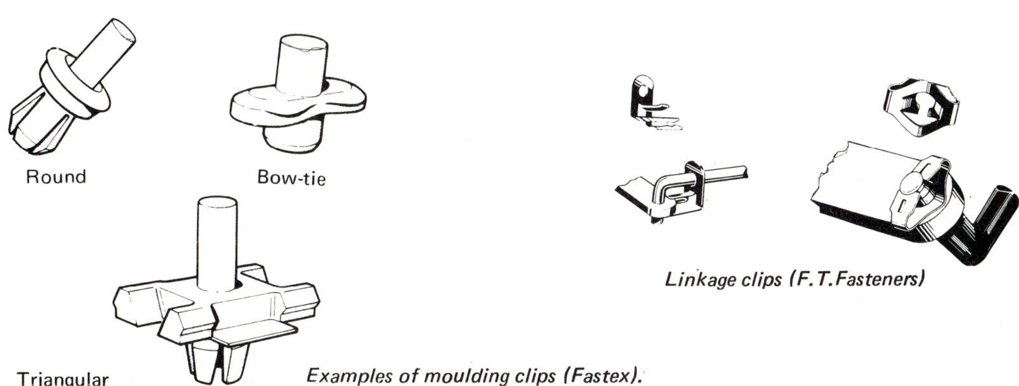

Examples of moulding clips (Fastex).

Linkage clips (F.T.Fasteners)

1
2
3
4
5
6
7
8
9

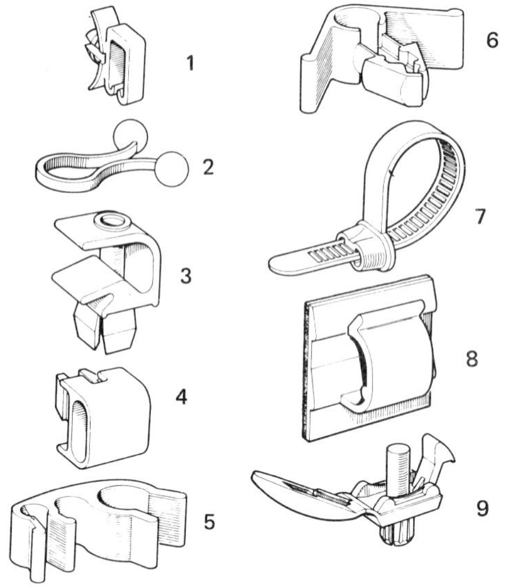

1—wire dressing clips
2—wire ties
3—rod clips
4—anti-rattle clips
5—pipe spacer clips
6—pipe retaining clips
7—ratchet/ cable straps
8—cord clips (adhesvie backed)
9—strain relief clips

Examples of miscellaneous plastic clips and straps. (Fastex).

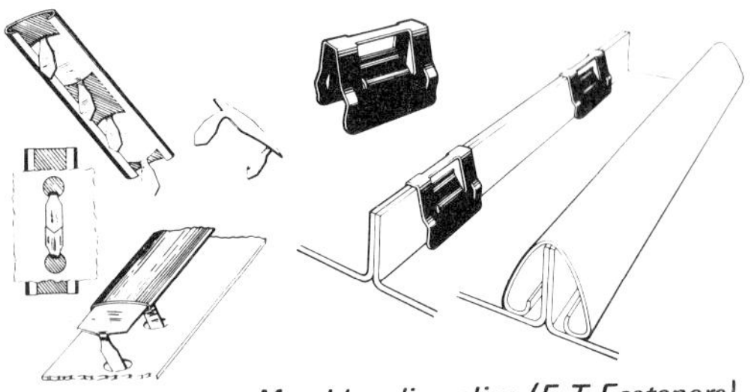

Metal beading clips (F.T.Fasteners).

1. Form a loop with the end of the band.
2. Pull loop through both sides of clip and hold it there.
3. Bring band twice around object to be fixed, each time passing it through clip. Cut band and insert the end into bolt slot and tighten bolt.

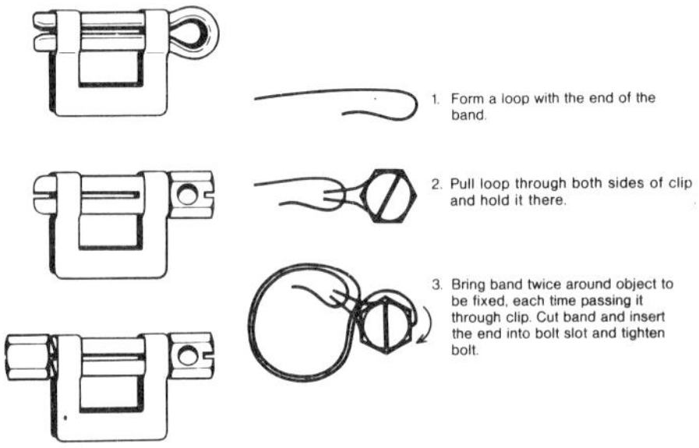

'Mirex' clamp fasteners for strap fastening.

Type 1
Type 2
Type 3
Beading
Panel

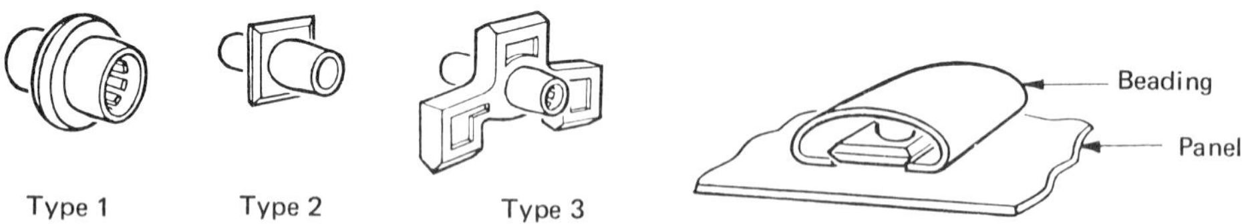

Plastic beading clips (F.T.Fasteners)

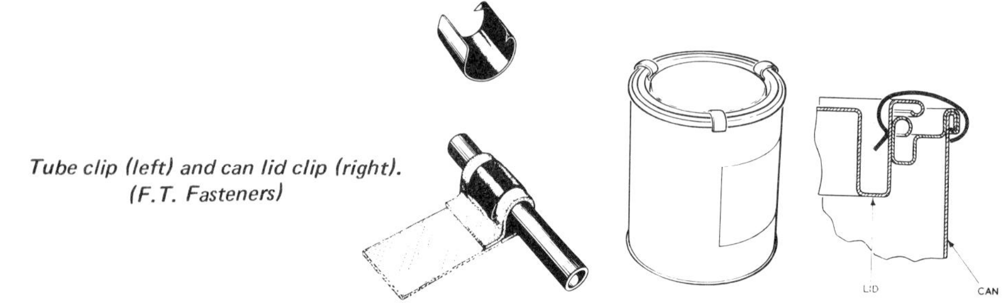

Tube clip (left) and can lid clip (right). (F.T. Fasteners)

LID
CAN

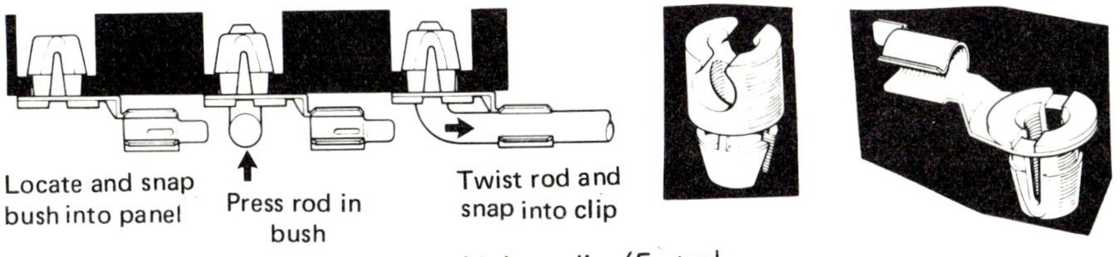
Locate and snap bush into panel
Press rod in bush
Twist rod and snap into clip
Linkage clips (Fastex)

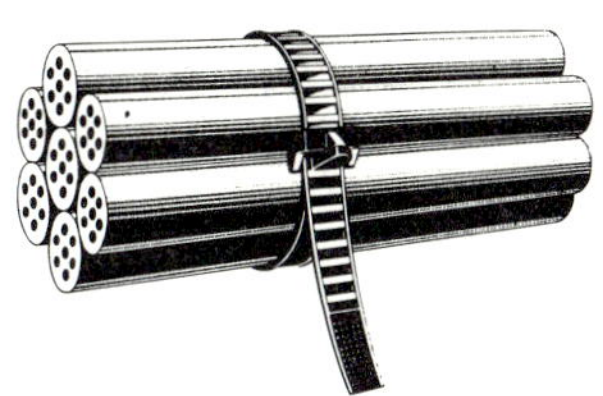
Cable strap (F.T.Fasteners).

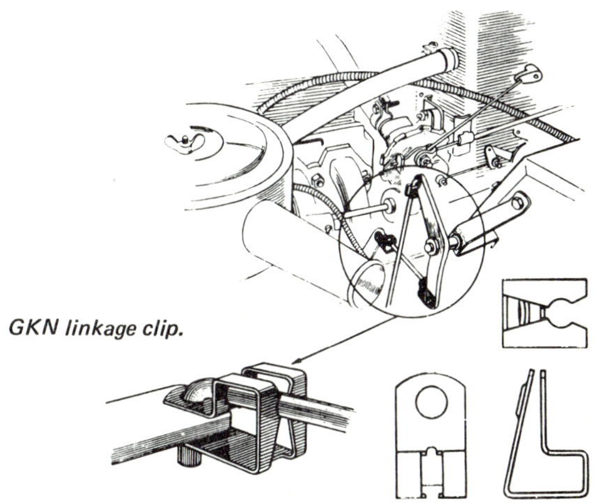
GKN linkage clip.

TYPES OF PLASTIC CLIPS AND STRAPS

Types of Clip	Description
Rod (Linkage clip)	Simple C-shape or U-shape clips for holding rods, etc. Various types and sizes.
Rod, anti-rattle (Linkage clips)	Locating or guiding for rods, wires, etc, to eliminate rattle. Various types and sizes.
Pipe	Similar to rod clips but in multiple as well as single forms, to accommodate standard pipe sizes. There are separate types for pipe spacing and pipe retaining.
Ties	Strap forms for binding with self-latching or self-locking geometry after completion of a turn. Large variety of types and sizes, both for general purpose and specific applications.
Straps	Similar to ties but with integral self-locking fastener feature (usually ratchet type) and provision for release for undoing strap.
Cable straps	Straps specifically designed for binding cables. Usually made in nylon or polyacetal.
Dressing clips (Cable clips)	Clips specially designed for dressing wires or cables, or as guides for control rods, flexible cable sheaths, etc.
Moulding clips	Self-locking or self-expanding clips designed to secure a variety of roll-formed, stamped or extruded sections of various jaw-gap sizes

Belt Fasteners

Modern flat transmission belts are of composite construction comprising a high tensile carcase of nylon or polyester with chamois leather or elastomeric facings for frictional grip. They are normally used as 'endless' belts, either produced to finished length or made endless after installing on the machine by means of a welded precision joint.

V-belts are normally truly endless, although there is a variety of patented methods joining continuous lengths of V-belting to simplify installation at fixed centres, etc. These include special forms of V-belt (eg 'link' belts). Such jointed or adjustable length belts may modify capacity or speed ratings, and may also require the use of matching pulleys to ensure adequate clearance for the fasteners used.

Traditional flat belting in leather, multi-ply fabric, multi-ply rubber, cotton or Balata continues to be used, cut to length and jointed by lacing. The lacing may take the form of individual wire 'stitches' or belt hooks clenched in position by hand tools or special machines, various proprietary forms of steel belt lacing or jump-joint fasteners.

Simple wire lacing is suitable for lighter drives. To facilitate assembly the individual hooks can be supplied mounted on a paper frame which holds the hooks securely in position until clamped onto the belt end – Figs 1A and 1B.

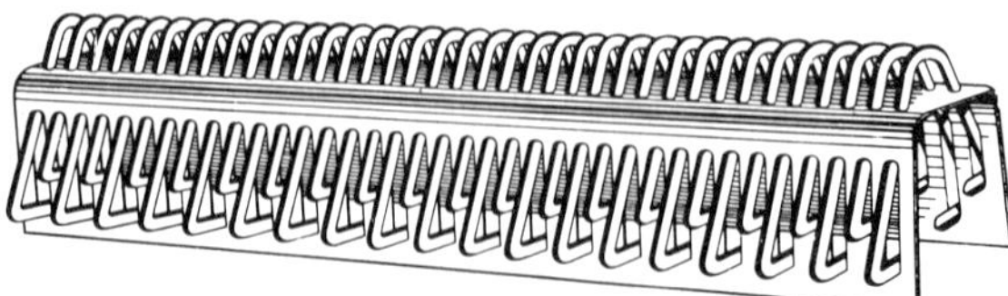

Fig 1A 'Hawk' carded belt hooks.

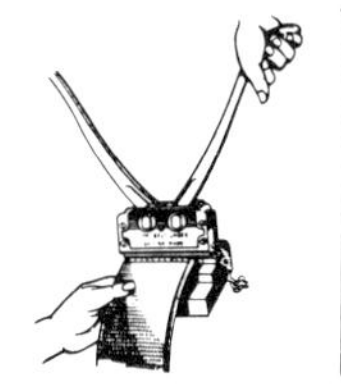

Fig 1B Alternative methods of assembling belt hooks – by hammering, by belt lacing machine, and with a vice tool.

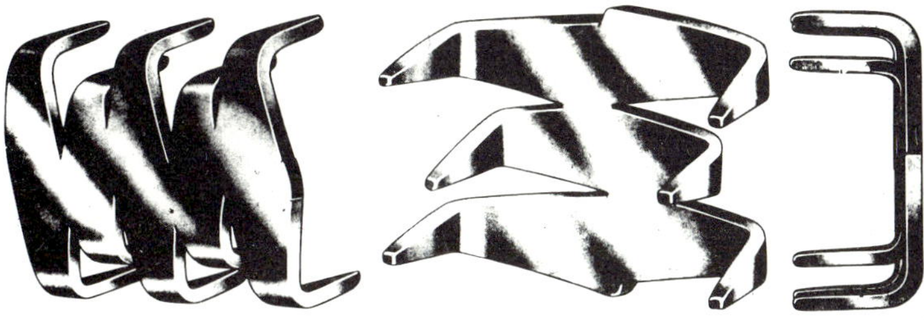

Fig 2 Jackson's Bristol steel belt lacing.

Zig-zag multi-point steel laces are also widely used, a proprietary type being shown in Fig 2. The points of the laces are staggered so that the prongs enter in two separate rows to provide maximum grip with minimum tendency to cut the fibre or substance of the belting. They are applicable to all types of leather, rubber, balata and fabric belting.

Higher strength is provided by continuous steel lacing designed to provide a piano-type hinge running across the complete width of the belt – see Fig 3. Again, this can be used for all types of flat belting, with the size selected specifically to match the thickness and constructional features of the belting.

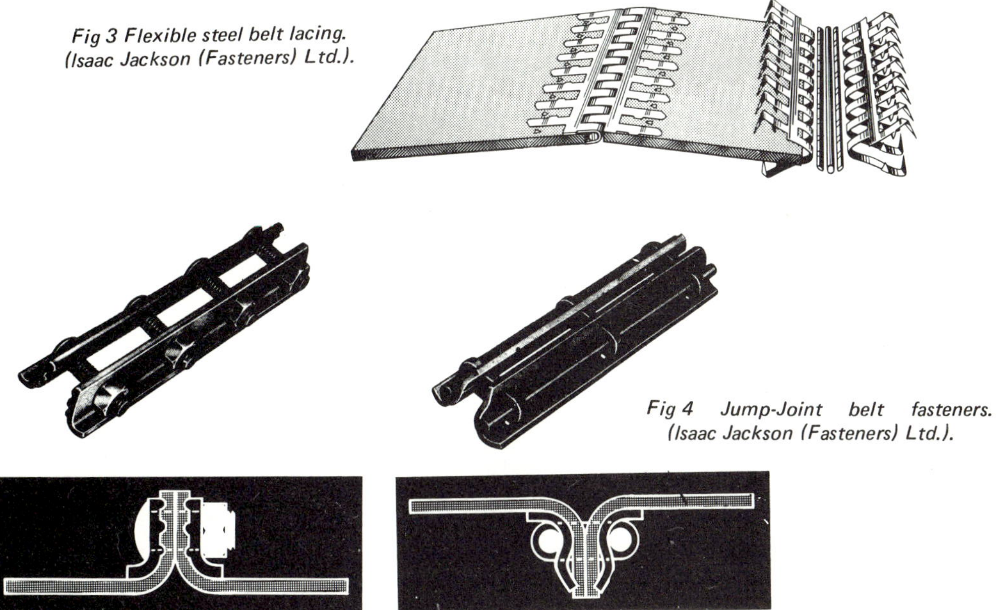

Fig 3 Flexible steel belt lacing. (Isaac Jackson (Fasteners) Ltd.).

Fig 4 Jump-Joint belt fasteners. (Isaac Jackson (Fasteners) Ltd.).

Jump-joint belt fasteners are designed to produce a clamped joint on turned up ends of the belt, the fastener coming on the outside of the run – Fig 4. For belts running over crowned pulleys it is advisable to use two or more fasteners of this type as this will minimise strain on the joint and give smoother running and better pulley contact. Jump-joints are probably the strongest and most secure of all types of belt fastener, but their application is obviously limited to belts with only one 'running' surface.

Fig 5 Plate (left), button plate and button type belt fasteners.

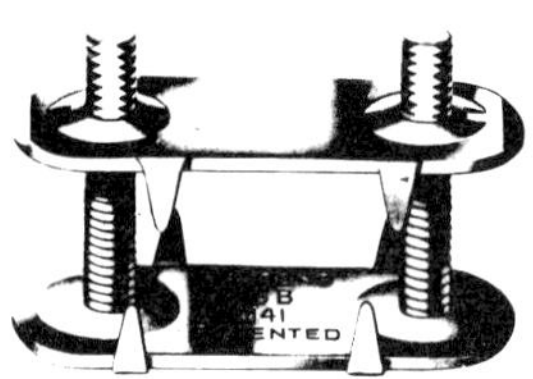

Fig 6 Conveyor belt fasteners. (Isaac Jackson (Fasteners) Ltd.).

Plate-type and button-type fasteners may also be used for jointing flat belts (Fig 5). Belt ends are cut square and butted firmly together and the plates are spaced equidistantly across the belt exactly halved by the belt ends and parallel to the belt edges.

Conveyor belt fasteners are usually of pinned, bolted or riveted plate type, the size and type being chosen with regard to the diameter of the pulleys over which the belt runs. If the plates are too large the belt may not conform with the pulley when the joint passes over it. If the centres of the plates are too short the fastenings will be placed too near the ends of the belt with a risk of pulling out. Examples of plate-type fasteners for conveyor belting are shown in Fig 6.

***TABLE I – GENERAL PERFORMANCE DATA OF BONDED JOINTED FLAT BELTS**

Belt Type and Material		Construction	Power Range HP	Speed Range Ft/Min	Tensile Strength	Thickness Range t	Min. Pulley Dia.	Temperature Range Deg F (Deg C)
JOINTED BELTS	Leather	Single	up to 1 000	up to 8 000	200 to 400 lb per $\frac{1}{16}$ in thickness	$\frac{5}{32}$ in to $\frac{5}{16}$ in	30 x t	–5 to + 120 (–20 to + 50)
		Double	up to 2 000	up to 8 000	250 to 450 lb per $\frac{1}{16}$ in thickness	$\frac{5}{16}$ in to $\frac{9}{16}$ in	30 x t	– 5 to + 120 (–20 to + 50)
		Three-Ply		up to 6 000			30 x t	
	Cotton	Woven Cotton Canvas	–	–	350 lb per ply (typical)	–	30 x t	–40 to + 300 (–40 to + 150)
	Hair	Woven	–	–	300 lb per $\frac{1}{16}$ in thickness	–	40 to 90 x t	
	Balata	3 ply to 10 ply	See Table III	500– 9 000	400–500 lb per ply	–	3 in to 30 in	0–100 (–20 to + 40)
	Composite	Nylon with Leather or Elastomer Face(s)	0.5–5 000	500 – 12 000	80 000 psi	0.03 in to 0.24 in	40 to 90 x t	–40 to + 175 (–40 to + 80)
TRULY ENDLESS BELTS	Cotton	Multiple Ply	–	–	–	$\frac{1}{16}$ in to $\frac{1}{4}$ in	–	–15 to + 140 (–25 to + 60)
	Composite	Nylon with Leather or Elastomer Face(s)	0.25–400	500–2 400	120 000 psi	0.05 in to 0.20 in	10 to 18.5 x t	–40 to + 175 (–40 to + 80)
		Polyester with Leather or Elastomer Face(s)	0.25–400	500–2 400	135 000 psi			
	Polyester Film	Cut Ring	Fractional	–	20 000 psi	0.0005 in to 0.015 in	100 xt	–40 to + 175 (–40 to + 80)
		Welded	Fractional	–	20 000 psi	0.0005 in upwards	100 xt	–40 to + 175 (–40 to + 80)
	Rubber Cord	Single Ply and Multiple Ply Woven Elastomers	–	–	–	0.020 in to 0.050 in	–	–20 to + 150 (–30 to +65)

*(Power Drives Handbook – Published by Trade & Technical Press Ltd)

Materials

Black bolts – the cheapest form of production – are manufactured from carbon steel with an average tensile strength of the order of 22–28tonf/in^2, (35–44kg/cm^2). Bright bolts are produced from carbon steel with an average tensile strength of 25tonf/in^2 (40kg/cm^2). Two other general qualities of (bulk) production are:-

'A' Quality – 25–28tonf/in^2 (40–44kg/mm^2), not stress relieved

'B' Quality – 28tonf/in^2 (44kg/mm^2), stress relieved.

The following BS specifications for materials apply to high tensile bolts for precision engineering with a minimum tensile strength of 45tonf/in^2 (71kg/mm^2).

	Minimum ultimate tensile strength	**Minimum yield stress**
Grade P (BS1768 and BS1083)	35tonf/in^2	21tonf/in^2
Grade R (BS1083)	45tonf/in^2	34tonf/in^2
Grade S (BS1768)	50tonf/in^2	40tonf/in^2
Grade T (BS1768)	55tonf/in^2	41tonf/in^2
Grade V (BS1768 and BS1083)	65tonf/in^2	52tonf/in^2
Grade X (BS1768 and BS1083)	75tonf/in^2	63tonf/in^2

These material specifications apply to ISO Unified and Whitworth bolts.

Grade S is also directly related to metric bolt sizes (Metric 8.8).

Strength gradings for ISO metric bolts and screws are designated by numbers separated by a decimal point. The first number specifies the ultimate tensile strength in kg/cm^2 x 10 and the number after the decimal point the factor to give yield stress of the material – eg 8.8 = 80kg/cm^2 tensile strength and 0.8 x 8 x 10 = 64kg/cm^2. The full range is shown in Table on page 453.

It will be noted that there is no near ISO equivalent to BS grade T; and no BS grade equivalent to ISO 14.9.

	Ultimate Tensile Strength		Yield Stress	
	kg/mm^2	tonf/in^2	kg/mm^2	tonf/in^2
Grade 4.6	40	25.9	24	15.2
Grade 4.8	40	25.9	32	20.3
Grade 5.6	50	31.7	30	19.0
Grade 6.6	60	38.1	36	22.9
Grade 8.8	80	50.8	64	40.6
Grade 10.9	100	63.5	90	57.2
Grade 12.9	120	76.2	108	68.8
Grade 14.9	140	88.9	126	80.0

Grading for Nuts

Imperial strength gradings for nuts are 0, 1, 3 and 5 (BS1768), and A, P, R and T (BS1083). Only A and P grades are in regular production, with recommendations as follows:-

Grade A – for use with Grade R and S high tensile bolts.

Grade P – for use with Grade V high tensile bolts.

ISO metric nut grades are 4, 5, 6, 8, 12 and 14 with corresponding proof load stress values of 40, 50, 60, 80 and 120kg/mm^2 respectively. Recommended bolt and nut combinations are:-

Bolt Grade	Nut Grade
4.6 or 4.8	4
5.6	5
6.6 or 6.8	6
8.8	8
10.9 or 12.9	12
14.9	14

As a general rule the strength of the nut material can be lower than that of the bolt although nuts are designed so that when used with the correct bolt the bolt should break before the bolt or nut threads strip.

In American practice the following steel specifications are used:-

Carbon and Alloy Steels

AISI 1010 – carriage bolts and machine screws, etc, without critical strength requirements.

AISI 1018, 1020, 1021 – bright cap screws and special uses.

AISI 1038 – high strength bolts and cap screws

AISI 1041, 1045, 1330, 1340, 4135 and 8637 – for special requirements.

IOB21, IOB22 and IOB23 – boron treated steels for high strength bolts and cap screws.

AISI 1100 series – general production of nuts.

AISI 1016, 1038 – nuts.

AISI 1045 – hot-formed nuts

Stainless Steels

AISI 410, 416 and 431 – most commonly used martensitic steels.

AISI 430 and 450F – most commonly used ferritic steels

304 and 305 – most commonly used austenitic steels

316, 321 and 347 – austenitic steels for special fasteners

Non-Ferrous Metals

The following brasses are commonly used for the production of fasteners:-

Free-cutting brass – milled production of screws and nuts.

Yellow brass – cold-headed bolts, screws and nuts.

Cold heading brass – cold-headed bolts, screws and nuts.

Leaded brass – milled screws and nuts.

Naval brass – various alloys for hot forging and cold heading.

Copper alloys used include the following types:-

Commercial bronze – cold-headed bolts, screws and nuts and milled nuts.

Naval bronze – various compositions for cold forging and hot forging bolts and nuts; also milled bolts and nuts.

Silicon bronze – for marine fasteners.

Cupro-nickel – superior strength and erosion and corrosion resistance (70/30 copper/nickel) for marine fasteners.

Aluminium alloys are used for the production of lightweight fasteners and also have good resistance to atmosphere and chemical corrosion. Typical alloys used are:-

2024-T4 and 7075-T73 heat treatable alloys (ASTM B316) – for male threaded fasteners, threaded after heat treatment

2024-T4, 6061-T6 and 6262-T9 (ASTM B211 and ASTM B316) – for nuts

2024-T4 (ASTM B211) – for machine screw nuts

ASTM B316 – non heat-treatable alloys for cold rivets; heat treatable alloys for hot rivets.

Nickel alloys may be used where strength has to be maintained at high temperatures, also for immunity from discolouration and corrosion in difficult conditions. Various nickel-copper alloys may be used for the production of bolts, screws, nuts and rivets; or, for greater strength, nickel-copper-aluminium alloy may be used. Both types of alloy are tough and difficult to machine.

Non-Metallic Materials

Nylon is the chief non-metallic material used for fasteners, in a variety of grades. The grade should be selected according to the service requirements – eg resistance to sunlight, ozone or chemicals, and dimensional stability. Usually nylon 6/10 is chosen for maximum resistance to ultravoilet rays. UV resistance can also be provided by dyeing nylon black or incorporating UV filters.

Polycarbonate is an alternative choice to nylon with better dimensional stability and impact performance. It is suitable for riveting.

Polyacetal – is another alternative to nylon with even better dimensional stability. Apart from that, it has no advantages over nylon.

Polythene (high density) is suitable for low cost, low to moderate strength plastic fasteners. It has particular virtue for electrical work because of its excellent insulating properties.

Polystyrene is another low-cost material with excellent electrical insulating properties. Strength is low, however, and in unmodified form polystyrene is brittle and sensitive to impact or shear.

PVC in its rigid form is an attractive material for low to moderate strength fasteners exposed to acid or corrosive surroundings.

Vinyl is a flexible material with good resistance to chemical attack. Its main use in fasteners is as a sealant or flexible element in cup type fasteners, etc.

Fluorocarbons, (PTFE, TFE, FPTFE) offer extreme resistance to chemical attack and higher temperature resistance than other common plastics. Tensile strength is poor, however.

Corrosion Resistant Fasteners

Alternative approaches where fasteners are to be resistant to staining or corrosion are:-

(i) Protective treatment in assembly and/or in situ.

(ii) Use of fasteners with protective coatings.

(iii) Use of corrosion resistant materials.

The choice of method adopted depends on the environment and duty involved, and also on the cost. The environment will govern the likelihood of corrosion conditions being present and their probably severity. If severe, then the use of corrosion-resistant fastener materials may be dictated, especially if the strength of the fastener is to be maintained. The duty involved may affect factors such as strength, appearance and the possibility of staining by 'weeping'. Cost usually dictates a compromise between a high degree of immunity to corrosion and minimal resistance. The optimum compromise solution would be where the fastener had slightly better resistance to the effects of corrosion than the adjacent components or materials.

Many corrosion problems arise from the fact that two dissimilar metals are in contact, leading to electro-chemical activity in damp conditions. This can be reduced or eliminated by controlling the ambience (eg particularly limiting the humidity), or isolating dissimilar metals by insulating compounds, washers, etc. Further protection can be given by applying coatings of grease, paint, waxes, etc.

Protective coatings applied to the fasteners themselves are usually electro-deposited — ie plating with cadmium, zinc, nickel, copper or chromium. The basic requirements are that:-

(i) the plating metal has the necessary resistance to the environment and satisfactory electro-chemical behaviour as regards the base metal to which it is applied.

(ii) sufficient thickness of coating is applied to give the necessary protection.

(iii) coating does not introduce dimensional problems (in general, practical limits of coating thickness for standard screw threads is as shown in Table I.)

(iv) anodic coatings are preferable on steel (eg cadmium and zinc) as these have sacrificial properties and prevent 'bleeding' of rust in the event of damage to, or porosity developing in, the coating. Unfortunately these are less attractive than cathodic coatings (eg nickel and chromium). Whilst bright coatings can be produced with zinc, they tend to deteriorate easily.

TABLE I – ELECTROPLATED COATING THICKNESSES RECOMMENDED FOR ISO INCH AND ISO METRIC THREADS

Screw thread diameter		**Batch average plating thickness**	
Over	*Up to and including*	*Minimum*	*Maximum*
0·060in	0·126in	0·00015in	0·00020in
0·126in	0·250in	0·00020in	0·00025in
0·250in	0·500in	0·00025in	0·00030in
0·500in	0·750in	0·00030in	0·00035in
M1·5	M3	0·0035mm	0·0050mm
M3	M6	0·0050mm	0·0065mm
M6	M12	0·0065mm	0·0080mm
M12	M18	0·0075mm	0·0090mm

TABLE IA – PERFORMANCE OF PLATED COATINGS ON STEEL (GKN)
(Time in years for first appearance of rusting with 0.001in coatings)

Environment	ZINC		CADMIUM	
	As plated	Passivated	As plated	Passivated
Industrial	2.1	2.6	0.9	1.5
Urban	3.0	3.8	2.1	3.0
Rural	4.8	4.8	7.5	15.0
Marine	2.1	2.1	2.0	3.0

Maximum protection for mild steel fasteners is given by bright zinc coatings with a supplementary finish – eg chromate passification, silicate finish or epoxy lacquer. Without supplementary treatment, zinc coating is unsuitable for resistance to high humidity, condensation,salt spray, etc, where cadmium plating is better. Some comparative data are given in Table II.

Electro-plated coatings present specific problems, notably the likelihood of unevenness of the deposit. The thickness of this coating tends to build up at sharp corners, and also to be thicker at the ends and thinnest in the centre of a long piece (eg a bolt). Plating thickness will also tend to be least in recessed areas; this can modify the thread angle in opposite directions on external and internal threads, although the effect is generally negligible. In critical cases, and especially if heavy coatings are called for, both the effect of plating thickness on clearances and the likely variation in local thickness of the plating may have to be considered carefully.

Where long term protection is required, the use of a corrosion resistant material instead of a protective coating would appear more logical but the cost penalty involved is usually high. Also 'resistant' materials are not necessarily fully resistant to corrosion or tarnish. Brass can tarnish readily in a salt atmosphere (as well as being a weaker fastener material), and many stainless steels will develop rust staining and surface corrosion in particular ambiences. Some comparative cost figures are given in Table III.

TABLE II – GENERAL RECOMMENDATIONS : SCREW MATERIALS AND FINISHES

Screw Material and Finish	Appearance	Typical Applications
STEEL	Plain metal, may turn rusty	General use
STEEL with:		
Bright zinc plate	Bright attractive protective coating	All dry interior applications and where a paint finish is applied, indoor or outdoor
Sherardised (zinc)	Dull grey protective coating. May turn brown unless painted	Most exterior fasteners for buildings. A good surface for painting.
Nickel plate	Bright reflective finish – may tarnish	Dry interior fasteners, eg shelves, heaters
Chromium plate	Attractive bright reflective finish	Fairly dry interior work. Kitchens, most domestic appliances.
Brass plate (electro-brass)	Reflective bright yellow finish	Cupboards and furniture for matching against brass. Dry interior work only.
Bronze metal antique	Dark brown finish	For interior use with oxidised copper fittings.
Dark florentine bronze	Near black finish	For interior use with oxidised copper fittings.
Antique copper bronze	Uniform bronze colour	For interior use with copper, bronze and matching timber finishes.
Black japanned	Overall black enamel finish	General interior use; repainting necessary outdoors for protection.
Berlin blacked	Overall black enamel finish duller than japanned	General interior use; repainting necessary outdoors for protection.
Blued and oiled	Dark blue/black oxide coating, protective lubricating oil finish	Temporary protection only. Requires painting in most applications
BRASS	Uniform bright yellow. Does not rust but may discolour.	Timber fastenings, brass hinges and door furniture. All marine constructions in timber.
BRASS with:		
Chromium plate	Brilliant 'polished' finish	With all chromium plated domestic goods.
SILICON BRONZE	Uniform dark brown colour	All exterior timber fastenings including boat building screws. Screws for copper and bronze components.
ALUMINIUM ALLOY (anodised and lubricated with lanolin)	Matt silver-grey finish	All fasteners for aluminium articles, eg door furniture, bathroom fittings
STAINLESS STEEL 18/8 type	Bright attractive finish	All construction applications where long term durability and freedom from rust staining is essential. May be used in contact with aluminium.

COMPATIBILITY TABLE – METAL-TO-METAL CONTACT

	Aluminium	Aluminium Bronze	Cast Iron	Copper	Manganese Bronze	Mild Steel	Stainless Steel	Phosphor Bronze
Aluminium	√	X	X	X	X	X	£	X
Aluminium bronze	X	√	X	√	O	X	£	√
Cast iron	X	X	√	X	X	√	X	X
Copper	X	√	X	√	O	X	O	√
Manganese bronze	X	O	X	O	√	X	O	O
Mild steel	X	X	√	X	X	√	X	X
Stainless steel	£	O	X	O	O	X	√	O
Phosphor bronze	X	√	X	√	O	X	O	√

√ Can be used together under all conditions
O Only safe to use together in dry conditions
£ Limited compatibility – may or may not prove suitable under all conditions
X Not compatible – should not be used together under any conditions

TABLE III – RELATIVE COST OF CORROSION RESISTANT TREATMENT/MATERIALS*

Bolt Material	Treatment	Relative Cost
Mild steel	Untreated	1.00
	0.003in zinc plated	1.40
	0.003in cadmium plated	1.53
	nickel + chromium plating	2.36
	heavy nickel + chromium plating	3.33
Brass	Untreated	5.07
18/8 stainless steel	Untreated	6.25

* Based on 1975 figures supplied by GKN

Increasing raw material prices, notably for cadmium, copper and zinc, are tending to close the margin between the cost of plated and stainless steel fasteners. An additional factor in favour of stainless steel is its high strength, which can permit the use of a smaller size of fastener for a specific duty.

TABLE IV – PRINCIPAL TYPES OF STAINLESS STEEL USED FOR FASTENER MANUFACTURE

Classification	Steel type description	Specification number UK	Germany (Werkstoff No.)	France (Afnor)	Italy (Uni 4047)
Austenitic 18/8 type	18/8 free machining	303 S21 (or S41)	4305	Z10 CNF 18–10	Z15 CNF 1808
	18 Cr, 10 Ni	304 S15	4301	Z6 CN 18–10	X6 CN 1911
	18 Cr, 12 Ni	305 S19	—	—	Z12 CN 1811
	18/10 Nb stabilised	347 S17	4550	Z10 CN Nb 18–10	X8 CN Nb 1811
	18/10 Ti stabilised	321 S12	4541	Z10 CNT 18–10	X8 CNT 1810
Austenitic 18/10/3 type	18 Cr, 10 Ni 3 Mo	316 S16	4401	Z6 CND 18–12	X8 CND 1712
Ferritic %	17 Cr steel	430 S15	4016	Z8 C17	X12 C17
Martensitic %	12 Cr steel	410 S21	4006	Z12 C13	X15 C13
%	12 Cr steel (free machining)	416 S21	4021	Z12 CF13	X16 CF 13
% %	17 Cr, 2 Ni	431 S29	4057	Z15 CN 16–2	X20 CN 16

Stainless steel is, however, a generic description of a range of materials with differing properties. The principal types used by European manufacturers are summarised in Table IV. Of these only 18/10/3 can be regarded as fully rust-resistant under virtually all conditions. Other types can show distinctly limited performance in this respect in certain conditions, but may prove quote satisfactory in others. 18/8 type remains the one in general use. The use of stainless steel fasteners has become widely established in America and Sweden, and material supplies are now generally available from EEC countries.

Brass is generally resistant to corrosion attack, although corrosion is readily developed by salt water or salt atmospheres. Certain brasses may be subject to definite corrosive attack whereby the zinc content is leached out. This is known as 'degrafication' and is most likely to occur with high zinc content brasses. The onset of degrafication can be seen by a reddening of the surface of the brass. The attack may be of the layer type, in which case it is confined to the surface of the fastener, or it may be localised and accompanied by pitting. This can increase in severity, resulting in a weakening of the fastener.

In general, however, brass and bronze are widely used for marine fasteners (bronze for higher strength), as well as more exotic alloys such as silicon bronze and aluminium bronze.

Resistance to High Temperature

Copper (and aluminium) based alloys are generally unsuitable for use at elevated temperatures, showing a substantial loss of strength at relatively moderate temperatures. Steels are more favourable, but become increasingly prone to oxidation above a particular 'threshold' temperature. This can be raised by including alloying elements such as chromium, silicon and aluminium.

Suitable materials, in ascending temperature order, are listed in Table V. It is an interesting point that the temperature for continuous exposure is higher than that for intermittent heating. This is due to the fact that intermittent heating tends to produce loose scale films, subjecting more of the base metal to oxidation attack in repeated heat cycles.

TABLE V – MAXIMUM SERVICE TEMPERATURES FOR FASTENER MATERIALS

Material	Temperature For	
	Continuous Exposure	Intermittent Exposure
	°C	°C
Mild steel, carbon steel	up to 500	450
Low alloy steel (1% chrome molybdenum)	600	550
17% chrome steels	750	700
18/10/3 stainless steel	850	800
25/25 nickel/chrome steel	900	850
Nickel alloy (Nimonic 80A)	1 050	1 000
25/20 chrome/nickel steel	1 100	1 050

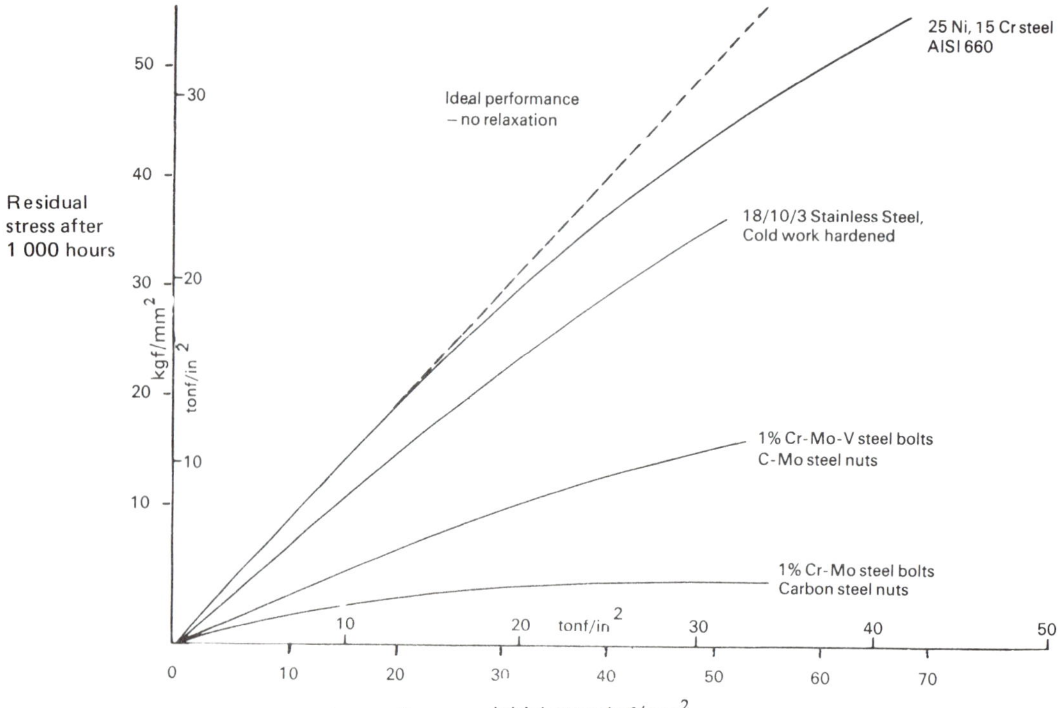

Fig 1 Stress relaxation of fasteners at 500° C. (GKN)

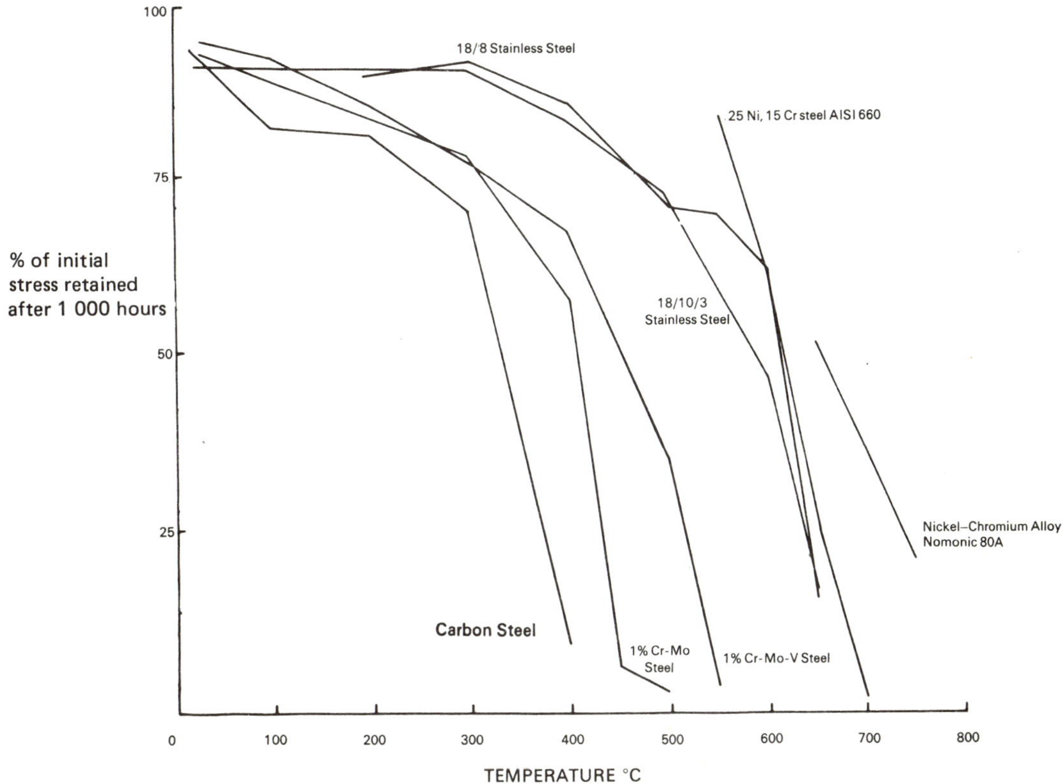

Fig 2 Stress relaxation of bolting steels and alloys. (GKN).

Other factors involved in the selection of a suitable material for high temperature fasteners are differential thermal expansion, resistance to stress relaxation and prevention of thread seizure. Stress relaxation can lead to creep and loss of internal tension – see Figs 1 and 2. Thread seizure is due to the growth of oxidation or corrosion films which can make it difficult to release fasteners. If necessary this can be relieved by the use of oxidation-inhibition coatings or high temperature lubricant films and relatively easy initial fits.

In general, thread seizure is least likely to occur with high alloy materials, and not likely to be unduly troublesome using fastener materials within the temperature ranges given in Table V. As a further safeguard there is the possibility of using a harder material for the nut than the bolt (eg stainless steel nuts on low alloy steel bolts).

Chemical Resistance

Where fasteners are subject to ambiences consisting of chemical solutions the compatibility of the fastener material needs to be established under typical screw conditions. This can differ from the results obtained by laboratory-type compatibility tests, so service experience is an invaluable guide. The majority of fastener requirements for ambiences which are active chemically can usually be met by a suitable grade of stainless steel, but there are exceptions. In extreme cases it may be necessary to resort to alternative forms of fastening or fabrication, or to investigate the possibility of using non-metallic fasteners.

Tools

The characteristics of the various types of clutch used on powered wrenches or powered screw-drivers are:-

(i) *Clutchless* (direct drive) – the fastener is driven until the motor stalls. This is the simplest method for a pneumatic driver (where an air motor can be stalled without harm) and some suitable form of pressure regulator can be incorporated to adjust the stall torque. Final torque achieved can, however, vary with air pressure, air motor condition, faulty lubrication and deterioration of the tool. With a hand-held tool there is also an objectionable 'kick' when the driver stalls.

(ii) *Positive clutch* – is usually of the ratchet-type, spring-loaded to be normally open. The clutch is engaged by pressure on the fastener. The driver then operates in direct drive until the torque reaches a sufficient level to throw the clutch faces out of engagement against the pressure applied by the operator. Final torque is dependent on the operator (ie pressure applied to the tool) and may vary considerably. Clutch engagement can also tend to throw the driver off the fastener head, unless this has a socket fit.

(iii) *Double slip clutch* – basically this comprises a spring-ratchet type clutch with a light spring and a second clutch with sloping jaw faces held together with a heavy spring. Operator's pressure engages the first clutch. When a predetermined torque has built up the jaws of the second clutch slip and 'ratchet' against spring pressure, but drive torque will still continue to rise. Thus slip torque is adjustable and normally set just below the desired torque. The slipping of the second clutch indicates to the operator that the required torque has nearly been reached. He can then judge how much longer to hold the driver in position. The performance achieved with this type of clutch is dependent on the operator and likely to vary with fatigue, etc. It is also noisy and vibrates when the second clutch ratchets.

(iv) *One-shot double-clutch* – a similar arrangement with two clutches but with the second (spring closed) clutch comprising a pair of jaws with pockets containing steel balls. Slipping torque is adjustable via spring pressure. When a predetermined torque is reached the balls tend to roll out of the pockets, forcing the jaws apart to disengage the drive. Final torque depends only on adjustment (spring pressure) and should be quite consistent. Also the clutch action is quiet and free from vibration. It is less suitable for driving duties where variable or different torques are required (eg self-tapping screws) as two clutch adjustments would be necessary for each operation.

Four from THOR
Above left high-power super-quality drills.
Below right high-power super-quality screwdrivers.
..And many more!
- the above represent a fraction of the world famous THOR range of drills, screwdrivers, nutrunners and wrenches with varying degrees of speed and torque to handle most industrial fastening jobs PLUS grinders, sanders and polishers for finishing operations and super-safety tool balancers and hoists for ease and convenience of handling . .
Write, phone or telex us for demonstration/literature
THOR TOOLS LIMITED
Tynemouth, Tyne & Wear NE29 7UE
Tel: North Shields (08945) 73181
Telegrams: Thortools, Tynemouth. Telex: North Shields 53602. Answer Back Thor N. Shields
SYMBOL OF
SW
EXCELLENCE
A Member of the Stewart Warner Group of Companies

Operation	Clutchless	Positive-clutch	Double-slip clutch	One-shot double clutch	Low-impact clutch
Bolt and screw tightening — rigid assemblies (nut-wrenching)	Medium or large size threads only	Only where close control of tightening torque is not required (all sizes)	Only where close control of tightening torque is not required.	Generally excellent for all except small sizes	Generally excellent for all sizes
Bolt and screw tightening — resilient materials (nut-wrenching)	Medium or large size nuts. Suitable for smaller sizes at lower speeds	Suitable for small and medium size screws. With shank screws can be fatiguing as considerable operating pressure may be required	Suitable for all sizes where close control of tightening torque is not required. Can be tiring on large screws	Generally excellent for all types	Generally excellent for all types
Nut-wrenching of stiff nuts	Medium or large screws Suitable for smaller sizes at lower speeds	Fair performance with all sizes	Good performance with all sizes, but needs close torque control	Generally excellent for all sizes	Generally excellent for all sizes
Self-tapping (thread-forming type in sheet metal)	Not usually suitable	Generally good for all sizes where stripping torque exceeds tapping torque by a generous margin	Generally good for all sizes with suitable operator technique	Generally good for all sizes	Generally good for all sizes
Self-tapping (thread-cutting types in thicker materials)	Not usually suitable	Generally good for all types and sizes when stripping torque exceeds tapping torque by a generous margin	Generally good for all types and sizes with suitable operator technique	Can be used for all types and sizes	Can be used for all types and sizes
Wood screws	Medium and large screws. Small screws need lower speeds. Not generally suitable for slotted heads	Generally excellent for all sizes	Generally good for all sizes	Generally good for all sizes	Generally good for all sizes

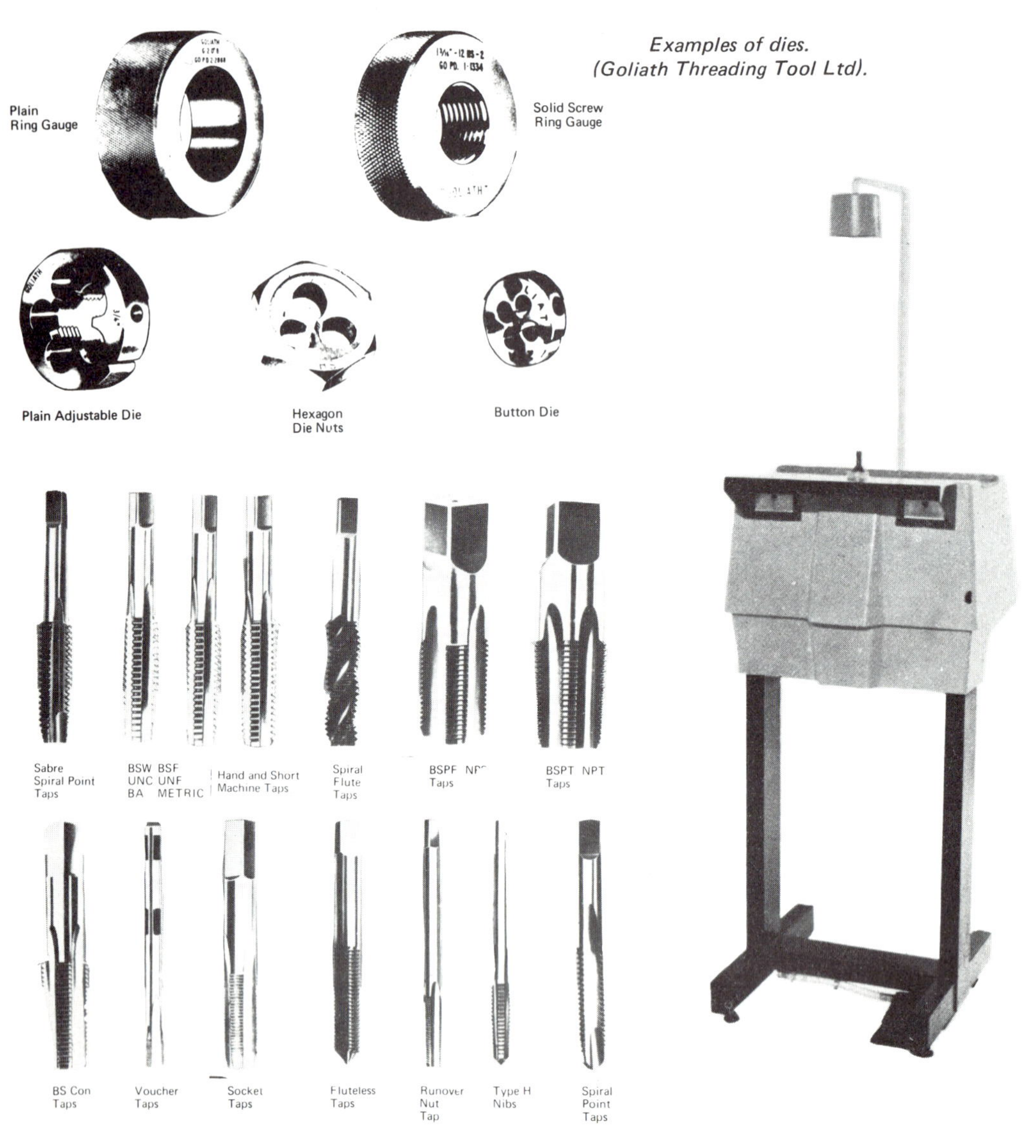

Examples of dies. (Goliath Threading Tool Ltd).

Examples of hand and machine taps. (Goliath Threading Tools Ltd).

Brivmatic high speed automatic riveter, speeds up to 3 500cycles/hour. (Advel Ltd).

(v) *Low-impact clutch* – a single clutch comprising a pair of spring-loaded jaws with pockets containing steel balls (like the second clutch mentioned in (iv)). Torque setting depends on spring pressure. Once the pre-set torque figure has been reached, the drive slips with negligible, or very slight, build-up in torque. It is quiet and vibration-free in operation. This type of clutch lends itself well to automatic start and stop, or automatic shut-off.

Characteristics of these types of drive allied to different duties are summarised in the Table.

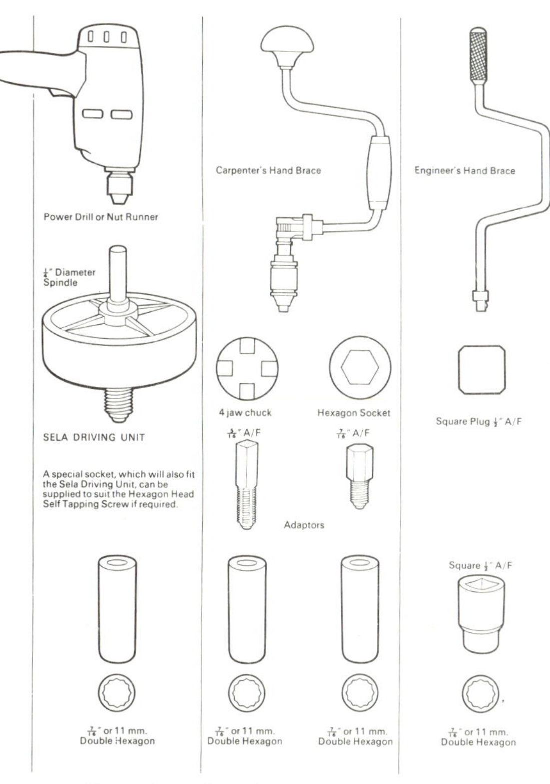

Examples of socket drivers for 'Sela' double hexagon headed screws. (The British Screw Co Ltd).

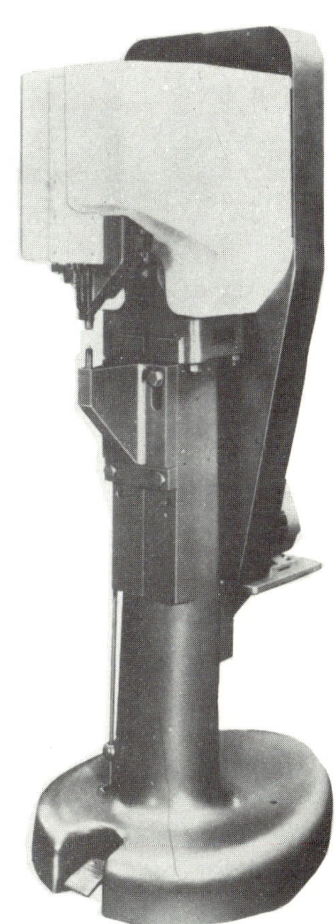

An Aylesbury 101 automatic feed power operated rivet setting machine. This machine is now regularly used for self-piercing rivet applications joining steel sheets up to 3/16in thickness. (The Bifurcated & Tubular Rivet Co Ltd).

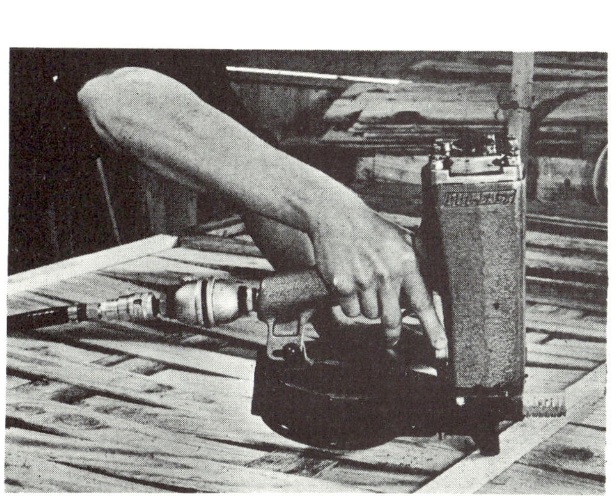

B.I.F. coil-feed nailer.

POWER TOOLS (PNEUMATIC AND ELECTRIC)

Tool	Remarks	Manufacturers
Screwdrivers	May or may not be fitted with clutch. Normally accept a variety of bits. The majority of designs can also be used as nut-runners, usually with standard ½in (13mm) square drive for socket wrench.	AEG Atlas Copco (GB) Ltd Black & Decker Ltd Bosch Ltd Broom & Wade Ltd Cleco Air Tools Ltd Consolidated Pneumatic Tool Co Ltd Desoutter Bros Ltd Gardner Denver Ltd Ingersoll-Rand Co Ltd Millers Rolls Co Ltd Samco-Strong Ltd Skil (GB) Ltd Stanley Bridges Ltd Thor Tools Ltd
Nut-runners (impact type)	Also known as impact wrenches	AEG Atlas Copco (GB) Ltd Black & Decker Ltd Bosch Ltd Cleco Air Tools Ltd Consolidated Pneumatic Tool Co Ltd Desoutter Bros Ltd Gardner Denver Ltd Ingersoll-Rand Co Ltd Millers Rolls Co Ltd Stanley Bridges Ltd Thor Tools Ltd Wolf Electric Tools Ltd
Torque wrenches	Usually pneumatic with adjustable pre-set torque. Torque multipliers may be fitted for higher torques (eg for unscrewing)	John Bedford & Sons Ltd Britool Ltd Cleco Air Tools Ltd Desoutter Bros Ltd MHH Engineering Co Ltd North Bar Tool Co

Rotary Drills	High speed pneumatic drills for drilling metals, timber, plastics, etc. Electric drills may have variable speeds or reduction gearboxes for slow speed drilling	Peter Abbott & Co Ltd AEG Atlas Copco (GB) Ltd Black & Decker Ltd Bosch Ltd Broome & Wade Ltd Cleco Air Tools Ltd Consolidated Pneumatic Tool Co Ltd Desoutter Bros Ltd DOM Products (Royston) Ltd Explosive Power Tools Ltd Gardner Denver Ltd Ingersoll-Rand Co Ltd Millers Rolls Co Ltd Rockwell International SA Skil (GB) Ltd Stanley Bridges Ltd Thor Tools Ltd Wolf Electric Tools Ltd
Vibratory/ Percussive Drills	Two basic types for drilling harder materials such as engineering brickwork, granite, concrete, etc. (i) percussive (ii) vibratory, combining rotary and percussive action. Some tools of type (ii) may also be switched to rotary action only	Type (i) Explosive Power Tools Ltd DOM Products (Royston) Ltd Hilti (GB) Ltd Tornado Fixings Ltd Ucan Developments Ltd Victor Products (Wallsea) Ltd Type (ii) See under *Rotary Drills*
Nailers and Staplers	Nailers are usually pneumatic, hopper fed and may have interchangeable magazines. Staplers are usually pneumatic, hopper or magazine fed	Gordian Strapping Ltd Paslode Co Samco-Strong Ltd Senco Pneumatics Ltd Spot Nails Ltd
Riveters	Rivet-setting machines are usually designed to accommodate a wide range of solid, bifurcated or semi-solid rivets. Blind riveting tools are usually designed for use with a specific type of blind rivet. Automatic riveters are designed for high speed repetitive riveting.	Ardel Ltd Bifurcated & Tubular Rivet Co Ltd Broome & Wade Ltd Gesipa Fasteners Ltd Tucker Fasteners Ltd

FREE-FLIGHT CARTRIDGE OPERATED TOOLS

Name	Details	Manufacturer
Bossong	0.22 calibre, ¼in, 5/16in and 3/8in sizes. Automatic cartridge ejection	Bossong Italiana SpA
Hilti	Two hammer actuated models	Hilti (GB) Ltd
Ramset	Duo-Jobmaster for ¼in and 3/8in fasteners. (0.22 calibre). Super Power Jobmaster for 3/8in and ½in fasteners (0.35 calibre)	Ram Set Fasteners Ltd
Rapid	Major (¼in and 3/8in) and Classic (¼in) breech loading	Explosive Power Tools Ltd
Spitmatic	Positive ejection of cartridge	SPIT International Ltd
Tornado	0.25 or 0.38 calibre with four hammer sizes ¼in, 5/16in and 3/8in. Breech loading	Tornado Fixings Ltd

PISTON-TYPE CARTRIDGE TOOLS

Name	Details	Manufacturer
Bonded	220 (0.22 calibre); 600 and 660 automatic ejection (0.25 calibre)	Bonded Direct Fixings Ltd
Hilti	5 piston actuated models	Hilti (GB) Ltd
Obo	Model K5 for nails and studs	Betterman Electro
Sela	Sliding breech, automatic cartridge ejection for use with Sela screws, pins and studs	British Screw Co Ltd

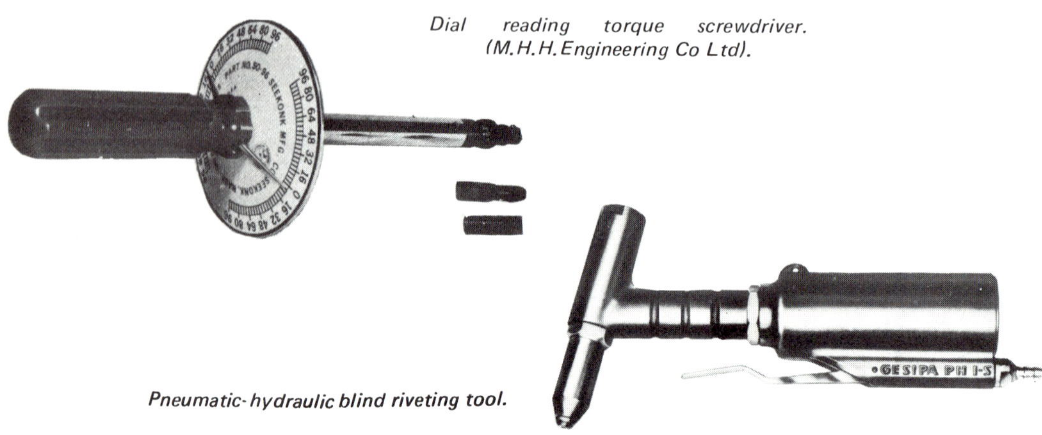

Dial reading torque screwdriver. (M.H.H. Engineering Co Ltd).

Pneumatic-hydraulic blind riveting tool.

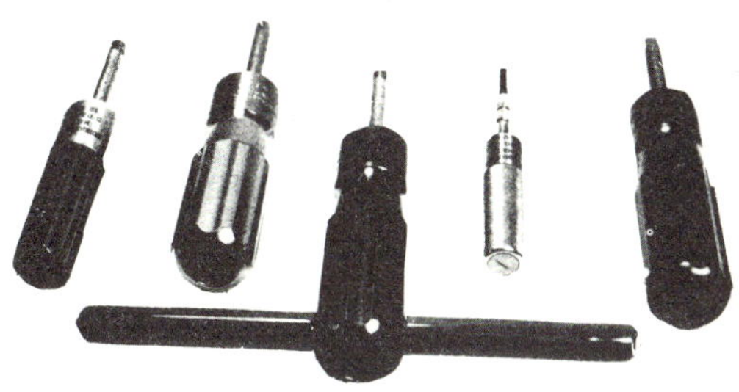

Torque limiting screwdrivers.
(M.H.H.Engineering Co.Ltd)

Air-operated stapler (British Industrial Fasteners Ltd).

Dial indicating torque wrench.
(M.H.H. Engineering Co Ltd).

A standard Aylesbury 300 power operated automatic feed rivet setting machine. The guard and fixture are not standard fittings. Standard guards are available.
(The Bifurcated & Tubular Rivet Co Ltd).

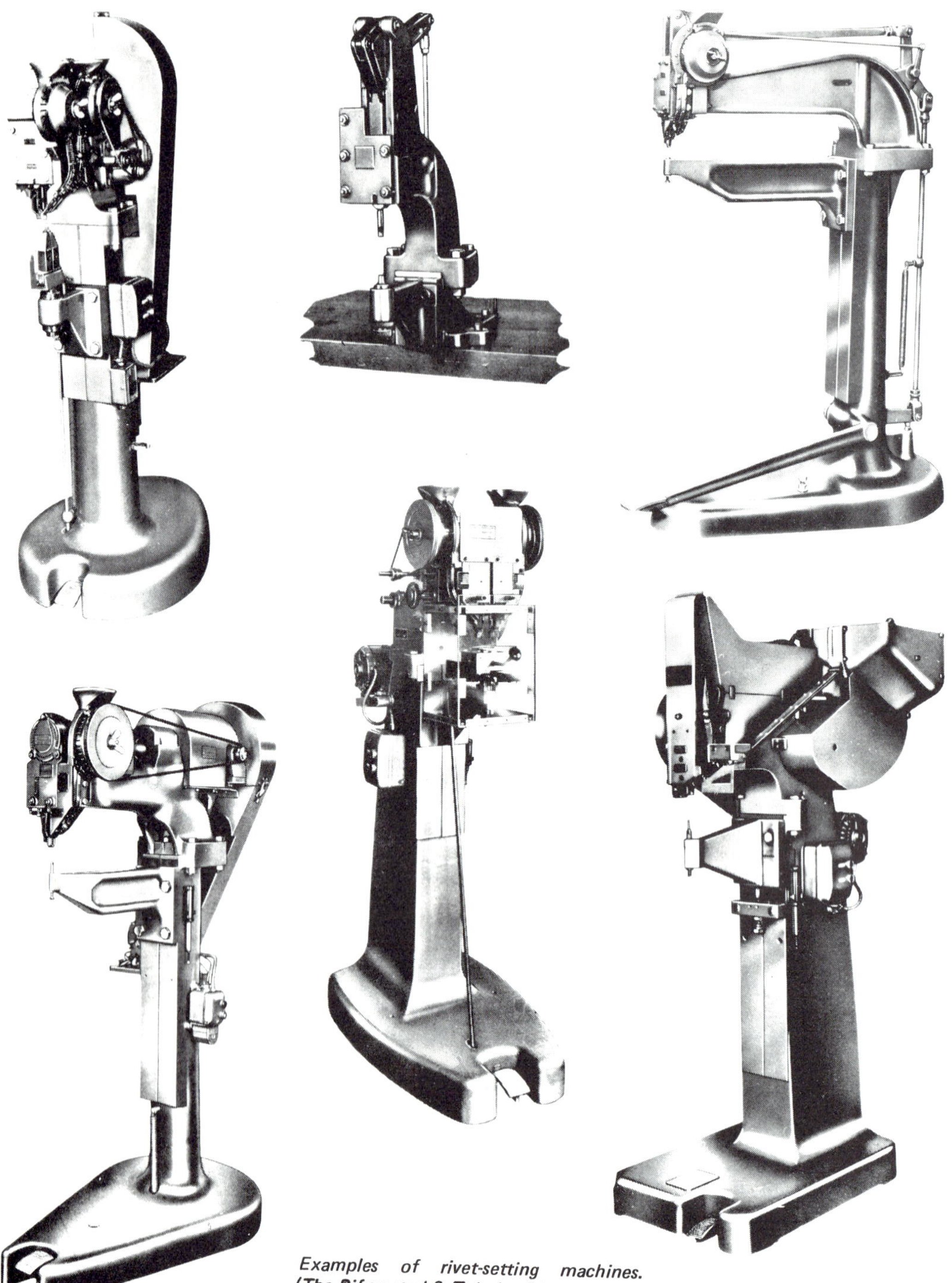

Examples of rivet-setting machines. (The Bifurcated & Tubular Rivet Co Ltd).

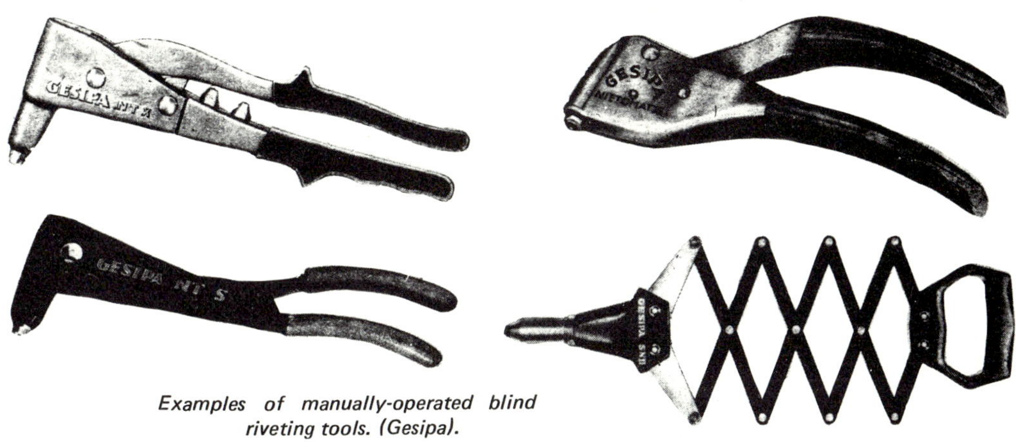

Examples of manually-operated blind riveting tools. (Gesipa).

Broad-crown air tackers. (British Industrial Fastener Ltd).

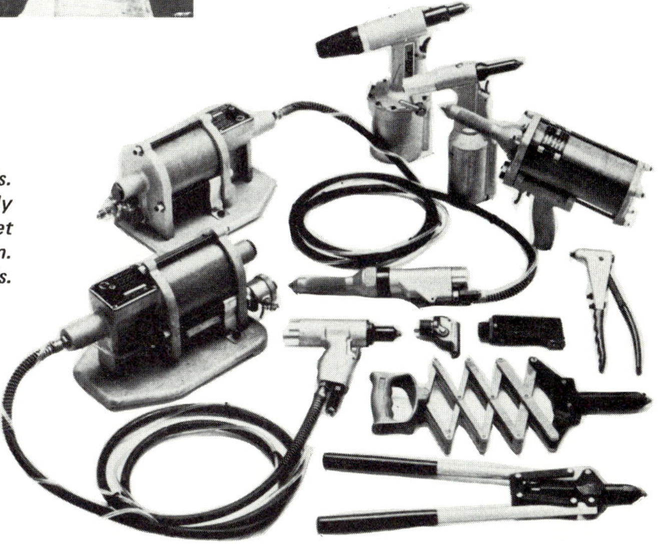

Typical range of blind riveting tools. 1–manually operated pliers. 2–manually operated lazytongs. 3–pneumatic rivet gun. 4–pneumatic-hydraulic rivet gun. 5–pneumatic operated stand tools. (Tucker Fasteners Ltd).

Specifications and Standards

International (ISO)

R68	– Screw Threads
R225	– Bolts, Screws, Studs, Dimensioning
R261	– ISO Metric Screw Threads, General Plan
R262	– ISO Metric Screw Threads for Screws, Bolts and Nuts
R263	– ISO Inch Screw Threads, General Plan and Selection for Screws, Bolts and Nuts
R272	– Hexagon Bolts and Nuts, Widths Across Flats, Heights of Heads, Thicknesses of Nuts
R273	– Clearance Holes for Metric Bolts
R288	– Slotted and Castle Nuts with Metric Thread

AAR (USA)

Association of American Railroads, 59, East Van Buren Street, Chicago, Illinois 60605.

M110	– Rivet Steel and Rivets
M125	– Machine Bolts and Nuts. Heat-treated Carbon-Steel and Alloy-Steel Track Bolts
M922	– Self-Locking Nuts and Self-Locking Cap Screws
L28	– Cap Screws and Finished Hexagon Head Bolts for Journal Roller Bearing Boxes

ABS (USA)

American Bureau of Shipping, 45 Broad Street, New York, NY10004.

Rules for Building and Classing Steel Vessels (covering dimensional material, and design requirements for steel rivets).

API (USA)

American Petroleum Institute Dv. of Production, 300 Corrigan Tower Building, Dallas, Texas 75201.

12A — Oil Storage Tanks with Riveted Shells
12B — Bolted Production Tanks
12E — Wooden Production Tanks

ASA (USA)

American Society of Mechanical Engineers, United Engineering Center, 345 East 47th Street, New York, NY10017.

B1.1 — Unified Screw Threads
B1.2 — Screw Thread Gages and Gaging
B1.5 — Acme Screw Threads
B1.7 — Nomenclature, Definitions, and Letter Symbols for Screw Threads
B1.8 — Stub-Acme Screw Threads
B1.9 — Buttress Screw Threads
B1.10 — Unified Miniature Screw Threads
B1.12 — Class 5 Interference-Fit Thread
B2.1 — Pipe Threads (Except Dryseal)
B2.2 — Dryseal Pipe Threads
B5.1 — T-Slots — Their Bolts, Nuts, Tonques and Cutters
B5.4 — Taps, Cut and Ground Thread
B5.20 — Machine Pins
B16.5 — Steel Pipe Flanges and Flanged Fittings
B18.1 — Small Solid Rivets
B18.2.1 — Square and Hex Bolts and Screws
B18.2.2 — Square and Hex Nuts
B18.3 — Socket Cap, Shoulder and Set Screws
B18.4 — Larqe Rivets
B18.6.1 — Slotted and Recessed Head Wood Screws
B18.6.2 — Cap and Set Screws
B18.6.3 — Slotted and Recessed Head Machine Screws, Nuts
B18.6.4 — Slotted and Recessed Head Tapping Screws and Metallic Drive Screws
B18.8 — High Strength, High Temperature Internal Wrenching Bolts
B18.9 — Plough Bolts
B18.10 — Track Bolts and Nuts
B18.11 — Miniature Screws
B18.12 — Glossary of Terms for Mechanical Fasteners

B18.13 — Screw and Washer Assemblies — Sems
B27.1 — Lock Washers
B27.2 — Plain Washers
Y14.6 — Drafting Manual for Screw Threads

ASTM (USA)

American Society for Testing and Materials, 1916 Race Street, Philadelphia, Pennsylvania 19103.

A31 — Boiler Rivet Steel and Rivets
A66 — Steel Screw Spikes
A76 — Low Carbon-Steel Track Bolts and Nuts
A84 — Staybolt Wrought-Iron, Solid
A152 — Wrought-Iron Rivets and Rivet Rounds
A183 — Heat-Treated Carbon- and Alloy-Steel Track Bolts and Carbon-Steel Nuts
A193 — Alloy-Steel Bolting Materials for High-Temperature Service
A194 — Carbon- and Alloy-Steel Nuts for Bolts for High-Pressure and High-Temperature Service
A195 — High-Strength Structural Rivet Steel
A307 — Low Carbon-Steel Externally and Internally Threaded Standard Fasteners
A320 — Alloy-Steel Bolting Materials for Low-Temperature Service
A325 — High Strength Carbon-Steel Bolts for Structural Steel Joints, Including Suitable Nuts and Plain Hardened Washers
A354 — Quenched and Tempered Alloy-Steel Bolts and Studs with Suitable Nuts
A394 — Galvanized Steel Transmission Tower Bolts and Nuts
A437 — Alloy-Steel Turbine-Type Bolting Material Specially Heat Treated for High-Temperature Service
A449 — Quenched and Tempered Steel Bolts and Studs
A453 — High Strength, High Temperature Bolting Materials with Expansion Coefficients Comparable to Austenitic Steels
A489 — Carbon-Steel Eyebolts
A490 — Quenched and Tempered Alloy-Steel Bolts for Structural Steel Joints
A502 — Specifications for Steel Structural Rivets
A540 — Alloy-Steel Bolting Materials for Special Applications

British Standards (UK)

British Standards Institution, 2, Park Street, London W1A 2BS.

Standards for Metric Threads and Fasteners

BS6443 — Screw Threads
BS3692 — Precision Hexagon Bolts, Screws and Nuts
BS4168 — Hexagon Solid Screws and Wood Keys
BS4183 — Machine Screws and Machine Screw Nuts

BS4190 – Black Hexagon Bolts, Screws and Nuts
BS4219 – Slotted Grub Screws
BS4320 – Metal Washers for General Engineering Purposes
BS4439 – Screwed Studs
BS4395 – High Strength Friction Grip Nuts
BS4186 – Clearance Holes
BS4278 – Eye Bolts
BS4463 – Crinkle Washers
BS4464 – Spring Washers

EEI (USA)

Edison Electric Institute, 750 Third Avenue, New York, NY10017

TD-1 – Steel Bolts and Nuts
TDJ-3 – Leg Screws
TD-5 – Eye Bolts and Eyelets
TDJ-23 – Steel and Malleable Iron Guy Clamps

NSA (USA)

National Standards Association, 1321 Fourteenth Street, NS Washington 5, DC

NAS159 – Bolt, Internal Wrenching, Aircraft
NAS190 – Dimensioning, Decimal Usage for
NAS353 – Nut, Aircraft Self-Locking – for 160 000 to 180 000 PSI Tension Bolts
NAS496 – Bolts, Internal Wrenching
NAS498 – Bolts, Shear
NAS499 – Fasteners, Panel, Fast Operating
NAS547 – Fastener, Rotary, Quick-Operating, High-Strength
NAS597 – Fittings, Rigid Tube Connector
NAS618 – Fastener – Recommended Shank, Hole and Head-to-Shank Fillet Radius Limits for
NAS621 – Fasteners, Titanium Alloy-Procurement Specification
NAS672 – Plating-High Strength Steels – Cadmium
NAS725 – Miniature Screws
NAS1058 – Fasteners, Blind
NAS1289 – Threaded Shear Fasteners, Flush and Protruding Type for Applications up to 900°F max – 5% CR Ultra-High Strength Alloy Steel (H-1) – Procurement Specification
NAS1290 – Threaded Shear Fastener, Flush and Protruding Type for Applications up to 1 400°F. Corrosion and Heat Resistant Rene 41 Nickel Base Alloy – Procurement Specification
NAS1400 – Rivet, Blind, Self-Plugging, Mechanically Locked Spindle
NAS1413 – Lockbolts and Collars, Pull Type, Stump Type, Flush Head, Protruding Head, Aluminium Alloy and Steel

NAS1589 – Fasteners, Threaded, Corrosion and Heat Resistant 1 200°F – Procurement Specification

NAS1597 – Steel, Corrosion and Heat Resistant, Bar, Rod, and Forging

NAS3350 – Nuts, Self-Locking, High Quality

NBS (USA)

US Department of Commerce, National Bureau of Standards, Supt of Documents, US Government Printing Office, Washington 25, DC

Handbook H28 – Screw-Thread Standards for Federal Services

SAE (USA)

Society of Automotive Engineers, 485 Lexington Avenue, New York, NY10017.

The SAE Handbook contains standards and recommendations covering the following:-

- Square and Hexagon Bolts and Nuts
- Round Head Bolts
- Socket Cap Screws
- Set Screws
- Machine Screws
- Tapping Screws
- Machine Pins
- Plain Washers
- Lock Washers
- Hi-Head Finished Hexagon Bolts
- High Nuts
- Crown Nuts
- Conical Spring Washers
- Ball Joints
- Ball Studs and Tie Rod Sockets
- Rivets and Riveting
- Rod Ends and Clevis Pins
- Mechanical and Quality Requirements for Threaded Fasteners
- Mechanical and Quality Requirements for Non-threaded Fasteners
- Mechanical and Quality Requirements for Tapping Screws

Federal (USA)

Business Service Center, General Services Administration, Washington 25, DC

GGG-G-61 – Gauges; Plug and Ring, Plain and Thread

GGG-T-70 – Tap, Thread Cutting

FF-W-84 – Washers, Lock (Spring)

FF-S-85 – Screws, Cap, Slotted and Hexagon Head

FF-S-86 – Screws, Cap, Socket Head

FF-S-92 – Screws, Machine: Slotted or Cross-Recessed

FF-W-92 — Washers, Metal, Flat (Plain)
FF-W-100 — Washers, Lock (Tooth)
FF-S-103 — Screws: Set
FF-S-107 — Screws, Tapping, Slotted and Plain Head (Sheet Metal, Machine and Drive)
FF-S-11 — Screws, Wood, Slotted-Head
QQ-M-151 — Metals; General Specification for Inspection of
FF-S-200 — Set Screws; Hexagon Socket and Spline Sockets, Headless
FF-S-210 — Set Screws: Square Head and Slotted Headless
QQ-N-290 — Nickel Plating (Electro-deposited)
FF-T-305 — Thumbscrews
QQ-C-320 — Chromium Plating (Electro-deposited)
QQ-Z-325 — Zinc Plating (Electro-deposited)
FF-P-386 — Pins, Cotter (Split)
QQ-P-416 — Plating, Cadmium (Electro-deposited)
FF-R-556 — Rivets, Burrs, and Caps; Copper and Brass
FF-B-561 — Bolts, Lag
FF-B-571 — Bolts, Nuts Studs and Tap-Rivets
FF-B-575 — Bolts, Hexagon and Square
FF-B-584 — Bolts, (Square Neck, Machine Ribbed Neck, Finned Neck, Tee Head, Plow) (Round Head)
FF-B-588 — Bolts, Toggle
FF-S-611 — Spikes, Track, Square-Shank
QQ-S-701 — Steel, Staybolt
FF-T-791 — Turnbuckles
FF-N-836 — Nuts, Hexagon and Square
FF-N-845 — Nuts, Plain, Wing
D075 — (COM-BDSA) — Glossary of Packaging Terms
FF-S-00109 — (COM-NBS) — Screws, Wood; Cross-Recessed Head
DGG-K-00275 — (GSA—FSS)— Key, Socket Head Screws; and Key Set, Socket Head Screw

MIL (USA)

Commanding Officer, Naval Aviation Supply Depot, 700 Robins Avenue, Philadelphia, Pennsylvania 19111. Attention: CDS

MIL-STD-9 — Screw Thread Conventions and Methods of Specifying
MIL-STD-105 — Sampling Procedures and Tables for Inspection of Attributes
MIL-STD-403 — Riveting, Rocket and Guided Missile Structural, Aluminium and Aluminium Alloy Solid Rivets, Non-Flush and Countersunk Type
MIL-STD-700 — Locking Devices for Screw Threaded Fasteners and Other Similar Applications in Nuclear Power Plants
MIL-B-857 — (Ships) — Bolts, Nuts and Studs
MIL-S-933 — Screws, Machine, Cap and Set; and Nuts

MIL-S-971 — (Ships) — Screws, Wood
JAN-R-1127 — Rivets, Belt, Copper; and Burrs, Copper
MIL-R-1150 — Rivets, Solid (Aluminium Alloy) and Aluminium Alloy Rivet Wire and Rod
MIL-R-1166 — Rivets for Sheet Metal Work
MIL-S-1222 — Studs, Threaded (Bolt-Stud); Nut, Plain Hexagon; and Steel Bar, Round; High Temperature Service
MIL-R-1223 — Rivets and Rivet Steel Bars (for Hull Construction)
MIL-F-1824 — Fastener — Externally Threaded 250°F, Self-Locking Element for (ASG)
MIL-R-2582 — Rivets, Tubular and Cap
MIL-R-2583 — Rivets, Belt and Burrs
MIL-B-2677 — Bolts and Clips (Alloy-Steel) and Nuts (Carbon-Steel)
MIL-R-2890 — Rivets and Tap-Rivets, Non-Ferrous (for Hull Construction)
MIL-B-2938 — Bolts and Nuts, Deck
MIL-N-3336 — Nut, Self-Locking, Instrument Mounting
MIL-N-3337 — Nut; Sheet Spring
MIL-B-3964 — Bolts and Nuts, Track
MIL-F-4209 — Fastener, Metal, Nut-Sleeve Assembly
MIL-A-5070 — Adapter, Hose-to-Tube, Pipe and Flange, Re-usable, Hydraulic, Pneumatic Fuel and Oil Lines
MIL-F-5509 — Fittings, Fluid Connection
MIL-P-5673 — Pins; Flat Head Aircraft
MIL-R-5674 — Rivets; Aluminium and Aluminium Alloy
MIL-S-6033 — Screws; Self-Tapping Steel, Aircraft
MIL-N-6034 — Nuts. Hexagonal
MIL-B-6461 (USAF) — Bolts; Aircraft Engine and Propeller
MIL-B-6812 — Bolts; Aircraft
MIL-S-7742 — Screw Threads; Standard, Aeronautical
MIL-B-7838 — Bolts, Internal Wrenching 160 000lb/in^2
MIL-S-7839 — Screws, Structural, Aircraft
MIL-N-7873 — Nut, Self-Locking, 1 200°F
MIL-B-7874 — Bolt, Machine 1 200°F
MIL-R-7885 — Rivets, Blind Aluminium Alloy
MIL-N-8065 — Nuts, Self-Locking, Free-Spinning 1 200°F
MIL-R-8814 (ASG) — Rivets, Blind, Non-Structural Type
MIL-B-8831 — Bolt 12 Point, External Wrenching 180 000lb/in^2
MIL-S-8879 — Screw Threads, Controlled Radius Root with increased Minor Diameter
MIL-F-10884 — Fasteners, Snap
MIL-R-12221 — Rivet, Solid Aluminium Alloy, Grade 7277 Tempered
MIL-P-12932 — Packaging of Bolts and Nuts
MIL-R-16503 (BuOrd) — Rivet Rods and Rivets, Steel (for Torpedo Construction)
MIL-R-16759 (BuOrd) — Rivets, Brass and Copper (for Torpedo Construction)

MIL-F-18240 – Fastener, Externally Threaded 250°F, Self-Locking Element for (ASG)
MIL-S-18241 (AER) – Screws, Self-Locking, 250°F
MIL-S-18247 – Studs, Plain, Steel, General Purpose
MIL-B-18695 (Navy) – Bolts and Nuts; Plow, Pole Line, Hook and Shoulder
MIL-P-20700 – Pins, Grooved, Headless, Longitudinal Groove
MIL-N-21337 – Nuts, Plain, Round, Retaining, Ball and Roller Bearings
MIL-S-21472 – Screws, Shoulder, Socket Head, Alloy-Steel
MIL-T-22745 – Turnbuckles, Forged Steel, Galvanized, Weldless
MIL-F-22978 – Fastener, Rotary, Quick Operating, High Strength
MIL-B-23470 – Bolt and Nut, Torque-Controlled, Pre-Stress
MIL-N-25027 (ASG) – Nut, Self-Locking, 250°F, 550°F and 800°F
MIL-P-27235 – Pins, Straight, Headed (Clevis Pins)
MIL-F-27272 – Fitting, Hose, High Temperature, Medium Pressure
MIL-R-27384 – Rivets, Blind, Drive Type
MIL-B-52209 – Bolt, Hook, Tie, with Hexagon Nut and Triangular Washer

EUROPEAN STANDARDS

The German DIN standards are listed first since these tend to be more numerous, individual standards being issued for different heads and points, etc.

Subject	German DIN	British BS	French NF	Netherlands NEN	Italian UNI
Machine screws:					
cheese head	84	4183	E27–115	1602	6107
pan head	85		E27–116	1603	6108
Steel cotter pins	94	1574		182	
Steel flat washers	125	4320		2269	
Block washers	126	4320		2268	
Spring washers	127	4464		1197	
Steel hexagon locknuts	439B	3692			5589–65
Black hexagon nuts	555			697	
Black hexagon setscrews	558	4190	E27–310	2335	5739–65
Black hexagon bolts	601		E27–315	2230	5727
Carriage bolts and nuts	603/555	4933		303–697	
Socket cap screws	912		E27–161	1241	5931
Socket setscrews,					
flat point	913	4168		2341	5923
cone point	914		E27–162	2342	5927
dog point	915			2344	5925
cup point	916			2343	5929
Socket countersunk head screws	7991	(4168)	E27–160	2359	5933
High strength friction grip bolts	6914	4395		5511	5712

BRITISH AND CORRESPONDING GERMAN STANDARDS FOR FASTENERS

Product	B.S.	DIN	Remarks
Precision Hexagon Head Bolts	3692	931	Completely interchangeable
Precision Hexagon Head Screws	3692	933	Completely interchangeable
Precision Hexagon Nuts	3692	934	Completely interchangeable
Black Hexagon Head Bolts	4190	601	Completely interchangeable
Black Hexagon Head Screws	4190	558	Completely interchangeable
Black Hexagon Nuts	4190	555	Completely interchangeable
Cup Square Bolts	4933	603	Bolts to DIN 603 have larger heads
High Strength Bolts (general grade)	4395 Part 1	—	No German equivalent
High Strength Bolts (higher grade)	4395 Part 2	6914	Bolts to DIN 6914 have shorter length of thread

Weight Data

WEIGHTS OF SLOTTED MACHINE SCREWS*
(Thomas Haddon & Stokes)

Metric Weights lb/1 000

Screw Dia mm	Weight 1.0mm	Head Weight			
		Csk	Rsd	Pan	Cheese
2.0	0.0399	0.08	0.14	0.21	0.23
3.0	0.0952	0.18	0.26	0.62	0.74
3.5	0.1299	0.40	0.71	0.80	0.93
4.0	0.1682	0.69	0.87	1.30	1.49
5.0	0.2685	0.96	1.20	2.48	2.84
6.0	0.3820	1.90	2.30	4.05	4.76
8.0	0.6880	3.64	4.94	8.80	12.54
10.0	1.1010	7.70	9.40	16.20	24.10

BA Weights lb/1 000

Screw	Weight 1.0mm	Csk	Rsd	Pan	Cheese
8	1.253	0.08	0.13	0.15	0.29
7	1.630	0.10	0.17	0.19	0.37
6	2.040	0.16	0.25	0.38	0.60
5	2.697	0.24	0.37	0.58	0.87
4	3.425	0.32	0.54	0.83	1.27
3	4.460	0,45	0.74	1.19	1.82
2	5.975	0.58	0.95	1.73	2.64
1	7.625	0.89	1.46	2.55	3.88
0	9.840	1.55	2.15	3.76	5.68

*Heights of Pozidriv machine screws are approximately the same as slotted head screws

cont...

WEIGHTS OF SLOTTED MACHINE SCREWS* (contd.)
(Thomas Haddon & Stokes)

Unified Weights lb/1 000

Screw Dia	Weight/ Inch	Head Weight			
		Csk	Rsd	Pan	Cheese
2	1.223	0.07	0.16	0.19	0.22
3	1.590	0.12	0.25	0.30	0.34
4	1.990	0.16	0.34	0.40	0.48
5	2.575	0.23	0.49	0.58	0.70
6	3.045	0.32	0.66	0.77	0.94
8	4.550	0.51	1.06	1.25	1.58
10	5.860	0.99	1.86	2.11	2.63
12	7.825	1.46	2.60	3.11	3.93
1/4in	10.380	2.32	4.33	4.83	5.21
5/16in	16.830	5.32	9.31	10.30	13.10
3/8in	24.650	8.61	12.80	16.87	22.06

BSW Weights lb/1 000

Screw Dia in	Weight/ Inch	Head Weight			
		Csk	Rsd	Round	Cheese
1/8	2.575	0.21	0.37	0.55	0.71
5/32	4.040	0.39	0.66	1.05	1.37
3/16	5.640	0.68	1.13	1.87	2.40
1/4	10.400	1.45	2.51	4.40	5.64
5/16	16.780	3.07	5.23	8.92	11.30
3/8	24.650	4.91	8.71	15.09	19.24
7/16	33.650	8.71		17.55	25.19
1/2	43.800	11.96		26.20	37.77

*Heights of Pozidriv machine screws are approximately the same as slotted head screws

GKN CAP SCREWS : ISO METRIC THREADS (Kilogrammes per 100*)

Length	Diameter mm									
mm	M3	M4	M5	M6	M8	M10	M12	M16	M20	M24
5	0·073	0·14								
6	0·077	0·15	0·23							
8	0·086	0·16	0·26	0·41						
10	0·095	0·18	0·29	0·44	0·95					
12	0·104	0·20	0·31	0·47	1·01	1·73				
16	0·13	0·23	0·35	0·54	1·14	1·93				
20	0·16	0·27	0·41	0·61	1·26	2·12	3·17			
25	0·20	0·32	0·48	0·70	1·44	2·37	3·51			
30	0·24	0·37	0·55	0·81	1·63	2·66	3·87	8·3		
35	0·27	0·42	0·63	0·91	1·82	2·96	4·33	8·9		
40	0·31	0·47	0·70	1·03	2·00	3·27	4·80	9·7	17·0	
45		0·52	0·77	1·13	2·20	3·57	5·30	10·35	18·0	
50		0·57	0·84	1·24	2·40	3·87	5·70	10·95	19·0	30·5
55			0·91	1·35	2·58	4·17	6·30	11·9	20·2	32·0
60			0·99	1·45	2·77	4·47	6·62	12·7	21·4	33·6
65				1·56	2·96	4·75	7·10	13·4	22·7	35·2
70				1·66	3·15	5·10	7·54	14·2	23·9	37·0
75				1·77	3·34	5·40	8·00	15·0	25·0	38·7
80				1·88			8·44	15·7	26·2	40·5
90								17·3	28·7	44·0
100								18·8	31·0	47·5
110								20·4	33·3	51·0
120								21·9	36·0	54·3
130									38·0	58·0
140									40·4	61·3
150									42·7	65·0

GKN COUNTERSUNK HEAD SCREWS : ISO METRIC THREADS (Kilogrammes per 100*)

Length	Diameter mm								
mm	M3	M4	M5	M6	M8	M10	M12	M16	M20
8	0·041	0·097	0·126	0·23					
10	0·048	0·106	0·15	0·27	0·51	0·86			
12	0·060	0·115	0·18	0·30	0·57	0·96			
16	0·073	0·143	0·22	0·37	0·70	1·17			
20	0·092	0·20	0·27	0·44	0·81	1·35	2·10		
25		0·29	0·45	0·63	0·97	1·60	2·43		
30		0·38	0·88	0·88	1·22	1·85	2·96		
35				1·12	1·50	2·10	3·14	5·79	9·13
40				1·38	1·76	2·40	3·48	6·46	10·16
45						2·70	3·95	7·12	11·20
50						3·00	4·41	7·87	12·24
60								9·84	14·45
70									16·95
80									19·38
90									20·58
100									24·28

*To obtain the weight in lbs multiply by 2.205

GKN BUTTON HEAD SCREWS : ISO METRIC THREADS (Kilogrammes per 100*)

Length mm	Diameter mm						
	M3	M4	M5	M6	M8	M10	M12
8	0·047	0·105	0·193	0·26	0·48		
10	0·055	0·125	0·222	0·31	0·55		
12	0·064	0·145	0·25	0·35	0·60		
16		0·186	0·31	0·44	0·73	1·27	
20			0·38	0·53	0·86	1·47	2·15
25			0·45	0·64	1·02	1·72	2·51
30			0·53	0·75	1·18	1·97	2·87
40					1·49	2·47	3·60
50							4·32

GKN SHOULDER SCREWS : ISO METRIC THREADS (Kilogrammes per 100*)

Shoulder Length mm	Shoulder Diameter mm				
	6	8	10	12	16
12	0·67				
16	0·76	1·43			
20	0·84	1·60	2·71	4·33	
25	0·96	1·78	3·02	4·77	
30	1·07	1·98	3·32	5·21	9·80
40	1·29	2·37	3·93	6·08	11·36
50	1·51	2·76	4·54	6·96	12·9
60			5·15	7·84	14·5
70			5·76	8·72	16·0
80			6·37	9·59	17·6
90					19·2
100					20·7

GKN SET SCREWS – ALL POINTS : ISO METRIC THREADS (Kilogrammes per 100*)

Length mm	Diameter mm									
	M3	M4	M5	M6	M8	M10	M12	M16	M20	M24
5	0·016	0·029	0·038							
6	0·020	0·036	0·052	0·073	0·10					
8	0·028	0·051	0·075	0·109	0·17					
10	0·038	0·065	0·104	0·14	0·23	0·33	0·45			
12	0·045	0·086	0·122	0·18	0·29	0·42	0·59			
16		0·109	0·172	0·25	0·42	0·62	0·87	1·43		
20		0·140	0·22	0·32	0·54	0·82	1·16	1·95	2·60	3·72
25			0·29	0·41	0·70	1·07	1·52	2·63	3·63	5·21
30				0·50	0·85	1·32	1·89	3·28	4·66	6·70
35				0·59	1·00	1·57	2·24	3·82	5·69	8·17
40				0·68	1·26	1·82	2·60	4·59	6·71	9·67
45					1·32	2·07	2·95	5·22	7·75	11·16
50						2·33	3·32	5·99	8·79	12·66
55								6·62	9·79	14·11
60								7·32	10·83	15·61

To obtain the weight in lbs multiply by 2.205

GKN CAP SCREWS : ISO INCH (UNIFIED) : BSW AND BSF THREADS (Kilogrammes per 100*)

Length mm	Diameter mm															
	UNC/UNF					BSW	BSW BSF	UNC/UNF/BSW/BSF								
in	No. 4	No. 5	No. 6	No. 8	No. 10	1/8	3/16	1/4	5/16	3/8	7/16	1/2	5/8	3/4	7/8	1
3/16	0·051		0·90													
1/4	0·057	0·084	0·104	0·155	0·218		0·218	0·42								
5/16	0·062		0·115	0·168	0·236											
3/8	0·068	0·100	0·127	0·180	0·254	0·104	0·254	0·48	0·81							
1/2	0·079	0·116	0·150	0·206	0·286	0·118	0·286	0·54	0·91	1·56						
5/8	0·091		0·170	0·231	0·32	0·136	0·32	0·60	1·00	1·70						
3/4	0·102		0·193	0·258	0·35	0·150	0·35	0·69	1·10	1·84	2·71	3·63				
7/8	0·117			0·282	0·39		0·39	0·72	1·19	1·98		4·02				
1	0·132		0·240	0·32	0·44	0·182	0·44	0·79	1·28	2·12	3·07	4·28	7·13			
1¼				0·38	0·52		0·52	0·94	1·50	2·41	3·47	4·86	7·95	12·1		
1½					0·61		0·61	1·09	1·73	2·74	3·91	5·36	8·54	13·4		28·4
1¾					0·70		0·70	1·24	1·95	3·08	4·39	6·00	9·63	14·8	21·6	30·8
2					0·79		0·79	1·41	2·18	3·41	4·86	6·58	10·35	16·2	23·4	33·2
2¼								1·56	2·41	3·75		7·22	11·3	17·5	25·2	35·7
2½							0·97	1·69	2·63	4·07	5·81	7·81	12·1	18·9	27·1	38·2
2¾								1·86	2·86	4·39		8·49	13·2	20·0		
3								2·03	3·08	4·77	6·77	9·03	14·0	21·6	30·8	43·2
3¼								2·18	3·31	5·04	7·04	9·72	15·1	22·7		
3½								2·33	3·53	5·45	7·72	10·26	15·9	24·3	34·5	48·1
4								2·64	3·98	6·08		11·5	17·5	27·1	38·3	53·1
4½									4·40	6·81		12·8	19·6	29·9	42·0	58·1
5									4·86	7·45		14·0	21·3	32·6	45·4	63·1
5½									5·31	8·13		15·3	23·3	35·4		
6									5·77	8·81		16·5	25·0	38·2	53·1	73·1
6½													27·0	40·9		
7													28·9	43·7	60·4	83·1
8													31·8	49·0		93·1
9														54·5		103·1

GKN CAP SCREWS : BA THREADS (Kilogrammes per 100*)

Length mm	Diameter mm							
	8BA	6BA	5BA	4BA	3BA	2BA	1BA	0BA
1/8	0·023							
3/16	0·026	0·051		0·095				
1/4	0·030	0·057	0·084	0·104		0·218		
5/16	0·034	0·062		0·115		0·236		
3/8	0·037	0·068	0·100	0·127	0·159	0·254		0·454
7/16		0·074		0·138		0·272		
1/2	0·045	0·079	0·116	0·149	0·186	0·291	0·350	0·508
5/8		0·091		0·171	0·209	0·322	0·386	0·568
3/4	0·061	0·102		0·193	0·231	0·359	0·431	0·622
7/8		0·117		0·216		0·395		0·676
1	0·078	0·132		0·240	0·295	0·436	0·513	0·731
1¼		0·161		0·289		0·522	0·622	0·872
1½				0·338		0·608		1·01
1¾				0·386		0·695		1·14
2				0·438		0·776		1·28
2¼						0·863		
2½						0·954		

*To obtain the weight in lbs multiply by 2.205

GKN COUNTERSUNK HEAD SCREWS : ISO INCH (UNIFIED), BSW, BSF AND BA THREADS

Length	Diameter in													
						UNC/UNF				BSW	UNC/UNF/BSW/BSF			
in	6BA	4BA	3BA	2BA	0BA	No. 4	No. 6	No. 8	No.10	3/16	1/4	5/16	3/8	1/2
1/4	0·032	0·052		0·113		0·041	0·059			0·127				
5/16	0·036	0·064		0·132										
3/8	0·041	0·073	0·091	0·145	0·209	0·048	0·077	0·091	0·154	0·154	0·213			
7/16				0·159										
1/2	0·054	0·095	0·113	0·177	0·263	0·059	0·095	0·109	0·186	0·186	0·318	0·536		
5/8	0·064	0·113		0·209	0·322		0·113	0·127	0·218	0·218	0·381	0·636	0·976	
3/4	0·077	0·132	0·168	0·236	0·377			0·145	0·245	0·245	0·445	0·731	1·12	2·18
7/8				0·268						0·277		0·831	1·26	
1	0·104	0·172	0·218	0·300	0·490					0·309	0·567	0·931	1·41	2·70
1 1/4				0·386	0·599					0·390	0·690	1·14	1·69	3·18
1 1/2				0·472						0·477	0·853	1·33	2·04	3·72
1 3/4											1·02	1·56	2·37	4·22
2											1·18	1·80	2·72	4·72
2 1/2											1·50	2·27	3·40	5·77
3													4·09	6·85

GKN SET SCREWS – ALL POINTS: ISO INCH (UNIFIED): BSW AND BSF THREADS (Kilogrammes per 100*)

Length	Diameter in													
	UNC/UNF					BSW	BSW BSF	UNC/UNF/BSW/BSF						
in	No. 4	No. 5	No. 6	No. 8	No.10	1/8	3/16	1/4	5/16	3/8	7/16	1/2	5/8	3/4
1/8	0·0091	0·011	0·013	0·024	0·025	0·011								
3/16	0·015	0·019	0·022	0·035	0·039	0·019	0·039	0·055						
1/4	0·021	0·026	0·031	0·045	0·059	0·026	0·059	0·082	0·141					
5/16	0·027		0·038	0·059	0·077	0·033	0·077	0·118	0·191	0·241		0·404		
3/8	0·033	0·042	0·045	0·068	0·091	0·042	0·091	0·141	0·241	0·318	0·50	0·531		
7/16								0·173	0·295	0·340	0·60	0·658		
1/2			0·064	0·100	0·127	0·059	0·127	0·209	0·345	0·463	0·70	0·785	1·06	
5/8			0·082		0·159		0·159	0·272	0·445	0·608	0·90	1·05	1·49	2·95
3/4			0·100		0·195		0·195	0·327	0·549	0·758	1·10	1·31	1·92	3·67
7/8								0·390	0·645	0·908		1·58		
1					0·268		0·268	0·454	0·754	1·06	1·51	1·84	2·76	5·11
1 1/4							0·335	0·572	0·953	1·35		2·37	3·65	6·54
1 1/2								0·700	1·16	1·64	2·31	2·91	4·47	7·98
1 3/4									1·36	1·94			5·33	
2								0·940	1·56	2·23		3·97	6·19	10·85
2 1/2												5·04		
3													9·64	

*To obtain the weight in lbs multiply by 2.205

GKN SET SCREWS – ALL POINTS : BA THREADS (Kilogrammes per 100*)

Length	Diameter in						
in	8BA	6BA	5BA	4BA	3BA	2BA	0BA
$\frac{1}{8}$	0·0040	0·0109	0·017	0·015		0·022	
$\frac{3}{16}$	0·0080	0·016	0·024	0·025	0·034	0·039	0·059
$\frac{1}{4}$	0·0109	0·022	0·033	0·031	0·043	0·057	0·086
$\frac{5}{16}$		0·027		0·045		0·071	0·114
$\frac{3}{8}$		0·034	0·046	0·057	0·066	0·091	0·141
$\frac{7}{16}$		0·039		0·066		0·109	
$\frac{1}{2}$		0·045		0·075	0·091	0·127	0·195
$\frac{5}{8}$		0·057		0·094		0·159	0·250
$\frac{3}{4}$		0·068		0·116		0·195	0·304
1				0·154		0·263	
$1\frac{1}{4}$						0·331	

GKN BUTTON HEAD SCREWS : ISO INCH (UNIFIED) BSW, BSF AND BA THREADS (Kilogrammes per 100*)

Length	Diameter in								
	UNC/UNF				BSW/BSF	UNC/UNF/BSW/BSF			
in	2BA	No. 6	No. 8	No. 10	$\frac{3}{16}$	$\frac{1}{4}$	$\frac{5}{16}$	$\frac{3}{8}$	$\frac{1}{2}$
$\frac{1}{4}$	0·136	0·045	0·064	0·136	0·136				
$\frac{3}{8}$	0·168	0·064	0·095	0·168	0·168	0·295			
$\frac{1}{2}$	0·209	0·086	0·123	0·209	0·209	0·359	0·622	0·967	
$\frac{5}{8}$	0·241		0·154	0·241	0·241	0·422	0·722		
$\frac{3}{4}$	0·277			0·277	0·277	0·481	0·822	1·27	
1				0·345		0·608	1·02	1·56	3·09
$1\frac{1}{4}$								1·86	3·65
$1\frac{1}{2}$								2·16	4·13

GKN SHOULDER SCREWS : BSW AND BSF THREADS (Kilogrammes per 100*)

Shoulder Length	Shoulder Diameter in					
in	$\frac{1}{4}$	$\frac{5}{16}$	$\frac{3}{8}$	$\frac{1}{2}$	$\frac{5}{8}$	$\frac{3}{4}$
$\frac{3}{4}$	0·804		2·25			
1	0·953	1·77	2·60	4·95		
$1\frac{1}{4}$	1·11	2·00	2·94	5·58	9·4	
$1\frac{1}{2}$	1·26	2·26	3·29	6·17	10·4	16·4
$1\frac{3}{4}$		2·50	3·64	6·81	11·4	17·9
2	1·57	2·75	3·99	7·45	12·4	19·3
$2\frac{1}{4}$		2·99	4·34	8·08	13·4	20·8
$2\frac{1}{2}$		3·23	4·68	8·67	14·4	22·2
$2\frac{3}{4}$			5·04	9·31	15·4	23·6
3			5·40	9·94	16·4	25·0
$3\frac{1}{4}$				10·58	17·5	26·4
$3\frac{1}{2}$			6·08	11·17	18·5	27·9
$3\frac{3}{4}$					19·5	
4				12·44	20·5	30·8
$4\frac{1}{2}$					22·6	33·6

*To obtain the weight in lbs multiply by 2.205

WEIGHT OF STEEL ROUND HEAD WOOD SCREWS (Weight in pounds per gross*)

Length in	Screw Gauge																
	0	1	2	3	4	5	6	7	8	9	10	11	12	14	16	18	20	24
¼	·036	·043	·065	·096	·120	·176	·234		·394									
⅜	·050	·055	·080	·114	·154	·217	·285	·347	·468		·706							
½	·060	·068	·099	·140	·187	·253	·342	·410	·542	·611	·814		1·06					
⅝	·074	·082	·120	·165	·219	·299	·397	·485	·616	·693	·921	1·06	1·21	1·63				
¾		·096	·142	·188	·251	·335	·453	·545	·690	·799	1·03	1·15	1·36	1·80	2·46			
⅞				·211	·282	·380	·503	·623	·764	·884	1·13	1·31	1·50	1·95	2·67			
1			·186	·236	·318	·431	·562	·692	·838	·987	1·24	1·42	1·61	2·14	2·88	3·47	4·30	
1¼				·294	·378	·508	·672	·830	·988	1·17	1·45	1·66	1·88	2·49	3·30	4·01	4·94	
1½					·451	·591	·782	·968	1·13	1·36	1·67	1·90	2·18	2·82	3·72	4·56	5·58	8·06
1¾					·516	·681	·892	1·10	1·28	1·55	1·88	2·13	2·47	3·19	4·14	5·10	6·23	
2					·590	·757	1·00	1·24	1·43	1·73	2·09	2·40	2·75	3·54	4·56	5·60	6·88	9·82
2¼							1·11	1·38	1·57	1·92	2·31	2·65	3·02	3·90	4·98		7·58	
2½					·717		1·22	1·52	1·73	2·07	2·52	2·90	3·30	4·26	5·40	6·74	8·13	11·60
2¾									1·87		2·74		3·58	4·62				
3							1·44		2·02		2·95		3·87	4·98	6·21	7·83	9·48	13·37
3½									2·32		3·38		4·44	5·70	7·09	8·92		
4									2·62		3·81		5·02	6·42	7·94	10·00	12·08	
4½													5·60	7·14				
5														7·86	9·62			

To obtain the weight of screws in other metals, multiply the above weights by the following factors:

Brass	1.072	Light Alloy	0.357
Copper	1.145	Stainless Steel 1.	1.012
Gunmetal	1.138		

*To obtain weights in kilograms per 100, multiply table figures by 0.315

WEIGHT OF BAR STEEL STOCK (pounds per Lineal Foot) *

in	Round Bars lb	Square Bars lb	Hex Bars (A/F) lb
3/16	0·094	0·120	0·103
1/4	0·167	0·213	0·184
5/16	0·261	0·332	0·287
3/8	0·376	0·478	0·414
7/16	0·511	0·651	0·564
1/2	0·668	0·849	0·735
9/16	0·845	1·076	0·930
5/8	1·043	1·328	1·148
11/16	1·262	1·607	1·390
3/4	1·502	1·912	1·650
13/16	1·763	2·245	1·943
7/8	2·044	2·603	2·250
15/16	2·347	2·988	2·585
1	2·670	3·400	2·940
1 1/8	3·380	4·303	3·720
1 1/4	4·172	5·312	4·60
1 3/8	5·049	6·428	5·56
1 1/2	6·008	7·650	6·62
1 5/8	7·051	8·978	7·80
1 3/4	8·178	10·412	9·01
1 7/8	9·388	11·953	10·35
2	10·681	13·60	11·78
2 1/8	12·06	15·35	13·30
2 1/4	13·52	17·21	14·91
2 3/8	15·06	19·18	16·62
2 1/2	16·69	21·25	18·41
2 5/8	18·40	23·43	20·30
2 3/4	20·19	25·71	22·27
2 7/8	22·07	28·10	24·33
3	24·03	30·60	26·51

*For weight in kilogrammes per lineal metre, multiply table figures by 1.5

NAIL COUNT – APPROXIMATE NUMBER OF NAILS PER POUND*

Nail Type	Length – Inches (mm)													
	½in (12.5)	¾in (19)	1in (25)	1¼in (31)	1½in (37.5)	1¾in (44)	2in (50)	2¼in (55)	2½in (62.5)	3in (75)	3½in (87.5)	4in (110)	5in (125)	6in (150)
Round Wire		1 765	800	585	350	275	165	125	125	70	45	35	25	15
Oval Wire	3 360	2 000	1 150	670	425	300	200	150	100	55	40	30	20	15
Cut Clasp			620	390	280	200	150	100	90	55	35	25	15	10
Lost Head	3 740	2 070	1 180	680	465	295	215	150	100	60	45	35	20	15
Clout	1 050	500	340	320	250	150	120							
Panel Pin	4 200	2 100	1 290	985	670		350							

*For number per kilogramme, multiply by 2.205

APPROXIMATE WEIGHT OF STEEL SNAP HEAD RIVETS *
(Weight in lb per 1 000 rivets)

Shank Length	Shank Diameter										
	1/16	3/32	1/8	5/32	3/16	7/32	¼	5/16	3/8	7/16	½
1/16	·1338										
1/8	·1889	·508	1·061	1·903							
3/16	·2440	·631	1·279	2·245	3·586						
1/4	·2991	·754	1·497	2·586	4·078	6·03	8·47	15·21	24·71		
5/16	·3542	·877	1·716	2·928	4·570	6·70	9·34	16·57	26·67		
3/8	·4093	1·000	1·934	3·270	5·062	7·37	10·21	17·94	28·64		
7/16	·4644	1·123	2·152	3·611	5·554	8·04	11·08	19·30	30·60		
1/2	·5195	1·246	2·370	3·953	6·046	8·71	11·96	20·67	32·57	48·16	67·82
9/16	·5746	1·369	2·588	4·295	6·538	9·38	12·83	22·03	34·53	50·83	71·31
5/8	·6297	1·492	2·807	4·637	7·030	10·05	13·70	23·40	36·50	53·50	74·80
11/16	·6848	1·615	3·025	4·978	7·522	10·72	14·57	24·76	38·46	56·17	78·29
3/4	·7399	1·738	3·243	5·320	8·014	11·39	15·45	26·13	40·43	58·84	81·78
13/16	·7950	1·861	3·461	5·662	8·506	12·06	16·32	27·49	42·39	61·51	85·27
7/8	·8501	1·984	3·679	6·003	8·998	12·73	17·19	28·86	44·36	64·18	88·76
15/16	·9052	2·107	3·898	6·345	9·490	13·40	18·06	30·22	46·32	66·85	92·25
1	·9603	2·230	4·116	6·687	9·982	14·07	18·94	31·59	48·29	69·52	95·74
1 1/8			4·552	7·370	10·96	15·41	20·68	34·32	52·22	74·86	102·7
1 1/4			4·989	8·054	11·95	16·75	22·43	37·05	56·15	80·20	109·7
1 3/8			5·425	8·737	12·93	18·09	24·17	39·78	60·08	85·54	116·6
1 1/2			5·861	9·420	13·91	19·43	25·92	42·51	64·01	90·88	123·6
1 5/8			6·298	10·10	14·90	20·77	27·66	45·24	67·94	96·22	130·6
1 3/4			6·734	10·78	15·88	22·11	29·41	47·97	71·87	101·5	137·6
1 7/8			7·171	11·47	16·87	23·45	31·15	50·70	75·80	106·9	144·6
2			7·607	12·15	17·85	24·79	32·90	53·43	79·73	112·2	151·5
2 1/4					19·82	27·47	36·39	58·89	87·59	122·9	165·5
2 1/2					21·79	30·15	39·88	64·35	95·45	133·6	179·5
2 3/4					23·75	32·83	43·37	69·81	103·3	144·2	193·4
3					25·72	35·51	46·86	75·27	111·1	154·9	207·4
3 1/4					27·69	38·19	50·35	80·73	119·0	165·6	221·3
3 1/2					29·66	40·87	53·84	86·19	126·8	176·3	235·3
3 3/4					31·63	43·55	57·33	91·65	134·7	187·0	249·3
4					33·59	46·23	60·82	97·11	142·6	197·6	263·2

*Baxters (Bolts, Screws and Rivets Ltd).

APPROXIMATE WEIGHT OF STEEL SNAP HEAD RIVETS (contd.)

Shank Length	Shank Diameter										
	1/16	3/32	1/8	5/32	3/16	7/32	1/4	5/16	3/8	7/16	1/2
1/16	·1281										
1/8	·1832	·491	1·019	1·823							
3/16	·2383	·614	1·237	2·165	3·436						
1/4	·2934	·737	1·455	2·506	3·928	5·80	8·130	14·54	23·54		
5/16	·3485	·860	1·674	2·848	4·420	6·47	9·002	15·90	25·50		
3/8	·4036	·983	1·892	3·190	4·912	7·14	9·875	17·27	27·47		
7/16	·4587	1·106	2·110	3·531	5·404	7·81	10·74	18·63	29·43		
1/2	·5138	1·229	2·328	3·873	5·896	8·48	11·62	20·00	31·40	46·22	65·05
9/16	·5689	1·352	2·546	4·215	6·388	9·15	12·49	21·36	33·36	48·89	68·54
5/8	·6240	1·475	2·765	4·557	6·880	9·82	13·36	22·73	35·33	51·56	72·03
11/16	·6791	1·598	2·983	4·898	7·372	10·49	14·23	24·09	37·29	54·23	75·52
3/4	·7342	1·721	3·201	5·240	7·864	11·16	15·11	25·46	39·26	56·90	79·01
13/16	·7893	1·844	3·419	5·582	8·356	11·83	15·98	26·82	41·22	59·57	82·50
7/8	·8444	1·967	3·637	5·923	8·848	12·50	16·85	28·19	43·19	62·24	85·99
15/16	·8995	2·090	3·856	6·265	9·340	13·17	17·72	29·55	45·15	64·91	89·48
1	·9546	2·213	4·074	6·607	9·832	13·84	18·60	30·92	47·12	67·58	92·97
1 1/8			4·510	7·290	10·81	15·18	20·34	33·65	51·05	72·92	99·95
1 1/4			4·947	7·974	11·80	16·52	22·09	36·38	54·98	78·26	106·9
1 3/8			5·383	8·657	12·78	17·86	23·83	39·11	58·91	83·60	113·9
1 1/2			5·819	9·340	13·76	19·20	25·58	41·84	62·84	88·94	120·8
1 5/8			6·256	10·02	14·75	20·54	27·32	44·57	66·77	94·28	127·8
1 3/4			6·692	10·70	15·73	21·88	29·07	47·30	70·70	99·62	134·8
1 7/8			7·129	11·39	16·72	23·22	30·81	50·03	74·63	104·9	141·8
2			7·565	12·07	17·70	24·56	32·56	52·76	78·56	110·3	148·8
2 1/4					19·67	27·24	36·05	58·22	86·42	120·9	162·7
2 1/2					21·64	29·92	39·54	63·68	94·28	131·6	176·7
2 3/4					23·60	32·60	43·03	69·14	102·1	142·3	190·6
3					25·57	35·28	46·52	74·60	110·0	153·0	204·6
3 1/4					27·54	37·96	50·01	80·06	117·8	163·7	218·6
3 1/2					29·51	40·64	53·50	85·52	125·7	174·3	232·5
3 3/4					31·48	43·32	56·99	90·98	133·5	185·0	246·5
4					33·44	46·00	60·48	96·44	141·4	195·7	260·4

Note: Weights are generally similar, regardless of the head.
However the following factors can be used for correcting table weights for difficult head forms:

Pan or Mushroom head — factor by 0.99
Flat head — factor by 0.98
90° Countersunk head — factor by 0.97
120° Countersunk head — factor by 0.95

For weights of Brass rivets multiply Steel weights by 1.064
For weights of Copper rivets multiply Steel weights by 1.156
For weights of Aluminium rivets multiply Steel weights by 0.348

Conversion Tables

STANDARD WIRE GAUGE

swg	Inches	Millimetres
8	0.160	4.064
9	0.144	3.658
10	0.128	3.251
11	0.116	2.946
12	0.104	2.642
13	0.092	2.337
14	0.080	2.032
15	0.072	1.829
16	0.064	1.626
17	0.056	1.422
18	0.048	1.219
19	0.040	1.016
20	0.036	0.914
21	0.032	0.813
22	0.028	0.711
23	0.024	0.610
24	0.022	0.559
25	0.020	0.508
26	0.018	0.457
27	0.0164	0.4166
28	0.0148	0.3759
29	0.0136	0.3454
30	0.0124	0.3150
31	0.0116	0.2946
32	0.0108	0.2743

GERMAN DIN GAUGE FOR SHEET

Number	Thickness mm
1	5.50
2	5.00
3	4.50
4	4.25
5	4.00
6	3.75
7	3.50
8	3.25
9	3.00
10	2.75
11	2.50
12	2.25
13	2.00
14	1.75
15	1.50
16	1.38
17	1.25
18	1.13
19	1.00
20	0.88
21	0.75
22	0.63
23	0.56
24	0.50
25	0.44
26	0.38
27	0.32
28	0.28
29	0.24
30	0.22
31	0.20
32	0.18

INCHES – MILLIMETRES – WOODSCREW GAUGE – SWG

INCHES Decimals	INCHES Fractions	mm	Wood Screw Gauge	Imperial Standard Wire Gauge
0.0124		0.314		30
0.0136		0.345		29
0.0148		0.375		28
0.0156	1/64	0.396		
0.0164		0.416		27
0.018		0.457		26
0.020		0.508		25
0.022		0.558		24
0.024		0.609		23
0.028		0.711		22
0.031	1/32	0.793		
0.032		0.812		21
0.036		0.914		20
0.039		1.000		
0.040		1.016		19
0.046	3/64	1.190		
0.048		1.219		18
0.056		1.422		17
0.060		1.520	0	
0.062	1/16	1.587		
0.064		1.625		16
0.070		1.780	1	
0.072		1.828		15
0.078	5/64	2.000		
0.080		2.032		14
0.082		2.080	2	
0.092		2.336		13
0.093	3/32	2.381		
0.094		2.390	3	
0.104		2.640		12
0.108		2.740	4	
0.109	7/64	2.778		

INCHES Decimals	INCHES Fractions	mm	Wood Screw Gauge	Imperial Standard Wire Gauge
0.116		2.946		11
0.118		3.000		
0.122		3.100	5	
0.125	1/8	3.175		
0.128		3.251		10
0.136		3.450	6	
0.140	9/64	3.571		
0.144		3.657		9
0.150		3.810	7	
0.156	5/32	3.968		
0.157		4.000		
0.160		4.064		8
0.164		4.170	8	
0.171	11/64	4.365		
0.176		4.470		7
0.178		4.520	9	
0.187	3/16	4.762		
0.192		4.880	10	6
0.196		5.000		
0.203	13/64	5.159		
0.212		5.384		5
0.218	7/32	5.556		
0.220		5.590	12	
0.232		5.892		4
0.234	15/64	5.953		
0.236		6.000		
0.248		6.300	14	
0.250	1/4	6.350		
0.252		6.400		3
0.265	17/64	6.746		
0.275		7.000		
0.276		7.010	16	2

INCHES Decimals	INCHES Fractions	mm	Wood Screw Gauge	Imperial Standard Wire Gauge
0.281	9/32	7.143		
0.296	19/64	7.540		
0.300		7.620		1
0.304		7.720	18	
0.312	5/16	7.937		
0.315		8.000		
0.324		8.229		0
0.328	21/64	8.334		
0.332		8.430	20	
0.343	11/32	8.731		
0.348		8.839		2/0
0.354		9.000		
0.359	23/64	9.128		
0.360		9.140	22	
0.372		9.448		3/0
0.375	3/8	9.525		
0.390	25/64	9.921		
0.393		10.000		
0.400		10.160		4/0
0.406	13/32	10.318		
0.421	27/64	10.715		
0.432		10.970		5/0
0.433		11.000		
0.437	7/16	11.112		
0.453	29/64	11.509		
0.464		11.800		6/0
0.468	15/32	11.906		
0.472		12.000		
0.484	31/64	12.303		
0.500	1/2	12.700		7/0

cont...

INCHES – MILLIMETRES – WOODSCREW GAUGE – SWG (contd.)

Decimals	Fractions	mm
0.511		13.00
0.515	33/64	13.09
0.531	17/32	13.49
0.546	35/64	13.89
0.551		14.00
0.562	9/16	14.28
0.578	37/64	14.68
0.590		15.00
0.593	19/32	15.08
0.609	39/64	15.47
0.625	5/8	15.87
0.629		16.00
0.640	41/64	16.27
0.656	21/32	16.66
0.669		17.00
0.671	43/64	17.06
0.687	11/16	17.46
0.703	45/64	17.85
0.708		18.00
0.718	23/32	18.25
0.734	47/64	18.65
0.748		19.00

Decimals	Fractions	mm
0.750	3/4	19.05
0.765	49/64	19.44
0.781	25/32	19.84
0.787		20.00
0.796	51/64	20.24
0.812	13/16	20.63
0.826		21.00
0.828	53/64	21.03
0.843	27/32	21.43
0.859	55/64	21.82
0.866		22.00
0.875	7/8	22.22
0.890	57/64	22.62
0.905		23.00
0.906	29/32	23.01
0.921	59/64	23.41
0.937	15/16	23.81
0.944		24.00
0.953	61/64	24.20
0.968	31/32	24.60
0.984	63/64	25.00
1.000	1	25.40

METRIC DIAMETERS AND LENGTHS COMPARED WITH INCH SIZES

Standard Diameters

Nearest recommended metric diameters to replace inch diameters

Metric Diameter		Inch Diameter	
	in		in
M2.5	0.098	7BA	0.098
M3	0.118	6BA	0.110
M4	0.158	4BA	0.142
M5	0.197	2BA	0.185
		3/16in	0.187
M6	0.2362	1/4in	0.2500
M8	0.3149	5/16in	0.3125
M10	0.3937	3/8in	0.3750
M12	0.4724	7/16in	0.4375
		1/2in	0.5000
M16	0.6299	5/8in	0.6250
M20	0.7874	3/4in	0.7500

METRIC AND LENGTHS COMPARED WITH INCH SIZES

Standard Lengths of Bolts

Nearest recommended lengths to replace existing inch lengths

Metric Length		Nearest Inch length	
mm	in *		*
12	0.4724	½	0.5000
14	0.5512	9/16	0.5625
16	0.6299	5/8	0.6250
18	0.7086		
20	0.7874	¾	0.7500
25	0.9842	1	1.0000
30	1.1811	1¼	1.2500
35	1.3779	1.3/8	1.3750
40	1.5748	1½	1.5000
45	1.7716	1¾	1.7500
50	1.9685	2	2.0000
55	2.1654	2¼	2.2500
60	2.3622		
65	2.5590	2½	2.5000
70	2.7559	2¾	2.7500
75	2.9527	3	3.0000
80	3.1496	3¼	3.2500
85	3.3464		
90	3.5433	3½	3.5000
95	3.7401	3¾	3.7500
100	3.9370	4	4.0000
110	4.3307	4½	4.5000
120	4.7244	4¾	4.7500
130	5.1181	5	5.0000
140	5.5118	5½	5.5000
150	5.9055	6	6.0000

* These figures are inch equivalents

NUMBER OF THREADS PER INCH IN ALL THREAD FORMS COMPARED

Metric Diameter	TPI		Other Threads Threads per inch					
ISO Metric Coarse		Approx	UNC	UNF	BSW	BSF	BA	Inch Diameter
	in *							in
M2.5	0.098	56.4					52.9	7BA (0.098)
M3	0.118	51.0					47.9	6BA (0.110)
M4	0.158	36.3					38.5	4BA (0.142)
M5	0.197	31.8			24	32	31.4	3/16 (0.187) and 2BA (0.185)
M6	0.236	25.4	20	28	20	26		1/4 (0.250)
M8	0.315	20.3	18	24	18	22		5/16 (0.312)
M10	0.394	16.9	16	24	16	20		3/8 (0.375)
M12	0.472	14.5	13	20	12	16		1/2 (0.500)
M16	0.630	12.7	11	18	11	14		5/8 (0.625)
M20	0.787	10.2	10	16	10	12		3/4 (0.750)

* These figures are inch equivalents

Figures in brackets are inch equivalents

BSI APPROVED DRILL SIZES AND OLD NUMBER AND LETTER EQUIVALENTS

Diameter			Old Equivalent
in	in	mm	
0.0079		0.20	
0.0087		0.22	
0.0099		0.25	
0.0110		0.28	
0.0118		0.30	
0.0126		0.32	
0.0138		0.35	80
0.0150		0.38	79
	1/64		
0.0157		0.40	78
0.0165		0.42	
0.0177		0.45	77
0.0189		0.48	
0.0197		0.50	76
0.0205		0.52	75
0.0216		0.55	
0.0228		0.58	74
0.0236		0.60	73
0.0244		0.62	
0.0256		0.65	72 and 71
0.0268		0.68	
0.0276		0.70	70
0.0283		0.72	
0.0295		0.75	69
0.0307		0.78	
0.0312	1/32		68
0.0315		0.80	
0.0323		0.82	67
0.0335		0.85	66
0.0346		0.88	
0.0354		0.90	65
0.0362		0.92	64
0.0374		0.95	63
0.0386		0.98	62
0.0394		1.00	61 and 60
0.0413		1.05	59 and 58
0.0433		1.10	57
0.0453		1.15	
0.0469	3/64		56
0.0472		1.20	
0.0492		1.25	
0.0512		1.30	55
0.0532		1.35	
0.0551		1.40	54
0.0571		1.45	
0.0590		1.50	53
0.0610		1.55	
0.0625	1/16		
0.0630		1.60	52
0.0650		1.65	
0.0669		1.70	51
0.0689		1.75	
0.0709		1.80	50
0.0728		1.85	49
0.0748		1.90	
0.0768		1.95	48
0.0781	5/64		
0.0787		2.00	47
0.0807		2.05	46
0.0827		2.10	45
0.0846		2.15	
0.0866		2.20	44
0.0886		2.25	43
0.0906		2.30	
0.0925		2.35	
0.0938	3/32		42
0.0945		2.40	
0.0965		2.45	41
0.0984		2.50	40
0.1004		2.55	39
0.1024		2.60	38
0.1043		2.65	37
0.1063		2.70	36
0.1083		2.75	
0.1094	7/64		
0.1102		2.80	35 and 34
0.1122		2.85	33
0.1142		2.90	
0.1161		2.95	32
0.1181		3.00	31
0.1220		3.10	
0.1250	1/8		
0.1260		3.20	
0.1299		3.30	30
0.1339		3.40	

cont...

BSI APPROVED DRILL SIZES AND OLD NUMBER AND LETTER EQUIVALENTS (contd.)

in	Diameter in	mm	Old Equivalent	in	Diameter in	mm	Old Equivalent
0.1378		3.50	29	0.2795		7.10	
0.1406	9/64		28	0.2812	9/32		K
0.1417		3.60		0.2835		7.20	
0.1457		3.70	27 and 26	0.2874		7.30	
0.1496		3.80	25	0.2913		7.40	L
0.1535		3.90	24 and 23	0.2953		7.50	M
0.1562	5/32			0.2969	19/64		
0.1575		4.00	22 and 21	0.2992		7.60	
0.1614		4.10	20	0.3031		7.70	N
0.1654		4.20	19	0.3071		7.80	
0.1693		4.30	18	0.3110		7.90	
0.1719	11/64			0.3125	5/16		
0.1732		4.40	17	0.3150		8.00	O
0.1772		4.50	16	0.3189		8.10	
0.1811		4.60	15 and 14	0.3228		8.20	P
0.1850		4.70	13	0.3268		8.30	
0.1875	3/16			0.3281			
0.1890		4.80	12	0.3307		8.40	
0.1929		4.90	11 and 10	0.3346		8.50	Q
0.1968		5.00	9	0.3386		8.60	R
0.2008		5.10	8 and 7	0.3425		8.70	
0.2031	13/64			0.3438			
0.2047		5.20	6 and 5	0.3465		8.80	S
0.2087		5.30	4	0.3504		8.90	
0.2126		5.40	3	0.3543		9.00	
0.2165		5.50		0.3583		9.10	T
0.2188	7/32			0.3594	23/64		
0.2205		5.60		0.3622		9.20	
0.2244		5.70		0.3661		9.30	U
0.2283		5.80		0.3701		9.40	
0.2323		5.90					
0.2349	15/64			0.3740		9.50	
0.2362		6.00	A and B	0.3750	3/8		V
				0.3780		9.60	
0.2402		6.10	C	0.3819		9.70	
0.2441		6.20	D	0.3858		9.80	W
0.2480		6.30					
0.2500	1/4		E	0.3898		9.90	
0.2520		6.40		0.3906	25/64		
				0.3937		10.00	
0.2559		6.50	F	0.3976		10.10	X
0.2598		6.60	G	0.4016		10.20	
0.2638		6.70					
0.2656	17/64		H	0.4055		10.30	Y
0.2677		6.80		0.4062	13/32		
				0.4094		10.40	
0.2717		6.90	I	0.4134		10.50	Z
0.2756		7.00	J				

CONVERSION : INCHES AND FRACTIONS TO MILLIMETRES

Fraction	Whole Inches 0	1	2	3	4	5	6	7	8	9	10	11
0		25.400	50.800	76.200	101.60	127.00	152.40	177.80	203.20	228.60	254.00	279.40
1/32	0.794	26.194	51.594	76.994	102.39	127.79	153.19	178.59	203.99	229.39	254.79	280.19
1/16	1.588	26.988	52.388	77.788	103.19	128.59	153.99	179.39	204.79	230.19	255.59	280.99
3/32	2.381	27.781	53.181	78.581	103.98	129.38	154.78	180.18	205.58	230.98	256.38	281.78
1/8	3.175	28.575	53.975	79.375	104.78	130.18	155.58	180.98	206.38	231.78	257.18	282.58
5/32	3.969	29.369	54.769	80.169	105.57	130.97	156.37	181.77	207.17	232.57	257.97	283.37
3/16	4.762	30.162	55.562	80.962	106.36	131.76	157.16	182.56	207.96	233.36	258.76	284.16
7/32	5.556	30.956	56.356	81.756	107.16	132.56	157.96	183.36	208.76	234.16	259.56	284.96
1/4	6.350	31.750	57.150	82.550	107.95	133.35	158.75	184.15	209.55	234.95	260.35	285.75
9/32	7.144	32.544	57.944	83.344	108.74	134.14	159.54	184.94	210.34	235.74	261.14	286.54
5/16	7.938	33.338	58.738	84.138	109.54	134.94	160.34	185.74	211.14	236.54	261.94	287.34
11/32	8.731	34.131	59.531	84.931	110.33	135.73	161.13	186.53	211.93	237.33	262.73	288.13
3/8	9.525	34.925	60.325	85.725	111.13	136.53	161.93	187.33	212.73	238.13	263.53	288.93
13/32	10.319	35.719	61.119	86.519	111.92	137.32	162.72	188.12	213.52	238.92	264.32	289.72
7/16	11.112	36.512	61.912	87.312	112.71	138.11	163.51	188.91	214.31	239.71	265.11	290.51
15/32	11.906	37.306	62.706	88.106	113.51	138.91	164.31	189.71	215.11	240.51	265.91	291.31
1/2	12.700	38.100	63.500	88.900	114.30	139.70	165.10	190.50	215.90	241.30	266.70	292.10
17/32	13.494	38.894	64.294	89.694	115.09	140.49	165.89	191.29	216.69	242.09	267.49	292.89
9/16	14.288	39.688	65.088	90.488	115.89	141.29	166.69	192.09	217.49	242.89	268.29	293.69
19/32	15.081	40.481	65.881	91.281	116.68	142.08	167.48	192.88	218.28	243.68	269.08	294.48
5/8	15.875	41.275	66.675	92.075	117.48	142.88	168.28	193.68	219.08	244.48	269.88	295.28
21/32	16.669	42.069	67.469	92.869	118.27	143.67	169.07	194.47	219.87	245.27	270.67	296.07
11/16	17.462	42.862	68.262	93.662	119.06	144.46	169.86	195.26	220.66	246.06	271.46	296.86
23/32	18.256	43.656	69.056	94.456	119.86	145.26	170.66	196.06	221.46	246.86	272.26	297.66
3/4	19.050	44.450	69.850	95.250	120.65	146.05	171.45	196.85	222.25	247.65	273.05	298.45
25/32	19.844	45.244	70.644	96.044	121.44	146.84	172.24	197.64	223.04	248.44	273.84	299.24
13/16	20.638	46.038	71.438	96.835	122.24	147.64	173.04	198.44	223.84	249.24	274.64	300.04
27/32	21.431	46.831	72.231	97.631	123.03	148.43	173.83	199.23	224.63	250.03	275.43	300.83
7/8	22.225	47.625	73.025	98.425	123.83	149.23	174.63	200.03	225.43	250.83	276.23	301.63
29/32	23.019	48.419	73.819	99.219	124.62	150.02	175.42	200.82	226.22	251.62	277.02	302.42
15/16	23.812	49.212	74.612	100.01	125.41	150.81	176.21	201.61	227.01	252.41	277.81	303.21
31/32	24.606	50.006	75.406	100.81	126.21	151.61	177.01	202.41	227.81	253.21	278.61	304.01

COMPARISON OF INCH AND ISO METRIC THREADS

Diameter	BS Whitworth			BS Fine				ISO Metric Coarse		
	TPI	Stress Area in^2	Basic Minor Diameter	TPI	Stress Area in^2	Basic Minor Diameter	Metric Diameter	TPI (approx)	Stress Area mm^2 (in^2)	Basic Minor
in			in			in	mm (in)			mm (in)
1/4 (0.2500)	20	0.0320	0.1860	26	0.0356	0.2008	6 (0.2362)	25.4	20.1 (0.0312)	4.773 (0.1879)
5/16 (0.3125)	18	0.0527	0.2413	22	0.0567	0.2543	8 (0.3149)	20.3	36.6 (0.0567)	6.466 (0.2546)
3/8 (0.3750)	16	0.0779	0.2950	20	0.0839	0.3110	10 (0.3937)	17.0	58.0 (0.0899)	8.160 (0.3212)
7/16 (0.4375)	14	0.1069	0.3461	18	0.1158	0.3663				
1/2 (0.5000)	12	0.1385	0.3932	16	0.1520	0.4200	12 (0.4724)	14.5	84.3 (0.1307)	9.853 (0.3879)
5/8 (0.6250)	11	0.227	0.5086	14	0.2430	0.5336	16 (0.6299)	12.7	157 (0.2433)	13.546 (0.5333)
3/4 (0.7500)	10	0.336	0.6220	12	0.3520	0.6432	20 (0.7874)	10.1	245 (0.3798)	16.933 (0.6666)
7/8 (0.8750)	9	0.463	0.7328	11	0.4870	0.7586	22 (0.8661)	10.1	303 (0.4697)	18.933 (0.7454)
1 (1.000)	8	0.608	0.8400	10	0.6420	0.8720	24 (0.9448)	8.5	353 (0.5472)	20.319 (0.7999)
1.1/8 (1.125)	7	0.767	0.9420	9	0.8147	0.9828	M27 (1.0629)	8.5	459 (0.7114)	23.319 (0.9181)
							M30 (1.1811)	7.3	561 (0.8695)	25.706 (1.0120)
1.1/4 (1.250)	7	0.973	1.0670	9	1.0270	1.0780	M33 (1.2992)	7.3	694 (1.0757)	28.706 (1.1302)
1.3/8 (1.375)	No specification for 1.3/8in Whitworth threads			8	1.2370	1.2150	M36 (1.4173)	6.4	817 (1.2663)	31.093 (1.2241)
1.1/2 (1.500)	6	1.409	1.2866	8	1.4960	1.3400	M39 (1.5354)	6.4	976 (1.5128)	34.093 (1.3422)

MM TO INCHES, FRACTIONS AND BA SIZES

mm	ins	mm	ins	mm	ins	mm	ins
·01	·00039	·26	·01024	·51	·02008	·76	·02992
·02	·00079	·27	·01063	·52	·02047	·77	·03032
·03	·00118	·28	·01102	·53	·02087	·78	·03071
·04	·00157	·29	·01142	·54	·02126	·79	·03110
·05	·00197	·30	·01181	·55	·02165	·80	·03150
·06	·00236	·31	·01220	·56	·02205	·81	·03189
·07	·00276	·32	·01260	·57	·02244	·82	·03228
·08	·00315	·33	·01299	·58	·02283	·83	·03268
·09	·00354	·34	·01339	·59	·02323	·84	·03307
·10	·00394	·35	·01378	·60	·02362	·85	·03346
·11	·00433	·36	·01417	·61	·02402	·86	·03386
·12	·00472	·37	·01457	·62	·02441	·87	·03425
·13	·00512	·38	·01496	·63	·02480	·88	·03465
·14	·00551	·39	·01535	·64	·02520	·89	·03504
·15	·00591	·40	·01575	·65	·02559	·90	·03543
·16	·00630	·41	·01614	·66	·02598	·91	·03583
·17	·00669	·42	·01654	·67	·02638	·92	·03622
·18	·00709	·43	·01693	·68	·02677	·93	·03661
·19	·00748	·44	·01732	·69	·02717	·94	·03701
·20	·00787	·45	·01772	·70	·02756	·95	·03740
·21	·00827	·46	·01811	·71	·02795	·96	·03780
·22	·00866	·47	·01850	·72	·02835	·97	·03819
·23	·00906	·48	·01890	·73	·02874	·98	·03858
·24	·00945	·49	·01929	·74	·02913	·99	·03898
·25	·00984	·50	·01969	·75	·02953	1	·0394

mm	ins	Fractions	BA Equiv	mm	ins	Fractions
1·59	0·062	$\frac{1}{16}$	—	6·35	0·250	$\frac{1}{4}$
2·00	0·079	—	—	7·00	0·276	—
2·20	0·087	—	8	7·14	0·281	$\frac{9}{32}$
2·38	0·094	$\frac{3}{32}$	—	7·94	0·312	$\frac{5}{16}$
2·50	0·098	—	7	8·00	0·315	—
2·80	0·110	—	6	8·73	0·344	$\frac{11}{32}$
3·00	0·118	—	—	9·00	0·354	—
3·17	0·125	$\frac{1}{8}$	—	9·52	0·375	$\frac{3}{8}$
3·20	0·126	—	5	10·00	0·394	—
3·50	0·138	—	—	10·32	0·406	$\frac{13}{32}$
3·60	0·142	—	4	11·00	0·433	—
3·97	0·156	$\frac{5}{32}$	—	11·11	0·438	$\frac{7}{16}$
4·00	0·158	—	—	11·91	0·468	$\frac{15}{32}$
4·10	0·161	—	3	12·00	0·472	—
4·50	0·177	—	—	12·70	0·500	$\frac{1}{2}$
4·70	0·185	—	2	13·00	0·512	—
4·76	0·187	$\frac{3}{16}$	—	14·00	0·551	—
5·00	0·197	—	—	14·29	0·563	$\frac{9}{16}$
5·30	0·209	—	1	15·00	0·591	—
5·56	0·219	$\frac{7}{32}$	—	15·87	0·625	$\frac{5}{8}$
6·00	0·236	—	0	16·00	0·630	—

mm	ins	Fractions
17·00	0·670	—
17·46	0·688	$\frac{11}{16}$
18·00	0·709	—
19·00	0·748	—
19·05	0·750	$\frac{3}{4}$
20·00	0·787	—
20·64	0·813	$\frac{13}{16}$
22·22	0·875	$\frac{7}{8}$
23·81	0·938	$\frac{15}{16}$
25·00	0·984	—
25·40	1·000	1

HARDNESS EQUIVALENTS

Vickers Hardness Number	Rockwell A Scale	Rockwell B Scale	Rockwell C Scale	Brinell dia mm 10/3 000kg	Brinell Hardness Number	Equivalent Strength * Tensile Strength 100lb/in^2	Tensile Strength tons/in^2	Tensile Strength kg/mm^2
717	80	—	59	2·50	601	2979	133	210
670	79	—	57	2·55	578	2845	127	200
633	78	—	56	2·60	555	2755	123	194
599	77	—	54	2·65	534	2666	119	187
570	77	—	52	2·70	514	2576	115	181
540	76	—	50	2·75	495	2464	110	173
520	75	—	49	2·80	477	2374	106	167
495	75	—	48	2·85	461	2285	102	161
476	74	—	47	2·90	444	2218	99	156
456	73	—	45	2·95	429	2128	95	150
440	73	—	44	3·00	415	2038	91	143
420	72	—	43	3·05	401	1971	88	139
408	72	—	42	3·10	388	1904	85	134
390	71	—	40	3·15	375	1837	82	129
380	70	—	39	3·20	363	1770	79	124
363	69	—	38	3·25	352	1725	77	121
358	69	—	36	3·30	341	1658	74	117
344	68	—	35	3·35	331	1613	72	113
332	68	—	34	3·40	321	1568	70	110
320	67	—	32	3·45	311	1523	68	107
305	66	—	31	3·50	302	1478	66	104
297	66	—	30	3·55	293	1434	64	101
288	65	—	29	3·60	285	1389	62	98
280	64	—	27	3·65	277	1344	60	95
270	64	—	26	3·70	269	1322	59	93
265	63	—	25	3·75	262	1277	57	90
258	63	—	24	3·80	255	1254	56	88
251	63	—	23	3·85	248	1232	55	87
244	62	—	22	3·90	241	1187	53	84
238	61	100	20	3·95	235	1165	52	82
232	61	99	—	4·00	229	1120	50	79
225	60	98	—	4·05	223	1098	49	77
220	60	97	—	4·10	217	1075	48	76
215	60	97	—	4·15	212	1030	46	72
210	59	95	—	4·20	207	1008	45	71
200	59	94	—	4·30	197	963	43	68
190	58	92	—	4·40	187	918	41	65
182	56	90	—	4·50	179	874	39	61
173	56	88	—	4·60	170	829	37	58
166	54	86	—	4·70	163	806	36	57
159	53	83	—	4·80	156	762	34	54
152	52	82	—	4·90	149	739	33	52
146	51	80	—	5·00	143	717	32	51
140	50	77	—	5·10	137	694	31	49
134	49	74	—	5·20	131	672	30	48
129	48	71	—	5·30	126	650	29	46
124	47	69	—	5·40	121	627	28	45

*Approximate only

WEIGHT CONVERSION TABLE

Ounces		Grammes
0.035	1	28.35
0.070	2	56.70
0.106	3	85.05
0.141	4	113.40
0.176	5	141.75
0.212	6	170.10
0.247	7	198.45
0.282	8	226.80
0.317	9	255.15
0.353	10	283.49
	11	311.84
	12	341.20
	13	368.54
	14	396.89
	15	425.24
	16	453.59

lb		kg
2.20	1	0.45
4.40	2	0.91
6.61	3	1.36
8.82	4	1.81
11.02	5	2.27
13.23	6	2.72
15.43	7	3.17
17.64	8	3.63
19.84	9	4.08
22.05	10	4.53
44.09	20	9.07
66.14	30	13.61
88.18	40	18.14
110.23	50	22.68
132.28	60	27.22
154.32	70	31.75
176.37	80	36.29
198.42	90	40.82
220.46	100	45.36
242.51	110	49.89
264.55	120	54.43
286.60	130	58.97
308.64	140	63.50
330.69	150	68.04
352.74	160	72.58
374.78	170	77.11
396.83	180	81.65
418.88	190	86.18
440.92	200	90.72
462.97	210	95.25
485.01	220	99.79

lb		kg
507.06	230	104.33
529.10	240	108.86
551.15	250	113.40
573.20	260	117.94
595.24	270	122.47
617.29	280	127.01
639.34	290	131.54
661.38	300	136.18
683.43	310	141.61
705.47	320	145.15
727.52	330	149.99
749.56	340	154.22
771.61	350	158.76
793.66	360	163.30
815.60	370	167.83
837.75	380	172.37
859.80	390	176.90
881.84	400	181.54
903.89	410	186.97
925.93	420	190.51
947.98	430	195.35
970.02	440	199.58
992.07	450	204.12
1014.12	460	208.66
1036.06	470	213.19
1058.21	480	217.73
1080.26	490	222.26
1102.30	500	226.90

DECIMAL/INCH EQUIVALENTS OF AMERICAN NUMBERED DRILLS

Number	Diameter
37	0.1040
38	0.1015
39	0.0995
40	0.0980
41	0.0960
42	0.0935
43	0.0890
44	0.0860
45	0.0820
46	0.0810
47	0.0785
48	0.0760
49	0.0730
50	0.0700
51	0.0670
52	0.0635
53	0.0595
54	0.0550
55	0.0520
56	0.0465
57	0.0430
58	0.0420
59	0.0410
60	0.0400
61	0.0390
62	0.0380
63	0.0370
64	0.0360
65	0.0350
66	0.0330
67	0.0320

Number	Diameter
68	0.0310
69	0.0292
70	0.0280
71	0.0260
72	0.0250
73	0.0240
74	0.0225
75	0.0210
76	0.0200
77	0.0180
78	0.0160
79	0.0145
80	0.0135
81	0.0130
82	0.0125
83	0.0120
84	0.0115
85	0.0110
86	0.0105
87	0.0100
88	0.0095
89	0.0091
90	0.0087
91	0.0083
92	0.0079
93	0.0075
94	0.0071
95	0.0067
96	0.0063
97	0.0059

Fasteners Glossary

AMS – American Material Specification, originated by the SAE Aeronautics Committee.

AN – American dimensional standards for aircraft fasteners

Acorn Nut – Acorn-shaped blind nut.

Allowance – The geometric difference permitted between the maximum and minimum sizes of mating components.

Angularity – the angle between the axes of two faces of a fastener.

Axis – specifically the axis of the pitch cylinder or core of a screw thread.

Bearing Surface – the underside of the head of a fastener, nut or washer.

Blinding (Head) – a head with rounded top surfaces, a flat bottom surface and slightly tapering sides.

Binding Post – fastener for electrical conductors.

Blank – plain or intermediate form of a fastener.

Blind Fastener – a fastener which can be assembled and tightened from one side of the work.

Bound Body -- member, fastener or component with an interference fit.

Brinell Hardness (Number) – a measure of the hardness of a material.

Chamfer -- a bevel or bevel edge.

Chamfer point – truncated cone point

Chasing – screw cutting by moving a tool parallel to the axis of the fastener being threaded.

Class (of thread) – specification of the tolerance and allowance on screw threads.

Clevis – U-shaped joint member fastened by a pin.

Cold Heading – cold forming to yield thicker sections or head a blank.

Coining – cold forming metal by pressing between dies.

Collar – ring or flange section.

Comparator – an instrument for inspecting screw threads in enlarged detail to compare with standard charts.

Counterbore – enlargement of the diameter of an original hole.

Countersink – internal chamfer.

Countersunk (Head) – headed with a chamfered or bevelled underside.

Crest – the surface of a thread which connects adjacent flanks at the top of the ridge.

Cup Point – conical point or conical depression in the point of a screw.

Cut Thread – thread produced by machining.

Dog Point – cylindrical point with a diameter less than that of the thread.

External Thread – a thread formed on the external surface of a fastener (eg as on a bolt).

Fastener – mechanical component designed to hold two (or more) separate parts together.

Female Thread – an internal thread.

Fillet – a concave junction between two otherwise straight intersecting surfaces.

Fillister (Head) – cylindrical head with rounded top.

Fin – a small protrusion formed under the head of a fastener.

Fit – general description of the degree or range of geometric compatibility of a fastener with its mating holes.

Flash – thin fin or ridge of surplus material produced when forming in dies, etc.

Form – the shape of one complete thread profile in section in the plane of the axis.

Gimlet Point – threaded conical point.

Ground Thread – thread formed or finished on the flanks by grinding.

Head Angle – the included angle of the bearing surface of the head.

Head Width – width across flats, or across the narrowest axis.

Header Point – chamfered point produced during heading operation.

Headless fastener – fastener with no enlargement on the end.

Incomplete Thread – that portion of the thread which is incomplete due to a chamfer, etc.

Included Angle – angle between the flanks of a thread, measured in section in an axial plane.

Interference Fit – mating threads where the internal thread is larger than the external thread.

Internal Thread – thread formed on the internal surface (eg as in a nut or a tapped hole).

Jam Nut – second nut used as a lock nut.

Lead – the reciprocal of number of threads per unit length.

Left Hand Thread – a thread which has to be rotated in an anti-clockwise direction to assemble with a mating thread.

Length of Engagement – the distance in the plane of the axis between the external points of contact of mated threads.

Lentil Head – oval head.

Lock Nut – a nut used as a jam nut, or capable of providing self-locking action by its design.

Male Thread – an external thread.

Major Cylinder – an imaginary cylindrical surface which just touches the crests of an external thread or the roots of an internal thread.

Minor Cylinder – an imaginary cylindrical surface which just touches the roots of an external thread or the crests of an internal thread.

Major Diameter – the diameter of the major cylinder of a parallel thread or of the major cone of a taper thread, in a specified plane normal to the axis.

Minor Diameter – the diameter of the minor cylinder of a parallel thread or of the minor cone of a taper thread, in a specified plane normal to the axis.

NAS – American aircraft fastener standards originated by the National Aircraft Standards Committee.

Nail Point – sharp pyramidal point.

Neck – a 'waist' generated under the head of a fastener.

Needle Point – long, fully tapered conical point.

Nominal Size – specified size for general identification, the actual size being subject to tolerances.

Oiled – treatment applied to fasteners for corrosion resistance (ie a coating of oil).

Oval Head – head with rounded top surface.

Oval Point – point with a radiused end.

Pan Head – head with a flat top radiused into cylindrical sides.

Parallel Thread – thread formed on the surface of a cylinder.

Pilot Point – cylindrical point formed with a positive diameter.

Pin – unthreaded fastener of cylindrical section.

Pinch point – short cone point.

Pipe Thread – thread forms specially developed for the external threading of pipes.

Pitch – the distance, measured parallel to the axis, between corresponding points on adjacent thread forms in the same axial plane section and on the same side of the axis. The Pitch (in inches) is the reciprocal of the number of threads per inch.

Pitch Cylinder – an imaginary cylinder, co-axial with the thread, which intersects the surface of a parallel thread in such a manner that the intercept on a generator of the cylinder between the points where it meets the opposite flanks of the thread groove is equal to half the basic pitch of the thread.

Pitch Line – the generator of the pitch cylinder or cone.

Pitch Point – the point where the pitch line intersects the flank of the thread.

Point – the shape of the end of the fastener opposite to the head.

Point Diameter – the diameter of the extreme 'point' end of a fastener.

Point Radius – the spherical radius of a point.

Proof Load – specified test load to establish or specify the strength of a fastener.

Recess Head – head with a recess form to match a specific type of driver, eg Philips recess, Pozidriv recess, etc.

Reduced Body – bolt or screw with part of the body having a diameter smaller than the major diameter of the thread.

Reference Dimension – normal size or dimension.

Rib – small ridges or keys formed around the shank of a fastener.

Right-hand Thread – a thread which must be rotated in a clockwise direction to engage with a mating thread.

Rockwell Hardness (Number) – a measure of the hardness of a material.

Rolled Point – cupped form of point produced by thread rolling.

Rolled Thread – process of providing a thread by rolling a blank between dies.

Root – the surface forming the bottoms of the flanks of a thread.

Runout – the amount by which the outside surface of one component runs out with respect to that of another component, expressed in terms of TIR (Total Indication Reading).

SAE – specification originated by the (American) Society of Automotive Engineers.

Screw Thread – see Thread.

Series – preassembled screws and washers.

Shank – the part of a fastener between the head and the point.

Shim – a thin spacer.

Shoulder – enlarged portion on the body or shank

Socket Head – recessed head to match a socket driver or key.

Square Neck – a square shoulder formed under the head of a bolt.

Stud – parallel (cylindrical) headless fastener threaded over part of the whole of its length.

Swell Neck – tapered neck.

Tap – tool for cutting a screw thread by rotation or the formation of a screw thread with such a tool.

Tapped hole – internally threaded hole.

Thread – the ridge produced by forming, on the surface of a cylinder or cone, a continuous helical or spiral groove of uniform section such that the distance measured parallel to the axis between two corresponding points on its contour is proportional to their relative angular displacement about the axis.

Thread Rolling – the process of producing a screw thread by cold offset forming or rolling between dies.

Tolerance – permissible limits on dimensions.

Tolerance Limits – maximum permitted deviation from a specified dimension.

Truss Head – shallow head with a rounded top surface and flat under surface.

Undercut Head – head undercut or reduced to give a greater length of thread.

Washer Face – circular boss on the face of a bolt head or nut.

Wrenching Head – a head form designed for turning with a wrench by application of the wrench either to the outside of the head (external wrenching) or in a recess (internal wrenching).

English-French-German-Spanish Glossary

HEADS

ENGLISH	FRENCH	GERMAN	SPANISH
binding head	tète cylindrique bombée	Linsenflachkopf	cabeza cilindrica reducida y bombeada
countersunk head	tête fraisée	Senkkopf	cabeza avellanada
cross recess	empreinte cruciforme	Kreuzschlitz	ranura en cruz
fillister head	tête cylindrique bombée	Linsenzylinderkopf	cabeza cilindrica y bombeada
hexagon head	tête hexagonale	Sechskantkopf	cabeza hexagonal exterior
hexagon socket	six pans creux	Innensechskant	hexagono interior
mushroom head	tête goutte du suif	Flachrundkopf	cabeza gota de sebo
oval head	tête fraisée bombéer	Linsensenkkopf	cabeza avellanada y bombeada
pan head	tête cylindre avec déponille	Linsenkopf	cabeza bombeada plana
Phillips	Phillips	Phillips	Phillips
Pozidriv	Pozidriv	Pozidriv	Pozidriv
round head	tête ronde	Halbrundkopf	cabeza redonda
square head	tête carrée	Vierkantkopf	cabeza cuadrada
slot	fente	Schlitz	ranura recta

POINTS

ENGLISH	FRENCH	GERMAN	SPANISH
chamfered	à bout plat chanfreiné	Kegelkuppe	extremo achaflanado
cone	à bout pointu	Spitze	punta cónica
cup	extrémité à cuvette	Ringschneide	extremo achaflanado con cono embutido
full dog	extrémité à téton long	Zapfen	punta con largo teton
half dog	extrémité à téton court	Kernansatz	punta con corto teton
plain end	à bout plat	ohne Kuppe	extremo plano
rounded end	extrémité bombée	Linsenkuppe	extremo bombeado

TYPES OF FASTENERS

ENGLISH	FRENCH	GERMAN	SPANISH
anchor bolt	boulon queue de carpe	Ankerbolzen	perno de anclaje
cap nut	écrou borgne bas	Hutmutter biedrige Form	tuerca ciega baja
eye bolt	boulon basculant	Augenschraube	tornillo con ojo
helical spring washer	rondelle Grower	glatter Federring	arandela espiral
plain washer	rondelle plate sans chanfrein	Scheibe	arandela plana
hexagon bolt	vis a tete hexagonale (ou six pans)	Sechskantschraube	tornillo de cabeza hexagonal
hexagon castle nut	écrou à créneaux dégagés	Kronenmutter	tuerca hexagonal almenada alta, con collarete
hexagon lock	écrou hexagonal bas	flache Sechskantmutter	tuerca hexagonal baja
hexagon nut	ecrou hexagonal	Sechskantmutter	tuerca hexagonal
hexagon slotted nut	écrou a creneaux	Kronenmutter	tuerca hexagonal almenada lisa
hexagon socket cap screw	vis à tête cylindrique à six pans creux	Zylinderschraube mit Innensechskant	tornillo de cabeza cilindrica con hexagono interior
hook bolt	vis à crochet	Hakenbolzen	perno de gancho
J-bolt	Boulon d'encrage a crochet	J-förmiger Bolzen	perno J
knurled nut	ecrou molete bas	flache Randelmutter	tuerca moleteada baja
knurled thumb screw	vis à tête molétee	Randelschraube	tornillo de cabeza moleteade
self-tapping screw	vis à tôle	Blechschrauben	tornillos de rosca cortante
square head bolt	vis a tete carree	Vierkantschraube	tornillo de cabeza cuadrada
square nut	ecrou carre	Vierkantmutter	tuerca cuadrada
stud	goujon	Stifschraube	espárrago liso
toothed washer	rondelle a dents	federnde Zahnscheibe	arandela elástica dentada
U-bolt	étrier fileté	Bugelschraube	Horquilla roscada
wing screw	vis à oreilles	Flugelschraube	tornillo con cabeza de mariposa
wing nut	ecrou a oreilles	Flügelmutter	tuerca de mariposa

TYPES OF THREAD

ENGLISH	FRENCH	GERMAN	SPANISH
inch thread	filetage en inches	Zollgewinde	rosca en pulgadas
metric thread	filetage métrique	Metrisches Gewinde	rosca métrica rosca gas
pipe thread	filetage au pas du gaz	Rohrgewinde	rosca gas
screw thread	filetage	Schraubengewinde	rosca cilindrica
tapping screw thread	filetage de vis à tôle	Blechschraubengewinde	rosca cortante
external thread (male thread)	filetage externe (filetage mâle)	Aussengewinde	rosca exterior (rosca macho)
internal thread (female thread)	filetage interne (filetage femelle)	Innengewinde	rosca interior (rosca hembra)
right-hand thread	filetage à droite	Rechtsgewinde	rosca a derechas
left-hand thread	filetage à gauche	Linksgewinde	rosca a izguierdas
parallel thread	filetage parallele	zylindrisches gewinde	rosca cilindrica
taper thread	filetage conique	Konisches Gewinde	rosca cónica
axis	aze	Achse	eje
pitch	pas de vis	Gewindesteigung	paso
pitch cylinder	cylindre de pas	Teilkreiszyclinder	cilindro primitivo
pitch line	diametre a flanc de filet	Teilkreis	circulo primitivo
pitch point	sommet de filet	Walzpunkt	punto de contacto de los circulos primitivos
major cylinder	cylindre principal	Aussenzylinder	cilindro principal
minor cylinder	cylindre secondaire	Kernzylinder	cilindro secundario
major diameter	diamètre extérieur	Aussengewinde-durchmesser	diámetro principal
minor diameter	diamètre intérieur	Kerndurchmesser	diámetro secundario
chamfer angle	angle du chanfrein	Fasenwinkel am Kopf	angulo de chaflan
effective (pitch) diameter	diamètre sur flancs de filet	Flankendurchmesser	diámetro del núcleo
effective length	longueur de serrage	Klemmlänge	longitud de apriete del perno
head angle	angle de fraisure	Senkwinkel	ángulo de la avellanado
head diameter	diamètre de la tête	Kopfdurchmesser	diametro de la cabeza
head height	hauteur de tête	Kopfhöhe	altura de la cabeza
nominal diameter	diamètre nominal	Nenndurchmesser	diámetro nominal
shank diameter	diamètre de la partie lisse	Schaftdurchmesser	diámetro de la espiga
shank length	longueur de la partie lisse	Lange des gewindefreien Schaftteiles	longitud de la parte no roscada
thread diameter	diamètre du filetage	Gewindedurchmesser	diámetro nominal de la rosca
thread length	longueur filetee	Gewindelange	longitud roscada
width across flats	dimension sur plats	Schlusselweite des Sechskantes	ancho entre caras

SECTION 1

Sub-section B

Buyers' Guide

Sub-section b(I)

TRADE NAMES INDEX

ADMIRAL – Bolts and nuts – Black high tensile specials – Thomas William Lench Ltd
AEROTIGHT – Stiff nuts – GKN Screws & Fasteners Ltd (GKN Socket Screws)
A-LOC – Lock nut – Suko-Sim GmbH
ANCHOR RIVET BUSHES – Threaded rivet bush as for sheet metal – P.S.M.Fasteners Ltd
ANCHOR-SERT – Blind fitting cage nuts – P.S.M.Fasteners Ltd
ARDOWER – Toggle latch – Londesborough Oil & Marine Engineering Ltd
ARMALOK – Nylon inserted locknut – Armstrong Fastenings Ltd
ARMASERT – Heat inserted threaded bush for thermoplastics – Armstrong Fastenings Ltd
AVDEL RIVET – Quick release pin – Avdel Ltd
AVDELOK – Vibration proof bolts – Avdel Ltd
AVEX – Break stem blind rivets – Avdel Ltd
AVLUG – Blind self-plugging rivet – Avdel Ltd
AYLESBURY – Rivets – The Bifurcated & Tubular Rivet Co Ltd
AYLESBURY – Rivet setting machines – The Bifurcated & Tubular Rivet Co Ltd

BALL-LOK – Self-piercing nuts – Avdel Ltd
BALL-LOK – Self-locking nut – Deltight Industries Ltd
BANK-LOK INSERTS – 'push-in' threaded self-locking inserts for plastic – P.S.M.Fasteners Ltd
BANC-LOK TAPPED HOLES – 'Push-in' Inserts for sheet metal – P.S.M.Fasteners Ltd
BEAR-NUT – Blind screw anchors – Tucker Fasteners Ltd
BIERBACH – Special nails and fasteners – Ernst Bierbach KG
BIGHEADS – Perforated headed bonding and load-spreading steel fasteners – Bighead Bonding Fasteners Ltd

BREWER RAPID RIVETER – Pneumatically operated automatic blind riveting machine
Brewer Rapid Riveting Systems (Machinery)

BRIV – High clench blind rivet – Avdel Ltd

BULTE SNAP RINGS – Plastic locking rings, gaskets, packing rings and washers – N.Bulte KG

CAMLOC – Quick-release fastener – Camloc Industrial Fixings (UK) Ltd

CHOBERT – Blind rivets – Avdel Ltd

CIRCLEX QCC – Worm drive hose clips heavy zinc plated, stainless steel worm drive hose clips, pipe and cable clamping devices – Circlex Ltd

CLEVELOK – All-metal self-locking nuts – Peter Abbott & Co Ltd

CLEVELOC – Self-locking nuts – Firth Cleveland Fastenings Ltd

CLIC-RIVET – Plastic rivet – Instrument Screw Co Ltd

CONELOK – Prevailing torque self-locking nut – Standard Pressed Steel International

CONELOK – All metal stiff nut – Armstrong Fastenings Ltd

'CORONET' BRAND – High strength bolts, load indicator washers, rivets, split cotter pins – Cooper & Turner Ltd

DELCLIP – Hose clip – Deltight Industries Ltd

DELRON – Inserts for honeycomb and other panels – Camloc Industrial Fixings (UK) Ltd

DELTIGHT – Free-running, self-locking nut – Deltight Industries Ltd

DODGE – Brass expansion insert for plastic materials – Armstrong Fastenings Ltd

DYNABOLT – Masonry anchor – Ramset Fasteners Ltd

DZUS –DART – Plastic-acetal resin quarter turn fastener – Dzus Fastener Europe Ltd

DZUS-SABRE – Blind hole fastener – Dzus Fastener Europe Ltd

DZUS – Quick-release fasteners – Dzus Fastener Europe Ltd

DZUS-SUPERSONIC – Quarter-turn fastener for high tensile and shear application – Dzus Fastener Europe Ltd

DZUS – Quick-release fasteners – Borstlap b.v.

FABORY – Bolts, nuts, screws, etc, – Blind rivets – Borstlap b.v.

FAN DISK – Locking washer – John Bradley & Co Ltd

FASTEX – Plastic fasteners, components and assemblies – Fastex A Division of ITW Ltd

FERO – Blind rivets – Borstlap b.v.

FIRTH CLEVELAND – Nyloc nuts – Balcombe Engineering Ltd. (Suppliers)

FIXT-NUTS – High torque resistant threaded clinched in anchors for sheet metal – Barton Cold-form Ltd

FLEXLOC – Prevailing torque self-locking nut – Standard Pressed Steel International

FLEXITHREAD – Self-swaging fastener – sheet metal – Standard Pressed Steel International

FLUSHNUTS – Perforated steel female fasteners for flush-bonding and load-spreading applications – Bighead Bonding Fasteners Ltd

FLUSHLOK BOLT – Threaded actuated blind fastener – Avdel Ltd
FLUSHLOK RIVETS – Quick-release pin – Avdel Ltd
F.T.FASTENERS – Special purpose fasteners – United-Carr Ltd
F.T.FASTENERS – Fasteners and mechanisms – United-Carr Ltd

GESIPA FASTENERS – Blind riveting tools and blind rivets – Gesipa Blindniettechnik GmbH
GIPLA – Gypsum Plasterboard screws – Borstlap b.v.
G.K.N. – Self-tapping screws, coach bolts and nuts – Balcombe Engineering Ltd (Suppliers)
GKN – Self-tapping screws, hexagon socket screws, thumb screws and wing nuts, hardened fixing pins – GKN Screws & Fasteners Ltd (GKN Wood Screws)
GROVIT – Blind rivet – Avdel Ltd

'HANK' – Rivet bushes – GKN Screws & Fasteners Ltd (GKN Socket Screws)
HAWK SUPER SPLICE – Conveyor belt fastener – Isaac Jackson (Fasteners) Ltd
HELI-COIL INSERTS – Wire thread inserts – Peter Abbott & Co Ltd
HELI-COIL – Wire thread inserts – Armstrong Fastenings Ltd
HIT-SERT – Thread inserts – Metric Screws (Fastenings) Ltd.
HEYCO – Range of moulded nylon bushings – Heyco Manufacturing Co Ltd
H.S.B. – Conveyor belt fastener – Isaac Jackson (Fasteners) Ltd

INBUS – Socket screw – Bauer & Schuarte
INBUS PLUS – Self-locket socket screw – Bauer & Schuarte
INFAST – Threaded fastener – Industrial Fasteners Ltd
I.V.I. – Wire and cut hand tacks and a full range of shoe machinery tacks and nails – The British United Shoe Machinery Co Ltd

JACK NUTS – Blind screw anchors – Tucker Fasteners Ltd
JO-BOLT – Self-plugging blind rivets – Avdel Ltd

KALEI – Pressnuts – Borstlap b.v.
KEIL-STOP – Self-locking screw – Suko-Sim GmbH

'L' – Impact nut, weld nut (sheet metal nut) – Karl Limbach & Cie KG
LIMPET – Rivet bush – RT. R. Fastenings Ltd
LONDESBOROUGH – Toggle latch – Londesborough Oil & Marine Engineering Ltd

MALESERT – Self-tapping stud type threaded insert – Tappex Threaded Inserts Ltd
MEPLAG – Metal/plastic fasteners – Metric Screws (Fastenings) Ltd

MOLLY – Blind screw anchors – Tućker Fasteners Ltd
MULTICLIP – Toggle latch – Londesborough Oil & Marine Engineering Ltd
MULTIFORM – Special fcold-formed parts – The Bifurcated & Tubular Rivet Co
MULTISERT – Threaded insert installable by three methods – Tappex Threaded Inserts Ltd

NETTLEFOLDS – Self-tapping screws – Stenman Holland B.V.
NETTLEFOLDS – Wood screws and coach screws – GKN Screws & Fasteners Ltd (GKN Wood Screws)
NUTSERTS – Threaded actuated blind fastener – Avdel Ltd
NYLOC – Self-locking nuts – Firth Cleveland Fastenings Ltd
NYLOC NUTS – Self-locking nuts – Peter Abbott & Co Ltd
NYLTITE – Nylon locking seal for screws – P.S.M.Fastenings Ltd

PARABOLT – Anchor bolts – Tucker Fasteners Ltd
PEINE – High strength structural fasteners – Borstlap b.v.
P.F.G. – Expansion bolts – Van Thiel United B.V.
PIP-PIN – Quick-release panel fastener – Avdel Ltd
PASTIDOME – Plastic caps for screws – GKN Screws & Fasteners Ltd
PULSERT – Blind fitting rivet nuts – P.S.M.Fastenings Ltd
POLY-STOP – Lock nut – Suko-Sim GmbH
POP – Blind rivets – Tucker Fasteners Ltd
POP – Blind riveting systems – Tucker Industries Pty Ltd
POZIDRIV – Recess head wood screws – GKN Screws & Fasteners Ltd (GKN Wood Screws)
POZITOPS – Plastic cover caps for Pozidriv screws – GKN Screws & Fasteners Ltd (GKN Wood Screws)
PUSHTURN – Vibration-proof bolts – Avdel Ltd
PUSHSERT – Threaded inserts press-in type – Tappex Threaded Inserts Ltd

RAMPA – Sleeves, bushes and sockets – H. & H. Brugmann jun., Schraubenfabrik
RAMSET – Cartridge-operated fixings and fixing tools – Ramset Fasteners Ltd
RIVEKLE – Blind riveting nuts – Borstlap b.v.
ROSAN PRESS NUT – Tapped self-clinching bush for sheet metal – Instrument Screw Co Ltd
ROSAN PRESS SCREW – Tapped self-clinching screw for sheet metal – Instrument Screw Co Ltd

SAAT SYSTEM – Self-piercing and tapping screws system – GKN Screws & Fasteners Ltd (GKN Self Tapping Screws)
SCHEIBEN – Mechanical connecting elements – Teckentrup KG
SELF-CLINCHING FASTENERS – Threaded bushes pressed into sheet metal – P.S.M.Fasteners Ltd
SEMS – Screws with pre-assembled washers – GKN Screws & Fasteners Ltd (GKN Self-tapping Screws)
SHEARSERT – Threaded insert – Tappex Threaded Inserts Ltd

SLIMSERT – Threaded insert for castings – Instrument Screw Co Ltd
SNEP – Locknuts – Borstlap b.v.
SONIC-LOK – Weld-in threaded inserts for plastic – P.S.M.Fastenings Ltd
SONICSERT – Thread insert installed ultrasonically – Tappex Threaded Inserts Ltd
SOPRAL – Aluminium alloy fasteners – Borstlap b.v.
SPAC NUTS – Self-piercing and clinching threaded bushes – P.S.M.Fastenings Ltd
SPIRE FASTENERS – Spring steel fasteners – Firth Cleveland Fastenings Ltd
SPIRO – 'Push-in' threaded inserts for plastics – P.S.M.Fastenings Ltd
SPIROLOX – Retaining rings – Wellworthy Ltd
SPOTNAILS – 'T' nails – Spotnails Ltd
SPRED-SERT – Thread inserts – Metric Screws (Fastenings) Ltd
SPRING-STOP – Lock nut – Suko-Sim GmbH
SPS HITEK – Nuts – Standard Pressed Steel International
STANDARD – High tensile setscrews and bolts – Peter Abbott & Co Ltd
STILETTO – Full range of nails and tacks for hand and machine applications – The British United Shoe Machinery Co Ltd
'STILETTO' – Masonry nails, machine nails, panel pins, deep drive panel pins, annular ring shanked nails, decorative nails – Whitfield Hodgsons & Brough Ltd
SUPERTEKS – Self-drilling screws – Borstlap b.v.
SURBUS – Fasteners with pre-assembled washers – Bauer & Shuarte

TAPIT – Nylon and steel fixing device – Deltight Industries Ltd
TAPPEX – threaded inserts and blind fasteners – Tappex Thread Inserts Ltd
TAPTITE – Thread forming screws – Stenman Holland B.V.
TAPTITE – Thread forming screws – Borstlap b.v.
TAPTITE – Thread forming screws – GKN Screws & Fasteners Ltd
(GKN Self-Tapping Screws)
TERRIER – Self-drilling masonry anchors, non-drill masonry anchors – Ramset Fasteners Ltd
THERMAG – Lock nut – Suko-Sim GmbH
THIEL – High tensile bolts – Van Thiel United B.V.
THOR – Portable pneumatic power tools for chemical, civil and contractor plant hire industries –
Thor Tools Ltd
THREADRIV – Power and hand tool drive bits – Peter Abbott & Co Ltd
TORQLOK – Nylon insert self-locking nut – Deltight Industries Ltd
TRISERT – Self-tapping threaded insert – Tappex Threaded Inserts Ltd
TRIVET – Rivet bush – T.R.Fastenings Ltd.
TUCKER – Blind riveting tools, eyelets and metal pressings – Tucker Fasteners Ltd
TUCKER – Eyelets – Tucker Industries Pty Ltd
TUCKER DRIVLOK – Spring pins – Tucker Industries Pty Ltd
TUCKER/MOLLY – Speciality blind fasteners – Tucker Industries Pty Ltd
TUCKER/PARABOLT – Masonry anchors – Tucker Industries Pty Ltd
TUCKER/WARREN – Stud welding systems – Tucker Industries Pty Ltd
TUK-LOK – RPrevailing torque locking fasteners – Tucker Industrial Pty Ltd
TWINFAST – Wood screws – GKN Screws & Fasteners Ltd

UNBRAKO – Socket screws – Balcombe Engineering Ltd (Suppliers)
UNEX – Hose clip – Deltight Industries Ltd
UNI-STOP – Lock nut – Suko-Sim GmbH
UNITEC – Nylon fasteners – Metric Screws (Fastenings) Ltd

VERBUS – Hexagon screws – Bauer & Schuarte
VERBUS PLUS – Self-locking hexagon screws – Bauer & Schuarte
VERBUS-TENSILOCK – Self-locking screws and nuts – Bauer & Schuarte
VIBREX – Quick-release fasteners – Peter Abbott & Co Ltd
VIBREX – Quick-operating fasteners – Metric Screws (Fastenings) Ltd

'W' POINT – Hexagon socket set screws – GKN Screws & Fasteners Ltd
(GKN Socket Screws)
WHIZTITE – Free spinning self-locking screws – Glynwed Fastenings
WHIZ-TITE – Self-lcoking screw – The Ormond Engineering Co Ltd

Sub-section b(2)

CLASSIFIED INDEX TO FASTENER MANUFACTURERS, DISTRIBUTORS AND MACHINERY SUPPLIERS

Alloy Steel Studbolts and Nuts

R. S. Rowlands Ltd

Anchor Bolts

Tucker Fasteners Ltd

Anchor Nuts

Tucker Industries Pty Ltd

Automotive and Truck Wheel Nuts

Firth Cleveland Fastenings Ltd

Ball Joints

Springfix Ltd

Belt Fasteners

Prestwich Parker Ltd

Black Bolts and Nuts

Industrial Fasteners Ltd
Benjamin Priest & Sons Ltd

Black Fasteners

The Allthread Group

Blind Captive Nuts

Tappex Thread Inserts Ltd

Blind Hole Fasteners

Dzus Fasteners Europe Ltd

Blind Rivets

Peter Abbott & Co Ltd
Adams & Hann Ltd
Avdel Ltd
Balcombe Engineering Ltd
Brewer Rapid Riveting Systems (Machinery)
Gesipa Blindniettechnik GmbH
Borstlap b.v.
Deltight Industries Ltd
Essanbee Products Ltd
Fastex A Division of ITW Ltd
GKN Fasteners & Hardware Distributors Ltd
Thos Haddon & Stokes Ltd
Industrial Fasteners Ltd
Instrument Screw Co Ltd
Metric Screws (Fastenings) Ltd
Miller Bridges Fastenings, Glynwed Distribution Limited
A. Raymond, Manufacturer de Boutons et d'articles Metalliques
Tucker Fasteners Ltd
Tucker Industries Pty Ltd
The Walford Manufacturing Co Ltd

Blind Screw Anchors

Tucker Fasteners Ltd

Blind Threaded Inserts

Avdel Ltd

Bolts, Hook 'J' and 'U'

Entwistle (Oldham) Ltd

Bolts and Nuts (See also under "Stainless Steel")
Entwistle (Oldham) Ltd

Bonding
Bighead Bonding Fasteners Ltd

Brass Bolts and Nuts
Essanbee Products Ltd

Bright Steel Bolts and Nuts
H. J. Barlow & Co Ltd

Building Construction Fasteners
Ernst Bierbach KG
T. H. Dilkes & Co Ltd

Button Fasteners
United-Carr Ltd

Cable Clips
United-Carr Ltd
A. Raymond, Manufacturer de Boutons et D'articles Metalliques

Cable and Pipe Fasteners
United-Carr Ltd

Cable Straps
Fastex (Division of ITW Ltd)

Captive Fasteners for Sheet Metal
P.S.M. Fasteners Ltd

Carriage Bolts and Nuts
The Bolt & Nut Co (Tipton) Ltd
Hasselfors (UK) Ltd
Benjamin Priest & Sons Ltd

Castor Pegs
Aston Screw & Rivet Co Ltd

Circlips and Retaining Rings
Adams & Hann Ltd
The Allthread Group
Balcombe Engineering Ltd
Borstlap b.v.
John Bradley & Co Ltd
N. Bulte Kg
Circlex Ltd
Firth Cleveland Fastenings Ltd
Hasselfors (UK) Ltd
Jesse Haywood & Co Ltd
Londesborough Oil & Marine Engineering Ltd
Metric Screw (Fastenings) Ltd
Morlock Industries Ltd
Prestwich Parker Ltd
Springfix Ltd
The Walford Manufacturing Co Ltd
Wellworthy Ltd

Clamping Discs
Teckentrup KG

Claw-Type Clamping Discs
Teckentrup KG

Clevises and Pivot Fasteners
Adams & Hann Ltd
Aston Screw & Rivet Co Ltd
Balcombe Engineering Ltd
Barton Cold-form Ltd
The Bifurcated & Tubular Rivet Co Ltd
Black & Luff Ltd
John Bradley & Co Ltd
Clevedon Rivets & Tools Ltd
Cooper & Turner Ltd
Firth Cleveland Fastenings Ltd
Hasselfors (UK) Ltd
Jesse Haywood & Co Ltd
Metric Screws (Fastenings) Ltd
G. Pearson & W.P.Beck Ltd
Springfix Ltd
Tucker Industries Pty Ltd
The Walford Manufacturing Co Ltd

Clevis Pins
Deltight Industries Ltd

Clinch Nuts (Anchor Nuts)
Barton Cold-form Ltd

Coach Screws
The Bolt & Nut Co (Tipton) Ltd
GKN Screws & Fasteners Ltd (GKN Wood Screws)
Hasselfors (UK) Ltd

Cold Forged Nuts
Crane/Midland Screw Co Ltd

Cold Formed Steel Nuts
Industrial Fasteners Ltd

Colliery Fasteners
George Cooper (Sheffield) Ltd

Compression Rings
Teckentrup KG

Compression Spring Washers
Industrial Trading Co Ltd

Connecting Rods
Springfix Ltd

Corrosion Resistant Fastenings
Southern Marine Fastening Co Ltd

Corrugated Fasteners
Spotnails Ltd

Cotter Pins
The Allthread Group

Cotter Pins, Stainless Steel
Everbright Fasteners Ltd

Couplers for Tube Scaffolds
Van Thiel United BV

Cover Plates and Plugs
A. Raymond, Manufacturer de Boutons et D'articles Metalliques

Cup and Countersunk Bolts and Nuts
Benjamin Priest & Sons Ltd

Cup Washers
GKN Screws & Fasteners Ltd (GKN Wood Screws)

Decorative Caps
John Bradley & Co Ltd

Dial Indicating Torque Wrenches
M H H Engineering Co Ltd (Tools)

Drive Fasteners
Fastex (Division of ITW Ltd)

Electrical Tags and Terminals
Tucker Fasteners Ltd

Elevator Bucket Bolts
Isaac Jackson (Fasteners) Ltd

Expansion Bolts
Van Thiel United BV

Fork Head Assemblies
Springfix Ltd

Free Spinning, Self-Locking Fasteners
Glynwed Fastenings

Eyelets
Adams & Hann Ltd
The Allthread Group
A. Raymond Manufacturer de Boutons et D'articles Metalliques
Tucker Fasteners Ltd
Tucker Industries Pty Ltd

Fabric Fasteners
Adams & Hann Ltd
Balcombe Engineering Ltd
A. Raymond Manufacturer de Boutons et D'articles Metalliques
Tucker Fasteners Ltd
Tucker Industries Pty Ltd

Fabco Roofing Fasteners
Deltight Industries Ltd

Flange Nuts
Maclean – Fogg Lock Nut Co

Flanges
Entwistle (Oldham) Ltd

Flat Washers, Stainless Steel
Everbright Fasteners Ltd

Fluid Power Couplings
GKN Fasteners & Distributors Ltd

Fork Head Assemblies
Springfix Ltd

Free Spinning, Self-Locking Fasteners
Glynwed Fastenings

Friction Bushes
United-Carr Ltd

Hand Riveting Tools
Gesipa Blindniettechnik GmbH

Hank Bushes Stainless Steel
Everbright Fasteners Ltd

Heli-Coil Inserts
Miller Bridges Fastenings, Glynwed Distribution Limited

Hexagon Bolts and Set Screws
Carbo Engineering Co Ltd

Hexagon Socket Screws
GKN Screws & Fastener Ltd
(GKN Socket Screws)

Hexagon Wrenches
Unbrako Distribution

High Strength Bolts
Cooper & Turner Ltd
R. S. Rowlands Ltd

High Tensile Bolts and Nuts
Bauer & Schaurte
Crane's Screw & Colgryp Castor Co Ltd
Glynwed Fastenings
The Allthread Group

High Tensile Bolts and Setscrews
Essanbee Products Ltd

High Tensile Steel Hexagon Bolts
Industrial Fasteners Ltd

High-Tensile Hexagon Head Bolts and Screws
H. K. Westendorff

High-Tensile Hexagonal Socket-Head Bolts
H. K. Westendorff

Hook Bolts
J. E. Woodall & Co Ltd
Yarwood Ingram Co, Glynwed Screws & Fasteners Ltd

Hose Clamps
Firth Cleveland Fastenings Ltd

Hose Clips
Carbo Engineering Co Ltd

Hose Clips, Stainless Steel
Everbright Fasteners Ltd

Huckrimp Fastening System
Thomas William Lench Ltd

Industrial Belt Fasteners
Isaac Jackson (Fasteners) Ltd

Inglefield Clips
Londesborough Oil & Marine Engineering Ltd

Inserts for Plastics
P.S.M. Fasteners Ltd

Keys and Keyways
Adams & Hann Ltd
Borstlap b.v.
Metric Screws (Fastenings) Ltd

Knob Clips
United-Carr Ltd

Latches and Strikes
Fastex A Division of ITW Ltd

Lens Clips
Fastex A Division of ITW Ltd

Lester Star Self-Locking Nut
Thomas William Lench Ltd

Linkage Clips
Fastex A Division of ITW Ltd
United-Carr Ltd

Linkage Systems
Springfix Ltd

Load-Spreading
Bighead Bonding Fasteners Ltd

Lock Washers
Teckentrup KG

Locking Screws
Maclean-Fogg Lock Nut Co

Locknuts, Stainless Steel
Everbright Fasteners Ltd

Lokut Nuts
Fastex A Division of ITW Ltd

Lock Nuts and Nut Retainers
Peter Abbott & Co Ltd
Adams & Hann Ltd
The Allthread Group
Armstrong Fasteners Ltd
Avdel Ltd
Balcombe Engineering Ltd
Carbo Engineering Co Ltd
Firth Cleveland Fastenings Ltd
Gesipa Blindniet technik GmbH
Borstlap b.v.
T. Chatani & Co Ltd
Firth Cleveland Fastenings Ltd
(Self-Locking Nut Division)

George Cooper (Sheffield) Ltd
Deltight Industries Ltd
T. H. Dilkes & Co Ltd
Entwistle (Oldham) Ltd
GKN Fasteners & Hardware Distributors Ltd
Hasselfors (UK) Ltd
Industrial Fasteners Ltd
Industrial Trading Co Ltd
Thomas William Lench Ltd
MHH Engineering Co Ltd(Tools)
Maclean-Fogg Lock Nut Company
Macnays Ltd
Metric Screws (Fastenings) Ltd
Miller Bridges Fastenings, Glynwed Distribution Ltd
Rex Nichols & Co Ltd
Nylon & Alloys
A. Raymond Manufacturer de Boutons et D'articles Metalliques
R. S. Rowlands Ltd
SpS Hitek
Sandiacre Screw Co Ltd
Frank Stacey (Glynwed Distribution) Ltd
Suko-Sim GmbH
T. R. Fastenings Ltd
Tucker Industries Pty Ltd
Unbraco Distribution
The Walford Manufacturing Co Ltd

Machine Screws Nuts and Bolts

Peter Abbott & Co Ltd
Adams & Hann Ltd
The Allthread Group
Armstrong Fasteners Ltd
Aston Screw & Rivet Co Ltd
Balcombe Engineering Ltd
H. J. Barlow & Co Ltd
The Bolt & Nut Co (Tipton) Ltd
Borstlap b.v.
John Bradley & Co Ltd
Carbo Engineering Co Ltd
George Cooper (Sheffield) Ltd
Cooper & Turner Ltd
Crane's Screw & Colgryp Castor Co Ltd
Deltight Industries Ltd
T. H. Dilkes & Co Ltd
Thomas Eaves Limited
Embassy Machine & Tool Co Ltd (Machinery Suppliers)
Essanbee Products Ltd
E. J. Francois Ltd
GKN Fasteners & Hardware Distributors Ltd
Thos. Haddon & Stokes Ltd
Hasselfors (UK) Ltd
Hollinwood Screw Ltd
Industrial Fasteners Ltd

Isaac Jackson (Fasteners) Ltd
Karl Limbach & Cie Kg
Thomas William Lench Ltd
MHH Engineering Co Ltd (Tools)
Macnays Ltd
Metal Products (Cork) Ltd
Metric Screws (Fastenings) Ltd
Miller Bridges Fastenings, Glynwed Distribution Ltd
Rex Nichols & Co Ltd
Nylon & Alloys
The Ormond Engineering Co Ltd
Plastic Screws Ltd
Prestwich Parker Ltd
R. S. Rowlands Ltd
Sandiacre Screw Co Ltd
Wilhelm Schumacher
Southern Marine Fastening Co Ltd
Frank Stacey (Glynwed Distribution) Ltd
Standard Pressed Steel International
T. R. Fastenings Ltd
Unbrako Distribution
Van Thiel United BV
The Walford Manufacturing Co Ltd
H. K. Westendorff

Machine Screws, Nuts and Bolts, Stainless Steel

Everbright Fasteners Ltd

Magnetic Catches

Fastex A Division of ITW Ltd

Masonry Anchors

Ramset Fasteners Ltd
Tucker Industries Pty Ltd

Metric Fasteners

Rex Nichols & Co Ltd
The Walford Manufacturing Co Ltd

Metric Screws, Nuts and Washers in Nylon

Plastic Screws Ltd

Metric and Imperial High Tensile Fastenings

Miller Bridges Fastenings, Glynwed Distribution Ltd

Moulding Clips

Fastex A Division of ITW Ltd

Nails (including Hammer-Driven Pins, etc)

Adams & Hann Ltd
The Allthread Group
Balcombe Engineering Ltd
Ernst Bierbach KG
The British United Shoe Machinery Co Ltd
T. H. Dilkes & Co Ltd
Embassy Machine & Tool Co Ltd (Machinery Supplies)
GKN Fasteners & Hardware Distributors Ltd
Metric Screws (Fastenings) Ltd
Rex Nichols & Co Ltd
Prestwich Parker Ltd
Ramset Fasteners Ltd
Spotnails Ltd
Tower Manufacturing, Glynwed Screws & Fastenings Ltd
The Walford Manufacturing Co Ltd
Whitfield Hodgsons & Brough Ltd

Nail Type Fasteners

Adams & Hann Ltd
Balcombe Engineering Ltd
Ernst Bierbach KG
Borstlap b.v.
The British United Shoe Machinery Co Ltd
Metric Screws (Fastenings) Ltd
Spotnails Ltd
Frank Stacey (Glynwed Distribution) Ltd
T. Fastenings Ltd
Whitfield Hodgsons & Brough Ltd

Nut Runners, Pneumatic

Thor Tools Ltd

Paint Clearing Screws

Glynwed Fastenings

Panel Clamps

Camloc Industrial Fixings (UK) Ltd

Patented and Special Design Fasteners

Peter Abbott & Co Ltd
Adams & Hann Ltd
Avdel Ltd
Balcombe Engineering Ltd
Barton Cold-form Ltd
Bauer & Schuarte
The Bifurcated and Tubular Rivet Co Ltd
Bighead Bonding Fasteners Ltd
Black & Luff Ltd
John Bradley & Co Ltd
Borstlap b.v.

H. & H. Brugmann
N. Bulte KG
Clevedon Rivets & Tools Ltd
Cooper & Turner Ltd
Dzus Fastener Europe Ltd
GKN Fasteners & Hardware Distributors Ltd
Fastex A Division of ITW Ltd
E. J. Francois Ltd
Firth Cleveland Fastenings Ltd
Gesipa Blindniettechnik GmbH
Glynwed Fastenings
Jesse Haywood & Co Ltd
Heyco Manufacturing Co Ltd
Instrument Screw Co Ltd
Karl Limbach & Cie KG
Londesborough Oil & Marine Engineering Ltd
Thomas William Lench Ltd
Metric Screws (Fastenings) Ltd
Nylon & Alloys
A. Raymond Manufacture de Boutons et D'articles Metalliques
R. S. Rowlands Ltd
Wilhelm Schumacher
Springfix Ltd
Frank Stacey (Glynwed Distribution) Ltd
Suko-Sim GmbH
Tucker Fasteners Ltd
Tucker Industries Pty Ltd
Unbrako Distribution
United-Carr Ltd
Yarwood Ingram Co, Glynwed Screws & Fastenings Ltd
The Walford Manufacturing Co Ltd

Pin Type Fasteners

Adams & Hann Ltd
Avdel Ltd
Balcombe Engineers Ltd
Borstlap b.v.
Carbo Engineering Co Ltd
GKN Fastener & Hardware Distributors Ltd
Metric Screws (Fastenings) Ltd
A. Raymond Manufacturer de Boutons et D'articles Metalliques
Spotnails Ltd
Tucker Industries Pty Ltd
Unbrako Distribution
The Walford Manufacturing Co Ltd
Whitfield Hodgsons & Brough Ltd

Pipe and Cable Clamps

Circlex Limited

Pipe Fittings (Screwed)

Entwistle (Oldham) Ltd

Pipe and Wire Fasteners

Fastex A Division of ITW Ltd

Plastics

Instrument Screw Co Ltd
Tucker Fasteners
United-Carr Ltd

Plastic Quick-Release Fasteners

Dzus Fasteners Europe Ltd

Plugs and Buttons

United-Carr Ltd

Pneumatic/Hydraulic Riveting Tools

Gesipa Blindniettechnik GmbH

Pneumatic/Mechanical Riveting Tools

Gesipa Blindniettechnik GmbH

Pre-Assembled Screws and Washers

Glynwed Fastenings

Pressure Plugs Gas and NPTF

Unbrako Distribution

Prevailing Torque Fasteners

Glynwed Fastenings

Push Fasteners

United Carr Ltd

Quarter-Turn Fastener

United-Carr Ltd

Quick-Clamping Discs

Teckentrup KG

Quick-Clamping Nuts

Teckentrup KG

Quick-Operating Fasteners

Peter Abbott & Co Ltd
Adams & Hann Ltd
Avdel Ltd
Balcombe Engineering Ltd
The Bifurcated & Tubular Rivet Co Ltd
Borstlap b.v.
Camloc Industrial Fixings (UK) Ltd
Circlex Ltd
Dzus Fastener Europe Ltd
Fastex A Division of ITW Ltd
Firth Cleveland Fastenings Ltd
Instrument Screw Co Ltd

Londesborough Oil & Marine Engineering Ltd
Metric Screws (Fastenings) Ltd
A. Raymond Manufacturer de Boutons et D'articles Metalliques
Springfix Ltd
Teckentrup KG
Unbrako Distribution
United-Carr Ltd

Railway Fasteners

George Cooper (Sheffield) Ltd

Retaining Rings

Fastex A Division of ITW Ltd
Teckentrup KG

Rivets

Peter Abbott & Co Ltd
The Allthread Group
Adams & Hann Ltd
Aston Screw & Rivet Co Ltd
Avdel Ltd
Barton Cold-form Ltd
The Bifurcated & Tubular Rivet Co Ltd
Black & Luff Ltd
Gesipa Blindniettechnik GmbH
Borstlap n.v.
John Bradley & Co Ltd
Clevedon Rivets & Tools Ltd
Cooper & Turner Ltd
Deltight Industries Ltd
Embassy Machine & Tool Co Ltd (Machinery Suppliers)
Fastex A Division of ITW Ltd
GKN Fasteners & Hardware Distributors Ltd
Jesse Haywood & Co Ltd
Metric Screws (Fastenings) Ltd
G. Pearson & W.P.Beck Ltd
A. Raymond Manufacturer de Boutons et D'articles Metalliques
Frank Stacey (Glynwed Distribution) Ltd
Tower Manufacturing, Glynwed Screws & Fastenings Ltd
Tucker Industries Pty Ltd
Van Thiel United BV
The Walford Manufacturing Co Ltd

Rivet Bushes

Deltight Industries Ltd
E. J. Francois Ltd
T. R. Fastenings Ltd
GKN Screws & Fasteners Ltd (GKN Socket Screws)

Riveted Discs

Teckentrupp KG

Rivet Setting Machines

The Bifurcated & Tubular Rivet Co Ltd

Rivet Squeezers

Thor Tools Ltd

Rod Ends

Springfix Ltd

Roofing Bolts

J. E. Woodall & Co Ltd

Roll Pins

Firth Cleveland Fastenings Ltd

Screwdrivers

Thor Tools Ltd

Screwdriving Bits (Tools)

Gardner-Denver (UK) Ltd

Screw Caps

GKN Screws & Fasteners Ltd (GKN Wood Screws)

Screwed In Nuts to DIN 7965

H. & H. Brugmann

Screw/Washer Pre-Assembled

Bauer & Schaurte

Screws with Pre-Assembled Washers (Sems)

GKN Screws & Fasteners Ltd (GKN Self-Tapping Screws)

Sealing Rings, Gaskets

Teckentrup KG

Self-Drilling Masonry Anchors

Ramset Fasteners Ltd

Self-Locking Nuts

The Allthread Group
Crane's Screw & Colgryp Castor Co Ltd
Deltight Industries Ltd
Miller Bridges Fastenings, Glynwed Distribution Ltd
Prestwich Parker Ltd

Self-Locking Nuts, Stainless Steel

Everbright Fasteners Ltd

Self-Piercing Nuts

Avdel Ltd

Self-Tapping Screws

Adams & Hann Ltd
The Allthread Group
Balcombe Engineering Ltd
Borstlap b.v.
Carbo Engineering Co Ltd
Crane's Screw & Colgryp Castor Co Ltd
Deltight Industries Ltd
T. H. Dilkes & Co Ltd
Embassy Machine & Tool Co Ltd (Machinery Suppliers)
Essanbee Products Ltd
E. J. Francois Ltd
GKN Fasteners & Hardware Distributors Ltd
Hasselfors (UK) Ltd
Industrial Fasteners Ltd
MHH Engineering Co Ltd (Tools)
Macnays Ltd
Metric Screws (Fastenings) Ltd
Miller Bridges Fastenings, Glynwed Distributors Ltd
Rex Nichols & Co Ltd
The Ormond Engineering Co Ltd
Prestwich Parker Ltd
A. Raymond Manufacturer de Boutons et D'articles Metalliques
R. S. Rowlands Ltd
Sandiacre Screw Co Ltd
Wilhelm Schumacher
Southern Marine Fastening Co Ltd
Frank Stacey (Glynwed Distribution) Ltd
Stenman Holland B.V.
T. R. Fastenings Ltd
The Walford Manufacturing Co Ltd
H. K. Westendorff

Self-Tapping Screws, Stainless Steel

Everbright Fasteners Ltd

Self-Retaining and Spring Fasteners

Adams & Hann Ltd
The Allthread Group
Avdel Ltd
Balcombe Engineering Ltd
Embassy Machine & Tool Co Ltd (Machinery Suppliers)
Carbo Engineering Co Ltd
GKN Fasteners & Hardware Distributors Ltd
Firth Cleveland Fastenings Ltd
Metric Screws (Fastenings) Ltd
Miller Bridges Fastenings, Glynwed Distributors Ltd
A. Raymond Manufacturer de Boutons et D'articles Metalliques
Springfix Ltd
Frank Stacey (Glynwed Distributors) Ltd
Tucker Fasteners Ltd
Tucker Industries Pty Ltd
Unbrako Distribution
H. K. Westendorff

Sheet Metal Fixings

Standard Pressed Steel International

Slotted Screws

H. K. Westendorff

Snap Nuts

Teckentrup KG

Socket Screws

Carbo Engineering Co Ltd
Crane/Midland Screw Co Ltd
Crane's Screw & Colgryp Castor Co Ltd
Industrial Fasteners Ltd
Miller Bridges Fastenings, Glynwed Distributors Ltd
Spensall Engineering Co Ltd
Supacraft Engineering Products Ltd
T. R. Fastenings Ltd
Unbrako Distribution

Socket Screws, Stainless Steel
Everbright TFasteners Ltd

Sockets (Tools)
Gardner-Denver (UK) Ltd

Special Cold-Formed Parts
The Bifurcated & Tubular Rivet Co Ltd
Black & Luff Ltd
Clevedon Rivets & Tools Ltd

Special Bolts and Nuts
Benjamin Priest & Sons Ltd

Special Fasteners
Ernst Bierbach KG
Everbright Fasteners Ltd
H. K. Westendorff

Special Nails
Ernst Bierbach KG

Special Nuts
Maclean – Fogg Lock Nut Co

Special Purpose Fasteners for Automobile, Domestic Appliance and TV Industries
United-Carr Ltd

Special Screws
Ernst Bierbach KG
Crane's Screw & Colgryp Co Ltd

Spring Nuts
Teckentrup KG

Spring and Lock Washers
Adams & Hann Ltd
The Allthread Group
Balcombe Engineering Ltd
Borstlap b.v.
John Bradley & Co Ltd
Carbo Engineering Co Ltd
Deltight Industries Ltd
T. H. Dilkes & Co Ltd
Essanbee Products Ltd
GKN Fasteners & Hardware Distributors Ltd
Thos. Haddon & Stokes Ltd
Hasselfors (UK) Ltd
Industrial Fasteners Ltd
Industrial Trading Co Ltd
Macnays Ltd

Metric Screws (Fastenings) Ltd
Morlock Industries Ltd
Miller Bridges Fastenings,
Glynwed Distribution Ltd
Rex Nichols & Co Ltd
Prestwich Parker Ltd
R. S. Rowlands Ltd
Sandiacre Screw Co Ltd
Johann Schürholz
Frank Stacey (Glynwed Distribution) Ltd
T. R. Fastenings Ltd.
Vossloh – Werke GmbH
The Walford Manufacturing Co Ltd

Spring and Lock Washers, Stainless Steel

Everbright Fasteners Ltd

Spring Pins

Camloc Industrial Fixings (UK) Ltd

Spring Pins, Stainless Steel

Everbright Fasteners Ltd

Spring Washers

Teckentrup KG

Square Weld Nuts

Deltight Industries Ltd

Stainless Steel Bolts, Nuts and Turned Parts

Crew & Sons Ltd

Stainless Steel Fastenings

The Allthread Group
Deltight Industries Ltd
Miller Bridges Fastenings,
Glynwed Distribution Ltd

Stainless Steel Screws, Nuts and Bolts

Essanbee Products Ltd

Steel Collars

Unbrako Distribution

Staples

Adams & Hann Ltd
Embassy Machine Tool Co Ltd
(Machinery Suppliers)
Spotnails Ltd
Tower Manufacturing,
Glynwed Screw & Fastenings Ltd

Stiff Nuts

GKN Screws & Fasteners Ltd
(GKN Socket Screws)

Strap Fasteners

Adams & Hann Ltd
T. H. Dilkes & Co Ltd
Heyco Manufacturing Co Ltd
A. Raymond Manufacturer de Boutons et
D'articles Metalliques
Tucker Fasteners Ltd
Tucker Industries Pty Ltd
United-Carr Ltd
Yarwood Ingram Company,
Glynwed Screws & Fastenings Ltd

Studs – Double-Ended

Yarwood Ingram Company,
Glynwed Screws & Fastenings Ltd

Studding, Stainless Steel

Everbright Fasteners Ltd

Studs, Studding and Stud Bolts

The Allthread Group
Entwistle (Oldham) Ltd
Hollinwood Screw Ltd
Thomas William Lench Ltd
Spensall Engineering Co Ltd
The Walford Manufacturing Co Ltd
H. K. Westendorff

Stud Welding Equipment

Tucker Fasteners Ltd

Stud Welding Systems

Tucker Industries Pty Ltd

Surgical Bone Screws

Adams & Hann Ltd

Tacks

The British Shoe Machinery Co Ltd

T-Bolts

Carbo Engineering Co Ltd

Tension Latches

Camloc Industrial Fixings (UK) Ltd

Threaded Bars and Rods

H. K. Westendorff

Threaded Inserts

Tappex Thread Inserts Ltd

Threaded Inserts for Plastic Materials

Armstrong Fastenings Ltd

Thread Locking Systems

Peter Abbott & Co Ltd
Adams & Hann Ltd
Balcombe Engineering Ltd
Bauer & Schuarte
Borstlap b.v.
N. Bulte KG
Industrial Fasteners Ltd
Indstrument Screw Co Ltd
MHH Engineering Co Ltd (Tools)
Metric Screws (Fastenings) Ltd
P. S. M. Fasteners Ltd
Wilhelm Schumacher
Suko-Sim GmbH
T. R. Fastenings Ltd
Tucker Industries Pty Ltd
Unbrako Distribution
The Walford Manufacturing Co Ltd

Thread Rolling

Aston Screw & Rivet Co Ltd

Threaded Studs

Hasselfors (UK) Ltd

H. K. Westendorff

Thread-Forming Screws

Crane's Screw Colgryp Castor Co Ltd
Stenman Holland BV

Thread Sealing System

P. S. M. Fasteners Ltd

Thumb Screws

GKN Screws & Fasteners Ltd
(GKN Socket Screws)

Tie Rods

G. Pearson & W. P. Beck Ltd
J. E. Woodall & Co Ltd

Toggle Latches

Dzus Fasteners Europe Ltd

Tool Drive Bits

Peter Abbott & Co Ltd

Tools, Fastening

Gardner-Denver (UK) Ltd

Tooling

Thor Tools Ltd

Toothed Lockwashers

Teckentrup KG

Torque-Limiting Screwdrivers

H MHH Engineering Co Ltd (Tools)

Torque Slipping Wrenches

MHH Engineering Co Ltd (Tools)

Trim Fasteners

United-Carr Limited

Trim Pad Clips

A. Raymond, Manufacturer de Boutons et
D'articles Metalliques

Turnbuckles

Hasselfors (UK) Ltd

'U' Bolts

J. E. Woodall & Co Ltd
Yarwood Ingram Co
Glynwed Screws & Fasteners Ltd

Universal Wrenches (Tools)

Gardner-Denver (UK) Ltd

Vibration Proof Bolts

Avdel Ltd

Washers

Essanbee Products Ltd
Entwistle (Oldham) Ltd
Sandiacre Screw Co Ltd
The Walford Manufacturing Co Ltd

Washers, Metal

Charles (Wednesbury) Ltd

Washers and Shims

Teckentrup KG

Wedglok Fasteners

Adams & Hann Ltd

Weld Nuts

Bauer & Schaurte
Maclean – Fogg Lock Nut Co
G. Pearson & W.P.Beck Ltd

Wing Nuts

Crane/Midland Screw Co Ltd
GKN Screws & Fasteners Ltd
(GKN Socket Screws)

Wire Thread Inserts

Peter Abbott & Co Ltd
Armstrong Fastenings Ltd

Woodscrews

The Allthread Group
Ernst Bierbach KG
Essanbee Products Ltd
GKN Screws & Fasteners Ltd
(GKN Wood Screws)
Hasselfors (UK) Ltd
Miller Bridges Fastenings,
Glynwed Distribution Ltd
The Walford Manufacturing Co Ltd

Woodscrews, Stainless Steel

Everbright Fasteners Ltd

Wrenches, Pneumatic

Thor Tools Ltd

Sub-section b(3)

Alphabetical List of Manufacturers with addresses, telephone numbers telegram addresses of Head Office, Works and Branches

Peter Abbott & Co Ltd., Bridge Close Industrial Estate, Romford, Essex.
Telephone : Romford 25111, Telex 897072, Grams : Locknuts Romford.
Branches : Peter Abbott & Co (Central) Ltd.,
Morley Street, Daybrook, Nottingham.
Telephone : 0602–264222, Telex 377216, Grams : Abbottco Nott'm.
Peter Abbott & Co (Northern) Ltd.,
Beza St Industrial Estate, Hunslett, Leeds.
Telephone : 0532–700681, Telex 557145, Grams : Abbottco Leeds.
Peter Abbott & Co (Western) Ltd.,
Manor Road, Marston Trading Estate, Frome, Somerset.

Adams & Hann Ltd., London Road, Barking, Essex.
Telephone : 01–594 2666, Telex 21772, Grams: Adan Barking.
Branches : Trafalgar Close, Chandler's Ford Industrial Estate, Eastleigh, Hants.
Telephone : Chandler's Ford 69828/30
Fernhill House, Fernhill, Horley, Surrey.
Telephone : Horley 6778, Telex 477016
26–30 John Street, Luton, Beds.
Telephone : Luton 23312

The Allthread Group, Allthread House, Wellington Road, High Wycombe, Bucks.
Telephone : 0494 36411, Telex 837168
Branches : Allthread International Ltd., 39 Queens Road, High Wycombe, Bucks.
Telephone : (0494) 36411, Telex 837575.
Allthread Distributors Scotland Company, 10 Loanbank Quadrant, Govan, Glasgow; Telephone : (041–445) 3277. Telex 778302

Allthread Distributors North East Company, Unit 12, Brough Park Trading Estate, The Fossway, Newcastle NE6 2XJ. Telephone : (0632) 658261, Telex 53442
Allthread Distributors Midland Company, Unit 6, Dunstall Trading Estate, Gorsebrook Road, Wolverhampton. Telephone : (0902) 26427-9
Allthread Distributors South West Company, Unit 4A, Severnside Trading Estate, Bristol. Telephone : Avonmouth (02752) 7231
Allthread Distributors Southern Company, Ferndown Industrial Estate, Ferndown, Dorset. Telephone : (020 17) 4045
Allthread Distributors Northern Company, Unit 1, Lowmills Road, Leeds 12. Telephone : (0532) 790066, Telex 556469
Allthread Distributors Central Company, Wellington Road, Cressex Industrial Estate, High Wycombe, Bucks. Telephone : (0494) 36411, Telex 837168
Allthread Distributors Anglia Company, South Street, Braintree, Essex. Telephone : Braintree 22766/22716
Allthread Distributors South East Company, Unit 3, Vale Road, Tonbridge,Kent. Telephone : (073 22) 65188 Telex 95410
Allthread Distributors Southern Company, Unit D, Southampton Airport, Southampton, Hants Telephone : (042 126) 2547-2295-7121. Telex 477406

Armstrong Fastenings Ltd., Gibson Lane, Melton, North Ferriby, North NHumberside.
Telephone : 0482 633311, Telex 52164, Grams: Armstrongs North Ferriby

Aston Screw & Rivet Co.Ltd., 50 Cardigan Street, Birmingham B4 7RT.
Telephone : 021–359 3177

Avdel Ltd., Black Fen Road, Welwyn Garden City, Herts.
Telephone : W.G.28161, Telex 24254, Grams: Avidev Welwyn Garden

Branches : Avdel Ltd., Hardwick Grange, Woolston, Warrington, Lancs, WA1 4RF
Telephone : 0925 811243
Avdel Pty Ltd., 18 Ellis Street, South Yarra 3141, Victoria, Australia.
Telephone : 24 7556, Telex 207.71 32377, Grams : Avdaust
Avdel Ltd., 2782 Slough Street, Mississauga, Ontario, Canada.
Telephone : 416 677 9400, Telex 207.21.6968757, Grams : Adcan
Avdel S.A., 66 Rue David D'Angers, Paris 19, France.
Telephone : 202 39 15, Telex 2022.23690, Grams : Avparis
Avdel GmbH, Klusriede 14-16, 3012 Langenhagen, Hanover, Germany.
Telephone : 0511 73 60 23, Telex 203.924600, Grams : Aviland
Avdel (India) Private Ltd., 409 Himalaya House, Palton Road, Bombay 1,India.
Telephone : 26 72 21. Telex 011.2802, Grams : Avdelind
Avdel S.p.A., Via G. Di Vittorio 307/10, 20099 Sesto S.Giovanni,Milan,Italy.
Telephone : 24 31 55, Telex 2023.35174, Grams : Avitalia
Avdel K.K., Shin-Nakanoshima Building, 49 Hinoue-Cho,Kita-ku Osaka, Japan
Telephone : 363 1876, Telex 207.726.3501 Grams : Expanex Osaka
Avdel S.A. De. C.V.,Ave, Urbina No.71, Parque Industrial Naucalpan, Naucalpan De Juarez, Edo. De Mexico. Telephone
Telephone : 576 70 43, Grams : Avimex
Avdel Burnside (Pty) Ltd., P.O..Box 2276, Pretoria, South Africa
Telephone : 492110, Grams Avoria
Avdel Corp., 10 Henry Street, Teterboro,New Jersey.U.S.A.

Telephone : 201 288 0500, Telex 207.255.7109906111, Grams : Aviboro
Avdel International, 6 Rue Du Simplon, 1207 Geneva, Switzerland.
Telephone : 022 395960, Telex 2064 22419

Balcombe Engineering Ltd., (Supplier), 19 Compton Terrace, London, N.1.
Telephone : 01–226 0292/3.
Works : Beeching Road, Bexhill-on-Sea, Sussex. Telephone : 0424-21-7636

H. J. Barlow & Co Ltd., Mounts Works, Bridge Street, Wednesbury, Staffs.
Telephone : 021-556 1910, Telex 33246
Branches : H. J. Barlow (Ogden) Ltd., Telephone : Watford 33713, Telex 923487
H. J. Barlow (Smedley) Ltd., Telephone : Poole 5741
John Bullough Ltd., Telephone : Atherton 4151, Telex 67693

Barton Cold-form Ltd., Kidderminster Road, Droitwich, Worcestershire.
Telephone : Droitwich 2021 (3 lines)

Bauer & Schaurte, Further Strasse 24-26, 404 Neuss-Rhein, Fed.Rep.Germany.
Telephone : 02101/5221, Telex 08517861, Grams : Bauschan Neuss

Ernst Bierbach KG, D–475 Unna, Postfach 39.
Telephone : 02303/1931, Telex 8229288, Grams : Nadelbierbach

The Bifurcated & Tubular Rivet Co Ltd., P.O.Box 2, Mandeville Road, Aylesbury, Bucks HP21 8AB
Telephone : 0296 5911, Telex 83210
Branches : Bifurcated Engineering GmbH, 2000 Norderstedt 3, Gutenbergring 31A,
Postfach 1109, West Germany. Telephone :(0411)5232161, Telex 02174247
The South African Rivet Manufacturing (Pty) Co Ltd., P.O.Box 12175,
Jacobs 4026, Natal, South Africa. Telephone : 476806, Grams:Tubular Durban

Bighead Bonding Fasteners Limited, Southern Road, Aylesbury, Bucks.
Telephone : Aylesbury 81663 (STD 0296)

Black & Luff Ltd., Birmingham Factory Centre, Pershore Road South, Birmingham B30 3HQ
Telephone : 021–459 2281, Telex 338798

The Bolt & Nut Co. (Tipton) Ltd., Park Lane East, Tipton, West Midlands DY4 8RF.
Telephone : 021 557 4731

Borstlap b.v., Zevenheuvelenweg 44, Tilburg, Holland.
Telephone : 013-678445, Telex 52155

John Bradley & Co Ltd., Hollow Mead, Birmingham B1 1QU.
Telephone : 021–643 4781, Grams : Bribolt

Brewer Rapid Riveting Systems, 10 Lagoon Road, Lilliput, Poole, Dorset.
Telephone : Canford Cliffs 708170

The British United Shoe Machinery Co Ltd., P.O.Box 88, Belgrave Road, Leicester.
Telephone : Leics.61551, Telex 34445, Grams : Prominent Leicester.
Branches : Whitfield, Hodgsons & Brough Ltd., Lawson St,Kettering,Northants.
Telephone : Kett.4491/2, Grams : Whitfields Ltd,Kettering.

H. & H. Brugmann jun., Schraubenfabrik, 2o53 Schwarzenbek, Grabauer Strasse 35.
Telephone : o4151/2o74, Telex o2-189 408

N. Bulte KG., D4710 Ludinghausen, Sendenerstrasse 14-16
Telephone : 02591/4025, Telex 089803, Grams : Bulte
Branches : F68000 Zimmerbach, Rue de Ecole

Camloc Industrial Fixings (UK) Ltd., 51 Highmeres Road, Leicester LE4 7LZ.
Telephone : Leicester 769371, Telex 34488.
Branches : Camloc GmbH, 6233 Kelkheim/Taunus, W.Germany.
Telephone : 61952001, Telex 10516
Camloc S.A.R.L., 60 Bis Rue du Depart, Enghien les Bains 95880
Telephone : 9896408, Telex 60974

Carbo Engineering Co Ltd., 20 Malthouse Lane, Washwood Heath, Birmingham B8 1SP,
Telephone : 021-327 6366, 4027, 4028, Telex 337912

Charles (Wednesbury) Ltd., Bridge Works, Wednesbury, Staffs.
Telephone : 021–556 1921, Telex 338083, Grams : Chamfer

T. Chatani & Co Ltd., P.O.Box Higashi 59, Osaka, Japan.
Telephone : 06–203 5331, Telex J63364
Branches : T. Chatani & Co Ltd., 4 Dusseldorf 1 Konigsallee 24, F.R.Germany
Telephone : (0211) 323924, Telex 8588207, Grams : Chatanico Dusseldorf
Chatani America Inc., 180 Madison Avenue, New York N.Y.10016,U.S.A.
Telephone : (212)532-5466,5437, Telex Chatani 224667, Grams : Terchat,N.Y.
Chatani Australia Pty Ltd.,P.O.Box 672, North Sydney, 2060 Australia.
Telephone : 922 1266, Telex AA 20799 Sydney, Grams : Chatanico,Sydney.

Circlex Limited, 451 London Road, Camberley, Surrey.
Telephone : Camberley 62461
Works : Station Approach, Fleet, Hampshire, Telephone : Fleet 21619

Clevedon Rivets & Tools Ltd., Reddicap Trading Estate, Sutton Coldfield, West Midlands B75 7DG
Telephone : 021–354 5238, Telex 339294

George Cooper (Sheffield) Ltd.., Sheffield Road, Tinsley, Sheffield S9 1RF.
Telephone : 41026, Telex 547092

Cooper & Turner Ltd., Vulcan Works, Vulcan Road, Sheffield S9 2FW.
Telephone : 43771, Telex 54607, Grams : Rivets Sheffield

Crane/Midland Screw Co Ltd., Floodgate Street, Birmingham B5 5SH.
Telephone : 021 772 3274. Telex 33472

Crane's Screw & Colgryp Castor Co Ltd., Floodgate Street, Birmingham B5 5SH
Telephone : 021 772 3274, Telex 33472

Crew & Sons Limited, Dibdale Street, Dudley, West Midlands D71 2SD
Telephone : Dudley 57231-5, Telex 339764, grams : Crewshyne Dudley

Deltight Industries Ltd., Wandle Way, Mitcham, Surrey CR4 4NB.
Telephone : 01–640 3261 (10 lines), Telex 946324
Branches : Solent Industrial Estate, Shamblehurst Lane, Hedge End,Southampton.
Telephone : 0489-2-3413, Telex 47284
Farrow Road, Rodney Way, Widford Industrial Estate,Chelmsford,Essex.
Telephone : 0245 60161, Telex 995173
Midland Deltight Ltd., Fryers Close, Walsall, WS3 2XQ.
Telephone : 0922-78471, Telex 338868
Whale Wharf, Littleton-on-Severn, Thornbury, Avon.
Telephone : 0454-416414
Bintcliffe Turner Ltd., Buckhurst Lane, Sevenoaks, Kent, TN13 1LY
Telephone : 53858, Telex 957125

T. H. Dilkes & Co.Ltd., 74 Lower Dartmouth Street, Bordesley Green, Birmingham B9 4NP
Telephone : 021 773 5451.
Branches : Old Wolverton Road, Old Wolverton, Milton Keynes MK12 5QA.
Telephone : Wolverton 315531
32/34 Midland Street, Glasgow G1 4PS.
Telephone : 041–248 6114

Dzus Fastener Europe Ltd., Farnham Trading Estate, Farnham,Surrey GU9 9PL
Telephone : Farnham 4422, Telex 858201, Grams : Adzusfast

Thomas Eaves Limited, 56-60 Holloway Head, Birmingham B1 1NP.
Telephone : 021-692 1481, Telex 339385

Embassy Machine & Tool Co Ltd., (Machinery Suppliers), Embatool Works, 104 High Street, London Colney, St. Albans, Herts AL2 1QL.
Telephone : Bowmansgreen 23461, Telex 21366, Grams : Embatool, St.Albans.

Entwistle (Oldham) Limited, Townfield Street, Oldham, Lancs.
Telephone : 061-624 9771, Telex 667029

Essanbee Products Ltd, 16 Hanbury Street, London, E.1.

Everbright Fasteners Ltd.,Stainless House, 4/6 Edwin Road, Twickenham, Middlesex TW1 4JN
Telephone : 01-891 0111, Grams : Everfast

Fastex A Division of ITW Ltd., Viables Estate, Basingstoke, Hampshire RG22 4BW.
Telephone:(0256)61151, Telex 847872
Branches : Fastex Division Der ITW –Ateco GmbH, 2 Norderstedt 1, Stormarnstrasse 43-49, Hamburg, West Germany, Telephone : 040 5252001, Telex 02 12912

Firth Cleveland Fastenings Ltd., Self-Locking Nut Division, Treforest Industrial Estate, Pontypridd, Mid Glamorgan, CF37 5YG.
Telephone : Treforest 2633, Telex 49624, Grams : Fircleve Pontypridd

E. J. Francois Ltd., 62-68 Rosebery Avenue, London EC1R 4RT.
Telephone : 01-837 9157/8, Grams : Nosamfran,London,E.C.1.

Gardner-Denver (U.K.) Ltd., Brick Knoll Park, Ashley Road, St.Albans, Herts, AL1 5UB
Telephone : St.Albans 65517, Telex 23150, Grams : Gardair St.Albans
Works : Apex Machine & Tool Co., Dayton, Ohio, U.S.A.

Gesipa Blindniettechnik GmbH, 6083 Walldorf bei Frankfurt (Main), Nordendstrasse 13-39, Germany. Telephone :(06105)5045-5049, Telex 4185720, Grams : Gesi D
Works : Gesipa Fasteners Limited, Dalton Lane, Keighley, West Yorkshire.
Telephone : 05352 7844, Telex 51429, Grams : Gesipa Keighley
Branches : Gesipa Blindnietvertriebsgellschaft mbH, Grundsteingasse 63, 1160 Vienna, Austria. Telephone : 432117, Grams : Gesipaniet Wien
Gesipa Do Brasil, Rua Pernambuco 400, Via Raposo Tavares, Km 18,5, Jardim Arpoador C.P.6.193, Sao Paulo, Brazil.
Telephone : 2115214
Gesipa Fasteners (Canada) Ltd., Toronto, Canada.
Gesipa Del Ecuador, S.A., Quito, Equador.
Gesipa AG, Zurich, Switzerland.

GKN Screws & Fasteners Ltd., (GKN Socket Screws), Alexander Socket Screws
Guns Lane, West Bromwich, West Midlands, B70 9HF Telephone : 021-553 3151, Telex 33145

Works : P.O.Box 76, Cranford Street, Smethwick, Warley,West Midlands B66 2TA.
Telephone : 021-558 1441, Telex 336511, Grams : Nettlefords,B'ham
Branches · 8 Gate Street,Lincoln's Inn Fields, London, WC2A 3HX.
Telephone : 01–831 6641, Telex 27814,
19 North Claremont Road, Glasgow, G3 7NR.
Telephone : 041-332 7884, Telex 77641
Netherton House, 23/29 Marsh Street. Bristol BS1 4AQ
Telephone : 0272 25173, Telex 44135
21 Ashton Lane, Sale, Cheshire M33 1WR
Telephone : 061-962 4586, Telex 66770

GKN Screws & Fasteners Limited, (GKN Self-Tapping Screws), P.O.Box 78, Grove Lane, Smethwick, Warley, West Midlands B66 2ST.
Telephone : 021-558 1441, Telex 336511, Grams : Nettlefolds, Birmingham.
Branches : 8 Gate Street, Lincoln's Inn Fields, London, WC2A 3HX
Telephone : 01-831 6641, Telex 27814
19 North Claremont Road, Glasgow, G3 7NR
Telephone : 041-332 7884, Telex 77641
Netherton House, 23/29 Marsh Street, Bristol BS1 4AQ
Telephone : 0272 25173, Telex 44135
21 Ashton Lane, Sale, Cheshire M33 1WR
Telephone : 061-962 4586, Telex 66770

GKN Screws & Fasteners Limited, (GKN Wood Screws),
P.O.Box 77, Heath Street, Smethwick, Warley, West Midlands B66 2RA
Telephone : 021-558 1441, Telex 336511, Grams: Nettlefolds Birmingham
Branches : Cranford Street, Smethwick, Warley, West Midlands.
Telephone : 021 558 1441, Telex 336511
8 Gate Street, Lincoln's Inn Fields, London, WC2A 3HX
Telephone : 01-831 6641, Telex 27814
19 North Claremont Road, Glasgow, G3 7NR
Telephone : 041-332 7884, Telex 77641
Netherton House, 23/29 Marsh Street, Bristol, BS1 4AQ
Telephone : 0272-25173, Telex 44135
21 Ashton Lane, Sale, Cheshire M33 1WR
Telephone : 061-962 4586, Telex 66770

GKN Fasteners & Hardware Distributors Ltd., 39 The Green, Banbury, Oxon
Telephone : 0295 53281
Branches : Alder Miles Druce Ltd., Beaconsfield Road, Hayes, Middlesex UB4 OLP
Telephone : 01-573 7766, Telex 267909
Branches at Bristol, Dartford, Dundee, Glasgow, Hebburn, Kidderminster, Leeds, Manchester, Nottingham, Southampton
B.A.R.Fasteners Ltd., Potters Lane, Wednesbury, West Midlands WS10 8BH
Telephone : 021 556 0951
Branch also at Newport, Gwent.
Davis & Timmins Ltd., Lichfield Road Trading Estate, Tamworth, Staffordshire B79 7TF. Telephone : 0827 60321
Branches also in Belfast, Birmingham, Blackburn, Bovey Tracey, Bridgend,

Bristol, Burgess Hill, Chandlers Ford, Doncaster, Dublin, Dudley, Glasgow, Leeds, Liverpool, Loughborough, Manchester, Newcastle, Reading, Saffron Walden, Swanley.

W. Galloway & Co Ltd., Chain Bridge Road, Blaydon-on-Tyne, Tyne and Wear NE21 5SS. Telephone : Blaydon 2121

Macnays Fasteners, P.O.Box 13, 48/50 West Street, Middlesbrough, Cleveland TS2 1LY. Telephone : Middlesbrough 48144

Merry & Co. (Manchester) Ltd., Keymer Street, Manchester M11 3HA. Telephone : 061-273 4505.

Millerservice Ltd., P.O.Box 19, Trading Estate, Slough, Berkshire SL1 4SG Telephone : Slough 25511, Telex 847062

Nettlefold Engineering Distributors Ltd., 2-20 Prices Lane, Reigate, Surrey RH2 8AJ
Telephone : Reigate 40181, Telex 946608
Branches also at Hayes, Hull and Wednesbury

Glynwed Fastenings, Midland Road, Darlaston, Staffordshire WS10 8JN.
Telephone : Works 021 526 2951, Sales 021 526 2895 Telex 33508.

Thomas Haddon & Stokes Ltd., High Street, Deritend, Birmingham, B12 OLW
Telephone : 772 2312, Telex 338071, Grams : Screwriv Birmingham

Hasselfors (U.K.) Ltd., Newby Road, Hazel Grove, Cheshire, SK7 5EB
Telephone : 061 483 7836, Telex 669710

Jesse Haywood & Co Ltd., Foundry Lane, Smethwick, Warley West Midlands B66 2LW
Telephone : 021 558 3027

Heyco Manufacturing Company Ltd., Uddens Trading Estate, Nr.Wimborne,Dorset BH21 7NL
Telephone : Ferndown 71411/2/3, Telex 41408, Grams: Heycoman Wimborne

Hollinwood Screw Ltd., Hawksley Industrial Estate, Manchester Road, Hollinwood, Oldham.
Telephone : 061 624 1487

Industrial Fasteners Limited, Hempsted Lane, Gloucester, GL2 6JB.
Telephone : 0452 25171, Telex 43101
Branches : Willow Walk, Sutton, Surrey, SM3 9QH
Telephone : 01-644 1251, Telex 28777
Treforest, Pontypridd, Glamorgan CF37 5BN
Telephone : Treforest 2321, Telex 497343
Arnhem Road, Newbury, Berkshire RG14 5RU
Telephone : 0635 41667, Telex 848375
Oldbury Road Industrial Estate, West Bromwich, Staffs.
Telephone : 021 553 6451, Telex 339417
21B Green Lane, Ashton-on-Mersey, Sale, Cheshire
Telephone : 061 969 4622,061 962 1498 Telex 667870

Industrial Trading Co Ltd., P.O.Box 51, 23 Cumberland Street, Worcester WR1 1QE
Telephone : 0905 20373, 51337 Telex 339652, Grams : Intra Worcester

Instrument Screw Co Ltd., 206 Northolt Road, South Harrow, Middlesex HA2 0ET
Telephone : 01-864 6566, Telex 925842, Grams : Screw Harrow

Isaac Jackson (Fasteners) Ltd., P.O.Box 2, Glossop, Derbyshire.
Telephone : Glossop 2091, Telex 668650

Karl Limbach & Cie KG, D-5650 Solingen 19, Erbslöhstr, 16-Postfach 19 03 65
Telephone : 31 00 57, Telex 08-514718, Grams : Limbachmetall Solingenwald

Thomas William Lench Limited, P.O.Box 31, Rowley Regis, Warley, West Midlands
Telephone : 021 559 1530, Telex 338735, Grams : Lench Warley

Londesborough Oil & Marine Engineering Ltd., 200 Aston Brook Street, Aston,Birmingham
Telephone : 021 359 5595, Telex 336997, Grams : Londeinst . Birmingham 6

Maclean-Fogg Lock Nut Company, 1000 Allansson Road, Mundelein, Illinois 60060, U.S.A.
Telephone : 312/556-0010, Telex 25-4470, Grams : Mac Lock Mundelein
Branches : A.P.A. Pehrson, 10 Manor Way, Beckenham, Kent.
Telephone : 650 0445, Telex 21792, Grams : Pehrson Bromley

Macnays Ltd., P.O.Box 14, Middlesbrough, Cleveland.
Telephone : 48144, Telex 58514, Grams : Inventions
Branches : Coventry House, Moorgate, London, E.C.2.
Telephone : 628 6626

Metal Products (Cork) Ltd., Albert Street, Cork.
Telephone : 25091, Telex 6149

Metric Screws (Fastenings) Ltd., Pease Pottage, Crawley, Sussex.
Telephone : 0293 25811/4, telex 87206

MHH Engineering Co Ltd., Bramley, Guildford, Surrey.
Telephone : 048 647 2772, Telex 859387

Miller Bridges Fastenings, Glynwed Distribution Limited, Humpage Road, Bordesley Green, Birmingham B9 5HP
Telephone : 021 773 1222, Telex 338768
Branches : Portway Road, Wednesbury, West Midlands WS10 7ED
Telephone : 021 556 6321, Telex 337046
Middlemore Lane West, Aldridge, Walsall WS9 8DS
Telephone : Aldridge 55121, Telex 338784
Pym Street, Off Hunslet Road, Hunslet, Leeds, 10
Telephone : Leeds 448331, Telex 557125
Trecenydd Industrial Estate, Caerphilly, Glamorgan
Telephone : Caerphilly 868 411
Nelison Road, Gateshead, Tyne & Wear NE10 OEG
Telephone : Newcastle 782131
Castlebank Industrial Estate, South Street, Glasgow

Morlock Industries Ltd., P.O.Box 2, Bridgnorth Road, Wombourn, Wolverhampton WV5 8AU
Telephone : Wombourn 2431, Telex 33276

Rex Nichols & Co Ltd., Sowers Road Industrial Estate, Rugby CV22 7DH
Telephone : Rugby 71313, Telex 311237
Branches : Cambridge Avenue, Slough, Berkshire SL1 4QH
Telephone : Slough 38556, Telex 847062

Nylon & Alloys, Hymax Works, 74 Half Acre Road, Hanwell, W7 3JJ
Telephone : 01–579 5166

The Ormond Engineering Company Ltd., 189 Pentonville Road, London, N1 9NF
Telephone : 01-837 2888, Telex 261875, Grams : Ormondengi, London N1

Works : Holford Yard, Cruikshank Street, London, WC1
Park Street, Ammanford, Dyfed SA18 2ET, South Wales
Telephone : (0269) 3338, Ammanford (STD 0269) 3338

G. Pearson & W. P. Beck Ltd., Beckriv Works, Feckenham Road, Astwood Bank, Redditch,Worcs.
Telephone : Astwood Bank 2061

Plastic Screws Limited, Uddens Trading Estate, Nr.Wimborne, Dorset BH21 7NL
Telephone : Ferndown 71411/2/3, Telex 41408, Grams : Plascrew Wimborne

Prestwich Parker Limited, Victoria Works, Atherton, Manchester.
Telephone : 0234 2561, Telex 67-683, Grams : Progress Atherton
Branches · Potter Cowan & Co Ltd., 139 Reid Street, Glasgow
Telephone : 041 554 7131, Grams : Tireless Glasgow
Potter Cowan & Co Ltd., Back Hilton Road, Ashgrove,Aberdeen
Telephone : 0224 492931

Benjamin Priest & Sons Limited, P.O.Box 38, Old Hill Works, Priest Street, Cradley Heath, Warley, West Midlands B64 6JW.
Telephone : Cradley Heath 66501, Telex 339932, Grams : Bolts, Warley

P. S. M. Fasteners Limited, Longacres, Willenhall, West Midlands, WV13 2JS
Telephone : Willenhall 68011, Telex 338565, Grams : Anchor Willenhall

Ramset Fasteners Ltd., 67 Bideford Avenue, Greenford, Middlesex UB6 7PX
Telephone : 01-998 2245, Grams : Ramfast Greenford

A. Raymond, Manufacture de Boutons et d'Articles Metalliques,
113 Cours Berriat, 38028 Grenoble-Cedex. Telephone : 96-57-45, Telex 32751, Grams : Araymond –Grenoble
Branches : A.Raymond – Druckknopf –und Metallwarenfabrik D 7850
Lorrach Postfach 42, Telephone : 07261 88081
Telex 07-73520, Grams : Raymond-Lorrach

R. S. Rowlands Ltd., RSR House, Lionel Road, Brentford, Middlesex TW8 OJB
Telephone : 01-994 5322, Grams: RS Rowlands, Hounslow
Works : Pennypot Industrial Estate, Hythe, Kent.
Telephone : Hythe 66757

Sandiacre Screw Company Limited, Bradley Street, Sandiacre, Nr.Nottingham NG10 3HJ
Telephone : 0602 394646, Telex 37185, Grams : S.S.C.Sandiacre

Wilhelm Schumacher, 5912 Hilchenbach 1, Postfach 1280.
Telephone : 02733/836, Telex 0872805, Grams : Weshaschraube Hilchenbach

Johann Schürholz, Germany 5970 Plettenberg, Königstrasse 88
Telephone : 02391/1730

Southern Marine Fastening Co Ltd., 20/22 Station Hill, Eastleigh, Hampshire
Telephone : 042126 8844

Spensall Engineering Co Ltd., Great Wilson Street, Leeds LS11 5AP
Telephone : 34803/4, Telex 55257

Spotnails Limited, Bessemer Road, Basingstoke, Hampshire RG21 3NT
Telephone : 3141, Telex 858286, Grams : Spotnails, Basingstoke

Springfix Ltd., 35 Kentish Town Road, NW1 8NU
Telephone : 01-485 9401, Telex 262397

SpS Hitek, Northey Road, Coventry, West Midlands.
Telephone : 0203-661121-4, Telex 31634

Frank Stacey (Glynwed Distribution) Limited, Calthorpe House, Hagley Road, Edgbaston,B'ham
Telephone : 021-455 9820, Telex 338291, Grams : Paratus
Works : Pleck Road, Walsall, Staffs.
Telephone : 92 28930

Standard Pressed Steel International, Mervue, Galway, Republic of Ireland.
Telephone : 3141, Telex 8331.

Stenman Holland B.V., P.O.Box 47, Veenendaal, Holland.
Telephone : 08385-19106, Telex 45198, Grams : Stenman

Suko-Sim GmbH, Sicherheitsmuttern-Verbindungselemente, D7187 Schrozberg, Postfach.
Telephone : 07935/591 Telex 07 4210

Supacraft Engineering Products Ltd., 35 Kentish Town Road, London NW1 8NU
Telephone : 01-485 9401, Telex 262397

T. R. Fastenings Ltd., Uckfield, Sussex.
Telephone : 0825 4711 and 2191

Tappex Thread Inserts Ltd., Masons Road, Stratford on Avon, Warwickshire CV37 9NT
Telephone : 4081

Teckentrup KG, 597 Plettenberg, Wilhelmstrasse 28, Postfach 151.
Telephone : 02357/2011 Telex 8263421

Thor Tools Ltd., Tynemouth, Tyne & Wear NE29 7UE
Telephone : North Shields 08945 73181, Telex 53602, Grams : Thortools, Tynemouth, Tyne & Wear, Thor NShields

Tower Manufacturing, Glynwed Screws & Fastenings Ltd., Navigation Road, Diglis, Worcester WR5 3DE
Telephone : 0905 356012, Telex 38880, Grams : Nails, Phone Worcester

Tucker Fasteners Ltd., Walsall Road, Birmingham B42 1BP.
Telephone : 021 356 4811, Telex 33480, Grams : Tuckerfast B'ham
Branches : 62 Horn Lane, Acton, London W3 6AT
Telephone : 01 992 7851
96 Oxford Road, All Saints, Manchester M13 9RW
Telephone : 061 273 5983

Tucker Industries Pty Ltd., 636 Whitehouse Road, Mitcham, Victoria, Australia
Telephone : 874 7733, Telex AA33741, Grams : Tuckereye Melbourne
Branches : 79 Carnarvon Street, Auburn, N.S.W.
Telephone : 648 4022, Telex AA 26469, Grams : Tuckereye Sydney

Unbrako Distribution, Bannerly Road, Garretts Green, Birmingham 33
Telephone : 021 783 4066, 021 784 3937, Telex 338703
Works : Unbrako Ltd., Burnaby Road, Coventry, CV6 4AE
Telephone : 0203 88722, Telex 31 608

Branches : Power Road, Chiswick, London, W.4.
Telephone : 01-995 0011, Telex 934748
Earl Haig Road, Hillington, S.W.2. Glasgow
Telephone : 041-883 4781, Telex 778717
9 Longwood Road, Trafford Park 17 Manchester
Telephone : 061 872 5731, Telex 667573

United-Carr Limited, Wallingford Road, Uxbridge, Middlesex UB8 2SZ
Telephone : Uxbridge 38681, Grams : Efftee Uxbridge, Telex 261707
Works : Buckingham Road, Aylesbury, Bucks, HP19 3QA
Telephone : Aylesbury 3541, Telex 83476

Van Thiel United B.V., P.O.Box 8, Beek en Donk, Holland.
Telephone : 04929 1931, Telex 51161, Grams : Thielbout

Vossloh – Werke GmbH, D 598, Werdohl, Germany
Telephone : 2392 521, Telex 826 444, Grams : Vosslohwerke

The Walford Manufacturing Co Ltd., Excelsior Works, Great Lister Street, B'ham B7 4LR
Telephone : 021 359 2681, Grams : Altruist
Works : Tipton, Staffs.
Branches : Exhall, Coventry
Birmingham 1
Haigh Avenue, Whitehill Industrial Estate, Stockport, Cheshire.
Telephone : 061 480 2463

Wellworthy Limited, Lymington, Hampshire SO4 9YE.
Telephone : Lymington 5222, Telex 47647-8

H. K. Westendorff, D 4000 Dusseldorf, Hoherweg 277, West Germany
Telephone : 7700–1 Telex o8582211 hkw-d Grams : Westschraube, Dusseldorf

Whitfield Hodgsons & Brough Ltd., P.O.Box 3, Lawson Street, Kettering, Northants.
Telephone : Kettering 517070

J. E. Woodall & Co Limited, Cox's Lane, Cradley Heath, Warley, West Midlands B64 5QX
Telephone : Cradley Heath 66871

Yarwood Ingram Company, Glynwed Screws & Fastenings Ltd., Ledsam Street, Ladywood, Birmingham 16.
Telephone : 021 454 3607

SECTION 2

Sub-section A

Adhesives

Basic Adhesive Types

Acrylics

Acrylics are relatively low strength thermoplastic resins but have applications as specialised adhesives, eg non-structural metal-to-metal bonding. Acrylic emulsions are widely used for bonding plasticised PVC to concrete, cement, timber, etc (eg floor coverings), where they exhibit excellent dimensional stability. Acrylic emulsions are also used for bonding scrim to balsa to make flexible panels for core material for GRP mouldings, a particular advantage being that the adhesive is soluble in polyester resin. Acrylic copolymers are produced as lower cost adhesives for PVC tiles and floor coverings. Polyester-acrylic adhesives are used for bonding metals and hard glazed surfaces, and as thread locking adhesives.

Casein

Casein – the original resin-type woodworking glue – is currently used mainly for bonding asbestos. Latex/casein adhesives are used for foil/paper lamination.

Ceramics

Ceramic adhesives have been developed for high temperature applications requiring high strength, particularly metal-to-metal bonding. They are based on boro-silicate or other glasses compounded with alkaline earths and oxides of alkaline metals. Ceramic adhesives are set by firing at temperatures between 700 and 1 200°C and in this respect resemble ceramic glazes.

Cyanoacrylate

This is an anaerobic synthetic resin where polymerisation is brought about in a matter of seconds by pressure on a thin film of the adhesive between two surfaces. Cyanoacrylate will adhere with a high bond strength to most materials and surfaces. The bond is resistant to oils and many solvents. It is only moderately resistant to water and is broken down by steam. It is also a characteristic of cyanoacrylate adhesives that they set rapidly in the presence of water. Maximum service temperature rating is 80°C continuous or 100°C for limited exposure.

Particular advantage: extremely rapid setting.

Particular disadvantage: extremely high cost.

Epoxy

Epoxy resin adhesives are produced either as two-part mixtures (resin and hardener) for self-curing at room temperatures; or one-part resins for heat-curing. Two-part epoxy resins may have long setting times (up to 24 hours) or be formulated for rapid setting (eg setting time may be reduced to as little as 5 minutes). Originally developed primarily as metal-to-metal adhesives, epoxy adhesives are suitable for bonding most materials including glass, ceramics, wood, many rubbers and some plastics. They are also noted for their excellent resistance to oils and good resistance to water and most solvents.

Pure epoxy resin is solid at room temperature and becomes plastic at 40° to 50°C. Above this the resin undergoes 'curing', eventually becoming a solid mass. At the same time it fulfills the major requirements of a good adhesive. Once the temperature is raised to the flow point the resin readily 'wets' metals and other non-porous surfaces, whilst the film strength when cured is high and the film properties stable. Curing can be achieved simply by holding the assembly at the required temperature for the specified time, no pressure being needed. The curing temperature is largely influenced by the materials. Below about 180°C, however, the curing time is long.

Cold-setting adhesives employ an accelerator and/or hardener to promote curing at room temperature. The more viscous cold-setting resin-hardener mixtures are sometimes thinned with a solvent for spray application, after which up to one hour may be allowed for evaporation of the solvent before the joint faces are brought together and clamped. In general, the thicker the resin used for cold setting (ie the more viscous the resin) the stronger the bond produced. Heat-curing tends to promote a higher bond strength. For maximum bond strength the full heat-curing type is preferred. Further improvements can be had with two-part adhesives, one of the best known being a mixture of polyvinyl formal powder and a phenol formaldehyde liquid resin (Redux). This gives a metal-to-metal joint which is normally stronger than riveting or spot welding and appreciably better than that achieved with the epoxy resins. Joint strength is, however, temperature dependent and begins to show a marked decrease above approximately 80°C. For maximum strength at elevated temperatures the formulation may be modified to provide a maximum service temperature of up to 250°C for stressed applications. The normal process with Redux bonding is to coat both joint surfaces with the liquid resin, after which powdered formvar is sprinkled or dusted on or applied by dip. No reaction takes place until the two surfaces are brought together under heat and pressure. The amount of pressure required is not critical, with 100lb per sq in a typical figure, although up to twice this may be employed in specific applications and with distinct advantages. The time cycle for curing varies with temperature but is normally from four minutes at 180°C, to 20 minutes at 145°C.

Modified epoxy resins, notably epoxy-phenolic and epoxy-polyamide, all have enhanced properties. The addition of nylon improves both the shear and peel strengths, and it can be cold-cured under contact pressure, although far better and more rapid curing is produced by heating at 149°C to 177°C,the curing time then being about one hour. Adhesives of this type are in the form of non-woven mat, plain film or film reinforced with glass fibre or nylon cloth. This two-part adhesive is sensitive to humidity and is also costly. It has been used successfully for bonding stainless steel, aluminium and titanium in aircraft and in aerospace and cryogenic structures.

The addition of phenolic resin materially improves the service temperature range of a straight epoxy resin and epoxy-phenolic adhesives retain up to 50% of their room temperature strength at up to 260°C or more, as well as being suitable for short-term exposure to temperatures as high as 538°C. Besides maintaining relatively low bond strength, creep is also exceptionally low at high temperatures. The bond is brittle and the peel strength-low. Curing is under heat and pressure,

163 to 177°C at 100lb per sq in for 45 minutes to one hour. The mixture is active at room temperature, so the shelf life is limited to a few months, although it may be improved by refrigerated storage. The addition of polyamides to epoxy resins greatly improves flexibility of the bond and gives higher peel strength, although mechanical strength and creep resistance are both reduced. The adhesive is also sensitive to humidity and tends to become brittle at high temperatures. Maximum service temperature is only 66°C. The adhesive is cheap and simple to use. It can be a two-part solution or paste, or a cast or calendered film, and will set at contact pressure. At room temperature, setting time is 3 to 5 days, but only 3 to 5 minutes at 204°C. Such adhesives are finding increasing industrial application, notably in the automobile industry.

Epoxy-polysulphide was one of the original modified forms of epoxy with enhanced elasticity and peel strength. Typical shear strength is 3 000-4 000lb/in². It continues to be used extensively in America for bonding metal plates and beams to concrete.

Epoxy-silicone resins offer greater heat resistance than other epoxy-type adhesives with a maximum service temperature rating of up to 300°C continuous or 500°–550°C for intermittent exposure. Shear strength is only moderate, however, and of the order of 1 500–2 000lb/in²

Furane

Limited use is made of furane copolymers as an adhesive for thermoset plastics and laminated plastics.

Natural Rubber

Adhesives based on natural rubber are either of solvent or heat-curing type. The latter yields a vulcanised bond of improved strength and higher service temperature rating (90°C as opposed to 60°C typical of a solvent-set natural rubber adhesive).

Solvent-set rubber adhesives and solutions may also incorporate resin for improved bond strength and are useful general-purpose adhesives, although strictly limited for engineering and structural applications because of generally low bond strength and the thermoplastic nature of the base material. They have good resistance to water and mould growth but are attacked by oils, solvents and many chemicals.

Specific subtrates for which natural rubber adhesives are generally suitable are:-

natural rubber (not synthetic rubbers)
some plastics (eg acrylic, GRP, PTFE and ABS)
expanded rubbers – natural rubber and natural latex foam
expanded elastomers – polyurethane, polystyrene
metals – aluminium, aluminium alloys and iron and steel
fabrics – especially cotton, wool, glass fibre cloth, felt, rayon and asbestos cloth
cardboard and paper
wood, chipboard and hardboard
leather
glass
ceramic and glazed surfaces
concrete
brickwork.

Note: choice and specific performance of a natural rubber adhesive will be dependent on the type and formulation.

Neoprene

Neoprene based adhesives have good resistance to water, solvents, chemicals and oils (with the exception of strong oxidising agents and aromatic hydrocarbons). They can be formulated for solvent-setting or vulcanising (either by heat-curing or by the addition of a catalyst), with or without resin modifiers.

Specific subtrates for which neoprene-based adhesives are suitable include:-

rubbers – natural rubber, butyl*, neoprene, nitrile*, polyurethane*

plastics – PVC, polystyrene, cellulose plastics, GRP, polyethylene*, polypropylene*, PTFE* and polycarbonate*, and thermoset and laminated plastics

expanded rubbers – synthetic rubber foams

expanded plastics – rigid PVC

metals – most

fabrics – most, except polyester (terylene)

cardboard and paper

wood, chipboard and hardboard

leather

glass* and glazed surfaces*

concrete

brickwork.

**Depends on formulation; some neoprene adhesives may be unsuitable.*

Nitrile

Nitrile based adhesives are generally similar to neoprene adhesives in variety and application. Shear strengths of up to 1 000lb/in^2 are possible.

Specific subtrates for which nitrile-based adhesives are suitable are:-

rubbers – nitrile and polyurethane

plastics – ABS, PVC, polyester, nylon, acrylic, cellulose plastics, PTFE*, polycarbonate and thermoset plastics

expanded rubbers – polyurethane

expanded plastics – PVC

metals – most (except tinplate and galvanised sheet)

fabrics – most (except rayon)

cardboard and paper

wood, chipboard and hardboard

leather

glass* and glazed surfaces*

concrete*

brickwork*.

**Depends on formulation; some nitrile adhesives may be unsuitable.*

Polyimides

Polyimide adhesives are of thermosetting type, curing at temperatures of 500 to 700°F. post curing is also required to develop maximum bond strength. They have high bond strength with service temperatures of the order of 550 to 750°F. They are a recently introduced group of structural adhesives which will bond most materials, but cost is high.

Polybenzimiazoles

These are thermally stable adhesives with a maximum service temperature of the order of 361°C to 538°C for continuous rating (although some degradation or ageing effects may be apparent above 427°C). They require high temperature curing and post curing, so their application is relatively limited.

Polyurethane

Polyurethane adhesives produce flexible bonds with good pull strength and good resistance to shock and vibration. Bond strength is high at normal room temperatures but decreases rapidly with increasing temperature. Polyurethane has excellent resistance to oils, acids, alkalis and many solvents, but resistance to moisture and water tends to be poor.

Specific subtrates for which polyurethane adhesives are suitable are:-

rubbers – polyurethane

plastics – ABS, PVC (plasticised and unplasticised), polyester, polyamide*, acrylic, cellulose plastics, GRP, polyester, polycarbonate and laminated and thermoset plastics

expanded rubbers – polyurethane

expanded plastics – PVC

metals – aluminium, aluminium alloys, copper and copper alloys, iron and steel.

fabrics – most (except rayon)

cardboard and paper

wood, chipboard, hardboard

leather

concrete

brickwork.

**Depends on formulation; some polyurethane adhesives may be unsuitable.*

Fluorocarbon (Viton)

Adhesives based on fluorocarbon polymers are suitable for higher temperature duties, but are normally used only for special applications.

Suitable subtrates are:-

plastics – fluorocarbons

metals – most

fabrics – felt and asbestos cloth

glass and glazed surfaces

concrete

brickwork.

Polyester

Polyester resins have limited application as adhesives. They are not, for example, good adhesives to use with cured GRP (epoxy resins being preferred in this case), but specific sub-trates for which polyester adhesives are suitable include:-

plastics – PVC (plasticised and unplasticised), polyester films, polystyrene*

metals – copper and copper alloys

fabrics – most (except acrylic, asbestos and rayon).

**Depends on formulation; some polyester adhesives may be unsuitable.*

Polysulphide

Polysulphide mixtures have a particular application as sealing compounds – see section on *Joint Sealants.* They are also formulated as adhesives for the following subtrates :-

plastics – GRP mouldings and polycarbonate*

metals – aluminium and aluminium alloys, copper and copper alloys, iron and steel

fabrics – suitable for most (except acrylic, polyester and polyamide)

wood, chipboard and hardboard

leather

glass and glazed surfaces

concrete

brickwork.

**Depends on formulation; some polysulphide adhesives may be unsuitable.*

Polyamides

Whilst the main application of polyamides in adhesives is as modifying agents in epoxy resins and phenolic resins, they have a limited use in bonding metals and many plastics. They are produced as solutions, films and hot melts. Heat-sealing polyamides set immediately on cooling. Chemical resistance is similar to that of nylons.

Phenolic Resins

Thermosetting phenolic resins (PF resins) are major adhesives used in the woodworking industry, particularly in the manufacture of plywood. They provide excellent resistance to water, solvents and oils etc. Structural or engineering adhesives are also based on mixtures of phenolic and other synthetic resins, eg

Phenolic-neoprene – heat-curing adhesives for metal-to-metal or metal-to-wood joints, etc. Shear strengths of the order of 2 500–3 000lb/in^2 can be achieved.

Phenolic-nitrile – heat-curing adhesives with enhanced strength (shear strength up to 4 000lb/in^2) and improved maximum service temperature (up to 175°C continuous or up to 250°C for intermittent exposure). These can be regarded as special adhesives for metals, etc, and for bonding non-metallic components to metals (eg linings to brake shoes).

Phenolic-polyamide – generally used in the form of a (thermoplastic) polyamide film in conjunction with a (thermosetting) phenolic resin solution. Bond strengths of up to 5 000lb/in^2 are attainable, and good strength is maintained up to quite high temperatures (eg 200lb/in^2 at 150°C).

Phenolic-vinyl – generally used in the form of a (thermoplastic) vinyl powder dusted on to a liquid phenolic resin film, although one-part mixtures are also available. Shear strengths of up to 5 000lb/in^2 are obtainable, but strength degrades rapidly above 100°C. This type of adhesive is widely used for bonding honeycomb-sandwich constructions, metal-to-plastic bonding and bonding cyclised rubbers to metals.

Polyvinyl Acetate (PVA)

Thermoplastic resin or 'white glue' is widely used in the modern woodworking industry, but is also suitable for use with metals, many plastics, glass, ceramics, leather, etc. It has a high bond strength, is non-staining and has good resistance to oils and mould growth. It has very poor resistance to heat, however, and strictly limited resistance to water and moisture. PVA emulsions are widely used as a ceramic tile adhesive.

Until recently all PVA adhesives were slow setting, requiring joints to be clamped for up to 24 hours, but modern formulations are fast-setting.

Silicone

Basic silicone adhesives have excellent resistance to high temperatures but relatively low strength. They are normally formulated with stronger adhesives to provide high temperature stability with good mechanical strength. Epoxy-silicone adhesives may be rated for continuous service temperatures up to 315°C, and up to 482°C for intermittent exposure.

Sodium Silicate

Sodium silicate based adhesives are used for bonding asbestos cloth lagging in high temperature insulation.

Urea Formaldehyde (UF)

UF is a thermosetting resin widely used in woodworking. It is available as a one-part mixture activated by mixing with water, or as a two-part mixture of resin (or resin powder for mixing with water) and catalyst. It has good resistance to oils, solvents and water (but not boiling water).

In addition to woodworking joints it can also be used as an adhesive for phenolic, melamine and urea thermoset plastics.

Resorcinol Formaldehyde (RF)

Resorcinol formaldehyde is another thermosetting resin used mainly for woodwork joints, with a superior strength, water resistance and maximum service temperature to UF resins. It can also be used for bonding acrylics, nylons, phenolics and urea plastics.

TABLE I – SERVICE TEMPERATURES FOR ADHESIVES

Adhesive Base and Type	Minimum Temperature °C	Maximum Service Temperature	
		Continuous °C	Intermittent or Short Term °C
Cyanoacrylate		80	100
Epoxy	–	90	–
Epoxy-Phenolic		up to 200	up to 260
Epoxy-Polyamide	–	100	–
Epoxy-Polysulphide		90	–
Epoxy-Silicone	–	300	up to 550
Natural Rubber	–40	65	–
Natural Rubber (vulcanised)	30	90	–
Neoprene	–50	90	–
Nitrile	–50	up to 150	–
Polyurethane	–200	95–150	–
Fluorocarbon	–	200	
Polyester	–	150	–
Polybenzimiazole	–	370–540	–
Polysulphide	–	90	
Polyamide	–	100–160	–
Polyimide	–	260–370	–
Phenolic	–40	100	–
Phenolic-neoprene	–55	90	–
Phenolic-Nitrile	–50	up to 175	up to 250
Phenolic-Polyamide	–55	150	–
Phenolic-Vinyl	–50	100	–
PVA	5	50	–
Silicone	–	315	480
Urea Formaldehyde (UF)		60	–
Resorcinol Formaldehyde (RF)	–175	175	–

TABLE II – TYPICAL STRENGTH FIGURES FOR ADHESIVES

Adhesive	Representative Strength Figures			Remarks
	Shear kg/cm^2	Shear lb/in^2	Peel	
Epoxy (unmodified)	350	5000	–	–
Filled Epoxy	140–210	2000–3000	–	aluminium oxide filler
Epoxy Polyamide	245	3500	28 piw	on aluminium
Epoxy–Nylon	420	6000	20–100 piw	requires clamping pressure
Epoxy–Polysulphide	200–280	3000–4000	–	–
Epoxy–Silicone	105–140	1500–2000		–
Neoprene	20	1000	–	–
Nitrile	70	1000	–	–
Phenolic–Neoprene	140–200	2000–3000	high	creeps at high loads
Phenolic–Nitrile	up to 280	up to 4000	–	–
Phenolic–Polyamide	350	5000	–	–
Phenolic–Vinyl	350	5000	–	–
PVA	up to 200	up to 3000	–	metal-to-metal bond
Polyimide	140–175	2000–2500	–	stainless steel to stainless steel
Polyurethane	35–105	500–1500	20–40 piw	–
Silicone (unmodified)	up to 140	up to 2000	–	–

Adhesive Joint Design

The four basic types of stress to which a bonded joint may be subjected are (see also Fig 1).

(i) Tensile – or a direct separating force.

(ii) Shear – or a sliding apart of the bonded surfaces.

(iii) Cleavage – or a levering apart of the bonded surfaces.

(iv) Peel – or a stripping off of one of the bonded surfaces.

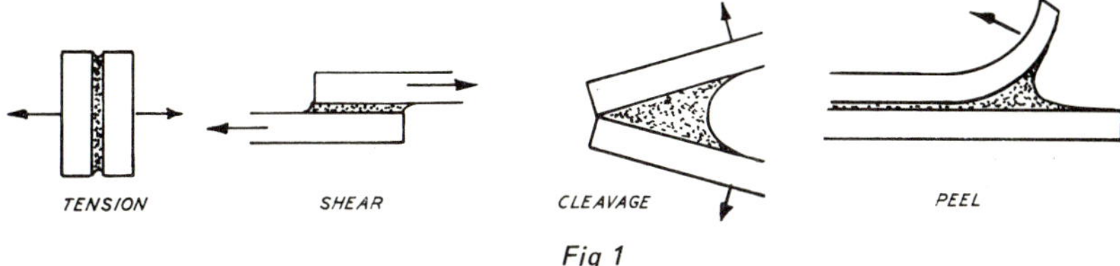

Fig 1

In the case of pure *tension* the forces throughout the joint are perpendicular to the plane of the loaded area and stress is distributed uniformly over this area. Thus all of the adhesive contributes to the strength of the joint.

In the case of pure *shear* the forces are again distributed uniformly, but this time parallel to the plane of the bonded area. Thus all the adhesive contributes to the strength of the joint.

In the case of *cleavage,* stress is concentrated in only part of the adhesive, with one end of the joint receiving maximum stress and the other end little or no stress. Thus only a proportion of the adhesive contributes to the joint strength.

In the case of *peel* the stress is concentrated in a very thin line of adhesive and so an even smaller proportion of the adhesive contributes to the joint strength.

Ideally, adhesive joints should be designed so that they are stressed in tension. In practice, stressing in shear is often a more realistic approach since this usually results in a simpler joint design. Joint designs which stress the load in cleavage and peel should be avoided; or cleavage and (especially) peel forces should be minimised where combined forces are present.

The most common form of adhesive joint is the *lap joint* where the load is normally stressed in shear. The simple lap joint suffers from the fact that the shear forces are not in line, which can set up undesirable cleavage and peel stresses – Fig 2. However, it is widely used because of its simplicity and can usually be given adequate strength by providing sufficient overlap (see later).

Alternative joint designs which reduce or eliminate this failing are shown in Fig 3. All suffer from the practical disadvantage that they require working or machining of the sheets to be jointed, adding to cost.

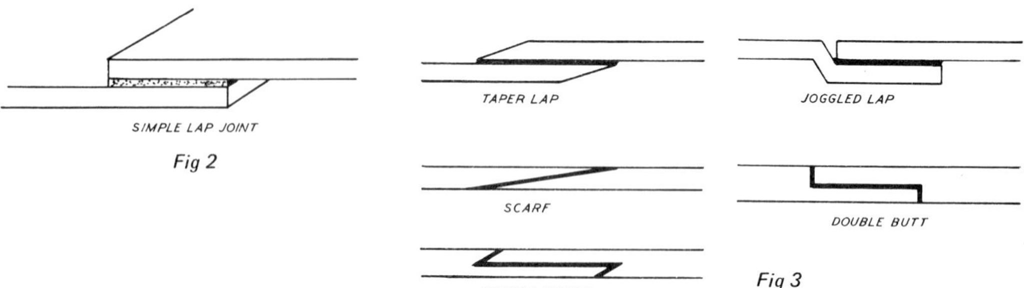

Fig 2

Fig 3

The *tapered* single lap reduces stress concentrations generated by cleavage and peel forces by allowing the edges of the joint to bend when distortion occurs under stress.

The *joggled* joint is more efficient since it provides stress alignment with the load bearing area in the same plane as the shear stress on the substrata. It also has the advantage that only one surface has to be formed.

The *scarf* is another joint which provides alignment of the forces providing shear stress, but may also introduce cleavage and peel stresses. Also the joint is difficult to form in thin materials and so is normally only used in wood or plywood.

The *double butt* is again limited to thicker materials, but is considerably superior to the scarf joint where bending forces are involved.

The *double scarf* is even better in resistance to bending forces, but even more difficult to machine.

Plain *butt joints* are normally weak since they are almost inevitably stressed in cleavage. Simple joints which offset this are the *inset butt* or the *braced butt* (Fig 4). Both are more applicable to woodworking than metals or sheet materials. In the latter case strong butt joints can be produced by an *angle* or *tee* configuration – see Fig 5.

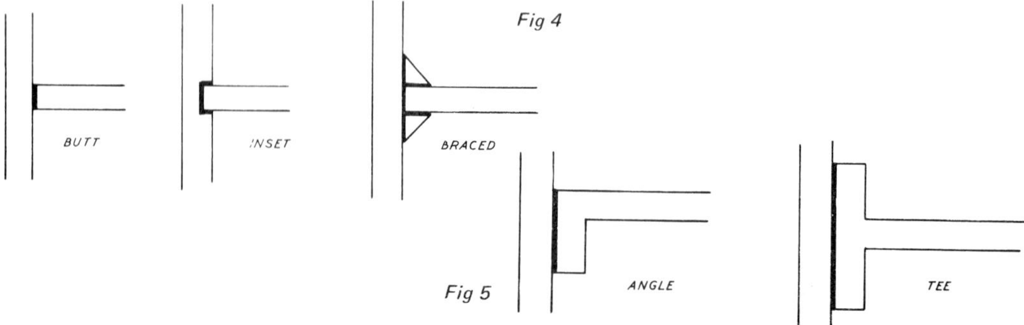

Fig 4

Fig 5

Corner joints and other types of joint used in woodworking should be chosen on the basis of providing adequate bonding area, and whether or not the adhesive requires clamping pressure while setting. It is easy to apply closing pressure to lap joints, for example, but difficult to apply uniform closing pressure over a mortise and tenon joint. Corner joints in sheet metal may be lapped, or assembled in corner fittings or attachments. Epoxy adhesives are normally used in both cases.

Bonded-on stiffeners applied to sheet metal etc represent a particular problem since flexing of the basic sheet will induce peel stresses in the adhesive. These will be a minimum if the flanges of the stiffening section can bend with the sheet.

Strength of bonded joints in shear

The strength of a simple lap joint stressed in shear is proportional to the length of the overlap, L – see Fig 6. It is more convenient to express joint strength as a joint factor, applicable to the specified shear strength of the adhesive, when the joint factor is calculated as the square root of the sheet thickness (T) divided by the length of lap, ie

$$\text{joint factor} = \frac{\sqrt{T}}{L}$$

The strength of the joint in shear = joint factor x shear strength of adhesive.

This simple analysis is valid with joint factors not exceeding 0.3. If larger joint factors are applicable, then the strength of the joint will be lower than that given by this formula. In practice, if the joint strength required is greater than that given by a joint factor of 0.3 with a simple butt joint, then either a stronger adhesive or a stronger type of joint should be used.

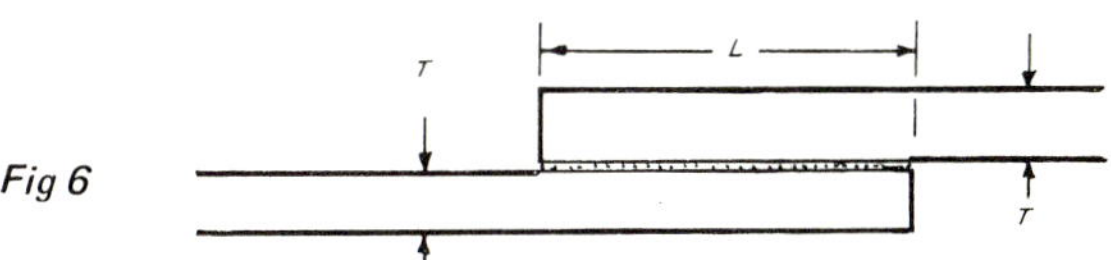

Fig 6

Strength of bonded joints in tension

The strength of bonded joints in pure tension follows from first principles, ie

$$\text{stress} = \frac{\text{tensile force applied}}{\text{bonded area}}$$

Thus:

strength in tension ≐maximum unit tensile strength of adhesive x bonded area.

The strength of bonded joints in cleavage and peel can only be determined empirically under specified loading conditions. Such tests may be included in material specifications – eg plywood manufacture – when a minimum figure would apply as a material property.

Specific tests are also applicable to determine tensile and shear strength of adhesives themselves; also bending stress, cleavage, peel, creep, fatigue, etc.

It must be borne in mind that stresses in bonded joints are seldom simple even when apparently loaded in pure shear or pure tension. Residual stresses may be induced in the adhesive itself through curing, and additional internal stresses be induced by differential expansion, etc. In the case of cycling stresses fatigue is not normally a problem, since most modern adhesives have high resistance to fatigue.

Surface Preparation

For non-critical work it is generally only necessary to ensure that the surfaces to be joined are free from grease and any loose surface contamination. Solvent cleaning is usually adequate, using Toluene, Xylene, or similar.

To obtain maximum bond strength more thorough degreasing and cleaning is called for together with surface treatment. The latter may involve abrasion or chemical treatment, depending on the materials involved. Specific cleaning/surface treatment procedures may also be specified by adhesive manufacturers and should be followed in such cases, particularly where a primer may be specified.

The following can be taken as a general guide to surface treatment required for various substrate materials. Procedures designated (N) are for normal applications; those designated (C) are for critical applications where maximum bond strength is required.

Substrate	Degrease	Surface Treatment
Aluminium	(N) solvent cleaner or (C) vapour degrease. Follow by (C) alkaline cleaner and rinse.	(N) none (C) use suitable etching solution or primer, followed by rinse in demineralised water
Aluminium Alloys	(N) as for aluminium (C) as for aluminium solvent cleaner	(N) sand surface (C) as for aluminium
Brass	solvent cleaner	(N) rub with sandpaper or emery and follow by (C) etch and rinse with demineralised water or (C) primer coating
Copper	as for brass	as for brass
Cast Iron and Steel	(N) solvent cleaner (C) vapour degrease	(N) lightly abrade surface to dull (C) sandpaper or emery surface and follow with etch or primer (if specified for adhesive)
Stainless Steel	(N) solvent cleaner (C) vapour degrease followed by alkaline cleaner.	(N) none (C) rinse and dry, follow with etch or primer (if specified for adhesive)

cont...

(contd.)

Substrate	Degrease	Surface Treatment
Tinplate	(N) solvent cleaner (C) vapour degrease	(N) none (C) primer coat (as appropriate to adhesive)
Enamelled Surface	(N) detergent (C) solvent cleaner	(N) wash well with detergent. Rinse and dry. (C) degrease with solvent cleaner. Rinse and dry.
Glass	(N) detergent (C) solvent cleaner	(N) wash with detergent or degrease with solvent cleaner. Rinse and dry. (C) abrade lightly, if recommended by adhesive manufacturer.
Wood (new surface)	(N) none (C) none	(N) remove dust (C) remove dust; use primer where specified for adhesive. *Note:* older surfaces should be sanded down to new wood.
Cement or Concrete	(N) none (C) none	(N) remove dust (C) remove dust, treat surface with primer if specified. *Note:* older concrete surfaces should be wire brushed down to new surface.
Rubber (Neoprene and nitrile)	(N) wash with detergent (C) degrease with solvent cleaner	(N) none (C) use primer, if specified for adhesive.
Silicone Rubber	(N) wash with detergent or solvent cleaner. (C) degrease with solvent cleaner.	(N) none (C) none

Plastic materials generally require individual treatment, but much depends on the type of adhesive used. In general, the basic requirements for preparation apply, ie the substrates should be

(i) free from grease.

(ii) free from dust, etc.

(iii) dry.

The following specific recommendations also apply

Plastic	Recommended Treatment
Thermoset Laminates	Degrease and dry. Ensure even spread of adhesive.
Thermoset Mouldings	Degrease, rinse and dry. Lightly abrade the surface if necessary (ie depending on type of adhesive used). Remove dust
Phenolics	Roughen surface or lightly abrade. Remove dust. Primer may be specified for certain types of adhesive.
Nylon	Degrease with solvent cleaner. Lightly abrade surface. Remove dust. Primer may be specified for certain adhesives.
Polyester and GRP	Degrease with solvent cleaner. Lightly abrade surface or use etching primer.
PVC	Degrease with solvent cleaner. Primer may be specified for certain adhesives.
PTFE	If bonded joints are required with PTFE, use PTFE which has been specially treated for bonding, or has a bondable backing.

Joint Sealants (Mastics)

Joint sealants are widely used in the building trade for bedding and sealing window and door frames, sealing expansion and movement points, sealing floor joints, sealing joints in prefabricated buildings, etc. They can be classified as general purpose sealants. Specially developed sealants are also used as caulking compounds for marine use, low temperature sealing in cold rooms and refrigerators and other specialised applications.

Joint sealants may be classified by chemical type – see Table I. Acrylic and polyurethane sealants have been introduced relatively recently and their long-term properties are not yet known.

TABLE I – TYPES OF JOINT SEALANT

Chemical Type	Base	Pretreatment	Remarks
Oleo-Resins	Drying and non-drying oils, resins and fillers.	Require primers to give adhesion to porous surfaces.	Best suited to lap joints, but can be tamped into butt joints. Available in strip form, or for knife or gun application
Rubber-Bitumen	Natural or synthetic rubber compounded with bitumen or pitch.	Movement accommodation of the order of 10%	Cold poured (set by solvent action) or hot poured.
Butyl	Butyl rubbers or degraded butyl rubbers, used alone or in combination with solvents, oils, extenders, etc.	Can be skin-forming, non-skinning, non-drying or curing (curing types cure by solvent evaporation).	Very wide range of compositions from soft, sticky compounds to hard rubber strips.
Polysulphide	Single or two-part mixtures based on polysulphide polymers. Two-part mixtures cure by chemical reaction – fast.	Primers required on most porous and friable surfaces.	Wide range of compositions from soft elastic materials to hard rubbers. Available in a wide range of colours. Usual base of caulking compounds.
Silicone	Single-part chemically curing silicone elastomer.	Require primers on most porous and friable surfaces.	Remain fully elastic. Available in a range of colours
Polyurethane	Two-part self-curing mixtures based on polyurethane polymers.	Primers required on most surfaces for adequate adhesion.	Remain fully elastic.
Acrylic	Mixtures of acrylic polymers.	–	Plastic in nature.

The truly elastic sealants – silicones, polyurethanes – may tend to produce adhesive failure under extended stress, although the amount of stress may be relatively small. A summary of the properties of sealing compounds is given in Table II. Tolerances for a range of proprietary sealants are given in Table III.

TABLE II – SUMMARY OF PROPERTIES OF SEALING COMPOUNDS (Bostik Ltd)

Type	Form	Skinning or Setting Time Hrs	Maximum Joint Width mm	Minimum Joint Depth mm	Butt Joint Movement Tolerance	Expected Life	Remarks
Oleo-resinous	Strip Knife grade Gun grade	-- 24 24	50 25 12 Special Grades 25	6 9 12	2% 5–15% 15%	2–10 yrs 2–10 yrs 2–10 yrs	Painting improves durability.
Butyl	Strip Knife grade Gun grade	– 24 24	50 25 12 Special Grades 25	6 6 6	2–10% 2–15% 5–15%	20 yrs 2–15 yrs 2–15 yrs	Durability is affected by nature of joint movement. Some sealants do not skin. Most collect dirt.
Acrylic	Gun grade	12	12	6	15%	Unknown possibly 15 yrs	Collects dirt.
Polysulphide single part	Gun grade	24	12	6	15–25%	20 yrs	Moderate cure.
Polysulphide 2 part	Pourable Gun grade	24 48	50 50	6 6	15–33% 15–33%	20 yrs 20 yrs	Fast curing
Silicone	Gun grade	2	12	6	12–25%	Unknown possibly 20 yrs	Single part. Moderately fast cure
Polyurethane	Pourable Gun grade	24 24	50 50	6 6	Unknown Unknown	Unknown possibly 15 yrs	Poor adhesion to most surfaces

TABLE III – TOLERANCES OF A RANGE OF PROPRIETARY SEALANTS

Title	Type	Maximum Movement Tensile %	Tolerance Shear %
Bostik Gun Mastic	Oil-based (oleo-resinous) mastic gun grade	15	40
Bostik Mastic 1566	Butyl rubber gun grade	20	50
Prestik 5913	Butyl rubber strip-form sealant. Available in a variety of sizes	N/A	20
Prestik 5925	Softer version of 5913	N/A	33
Bostik 1135	Black single part polysulphide	25	70
Bostik 1137	Grey single part polysulphide	25	70
Bostik 1138	White single part polysulphide	25	70
Bostik 2135	Black two part polysulphide	33	80
Bostik 2137	Grey two part polysulphide	33	80
Bostik 2138	White two part polysulphide	33	80
Bostik 1581	One part silicone (white, black and translucent)	25	60
Bostik Highway Sealant 2139	Two part polysulphide (black)	25	N/A

Back-up Materials

Back-up materials are used to fill the bulk of a joint, the sealing compound then being applied to seal the face of the joint. Fibreboard, rope and expanded plastics are commonly used back-up materials, but with rubber sealants it is important that the sealant is not restricted during movement. This requires that the sealant does not stick to the back-up material and also that the back-up material is deflected at least as easily as the rubber sealant. Expanded plastics are preferred in such cases.

Joint Design

Vertical and horizontal joint lines should link at intersections to form a continuous seal. It is also important to avoid regions of high stress as this can result in premature failure of the seal. Fig 1 shows 'right' and 'wrong' applications of back-up and sealant in typical joint designs.

It is also desirable as far as possible to provide protection against direct weathering to eliminate erosion, and direct sunlight, which has a degrading effect on most sealants. Examples of good joint design in these respects are shown in Fig 2.

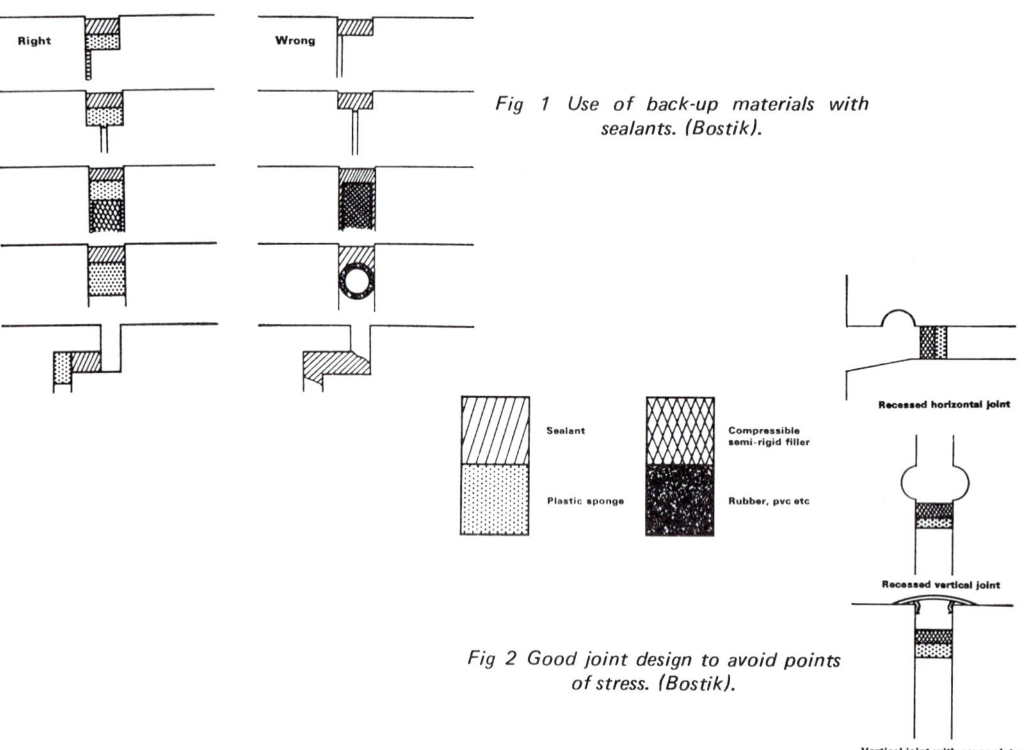

Fig 1 Use of back-up materials with sealants. (Bostik).

Fig 2 Good joint design to avoid points of stress. (Bostik).

Sealed Moving Joints

Movements can occur in building joints and the type and speed of movement will affect the behaviour of plastic and elastic sealants. Thus the design of a sealed moving joint must relate the expected joint movement to the movement accommodation of the sealant. Typical movements of joints in buildings are given in Table IV. The following design procedure is quoted from Bostik Ltd.

TABLE IV – MOVEMENT OF JOINTS (after Building Research Station data)

Joint In	Maximum Rate	Range of Movement mm	Approximate Number of Movements in One Year
Concrete roof kerb	0.6mm/hour	4.80	300
Concrete roof kerb	0.178mm/hour	0.76	40
Roof	0.331mm/hour	4.45	300
Upper brickwork	0.433mm/hour	4.27	300
Lower brickwork	0.280mm/hour	1.27	100
Aluminium mullion	3.048mm at least in one hour	2.08	200

In calculating the expected movement, consideration must be given to both thermal and moisture movement and the tolerances on the units. When the total movement has been determined the joint width is calculated as follows:

$$\text{joint width} = \frac{\text{expected movement}}{\text{movement tolerance of sealant}}$$

The depth of sealant is then determined:

a) For polysulphide sealants

for a butt joint: $\text{depth} = \frac{\text{width}}{2}$

for a lap joint: depth should be not less than width (see Fig 3).

b) For oil-based and butyl-based gun grade mastics depth should not be less than width for both butt joints and shear joints.

When polysulphide sealants are used care should be taken to avoid adhesion to the bottom of the joint. This is usually done by backing the joint with polyethylene foam.

The following example illustrates the method described above:

Consider a building 50 metres long by 12.5 metres high having a pre-cast concrete frame erected to a tolerance plus or minus 6mm overall. The building is clad with natural coloured pre-cast concrete panels each approximately 5 metres long by 2 metres high cast to a tolerance of plus or minus 3mm. It is to have butt joints sealed with a polysulphide sealant.

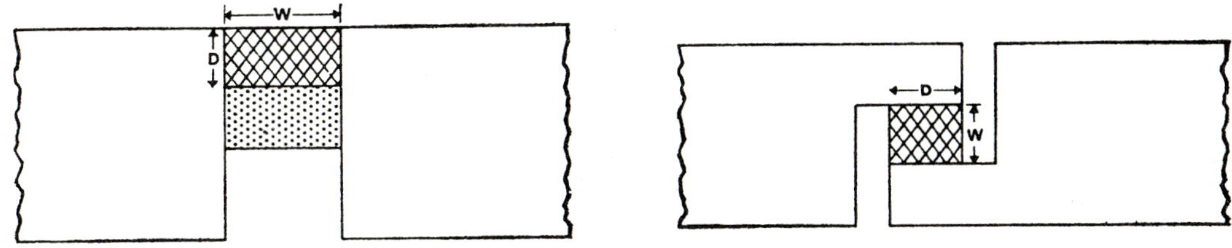

Fig 3 Butt joint (left) and lap joint (right).

Moisture Movement

Irreversible moisture movement of concrete = 0.08%

Reversible moisture movement of concrete (see Table V) = 0.05%.

The panels are pre-cast, and, provided they are adequately cured, irreversible movement should be complete.

$$\text{Reversible longitudinal movement of panels} = 5 \times \frac{0.05}{100} = 2.5\text{mm}$$

$$\text{Reversible vertical movement of panels} = 2.5 \times \frac{0.5}{100} = 1.25\text{mm}$$

It is assumed that settlement movements will not affect this structure.

TABLE V – MOISTURE MOVEMENTS OF COMMON MATERIALS

Material	% Movement (Irreversible)	% Movement (Reversible)
Timber		**(1) 1.7 (radial)** **(2) 2.5 (tangential)** **(3) Negligible (longitudinal)**
Portland stone (limestone)		0.004
Darley Dale stone (sandstone)		0.015
A typical good clay facing brick		0.01
Sand-lime brick	0.001–0.05	0.001–0.05
Portland cement concrete (dense aggregate)	0.02–0.08	0.01–0.055
Fibre building board (insulating)		0.04–0.5
Laminated plastic board		0.1–0.5
Metals		Nil

Thermal Movement

The anticipated maximum surface temperature is 50°C.

The anticipated minimum surface temperature is –20°C (see Table VI).

Thus the anticipated surface temperature range is 70°C.

Thermal coefficient of expansion of concrete is 11.0×10^{-6} per °C (see Table VII).

Therefore, potential longitudinal thermal movement of panels = $5 \times 70 \times 11 \times 10^{-6}$ = 3.85mm

Potential vertical thermal movement of panels = $2.5 \times 70 \times 11 \times 10^{-6}$ = 1.93mm

Therefore:

Total longitudinal movement of panels 3.85mm + 2.50mm = 6.35mm

Total vertical movement of panels 1.93mm + 1.25mm = 3.18mm

These joints are to be sealed with a polysulphide sealant having a butt joint movement tolerance of 33%.

Therefore, the minimum width of vertical joint to accommodate horizontal movements between

$$\text{panels} = 6.35 \times \frac{100}{33} = 19.2\text{mm}$$

The minimum width of horizontal joint to accommodate vertical movement between panels

$$= 3.18 \times \frac{100}{33} = 9.64\text{mm}$$

TABLE VI – ESTIMATED TEMPERATURES ON BUILDINGS IN ENGLAND

Part of Construction	Temperature			
	Maximum		Minimum	
	°C	°F	°C	°F
Light-coloured masonry wall (outer 3 inches) exposed concrete eaves, edges of floor slab	50	120	–20	–5
Similar construction, but dark coloured	65	150	–20	–5
Black glass, ceramic tiles, or metal, insulated behind	80	180	–25	–15
White glass, ceramic tiles, or metal, insulated behind	60	140	–25	–15
Black metal tray, exposed behind clear glass and insulated behind	130	260	–10	+ 15
Clear glass in front of dark insulated background, such as tray above	80	180	–25	–15
Aluminium mullion in a curtain wall (natural colour or white)	50	120	–15	+5

Note: Figures are 'rounded off' and not exact Farenheit–Centigrade conversions

TABLE VII – THERMAL COEFFICIENT OF EXPANSION OF COMMON BUILDING MATERIALS

Material	Coefficient of thermal expansion per °C difference in temperature	Increase in 100 metres length for a 10°C change in temperature mm
Aluminium	23.4×10^{-6}	23.4
Brass	18.9×10^{-6}	18.9
Bronze	18.4×10^{-6}	18.4
Copper	16.1×10^{-6}	16.1
Steel	$10.4 - 17.0 \times 10^{-6}$	10.4 – 17.0
Stainless steel	17.3×10^{-6}	17.3
Glass	8.9×10^{-6}	8.9
Concrete	11.0×10^{-6}	11
Brick	9.5×10^{-6}	9.5
Wood – cross grain	50×10^{-6} approx	50
Wood – along grain	7×10^{-6} approx	7
Rigid PVC	70×10^{-6}	70

The tolerances to be absorbed at the joints must be taken into account in calculating the nominal joint width. Tolerances on panels total ± 3mm ie ± 1.5mm each end.

Tolerances on frame ± 6mm. The longitudinal frame tolerance is taken over 10 panels, ie 9 joints.

Therefore the tolerance at each joint is ± 6mm x 1/9 = ± 0.67mm.

The vertical frame tolerance is taken up on 5 panels, ie 4 joints.

Therefore, the tolerance at each joint is ± 6mm x ¼ ± 1.50mm.

Total tolerance at the horizontal joints = 2(± 0.33mm) + (± 0.33mm) = ± 0.66mm

Total tolerance at vertical joints = 2(± 0.33mm) + (± 0.67mm) = ± 1mm.

Nominal joint width = minimum width + tolerance. Nominal vertical joint width = 19.24mm + 1mm = 20.24mm. Nominal horizontal joint width = 9.64mm + 0.66mm = 10.3mm.

Rounding these figures:

Nominal vertical joint width 20mm.

Nominal horizontal joint width 10.5mm.

Therefore, depth of vertical joint = 10mm Depth of horizontal joint = 5mm.

The joints should be backed by polyethylene foam. The cross section of the two joints is shown in Fig 4.

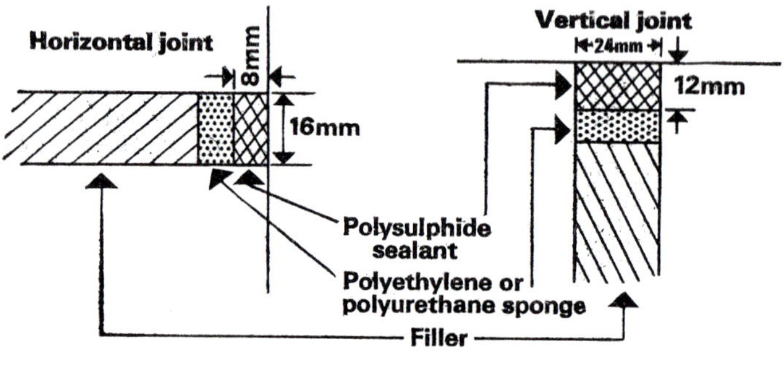

Fig 4

Thread Locking Adhesives

Thread locking adhesives provide a simple, economic and effective method of converting non-locking fasteners into locking types, by means of the adhesive coating. Significant factors are:-

(i) the strength of the adhesive, which governs the breakout torque.

(ii) the size of the fastener, which also affects the breakout torque.

(iii) the viscosity of the adhesive, which governs the gap-filling properties.

Low strength adhesives are normally used for locking screws and nuts where the components must be removable. Gap-filling requirements are met by using an adhesive with a low viscosity for small screws and nuts, eg 6mm (¼in) or less, and a medium viscosity for large sizes. Alternatively, since a slightly greater degree of gap-filling is required for locking female (nut) threads, than male (screw) threads of the same size, individual manufacturers may specify different viscosity adhesives for screws and nuts.

Higher strength adhesives are required for locking studs, and here again a medium viscosity is usually used for optimum gap-filling properties. High strength adhesives are used for permanent locking of threads (eg nuts and studs).

Adhesive types used may be:-

(i) One-part, self-setting containing solvent.

(ii) One-part, anaerobic.

(iii) Two-part, curing by the admixture of a catalyst or accelerator.

(iv) 'Passivated', pre-applied adhesive.

Curing or setting time depends on the size and nature of the joint, and the temperature – Fig 1. Setting time in the case of self-setting and anaerobic adhesives may be accelerated by the application of a catalyst, either as an additive, or directly to the surface to be coated prior to the application of an anaerobic adhesive – Fig 2.

The main practical requirement of all types of thread locking adhesive is that the surface to be coated should be degreased and clean. They are all suitable for use on ferrous metal surfaces without further treatment.

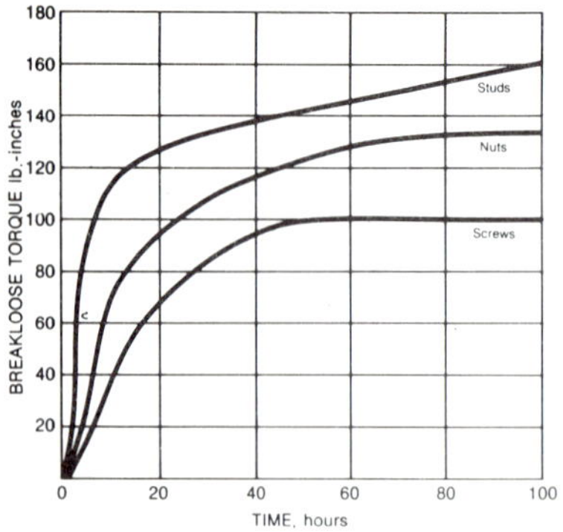

Fig 1 Effect of time versus breakloose torque on 5/16in heat treated steel fasteners cured at 20° (68°F). Fasteners were coated with adhesive for studs in top curve, adhesive for nuts in middle curve, and adhesive for screws in bottom curve. (Unbrako Ltd).

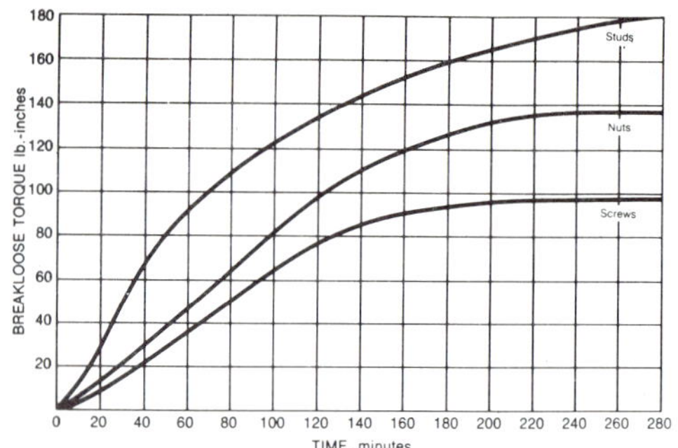

Fig 2 Effect of time versus breakloose torque on 5/16in heat treated steel fasteners coated with Unbrako accelerator. Fasteners were coated with adhesive for studs in top curve, adhesive for nuts in middle curve, and adhesive for screws in bottom curve. (Unbrako Ltd).

TABLE I – UNBRAKO ANAEROBIC LOCKING ADHESIVES DATA

UNBRAKO Adhesive	Colour	Specific Gravity (±0.05)	Viscosity at 20°C (68°F) cps	Gap Filling Ability (max) in	mm	Breakloose Torque† lb-inches	Newton-metres
for screws	purple	1.10	150–300	0.008*	0.20	50–100	5.6–11.3
for nuts	blue	1.10	500–800	0.010*	0.25	100–150	11.3–17.0
for studs	red	1.05	500–800	0.010*	0.25	150–200	11.4–23.0
						Shear Strength ϕ lb/in²	Newtons/mm²
for bearings	yellow	1.10	500–800	0.010	0.25	1500–2000	10.3–13.8
for shafts	green	1.10	600–1000	0.012	0.30	3000–4000	20.7–27.5

* *Accommodates class 1, 2, and 3 thread fits.*
† *Typical values when tested with 3/8–24 Grade 8 steel fasteners. Tested in unseated condition to MIL-S-22473D.*
ϕ *Shear strength ratings are based on tests using a ½" i.d. x ½" long steel bushing and ½" d. x 1" long pin with 0.002" diametral clearance and 63 microinch finish.*

Certain non-ferrous surfaces, hardened, plated or passivated surfaces may have an inhibiting effect on the curing of certain types of adhesive and require special treatment. Where necessary, this will be specified by the adhesive manufacturer.

Typical performance data for anaerobic thread-locking adhesives are given in Table I. Breakloose torque figures can be compared with those of primary torque fasteners – see appropriate chapter in Mechanical Fasteners section.

Most thread locking adhesives are fully resistant to vibration, oils, solvents and heat. They will, however, have a maximum service temperature above which degradation of the bond will occur. Most types have a limited shelf life and should be stored in a cool place, when a shelf life in excess of one year can be expected.

'Passivated' pre-applied adhesives are the latest development in thread locking adhesives, eg Scotch-Grip Adhesive 2353. This is a micro-encapsulated product which can be applied to screw threads, the adhesive then remaining dormant until the twisting action of applying the fastener causes the adhesive to cure. The adhesive begins to develop strength immediately at room temperature and achieves almost 90% of its ultimate strength in 24 hours. Complete curing is possible in 15 minutes at a temperature of 71°C (160°F), but would be dependent on the heating rate of the assembly. Breakloose torque figures obtained on test are summarised in Table II.

A specific advantage of this type of thread-locking adhesive is that precoated fasteners and threaded parts can be packed and shipped in normal fashion, and require no special storage. Shelf life of the adhesive is stated to be at least 48 months.

TABLE II – PERFORMANCE ON VARIOUS FINISHES AND SUBSTRATES

The adhesive performance on seven finishes and substrates was determined using 3/8in – 16 x 1¼in hex head cap screws with matching nuts. A cure time of 24 hours at 24°C (75°F) was employed for each use. Ten samples were used for each determination.

Grade	5	5	5	5	5	304	2024-T4
Finish	Plain	Phosphate and oil	Cadmium	Zinc	Chrome bright	Stainless steel	Aluminium
Adhesive weight grams/inch	0.1620	0.1434	0.1662	0.1753	0.1617	0.1655	0.1667
Break-away torque – ft/lbs Control – no adhesive	0	0	0	0	0	0	0
1st use with adhesive	20.8	23.2	22.6	18.8	23.0	8.6	8.8
2nd use with adhesive	28.4	23.4	19.2	17.4	13.3	9.8	10.0
3rd use with adhesive	14.0	8.6	7.8	8.6	3.5	3.6	5.0
Prevailing-out torque–ft/lbs Control – no adhesive	0	0	0	0	0	0	0
1st use with adhesive	9.6	13.4	10.6	12.2	13.5	2.8	5.0
2nd use with adhesive	17.6	13.8	9.6	8.6	7.5	5.6	4.8
3rd use with adhesive	8.8	7.4	5.6	5.6	2.5	2.4	3.8
Prevailing-in torque–in/lbs Control – no adhesive	0	0	0	0	0	0	0
1st use with adhesive	7.8	6.2	5.6	5.8	4.5	2.8	4.6

Adhesives for Non-Threaded Parts

Similar adhesives can be used for locking non-threaded parts – eg for fitting ball and roller bearings, collars, bushes, pulleys, gears, etc. These are normally medium or high strength adhesives, depending on the application, of medium viscosity having a gap-filling capability of the order of 0.25mm (0.010in) to 3mm (0.012in).

The particular advantage of such an adhesive is that it allows components to be assembled with a slip fit instead of a press fit, or eliminates the need for mechanical fastening via keys, splines, pins, etc. Also the gap-filling properties of the adhesive can provide satisfactory locking on worn or poorly machined parts, etc. Fig 3 shows comparisons of the push-out force of shafts in collars press-fitted and slip-fitted with adhesive.

Adhesive used may be *medium strength* to allow the component to be refitted – eg for mounting bearings into housings or on shafts, or fixing bushes, collars or pins; or *high strength* for permanent assembly – eg assembly of liners, gears, bushes, sleeves, splines, etc.

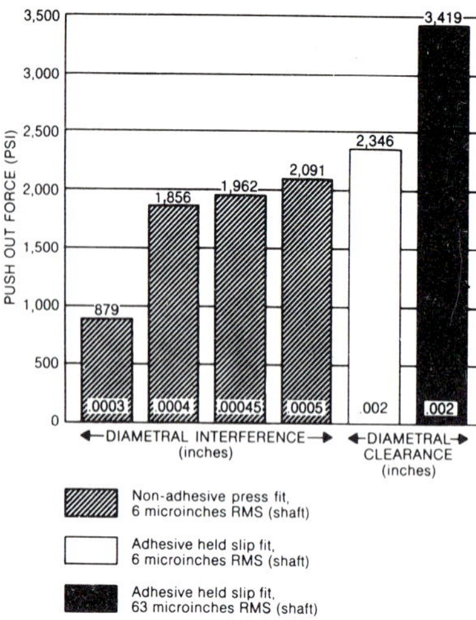

Fig 3 Comparison of push out force of shafts in collars. First four are for interference press fits without Unbrako anaerobic adhesive; last two are for slip fits with adhesive. Collar i.d. and shaft diameter were a nominal one-half inch. (Unbrako Ltd).

Thread Sealing Adhesives

The chief property required of a thread sealing adhesive is gap-filling ability to provide a gas- or liquid-tight seal. At the same time it will usually provide a varying degree of thread locking, depending on the type of adhesive used. It is more important, however, that the properties of the adhesive be tailored so that it will not exude and block the pipe when assembled, or extrude or disintegrate under pressure. It must also be fully compatible with the type of fluid carried by the pipe.

Thread sealants for general pipe sealing applications are usually medium strength, medium viscosity adhesives. Low strength, high viscosity adhesives may be used for higher temperature services (eg sealing piping on internal combustion engines).

Adhesive Tapes

Adhesive tapes — also known as self-adhesive tapes or pressure-sensitive tapes — consist of a tape backing coated on one, or both, sides with an adhesive which forms a bond when the tape is pressed against a surface with light pressure. Tape-backing materials include papers and treated papers, plastic films, fabrics, glass cloth and metal. The adhesive is usually a rubber-resin, but tapes designed for sticking to 'waxy' surfaces may have a synthetic adhesive (Table I).

TABLE I — MECHANICAL PROPERTIES OF TYPICAL SELF-ADHESIVE TAPES

Tape	Adhesive	Thickness		Tensile Strength		Elongation to Break	Initial Tear Resistance		Bursting Strength	
		in	mm	lb/in	kg/25mm	%	lb	kg	lb/in	kg/25mm
Paper	rubber-resin	0.006-0.016	0.15-0.40	10-80	4.5-36	2-12	–	–	–	–
Crepe paper	rubber-resin	0.0075-0.010	0.18-0.25	20	9	8-17	–	–	–	–
Crepe paper	phenolic	0.008	0.20	10	4.5	9	–	–	–	–
Cellulose acetate	resin-latex	0.003	0.08	20	9	up to 10	–	–	–	–
		0.005	0.125	40	18	–	11.5	5	96	44
PVC	latex	0.007	0.18	20-22	9-10	175	–	–	–	–
		0.009	0.23	25	11.5	170	–	–	–	–
Vinyl (unplasticised PVC)	resin-latex	0.003	0.08	20-22	9-10	25	18.5	8.5	36	16
Polythene	resin-latex	0.009	0.20	8	4	150	–	–	11	5
Polyester		0.002-0.0025	0.05-0.06	25	11.5	110	–	–	–	–
Cellulose acetate fibre	resin-latex	0.003	0.08	20	9	5	–	–	–	–
Filament tapes		0.007-0.010	0.18-0.25	240-500	108-225	3-15	–	–	–	–
Glass fibre	phenolic	0.008	0.20	130	60	6-15	60	27	130	60
Cellulose acetate cloth	phenolic	0.008	0.20	35	16	20	–	–	–	–
Cloth (Egyptian cloth)	rubber-resin	0.012	0.30	40	18	–	27	12	70	32
PVC impregnated cloth	zinc oxide based	0.012	0.30	60	27	9	41.5	19	130	60
Insulating tape (old type)	rubber solution	0.013	0.33	45	20	–	27.5	12.5	–	–
Aluminium foil	rubber-resin	0.005	0.125	30	14	5	–	–	–	–
Lead foil	rubber-resin	0.007	0.18	20	9	15	–	–	–	–

The important physical properties of an adhesive tape are:-

(i) Thickness – normally expressed in thousandths of an inch (mils).

(ii) Tensile strength – expressed as the force or pull to break a 25mm (1in) wide strip of tape.

(iii) Elongation to break (also known as Stretch Ratio) – or the amount which the tape can be stretched before breaking, expressed as a percentage of the original length.

(iv) Initial tear resistance – in terms of lbf or kgf pull.

(v) Bursting strength – in lb/in^2 or bar.

(vi) Adhesive strength – normally determined as the force required to pull a 25mm (1in) wide strip of tape from a clean aluminium plate (BS1133) or steel plate (German Federal test method STD No 147). (Table II).

TABLE II – ADHESIVE STRENGTH OF TYPICAL TAPES

Tape-Backing	Adhesive Strength BS1133*		Fed STD No 147†
	in	mm	oz
Cellulose acetate	0.1	2.5	25–35
Cellulose hydrate	0–0.5	0–1.25	–
Ethyl cellulose	4.5–5	115–125	–
Polythene	5–6	125–150	–
Vinyl	0.9–1.5	22–40	–
PVC	0.1	2.5	–
Plastic films (generally)	–	–	18–35
Polyester	–	–	25–40
Acetate fibre	–	–	25–35
Filament tapes	–	–	40
Aluminium foil	–	–	50
Lead foil	–	–	40
Paper masking tape	16–25	400–625	40 (typical)
Cloth masking tape	50–60	1250–1500	40

**1in width of tape stuck to a clean aluminium surface length of tape peeled off in 5 minutes with 30 ounce vertical pull.*

†Force to pull 1in wide strip of tape from a steel surface.

For electrical tapes the following properties are also significant (Table III).

(vii) Insulation resistance.

(viii) Breakdown voltage

(ix) pH value

(x) Sulphate content

(xi) Chloride content.

(xii) Moisture vapour permeability – Table IV.

TABLE III – ELECTRICAL PROPERTIES OF TYPICAL SELF-ADHESIVE TAPES

Tape	Insulation Resistance (megohms)	Breakdown Voltage	pH	Sulphate Content %	Chloride Content %
Cellulose hydrate	10	less than 500	7.5	0.05 (max)	0.01 (max)
Ethyl cellulose	1000000	5300	7.0	0.05 (max)	0.01 (max)
Acetate film (0.003in)	800000	4500	7–7.5	0.05 (max)	0.01 (max)
(0.005in)	1000000	4800	7.1	0.05 (max)	0.01 (max)
Acetate fibre	–	4000	6.5–7.5	–	0.0005
Acetate cloth	–	1700	6.5–7.5	–	0.0005
PVC (0.007in)	–	Up to 5000	6–6.5	–	0.0008
(0.009in)	5000000	4500	6.4	0.05 (max)	0.01 (max)
Vinyl	5000000	2000	7.5	0.05 (max)	0.01
Polythene	50000000	5800	6.2	0.05 (max)	0.02
Crepe paper (treated)	20000	1300	6.5–7.5	–	0.0025–0.004
PVC impregnated cloth	200000	2000	7.1	0.05 (max)	0.01 (max)
Insulating tape	2000	1500	8	0.02	0.01
Glass fibre	1000000	2300	7.1	less than 0.05	less than 0.01

TABLE IV – MOISTURE VAPOUR PERMEABILITY

Tape	Adhesive	Thickness in	Thickness mm	Permeability (grammes per sq metre)
Cellulose hydrate	rubber	0.003	0.08	36
Cellulose acetate	resin-latex	0.005	0.125	2.6
Ethyl cellulose	resin-latex	0.005	0.125	26
PVC	latex	0.009	0.23	4.0
Vinyl	resin-latex	0.003	0.08	7.0
Polythene	resin-latex	0.004	0.10	less than 1.0
Crepe paper	resin-rubber	0.008	0.20	31
Cloth	resin-rubber	0.012	0.30	150
Cloth (PVC impregnated)	zinc oxide based	0.012	0.30	11
Glass fibre	pressure-sensitive	0.008	0.20	7.0

Paper Tapes.

Paper tapes are made in a wide variety of thicknesses and strengths and are used mainly for holding, masking, labelling and simple sealing duties. Plain paper tapes have little or no stretch. Crepe paper tapes have moderate stretch and high conformability, and are thus particularly suitable for masking duties on curved surfaces.

Cellulose Tapes

Cellophane (cellulose acetate) tapes are the original type of plastic self adhesive tape developed for general and industrial use. There are numerous types – high-lock, low-lock, clear, coloured, etc, with a wide application. Other cellulose tape-backings include: (see also Table I)

cellulose hydrate – mainly used for sealing, holding and packaging applications.

ethyl cellulose – rather more flexible than cellulose hydrate tapes and more conformable to curved surfaces, etc.

acetate fibre tapes – with a backing of cellulose acetate film and tissue paper laminations to improve tear resistance.

Plastic Tapes

Plastic tape backings of plasticised vinyl, polyethylene and polypropylene produce stretchable tapes, with high conformability. They are widely used for holding, masking, weather protection, moisture barriers and as electrical insulating tapes (Table III). Unplasticised PVC tapes are clear and generally similar in appearance to cellophane tapes but with higher conformability and greater stability and resistance to moisture.

Polyester Tapes

Polyester tapes have excellent mechanical, electrical and chemical properties and are coming into wider use.

Filament Tapes

These are high-strength tapes consisting of continuous filaments of nylon or glass fibre yielding tensile strengths of 250 to 500lb per inch. They are virtually tearproof and are widely used for palletising, bundling, etc and other duties where high resistance to shock and impact is required.

Foam Tapes

Foam tapes are made in a wide variety of cellular elastomers in a range of different densities. Thickness can range from 0.030in to 0.250in. Such tapes may be single-coated or double-coated with adhesive. Higher density foam tapes with high shear strength are used for such duties as mounting relatively heavy components on rough surfaces. Lower density tapes may be designed for such duties as insulating, padding, cushioning, gasketing, draught-sealing, etc.

Double-coated Tapes

Double-coated tapes have high-lock adhesive on both sides, usually with a paper liner facing one side (see Table I). They are widely used for fixing and holding duties particularly in the printing, paper, electrical, textile and plastic industries, eg –

Printing Industry – fixing of blocks, metal and plastic plates in flat-bed and rotary printing. Rubber blocks in flexograph printing. Rubber blocks in heated carbon printing.

Paper Industry – Splicing. Securing the start of the reel. Automatic reel change. Manual reel change. Ready-made packing cut outs. Securing samples and posters. Securing book wrappers. Rendering sanitary towels and babies' napkins self-adhesive. Mounting of displays and decorative materials. Rendering visiting cards self-adhesive.

Electrical Industry – Rendering drilling jigs self-adhesive. Securing components on assembly line. Mounting radio dials. Securing parts during machining. Rendering insulating materials self-adhesive. Affixing decorative foils for lighting and on dashboards. Securing impulse tapes. Rendering soundproofing material self-adhesive. Securing refrigerator guard plates. Affixing high quality wood veneers. Securing wires before mounting. Affixing name plates in appliances. Rendering magnetic catches self-adhesive.

Textile Industry – Affixing brake material in shuttles. Affixing swatches on sample cards. Joining disposable clothing of non-woven textiles. Briefly affixing materials. Assembling folding cartons.

TABLE V – PROPERTIES OF TYPICAL DOUBLE-COATED TAPES*

Backing	Thickness mm	Tensile Strength kg/25mm	Elongation at Break %	Adhesive Power g/25mm
Cloth with resin adhesive	0.41	22.5	10	1200
Cotton cloth	0.45	18	10	1200
Tissue base	0.12	1.5	–	1500
Tissue base	0.09	1.5	–	800
Paper	0.23	25	6	1500
UPVC	0.11	10	40	1000
Rayon fabric	0.38	18	10	1200
Adhesive only	0.04	–	–	1000
Non-plasticised PVC	0.12	–	–	1100
Non-plasticised PVC	0.15	–	–	1100
Non-plasticised PVC	0.20	–	–	1100
Non-plasticised PVC	0.25	–	–	1100
Rayon fabric	0.38	–	–	1200
Rayon fabric	0.45	–	–	1200

**Tesa Tapes Ltd*

Rendering decorative ribbon self-adhesive. Cleaning lengths of material by the lint pick-up roll system. Securing material during screen printing. Securing lengths of material to centre cores. Affixing zip fasteners.

Leather and Shoe Industry – Folding of uppers. Rendering shoe inlays self-adhesive. Joining insoles and polyurethane foam soles. Joining fur and leather pieces. Affixing lining in leather clothing.

Plastics Industry – Rendering skirting and kicking plates self-adhesive. Affixing plastic floor tiles. Sticking on abrasive grinding discs. Splicing plastic foil. Rendering hooks and key racks self-adhesive. Securing mirrors and powder compacts. Securing plastic covers. Rendering plastic foam and rubber pieces self-adhesive. Securing plastic sheeting during chemical cleaning of facades. Rendering weather seals self-adhesive. Rendering pelmets self-adhesive.

Carpet Laying – Laying carpets, underlays and other floor coverings.

Building Industry – Mounting cable ducts. Simplifying the mounting of double glazing. Fixing facing slabs in concrete moulds. Affixing plaques to doors. Constructing models. Affixing insulating boards between inner construction and facade lining.

Furniture Industry – Fixing suspended, filling strips. Affixing materials to screens and display walls. Affixing masking materials. Packing furniture. Affixing covering materials. Sticking mirrors to wardrobes, etc. Securing cutlery in display boxes. Rendering letters self-adhesive.

Toy Industry – Assembling dolls' furniture, plastic pieces, etc. Affixing dolls' eyes and hair. Short-term securing of toys in display cartons. Rendering hair pieces, wigs and beards for fancy dress self-adhesive. Affixing dedications inscribed in plastic on presentation cups. Securing small pieces prior to drying in the oven.

Metal Tapes

Aluminium foil tapes are used for sealing, protection and similar applications. Thickness is usually 0.005in and tensile strength about 30lb per inch. Elongation is of the order of 5%.

Lead foil tapes are used for sealing and also for masking in hard chrome plating. Thickness is usually 0.006–0.007in and tensile strength about 20lb per inch. Elongation is of the order of 15%.

Transfer Tapes

Transfer tapes consist of thin adhesive film only and are usually of the order of 0.002in thick. They are used for metal-to-metal bonding, attaching plastic or metal nameplates to surfaces and similar duties.

Bonding Film Tapes

Bonding film tapes are similar to transfer tapes but generally thicker and stronger so that they can be cut to any specific shape required. They are used for metal-to-metal bonding of heavier parts, and similar applications.

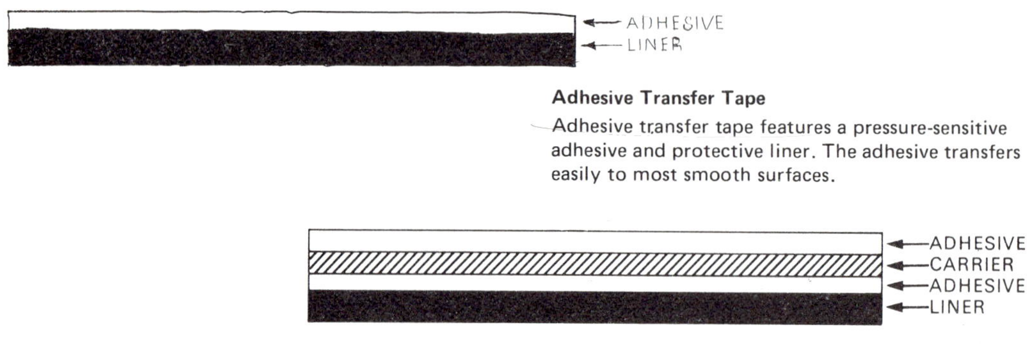

Adhesive Transfer Tape

Adhesive transfer tape features a pressure-sensitive adhesive and protective liner. The adhesive transfers easily to most smooth surfaces.

Double-Coated Tape

Double-coated tapes provide a protective liner, a layer of adhesive, a carrier made up of paper or film, and a second coating of adhesive. The thickness of the carrier allows some conformability to irregular surfaces and better removability than adhesive transfer tape.

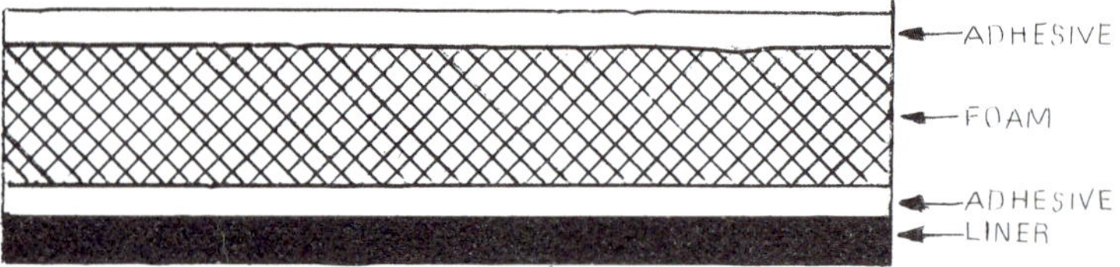

Foam Tape

These high strength foams are coated on both sides with a pressure-sensitive adhesive. Foam tapes readily conform to irregular surfaces. Available in urethane for heat resistance, vinyl for moisture resistance and self extinguishing properties, neoprene for internal strength and polyethylene for adhesion to plastics.

Fig 1 Three types of tape used for jointing (3M).

TABLE VI – EXAMPLES OF JOINTING TAPES (3M)

Adhesive Carrier	Solvent Resistance	Temperature Resistance	Product Characteristics	Density (lbs/cu ft)
Paper	Fair	65°C (150°F) ext 93°C (200°F) short	Excellent adhesion to rubber	
Paper	Good	107°C (225°F) ext 148°C (300°F) short	Good temp. and plasticiser resistance	
Saturated paper	Fair	65°C (150°F) ext 93°C (200°F) short	Easy-release liner	
Vinyl plastic	Good	43°C (110°F) ext 65°C (150°F) short	Good plasticiser resistance	
Polyester film	Fair	82°C (180°F) ext 121°C (250°F) short	High initial adhesion	
Vinyl film	Fair	43°C (110°F) ext 54°C (130°F) short	Large adhesive mass	
Tissue	Fair	82°C (180°F) ext 93°C (200°F) short	Good general-purpose bond	
Transfer	Fair	82°C (180°F) ext 121°C (250°F) short	High initial adhesion	
Adhesive	Fair	82°C (180°F) ext 121°C (250°F) short	High initial adhesion	
Urethane foam	Excellent	93°C (200°F) ext 190°C (375°F) short	Firm, high-shear strength	19
Urethane foam	Excellent	93°C (200°F) ext 190°C (375°F) short	Firm, high-shear strength	17
Urethane foam	Excellent	93°C (200°F) ext 190°C (375°F) short	Firm, high-shear strength	14
Urethane foam	Excellent	93°C (200°F) ext 190°C (375°F) short	Firm, high-shear strength	12
Vinyl foam	Excellent	65°C (150°F) ext 107°C (225°F) short	Firm, high-shear strength	15
Vinyl foam	Excellent	65°C (150°F) ext 107°C (225°F) short	Firm, high-shear strength	15
Vinyl foam	Excellent	65°C (150°F) ext 107°C (225°F) short	Firm, high-shear strength	15
Polyethylene	Fair	70°C (158°F)	High tack, excellent to plastics	6
Neoprene foam	Excellent	93°C (200°F) ext	Firm, high shear strength	35
Elastomeric foam	Excellent	65°C (150°F) ext	Firm, high-shear strength	40
Polypropylene foam	Excellent	93°C (200°F) ext	Firm, high-shear strength	30

**Temperature resistance shows average temperatures the system will withstand under typical loads. It does not describe permanent bonding under maximum loads at these temperatures.*

TABLE VII SELECTION GUIDE FOR JOINTING TAPES

Substrate ↓ To →	Paper	Metal	Glass & Ceramic	Rubber	Paint	Cloth	Vinyl	Polythene and Polypropylene	ABS and Acrylic	Polystyrene
Paper	T/ 2/E F/U	–	–	–	–	–	–	–	–	–
Metal	T/ 2/E F/U	T/ 2/E F/U	–	–	–	–	–	–	–	–
Glass & Ceramic	T/ 2/E	T/ 2/E	T/ 2/E	–	–	–	–	–	–	–
Rubber	T/ 2/V F/U	T/T 2/V F/U	T/T 2/V F/U	T/T 2/V F/U	–	–	–	–	–	–
Paint	T/ 2/P F/U	T/ 2/P F/U	T/ 2/P F/U	T/T 2/P F/U	T/ 2/P F/U	–	–	–	–	–
Cloth	T/ 2/V F/U	T/ 2/V F/U	T/ 2/V F/U	T/ 2/V F/U	T/ 2/V F/U	T/ 2/V F/U	–	–	–	–
Vinyl	T/T 2/P F/U	T/T 2/P F/U	T/T 2/P F/U	T/T 2/P F/U	T/T 2/P F/U	T/T 2/P F/U	T/T 2/P F/U	–	–	–
Polythene & Polypropylene	T/T 2/P F/F	2/P F/F	2/P F/F	2/P F/F	2/P F/F	2/P F/F	2/P F/F	2/P F/F	–	–
ABS and Acrylics	T/T F/F	– F/F	– F/F	T/T F/F	– F/F	– F/F	T/T F/F	– F/F	–	–
Polystyrene	T/ 2/E F/U	T/ 2/E F/U	T/ 2/E F/U	T/ 2/E F/U	T/ 2/E F/U	T/ 2/E F/U	T/T 2/P F/U	– – F/U	– – F/U	T/ 2/E F/U

KEY:

T/ Transfer tape
2/ Double sided tape
F/ Foam tape
/P Paper
/T Tissue
/E Polyester
/F Polythene
/N Neoprene
/U Urethane
/V Vinyl

Methods of Application

METHODS OF APPLYING ADHESIVE

Method	Suitability, etc
Direct from Tube or Applicator	Particularly suitable for sealants, mastics, etc. Gives poor spread over larger areas, requiring additional use of a spreader; includes tubes, squeeze bottles and syringe-type applicators.
Finger	Suitable for spreading small quantities of adhesive, but many adhesives may be difficult to remove from the finger.
Oilcan	Useful for 'spot' application of adhesives, also for applying adhesives to blind holes.
Brush	Suitable for most low to medium viscosity adhesives, the size of brush being chosen to suit the surface area involved. Stiffness of brush should be chosen to suit viscosity of adhesive. Brushes with metal ferrules may have to be avoided with certain resins. Brushes are often difficult to clean.
Capillary Applicators	Capillary applicators are suitable for use with thin water-based and other low viscosity adhesives. They consist, basically, of a wick dipping into a trough of adhesive. Capillary action draws the adhesive to the top edge of the wick over which the surface to be coated is passed. Capillary applicators may be used to apply water only to a pre-glued surface (eg gum strip)
Flow Brush	Special type of applicator for distributing and spreading adhesives. Only suitable for certain types of adhesive.
Roller	Gives faster and more uniform distribution than a brush and is therefore better for larger areas. It is also more suitable than a brush for applying more viscous adhesives, and/or thicker adhesive films. Several different types of roller are used, including those fed by a hose for production line work (generally called a *Flow Roller*).

cont...

METHODS OF APPLYING ADHESIVE

Method	Suitability, etc
Glue Gun	Types vary from small plunger-operated guns (ie syringes) to pressurised guns fed by a hose. They are particularly suitable for the spot application of adhesive or adhesive lines in production work. A variation is the *Glue Striper.*
Silk Screening	Particularly useful for repetitive runs, where adhesive has to be applied to limited areas or irregular shapes. Screen is used with a rubber squeegee.
Stencil	Useful for coating discrete or irregularly shaped areas. Adhesive may be applied by brush, roller, etc.
Rubber Stamp	Useful for coating specific shapes with adhesives too viscous to be applied easily by silk screening or stencilling.
Spray Gun	Particularly suitable for large scale application of adhesives with low viscosity, but requires relatively skilled operation.
Glue Roll	Many different types and sizes and well suited to repetitive work involving the coating of flat surfaces and flexible stock.
Gravure Roll	Similar in principle to the glue roll but applies a controlled amount of adhesive to the work.
Grooved Rolls	Used for applying wide bands of adhesive to a surface whilst retaining a clear margin between the adhesive areas.
Disc Rolls	Used for applying strips of adhesive to a surface.
Segmental Rolls	Used for applying adhesive lines in slots or interrupted strips. A further variation on this is the *Shoe Roller.*
Trough	Hopper fed mechanism used for edge gluing of stocks of flat sheets (eg paper).
Belt	Various types of belt mechanism are used for line application of adhesives; segmental belts may be used for for interrupted line application.
Spotters	Various mechanisms are used for applying spots of adhesive to surfaces in a predetermined pattern. The spotting heads may be tubes fed by gravity or pressure, or pins or dowels dipped into a trough containing adhesive and then raised vertically against the work surface.
Air Knife	The air knife is a wide, narrow nozzle from which high pressure air is ejected. The surface to be coated is drawn under the knife, and a surplus of adhesive allowed to flow on to the surface in front of the knife. The flow of air causes the adhesive to be distributed uniformly over the surface as it passes under the knife.

Trouble - Shooting Chart - Bonding Faults

TROUBLE-SHOOTING CHART – BONDING FAULTS

Fault	Possible Causes
GENERAL:	
Lack of adhesion	(i) Use of unsuitable adhesive (ii) Greasy substrates
Lack of bond strength	(i) Use of unsuitable adhesive (ii) Unsatisfactory surface preparation (iii) Insufficient time for adhesive to set (iv) Lack of closing pressure (v) Poor spread of adhesive
Blistering	(i) Insufficient adhesive (ii) Insufficient setting time (iii) Too low a temperature (iv) Insufficient closing pressure (v) Entrapped air, water or solvent vapour
THERMOSETTING ADHESIVES:	
Lack of adhesion or poor bond strength	(i) Insufficient curing time (ii) ambient temperature too low (iii) Incorrect proportions of resin and catalyst or hardener (iv) Unsatisfactory surface preparation (v) Insufficient spread of adhesive (vi) Lack of closing pressure
Bond failure under stress	(i) Wrong type of adhesive (ii) Unsatisfactory surface preparation (iii) Insufficient spread of adhesive (which should be obvious on examination of failed bond surfaces)
CONTACT ADHESIVES:	
Film remains tacky	(i) Entrapped solvent – insufficient time allowed for adhesive to set off before bringing substrates together. (ii) Migration of plasticiser from substrates into adhesive – ie unsuitable combination.
Lack of bond	(i) Unsuitable adhesive (ii) Unsatisfactory surface preparation
Lack of strength	(i) insufficient spread of adhesive (ii) Insufficient closure pressure
HOT MELT:	
Lack of bond	(i) Unsuitable adhesive (ii) Adhesive temperature too low when used (iii) Substrates too cool at time of assembly

General Guide to Adhesive Selections

(See also under *Adhesive Types*).

Metal Bonding – Structural Applications

Aluminium and aluminium alloys, copper and copper alloys, iron and steel

epoxy
epoxy-phenolic
epoxy-polyamide
epoxy-silicone
phenolic-neoprene
phenolic-nitrile
phenolic-polyamide
polyamide
polyurethane
cyanoacrylate
ceramic

Metal Bonding – Non-Structural

Aluminium and aluminium alloys, copper and copper alloys, iron and steel. (see under individual adhesive headings for limitations with plated or passivated surfaces, etc).

acrylic
epoxy
PVA
polyamide
rubber-based adhesives

Plastics – Solid or sheet form, or mouldings

(Note: 'cement' designates a solution of the same plastic in a solvent).

ABS – epoxy (abrade surface first)
– rubber based (impact adhesive)

Acrylics – cement ('perspex' cement)
– epoxy
– resorcinol formaldehyde (RF)
– rubber based

Cellulose acetate	— cement ('cellulose'cement) — epoxy — PVA
Cellulose nitrate	— cement ('cellulose' cement) — epoxy — resorcinol formaldehyde (RF)
Nylons	— epoxy — phenolics — resorcinol formaldehyde (RF)
Polystyrene	— cement ('plastic' cement) — epoxy
Polyethylene	— phenolics — silicone
PTFE	— silicone
PVC	— acrylics — neoprene rubber — nitrile rubber — phenolic-neoprene — phenolic-nitrile — phenolic-vinyl
GRP	— epoxy
Thermoset plastics	— epoxy — phenolic-neoprene — phenolic-nitrile — phenolic-vinyl — polyester — synthetic rubber-based — resorcinol formaldehyde (RF) — urea formaldehyde (UF)

Laminated Thermoset Plastics

— synthetic rubber based (impact adhesives)
— epoxy
— phenolic-neoprene
— phenolic nitrile
— phenolic vinyl
— polyester
— resorcinol formaldehyde (RF)
— urea formaldehyde (UF)

Proprietary Adhesives Selection Guide

ATLAS ADHESIVES & COATINGS FOR INSULATION MATERIAL

Insulant	Adhesive	Sealer	Coating	Mastic
Calcium silicate, 85% magnesia and asbestos	Fibrous 81–27	Heat Resistant Sealer 30–43 Foamseal 30–45	Sealfas 30–36 Lagtone	H.I. Mastic 90–07 Sealfas Mastic GPM Mastic 35 Range
Fibrous glass and mineral wool	Fire Resistive 81–33 Safetee Ductfas 81–99 Lagfas 81–42W Ductfas 81–22 Koldfas 82–08 Clipfas Neoprene 13–29 (with Clipfas Hangers)	Heat Resistive 30–43 Foamseal 30–45	Sealfas 30–36 Lagtone	H.I. Mastic 90–07 Sealfas Mastic White and Colours GPM Mastic 35 Range
Polyurethane	Fire Resistive 81–33 Safetee Ductfas 81–99 Koldfas 82–08 Ductfas 81–22 Flexfas 82–10 Two-part adhesive 81–80	Foamseal 30–45	Sealfas 30–36 Lagtone	C.I. Mastic 60–25 H.I. Mastic 90–07 Sealfas Mastic Fire Resistive 60–75 Monolar Mastic 60–36 GPM Mastic 35 Range
Cork	Koldfas 82–08 Safetee Ductfas 81–99 Fire Resistive 81–33 Two-part adhesive 81–80	Foamseal 30–45	Lagtone	Sealfas Mastic Fire Resistive 60–75 H.I. Mastic 90–07 C.I. Mastic 60–25 Monolar Mastic 60–36 GPM Mastic 35 Range
Polystyrene	Flexfas 82–10 Koldfas 82–08 Ductfas 81–22 Clipfas Neoprene 13–29 (with Clipfas Hangers) Two-part adhesive 81–80	Foamseal 30–45	Sealfas 30–36 Lagtone	Sealfas Mastic GPM Mastic 35 Range
Cellular glass	Flexfas 82–10 Koldfas 82–08 Anti abrasion 30–16 Two-part adhesive 81–80	Foamseal 30–45	Sealfas 30–36 Lagtone	C.I. Mastic 60–25 Fire Resistive 60–75 Sealfas Mastic Monolar Mastic 60–36 GPM Mastic 35 Range

BEXOL ADHESIVES – RECOMMENDED APPLICATIONS

Bexol No	Application
13	Acetate to acetate
21	Cellulose nitrate to wood
44	Xylonite to xylonite
65	Fabric to acetate
101	Polystyrene
531	ABS to ABS
600	Flexible PVC to flexible PVC Flexible PVC to wood
608	PVC
657	General purpose adhesive Cellulose acetate to paper and fabric PVC to paper and fabric Polystyrene to paper and fabric SAN, to paper and fabric ABS to paper and fabric Expanded polystyrene to rigid PVC, ABS, SAN, polystyrene, wood, paper and fabric
1122	Fast drying clear adhesive for bonding toughened polystyrene or ABS
1123	Slow drying clear adhesive for bonding toughened polystyrene or ABS
1129	Quick drying bodied cement for toughened polystyrene or ABS
1162	Quick drying polystyrene adhesive
1409	Cellulose acetate to paper
1505	Fast drying clear general purpose adhesive for rigid PVC, ABS, SAN to each other and to wood, metal and fabric
1528	General purpose contact adhesive for bonding PVC, ABS, polystyrene , rubber, wood, Wareite to non-metal substrates
1533	Vinyl metal adhesive for roller or curtain coating
1551 and 1564 (clear)	Pipe jointing adhesive for rigid PVC pipe joints
1563 and 1587 (clear)	Pipe jointing adhesive for bonding rigid PVC, ABS and C PVC extruded pipes and fittings
1606	Non-flam, Non-gap filling adhesive for bonding rigid PVC to itself, and for bonding rigid PVC, C PVC, ABS pipes and fittings
1607	Vinyl metal adhesive for electro-static deposition
1628	Pressure sensitive emulsion adhesive for film to paper and cardboard
1629	PVC pipe cement with gap filling properties Thicker version of 1606
1634	Roller coated vinyl metal adhesive
1637	PVC or ABS
1644	Low temperature heat seal adhesive
1649	VPS to chipboard
1650	PVC to hardboard for Thames Board Mills – Autoboard
1651	Vinyl metal adhesive for roller or curtain coating, Modified version of 1533

BLOOMINGDALE ADHESIVES AND APPLICATIONS (CYANAMID)

Adhesive	FILM ADHESIVES Curing Temperature Range – °F	Operating Temperature Range – °F	Intended Use	Applicable Specification (* = Qualified)
FM-24 adhesive film unsupported, modified epoxy	175 to 350	–67 to +180	Metal to metal & sandwich -- perforated skin applications	MMM-A-132, Type I Class 3, MIL-A-25463 Type I, Class 2
FM-25 adhesive film unsupported, modified epoxy	175 to 350	–67 to + 180	Same as FM-24 but non-reticulating	MMM-A-132, Type I Class 3, MIL-A-25463 Type I Class 2
FM-34 adhesive film polyimide on glass cloth	500 to 700	–400 to +700	Metal to metal & sandwich	MMM-A-132, Type IV MIL-A-25463 Type IV
FM-47 adhesive film, type I, vinyl-phenolic on glass cloth	300 to 350	–400 to +225	Metal to metal & sandwich	MMM-A-132, Type I Class 3 MIL-A-25463 Type I Class 2
FM-47 adhesive film, type II, vinyl-phenolic on glass cloth	300 to 350	–400 to +225	Metal to metal & sandwich	MMM-A-132, Type I Class 3* MIL-A-25463 Type I Class 2*
FM-53 adhesive film modified epoxy on nylon cloth	175 to 300	–67 to +180	Metal to metal & sandwich	MMM-A-132, Type I, Class 1 MIL-A-25463 Type I, Class 2
FM-53U adhesive film unsupported modified epoxy	175 to 300	–67 to +180	Metal to metal & sandwich	MMM-A-132, Type I, Class 2 MIL-A-25463 Type I, Class 2
FM-61 adhesive film, duplex nitrile – elastomer & modified epoxy on nylon cloth	320 to 350	–400 to +350	Metal to metal & sandwich	MMM-A-132, Type I, Class 3* MIL-A-25463 Type I, Class 2*
FM-73 adhesive film, supported modified epoxy	225 to 300	–67 to +250	Metal to metal & sandwich	MMM-A-132, Type I, Class 2 MIL-A-25463 Type I, Class 2
FM-96 adhesive film, modified epoxy on nylon carrier	320 to 350	–100 to +350	Metal to metal	MMM-A-132 Type I, Class 3* MIL-A-25463 Type I, Class 2*
FM-96U adhesive film, unsupported modified epoxy	320 to 350	–100 to +350	Same as FM-96 without carrier	MMM-A-132 Type I, Class 3 and Type II
FM-123-2 adhesive film, modified nitrile epoxy on dacron mat	225 to 250	–67 to +250	Metal to metal & sandwich	MMM-A-132, Type 1, Class 2* MIL-A-25463 Type I, Class 2*
FM-123-5 adhesive film modified nitrile epoxy on nylon carrier	200 to 250	–67 to +250	Metal to metal & sandwich	MMM-A-132, Type I, Class 2* MIL-A-25463 Type I, Class 2*
FM-137 adhesive film modified nitrile epoxy on dacron mat	200 to 250	–67 to +250	Metal to metal & sandwich	MMM-A-132, Type I, Class 2* MIL-A-25463 Type I, Class 2*
FM-150-2 adhesive film, modified epoxy on a nylon carrier	325 to 350	–67 to +350	Metal to metal & sandwich	MMM-A-132, Type II MIL-A-25463, Type Type II, Class 2

BLOOMINGDALE ADHESIVES AND APPLICATIONS (CYANAMID)

Adhesive	Curing Temperature Range – °F	Operating Temperature Range – °F	Intended Use	Applicable Specification (* = Qualified)
		FILM ADHESIVES		
FM-150-2U adhesive film unsupported modified epoxy	325 to 350	–67 to +350	Metal to metal & sandwich – perforated skin applications	MMM-A-132 Type II, MIL-A-25463 Type II, Class 2
FM-238 adhesive film, unsupported nitrile phenolic	325 to 350	–67 to +350	Metal to metal	MMM-A-132 Type I, Class 2*
FM-400 adhesive film, modified epoxy – aluminium filled on nylon carrier	330 to 350	–67 to +420	Metal to metal & sandwich	MMM-A-132, Type II, MIL-A-25463 Type II, Class 2
FM-400-6 adhesive film, unsupported modified epoxy, reticulating	330 to 350	–67 to +420	Metal to metal & sandwich – perforated skin applications	MMM-A-132, Type II, MIL-A-25463 Type II, Class 2
HT-424 adhesive film epoxy phenolic, aluminium filled on glass cloth	250 to 350	–400 to +500 & short time to 1000	Metal to metal & sandwich	MMM-A-132, Type II and III*, MIL-A-25463, Type II & III Class 2*
HT-435 adhesive film epoxy phenolic on glass cloth	325 to 350	–400 to +400	Same as HT-424 except without aluminium filler for electrical transparency	MMM-A-132 Type II
FM-1000 adhesive film unsupported polyamide-epoxy film	320 to 350	–423 to +200	Metal to metal & sandwich	MMM-A-132, Type I, Class 1*
FM-1000-EP-15 adhesive film, polyamide-epoxy on nylon carrier	320 to 350	–423 to +200	Metal to metal	MMM-A-132, Type I, Class 1*
FM-1015 adhesive film, polyamide-epoxy on a nylon carrier	325 to 350	–67 to +180	Metal to metal & sandwich	MMM-A-132 Type I, Class 1
FM-1015U adhesive film, unsupported polyamide-epoxy	325 to 350	–67 to +180	Metal to metal & sandwich	MMM-A-132 Type I, Class 1
FM-1044R adhesive film unsupported polyamide epoxy film	320 to 350	–67 to +180	Bonding copper, circuit boards 0.001 to 0.003in T	MIL-P-139498 Applied to copper clad laminates
FM-1045 adhesive film unsupported polyamide epoxy film	250 to 350	–67 to +180	Aircraft interior honeycomb applications	MMM-A-132 Type I, Class 2
		LIQUID AND PASTE ADHESIVES		
BR-34 primer polyimide in solvent blend	500 to 700	–400 to +700	Primer with FM-34 adhesive; also as metal to metal, esp. Ti adhesive	MMM-A-132 Type IV
BR-34B-18 adhesive polyimide in solvent blend	500 to 550	–400 to +700	Metal to metal and core beading, esp. non-metallic assemblies	MMM-A-132 Type IV
FM-47 adhesive, 20% solids vinyl-phenolic	300 to 350	–400 to +225	Metal to metal and as primer with FM-47 films	MMM-A-132 Type 1, Class 3*
BR-92, adhesive, modified epoxy paste, spreadable two-part system	75 to 350	–100 to +180	Metal to metal	MMM-A-132, Type I, Class 3* (with curing agent 'A')
HT-424F primer resins & aluminium in solvent blend	325 to 350	–400 to +500 & short time to 1000	Optional use with HT-424 film, (not mandatory). Also as load bearing liquid adhesive.	MMM-A-132 Type II and III MIL-A-25463 Type II and III Class 2

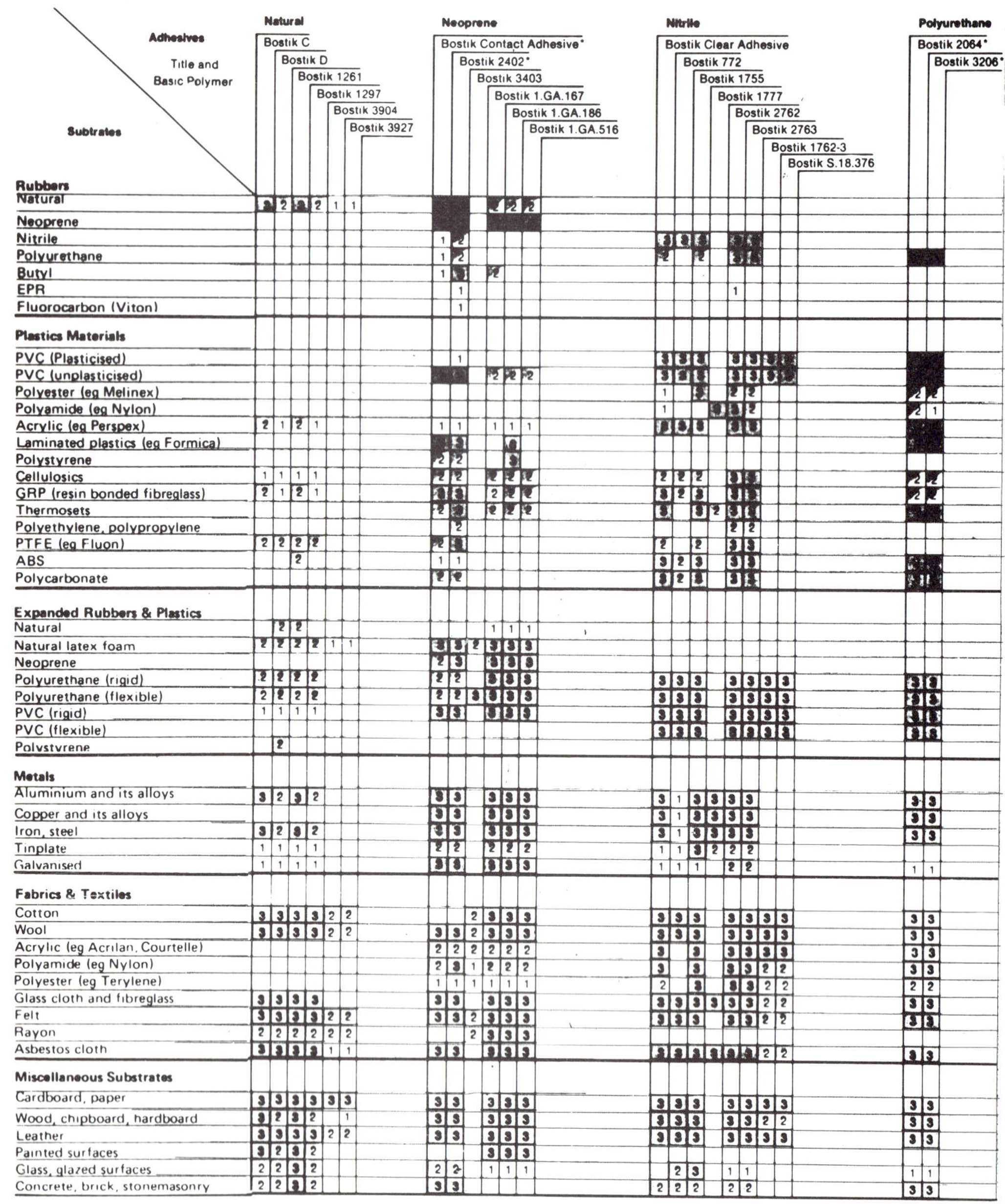

Adhesives: Title and Basic Polymer / Subtrates	Natural: Bostik C	Natural: Bostik D	Natural: Bostik 1261	Natural: Bostik 1297	Natural: Bostik 3904	Natural: Bostik 3927	Neoprene: Bostik Contact Adhesive*	Neoprene: Bostik 2402*	Neoprene: Bostik 3403	Neoprene: Bostik 1.GA.167	Neoprene: Bostik 1.GA.186	Neoprene: Bostik 1.GA.516	Nitrile: Bostik Clear Adhesive	Nitrile: Bostik 772	Nitrile: Bostik 1755	Nitrile: Bostik 1777	Nitrile: Bostik 2762	Nitrile: Bostik 2763	Nitrile: Bostik 1762-3	Nitrile: Bostik S.18.376	Polyurethane: Bostik 2064*	Polyurethane: Bostik 3206*
Rubbers																						
Natural	3	2	3	2	1	1	■	■		2	2	2										
Neoprene							■	■		■	■	■										
Nitrile							1	2					3	3	3		3	3				
Polyurethane							1	2					2		2		3	3			■	■
Butyl							1	3		2												
EPR								1									1					
Fluorocarbon (Viton)								1														
Plastics Materials																						
PVC (Plasticised)								1					3	3	3		3	3	3	3	■	■
PVC (unplasticised)							■	■		2	2	2	3	3	3		3	3	3	3	■	■
Polyester (eg Melinex)													1		3		2	2			2	2
Polyamide (eg Nylon)													1			3	3	2			2	1
Acrylic (eg Perspex)	2	1	2	1			1	1		1	1	1	3	3	3		3	3			■	■
Laminated plastics (eg Formica)							■	3			3										■	■
Polystyrene							2	2			3											
Cellulosics	1	1	1	1			2	2		2	2	2	2	2	2		3	3			2	2
GRP (resin bonded fibreglass)	2	1	2	1			3	3		2	2	2	3	2	3		3	3			2	2
Thermosets							2	3		2	2	2	3		3	2	3	3			■	■
Polyethylene, polypropylene								2									2	2				
PTFE (eg Fluon)	2	2	2	2			2	3					2		2		3	3				
ABS			2				1	1					3	2	3		3	3			■	■
Polycarbonate							2	2					3	2	3		3	3			■	■
Expanded Rubbers & Plastics																						
Natural		2	2							1	1	1										
Natural latex foam	2	2	2	2	1	1	3	3	2	3	3	3										
Neoprene							2	3		3	3	3										
Polyurethane (rigid)	2	2	2	2			2	2		3	3	3	3	3	3		3	3	3	3	3	3
Polyurethane (flexible)	2	2	2	2			2	2	3	3	3	3	3	3	3		3	3	3	3	3	3
PVC (rigid)	1	1	1	1			3	3		3	3	3	3	3	3		3	3	3	3	3	3
PVC (flexible)													3	3	3		3	3	3	3	3	3
Polystyrene		2																				
Metals																						
Aluminium and its alloys	3	2	3	2			3	3		3	3	3	3	1	3	3	3	3			3	3
Copper and its alloys							3	3		3	3	3	3	1	3	3	3	3			3	3
Iron, steel	3	2	3	2			3	3		3	3	3	3	1	3	3	3	3			3	3
Tinplate	1	1	1	1			2	2		2	2	2	1	1	3	2	2	2				
Galvanised	1	1	1	1			3	3		3	3	3	1	1	1		2	2			1	1
Fabrics & Textiles																						
Cotton	3	3	3	3	2	2			2	3	3	3	3	3	3		3	3	3	3	3	3
Wool	3	3	3	3	2	2	3	3	2	3	3	3	3	3	3		3	3	3	3	3	3
Acrylic (eg Acrilan, Courtelle)							2	2	2	2	2	2	3		3		3	3	3	3	3	3
Polyamide (eg Nylon)							2	3	1	2	2	2	3		3		3	3	2	2	3	3
Polyester (eg Terylene)							1	1	1	1	1	1	2		3		3	3	2	2	2	2
Glass cloth and fibreglass	3	3	3	3			3	3		3	3	3	3	3	3	3	3	3	2	2	3	3
Felt	3	3	3	3	2	2	3	3	2	3	3	3	3	3	3		3	3	2	2	3	3
Rayon	2	2	2	2	2	2			2	3	3	3										
Asbestos cloth	3	3	3	3	1	1	3	3		3	3	3	3	3	3	3	3	3	2	2	3	3
Miscellaneous Substrates																						
Cardboard, paper	3	3	3	3	3	3	3	3		3	3	3	3	3	3		3	3	3	3	3	3
Wood, chipboard, hardboard	3	2	3	2		1	3	3		3	3	3	3	3	3		3	3	2	2	3	3
Leather	3	3	3	3	2	2	3	3		3	3	3	3	3	3		3	3	3	3	3	3
Painted surfaces	3	2	3	2						3	3	3										
Glass, glazed surfaces	2	2	3	2			2	2		1	1	1		2	3		1	1			1	1
Concrete, brick, stonemasonry	2	2	3	2			3	3					2	2	2		2	2			3	3

KEY:

3 Most suitable   2 Good   1 Fair   ☐ Unsuitable

† These substrates need special pre-treatment

* These adhesives need an appropriate primer for best adhesion to metals, glass, masonry.

SELECTION CHART (See also page 592)

Polyester: Bostik 68.GA131	Viton: Bostik 2221*	Viton: Bostik 2222*	Polysulphide: Bostik 2114*	Polysulphide: Bostik 2115*	Polysulphide: Bostik 2135-7-8*	Epoxy: Bostik 2000	Epoxy: Bostik 2001	Epoxy Polysulphide: Bostik 2024	Natural Latex: Bostik 4600-1-2	Natural Latex: Bostik 4617	Synthetic Resin Emulsion: Bostik 4141-2	Synthetic Resin Emulsion: Bostik P.V.A.	Film Form Adhesives: Bostik 9100-1	Film Form Adhesives: Bostik 9105
									1	1				1
														2
													2	2
						2	1						1	
	3	3												
3											2		[illegible]	[illegible]
3						1	1				2		[illegible]	[illegible]
3											1		1	2
						1	1						1	2
						1	1				1		2	
						2	2							
						1	1				1			2
2													2	
			3	3	3	3	3	3			1		2	2
											1		3	
						1	1						3	3
					3								3	
									1	1				
						3	3	3		1	2		2	
						3	3	3			2		2	
						3	3	3		1	2		2	2
											2		2	2
						3	3	3		1	2	2		
	3	3	3	3	3	3	3	3		2	2		2	2
3	3	3	3	3	3								2	2
	3	3	3	3	3	3	3	3		2	2		2	3
	2	2								2	1		1	2
	2	2				3	3			2	1		1	
3			3	3	3	3	3	3	2	2	2	2	3	3
3			3	3	3	3	3	3	2	2	2	2	3	3
											1		3	
3											1		2	2
3											1		2	3
	3	3	3	3	3	3	3	3		2	2		3	
3			3	3	3	3	3	3	2	2	2	2	3	3
			3	3	3	3	3	3		2	2			
	3	3	3	3	3	3	3	3	2	2	2	2	3	3
3			3	3	3	3	3	3	2	3	3	3	3	3
3			3	3	3	3	3	3	1	3	3	3	3	3
3			3	3	3	3	3	3	3	3	3	3	3	3
			3	3	3					3	2			
	3	3	3	3	3	3	3	3		2	2	3		2
	3	3	3	3	3	3	3	3						

How to use the Bostik Adhesive Selector Chart

1. Find the materials to be joined in the 'Substrates' column.
2. Note the Bostik adhesives along the top of the guide. Vertically and opposite the material to be bonded are figures 1, 2 or 3. A figure 3 indicates the best adhesive, figures 2 and 1 are less suitable.

   Choose an adhesive which gives a maximum figure when the figures opposite the two particular adherends are added together.

   N.B. Bear in mind that 2 + 2 is a better choice than 3 + 1.

   For example, if bonding nitrile rubber to polyurethane rubber, Bostik 2762 and 2763 would be best (figure total of 3 + 3 = 6), with Bostik Clear and 1755 as second choices (figure total 3 + 2 = 5).

3M ADHESIVES

	Scotchgrip Adhesive	Description	Method of Application					Service† Temp Limits		Resistance*				Colour	Solvent
			Brush	Spray	Roller	Flow	Trowel	Low °C	High °C	Oil	Weathering	Water	Solvent		
A	Scotchgrip Industrial Adhesive 847	High-strength, fast-setting nitrile adhesive. Resistant to most migratory plasticisers.	X			X	X	-40	93	E	E	G	G	Dark Brown	Ketone
B	Scotchgrip Industrial Adhesive 1888V	Fast setting, high strength neoprene adhesive.	X		X	X	X	-30	100	G	E	E		Tan	Aromatic
C	Scotchgrip Industrial Adhesive 3075	High performance, thermosetting nitrile rubber adhesive for metal/metal bonding				X		-40	121	E	E	G	G	Dark	Ketone
D	Fastbond 10 Contact Cement	Neoprene contact adhesive with long bonding range for bonding plastic laminates.	X		X		X	-30	105	F	E	E		Yellow	Blend
E	Scotchgrip Insulation Adhesive 34	Fast-tacking, synthetic rubber adhesive. Sprayable through low cost, low-pressure spray equipment.	X	X				-35	70		G	G		Clear	Aliphatic
F	Scotchgrip Insulation Adhesive 7821	Water-dispersed, asphalt-based adhesive with above average wet strength.	X	X	X		X	-40	80		E	F		Black	Water
G	Scotchgrip Spray Adhesive 77	Aerosol-packed, fast tacking, synthetic rubber adhesive. For temporary or permanent bonds.	Aerosol					-35	60		G	G		Clear	Petroleum Naphtha
H	Sctochgrip Flexible Foam Adhesive 4500	Non-flammable, fast-tacking sprayable adhesive for bonding flexible foam.		X				-29	60		G	G		Translucent	Chlorinated
J	Scotchgrip Vinyl Adhesive 1099	Fast-drying, high-strength nitrile adhesive. Resistant to most migratory plasticisers.	X			X		-40	120	E	G	G	G	Cream	Ketone
K	Scotch Spraymount Artists Adhesive	Aerosol-packed, pressure-sensitive adhesive for temporary bonds	Aerosol					-35	60		G	G		Clear	Chlorinated
L	Scotch Photomount Adhesive	Aerosol-packed adhesive for permanent bonds.	Aerosol					-35	60		G	G		Clear	Petroleum Naphtha

M N	Scotchclad	coatings 776, 1034, 1490X
O P	Scotchseal	sealants 750C, 802

† Service temperature limits may vary according to the application.

*E: Excellent G: Good. F: Fair. "Blank". Not recommended.

SELECTION GUIDE

				FOAMS						PLASTICS											RUBBERS			
		Fabrics & Papers	Felts	Polystyrene	Polyurethane	PVC	Glass & Ceramics	Metals	ABS	Acetals	Nylons	Polyester (Film)	Polyethylene	Polypropylene	Hi-Impact Polystyrene	Unsupported Flexible PVC	Supported Flexible PVC	Rigid PVC	Plastic Laminates	GRP	Natural Rubber	Neoprenes	Nitriles	Wood
	Fabrics & Paper	E G K L	E G K L	E G	E G H	B D	E G K L	E G K L	G J K L	B D G K	A J	E G J M	E G K	E G K	B D E G	A J M	A B J	A J K L	D E G K	B D E G	B D E G	B D E G	A J	D G K L
	Felts		D E F G	E F G	E G H	D E G	E G K L	E F G K	B D E J	B D	A J	E G J M	E G K	E G K	B D E G	A J M	E G K L	A J M	D E G K	B D E G	B D E G	B D E G	A J M	B D E G
FOAMS	Polystyrene			E G	E G	E G	E G	E F G	E G	E G	E G	E G	E G	E G	E G		E G	E G	E G	E G	E F G	E F G	E G	E F G
	Polyurethane				H	E G H	E G H	E G H	E G H	E G H	E G	E G M	E G	E G	H		E G H	E G	E G	E G	F H	B D	J M	E G H
	PVC					A J M	A J M	A J M	B D E G	B D E G	J	J M	E G	E G	J M	A J M	A J M	A J M	B D J	B D J	B D	B D	A J M	B D J
	Glass & Ceramics						J N O P	J O N P	J N O P	B N O P	A J O P	J M	N	N	J O P	A J M	A J M	A J M P	D J O P	D J O P	B D P	B D P	J M P	O N P
	Metals							C N O P	C N O P	B N O P	J N O P	J M	N.	N	J M N	A J M	A J M	A J N	D N	O N P	B D	B D	A J M	O N P
	ABS								J O P	B O P	J O P	J M	N	N	B D P	A J	A B J	A J	B D	O P	B D P	B D P	A J P	B D O P
	Acetals									B D O P	B D O P	J M	N	N	B D O P	A J M	A J M	A J O P	B D O P	B D O P	B D P	B D P	A J P	D O P
	Nylons										J O P	J M	N	N	J P	A J	A J	A J P	A J	A J O P	A J	J P	J P	J P
	Polyester (Film)											J M	E G K	E G K	J M	J M	J M	J M E G	J M E G	J M E G	J	J	J M	J M E
PLASTICS	Polyethylene												E G N	E G N	E G N		E G	E G N	E G N	E G N	N	N	N	E G N
	Polypropylene													N	N			N	N	N	N	N	N	N
	Hi-Impact Polystyrene														B D J P	J M	B D J	B J P	D P	B D P	B D P	B D P	A J P	D J O P
	Unsupported Flexible PVC															J M	J M	A J M	A J M	A J M	J	J	A J M	A C J M
	Supported Flexible PVC																J M	A J M	A J M	A J M	J	J	A J M	A J M
	Rigid PVC																	A J O P	A J O P	A J O P	J P	D J P	A J M P	A J O P
	Plastic Laminates																		D	D	D	D	D J	D N
	GRP																			N O P	N O P	B D P	J P	D O N P
RUBBERS	Natural Rubber																				B D	B D	B J	B D
	Neoprenes																					B D	B J	B D
	Nitriles																						A J M	A J
	Wood																							D O N P

BOSTIK ADHESIVES AND SEALING COMPOUNDS (see also pages 588–589)

Bostik	Description	Base	Colour
BCA3	Synthetic rubber/resin adhesive	Polychloroprene	Buff
'C' adhesive	Reclaim rubber/resin adhesive	Reclaim	Black
Clear adhesive	General purpose synthetic rubber/resin adhesive	Nitrile	Clear
183C	Vapour seal adhesive	Bitumen rubber/emulsion	Black
247	Bostik Solbit	Bitumen	Black
281	Bostik Colset	Bitumen	Black
590	Chassis black	Asphalt/resin	Black
6	Reclaim rubber/resin Sealing compound	Reclaim	Black
772 771 in tubes	Synthetic rubber/resin adhesive and sealing compound	Nitrile	Blue-black
1297	Rubber/resin adhesive	Natural rubber	White
1311	Panel adhesive	Natural rubber	White
1530	Sealing compound	P.I.B.	Off-white
2000	Two part Epoxy resin adhesive	Epoxy	Brown
2024	Two part, cold-curing epoxy/polysulphide adhesive and sealant	Epoxy-polysulphide	Dark grey
2117	Two part polysulphide caulking compound	Polysulphide	Black
2135, 2137, 2138	Two part polysulphide sealants	Polysulphide	2135 Black, 2137 grey, 2138 white
2221, 2222, 2223, 2224	Two part Viton synthetic sealants	Viton	2221 black, 2223 black, 2222 white, 2224 white
2402	Two part cold-curing synthetic rubber/resin adhesive	Polychloroprene	Buff
3206	Synthetic rubber/resin adhesive	Polyurethane	Clear amber
4141, 4142	Synthetic polymer adhesives	Synthetic resin emulsion	White, colourless when dry
6360, 6365	Thermogrip hot melt adhesive	E.V.A.	Cream
5653 Prestik	Extruded sealing strip	Oil/asbestos	Cream
5686 Prestik	Extruded sealing strip	Oil/asbestos	Dark grey
5703 Prestik	Extruded sealing strip	P.I.B.	White
5913 Prestik	Extruded sealing strip	Butyl rubber	White
5925	Extruded sealing strip	Butyl rubber	Buff
9015	Asphalt tile adhesive	Bituminous	Black
9100	Film adhesive	Nitrile	Translucent film on white carrier
9105	Hot melt film adhesive	Polyester	White translucent
9410	Thermogrip hot melt adhesive	Polyethylene	Straw
6361, 6366	Thermogrip hot melt adhesive	E.V.A.	Cream

TRETOBOND BUILDING ADHESIVES SELECTION GUIDE

	Plywood	Hardboard	Plasterboard	Concrete	Brick	Plaster	Asbestos (flat)	Lightweight Concrete blocks	Expanded polystyrene	Sheet metal	Timber	Timber battens
Laminated plastics	**404** **468**	404 468	404 468	— —	— —	— —	**404** 468	— —	— 375	**404** 468	**404** **468**	**404** **468**
Decorative hardboard	**404** **425**	404 425	**404** **425**	404 425	— 780	404 425	404 425	— 780	— 375	404 425	404 425	404 425
Plywood wallboard	425 780	425 780	425 780	425 780	— 780	425 780	425 780	— 780	— 375	425 780	425 780	425 780
Asbestos wallboard	425 780	425 780	425 780	425 780	— 780	425 780	425 780	— 780	— 375	425 780	425 780	425 780
Fibreboard wallboard	425	425	425	425	425	425	425	425	375	425	425	425
Rigid urethane boards "Purlboard" "Lamithane" 'Plaschem"	425 780	425 780	425 780	425 780	425 780	425 780	425 780	425 780	— 375	425 780	425 780	425 780
Acoustic tiles	425 AT79	425 AT79	425 AT79	425 AT79	425 AT79	425 AT79	425 AT79	425 AT79	— —	425 AT79	425 AT79	425 AT79
Expanded polystyrene boards	**740** 375	740 375	740 375	740 375	740 375	740 375	740 375	740 375	740 375	375	375	375
Expanded polystyrene ceiling tiles and coving	282	282	282	282	282	282	282	282		375	282	375
Compressed asbestos sills	780	780	780	780	780	780	780	780	—	780	780	780
Rigid PVC, timber and metal skirtings	780	780	780	780	780	780	780	780	—	780	780	780
Rigid PVC architraves	780	780	780	780	780	780	780	780	—	780	780	780
Timber battens	425 780	425 780	425 780	425 780	425 780	425 780	425 780	780	375	425 780	425 780	425 780
Quarry tiles, rubber tiles. "Onazote" Felt	737	737	737	737	737	737	737	737	375	737	737	737
Timber (joinery work) general purpose adhesive work, priming, plastering screeding, cement rendering												

Universal Tretobond building adhesive
& general purpose bonding agent

SELECTION GUIDE: DUNLOP ADHESIVES

	Wood	Rubber	Rigid PVC	PVC Leathercloth (cotton backed)	Flexible PVC	Polyurethane Foam	Polythene	Paper, Card	Natural Fabrics	Natural Latex Foam	Metals	Leather	Hardboard	GRP	Cork	Aluminium Foil
Aluminium Foil	S 758	S 691	S 691	S 758	S 1115	S 758	S 957	S 758	S 758	S 758	S 758	S 758	S 758	S 758	S 758	S 758
Cork	S 889	S 758	S 758	S 758	S 1115	L 107	S 809	A 1020	A 1020	L 107	S 758	S 758	S 758	S 758	L 107	—
GRP	S 691	S 691	S 691	S 758	S 1115	S 758	S 957	S 758	S 758 & 888	S 758	S 691	S 691	S 691	S 758 & 691	—	—
Hardboard	S 758	S 758	S 758 & 691	S 758	S 1115	L 107	S 809	A 1020	A 1020	L 107	S 758	S 758	S 758	—	—	—
Leather	S 758	S 758	S 758	S 758	S 1115 & 1310	S 758	S 957	L 107	L 107	L 107	S 758	S 758	—	—	—	—
Metals	S 758	S 758	S 758	S 758	S 691 & 1115	S 758	S 809	S 758	S 758 & 888	S 834	S 758	—	—	—	—	—
Natural Latex Foam	L 107	S 758	S 758	S 758	S 1115	S 834	S 809	A 1020	A 1020	S 834	—	—	—	—	—	—
Natural Fabrics	S 758	S 758	S 758	S 1115	S 1115	A 1020 107	S 809	A 1020	A 1020	—	—	—	—	—	—	—

Paper, Card	L 107 S 758	S 758	S 758	S 1115	S 1115	A 1020	S 809	A 1020	—	—	—	—	—	—	—	—
Polythene	S 957	S 957	S 957	S 957	S 957	S 957	S 957	—	—	—	—	—	—	—	—	—
Polyurethane Foam	L 107 S 758	S 758	S 758	S 758	S 1115	S 834	—	—	—	—	—	—	—	—	—	—
Flexible PVC	S 1115	S 691	S 1115	S 1310 S 1115	S 1115	—	—	—	—	—	—	—	—	—	—	—
PVC Leathercloth (cotton backed)	S 758	S 691	S 691	S 758 & 1115	—	—	—	—	—	—	—	—	—	—	—	—
Rigid PVC	S 758 & 691	S 691	S 758 & 691	—	—	—	—	—	—	—	—	—	—	—	—	—
Rubber	S 691	S 758	—	—	—	—	—	—	—	—	—	—	—	—	—	—
Wood	wood working adhesive	—	—	—	—	—	—	—	—	—	—	—	—	—	—	—

Adhesive Types: A – Latex L – Natural Rubber S – Synthetic Rubber

SELECTION GUIDE TO 'EVO-STIK' ADHESIVES

(Building trades, light engineering, etc applications)

'IMPACT' SH.25 – one part synthetic rubber based adhesive designed for bonding floor-coverings of PVC, rubber and linoleum to a wide range of substrates including concrete, cement screeds, wood, metal, and a variety of boards.

'IMPACT' SH.64 – single part, synthetic rubber/resin adhesive specially developed for PVC bonding. Also suitable for bonding nylon, 'Terylene', cotton, leather, some rubber compositions and metal.

IMPACT' SH.100 – one part synthetic rubber based adhesive designed for bonding floor coverings of PVC, rubber and linoleum to a wide range of substrates including concrete, cement screeds, wood, metal and a variety of boards.

426 – a solvent-based bituminous adhesive designed for bonding all types of thermoplastic and vinyl-asbestos floor tiles to sub-floors of concrete, sand/cement, wood, or proprietary screeding compounds.

429 – dipping grade bitumen/rubber emulsion for bonding wood block flooring to sub-floors of concrete, sand/cement, wood, proprietary screeding compounds and asphalt.

432 – bitumen/rubber emulsion for bonding linoleum, vinyl/asbestos and felt-backed floor coverings to sub-floors of concrete, sand/cement, wood, and proprietary screeding compounds.

IMPACT' 523 – one part synthetic rubber/resin adhesive for bonding plasticised PVC, leather, and most types of board to wood, board, and metal surfaces.

'IMPACT' 524 – one part synthetic rubber/resin solution particularly suitable for bonding natural and most synthetic rubbers, irrespective of hardness, to themselves and to metal. It is also suitable for bonding leather, PVC and nitrocellulose leathercloths to various metals, including chromium plate.

IMPACT' 528 – general purpose one part synthetic rubber resin solution, for bonding laminated plastics and other rigid plastics (PVC, ABS, but excluding polyolefins) to wood, metal and all types of boards and dry screeds with the exception of bituminous types.

'IMPACT' 567 – a one part synthetic rubber/resin adhesive possessing high tack, particularly suitable for the bonding of micro-cellular and foam plastics and rubber (eg cellular PVC, 'Plasticell', but not expanded polystyrene) to a variety of building surfaces.

'IMPACT' 584 – single part synthetic rubber/resin solution specifically formulated for bonding expanded polystyrene and other rigid and semi-rigid cellular plastics to themselves and to other substrates.

'IMPACT' 584/S – low pressure sprayable, single part synthetic rubber/resin solution specially formulated for bonding expanded polystyrene and other rigid and semi-rigid cellular plastics to themselves and to other substrates.

SL 807 – one part acrylic resin emulsion adhesive which is suitable for bonding fabric backed or sheet PVC to porous surfaces, eg plaster or hardboard. It may also be used to bond absorbent materials such as leather, boards, soft foams and fabrics to themselves, to the substrates above, or to PVC, ABS and metals.

863 – heavy-bodied, synthetic resin emulsion adhesive for bonding expanded polystyrene products to vertical and horizontal absorbent surfaces such as plaster, cement, smooth concrete, hardboard, plasterboard, plywood and timber.

873 – synthetic resin/rubber latex blend which is suitable for bonding most types of PVC flooring materials (including felt-backed, foam-backed, hessian-backed and latex asbestos-backed types), nylon and polypropylene needleloom carpeting (except water-sensitive types), jute-backed carpeting, latex foam-backed carpeting, cork, linoleum, and hardened linoleum to sub-floors of concrete, sand/cement, latex cement, wood-chipboard, hardboard, plywood and timber.

917 – synthetic resin emulsion adhesive for bonding high vinyl content, translucent, and plastisol backed floorcoverings to sub-floors of concrete, sand/cement, wood and proprietary screeding compounds.

'IMPACT' 1001 – rubber solution for bonding most types of rubber flooring to sub-floors of concrete, sand/cement, wood and proprietary screeding compounds.

5007/2 and 5007/3 – one part formulations, based on a blend of nitrile synthetic rubber and resins, for bonding cellular plastics (eg cellular PVC, polyester and polyether urethane foams) to themselves or to other substrates such as PVC (rigid and flexible), wood, metal or cloth.

5008 – solvent based blend of nitrile synthetic rubber and resins for bonding PVC, metals, polyethylene terephthalate film (Mylar, Melinex), leather and wood. It is also available in film form as 'Nutrim' film.

5017/1 – one part synthetic rubber/resin gap filling adhesive/sealer intended for application by extrusion gun.

'IMPACT' 5080 – high solids, heavy duty adhesive based on reclaim rubber suitable for bonding rubber to metal or wood, and for lining wooden cases with laminated papers or felts.

5183 – general purpose, non-staining, one part synthetic rubber/resin solution, suitable for bonding most plastics (including PVC), metal, glass, pottery, fabrics, rubbers and wood to themselves and to other substrates.

5210– synthetic rubber/resin adhesive which is capable of cross-linking under the influence of heat. It gives excellent adhesion to many substrates including Melinex, metals, leather, wood and laminated plastics.

'IMPACT' 5390 – one part synthetic rubber/resin solution suitable for bonding wood veneers, laminated plastics, metals and rigid PVC to themselves or to seasoned timber and hardboard.

'IMPACT' 5696 – sprayable, general purpose, synthetic rubber/resin solution particularly suitable for bonding laminated plastics to wood, metals, and all types of boards and dry screeds. It is suitable for use by hot or cold spray techniques.

'IMPACT' 5736 and 5736/S – one part synthetic rubber/resin adhesives based on 'non-flam' solvents for general upholstery work. Both are suitable for bonding polyether and polyester foams, fabric backed PVC, plywood and hardboard to themselves and to each other.

'GUN-O-PRENE' – gun grade, high bond strength, synthetic rubber/resin adhesive with gap filling properties, for bonding most types of wall cladding direct to wall surfaces or timber battens provided that they are rigid in themselves and not subject to flexing between points of adhesive contact.

'THERMAFLO' 6820 – general purpose hot melt adhesive with good adhesion to a wide range of materials including cotton duck, wood, mild steel, aluminium, rubber, leather, polyolefins, rigid PVC, ABS, glass and acrylics.

7131 – two part epoxide adhesive which will cure at ambient or elevated temperatures.

'IMPACT' 8148 – single part polyurethane rubber adhesive for bonding PVC to itself, leather, wood, ABS, and most types of board.

9252 – acrylic, pressure sensitive, aqueous dispersion suitable for applications where pressure sensitive bonds of very high performance are required.

'IMPACT' 5655 – general purpose synthetic rubber/resin solution, for bonding laminated plastics and other rigid plastics (PVC, thermosets) to wood, metals, all types of board, cement renderings and plaster.

HERMETITE Double Bond Adhesives

These are two-part epoxide resin adhesives for bonding metals, plastics, rubbers, ceramics, glass, concrete, etc.

No 1 – solventless translucent amber gel

No 2 – solventless paste

No 3 – solventless low viscosity amber liquid

Specifications and Standards British

British

BS844	Methods of sampling and testing vegetable adhesives.
BS1203	Synthetic resin adhesives for plywood.
BS1204	Part 1: Gap filling adhesives.
	Part 2: Close contact adhesives.
BS1444	Cold setting casein adhesive particles for wood
BS3544	Methods of test for polyvinyl acetate adhesives for wood.
BS3940	Adhesives based on bitumen and coal tar.
BS4071	Polyvinyl acetate (PVA) emulsion adhesives for wood.
DTD775B	Adhesives suitable for joining metals
DTD861A	Adhesives for metals
DTD5577	Heat stable structural adhesives.

US Federal Standards

MMM-A-132	Adhesives, heat resistant, structural, metal to metal. Four temperature ranges, types I, II, III and IV.
MMM-A-131A	Adhesives, glass to metal for bonding optical elements. Type I liquid. Type II dry film.

US Military Standards

MIL-A-25463	Adhesives for honeycombs. Also in four temperature ranges and two classes. Class 1 metal facing to cores. Class 2 metal facing to cores, inserts and other metal parts.
MIL-A-81236	Adhesive, epoxy resin with polyamide curing agent.
MIL-A-81253	Adhesive, modified epoxy resin with polyamide curing agent.

Adhesive Glossary

A-Stage — an early stage in the curing of thermosetting resins where the material is still plastic, fusible and soluble in certain solvents.

Adhere — to stick.

Adherence — the strength of an adhesive bond.

Adhesion — the sticking together of two subjects by surface attachment.

Adhesive — a substance capable of promoting adhesion by surface attachment.

Adhesive Dispersion — a two phase system, one phase being suspended in a liquid.

Agressive Tack — a state of near-dry tackiness in a cure adhesive.

Ageing Time — the time between closing an adhesive joint and the joint achieving maximum bond strength.

Anaerobic — an adhesive which sets or cures in the absence of air.

B-Stage — an intermediate stage in the setting or curing of a thermosetting resin where some thermoplastic characteristics are still retained.

Batch — adhesives or mixtures, produced at the same time to a particular formulation.

Binder — component used in an adhesive to add body and which is primarily responsible for the strength of the adhesive film.

Blocking — unwanted sticking together of materials under moderate pressure — eg during storage.

Bond — the general description of a joint produced by an adhesive.

Bond Strength — the strength of an adhesive joint.

C-Stage — the final stage in the setting or curing of a thermosetting resin.

Catalyst — a substance which accelerates a setting or curing reaction of a thermosetting resin.

Cement — an adhesive which is a solution of the same substance to be glued, in a suitable solvent. Also used more loosely to describe other types of simple adhesives.

Closed Time — the time a made-up joint is kept closed under pressure.

Cohesion — the condition of holding together a substance under primary or secondary valency forces.

Cold Flow — creep or extrusion of a substance or adhesive film under pressure or load at normal temperatures.

Condensation – combination of two or more molecules into some other simple substance with the separation of water.

Consistency – the physical state of a liquid exhibited by its viscosity and plasticity, etc.

Copolymer – a substance produced by the polymerisation of two or more different monomers.

Creep – dimensional changes which occur with time under the influence of pressure or load and/or temperature.

Cure – an irreversible change in a substance by condensation, polymerisation, etc – eg the setting of a thermosetting resin to C-stage.

Curing Time – the time taken for a thermosetting resin to reach the C-stage.

Curing Temperature – the temperature to which an adhesive has to be raised to achieve setting or curing.

Delamination – separation of layers in a laminate, due to bond failure.

Diluent – any solvent, thinner, etc which reduces the consistency of a liquid adhesive

Doctor Blade – a scraper blade or mechanism used with coating rolls to regulate the amount of adhesive applied. Also called a Doctor Bar.

Dry Strength – the strength of an adhesive joint measured immediately after a specified dry period.

Dry Tack – the property of certain adhesives to adhere on contact when apparently dry (eg contact adhesives).

Drying Temperature – the temperature specified, or required, to dry or set an adhesive.

Extender – a substance added to an adhesive to improve body or consistency whilst reducing the amount of binder required.

Filler – a relatively inert or inert substance added to an adhesive to enhance its working properties.

Flow – movement of an adhesive when applied, or during assembly, and before setting.

Gel – semi-solid substance consisting of solids in continuous phase and the liquid content in discontinuous phase.

Gelation – the stage at which a liquid forms a gel or gel-like substance.

Glue – older term for adhesive, still in general use. In modern terminology 'glue' is generally reserved as descriptive of a gelatine-based or natural adhesive.

Gum – type of natural adhesive characterised as colloidal in substance.

Hardener – chemical substance which reacts with a thermosetting resin to promote curing or setting at room temperatures, or is used to control the curing process.

Inhibitor – a substance which inhibits or retards a chemical reaction.

Laminate – a laminar structure consisting of two or more surfaces bonded together.

Lamination – the process of producing a laminate.

Mechanical adhesion – a bond or adhesive joint formed by a mechanical interlocking action.

Modifier – a chemically inert substance added to an adhesive to change or enhance its properties. (ie basically similar to a filler).

Mucilage – an adhesive consisting of a solution of gum in water. Also more generally used for a low strength liquid adhesive.

Open Time – the time between spreading an adhesive on a joint and closing that joint under heat or pressure, etc.

Paste – the general description of a low-strength adhesive of a viscous nature and (usually) opaque appearance.

Plasticity – elasticity in a substance – ie capable of being deformed without shearing.

Plasticiser – an additive to increase the flexibility of a substance.

Polymer – a compound produced by the reaction of simple molecules (monomers).

Polymerisation – the chemical reaction whereby molecules of a simple substance are linked to produce a heavier molecule. Where different monomers are involved, the process is known as Copolymerisation.

Pot Life – the working life of an adhesive once prepared or mixed ready for use.

Primer – a solution applied to a surface prior to the application of an adhesive to improve the adhesive bond.

Resin – the generic term for a specific type of natural or synthetic organic material on which most adhesives are based.

Resol – the A-stage of a thermosetting resin.

Resistol – the B-stage of a thermosetting resin.

Resite – the C-stage of a thermosetting resin.

Retarder – an inhibitor.

Self-Curing – a thermosetting resin which sets or cures without the application of heat.

Set – the hardening or setting of a liquid adhesive into a solid state.

Shelf Life – storage or shelf life of an adhesive before it degrades.

Short (ness) – an adhesive that does not 'string'.

Specific Adhesion – the closure force produced between two surfaces bonded by valency forces.

Spread – the quantity or distribution of adhesive, specifically referred to in weight or quantity per unit area.

Storage Life – another term for shelf life.

String (ing) – the property exhibited by certain adhesives to develop filaments or 'strings' when the two surfaces are separated before setting, or when the adhesive applicator is withdrawn from a surface.

Substrate(s) – material(s) on which an adhesive is spread and subsequently bonded together.

Tack (Tackiness) – natural stickiness immediately apparent in some adhesives when spread, or when the substrates are brought together.

Tack Range – the time during which an adhesive remains tacky after spreading and/or joint closure.

Thermoplastic – materials which are softened by heat and show reversible characteristics in this respect.

Thermoset – materials which lose their plastic nature and become hard and rigid when set or cured by heat or chemical action.

Thinner – a volatile liquid which acts as a solvent for the substance concerned.

Thixotropy – liquids with non-sagging or non-running characteristics.

Viscosity – a measure of the 'thickness' or fluidity of an adhesive.

Vulcanisation – the process or reaction used to harden a rubber and increase its tensile strength.

Webbing – 'stringing'.

Working Life – the pot life of an adhesive.

THE BRITISH ADHESIVE MANUFACTURERS' ASSOCIATION

Most of the major adhesive manufacturers in the UK are members of the Association.

A major aim is to maintain and develop quality standards throughout the industry. To this end and as appropriate the Association's Technical Panel works on the development of test methods and performance standards, often in conjunction with the BSI. Also, the Association co-ordinates and expresses the views of the industry to Government, when such matters as EEC Directives are being considered.

Another important aspect of the Association's work is to encourage the correct use of adhesives by methods such as the presentation of seminars and the improvement of labelling.

Work is carried out on various other subjects of interest to Members, for example, statistics.

Members

Adcol Ltd.,
Airborne Industries Ltd..
Alfred Adams & Co Ltd.,
Anglo-Chemical Co (Leicester) Ltd.,
Apollo Chemicals Ltd.,
Associated Adhesives Ltd (Gloy),
Avdel Adhesives Ltd.,
F. Ball & Co Ltd.,
Borden (UK) Ltd.,
Bostik Ltd.,
Caswell & Co Ltd.,
Ciba-Geigy Ltd.,
Clam-Brummer Ltd.,
Cow Proofings Ltd.,
CPC (United Kingdom) Ltd.,
Crispin Chemical Co Ltd.,
Croda Glues Ltd.,
Dunlop Semtex Limited,
Evode Limited,
Henkel Chemicals Ltd.,
W. E. Howlett & Son Ltd.,
Idenden Adhesives Ltd.,
Industrial Adhesives Ltd.,
Samuel Jones & Co Ltd.,
Larkhill Soling Co Ltd.,
3M United Kingdom Ltd.,
Monarch Adhesives Ltd.,
National Adhesives & Resins Ltd..
Pafra Ltd.,
Pertec,
Plus Products Ltd.,
Radburne & Bennett (Rushden) Ltd.,
Roberts Smoothedge Ltd.,
Royal Sovereign Group,
Rubber Latex Ltd.,
Shell Composites Ltd.,
Sovereign Chemicals Ltd.,
Starch Products Ltd.,
Swift Chemical Company,
Tivoli Kay Adhesives Ltd.,
Tretobond Ltd.,
Union Glue & Gelatine Co Ltd.,
Vik Supplies Ltd.,
Williams Adhesives Ltd.,
Harold Wilson & Co Ltd.,
Yorkshire Chemicals Ltd.

Associate Members

Du Pont (UK) Ltd.,
Harlow Chemical Company Ltd.

SECTION 2

Sub-section (B)

Buyers' Guide

Sub-section b(I)

TRADE NAMES INDEX – ADHESIVE SECTION

AVDELBOND A – Surface activated acrylic-metal catalysed-oxtgen inhibited – Avdel Ltd
AVDELBOND C – Surface activated acrylic-moisture catalysed epoxt resin – Avdel Ltd
AVDELBOND E – Epoxt resin – Avdel Ltd

BIGBOND – Quick-setting synthetic rubber adhesive suitable for securing bigheads to most surfaces for retaining thermal and acoustic insulants – Bighead Bonding Fasteners Ltd

PRESSURE PACK – Self-contained adhesive applicator – Avdel Ltd
PULSE PACK – Pneumatic timed pulse adhesive applicator – Avdel Ltd

SELF-FORMING GASKET – Self-forming gasket material – Avdel Ltd

TRIMATIC – Epoxy resin handling machine systems – Avdel Ltd

UNBRAKO – Industrial and aerospace fasteners, anerbolic adhesives – Unbrako Distributors

Classified Index to Adhesive Manufacturers, Distributors and Suppliers

Acrylics

Avdel Ltd
Unbrako Distributors

Application

Avdel Adhesives Limited

Anaerobic

Avdel Adhesives Limited

Anaerobic Adhesives

Unbrako Ltd

Cyanoacrylate

Avdel Adhesives Limited

Epoxy Resins

Avdel Ltd

Industrial & Aerospace Fasteners

Unbrako Ltd

Proprietary Adhesives

Avdel Ltd

Synthetic Rubber

Bighead Bonding Fasteners Ltd.

Sub-section b(3)

Alphabetical List of Manufacturers with addresses, telephone numbers telegram addresses of Head Office, Works and Branches

Avdel Adhesives Limited, Woodside Road, Eastleigh, Hants, SO5 4EX.

Telephone : Eastleigh 7127/8

Branches: No.2 Factory, Thomas Road, North Baddesley, Hants.
Avdel Pty Ltd, 18 Ellis Street, South Yarra 3141, Victoria, Australia.
Avdel S.A., 66 Rue David D'Angers, Paris 19, France
Avdel KK, Shin Nakanoshima Building, 49 Hinoue-Cho, Kita-Ku, Osaka, Japan.
Avdel Ltd., 2782 Slough Street, Mississauga, Ontario, Canada.
Avdel GmbH, Klusriede 14-16, 3012 Langenhagen, Hanover, Germany.
Avdel Corp., 10 Henry Street, Teterboro, New Jersey, U.S.A.

Bighead Bonding Fasteners Limited, Southern Road, Aylesbury, Bucks.

Telephone : 0296 81663

Unbrako Distribution, Bannerly Road, Garretts Green, Birmingham, 33.

Telephone : 021–783 4066, 021–784 3937, Telex : 338 703

Branches : Burnaby Road, Coventry, CV6 4AE.
Telephone : 0203–88722, Telex: 31–608

Index

A

B

Index to Advertisers